T0145092

Lecture Notes in Computer Science 12683

More information about this subseries at http://www.springer.com/series/7409

Christian S. Jensen · Ee-Peng Lim ·
De-Nian Yang · Wang-Chien Lee ·
Vincent S. Tseng · Vana Kalogeraki ·
Jen-Wei Huang · Chih-Ya Shen (Eds.)

Database Systems
for Advanced Applications

26th International Conference, DASFAA 2021
Taipei, Taiwan, April 11–14, 2021
Proceedings, Part III

 Springer

Editors
Christian S. Jensen (iD)
Aalborg University
Aalborg, Denmark

De-Nian Yang
Academia Sinica
Taipei, Taiwan

Vincent S. Tseng
National Chiao Tung University
Hsinchu, Taiwan

Jen-Wei Huang (iD)
National Cheng Kung University
Tainan City, Taiwan

Ee-Peng Lim (iD)
Singapore Management University
Singapore, Singapore

Wang-Chien Lee
The Pennsylvania State University
University Park, PA, USA

Vana Kalogeraki
Athens University of Economics
and Business
Athens, Greece

Chih-Ya Shen
National Tsing Hua University
Hsinchu, Taiwan

ISSN 0302-9743   ISSN 1611-3349 (electronic)
Lecture Notes in Computer Science
ISBN 978-3-030-73199-1   ISBN 978-3-030-73200-4 (eBook)
https://doi.org/10.1007/978-3-030-73200-4

LNCS Sublibrary: SL3 – Information Systems and Applications, incl. Internet/Web, and HCI

This Springer imprint is published by the registered company Springer Nature Switzerland AG
The registered company address is: Gewerbestrasse 11, 6330 Cham, Switzerland

Preface

Welcome to DASFAA 2021, the 26th International Conference on Database Systems for Advanced Applications, held from April 11 to April 14, 2021! The conference was originally planned to be held in Taipei, Taiwan. Due to the outbreak of the COVID-19 pandemic and the consequent health concerns and restrictions on international travel all over the world, this prestigious event eventually happens on-line as a virtual conference, thanks to the tremendous effort made by the authors, participants, technical program committee, organization committee, and steering committee. While the traditional face-to-face research exchanges and social interactions in the DASFAA community are temporarily paused this year, the long and successful history of the events, which established DASFAA as a premier research conference in the database area, continues!

On behalf of the program committee, it is our great pleasure to present the proceedings of DASFAA 2021, which includes 131 papers in the research track, 8 papers in the industrial track, 8 demo papers, and 4 tutorials. In addition, the conference program included three keynote presentations by Prof. Beng Chin Ooi from National University of Singapore, Singapore, Prof. Jiawei Han from the University of Illinois at Urbana-Champaign, USA, and Dr. Eunice Chiu, Vice President of NVIDIA, Taiwan.

The highly selective papers in the DASFAA 2021 proceedings report the latest and most exciting research results from academia and industry in the general area of database systems for advanced applications. The quality of the accepted research papers at DASFAA 2021 is extremely high, owing to a robust and rigorous double-blind review process (supported by the Microsoft CMT system). This year, we received 490 excellent submissions, of which 98 full papers (acceptance ratio of 20%) and 33 short papers (acceptance ratio of 26.7%) were accepted. The selection process was competitive and thorough. Each paper received at least three reviews, with some papers receiving as many as four to five reviews, followed by a discussion, and then further evaluated by a senior program committee (SPC) member. We, the technical program committee (TPC) co-chairs, considered the recommendations from the SPC members and looked into each submission as well as the reviews and discussions to make the final decisions, which took into account multiple factors such as depth and novelty of technical content and relevance to the conference. The most popular topic areas for the selected papers include information retrieval and search, search and recommendation techniques; RDF, knowledge graphs, semantic web, and knowledge management; and spatial, temporal, sequence, and streaming data management, while the dominant keywords are network, recommendation, graph, learning, and model. These topic areas and keywords shed light on the direction in which the research in DASFAA is moving.

Five workshops are held in conjunction with DASFAA 2021: the 1st International Workshop on Machine Learning and Deep Learning for Data Security Applications (MLDLDSA 2021), the 6th International Workshop on Mobile Data Management,

Mining, and Computing on Social Networks (Mobisocial 2021), the 6th International Workshop on Big Data Quality Management (BDQM 2021), the 3rd International Workshop on Mobile Ubiquitous Systems and Technologies (MUST 2021), and the 5th International Workshop on Graph Data Management and Analysis (GDMA 2021). The workshop papers are included in a separate volume of the proceedings, also published by Springer in its Lecture Notes in Computer Science series.

We would like to express our sincere gratitude to all of the 43 senior program committee (SPC) members, the 278 program committee (PC) members, and the numerous external reviewers for their hard work in providing us with comprehensive and insightful reviews and recommendations. Many thanks to all the authors for submitting their papers, which contributed significantly to the technical program and the success of the conference. We are grateful to the general chairs, Christian S. Jensen, Ee-Peng Lim, and De-Nian Yang for their help. We wish to thank everyone who contributed to the proceedings, including Jianliang Xu, Chia-Hui Chang and Wen-Chih Peng (workshop chairs), Xing Xie and Shou-De Lin (industrial program chairs), Wenjie Zhang, Wook-Shin Han and Hung-Yu Kao (demonstration chairs), and Ying Zhang and Mi-Yen Yeh (tutorial chairs), as well as the organizers of the workshops, their respective PC members and reviewers.

We are also grateful to all the members of the Organizing Committee and the numerous volunteers for their tireless work before and during the conference. Also, we would like to express our sincere thanks to Chih-Ya Shen and Jen-Wei Huang (proceedings chairs) for working with the Springer team to produce the proceedings. Special thanks go to Xiaofang Zhou (DASFAA steering committee liaison) for his guidance. Lastly, we acknowledge the generous financial support from various industrial companies and academic institutes.

We hope that you will enjoy the DASFAA 2021 conference, its technical program and the proceedings!

February 2021   Wang-Chien Lee
                                                      Vincent S. Tseng
                                                      Vana Kalogeraki

Organization

Organizing Committee

Honorary Chairs

Philip S. Yu	University of Illinois at Chicago, USA
Ming-Syan Chen	National Taiwan University, Taiwan
Masaru Kitsuregawa	University of Tokyo, Japan

General Chairs

Christian S. Jensen	Aalborg University, Denmark
Ee-Peng Lim	Singapore Management University, Singapore
De-Nian Yang	Academia Sinica, Taiwan

Program Committee Chairs

Wang-Chien Lee	Pennsylvania State University, USA
Vincent S. Tseng	National Chiao Tung University, Taiwan
Vana Kalogeraki	Athens University of Economics and Business, Greece

Steering Committee

BongHee Hong	Pusan National University, Korea
Xiaofang Zhou	University of Queensland, Australia
Yasushi Sakurai	Osaka University, Japan
Lei Chen	Hong Kong University of Science and Technology, Hong Kong
Xiaoyong Du	Renmin University of China, China
Hong Gao	Harbin Institute of Technology, China
Kyuseok Shim	Seoul National University, Korea
Krishna Reddy	IIIT, India
Yunmook Nah	DKU, Korea
Wenjia Zhang	University of New South Wales, Australia
Guoliang Li	Tsinghua University, China
Sourav S. Bhowmick	Nanyang Technological University, Singapore
Atsuyuki Morishima	University of Tsukaba, Japan
Sang-Won Lee	SKKU, Korea

Industrial Program Chairs

Xing Xie	Microsoft Research Asia, China
Shou-De Lin	Appier, Taiwan

Demo Chairs

Wenjie Zhang	University of New South Wales, Australia
Wook-Shin Han	Pohang University of Science and Technology, Korea
Hung-Yu Kao	National Cheng Kung University, Taiwan

Tutorial Chairs

Ying Zhang	University of Technology Sydney, Australia
Mi-Yen Yeh	Academia Sinica, Taiwan

Workshop Chairs

Chia-Hui Chang	National Central University, Taiwan
Jianliang Xu	Hong Kong Baptist University, Hong Kong
Wen-Chih Peng	National Chiao Tung University, Taiwan

Panel Chairs

Zi Huang	The University of Queensland, Australia
Takahiro Hara	Osaka University, Japan
Shan-Hung Wu	National Tsing Hua University, Taiwan

Ph.D Consortium

Lydia Chen	Delft University of Technology, Netherlands
Kun-Ta Chuang	National Cheng Kung University, Taiwan

Publicity Chairs

Wen Hua	The University of Queensland, Australia
Yongxin Tong	Beihang University, China
Jiun-Long Huang	National Chiao Tung University, Taiwan

Proceedings Chairs

Jen-Wei Huang	National Cheng Kung University, Taiwan
Chih-Ya Shen	National Tsing Hua University, Taiwan

Registration Chairs

Chuan-Ju Wang	Academia Sinica, Taiwan
Hong-Han Shuai	National Chiao Tung University, Taiwan

Sponsor Chair

Chih-Hua Tai	National Taipei University, Taiwan

Web Chairs

Ya-Wen Teng	Academia Sinica, Taiwan
Yi-Cheng Chen	National Central University, Taiwan

Finance Chair

Yi-Ling Chen	National Taiwan University of Science and Technology, Taiwan

Local Arrangement Chairs

Chien-Chin Chen	National Taiwan University, Taiwan
Chih-Chieh Hung	National Chung Hsing University, Taiwan

DASFAA Steering Committee Liaison

Xiaofang Zhou	The Hong Kong University of Science and Technology, Hong Kong

Program Committee

Senior Program Committee Members

Zhifeng Bao	RMIT University, Vietnam
Sourav S. Bhowmick	Nanyang Technological University, Singapore
Nikos Bikakis	ATHENA Research Center, Greece
Kevin Chang	University of Illinois at Urbana-Champaign, USA
Lei Chen	Hong Kong University of Science and Technology, China
Bin Cui	Peking University, China
Xiaoyong Du	Renmin University of China, China
Hakan Ferhatosmanoglu	University of Warwick, UK
Avigdor Gal	Israel Institute of Technology, Israel
Hong Gao	Harbin Institute of Technology, China
Dimitrios Gunopulos	University of Athens, Greece
Bingsheng He	National University of Singapore, Singapore
Yoshiharu Ishikawa	Nagoya University, Japan

Nick Koudas	University of Toronto, Canada
Wei-Shinn Ku	Auburn University, USA
Dik-Lun Lee	Hong Kong University of Science and Technology, China
Dongwon Lee	Pennsylvania State University, USA
Guoliang Li	Tsinghua University, China
Ling Liu	Georgia Institute of Technology, USA
Chang-Tien Lu	Virginia Polytechnic Institute and State University, USA
Mohamed Mokbel	University of Minnesota Twin Cities, USA
Mario Nascimento	University of Alberta, Canada
Krishna Reddy P.	International Institute of Information Technology, India
Dimitris Papadias	The Hong Kong University of Science and Technology, China
Wen-Chih Peng	National Chiao Tung University, Taiwan
Evaggelia Pitoura	University of Ioannina, Greece
Cyrus Shahabi	University of Southern California, USA
Kyuseok Shim	Seoul National University, Korea
Kian-Lee Tan	National University of Singapore, Singapore
Yufei Tao	The Chinese University of Hong Kong, China
Vassilis Tsotras	University of California, Riverside, USA
Jianyong Wang	Tsinghua University, China
Matthias Weidlich	Humboldt-Universität zu Berlin, Germany
Xiaokui Xiao	National University of Singapore, Singapore
Jianliang Xu	Hong Kong Baptist University, China
Bin Yang	Aalborg University, Denmark
Jeffrey Xu Yu	The Chinese University of Hong Kong, China
Wenjie Zhang	University of New South Wales, Australia
Baihua Zheng	Singapore Management University, Singapore
Aoying Zhou	East China Normal University, China
Xiaofang Zhou	The University of Queensland, Australia
Roger Zimmermann	National University of Singapore, Singapore

Program Committee Members

Alberto Abelló	Universitat Politècnica de Catalunya, Spain
Marco Aldinucci	University of Torino, Italy
Toshiyuki Amagasa	University of Tsukuba, Japan
Ting Bai	Beijing University of Posts and Telecommunications, China
Spiridon Bakiras	Hamad Bin Khalifa University, Qatar
Wolf-Tilo Balke	Technische Universität Braunschweig, Germany
Ladjel Bellatreche	ISAE-ENSMA, France
Boualem Benatallah	University of New South Wales, Australia
Athman Bouguettaya	University of Sydney, Australia
Panagiotis Bouros	Johannes Gutenberg University Mainz, Germany

Stéphane Bressan	National University of Singapore, Singapore
Andrea Cali	Birkbeck University of London, UK
K. Selçuk Candan	Arizona State University, USA
Lei Cao	Massachusetts Institute of Technology, USA
Xin Cao	University of New South Wales, Australia
Yang Cao	Kyoto University, Japan
Sharma Chakravarthy	University of Texas at Arlington, USA
Tsz Nam Chan	Hong Kong Baptist University, China
Varun Chandola	University at Buffalo, USA
Lijun Chang	University of Sydney, Australia
Cindy Chen	University of Massachusetts Lowell, USA
Feng Chen	University of Texas at Dallas, USA
Huiyuan Chen	Case Western Reserve University, USA
Qun Chen	Northwestern Polytechnical University, China
Rui Chen	Samsung Research America, USA
Shimin Chen	Chinese Academy of Sciences, China
Yang Chen	Fudan University, China
Brian Chen	Columbia University, USA
Tzu-Ling Cheng	National Taiwan University, Taiwan
Meng-Fen Chiang	Auckland University, New Zealand
Theodoros Chondrogiannis	University of Konstanz, Germany
Chi-Yin Chow	City University of Hong Kong, China
Panos Chrysanthis	University of Pittsburgh, USA
Lingyang Chu	Huawei Technologies Canada, Canada
Kun-Ta Chuang	National Cheng Kung University, Taiwan
Jonghoon Chun	Myongji University, Korea
Antonio Corral	University of Almeria, Spain
Alfredo Cuzzocrea	Universitá della Calabria, Italy
Jian Dai	Alibaba Group, China
Maria Luisa Damiani	University of Milan, Italy
Lars Dannecker	SAP SE, Germany
Alex Delis	National and Kapodistrian University of Athens, Greece
Ting Deng	Beihang University, China
Bolin Ding	Alibaba Group, China
Carlotta Domeniconi	George Mason University, USA
Christos Doulkeridis	University of Piraeus, Greece
Eduard Dragut	Temple University, USA
Amr Ebaid	Purdue University, USA
Ahmed Eldawy	University of California, Riverside, USA
Sameh Elnikety	Microsoft Research, USA
Damiani Ernesto	University of Milan, Italy
Ju Fan	Renmin University of China, China
Yixiang Fang	University of New South Wales, Australia
Yuan Fang	Singapore Management University, Singapore
Tao-yang Fu	Penn State University, USA

Yi-Fu Fu	National Taiwan University, Taiwan
Jinyang Gao	Alibaba Group, China
Shi Gao	Google, USA
Wei Gao	Singapore Management University, Singapore
Xiaofeng Gao	Shanghai Jiaotong University, China
Xin Gao	King Abdullah University of Science and Technology, Saudi Arabia
Yunjun Gao	Zhejiang University, China
Jingyue Gao	Peking University, China
Neil Zhenqiang Gong	Iowa State University, USA
Vikram Goyal	Indraprastha Institute of Information Technology, Delhi, India
Chenjuan Guo	Aalborg University, Denmark
Rajeev Gupta	Microsoft India, India
Ralf Hartmut Güting	Fernuniversität in Hagen, Germany
Maria Halkidi	University of Pireaus, Greece
Takahiro Hara	Osaka University, Japan
Zhenying He	Fudan University, China
Yuan Hong	Illinois Institute of Technology, USA
Hsun-Ping Hsieh	National Cheng Kung University, Taiwan
Bay-Yuan Hsu	National Taipei University, Taiwan
Haibo Hu	Hong Kong Polytechnic University, China
Juhua Hu	University of Washington, USA
Wen Hua	The University of Queensland, Australia
Jiun-Long Huang	National Chiao Tung University, Taiwan
Xin Huang	Hong Kong Baptist University, China
Eenjun Hwang	Korea University, Korea
San-Yih Hwang	National Sun Yat-sen University, Taiwan
Saiful Islam	Griffith University, Australia
Mizuho Iwaihara	Waseda University, Japan
Jiawei Jiang	ETH Zurich, Switzerland
Bo Jin	Dalian University of Technology, China
Cheqing Jin	East China Normal University, China
Sungwon Jung	Sogang University, Korea
Panos Kalnis	King Abdullah University of Science and Technology, Saudi Arabia
Verena Kantere	National Technical University of Athens, Greece
Hung-Yu Kao	National Cheng Kung University, Taiwan
Katayama Kaoru	Tokyo Metropolitan University, Japan
Bojan Karlas	ETH Zurich, Switzerland
Ioannis Katakis	University of Nicosia, Cyprus
Norio Katayama	National Institute of Informatics, Japan
Chulyun Kim	Sookmyung Women's University, Korea
Donghyun Kim	Georgia State University, USA
Jinho Kim	Kangwon National University, Korea
Kyoung-Sook Kim	Artificial Intelligence Research Center, Japan

Seon Ho Kim	University of Southern California, USA
Younghoon Kim	HanYang University, Korea
Jia-Ling Koh	National Taiwan Normal University, Taiwan
Ioannis Konstantinou	National Technical University of Athens, Greece
Dimitrios Kotzinos	University of Cergy-Pontoise, France
Manolis Koubarakis	University of Athens, Greece
Peer Kröger	Ludwig-Maximilians-Universität München, Germany
Jae-Gil Lee	Korea Advanced Institute of Science and Technology, Korea
Mong Li Lee	National University of Singapore, Singapore
Wookey Lee	Inha University, Korea
Wang-Chien Lee	Pennsylvania State University, USA
Young-Koo Lee	Kyung Hee University, Korea
Cheng-Te Li	National Cheng Kung University, Taiwan
Cuiping Li	Renmin University of China, China
Hui Li	Xidian University, China
Jianxin Li	Deakin University, Australia
Ruiyuan Li	Xidian University, China
Xue Li	The University of Queensland, Australia
Yingshu Li	Georgia State University, USA
Zhixu Li	Soochow University, Taiwan
Xiang Lian	Kent State University, USA
Keng-Te Liao	National Taiwan University, Taiwan
Yusan Lin	Visa Research, USA
Sebastian Link	University of Auckland, New Zealand
Iouliana Litou	Athens University of Economics and Business, Greece
An Liu	Soochow University, Taiwan
Jinfei Liu	Emory University, USA
Qi Liu	University of Science and Technology of China, China
Danyang Liu	University of Science and Technology of China, China
Rafael Berlanga Llavori	Universitat Jaume I, Spain
Hung-Yi Lo	National Taiwan University, Taiwan
Woong-Kee Loh	Gachon University, Korea
Cheng Long	Nanyang Technological University, Singapore
Hsueh-Chan Lu	National Cheng Kung University, Taiwan
Hua Lu	Roskilde University, Denmark
Jiaheng Lu	University of Helsinki, Finland
Ping Lu	Beihang University, China
Qiong Luo	Hong Kong University of Science and Technology, China
Zhaojing Luo	National University of Singapore, Singapore
Sanjay Madria	Missouri University of Science & Technology, USA
Silviu Maniu	Universite Paris-Sud, France
Yannis Manolopoulos	Open University of Cyprus, Cyprus
Marco Mesiti	University of Milan, Italy
Jun-Ki Min	Korea University of Technology and Education, Korea

Jun Miyazaki	Tokyo Institute of Technology, Japan
Yang-Sae Moon	Kangwon National University, Korea
Yasuhiko Morimoto	Hiroshima University, Japan
Mirella Moro	Universidade Federal de Minas Gerais, Brazil
Parth Nagarkar	New Mexico State University, USA
Miyuki Nakano	Tsuda University, Japan
Raymond Ng	The University of British Columbia, Canada
Wilfred Ng	The Hong Kong University of Science and Technology, China
Quoc Viet Hung Nguyen	Griffith University, Australia
Kjetil Nørvåg	Norwegian University of Science and Technology, Norway
Nikos Ntarmos	University of Glasgow, UK
Werner Nutt	Free University of Bozen-Bolzano, Italy
Makoto Onizuka	Osaka University, Japan
Xiao Pan	Shijiazhuang Tiedao University, China
Panagiotis Papapetrou	Stockholm University, Sweden
Noseong Park	George Mason University, USA
Sanghyun Park	Yonsei University, Korea
Chanyoung Park	University of Illinois at Urbana-Champaign, USA
Dhaval Patel	IBM TJ Watson Research Center, USA
Yun Peng	Hong Kong Baptist University, China
Zhiyong Peng	Wuhan University, China
Ruggero Pensa	University of Torino, Italy
Dieter Pfoser	George Mason University, USA
Jianzhong Qi	The University of Melbourne, Australia
Zhengping Qian	Alibaba Group, China
Xiao Qin	IBM Research, USA
Karthik Ramachandra	Microsoft Research India, India
Weixiong Rao	Tongji University, China
Kui Ren	Zhejiang University, China
Chiara Renso	Institute of Information Science and Technologies, Italy
Oscar Romero	Universitat Politècnica de Catalunya, Spain
Olivier Ruas	Inria, France
Babak Salimi	University of California, Riverside, USA
Maria Luisa Sapino	University of Torino, Italy
Claudio Schifanella	University of Turin, Italy
Markus Schneider	University of Florida, USA
Xuequn Shang	Northwestern Polytechnical University, China
Zechao Shang	Univesity of Chicago, USA
Yingxia Shao	Beijing University of Posts and Telecommunications, China
Chih-Ya Shen	National Tsing Hua University, Taiwan
Yanyan Shen	Shanghai Jiao Tong University, China
Yan Shi	Shanghai Jiao Tong University, China
Junho Shim	Sookmyung Women's University, Korea

Hiroaki Shiokawa	University of Tsukuba, Japan
Hong-Han Shuai	National Chiao Tung University, Taiwan
Shaoxu Song	Tsinghua University, China
Anna Squicciarini	Pennsylvania State University, USA
Kostas Stefanidis	Tampere University, Finland
Kento Sugiura	Nagoya University, Japan
Aixin Sun	Nanyang Technological University, Singapore
Weiwei Sun	Fudan University, China
Nobutaka Suzuki	University of Tsukuba, Japan
Yu Suzuki	Nara Institute of Science and Technology, Japan
Atsuhiro Takasu	National Institute of Informatics, Japan
Jing Tang	National University of Singapore, Singapore
Lv-An Tang	NEC Labs America, USA
Tony Tang	National Taiwan University, Taiwan
Yong Tang	South China Normal University, China
Chao Tian	Alibaba Group, China
Yongxin Tong	Beihang University, China
Kristian Torp	Aalborg University, Denmark
Yun-Da Tsai	National Taiwan University, Taiwan
Goce Trajcevski	Iowa State University, USA
Efthymia Tsamoura	Samsung AI Research, Korea
Leong Hou U.	University of Macau, China
Athena Vakal	Aristotle University, Greece
Michalis Vazirgiannis	École Polytechnique, France
Sabrina De Capitani di Vimercati	Università degli Studi di Milano, Italy
Akrivi Vlachou	University of the Aegean, Greece
Bin Wang	Northeastern University, China
Changdong Wang	Sun Yat-sen University, China
Chaokun Wang	Tsinghua University, China
Chaoyue Wang	University of Sydney, Australia
Guoren Wang	Beijing Institute of Technology, China
Hongzhi Wang	Harbin Institute of Technology, China
Jie Wang	Indiana University, USA
Jin Wang	Megagon Labs, Japan
Li Wang	Taiyuan University of Technology, China
Peng Wang	Fudan University, China
Pinghui Wang	Xi'an Jiaotong University, China
Sen Wang	The University of Queensland, Australia
Sibo Wang	The Chinese University of Hong Kong, China
Wei Wang	University of New South Wales, Australia
Wei Wang	National University of Singapore, Singapore
Xiaoyang Wang	Zhejiang Gongshang University, China
Xin Wang	Tianjin University, China
Zeke Wang	Zhejiang University, China
Yiqi Wang	Michigan State University, USA

Raymond Chi-Wing Wong	Hong Kong University of Science and Technology, China
Kesheng Wu	Lawrence Berkeley National Laboratory, USA
Weili Wu	University of Texas at Dallas, USA
Chuhan Wu	Tsinghua University, China
Wush Wu	National Taiwan University, Taiwan
Chuan Xiao	Osaka University, Japan
Keli Xiao	Stony Brook University, USA
Yanghua Xiao	Fudan University, China
Dong Xie	Pennsylvania State University, USA
Xike Xie	University of Science and Technology of China, China
Jianqiu Xu	Nanjing University of Aeronautics and Astronautics, China
Fengli Xu	Tsinghua University, China
Tong Xu	University of Science and Technology of China, China
De-Nian Yang	Academia Sinica, Taiwan
Shiyu Yang	East China Normal University, China
Xiaochun Yang	Northeastern University, China
Yu Yang	City University of Hong Kong, China
Zhi Yang	Peking University, China
Chun-Pai Yang	National Taiwan University, Taiwan
Junhan Yang	University of Science and Technology of China, China
Bin Yao	Shanghai Jiaotong University, China
Junjie Yao	East China Normal University, China
Demetrios Zeinalipour Yazti	University of Cyprus, Turkey
Qingqing Ye	The Hong Kong Polytechnic University, China
Mi-Yen Yeh	Academia Sinica, Taiwan
Hongzhi Yin	The University of Queensland, Australia
Peifeng Yin	Pinterest, USA
Qiang Yin	Alibaba Group, China
Man Lung Yiu	Hong Kong Polytechnic University, China
Haruo Yokota	Tokyo Institute of Technology, Japan
Masatoshi Yoshikawa	Kyoto University, Japan
Baosheng Yu	University of Sydney, Australia
Ge Yu	Northeast University, China
Yi Yu	National Information Infrastructure Enterprise Promotion Association, Taiwan
Long Yuan	Nanjing University of Science and Technology, China
Kai Zeng	Alibaba Group, China
Fan Zhang	Guangzhou University, China
Jilian Zhang	Jinan University, China
Meihui Zhang	Beijing Institute of Technology, China
Xiaofei Zhang	University of Memphis, USA
Xiaowang Zhang	Tianjin University, China
Yan Zhang	Peking University, China
Zhongnan Zhang	Software School of Xiamen University, China

Pengpeng Zhao	Soochow University, Taiwan
Xiang Zhao	National University of Defence Technology, China
Bolong Zheng	Huazhong University of Science and Technology, China
Yudian Zheng	Twitter, USA
Jiaofei Zhong	California State University, East, USA
Rui Zhou	Swinburne University of Technology, Australia
Wenchao Zhou	Georgetown University, USA
Xiangmin Zhou	RMIT University, Vietnam
Yuanchun Zhou	Computer Network Information Center, Chinese Academy of Sciences, China
Lei Zhu	Shandong Normal Unversity, China
Qiang Zhu	University of Michigan-Dearborn, USA
Yuanyuan Zhu	Wuhan University, China
Yuqing Zhu	California State University, Los Angeles, USA
Andreas Züfle	George Mason University, USA

External Reviewers

Amani Abusafia
Ahmed Al-Baghdadi
Balsam Alkouz
Haris B. C.
Mohammed Bahutair
Elena Battaglia
Kovan Bavi
Aparna Bhat
Umme Billah
Livio Bioglio
Panagiotis Bozanis
Hangjia Ceng
Dipankar Chaki
Harry Kai-Ho Chan
Yanchuan Chang
Xiaocong Chen
Tianwen Chen
Zhi Chen
Lu Chen
Yuxing Chen
Xi Chen
Chen Chen
Guo Chen
Meng-Fen Chiang
Soteris Constantinou
Jian Dai

Sujatha Das Gollapalli
Panos Drakatos
Venkatesh Emani
Abir Farouzi
Chuanwen Feng
Jorge Galicia Auyon
Qiao Gao
Francisco Garcia-Garcia
Tingjian Ge
Harris Georgiou
Jinhua Guo
Surabhi Gupta
Yaowei Han
Yongjing Hao
Xiaotian Hao
Huajun He
Hanbin Hong
Xinting Huang
Maximilian Hünemörder
Omid Jafari
Zijing Ji
Yuli Jiang
Sunhwa Jo
Seungwon Jung
Seungmin Jung
Evangelos Karatzas

Enamul Karim
Humayun Kayesh
Jaeboum Kim
Min-Kyu Kim
Ranganath Kondapally
Deyu Kong
Andreas Konstantinidis
Gourav Kumar
Abdallah Lakhdari
Dihia Lanasri
Hieu Hanh Le
Suan Lee
Xiaofan Li
Xiao Li
Huan Li
Pengfei Li
Yan Li
Sizhuo Li
Yin-Hsiang Liao
Dandan Lin
Guanli Liu
Ruixuan Liu
Tiantian Liu
Kaijun Liu
Baozhu Liu
Xin Liu
Bingyu Liu
Andreas Lohrer
Yunkai Lou
Jin Lu
Rosni Lumbantoruan
Priya Mani
Shohei Matsugu
Yukai Miao
Paschalis Mpeis
Kiran Mukunda
Siwan No
Alex Ntoulas
Sungwoo Park
Daraksha Parveen
Raj Patel
Gang Qian
Jiangbo Qian
Gyeongjin Ra

Niranjan Rai
Weilong Ren
Matt Revelle
Qianxiong Ruan
Georgios Santipantakis
Abhishek Santra
Nadine Schüler
Bipasha Sen
Babar Shahzaad
Yuxin Shen
Gengyuan Shi
Toshiyuki Shimizu
Lorina Sinanaj
Longxu Sun
Panagiotis Tampakis
Eleftherios Tiakas
Valter Uotila
Michael Vassilakopoulos
Yaoshu Wang
Pei Wang
Kaixin Wang
Han Wang
Lan Wang
Lei Wang
Han Wang
Yuting Xie
Shangyu Xie
Zhewei Xu
Richeng Xuan
Kailun Yan
Shuyi Yang
Kai Yao
Fuqiang Yu
Feng (George) Yu
Changlong Yu
Zhuoxu Zhang
Liang Zhang
Shuxun Zhang
Liming Zhang
Jie Zhang
Shuyuan Zheng
Fan Zhou
Shaowen Zhou
Kai Zou

Contents – Part III

Recommendation

Emerging Applications

Industrial Papers

Demo Papers

Ph.D Consortium

Tutorials

Recommendation

Gated Sequential Recommendation System with Social and Textual Information Under Dynamic Contexts

Haoyu Geng, Shuodian Yu, and Xiaofeng Gao$^{(\boxtimes)}$

Shanghai Key Laboratory of Scalable Computing and Systems,
Department of Computer Science and Engineering, Shanghai Jiao Tong University,
Shanghai, China
{genghaoyu98,timplex233}@sjtu.edu.cn, gao-xf@cs.sjtu.edu.cn

Abstract. Recommendation systems are undergoing plentiful practices in research and industry to improve consumers' satisfaction. In recent years, many research papers leverage abundant data from heterogeneous information sources to grasp diverse preferences and improve overall accuracy. Some noticeable papers proposed to extract users' preference from information along with ratings such as reviews or social relations. However, their combinations are generally static and less expressive without considerations on dynamic contexts in users' purchases and choices.

In this paper, we propose *Heterogeneous Information Sequential Recom-mendation System (HISR)*, a dual-GRU structure that builds the sequential dynamics behind the customer behaviors, and combines preference features from review text and social attentional relations under dynamics contexts. A novel gating layer is applied to dynamically select and explicitly combine two views of data. Moreover, in social attention module, temporal textual information is brought in as a clue to dynamically select friends that are helpful for contextual purchase intentions as an implicit combination. We validate our proposed method on two large subsets of real-world local business dataset Yelp, and our method outperforms the state of the art methods on related tasks including social, sequential and heterogeneous recommendations.

Keywords: Recommendation systems · Sequential recommendation · Gating mechanism

1 Introduction

With the emergence of online service websites, recommendation systems have become a vital technology for companies. Recommendation systems

This work was supported by the National Key R&D Program of China [2020YFB1707903]; the National Natural Science Foundation of China [61872238, 61972254], the Huawei Cloud [TC20201127009], the CCF-Tencent Open Fund [RAGR20200105], and the Tencent Marketing Solution Rhino-Bird Focused Research Program [FR202001]. This paper is completed when Haoyu Geng is an undergraduate student in Advanced Network Laboratory supervised by Prof. Xiaofeng Gao.

C. S. Jensen et al. (Eds.): DASFAA 2021, LNCS 12683, pp. 3–19, 2021.
https://doi.org/10.1007/978-3-030-73200-4_1

provide better service for customers while promoting their products and also attracting more purchases. Particularly, there are more and more diverse forms that customers could interact with the products, such as click history, images, social relations, review texts, which provide traces to better grasp customers'preference for better recommendation accuracy. With increasingly abundant interaction modes and heterogeneous data, cross-domain recommendation becomes promising in many future applications, especially for large online service providers which provide multiple services and share the user profiles.

A common practice to achieve cross-domain recommendation with heterogeneous information is combining rating records with another. [3,25] leverage social relations to help with matrix factorization. [10] takes temporal information into user purchase history to build sequential dynamics of users' preferences. [12] combines more than two information sources for better performance. Moreover, multiple information source could alleviate the data sparsity problems, therefore the accuracy for side users are improved significantly.

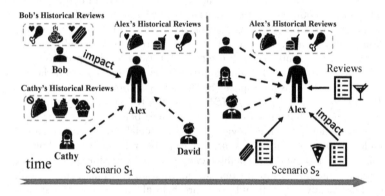

Fig. 1. Illustration of diverse online behaviors in scenario S_1 and S_2. Alex is prone to select restaurants by friends in S_1, while he would read reviews by himself in S_2. Alex is more likely to refer to Bob on the drumstick, while he would refer to Cathy on fruit.

However, most recent works failed to model the preference dynamics under varying purchase contexts of users. Figure 1 illustrates an example for such phenomenon. User Alex has different decision patterns in different scenarios S_1 and S_2. In S_1, he decides his purchase mostly on friends' recommendations; while in S_2, he would turn to crowd-sourcing reviews since none of his friends have bought his intended items before. Moreover, when referring to the friends, Alex will not take all the social connections equally. For example in S_1, Alex's friend Bob is familiar with local businesses selling drumstick, noodles and sand-witches, while his another friend Cathy comes to eat taco, fruit and cakes from time to time. Certainly, Alex may refer to Bob when looking for something like drumstick to eat, and turn to Cathy for fruit recommendations at another time. Both diversions of interests mentioned above come from evolution of temporal purchase

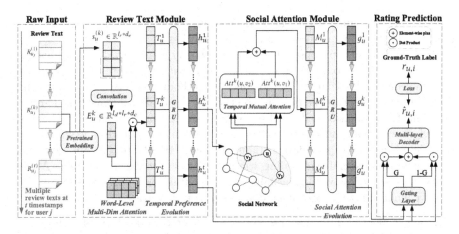

Fig. 2. Illustration of **HISR**. Textual and social information are organized in sequential view, and review text module and social attention module are aggregated by a novel gating layer to produce hybrid rating predictions.

intentions, and algorithms for better preference evolution modeling is in need for explicit and implicit combination of multiple views of data.

Based on the analysis above, we would address three limitations that hinder further exploitation of abundant information. Firstly, though recent research combines multiple information sources, the preference of users are dynamic with respect to contexts, rather than static. Secondly, some research works have noticed distinct attention to users' friends, while few of them find clues from textual reviews to implicitly enhance social attentions. Thirdly, for explicit module combination, previous methods usually take fixed strategies like concatenations or multiplications, which outdate in dynamical scenarios. Therefore, we would follow three lines of thoughts to address these issues respectively. We design user preference evolution in two views to tackle temporal contexts, extract temporal purchase intentions from textual clues to enhance social attention for implicit combination, and apply gating strategy that decides the extent and proportion for explicit combination.

In this paper, we propose **H**eterogeneous **I**nformation **S**equential **R**ecommendation System (**HISR**), a dual-GRU structure that combines review text module and social attention module by a novel gating mechanism, and models evolution of users' preference under dynamic contexts (as shown in Fig. 2). In the review text module, we would extract semantic information from texts by convolutional operations, and multi-dimension word attention layer reveals users' interests from multiple aspects. In the social attention module, it takes extracted textual representations as indicators, and measures the similarity score between users' current interests with his/her friends' for implicit combination into social attention. Both modules take advantages of GRU structures [5] for dynamic modeling, where long-term textual preferences are learnt in the former module, and short-term purchase intentions are utilized in the latter. Finally, the two modules are combined with a novel gating layer to select the informative parts dynamically for explicit context-aware combination.

We validate our proposed method on two large real-world local business datasets of Yelp[1]. Comparison of our methods with the state of the art baselines in sequential, social and heterogeneous methods respectively indicates that our results outperforms the baselines on three major evaluation metrics. Through ablation study, we also verify that our improvements come from effectively grasp two views of data, and the explicit gating combination strategy is superior to traditional ones. We provide a variant of model, HISR-SIM, as offline version with pretrained social embedding that are more effective in training. A case study is presented as well to visualize our sequential modeling results.

In conclusion, the major contributions of this paper are:

- Bring sequential dynamics into heterogeneous recommendations with textual and social information, which constructs long-term preferences and short-term intentions.
- Propose social attention module based on similarity measure of temporal purchase intentions to deal with dynamic contexts.
- Apply a gating layer for explicit combination for two views of data, which fits with heterogeneous dynamic model with more expressiveness.

2   Problem Formalization

Let $U = \{u_1, u_2, \ldots, u_m\}$ denote the user set of m users, and $I = \{i, i_2, \cdots, i_n\}$ denote the item set of n items. The set of reviews $R = \{R_{u_1}, R_{u_2}, \ldots, R_{u_m}\}$ is consist of reviews R_{u_j} for each user u_j. In $R_{u_j} = \left\{ R_{u_j}^{(1)}, R_{u_j}^{(2)}, \ldots, R_{u_j}^{(t)} \right\}$, $R_{u_j}^{(k)}$ is the review text for user j at timestamp k ($k \in \{1, 2, 3, \ldots, t\}$). The rating set is $r = \{r_{u_1}, r_{u_2}, \cdots, r_{u_m}\}$, where $r_{u_j} = \{r_{u_j, i_1}, r_{u_j, i_2} \ldots, r_{u_k, i_n}\}$ are the ratings for user u_j for items $i_1, i_2, \ldots, i_n$. Social relations among users are represented by matrix $S \in \mathbb{R}^{m \times m}$. $S_{uv} = 1$ indicates u trusts v or u regards v as his/her friend, otherwise $S_{uv} = 0$.

Definition 1. *Given user set U, item set I, review set R, social relation matrix S, and rating set r, our top-N recommendation task is to learn a scoring function $\phi : U \times I \times R \times r \times S \to \mathbb{R}$ from the dataset that replicates for each user its perfect ranking.*

3   Proposed Method

As is shown in Fig. 2, our proposed **HISR** consists of two components: review text module and social attention module. Both modules take advantages of GRU units to model the purchase dynamics and evolution, and two-side representations are combined by a novel gating layer as the final user representation for rating predictions and top-N recommendations.

---

[1] https://www.yelp.com/dataset.

3.1 Review Text Module

In the review text module, we aim to take advantage of reviews to learn long-term purchase preferences as well as short-term intentions behind user behaviors. Particularly, we hope to extract informative insights of user preferences, which is useful for implicit and explicit combinations with social attention module.

Word Embedding Lookup. In the lookup layer, we would transform the words into vector representations. Since it is not our main focus, we adopt pretrained word embedding model GloVe [15] as our word embedding. Suppose the single review text R_u^k for user u at timestamp k is consist of words $[\cdots, w_{i-1}, w_i, \cdots]$. We could look up in the pretrained word embedding for each word w_i, and concatenate them to form a word embedding matrix $s_u^{(t)} \in R^{l_r \times d_e}$ where l_r is the length of review text and d_e is the dimension of embedding word. Review texts with more than l_r words are clipped, and with less than l_r words are padded, so that each review contains exactly l_r words.

Convolutional Operations. Convolutional operations in papers [11,21,23] are utilized to extract contextual representations. Consecutive sentences in the review and consecutive words in sentences are highly related, therefore we use it to link the relevance among contextual words and produce the semantic information in user's review histories. The semantic extraction layer is put on each review to get w_i^c as the contextual representation for i-th word, denoting as:

$$w_{h,j}^c = \sigma(s_{h-\frac{d_c}{2}:h+\frac{d_c}{2}}), \tag{1}$$

where $s_{h-\frac{d_c}{2}:h+\frac{d_c}{2}}$ is word sequence slice centering at h with sliding window length d_c, $w_{h,j}^c \in R^1$ is contextual representation for j-th convolutional filter, and σ is the non-linear activation function. $w_i^c = [\cdots, w_{i,j-1}^c, w_{i,j}^c, w_{i,j+1}^c, \cdots]$, $w_i^c \in R^{d_c}$, d_c is the number of convolutional filters. Since different users have distinct pieces of reviews on their documents, for simplicity, we set each user have same number of reviews by clipping the latest reviews, and padding empty ones. The user review embedding is $E_u^k \in R^{l_d \times l_r \times d_c}$ at timestamp k, where l_d is the number of reviews in each user's document.

Word-Level Multi-Dimensional Attention Mechanism. Attention Mechanism has been proved effective in multiple tasks in recent years [1,17,18]. As we are inspired by the advances in review-related recommenders in [12,23], words in the reviews do not always contain the same amount of information. Moreover, the one-dimensional attentional weight could not grasp the diverse preference that is revealed in the textual reviews. Therefore, we adopt the multi-dimensional attention to avoid overly focus on one specific aspect of content. Specifically, the attention mechanism is put on each review embedding representation E_u^k, and the attention score $A_u^k \in R^{l_r \times d_a}$ is defined as:

$$A_u^k = \mathbf{softmax}\left(\tanh\left(W_c E_u^k + b_c\right) W_a + b_a\right), \tag{2}$$

where d_a is the dimension of attention, $W_c \in R^{d_c \times d_c}$ is the weight matrix for convolutional representation, and $b_a \in R^{d_c \times d_c}$ is the bias term. As the max pooling operation tends to lose information from the text when it overestimates the informative words and underestimates others, we takes the attentive sum of the word contextual embedding as the representation of each review. With the activation function tanh($\cdot$), multi-dimensional weight matrix $W_a \in R^{d_c \times d_a}$ is put on the review representations:

$$p_u^k = A_u^k E_u, \quad T_u^k = p_u^k W_{ag}, \tag{3}$$

where $p_u^k \in R^{d_c \times d_a}$ is the attentional representation of review at timestamp k, $W_{ag} \in R^{d_a}$ is the aggregation vector that combines attentions of multiple aspects into one vector. Therefore, $T_u^k \in R^{d_c}$ represents each review in the user u's document at timestamp k.

Temporal Preference Evolution. With the information we grasped through the text processing and modeling above, we further propose to build the preference evolution layer to dynamically grasp the long-term purchase habits as well as short-term purchase intentions. Recurrent neural networks are widely-used to build the sequential relations, specifically, we adopt Gated Recurrent Unit (GRU) [5] to construct the evolution for user text modeling:

$$h_u^k = GRU\left([h_u^{k-1}, T_u^k]\right). \tag{4}$$

The hidden state $h_u^k \in R^{d_h}$ at the last timestamp t will be regarded as text-view long-term user embedding as it preserves user preferences for all times, and the hidden state h_u^k ($k \in \{1, 2, ..., t\}$) at corresponding timestamps will be reckoned as short-term purchase intentions, which reflects fine-grained temporal preference of users.

3.2 Social Attention Module

While the review text module extracts purchase preferences, in some cases the sparsity of review texts and inactivity of user feed-backs on online services pose difficulties. We propose social attention module to alleviate this issue, moreover, to combine the short-term purchase intentions with temporal social reference as indications for attention strength. Additional information from social networks would be a promising complement for comprehensive user understanding.

Social Embedding Lookup. To characterize the preferences of each user in the social network, we introduce the long-term user preference representation in review text module h_u^t as user embedding of user u:

$$e_u = h_u^t. \tag{5}$$

Note that there is alternatives for lookup layer, pretrained network embedding, which will be further elaborated in model discussions.

Temporal Mutual Attention. In order to grasp the evolution of trust for different friends under dynamic scenarios, we need to bring in attention into temporal contexts. Recent trends in attention mechanism effectively assign distinct strengths to different friends. The attention would be bi-directional, therefore mutual attention indicates relevant attention between two people. We define $Att^k(u, v)$ as the mutual attention score from user u to user v at timestamp k, and it is measured by the similarity between convergence of users' long-term preference and short-term intention and that of their friends at the corresponding timestamp:

$$Att^k(u, v) = \cos\left(h_u^k \odot e_u, \quad h_u^k \odot e_v\right), \tag{6}$$

where $\odot$ is the element-wise production, which enhances users' temporal purchases and weaken his preference that is irrelevant to the context.

Hybrid User Embedding with Social Regularization. With the text-aware mutual attention, we could get hybrid user social embedding. With user v in friends $S(u)$ of user u, the hybrid user embedding M_u^k is the weighted sum of user embedding of user u at timestamp k, which aggregates the influence of social relations with distinct strength:

$$M_u^k = \sum_{v \in S(u)} Att^{(k)}(u, v)e_v. \tag{7}$$

Social Attention Evolution. The attentions towards friends with respect to time are not independent; instead, it is accumulative and based on previous knowledge. Therefore, we would build the dual structure of gated recurrent unit in social attention module alone with previous module to construct evolution on cognition and impressions of users towards friends. g_u^k for user u at timestamp k indicates hidden state of social attention module, which lies in the same space with h_u^k in review text module, defined as:

$$g_u^k = \mathrm{GRU}\left([g_u^{k-1}, M_u^k]\right), \tag{8}$$

where $g_u^k \in \mathbf{R}^{d_h}$, and g_u^t at the last timestamp t would be seen as comprehensive social representation as it aggregates the evolution of all times.

3.3   Gating Layer

Our both modules deal with dynamic contexts with sequential model, therefore, the combination of the two need to fit with the dynamic nature and effectively distill informative parts in two view of data. Instead of taking the prevalent practices for representation combination, we apply a gating layer for a novel combination of text-view and social-view user representations:

$$\begin{aligned} G &= \mathrm{sigmoid}\left(W_1 h_u^t + W_2 g_u^t + b\right), \\ z_u &= G \odot h_u^t + (1 - G) \odot g_u^t, \end{aligned} \tag{9}$$

where $W_1 \in \mathbf{R}^{d_h \times d_h}, W_2 \in \mathbf{R}^{d_h \times d_h}$ are weightings and $b \in \mathbf{R}^{d_h}$ is bias term.

Gating mechanism is widely used in many fields [9,12], and it has expressive power by deciding how much is carried out from input to output. However, it has not been fully exploited in heterogeneous recommendations. In our task, it is particularly attracting when gating layer dynamically decides the way and proportion to combine two views of information, especially in the dynamic contexts. As we expect it to grasp users' preference over social view and textual view, the gating mechanism outshines other simple combination strategies and we would further compare these in experiments.

3.4   Decode Layer for Rating Prediction

The decode layer is a multi-layer perceptron to decode the low-dimension latent vector to reconstruct the rating predictions:

$$r_u = \text{decode}(z_u), \tag{10}$$

where $z_u \in \boldsymbol{R}^n$ is the rating predictions over n items for user u. As the decode layer is a multi-layer neural network, it is to recover the user ratings from the dense user embedding, where there is a linear transformation and tanh activation for each layer respectively.

3.5   Training

The loss function is root-mean-square error with L2-norm regularization term:

$$J = \sum_{(u,i)\in U\times I} \mathbf{1}_{r_{u,i}\in r} \left(\hat{r}_{u,i} - r_{u,i}\right)^2 + \lambda_{\Theta}\|\Theta\|^2, \tag{11}$$

where Θ is collection of all parameters, and λ_{Θ} is the parameter of regularization.

Finally, we produce the top-N ranked list according to the rating predictions.

4   Experiments

4.1   Datasets

We adopt Yelp[2] as our dataset. Yelp is a large collection of local business reviews allowing users to rate items, browse/write reviews, and connect with friends. We extracted two subsets on Ontario province in Canada and North Carolina state in US. All the datasets are processed to ensure that all items and users have at least five interactions (ratings and reviews). The statistical details of these datasets are presented in Table 1.

---

[2] https://www.yelp.com/dataset.

Table 1. Statistics of Yelp datasets

Datasets	Yelp_ON	Yelp_NC
# of users	27625	14780
# of items	18840	7553
# of ratings	553045	235084
# of of Density (ratings)	0.1063%	0.2105%
# of reviews	553045	235084
# of Social Relations	173295	75408
# of Density (social relations)	0.0333%	0.0675%

4.2 Baselines

To evaluate the performance on the recommendation task, we compared our model with the following classical and state-of-the-art recommendation methods.

- **PMF** [13]: **Probabilistic Matrix Factorization** is a classical matrix factorization variant that performs well on the large, sparse, imbalanced dataset.
- **BPR** [16]: **Bayesian Personalized Ranking** learns personalized rankings from implicit feedback for recommendations.
- **SBPR** [25]: uses social connections to derive more accurate ranking-based models by Bayesian personalized ranking.
- **NCF** [8]: **Neural Collaborative Filtering** is a recently proposed deep learning based framework that combines matrix factorization (MF) with a multilayer perceptron (MLP) for top-N recommendation.
- **SASRec** [10]: **Self-Attentive Sequential Recommendation** is the state of the art self-attention based sequential model that captures long-term semantics. At each time step, SASRec seeks to identify which items are 'relevant' from a user's action history, and predict the next item accordingly.
- **SAMN** [3]: **Social Attentional Memory Network** is a state of the art method in social recommendation. The friend-level model based on an attention-based memory captures the varying aspect attentions to his different friends and adaptively select informative friends for user modeling.
- **GATE** [12]: **Gated Attentive-autoencoder** is state of the art method for heterogeneous recommendation. It learns hidden representations of item's contents and binary ratings through a neural gating structure, and exploits neighboring relations between items to infer users' preferences.

4.3 Experimental Setup

Evaluation Metrics: We set four metrics: $Precision@K$ measures what proportion of positive identifications is actually correct. $NDCG@K$ accounts for the position of hits by assigning higher scores to hits at top ranks, while F1 score measures precision and recall comprehensively.

Data Processing: We set our experiments under strong generalization: split all users into training/validation/testing sets, and recommend for unseen users in test stage. During testing, 70% of historical records are extracted for the held-out users to learn necessary representations of the model, and are used to recommend for the next 30% records. Strong generalization is more difficult than weak generalization taken in baseline methods, where proportions of historical records are held out for all users in testing.

Parameters: The parameter settings and hyper-parameters tuning of baselines follow their corresponding paper to get their optimal performance. The parameters of ours are set with batch size 128, learning rate 0.01 with dropping half every 5 epochs. The dimension of word embedding d_e is 50 by default. In review text processing, each sentence is set to length $l_r = 40$, and each user document contains $l_d = 50$ reviews. The number of convolutional filters (d_c) is 64, and the dimension of RNN hidden vector d_h is 100. The decoder layer is three layer fully connected neural networks, and the number of neurons is 600, 200 and 1. The activation function for the neural networks are all ReLU, and that for all softmax layers is tanh. The regularization parameter 0.9 in the objective function. The model is trained for 30 epochs by default.

4.4   Performance Comparison

Full results and comparisons are demonstrated in Table 2, and we have several observations regarding the result table.

Firstly, As our anticipation, the performance on NC subset is better than ON mainly because of lower sparsity in social relations and reviews. The density of yelp NC subset is twice as that of ON, while our model could address sparsity issues and narrows the gap between two datasets.

Table 2. Performance comparison with baselines

Method	Yelp_ON			Yelp_NC		
	F1	P@10	NDCG@10	F1	P@10	NDCG@10
PMF [13]	0.000379	0.000298	0.000346	0.000590	0.000414	0.000456
BPR [16]	0.011038	0.007716	0.015311	0.019033	0.012602	0.027625
SBPR [25]	0.007566	0.004781	0.012324	0.014571	0.008900	0.024261
NCF [8]	0.011732	0.008365	0.015314	0.015886	0.010643	0.021759
SASRec [10]	0.012565	0.008726	0.016371	<u>0.025737</u>	<u>0.019216</u>	0.031811
SAMN [3]	<u>0.013633</u>	0.008767	<u>0.019660</u>	0.021426	0.013217	<u>0.035317</u>
GATE [12]	0.014123	<u>0.011794</u>	0.012770	0.021861	0.014468	0.016295
HISR-SIM	0.018735	0.016894	0.017594	0.027747	0.026890	0.027989
HISR	0.019145	0.019392	0.020303	0.028181	0.025437	0.035620
Improvement	35.55%	64.42%	3.27%	9.49%	32.37%	0.85%

Note: Underline here represents the best result that the baseline achieves on the corresponding metric.

Secondly, as the heterogeneous method, we argue that our recommender is better than that for single information source. Moreover, our results are better than the recent heterogeneous methods GATE, which utilizes gating layer for information combination, but fails to model the contextual dynamics. Furthermore, our proposed similarity metric to build mutual social attention could better address the issues in text modules and makes them a deeper combination.

Thirdly, for social recommenders, SBPR leverages social connections, however, it fails to build the sequential dynamics. Even though SAMN captures social connections with attention mechanism, lack of social data for side users could limit its potential. SAMN has relatively good ranking quality with high NDCG, but comes with lower prediction accuracy.

Finally, as sequential recommender, SASRec has achieved great efficiency. Self attention structure is validated to be powerful in many tasks, however, empirical practices indicates that recurrent network still achieves comparable results for sequential dynamics modeling.

5   Model Discussions

5.1   Variant of the Model

Our proposed model takes output of textual review module as user embedding for social attention module. However, this embedding technique only leverages one-hop local connection As network embedding techniques could grasp highly non-linear global network structure, we adopt the work of SDNE [19] as pretrained social embedding of social attention module. The variant of the model is denoted as HISR-SIM. The pretrained user embedding is not necessarily on the same embedding space with the review module, so we apply a linear transformation.

As illustrated in Table 2, HISR-SIM is comparable to original HISR. Note that as user embeddings are pretrained, it saves much time and GPU memory during the training process, and the performance is more stable than HISR. HISR-SIM identifies the generalizability, and is more suitable for offline version.

5.2   Ablation Study

We set our ablation in two aspects and four variants. We argue that not only multiple information sources cover the side users for better recommendation accuracy, but the gating layer for explicit embedding combination improves the overall performance as well, as shown in Fig. 3.

Firstly, to validate the effectiveness of combination of multiple information sources, we would conduct split experiments on social component and review component, denoted as HISR-Text and HISR-Social, respectively. We block the one of the modules, and take another module to decode and recommend directly. Note that as the social attention component utilizes the representations in the review text module, there is no explicit combination of the two. As is shown, HISR-Text achieves better performance on four metrics than HISR-Social. Since

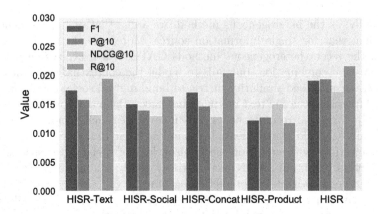

Fig. 3. Ablation study

the performance of both modules are lower than heterogeneous one, we argue that two modules address different aspects in users' preferences respectively, while combination of the two covers the side users for both modules.

Secondly, we would validate effectiveness of our gating layer for explicit combination. For comparison, we take two mostly adopted methods for vector combination: Hadamard product (element-wise multiplication) and concatenation. As is shown in Fig. 2, both the combination strategies fail to outperform our gating strategy. Concatenation presents more power than Hadamard product, especially on recall. However, neither of the alternative combination strategies could grasp the side preferences for side users as gating mechanism does.

5.3   Hyper-Parameter Sensitivity Analysis

To analyze reproducibility and generalization of HISR, we evaluate influence of hyper-parameters. We take four representative ones, word embedding size, regularization weight, hidden state size in RNN, and attention dimensions, for more experiments. The default values are 50, 0.9, 100 and 30, respectively.

Word Embedding Size. The pretrained word embedding comes from Wikipedia 2014 corpus[3]. In Fig. 4(a), the performance on four metrics drops slightly with the increasing embedding size. The choice of embedding size has relatively little influence on final results, so the short embedding is good enough while saving memory and training time in practice.

Regularization Term Weight. Empirical practices indicate that weight near 0.9 leads to good results. In Fig. 4(b), smaller values of regularization takes lower results, probably falls in overfitting. When the regularization increases, NDCG increases while other metrics decreases. for model perform stability, we need to set a medium regularization to avoid overfitting or underfitting.

---

[3] https://nlp.stanford.edu/projects/glove/.

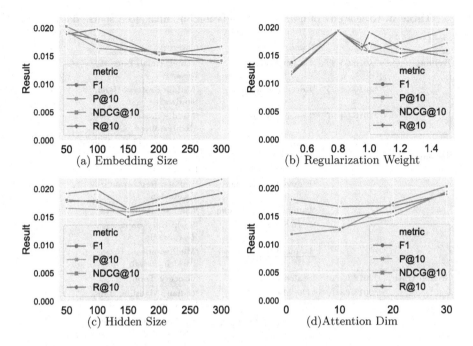

Fig. 4. Performance w.r.t different hyper-parameter settings

Hidden Size in Recurrent Neural Networks. Figure 4(c) compares results for the experiments. The result of recall increases as the hidden size increases, probably because higher dimension of user representation is more expressive and makes our model to retrieve desired business from candidate list.

Dimension in Word Attention. Figure 4(d) illustrates that NDCG increases significantly when the dimension increases, indicating that the desired items are ranked top. Precision is improved as well, showing probably that our model could better distinguish unfavorable businesses from favorable ones.

5.4   Case Study

To present our results more explicitly, we extract a recommendation scenario shown in Table 3. We list training part and ground truth for an anonymous user, namingly Alex in Yelp dataset. Alex is a representative user, as the other half, who has only a few pieces of reviews and lacks social contacts on online platforms. Scenarios like this would be generally difficult for recommenders, even impossible for pure social recommendations. We retrieve top 5 model outputs. Note that the business names and categories are all invisible to the model, and they are only for demonstration in the case study.

As is shown in Table 3, the samples in the dataset are put in the sequential order. We could guess from the information that Alex generally takes reviews on restaurants, and he maybe likes Asian food (and occasionally European food).

Table 3. Case study of user recommendation on imbalanced sparse data

	Date	Name	Category	Rate
Training	'2016-03-30'	'Burger Factory'	' Food, Halal, Burgers, Desserts'	5.0
	'2016-04-05'	'Reyan Restaurant'	'Middle Eastern, Halal, Sandwiches'	5.0
	'2016-04-07'	'Roman Zaman'	'Middle Eastern, Mediterranean, Greek'	5.0
	'2017-10-03'	'Braiseryy'	'Brazilian, Chicken Wings'	5.0
Recommended	–	'Schnitzel Queen'	'Sandwiches, German, Restaurants'	–
	–	'Blossom Vegetarian'	'Buffets, Restaurants, Vegetarian'	–
	–	'Dumplings&Szechuan's'	'Chinese, Restaurants'	–
	–	'Cece Sushi'	'Sushi Bars, Japanese, Restaurants'	–
	–	'Istanbul Kebab House'	'Mediterranean, Pizza, Turkish, Kebab'	–
Ground truth	'2018-03-10'	'Seafood City'	'Grocery, Food'	2.0
	'2018-03-10'	'The Maharaja'	'Indian, Nightlife, Bars, Pakistani'	5.0
	'2018-03-15'	'Istanbul Kebab House'	'Mediterranean, Pizza, Turkish, Kebab'	5.0

He puts full stars for the restaurants he went in 2016 and 2017. As is shown in our top-5 recommendations, Chinese and Japanese food are recommended, as well as German restaurants. There is a hit on Mediterranean's food, where Alex later comes to this restaurant and leaves full stars.

6   Related Work

Recommendation with Review Texts. With the advances in natural language processing, there are some works that take advantages of the review texts to learn user preferences. [23] proposes to take reviews to model the latent factor for user-item pairs instead of a static representation. [20] used a joint tensor factorization method of two tasks: user preference modeling for recommendation and opinionated content modeling for explanation. [7] extracts aspects from user reviews, and put attentive models to aspects and friends. [2,11,23] uses convolutional neural networks to extract semantic information from reviews. [20] uses sentiment analysis and joint tensor factorization for recommendations.

Social Recommendation. An influential work in social recommendations [25] leverages social correlations for recommendations, and the underlying theory is that users' behaviors are influenced or similar to his/her friends, proven by social correlation theories. We take the state of the art method [3] as our baseline, where it adopts memory network to indicate different strength and aspects when attending friends' preferences.

Sequential Recommendation. Sequential recommendation has achieved great success to model dynamic users' preference in recent work. [10] takes advantages of recent self-attention mechanism to build the sequential model. [4] uses memory network when building a memory matrix for each user to record their historical purchases or feature-level preferences. [24] employs variational inference to model the uncertainty in sequential recommendation.

Recommendation with Multiple Information Sources. Heterogeneous recommendation falls into two lines. In the first, the hybridization of algorithms integrates different recommendation techniques. [6] presents a joint optimization framework for the multi-behavior data for users. [14] designs a review encoder based on mutual self attention to extract the semantic features of users and items from their reviews. [22] is a recent work that combines reviews of encoder-decoder with user-item graph with attentive graph neural networks. [12] is the state of the art baseline that combines ratings and reviews with a gating layer, and finds the one-hop neighbors to enhance recommendations. In the second, hybridization of information combines information of heterogeneous data forms, and heterogeneous information networks (HIN) is the representative approach.

However, there is few work that model the sequential dynamics in users' preference with multiple information sources. While sequential model builds the rules of purchase habits, the sparsity problem in the dataset may harm the performance if there are cold start users and aspects of items. The problem mentioned above made the performance of previous recommendations constrained.

7   Conclusion

In this paper, we propose Heterogeneous Information Sequential Recommendation System (**HISR**) which tackles recommendations under dynamical contexts. The combination of multiple information sources lies in two parts: the gating layer for explicit infusion of two views of data, and the implicit combination in social attentional module. To our best knowledge, we are first to combine the sequential dynamics with the multiple information sources, and present the dual-GRU structure to model review texts and social attentions. Further, we evaluate our proposed method with real local business dataset. We also validate effectiveness of module components by ablation study and further discussions.

References

1. Bahdanau, D., Cho, K., Bengio, Y.: Neural machine translation by jointly learning to align and translate. arXiv preprint arXiv:1409.0473 (2014)
2. Chen, C., Zhang, M., Liu, Y., Ma, S.: Neural attentional rating regression with review-level explanations. In: The International World Wide Web Conference (WWW), pp. 1583–1592 (2018)
3. Chen, C., Zhang, M., Liu, Y., Ma, S.: Social attentional memory network: modeling aspect-and friend-level differences in recommendation. In: ACM International Conference on Web Search and Data Mining (WSDM), pp. 177–185 (2019)

4. Chen, X., et al.: Sequential recommendation with user memory networks. In: ACM International Conference on Web Search and Data Mining (WSDM), pp. 108–116 (2018)
5. Cho, K., et al.: Learning phrase representations using RNN encoder-decoder for statistical machine translation. arXiv preprint arXiv:1406.1078 (2014)
6. Gao, C., et al.: Neural multi-task recommendation from multi-behavior data. In: IEEE International Conference on Data Engineering (ICDE), pp. 1554–1557 (2019)
7. Guan, X., et al.: Attentive aspect modeling for review-aware recommendation. ACM Trans. Inf. Syst. (TOIS) **37**(3), 1–27 (2019)
8. He, X., Liao, L., Zhang, H., Nie, L., Hu, X., Chua, T.S.: Neural collaborative filtering. In: The International World Wide Web Conference (WWW), pp. 173–182 (2017)
9. Huang, T., She, Q., Wang, Z., Zhang, J.: GateNet: gating-enhanced deep network for click-through rate prediction. arXiv preprint arXiv:2007.03519 (2020)
10. Kang, W.C., McAuley, J.: Self-attentive sequential recommendation. In: IEEE International Conference on Data Mining (ICDM), pp. 197–206 (2018)
11. Liu, D., Li, J., Du, B., Chang, J., Gao, R.: DAML: dual attention mutual learning between ratings and reviews for item recommendation. In: ACM SIGKDD International Conference on Knowledge Discovery and Data Mining, New York, NY, USA, pp. 344–352 (2019)
12. Ma, C., Kang, P., Wu, B., Wang, Q., Liu, X.: Gated attentive-autoencoder for content-aware recommendation. In: ACM International Conference on Web Search and Data Mining (WSDM), New York, NY, USA, pp. 519–527 (2019)
13. Mnih, A., Salakhutdinov, R.R.: Probabilistic matrix factorization. In: Neural Information Processing Systems, pp. 1257–1264 (2008)
14. Peng, Q., Liu, H., Yu, Y., Xu, H., Dai, W., Jiao, P.: Mutual self attention recommendation with gated fusion between ratings and reviews. In: Nah, Y., Cui, B., Lee, S.-W., Yu, J.X., Moon, Y.-S., Whang, S.E. (eds.) DASFAA 2020. LNCS, vol. 12114, pp. 540–556. Springer, Cham (2020). https://doi.org/10.1007/978-3-030-59419-0_33
15. Pennington, J., Socher, R., Manning, C.: GloVe: global vectors for word representation. In: International Conference on Empirical Methods in Natural Language Processing (EMNLP), Doha, Qatar, pp. 1532–1543, October 2014
16. Rendle, S., Freudenthaler, C., Gantner, Z., Schmidt-Thieme, L.: BPR: Bayesian personalized ranking from implicit feedback. In: Conference on Uncertainty in Artificial Intelligence (UAI), pp. 452–461 (2009)
17. Wan, S., Lan, Y., Guo, J., Xu, J., Pang, L., Cheng, X.: A deep architecture for semantic matching with multiple positional sentence representations. In: AAAI Conference on Artificial Intelligence, pp. 2835–2841 (2016)
18. Wang, B., Liu, K., Zhao, J.: Inner attention based recurrent neural networks for answer selection. In: The Annual Meeting of the Association for Computational Linguistics (ACL), pp. 1288–1297 (2016)
19. Wang, D., Cui, P., Zhu, W.: Structural deep network embedding. In: ACM SIGKDD International Conference on Knowledge Discovery and Data Mining, New York, NY, USA, pp. 1225–1234 (2016)
20. Wang, N., Wang, H., Jia, Y., Yin, Y.: Explainable recommendation via multi-task learning in opinionated text data. In: International ACM SIGIR Conference on Research & Development in Information Retrieval, pp. 165–174 (2018)

21. Wu, C., Wu, F., Qi, T., Ge, S., Huang, Y., Xie, X.: Reviews meet graphs: enhancing user and item representations for recommendation with hierarchical attentive graph neural network. In: Conference on Empirical Methods in Natural Language Processing and International Joint Conference on Natural Language Processing (EMNLP-IJCNLP), Hong Kong, China, pp. 4884–4893 (2019)

22. Wu, C., Wu, F., Qi, T., Ge, S., Huang, Y., Xie, X.: Reviews meet graphs: enhancing user and item representations for recommendation with hierarchical attentive graph neural network. In: Conference on Empirical Methods in Natural Language Processing and International Joint Conference on Natural Language Processing (EMNLP-IJCNLP), pp. 4886–4895 (2019)

23. Wu, L., Quan, C., Li, C., Wang, Q., Zheng, B., Luo, X.: A context-aware user-item representation learning for item recommendation. ACM Trans. Inf. Syst. **37**(2), 1–29 (2019)

24. Zhao, J., Zhao, P., Liu, Y., Sheng, V.S., Li, Z., Zhao, L.: Hierarchical variational attention for sequential recommendation. In: Nah, Y., Cui, B., Lee, S.-W., Yu, J.X., Moon, Y.-S., Whang, S.E. (eds.) DASFAA 2020. LNCS, vol. 12114, pp. 523–539. Springer, Cham (2020). https://doi.org/10.1007/978-3-030-59419-0_32

25. Zhao, T., McAuley, J., King, I.: Leveraging social connections to improve personalized ranking for collaborative filtering. In: ACM International Conference on Information and Knowledge Management (CIKM), pp. 261–270 (2014)

SRecGAN: Pairwise Adversarial Training for Sequential Recommendation

Guangben Lu, Ziheng Zhao, Xiaofeng Gao$^{(\boxtimes)}$, and Guihai Chen

Shanghai Key Laboratory of Scalable Computing and Systems,
Department of Computer Science and Engineering,
Shanghai Jiao Tong University, Shanghai, China
{benlucas,Zhao_Ziheng}@sjtu.edu.cn
{gao-xf,gchen}@cs.sjtu.edu.cn

Abstract. Sequential recommendation is essentially a learning-to-rank task under special conditions. Bayesian Personalized Ranking (BPR) has been proved its effectiveness for such a task by maximizing the margin between observed and unobserved interactions. However, there exist unobserved positive items that are very likely to be selected in the future. Treating those items as negative leads astray and poses a limitation to further exploiting its potential. To alleviate such problem, we present a novel approach, Sequential Recommendation GAN (SRecGAN), which learns to capture latent users' interests and to predict the next item in a pairwise adversarial manner. It can be interpreted as playing a minimax game, where the generator would learn a similarity function and try to diminish the distance between the observed samples and its unobserved counterpart, whereas the discriminator would try to maximize their margin. This intense adversarial competition provides increasing learning difficulties and constantly pushes the boundaries of its performance. Extensive experiments on three real-world datasets demonstrate the superiority of our methods over some strong baselines and prove the effectiveness of adversarial training in sequential recommendation.

Keywords: Sequential recommendation · Generative adversarial networks · Pairwise comparison · Interest evolution

1 Introduction

In an era of data deluge, users are increasingly immersed in massive information and options, which calls for effective and powerful recommendation systems. Different from traditional methods such as Collaborative Filtering (CF), sequential recommendation has a distinct edge in capturing users' dynamic interests

This work was supported by the National Key R&D Program of China [2020YFB1707903]; the National Natural Science Foundation of China [61872238, 61972254], the Huawei Cloud [TC20201127009], the CCF-Tencent Open Fund [RAGR20200105], and the Tencent Marketing Solution Rhino-Bird Focused Research Program [FR202001].

C. S. Jensen et al. (Eds.): DASFAA 2021, LNCS 12683, pp. 20–35, 2021.
https://doi.org/10.1007/978-3-030-73200-4_2

and thus gains extensive attention in recent years. Research in this field have developed different schools of thinking. A classical method is to model users' behaviors through Markov Chain, which assumes that their feature behaviors depends on the last few ones [7,16]. Another method is based on deep learning approaches. This includes prevailing techniques such as recurrent neural network [9,10], self-attention mechanism [11], convolutional neural network [17] and dual structure [14,22].

The core of sequential recommendation can be formulated as creating a user-specific ranking for a set of items with user's recent behavioral history. Bayesian Personalized Ranking (BPR) [15] has been proved its effectiveness for such a task. It is a pairwise learning-to-rank method that maximizes the margin as much as possible between an observed interaction and its unobserved counterparts [8]. This behavior of BPR treats the unobserved positive items equally the same as those negative items. However, those unobserved positive items are very likely to be selected in the future and its margin with the observed items should be minimized instead of maximized naively. Thereby, in this work, we aim to explore an extended version of BPR which could wisely determine the margin to alleviate above problem.

Inspired by the success of applying GAN in recommendation system [2,19, 21], we concentrate upon the integration of adversarial training with BPR and propose a GAN-based Sequential Recommender, called SRecGAN. This combination enjoys merits from both sides: the adversarial training helps the model to find a certain margin instead of maximizing it naively, while the pairwise learning pushes the model to its limit. The formulation of SRecGAN can be seen as playing a minimax game: given the dynamic interests state from the sequential layer, the generator strives to learn a similarity function. With pairwise comparison, a score difference (i.e. margin) is obtained between an instance pair rated by the generator which preserves the preference information. On the other hand, the discriminator focuses on classification task and learns to judge the rationality of the generated margin. When the game reaches equilibrium, the generator can effectively capture user's dynamic interests and produce high-quality recommendations.

Extensive experiments have been carried out to verify the feasibility of our method over three real-world datasets compared with several competitive methods. The results show that SRecGAN achieves significant improvements on various evaluation metrics. Furthermore, a series of ablation studies further justify the rationality of our framework. The main contribution of our paper can be summarized as follows:

- We explore the integration of adversarial training with BPR and propose a novel framework called SRecGAN. Such combination helps to ease the issue that unobserved positive items are misleadingly treated.
- We design the sequential layer to effectively capture the user's interests as well as their evolving process and dynamics. Furthermore, we evaluate some sequential extraction methods realized under SRecGAN framework to verify the versatility of the module.

– We conduct extensive experiments and ablation studies on three real-world datasets. The results not only show the superiority of our method, but also justify the rationality of our framework.

The rest of the paper is organized as follows: Sect. 2 formalizes the problem and lists the notation. Section 3 describes the design of SRecGAN model and the proposed interest capturing method in details. Experimental results are presented in Sect. 4 and Sect. 5 introduces the related work to this paper. Finally, Sect. 6 concludes this paper.

2   Preliminary

In sequential recommendation problem, we denote a set of users as $\mathcal{U}$ and a set of items as $\mathcal{I}$. Each user u has a chronologically-ordered interaction sequence of items $\mathcal{S} = \{s_1, s_2, \ldots, s_{|\mathcal{S}|}\}$, where $s_k \in \mathcal{I}$. For fairness, we constrain the maximum length of the sequence that our model handles down to n and denote $s := \{(s_1, s_2, \ldots, s_n)\}$ as the subsequence set of entire sequence.

BPR is a generic method for personalized ranking and can be seamlessly transferred into sequential settings. Given a historical interaction sequence s of a user u, it assumes that u prefers the observed item $\mathcal{I}_{u,s}^+$ over all other non-observed items as the next item to act on. To this end, the margin between an observed interaction and its unobserved counterparts is maximized as much as possible. We create pairwise training instances $\mathcal{Q}$ by:

$$\mathcal{Q} := \{(u, s, i, j) \mid i \in \mathcal{I}_{u,s}^+ \wedge j \in \mathcal{I} \setminus \mathcal{I}_{u,s}^+\}$$

and the objective function of BPR to be minimized is

$$\mathcal{J}_{BPR}(\mathcal{Q} \mid \Theta) = \sum_{(u,s,i,j) \in \mathcal{Q}} - \ln \sigma \left(\hat{y}_{u,s,i}(\Theta) - \hat{y}_{u,s,j}(\Theta)\right) + \lambda_\Theta \|\Theta\|^2$$

where $\sigma(\cdot)$ is the sigmoid function, $\hat{y}_{u,s,i}(\Theta)$ is an arbitrary real-valued similarity function of the model parameters vector Θ, and λ_Θ are model specific regularization coefficient.

3   Methodology

In this section, we elaborate technical details for our proposed model SRecGAN. First, we introduce the generator and discriminator of SRecGAN as shown in Fig. 1. Then we highlight the techniques used for capturing users' interests evolution and detail the specifications of the sequential layer.

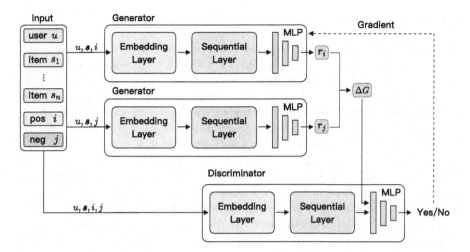

Fig. 1. The framework of SRecGAN. The generator (G) and the discriminator (D) play a minimax game with each other during the training. When the game reaches equilibrium, G can effectively capture user's dynamic interests and produce high-quality recommendations.

3.1 Generator

The generator is widely designed to learn the distribution of ground truth data with a fixed similarity function and then samples the most relevant items to a given user. However, back-propagation would be invalid since this sampling procedure is non-differentiable [23]. Distinctively, our generator strives to learn the user-item representations and the flexible similarity function simultaneously. Given a pairwise interaction quartet $q = (u, s, i, j) \in \mathcal{Q}$, we can obtain two triplets $\tau_i = (u, s, i)$ and $\tau_j = (u, s, j)$. As shown in Fig. 1, through sequential layer, the generator captures the dynamic interest patterns in the sequence s first and then estimates the preference score $r(\tau)$ that reflects the chance of item i being selected by user u as the next item. Concretely, we have

$$r(u, s, i) = \sigma(f_h(r_E^{user}(u), r_E^{item}(i), f_s(r_E^{item}(s))))\tag{1}$$

where $f_h(\cdot)$ is the fully-connected layer, $f_s(\cdot)$ is the sequential layer. $r_E^{user}(\cdot), r_E^{item}(\cdot)$ are the embedding layers for user and item respectively. Here we skip the details of the structure, which can be found in Sect. 3.3.

Normal generative models focus on the score itself, however, user bias causes distinct variation in rating values [13]. In other words, different users have different rating standards, some users have the tendency to give higher ratings while some rate conservatively. Based on the theory that the weighing of the scales reflects a user's preference, i.e. $r(\tau_i) > r(\tau_j)$ means user u with sequence s prefers item i to item j as the next item to click, we calculate the score difference between each pairwise comparison to preserve the valuable preference information.

$$\Delta G_{\theta, q} = r(u, s, i) - r(u, s, j)\tag{2}$$

3.2   Discriminator

The traditional discriminator in policy gradient based GAN is in fact a binary classier to predict the label of each item [19]. CFGAN [2] demonstrates that such a paradigm inevitably encounters the dilemma where the generator samples the items exactly the same as those in ground truth, causing a large proportion of contradictory labeled items in the training data. Therefore, the discriminator is confused and produces wrong feedback signals to the generator. This issue ends with the degradation of both generator and discriminator.

To exploit the potential of the generator, our discriminator forces the score difference of each comparison to approximate a critical margin which renders it improved labeling. Each instance of the discriminator includes the same pairwise interaction quartet $q = (u, s, i, j) \in \mathcal{Q}$, the score difference ΔG_q between item i and item j, and the label ℓ of the instance. All the instance whose score difference comes from the generator will be labeled as -1, whilst others will be labeled as 1 if it comes from the ground truth data. We equip those ground truth data with an ideal score difference ΔG_q^*, which is set to be the maximum of all possible values. Specifically, we have

$$d_\phi(\Delta G_q, u, s, i, \ell) = \sigma(\ell \cdot f_h(\Delta G_q, d_E^{user}(u), d_E^{item}(i), f_s(d_E^{item}(s)))) \quad (3)$$

where $\sigma(\cdot), f_h(\cdot), f_s(\cdot)$ is the same as the aforementioned, $d_E^{user}(\cdot), d_E^{item}(\cdot)$ are the embedding layers for user and item respectively.

$$\Delta G_q = \begin{cases} \Delta G_{\theta,q} & \text{if } \ell = -1 \\ \Delta G_q^* & \text{if } \ell = 1 \end{cases} \quad (4)$$

Optimally, the discriminator can not be fooled even when the generated score difference approaches the critical margin.

3.3   Layer Structure

Embedding Layer. Embedding is widely used to map discrete symbols into continuous embedding vectors that reflect their semantic meanings. To better encode user's and item's features, we create a user embedding matrix $\mathbf{M} \in \mathbb{R}^{|\mathcal{U}| \times d_u}$ and an item embedding matrix $\mathbf{N} \in \mathbb{R}^{|\mathcal{I}| \times d_i}$ where d_u, d_i are the latent dimensionality respectively. The k-th user u is first encoded as a one-hot vector where only the k-th value is 1 and other values are zero, then we retrieve the corresponding latent vector $e_u \in \mathbf{M}$ to represent the static preferences and interests. Similarly, we can retrieve $e_i \in \mathbf{N}$ for item i to characterize its static attribute and influence. For padding item, we set its embedding to be a constant zero vector $\mathbf{0}$. Note that in SRecGAN, the generator and discriminator are two self-contained models, so we adopt independent embedding parameters for them.

Sequential Layer. We leverage the sequential layer to capture users' imperceptible yet dynamics interests. Since the interest extraction is independent of

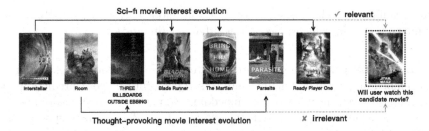

Fig. 2. Illustration of a user's interest evolution. Movies are aligned in chronological order. There are two streams of user's interest evolution, namely sci-fi and thought-provoking movie interest evolution.

the adversarial learning framework, we design the generator and discriminator to share the same architecture in sequential layer.

A significant feature of a user's behavior patterns is the constant evolution of his interests [26]. Using Fig. 2 as an example, after watching Interstellar, the user dove into exploring more sci-fi movies. What's more, there is another stream indicates that the user's interests in thought-provoking movie evolved. This figure vividly shows that the user's interests can not only evolve, but also drift from stream to stream. Capturing this interest evolution amplifies the representation of dynamic interests h_d with more relative history information and leads to a better prediction by following the interest evolution trend.

We follow the trend to combine GRU with attention mechanism [26]. Different from the inner attention-based GRU [18], we implement attention mechanism to activate relative items in the behavior sequence to target item first and then use GRU to learn the evolving interest. Here we apply additive attention [1] and the attention weight of candidate item i on clicked item j can be

$$a_{ij} = \frac{\exp(W_A\,[e_j, e_i] + b_A)}{\sum_{k=1}^{n} \exp(W_A\,[e_k, e_i] + b_A)} \tag{5}$$

where W_A is the weight matrix and b_A is the bias vector. Then, the t-th hidden states h_t of GRU with t-th item embedding e_t is computed by

$$u_t = \sigma\left(W^u a_{it} e_t + U^u h_{t-1} + b^u\right) \tag{6}$$
$$r_t = \sigma\left(W^r a_{it} e_t + U^r h_{t-1} + b^r\right) \tag{7}$$
$$\tilde{h}_t = \tanh\left(W^h a_{it} e_t + r_t \circ U^h h_{t-1} + b^h\right) \tag{8}$$
$$h_t = (1 - u_t) \circ h_{t-1} + u_t \circ \tilde{h}_t \tag{9}$$

where $\sigma(\cdot)$ is the sigmoid function, $\circ$ is element-wise product, W^u, W^r, W^h, U^u, U^r, U^h are the weight matrices and b^u, b^r, b^h are the bias vectors.

After reading the behavior sequence, the final hidden state of the GRU is the summary vector h_d which reflects user's dynamic interests.

Prediction Layer. In generator, we concatenate user embedding e_u, candidate item embedding e_i and dynamic interests h_d for future score prediction. In discriminator, we concatenate user embedding e_u, positive item embedding e_{pos}, negative item embedding e_{neg}, dynamic interests $h_{d_{pos}}$, $h_{d_{neg}}$ which represent the relevance between behavior sequence and the corresponding candidate items, the score difference together for future classification. We leverage an oft-used "tower" network structure for prediction where the bottom layer is the widest and each successive hidden layer reduces the number of units [3].

3.4 Training Algorithm

This subsection describes the training regime for our models. To learn the preference distribution, given a similarity function, the generator would try to diminish the distance between the observed samples and its unobserved counterpart, whereas the discriminator would try to maximize their margin. In other words, the generator and the discriminator play the following two-player minimax game with value function:

$$\min_{\theta} \max_{\phi} J(G, D) = \mathbb{E}_{\Delta G_q^* \sim \mathcal{P}_{\Delta G^*}} \left[\log D_\phi \left(\Delta G_q^* \right) \right] + \\ \mathbb{E}_{q \sim \mathcal{P}_{\mathcal{Q}}} \left[\log \left(1 - D_\phi \left(\Delta G_{\theta,q} \right) \right) \right] \tag{10}$$

where $q = (u, s, i, j) \in \mathcal{Q}$, $\mathcal{P}_{\mathcal{Q}}$ is the empirical preference distribution, and $\mathcal{P}_{\Delta G^*}$ is the pursued score difference distribution. Separately, our discriminator's objective function, denoted as J^D, is as follows[1]:

$$J^D = -\mathbb{E}_{\Delta G_q^* \sim \mathcal{P}_{\Delta G^*}} \left[\log D_\phi \left(\Delta G_q^* \right) \right] - \mathbb{E}_{q \sim \mathcal{P}_{\mathcal{Q}}} \left[\log \left(1 - D_\phi \left(\Delta G_{\theta,q} \right) \right) \right] \\ = -\sum_q \log D_\phi \left(\Delta G_q^* \right) - \sum_q \log \left(1 - D_\phi \left(\Delta G_{\theta,q} \right) \right) \tag{11}$$

And, similarly, that of our generator is:

$$J^G = \mathbb{E}_{q \sim \mathcal{P}_{\mathcal{Q}}} \left[\log \left(1 - D_\phi \left(\Delta G_{\theta,q} \right) \right) \right] \\ = \sum_q \log \left(1 - D_\phi \left(\Delta G_{\theta,q} \right) \right) \tag{12}$$

Here we provide the learning algorithm as shown in Algorithm 1. It is noteworthy that we initialize the generator using BPR (line 2). The reason is that during the implementation, we discover that the generator with randomly initialized parameters would possibly encounter mode collapse at early stage, where it finds a trivial solution to fool the discriminator: i.e. rating every instance simply the same without considering users' preference at all. This is due to the fact that the intense adversarial training in the early stage with meaningless initial

---

[1] For simplicity, we define both of the objective functions to be a minimization problem.

Algorithm 1. SRecGAN

Input: training quartet set $\mathcal{Q}$, learning rate μ^G for the generator (G) and μ^D for the discriminator (D), minibatch size M^G for G and M^D for D.

1: Initialize generator's parameters θ and discriminator's parameters ϕ.
2: Pre-train θ with BPR.
3: **for** number of training iterations **do**
4:    **for** D-steps **do**
5:       The generator generates score difference for each quartet (u, s, i, j) in $\mathcal{Q}$
6:       Update D by $\phi \leftarrow \phi - \dfrac{\mu^D}{M^D} \cdot \nabla_\phi J^D$
7:    **end for**
8:    **for** G-steps **do**
9:       Update G by $\theta \leftarrow \theta - \dfrac{\mu^G}{M^G} \cdot \nabla_\theta J^G$
10:    **end for**
11: **end for**

---

embeddings may easily lead astray. To prevent this, we pre-train the generator to obtain better parameters.

By fully exploiting the potential of GAN, our proposed framework produces the synergy effect between the generator and the discriminator and is able to catch the intricate relationships between users and items. When the adversarial training ends, our generator can provide high-quality recommendations.

3.5 Discussions

In this subsection, we point out the limitation of BPR, which motivates us to propose SRecGAN. As we have indicated before, BPR improves itself by maximizing the margin between observed and unobserved interactions. However, due to users' limited exposure to the data, the unobserved interactions are actually a mixture of both unobserved negative samples and positive ones. Unobserved negative samples refer to interactions that have not yet happened and would also not happen in the future, while unobserved positive samples are the potential interactions that are very likely to happen in the future but have not yet been exposed to users. Without prior knowledge, one can hardly distinguish those two kinds of samples. By ignoring the unobserved positive samples and naively maximizing its margin with the observed ones, BPR poses a limitation to further exploiting its potential.

Fortunately, in SRecGAN, the above problem can be alleviated to a certain degree. We draw an analogy in Fig. 3. There exist linking lines between observed positive items and unobserved positive items, indicating their potential future interactions. In each epoch of training, the generator tries to minimize the margin between the observed and unobserved samples. When their margin get closer to the critical point, the discriminator get confused. For the discriminator, its mission is to judge the rationality of the generated margin and tries to score

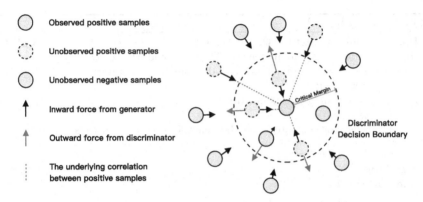

Fig. 3. An Illustration of SRecGAN training.

down the unobserved samples, to put it another way, maximizing the margin between observed and unobserved samples.

Eventually, when the game reaches a dynamic equilibrium, the generator can effectively estimate the preference distribution. The unobserved positive items would be assigned with higher scores and ranked before the negative ones since they are linked to those observed ones.

Table 1. Dataset statistics

Dataset	#User	#Item	#Actions	Sparsity
MovieLens-100K	943	1349	99287	92.20%
MovieLens-1M	6040	3416	999611	95.16%
Netflix	10000	5332	4937186	90.74%

4 Evaluation

This section comprehensively analyzes the results of our extensive experiments to show the effectiveness of our models. Our experiments are designed to answer the following research questions:

- **RQ1:** How effective is it to integrate BPR with adversarial training?
- **RQ2:** How do the key hyper-parameters affect the accuracy of SRecGAN?
- **RQ3:** How does SRecGAN perform compared with other strong methods?
- **RQ4:** Is sequential layer in SRecGAN a generic module?

4.1 Experimental Settings

In this subsection, we reveal our experimental details.

Dataset. We adopt three real-world datasets: ML-100K [6], ML-1M [6] and Netflix. For Netflix dataset, we randomly sample 10,000 users with their rated movies for the evaluation. Table 1 summarizes their detailed statistics. Following the experiment setting of [11], we treat users' ratings as implicit feedbacks and leave the last two most recent items as the testing set while the rest as the training set. Note that during testing, the input sequences contain training actions.

Comparison Methods. To show the effectiveness of our methods, we include four related mainstream sequential recommenders and two GAN-based recommenders, which are listed as follows:

- **GRU4Rec⁺** [9]: An improved version of GRU4Rec with different loss function and sampling strategy. Note that GRU4Rec is the pioneering work that applies RNNs in session-based recommendation.
- **DEEMS** [22]: A novel dual structure method that unifies sequential recommendation and information dissemination models via a two-player game. It leverages the dynamic patterns of both users and items for prediction. Note that DEEMS has two variants and we choose the stronger one, i.e. DEEMS-RNN, as the baseline.
- **DREAM** [24]: A RNN-based model for next basket recommendation. DREAM is able to enrich users' representation with both current interests and global sequential features for better prediction.
- **SASRec** [11]: A self-attention based model which adaptively learns the dependency between items. It achieves very competitive performance in sequential recommendation task.
- **IRGAN** [19]: A GAN-based method that combines generative and discriminative information retrieval via adversarial training, in which a simple matrix factorization is used. Note that IRGAN is the first to introduce GAN to recommendation system.
- **CFGAN** [2]: A novel collaborative filtering framework with vector-wise adversarial training.

Evaluation Metric. We adopt three popular top-K metrics to evaluate the performance: AUC, NDCG@K, P@K. The former two are ranking-based metric while the last one is accuracy-based. In the generated recommendation list, accuracy-based metrics focus on how many correct items are included while ranking-based metrics are sensitive about the ranked position of correct items. In this experiment we set K equal to 3.

Implementation Details. First and foremost, we set the number of hidden units in the generator as 128 ReLU $\rightarrow$ 64 ReLU $\rightarrow$ 32 ReLU $\rightarrow$ 8 ReLU and 32 ReLU $\rightarrow$ 8 ReLU in the discriminator. In the sequential layer, we set the units number of GRU hidden states to be 32, and use Xavier [4] initialization and orthogonal initialization for the input weights and recurrent weights of GRU respectively. For the biases, we initialize them as zero vectors. We choose Adam [12] optimizer for fast convergence and set the learning rate to be 0.005 for the generator and 0.001 for the discriminator. The mini-batch size is 256 and the maximum length of user's behavior sequence is 10. In the training stage, for each user, we see the real next behavior as positive instance and sample negative instance from item set except the clicked items. In order to keep the balance, the number of negative instance is the same as the positive instance.

4.2 RQ1: Effectiveness of Adversarial Training

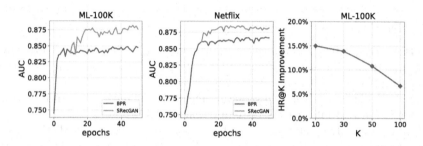

Fig. 4. Learning trend of SRecGAN and BPR on ML-100K and Netflix (left). Hit Ratio improvement when deliberately marks some observed items as unobserved ones (right).

We conduct two experiments to address following problems: (1) How is the effect of adversarial training? (2) Can unobserved positive items really be ranked before the negative ones? In so far as the first issue is concerned, we pre-train the generator with BPR for 10 epochs. After that, our experiment is divided into two branches, one of which keeps training the generator with BPR, and the other branch activates adversarial training, i.e. SRecGAN. Figure 4 show the performance of BPR and SRecGAN on ML-100K and Netflix respectively. As we can see, after 10 epochs, the BPR model can hardly elevate its accuracy in both datasets. By contrast, adversarial training can significantly boost the performance and gains around 4% and 1% improvements in ML-100K and Netflix. Then in terms of the second, since we do not know users' potential future interactions, we deliberately mark the last 10% observed items in the training set as unobserved positive items and view it as negative during the training. We employ hit ratio HR@K to measure how many above items are recollected in first K items, where K in [10, 30, 50, 100]. The right subgraph in Fig. 4 demonstrates that SRecGAN gains 15% HR@10 improvement compared to BPR on ML-100K, which shows the superiority of SRecGAN in rating those unobserved positive items.

Both experiments well confirm our assumption. Different from BPR method that strive to maximize the margin between positive and negative items according to a static optimization goal, SRecGAN utilizes pair-wise adversarial training to adjust the guidance for generator in a dynamic way. Such mechanism not only provides stricter supervisory signals to help adjust the generator efficiently, but also benefits the generator to rate those unobserved positive item properly.

4.3   RQ2: Influence of Hyper-parameters

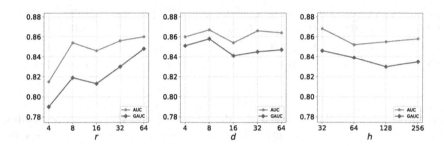

Fig. 5. Results of different hyper-parameter settings in SRecGAN.

We study the influence of some key hyper-parameters in SRecGAN on ML-100K in this section as demonstrated in Fig. 5.

Influence of Embedding Size. As aforementioned, we adopt independent embedding parameters for the generator and discriminator. We denote them as r and d respectively. Figure 5 shows how the AUC and GAUC change as the embedding size varies from 4 to 64. To achieve the best performance, we set embedding size as 64 and 8 for generator and discriminatory respectively in the following experiments. This size discrepancy mainly due to the different task of generator and discriminator.

Influence of Hidden Units Size in Sequential Layer. We change the size of hidden units h from 32 to 256 to explore how it affect the model's performance. From Fig. 5, we observe that both AUC and GAUC reach the peak when the size is 32. A large hidden layer may result in overfitting and heavy computational cost. Therefore, we set the hidden layer size as 32 in the following experiments.

4.4   RQ3: Performance Comparison

By analyzing Tables 2, we can find that our proposed model SRecGAN have superior recommendation accuracy than all the competitors on three evaluation metrics. For example, SRecGAN outperforms the strongest baseline SASRec by 1.61%, 1.38%, 2.01% in terms of AUC, P, NDCG on Netflix, respectively. The main reasons are two-folds. First, by leveraging a neural network, our generator is capable of handling the complicated interactions between users and items. Therefore, more reasonable prediction scores can be obtained. Second, the improvement mainly comes from the difference in the training approach.

Table 2. Experiment results

	ML-100K			ML-1M			Netflix		
	AUC	P	NDCG	AUC	P	NDCG	AUC	P	NDCG
IRGAN	0.739	0.505	0.698	0.799	0.542	0.760	0.755	0.512	0.719
CFGAN	0.772	0.525	0.750	0.848	0.578	0.829	0.768	0.519	0.738
DEEMS	0.778	0.532	0.737	0.815	0.554	0.783	0.814	0.554	0.785
DREAM	0.798	0.549	0.762	0.837	0.568	0.807	0.822	0.560	0.796
GRU4Rec+	0.803	0.544	0.772	0.860	0.579	0.840	0.755	0.506	0.716
SASRec	<u>0.863</u>	**0.589**	<u>0.848</u>	<u>0.881</u>	<u>0.599</u>	<u>0.864</u>	<u>0.871</u>	<u>0.575</u>	<u>0.845</u>
SRecGAN	**0.882**	<u>0.587</u>	**0.862**	**0.893**	**0.606**	**0.876**	**0.885**	**0.583**	**0.862**

Our intense competition between the discriminator and the generator results in mutual progress. Consequently, the method can achieve higher accuracy in recommendation.

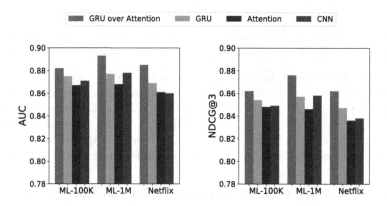

Fig. 6. Performance comparison of different sequential layer settings.

4.5   RQ4: Ablation Study on Sequential Layer

We design the sequential layer as a generic module to capture user's dynamic interests. In addition to our GRU over attention methods, more methods can be implemented in this layer, such as GRU [9], Attention [11] and CNN [17]. We rebuild the sequential layer with these oft-used techniques and investigate their perform difference. Figure 6 shows the feasibility of these techniques and verifies the versatility of sequential layer. Furthermore, we observe that our method provides recommendation accuracy consistently and universally higher than other methods.

5 Related Work

5.1 Sequential Recommendation

Sequential behaviors usually reveal a user's preferences and thus, mining sequential patterns is critical for recommendation systems to make more accurate predictions. Some sequential recommendation models base their work on Markov Chain (MC) method. FPMC [16] is a representative baseline that integrates matrix factorization (MF) with first order MC to model the transition between adjacent items. Later on, FOSSIL [7] improved the performance by fusing similarity-based method with high-order MC, which can better model both long-term and short-term dynamics.

Another line of work focus on deep learning techniques. DIN [27] adds a local activation unit to multilayer perceptron (MLP), which can adaptively learn the representation for different advertisements. [10] proposed GRU4Rec as the first model applying recurrent neural network (RNN) in sequential recommendation. In this model, multiple GRU layers are used to learn the patterns within a session (behavior sequence). As an improved version of [10], GRU4Rec$^+$[9] modifies the sampling strategy and makes a great improvement. Based on RNN, [24] proposed another dynamic recurrent basket model called DREAM to learn both the user's current interests and the sequential features. Moreover, [17] proposed Caser, a model that considers history behaviors embeddings as an "image" and adopts convolutional neural network (CNN) in learning. In addition, SASRec [11] utilizes self-attention mechanism to model user's sequence and discover the relevance between items. Recently, DEEMS [22] and SCoRe [14] construct dual models to capture temporal dynamics from both user and item sequences and prove their advancement in recommendation accuracy.

5.2 Generative Adversarial Nets

Generative Adversarial Nets was originally proposed by Goodfellow et al. [5] to generate realistic images. Recently, more and more studies focus on its application in Recommendation Systems. IRGAN [19] is the pioneering work that applies GANs in Recommendation Systems. It proposed an adversarial framework for information retrieval: the generator predicts relevant documents for a given query, and the discriminator distinguishes the generated document-query pairs and the real ones. It solves the problem that traditional GANs can only generate continuous data by leveraging policy gradient instead of gradient descent. However, the discriminator may be confused as the a document-query pair can be labeled as both real and fake (generated), so as to affect the optimization of the model. Accordingly, CFGAN [2] fixed this problem by adopting a generator which generates purchase vectors instead of item indexes. Besides, [21] proposed an adversarial framework with pairwise comparisons called CRGAN, where the generator learns a continuous score function on items. Nevertheless, the above mentioned approaches are aimed for the general recommendation, failing to consider the temporal information or capture the sequential pattern in historical data, thereby leaving it an open literature.

The existing work on GAN's application in sequential recommendation are actually quite few. PLASTIC [25] combines RNN and MF in an adversarial framework to learn user's sequential preference, and yet the fixed score function in MF is too simple to learn a sufficient expression. Work in [20] adopts a neural memory network to capture both long-term and short-term interests of users in stream recommendation. However, it lacks capturing dependency between items.

6 Conclusion

In this paper, we identify the limitation of existing BPR method that naively maximizes the margin between an observed sample and its unobserved counterparts. In order to address this limitation, we suggest integrating adversarial training with BPR to distinguish unobserved positive and negative items. Through the adversarial training process, the generator consistently pushes the discriminator to its limits by diminishing the distance between observed and unobserved items. The discriminator also consistently provides strict yet beneficial feedbacks to the generator which renders it improved scoring. Extensive experiments and sufficient ablation studies on three datasets show both the superiority and rationality of our method.

For future study, we will conduct further experiments on more real-world datasets. We plan to improve the stability of SRecGAN by exploring some recent state-of-the-art GANs, e.g. adopting the Wasserstein GAN framework by mapping the score difference into a Wasserstein distance. We also plan to leverage item side information to model more reasonable interest representations.

References

1. Bahdanau, D., Cho, K., Bengio, Y.: Neural machine translation by jointly learning to align and translate. In: ICLR (2015)
2. Chae, D., Kang, J., Kim, S., Lee, J.: CFGAN: a generic collaborative filtering framework based on generative adversarial networks. In: CIKM, pp. 137–146. ACM (2018)
3. Covington, P., Adams, J., Sargin, E.: Deep neural networks for Youtube recommendations. In: RecSys, pp. 191–198. ACM (2016)
4. Glorot, X., Bengio, Y.: Understanding the difficulty of training deep feedforward neural networks. In: AISTATS. JMLR Proceedings, vol. 9, pp. 249–256. JMLR.org (2010)
5. Goodfellow, I.J., et al.: Generative adversarial nets. In: NIPS, pp. 2672–2680 (2014)
6. Harper, F.M., Konstan, J.A.: The MovieLens datasets: History and context. ACM Trans. Interact. Intell. Syst. 5(4), 19:1–19:19 (2016)
7. He, R., McAuley, J.J.: Fusing similarity models with Markov chains for sparse sequential recommendation. In: ICDM, pp. 191–200. IEEE Computer Society (2016)
8. He, X., He, Z., Du, X., Chua, T.: Adversarial personalized ranking for recommendation. In: SIGIR, pp. 355–364. ACM (2018)
9. Hidasi, B., Karatzoglou, A.: Recurrent neural networks with top-k gains for session-based recommendations. In: CIKM, pp. 843–852. ACM (2018)

10. Hidasi, B., Karatzoglou, A., Baltrunas, L., Tikk, D.: Session-based recommendations with recurrent neural networks. In: ICLR (Poster) (2016)
11. Kang, W., McAuley, J.J.: Self-attentive sequential recommendation. In: ICDM, pp. 197–206. IEEE Computer Society (2018)
12. Kingma, D.P., Ba, J.: Adam: a method for stochastic optimization. In: ICLR (Poster) (2015)
13. Koren, Y., Bell, R.M., Volinsky, C.: Matrix factorization techniques for recommender systems. Computer **42**(8), 30–37 (2009)
14. Qin, J., Ren, K., Fang, Y., Zhang, W., Yu, Y.: Sequential recommendation with dual side neighbor-based collaborative relation modeling. In: WSDM, pp. 465–473. ACM (2020)
15. Rendle, S., Freudenthaler, C., Gantner, Z., Schmidt-Thieme, L.: BPR: Bayesian personalized ranking from implicit feedback. In: UAI, pp. 452–461. AUAI Press (2009)
16. Rendle, S., Freudenthaler, C., Schmidt-Thieme, L.: Factorizing personalized Markov chains for next-basket recommendation. In: WWW, pp. 811–820. ACM (2010)
17. Tang, J., Wang, K.: Personalized top-n sequential recommendation via convolutional sequence embedding. In: WSDM, pp. 565–573. ACM (2018)
18. Wang, B., Liu, K., Zhao, J.: Inner attention based recurrent neural networks for answer selection. In: ACL (1). The Association for Computer Linguistics (2016)
19. Wang, J., et al.: IRGAN: a minimax game for unifying generative and discriminative information retrieval models. In: SIGIR, pp. 515–524. ACM (2017)
20. Wang, Q., Yin, H., Hu, Z., Lian, D., Wang, H., Huang, Z.: Neural memory streaming recommender networks with adversarial training. In: KDD, pp. 2467–2475. ACM (2018)
21. Wang, Z., Xu, Q., Ma, K., Jiang, Y., Cao, X., Huang, Q.: Adversarial preference learning with pairwise comparisons. In: ACM Multimedia, pp. 656–664. ACM (2019)
22. Wu, Q., Gao, Y., Gao, X., Weng, P., Chen, G.: Dual sequential prediction models linking sequential recommendation and information dissemination. In: KDD, pp. 447–457. ACM (2019)
23. Xu, Z., et al.: Neural response generation via GAN with an approximate embedding layer. In: EMNLP, pp. 617–626. Association for Computational Linguistics (2017)
24. Yu, F., Liu, Q., Wu, S., Wang, L., Tan, T.: A dynamic recurrent model for next basket recommendation. In: SIGIR, pp. 729–732. ACM (2016)
25. Zhao, W., Wang, B., Ye, J., Gao, Y., Yang, M., Chen, X.: PLASTIC: prioritize long and short-term information in top-n recommendation using adversarial training. In: IJCAI, pp. 3676–3682. ijcai.org (2018)
26. Zhou, G., et al.: Deep interest evolution network for click-through rate prediction. In: AAAI, pp. 5941–5948. AAAI Press (2019)
27. Zhou, G., et al.: Deep interest network for click-through rate prediction. In: KDD, pp. 1059–1068. ACM (2018)

SSRGAN: A Generative Adversarial Network for Streaming Sequential Recommendation

Yao Lv[1,2], Jiajie Xu[1,2(✉)], Rui Zhou[3], Junhua Fang[1], and Chengfei Liu[3]

[1] Institute of Artificial Intelligence, School of Computer Science and Technology,
Soochow University, Suzhou, China
20194227033@stu.suda.edu.cn,{xujj,jhfang}@suda.edu.cn
[2] Neusoft Corporation, Shenyang, China
[3] Swinburne University of Technology, Melbourne, Australia
{rzhou,cliu}@swin.edu.au

Abstract. Studying the sequential recommendation in streaming settings becomes meaningful because large volumes of user-item interactions are generated in a chronological order. Although a few streaming update strategies have been developed, they cannot be applied in sequential recommendation, because they can hardly capture the long-term user preference only by updating the model with random sampled new instances. Besides, some latent information is ignored because the existing streaming update strategies are designed for individual interactions, without considering the interaction subsequence. In this paper, we propose a Streaming Sequential Recommendation with Generative Adversarial Network (SSRGAN) to solve the streaming sequential recommendation problem. To maintain the long-term memory and keep sequential information, we use the reservoir-based streaming storage mechanism and exploit an active subsequence selection strategy to update model. Moreover, to improve the effectiveness and efficiency of online model training, we propose a novel negative sampling strategy based on GAN to generate the most informative negative samples and use Gumble-Softmax to overcome the gradient block problem. We conduct extensive experiments on two real-world datasets and the results shows the superiority of our approaches in streaming sequential recommendation.

Keywords: Streaming recommendation · Sequential recommendation · Generative Adversarial Network

1 Introduction

With the rapid development of mobile devices and Internet services, recommender systems play a more and more important role in solving the problem of information overload and satisfying the diverse needs of users [12,19]. On the other hand, large E-commerce platforms such as Tmall and Amazon generate

© Springer Nature Switzerland AG 2021
C. S. Jensen et al. (Eds.): DASFAA 2021, LNCS 12683, pp. 36–52, 2021.
https://doi.org/10.1007/978-3-030-73200-4_3

tremendous amounts of interaction data at a high speed. For instance, up to 583,000 transactions were generated in Tmall per second during its shopping event on 11, November 2020. Such streaming data is continuous, time-varying and rapid, which is quite different from the static data in an offline setting. Therefore, it is more reasonable to study the recommender systems under a data streaming scenario.

Several methods has been proposed to solve the streaming challenge in conventional recommender systems. One kind of solution is memory-based algorithms based on users' historical activities. [4,13] adopt neighborhood-based methods to calculate the similarity of users and then recommend items for user based on the current preference of most similar users calculated by historical data. Another kind of solution is model-based algorithms such as online learning. This method updates the model only by using new arrival date to capture the drift of users' preference [25]. Furthermore, the random sampling strategy with reservoir [6] was employed to solve the long-term challenge. However, these methods can only be applied in non-sequential patterns and cannot capture hidden information in sequential behaviors. Different from conventional recommender systems, sequential recommender systems try to model the evolution of user preference and item popularity over time in user-item interaction sequences. Some recommender systems adopt MCs [9], CNNs [26], GRU [11] or Self-Attention [17] to tackle the sequential recommendation problem. Unfortunately, they can only be applied in offline settings.

Although many existing methods have achieved good results in both sequential and streaming recommendation, we found that there are still several challenges in solving the sequential recommendation problem in a streaming setting. Firstly, existing streaming update strategies [15,21] cannot be applied directly in sequential recommendation because we can only update the model using the newest coming data and this solution may cause the loss of long-term memory. Also, due to the high speed of data input, it takes a certain amount of time to update the model. A good model should have the ability to respond to the data instantly, e.g., not trying to update with all the new data. Secondly, existing streaming update strategies [8,25] fail to capture the drift of users' preferences and item popularity over time because sequence information is not considered when data is updated. Thirdly, in the period of online streaming model optimization, popularity-biased or random sampling strategies [2] are used in most recommender systems to generate negative samples. However, these negative samples contribute little to the model updating because most of them could be discriminated from positive samples without difficulty.

To address the above limitations of existing works, we propose a novel model, SSRGAN, to solve the sequential recommendation in a streaming setting. The model consists of a reservoir-based model update mechanism and an online adversarial sequential recommendation module. More specifically, to tackle the challenge of long-term memory, we use the reservoir-based streaming storage technique in the model updating component and exploit an active subsequence selection strategy, where subsequences that can change the sequential model most are chosen to update the model. It is worth noting that we try to keep the sequence information during the sampling process. In online model updating process, to speed up the training effectiveness and efficiency, we develop a negative sampler based on the Generative Adversarial Network (GAN) [7] and introduce the Gumbel-Softmax approximation [14] to tackle the gradient block problem in discrete sampling step. Moreover, we choose a hierarchical gating network to do the recommendation task due to its high accuracy and efficiency in sequential recommendation.

The main contributions of this study are summarized as follows:

- We define the sequential recommendation problem under a streaming scenario, and propose a streaming sequential recommendation model with generative adversarial network to solve it. It can ensure effectiveness and efficiency of streaming sequential recommendation.
- We introduce a reservoir-based data storage mechanism to solve the long-term memory challenge in streaming settings and propose an active sampling strategy which selects the most informative subsequences to update model with maintaining the sequential information.
- We design a novel negative sampling strategy to generate adversarial negative samples for model optimization in streaming settings, which greatly improves the training effectiveness of online model updating. We introduce the Gumbel-Softmax approximation to overcome the gradient block problem, which has not been studied in sequential recommender system.
- We conduct a comprehensive series of experiments on two real-world datasets, which shows the effectiveness of our proposed SSRGAN model in streaming scenario by comparing with the state-of-the-art methods.

2   Related Work

2.1   Streaming Recommendation

Making recommendation in the continuous streaming data has been widely studied recently. There are two major kinds of streaming recommender systems.

Memory-Based. The first category is memory-based algorithms based on users' historical activities. [4,13] adopt neighborhood-based methods to calculate the similarities of users in an offline phase and then recommend items for a user according to the current preference of most similar users calculated by historical data. All similarities need to be re-computed in the period of model updating in these methods, which is very inefficient for streaming recommendation. Another problem is that we cannot use all the historical activities at a time. Min-hash technique has been proposed to solve this problem [22]. By tracking the related users of each item in the minimum hash index, the similarity between users can be approximately calculated. They use hash functions to reduce the memory consumption instead of cutting down user-items interactions.

Model-Based. Another category is model-based algorithms, which try to update the pre-trained model based on new coming user-item interactions. [25] try to capture the evolution of users' interests by updating the model only by new arrival data. However, they fail to capture users' long-term preferences to some degree. Then the random sampling strategy with a reservoir [6] was proposed to maintain the long-term memory. Furthermore, [21] proposes an online-update algorithm to solve the overload problem by selecting new data instances. In our work, we aim to both maintain long-term user preferences and capture short-term user preferences in streaming recommendation.

2.2 Sequential Recommendation

Sequential recommendation has shown better performance than traditional recommender systems in many tasks. Many sequential recommender systems try to model item-item transition term to capture sequential patterns in successive items. For instance, [20] uses first-order Markov Chains (MC) to capture long-term preferences and short-term transitions respectively. Recently, methods based on deep neural network were proposed to learn the sequential dynamics. [11] firstly introduces Gated Recurrent Units (GRU) in order to model the sequential patterns for the session-based recommendation, and an developed method [10] was proposed to improve its Top-k recommendation performance. Another line of work [23] adopts convolutional neural network (CNN) to process the item embedding sequence to extract item transitions for future prediction. Moreover, following the idea in natural language processing, the Self-Attention mechanism is applied to sequential recommendation problem, which can adaptively consider the interaction preferences between items [16]. However, these methods are somewhat limited because they adopt offline batch training and cannot be applied in online streaming settings directly.

3 Problem Definition

We formulate the sequential recommendation problem and then extend it to the streaming setting. We take sequential implicit feedback as the input data for sequential recommendation in this paper. The number of users is M and the number of items is N. For each user, the preference is represented by a time ordered user-item interaction sequence $\mathcal{S}^i = \left(\mathcal{S}_1^i, \mathcal{S}_2^i, \ldots, \mathcal{S}_{|\mathcal{S}^i|}^i\right)$, where $\mathcal{S}_j^i$ is an item j that user i has interacted with. At time step t, given the earlier $|L|$ successive items of a user, the aim is to predict the next item that the user is likely to interact with at $t + 1$ time step. When it comes to streaming sequential recommendation, the difference is that the given subsequence is in a chronologically ordered and continuous manner.

4 Proposed Method

4.1 Overview of SSRGAN

In this paper, we propose a novel SSRGAN model to solve the streaming sequential recommendation problem. The architecture of SSRGAN model is shown in Fig. 1. We first introduce our offline HGN model (in Sect. 4.2) and then extend it to online streaming setting. There are two key components in streaming setting: the reservoir-based sequential updating module (in Sect. 4.3) and the adversarial sequential recommendation module (in Sect. 4.4).

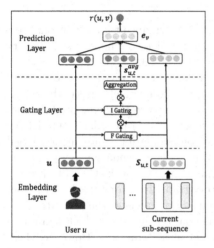

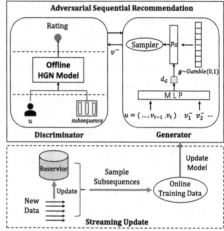

(a) offline HGN model   (b) online SSRGAN model

Fig. 1. The overview of the SSRGAN model. (a) An offline sequential recommendation model named Hierarchical Gating Networks (HGN). The HGN model can be updated in a short time with high accuracy. (b) The online model for streaming sequential recommendation. It consists of two parts: the reservoir-based sequential updating module and the adversarial sequential recommendation module.

4.2 Hierarchical Gating Networks for Sequential Recommendation

Follow the idea in [18], we introduce the HGN model to do the sequential recommendation task. HGN is a state-of-the-art sequential recommendation method with high updating efficiency. It captures the users' long and short-term preference from feature and instance level without using complex recurrent or convolutional neural networks. For each user u, we extract every $|L|$ successive items as input and their next one item as the target output.

4.2.1   Gating Layers for Group-Level Influence

In a user's interaction sequence, a group of previous items may be closely related to the items to be interacted in the near future. Therefore, it is crucial to model the interactions in group-level. A hierarchical gating network is proposed to model the user-item interactions from group-level, which consists of two different layers: a feature gating layer and an instance gating layer. These two layers select effective latent features and relevant items from different aspects respectively.

Feature Gating. There are many hidden connections between the latent features of different items. Thus, it is crucial to capture the sequence features in the latent space based on users' long-term preferences.

We feed the embeddings of previous L items into our model at time step t. Then we convert the item index v to a low-dimensional vector $\mathbf{v}$ by the item embedding matrix $Q \in \mathbb{R}^{d \times N}$. Here d is the number of latent dimensions. As a result, the subsequence at time t are embedded to $S_{u,t} = (\mathbf{v}_{t-L}, \ldots, \mathbf{v}_{t-2}, \mathbf{v}_{t-1})$.

Then, inspired by the gated linear unit (GLU) [5], we adopt the similar idea to select the relevant features to predict future items. Moreover, different users have different preferences on items, so the GLU should be improved to be user-specific. We use the inner product instead of the convolution operation to reduce the number of learnable parameters, and the operation is as follows:

$$S_{u,t}^{F} = S_{u,t} \otimes \sigma \left(W_{g_1} \cdot S_{u,t} + W_{g_2} \cdot \mathbf{u} + \mathbf{b}_g \right) \tag{1}$$

where $S_{u,t}^{F} \in \mathbb{R}^{d \times L}$ is the subsequence embedding after the feature gating, σ represents the sigmoid function, $\mathbf{u} \in \mathbb{R}^d$ is the embedding of user u, $W_{g_1}, W_{g_2} \in \mathbb{R}^{d \times d}$ and $\mathbf{b}_g \in \mathbb{R}^d$ are the learnable parameters in GLU, and $\otimes$ means the element-wise product between matrices. By applying the feature gating, different latent features of items can be chosen to the next layer.

Instance Gating. Given L successive items, it is likely that some items are more informative in the subsequence to predict the next one item. We use an instance-level gating layer to chosen relevant items which contribute more to do recommendation according to the preferences of user:

$$S_{u,t}^{I} = S_{u,t}^{F} \otimes \sigma \left(\mathbf{w}_{g_3}^{\top} \cdot S_{u,t}^{F} + \mathbf{u}^{\top} \cdot W_{g_4} \right) \tag{2}$$

where $\mathbf{w}_{g_3} \in \mathbb{R}^d, W_{g_4} \in \mathbb{R}^{d \times |L|}$ are learnable parameters and $S_{u,t}^{I} \in \mathbb{R}^{d \times L}$ is the embedded subsequence after the instance gating. By doing so, the more relevant items will be more useful to predict the items in the near future and other items could be ignored.

We use the average pooling on $S_{u,t}^{I}$ to transfer the subsequence embeddings $S_{u,t}^{I}$ into a group-level latent representation:

$$s_{u,t}^{avg} = \text{avg} - \text{pooling} \left(S_{u,t}^{I} \right) \tag{3}$$

where $s_{u,t}^{avg} \in \mathbb{R}^d$. Since informative features and items are chosen after the feature and instance-level layers, the hidden information can be accumulated by applying the average pooling.

4.2.2   Item-Item Product

In the recommendation task, we should not only consider the group-level information for sequential recommendation, but also try to capture relations between items explicitly. We can find out that the items in the input L items and the target one are closely related. As a result, we learn to aggregate the inner product between the input and output item embeddings to capture the item relations between L and the target item: $\sum_{v \in S_{u,t}} \mathbf{v}^\top \cdot E$, where $E \in \mathbb{R}^{d \times N}$ is the target item embedding. Then the latent information between items can be accumulated by the aggregation operation.

4.2.3   Prediction and Training

We use the matrix factorization (MF) to capture the global and long-term preferences of users. For user u, given the subsequence at time t as the input, the model outputs the score of item v:

$$\hat{r}(u, v) = \mathbf{u}^\top \cdot \mathbf{e}_v + \mathbf{s}_{u,t}^{avg\top} \cdot \mathbf{e}_v + \sum_{v \in S_{u,t}} \mathbf{v}^\top \cdot \mathbf{e}_v \qquad (4)$$

where $\mathbf{e}_v \in \mathbb{R}^d$ is the v-th output item embedding.

In prediction layer, different parts capture different latent information. The matrix factorization part captures the long-term preference of users and the hierarchical gating network module models the short-term user interests. In the third part, the item-item product represents the relations between various items.

Inspired by [19], we adopt the Bayesian Personalized Ranking objective to optimize our proposed model by the pairwise ranking between the interacted and unobserved items. The pairwise loss function is:

$$\mathcal{L} = \sum_{(u, L_u, v, v^-) \in \mathcal{D}} -\log \sigma \left(\hat{r}(u, v) - \hat{r}(u, v^-) \right) + \lambda \left(\|\Phi\|^2 \right) \qquad (5)$$

where L_u is one of the subsequences of user u, v is the target item that u will interact, and v^- denotes the unobserved negative item, Φ is the model parameters, λ is the regularization parameter.

4.3   Streaming Sequential Model Updating Method

In this section, we extend our offline model to a streaming setting. Our aim is to update our model with the new arrival data while keeping the knowledge learned from the historical sequences.

To keep the long-memory of the historical data, the reservoir technique [25] is widely used in the streaming database management systems. We conduct a random sampling strategy to decide the data stored in the reservoir.

For sequential recommendation, we need to note that the items come one by one in a streaming setting. However, we cannot view the items as individuals because each item is closely related to the items in its neighbourhood. To keep the sequence nature of coming data, we consider the L successive items together.

We maintain the l-th successive L items in a user's interaction history as a subsequence $S_{u,l}$. It is worth noting that when a new item comes over at time t, we combine it with the $(L-1)$ previous items in a buffer to form the new subsequence $S_{u,t}$.

The subsequences stored in the reservoir are denoted as C, and the data instance arrives at time step t. When $|C| < t$, we will store $S_{u,t}$ in the reservoir with the probability:

$$p_{\text{store}} = \frac{|C|}{t} \tag{6}$$

and uniformly replace a random subsequence already in the reservoir. Moreover, to capture the information contained in the latest generated data, we update the pre-trained model by training on the union of the reservoir C [8] and new data S^{new}.

The reservoir sampling strategy above enables the model to continually update according to the new and old data. However, in the streaming setting, the limited computing resources decide that few data instances can be utilized to update the model. Actually, it is more worthwhile to update the model with the subsequence which can change the model most. Such a subsequence is called an informative subsequence. Therefore, we propose an active sequential sampling strategy that samples the most informative subsequence.

We use the prediction function MF, i.e., the inner product of the latent factors, with the current parameters to compute the informativeness of a subsequence. Let $p_u \in \mathbb{R}^{1 \times d}$ and $q_v \in \mathbb{R}^{1 \times d}$ be the latent factors of user u and item v respectively. Then, the inner product of $p_u \in \mathbb{R}^{1 \times d}$ and $q_v \in \mathbb{R}^{1 \times d}$ is denoted as follows:

$$s_{u,v} = <p_u, q_v> = p_u \otimes q_v \tag{7}$$

The value of $<p_u, q_v>$ shows how closely p_u is related to q_v and denotes how much user u likes item v.

Traditional active learning strategy only samples one item at a time. However, in sequential recommendation, each subsequence is composed of several successive items. As a result, we use the numerical sum of individual items to measure the informative score of the subsequence. Given the i-th subsequence $S_i = (v_1, \ldots, v_{L-1}, v_L)$, the informative score S_{score} of the subsequence is computed as:

$$S_i^{score} = \frac{\sum_{k=1}^{L} s_{u,k}}{L} \tag{8}$$

Then we rank all the subsequences by their scores in a rank list in a descending order. If a subsequence gets a low score from the prediction model, it means this subsequence is worthwhile to be taken into consideration because it contains the latest preference of a user that the current model cannot learn well. In other words, the lower the rank is, the more likely it is to be selected. For subsequence S_i with $rank_i$, the sampling probability is calculated as follows:

$$p(S_i) = \frac{rank_i}{\sum_{s_i \in C \cup S^{new}} rank_i} \tag{9}$$

Then we update our current model with new sampled subsequences. The detailed construction of online training method is shown in Algorithm 1.

4.4   Adaptive Adversarial Negative Sampler

During the period of optimization, we sample the items that user u has never interacted with as the negative samples. However, most existing negative sampling strategies use popularity-biased sampling strategies [2] or random sampling [1] to generate negative samples. These methods are not informative enough and cannot capture complicated latent user or item representations. In addition, the existing negative sampling strategies cannot capture the user interests drift in streaming sequential recommendation, which greatly affects the effectiveness of online training.

Algorithm 1: Online Update Method

Input:
   the current model M, the current reservoir C, the new subsequences S^{new}
Output:
   the updated model M' and the updated reservoir C';
1: **for** each subsequence S_i in $C \cup S^{new}$ **do**
2:    Compute the subsequence score S_i^{score} by Eq. 8;
3: **end for**
4: Compute the sample probability $p(S_i)$;
5: Sample the subsequence set S from $C \cup S^{new}$ according to Eq. 9;
6: Update current model M with S to M';
7: **for** each subsequence S_i in S^{new} **do**
8:    Update the reservior C to C' with S_i according to Eq. 6;
9: **end for**

In this section, to overcome the shortcomings of existing sampling methods, we propose an adversarial training framework based on GAN in streaming sequential recommendation. There are two components in this framework: the discriminator and the generator. The discriminator is the sequential recommendation model and the target is to distinguish the true items from false items generated by the generator. The aim of generator is to generate adversarial negative items which can confuse the discriminator.

4.4.1   Discriminator

Let D_θ denote the discriminator component, where θ denotes the set of all parameters in G. The loss function of D is:

$$\mathcal{L}_D = \sum_{(u, L_u, v, v^-) \in \mathcal{D}} -\log \sigma \left(\hat{r}_D (u, v) - \hat{r}_D (u, v^-) \right) + \lambda \left(\|\Phi\|^2 \right)$$

$$\left(u, v^- \right) \sim P_G \left(u, v^- \mid u, v \right)$$

(10)

Given a positive user-item interaction (u, v), the probability distribution for generating negative interaction (u, v^-) by the generator is $P_G(u, v^- \mid u, v)$; $\hat{r}_D$ denotes the ratings of items calculated by the discriminator, which is the same as $\hat{r}$ defined in Eq. 5. We have described the optimization procedures in Sect. 4.2.3. Equation 10 is different from Eq. 5 in that the negative item v^- in Eq. 10 is generated by the generator.

4.4.2 Generator

The generator aims to generate adversarial negative items and use them to deceive the discriminator. We employ the metric learning approach to learn a metric space to find the noise items which are most similar to the positive items. The similarity between users and items is measured by Euclidean distance.

To enable the generator have the ability to generate more plausible items to deceive the discriminator, the objective function maximizes the expectation of $-d_D(u, v^-)$:

$$\mathcal{L}_G = \sum_{(u,v) \in \mathcal{S}} \mathbb{E}\left[-d_D\left(u, v^-\right)\right]$$
$$\left(u, v^-\right) \sim P_G\left(u, v^- \mid u, v\right) \tag{11}$$

The probability distribution $P_G(u, v^- \mid u, v)$ is as follows:

$$P_G\left(u, v^- \mid u, v\right) = \frac{\exp\left(-d_G\left(u, v^-\right)\right)}{\sum_{v^- \in \mathcal{V}_u^-} \exp\left(-d_G\left(u, v^-\right)\right)} \tag{12}$$

where $d_G(u, v)$ is the Euclidean distance between user u and item v measured by the generator through a neural network.

In particular, in order to maintain the characteristic of sequential recommendation, we use the latest successive L items $\mathcal{S}_{u,t}$ that a user has interacted with to represent the current user's preferences.

We feed the embedded vectors into a multilayer perceptron (MLP) separately so that the user u and item v can be represented in the same latent space. Then we measure the distance d_G between the output user and item vectors. $\mathcal{V}_u^-$ is a large set including all items which user u has not interacted with. In order to improve the training efficiency, we only select T items to form a small subset $\mathcal{V}_u^-$ from the whole item set $\mathcal{V}_u^-$.

However, the original GAN and their variants are usually used to generate continuous data in image processing domain. The gradient descent cannot be directly applied to solve the GAN formulation because sampling from a multinomial distribution is a discrete process. To tackle this challenge, some works [24] adopt a policy gradient strategy such as reinforcement learning, which utilizes a policy strategy to estimate the gradients of generator. However, this method often suffers from high variance.

We adopt the Gumbel-Softmax distribution [14] in optimization process to improve the stability of training. It is a continuous approximation to a multinomial distribution parameterized in terms of the softmax function. Let $\mathbf{g}$ be a K-dimensional noise vector sampled from $Gumbel(0,1)$. Then we obtain the sampled item v in an approximate one-hot representation:

$$v_i = \frac{\exp\left(\left(\log P_G\left(v_i\right) + g_i\right)/\tau\right)}{\sum_{j=1}^{K} \exp\left(\left(\log P_G\left(v_j\right) + g_j\right)/\tau\right)} \quad \text{for } i = 1,\ldots,K \tag{13}$$

where τ is an inverse temperature parameter. When τ approaches 0, samples become one-hot and the Gumbel-Softmax distribution is the same as the multinomial distribution P_G. When τ approaches positive infinity, the samples are the uniform probability vector. In this way, the whole process can be differentiable, and the GAN on discrete data can be trained by using τ approaches a standard backpropagation algorithm such as Adagrad. The detailed training process is shown in Algorithm 2.

Algorithm 2: The Adversarial Training Algorithm

Input:
    Training data S, number of epochs and batchsize;
Output:
    The diacriminator D, the generator G;
1: Initialize the parameters θ_G and θ_D for D and G
2: **for** each epoch **do**
3:     Sample a mini-batch $s_{batch} \in S$
4:     **for** each u in s_{batch} **do**
5:         Randomly sample T negative items: V_u^-;
6:         Calculate the sampling probability distribution based on Eq. 12;
7:         Sample a K-dimension $g \sim Gumble(0,1)$;
8:         Obtain a negative item v^- based on Eq. 13;
9:     **end for**
10:    Update θ_G and θ_D;
11: **end for**

---

5 Experiments

In this section, we first introduce our experimental setup. Then we show the results and compare with other methods to evaluate the effectiveness of our proposed steaming sequential recommendation.

5.1 Datasets

We use two datasets *MovieLens-20M* and *Amazon-CDs* from different domains as our experimental data source. *MovieLens-20M*[1] is a widely used dataset with

---

[1] https://grouplens.org/datasets/movielens/20m/.

20 million user-movie interactions for evaluating recommender systems. *Amazon-CDs*[2] dataset is a Amazon consumer review dataset, which includes a huge amount of user-item interactions, e.g., user ratings and reviews. We delete all user-item interactions with ratings less than four because these interactions cannot be considered as implicit feedbacks. We only retain the users with more than nine ratings and the items with more than four ratings to prevent noisy data. The statistics of two datasets after preprocessing are summarized in Table 1.

Table 1. Statistics of two datasets.

Dataset	Users	Items	Interactions	Density
MovieLens-20M	129,797	13,663	9,926,630	0.560%
Amazon-CDs	17,052	35,118	472,265	0.079%

5.2 Evaluation Setup

For sequential streaming recommendation, to mimic a real streaming recommendation scenario, we follow the dataset splitting strategy used in [24]. We order all user-item interactions records in time order and then split them into two parts (60% and 40%). The first part forms an initial training set which denoted as D^{train}. The remaining part is divided into four equal test sets denoted as D_1^{test} ... D_4^{test} to simulate the streaming settings. First of all, we train our model on first partition D^{train} to decide the initial parameters and the best hyper parameters. After the initial training, the four test sets are predicted one by one. We use the previous test set D_{i-1}^{test} to update the model, and then use the newly updated model to predicts the current test set D_i^{test}. After testing on D_i^{test}, the model will be updated by the next test set D_{i+1}^{test}.

To evaluate the performance of our model, we utilize two popular evaluation metrics commonly used in previous works: *Recall@10* and top-k Normalized Discounted Cumulative Gain (*NDCG@10*). *Recall@k* shows the percentage of correctly predicted samples in all positive samples. *NDCG@k* represents the position information of correctly predicted samples.

In our experiments, we set $d = 50$ for the latent dimension of all the methods. For our proposed model, we set $L = 5$. The batch size is set to 1024. The learning rate is set to 0.001 for *MovieLens-20M* dataset, and 0.0001 for *Amazon-CDs*.

---

[2] http://jmcauley.ucsd.edu/data/amazon/.

5.3   Baseline Methods

We compare our proposed model with the following sequential recommendation baselines in streaming settings:

- **GRU4Rec** [11]: This method uses gated recurrent unit for recommendation. It models user-items interactions by using recurrent neural networks to solve the session-based recommendation problem.
- **Caser** [23]: It is a sequential recommendation model uses convolutional neural network. It embeds the recent $|L|$ items to latent spaces and learn sequential patterns using convolutional filters.
- **SASRec** [16]: This method introduces Self-Attention to sequential recommendation. SASRec models the user-item interactions and can consider the interacted items adaptively.
- **HGN** [18]: Hierarchical gating networks. It captures long-term and short-term user interests from feature and instance level without utilizing complicated convolutional or recurrent neural networks.

To evaluate the streaming strategy in sequential recommendation system, we further apply our SSRGAN model in four various streaming settings. Since all existing sequential recommendation methods are not applied in streaming scenario, we implement these streaming strategies on *MovieLens-20M* dataset with huge amount of interactions, and compare SSRGAN with the following variant versions:

- **SSRGAN-Static:** For this method, we simply use the offline pre-trained model and eliminate the online update part.
- **SSRGAN-New** [15]: This method retrain the offline pre-trained model simply with only new coming events.
- **SSRGAN-Item** [6]: This method performs active item sampling strategy on the union set of new arrival and the current reservoir.
- **SSRGAN-Seq:** This method samples the most informative subsequence from the union set of the current reservoir and new arrival data actively. It keeps the sequentiality of data at the same time.

5.4   Experimental Results

5.4.1   Streaming Sequential Recommendation Results

The general experimental results are shown in Fig. 2.

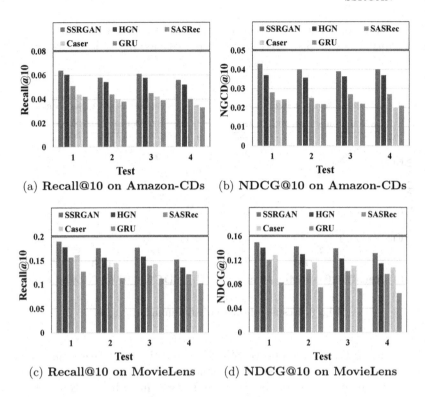

(a) Recall@10 on Amazon-CDs   (b) NDCG@10 on Amazon-CDs

(c) Recall@10 on MovieLens   (d) NDCG@10 on MovieLens

Fig. 2. Results of streaming sequential recommendation.

We can see that our proposed model performs better than other baselines in streaming settings on both datasets. Our SSRGAN obtains better results than the state-of-the-art HGN model because we adopt an adversarial architecture to train our model, which shows that it is important to generate informative negative items to improve the recommendation results. The RNN-based method GRU4Rec cannot perform well may because it is firstly designed for session-based recommendation and ignores the long-term user interest. Caser and SASRec both model the user-item and item-item to capture the long and short-term preference of users. This could be a possible reason for their better performance than GRU4Rec. HGN model performs better than other baselines because it applies both instance and feature-level selection, which plays an important role in modeling users preference. Also, the item-item relations between relevant items was captured in HGN.

We can also notice that our SSRGAN method perform better in *MovieLens-20M* than in *Amazon-CDs*. The main reason that is *Amazon-CDs* is sparaser than *MovieLens-20M*. In *Amazon-CDs*, a user may has a small amount of reviews, which is difficult to form subsequence of sufficient length.

5.4.2   Effect of Different Streaming Update Strategies

The performances of SSRGAN with various streaming update strategies are shown in Fig. 3.

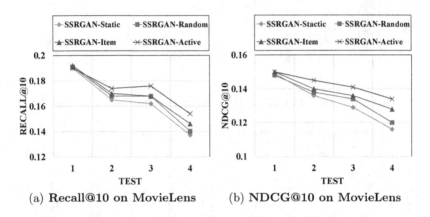

(a) Recall@10 on MovieLens   (b) NDCG@10 on MovieLens

Fig. 3. Results of different streaming strategies.

We can see that our proposed SSRGAN model with active sequence sampling strategy achieves the best performance in all situations. As time goes by, the recommendation results get worse and worse in SSRGAN-Static because users preference drifts in some degree. An important conclusion we can easily get is that we should update the model in time because the static model cannot capture the latent feature in new streaming data. The SSRGAN-Random method has a improvement over the static method. However, this method does not bring satisfactory results because randomly selected samples may not contribute enough to the model update. Furthermore, the SSRGAN-Item method cannot perform as well as our model mainly because it fails to maintain the sequential information in active sampling. Our active sampling method which chooses the most informative subsequences is proved to be effective in dealing with the continuous, high speed streaming data.

5.4.3   Effect of Adversarial Training

Since other sequential recommendation baselines are not aimed for streaming recommendation, We also conduct our experiments in offline settings. We eliminate the online streaming update part and retain the adversarial training part to prove the effectiveness of GAN. Following [3], we use the 70% of data as the training set and use the next 10% of interactions as the validation set to choose the best hyper-parameter for all methods. The remaining 20% of data in each user sequences are used as the test.

The performance comparison results are shown in Table 2. It can be observed that: (1) Our SRGAN model is consistently better than these baselines by a large margin in offline settings. (2) All results of all those offline methods are not as good as those methods in online settings. The most likely reason is that the datasets are divided differently in online/offline settings.

Table 2. The offline performance comparison of all sequential recommendation methods.

	MovieLens-20M		Amazon-CDs	
	Recall@10	NDCG@10	Recall@10	NDCG@10
GRU4Rec	0.0814	0.0826	0.0322	0.0244
Caser	0.1170	0.1293	0.0316	0.0242
SASRec	0.1069	0.1214	0.0371	0.0293
HGN	0.1224	0.1418	0.0462	0.0398
SRGAN	**0.1290**	**0.1507**	**0.0505**	**0.0436**

6 Conclusion

In this paper, we investigate the sequential recommendation problem in a streaming setting and propose a novel SSRGAN model to tackle the encountered challenges. To deal with the continuous streaming sequential data, we introduce a reservoir-based subsequence storage mechanism and an active subsequence sampling strategy to update the model. Moreover, to improve the effectiveness and efficiency of online model training, we propose a negative sampling strategy based on GAN to generate the most informative negative samples and use Gumble-Softmax to overcome the gradient block problem. We conduct extensive experiments on two real-world datasets and the results show the effectiveness of our approaches in streaming sequential recommendation task.

Acknowledgements. This work was supported by the National Natural Science Foundation of China under Grant Nos. 61872258, 61802273, Major project of natural science research in Universities of Jiangsu Province under grant number 20KJA520005.

References

1. Bansal, T., Belanger, D., McCallum, A.: Ask the GRU: multi-task learning for deep text recommendations. In: RecSys 2016, pp. 107–114 (2016)
2. Chen, T., Sun, Y., Shi, Y., Hong, L.: On sampling strategies for neural network-based collaborative filtering. In: SIGKDD 2017, pp. 767–776 (2017)
3. Chen, X., Xu, J., Zhou, R., Zhao, P., Zhao, L.: S^2R-tree: a pivot-based indexing structure for semantic-aware spatial keyword search. GeoInformatica **24**(11), 3–25 (2019). https://doi.org/10.1007/s10707-019-00372-z
4. Das, A., Datar, M., Garg, A., Rajaram, S.: Google news personalization: scalable online collaborative filtering. In: WWW 2007, pp. 271–280 (2007)
5. Dauphin, Y.N., Fan, A., Auli, M., Grangier, D.: Language modeling with gated convolutional networks. In: ICML 2017, pp. 933–941 (2017)

6. Diaz-Aviles, E., Drumond, L., Schmidt-Thieme, L., Nejdl, W.: Real-time top-n recommendation in social streams. In: RecSys 2012, pp. 59–66 (2012)
7. Goodfellow, I.J., et al.: In: NIPS 2014, pp. 2672–2680 (2014)
8. Guo, L., Yin, H., Wang, Q., Chen, T., Zhou, A., Hung, N.Q.V.: Streaming session-based recommendation. In: SIGKDD 2019, pp. 1569–1577 (2019)
9. He, R., McAuley, J.J.: Fusing similarity models with Markov chains for sparse sequential recommendation. In: ICDM 2016, pp. 191–200 (2016)
10. Hidasi, B., Karatzoglou, A.: Recurrent neural networks with top-k gains for session-based recommendations. In: CIKM 2018, pp. 843–852 (2018)
11. Hidasi, B., Karatzoglou, A., Baltrunas, L., Tikk, D.: Session-based recommendations with recurrent neural networks. In: ICLR 2016 (2016)
12. Hu, X., Xu, J., Wang, W., Li, Z., Liu, A.: A graph embedding based model for fine-grained POI recommendation. Neurocomputing 428, 376–384 (2020)
13. Huang, Y., Cui, B., Zhang, W., Jiang, J., Xu, Y.: TencentRec: real-time stream recommendation in practice. In: SIGMOD 2015, pp. 227–238 (2015)
14. Jang, E., Gu, S., Poole, B.: Categorical reparameterization with gumbel-softmax. In: ICLR 2017 (2017)
15. Jugovac, M., Jannach, D., Karimi, M.: Streamingrec: a framework for benchmarking stream-based news recommenders. In: RecSys 2018, pp. 269–273 (2018)
16. Kang, W., McAuley, J.J.: Self-attentive sequential recommendation. In: ICDM 2018, pp. 197–206 (2018)
17. Li, Y., Xu, J., Zhao, P., Fang, J., Chen, W., Zhao, L.: ATLRec: an attentional adversarial transfer learning network for cross-domain recommendation. J. Comput. Sci. Technol. 35(4), 794–808 (2020). https://doi.org/10.1007/s11390-020-0314-8
18. Ma, C., Kang, P., Liu, X.: Hierarchical gating networks for sequential recommendation. In: SIGKDD 2019, pp. 825–833 (2019)
19. Rendle, S., Freudenthaler, C., Gantner, Z., Schmidt-Thieme, L.: BPR: Bayesian personalized ranking from implicit feedback. In: UAI 2009, pp. 452–461 (2009)
20. Rendle, S., Freudenthaler, C., Schmidt-Thieme, L.: Factorizing personalized Markov chains for next-basket recommendation. In: WWW 2010, pp. 811–820 (2010)
21. Rendle, S., Schmidt-Thieme, L.: Online-updating regularized kernel matrix factorization models for large-scale recommender systems. In: RecSys 2008, pp. 251–258 (2008)
22. Subbian, K., Aggarwal, C.C., Hegde, K.: Recommendations for streaming data. In: CIKM 2016, pp. 2185–2190 (2016)
23. Tang, J., Wang, K.: Personalized top-n sequential recommendation via convolutional sequence embedding. In: WSDM 2018, pp. 565–573 (2018)
24. Wang, Q., Yin, H., Hu, Z., Lian, D., Wang, H., Huang, Z.: Neural memory streaming recommender networks with adversarial training. In: SIGKDD 2018, pp. 2467–2475 (2018)
25. Wang, W., Yin, H., Huang, Z., Wang, Q., Du, X., Nguyen, Q.V.H.: Streaming ranking based recommender systems. In: SIGIR 2018, pp. 525–534 (2018)
26. Xu, C., et al.: Recurrent convolutional neural network for sequential recommendation. In: WWW 2019, pp. 3398–3404 (2019)

Topological Interpretable Multi-scale Sequential Recommendation

Tao Yuan[1,2], Shuzi Niu[2], and Huiyuan Li[2(✉)]

[1] University of Chinese Academy of Sciences, Beijing, China
[2] Institute of Software, Chinese Academy of Sciences, Beijing, China
{yuantao,shuzi,huiyuan}@iscas.ac.cn

Abstract. Sequential recommendation attempts to predict next items based on user historical sequences. However, items to be predicted next depend on user's long, short or mid-term interest. The multi-scale modeling of user interest in an interpretable way poses a great challenge in sequential recommendation. Hence, we propose a topological data analysis based framework to model target items' explicit dependency on previous items or item chunks with different time scales, which are easily changed into sequential patterns. First, we propose a topological transformation layer to map each user interaction sequence into persistent homology organized in a multi-scale interest tree. Then, this multi-scale interest tree is encoded to represent natural inclusion relations across scales through an recurrent aggregation process, namely tree aggregation block. Next, we add this block to the vanilla transformer, referred to as recurrent tree transformer, and utilize this new transformer to generate a unified user interest representation. The last fully connected layer is utilized to model the interaction between this unified representation and item embedding. Comprehensive experiments are conducted on two public benchmark datasets. Performance improvement on both datasets is averagely 5% over state-of-the-art baselines.

Keywords: Multi-scale modeling · Sequential recommendation · Topological data analysis

1 Introduction

Sequential recommendation aims to predict the next items by treating user historical interactions as sequences. Traditional multi-scale approaches aim to obtain the latent user interest representation and make accurate predictions. Here we put more emphasis on why a user interacted with an item at certain time.

It is well known that long and short-term user interests play a critical role in the next item prediction task. A sequence of rated movies from IMDB before time t is depicted at different time scales in Fig. 1. We simply use its IMDB categories to describe each movie. Long-term user interest represents the stable information

© Springer Nature Switzerland AG 2021
C. S. Jensen et al. (Eds.): DASFAA 2021, LNCS 12683, pp. 53–68, 2021.
https://doi.org/10.1007/978-3-030-73200-4_4

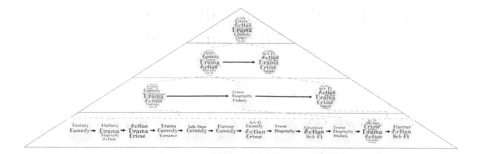

Fig. 1. Illustration of multi-scale user interests on movies in MovieLens-100k

of the sequence shown as the top level in Fig. 1. Short-term user interest changes dynamically along with time as the bottom level in Fig. 1. Only through the long-term user interest, Eve's Bayou (1997) labeled with "Drama" will be easily predicted at time t. For movie "Gattaca(1997)" labeled with "Drama, Sci-Fi, Thriller", it is hard to be predicted only by the top and bottom level in Fig. 1. Taking two middle levels into consideration, we can see that this example user is obviously interested in "Drama, Sci-Fi, Thriller" as shown in the last category cloud from each mid-term scale. It is necessary to incorporate multiple scales user interests into the next item prediction task.

Though some studies [14,23] take the multi-scale nature into consideration, user interests at different scales are only represented in a latent space and where are they from remains unknown. Therefore, how to model multi-scale user interests in an explicit and explainable way becomes a great challenge. To tackle this challenge, we propose a topological data analysis based framework. To explicitly capture multiple scales user interests, we introduce persistent homology in a topological space. To understand which patterns contribute to the target item, we propose a recurrent tree transformer to obtain a unified user interest representation by weighted average over different scales and drive each pattern's support value. Thus **T**opological **I**nterpretable **M**ulti-scale sequential recommendation is proposed, referred to as TIMe.

Specifically, we first formalize interpretable multi-scale sequential recommendation problem in terms of sequential pattern mining. Then, a topological transformation layer is proposed to map the user interaction sequence into item complex sequences with different diameters. Next, multi-scale interest tree is derived from those item complex sequences to describe the inclusion relation between complexes of different scales. Correspondingly, a tree transformer is proposed to obtain a unified latent representation by recursively aggregating latent representations of units in this user interest tree in a bottom-up manner. Attention matrix from this transformer will tell us which units play an essential role in next item prediction. Finally, a fully connected layer is introduced to model the interaction between this unified representation of user interest and item embedding. We conduct comprehensive experiments on two public benchmark datasets,

MovieLens-100k and Amazon Movies & TV. Experimental results demonstrate that performance improvement is averagely 5% over state-of-the-art baselines.

Our main contributions lie as follows. (1) We formalize the interpretable multi-scale sequential recommendation problem as a sequential pattern mining problem. (2) Topological transformation layer is proposed to explicitly model multi-scale user interests as candidate causal patterns. (3) Tree aggregation block is proposed to learn tree node embedding keeping parent-child relations in a recurrent aggregation process. (4) Recurrent tree transformer is designed to obtain a unified user interest representation and each pattern's support value from attention matrices.

2   Related Work

We briefly review the related studies on the sequential recommendation methods and the topological data analysis methods.

2.1   Sequential Recommendation

General recommendation algorithms mainly take the user-item interaction history as a whole set by ignoring the time effect. Different from that, sequential recommendation aims at identifying the sequential pattern from user historical sequences and predict the next items according to these patterns [13]. Intuitively, the time interval length of the pattern from the prediction point has effect on the prediction task. According to different scales of focused patterns' time interval length, existing approaches usually fall into the following categories.

Long-Term Sequence Model. For the long-term sequence model, the whole historical sequence is supposed to be significant to the next item prediction. It adopts the global sequence to capture the user long-term interests, such as DREAM [21], SASRec [9] and BST [1]. Dynamic Recurrent Basket Model (DREAM) learns a user dynamic representation and global sequential patterns among baskets through recurrent neural network. While Self-Attention based Sequential model (SASRec) and Behavior Sequence Transformer(BST) capture user's long-term interests using an self-attention mechanism and powerful Transformer model respectively.

Short-Term Sequence Model. Considering the time sensitivity of user interest, short-term sequence model tends to emphasize the effect of the last interaction or at the near time on the prediction. It employs the local user sequence to model user short-term interests, such as STAMP [11] and GRU4REC [7]. Short Term Attention Memory Priority model (STAMP) explicitly takes the effect of user current actions on the next item prediction into consideration from the short-term memory of last items. GRU4REC captures the sequential click pattern in a session by GRU [2] and predicts the next click by optimizing ranking based measures.

Long-Short-Term Sequence Model. Long-short-term sequence model aims at modeling both long-term and short-term effect on the next item prediction, such as HRM [20], LSIC [23] and MARank [22]. Hierarchical representation model (HRM) represents global interest and local interest in a two-layer network and combine them with non-linear operators, so it's able to subsume some existing models theoretically. Long-term information changes slowly across time while short-term information sensitive to time. Long and Short-Term Information in Context aware recommendation (LSIC) is leveraged by adversarial training, where long and short-term information are learned through matrix factorization and recurrent neural network respectively. Suppose different levels of transition dependencies among items provide different important information for the next item prediction, Multi-order Attentive Ranking Model (MARank) is to unify individual-level and union-level item dependency for preference inference.

Multi-scale Sequence Model. In fact, each user has static interest in nature and also changes her personal interests. The static interest is viewed as the long-term effect, but the changing interest is not only corresponding to the short-term effect. In fact, there are multiple levels of dynamical interacting factors that have influence on the next item prediction, such as MARank [22] and HPMN [14]. MARank introduces two levels of item transition dependency for prediction, but the union-level item transition dependency is not explicitly modeled. To capture the multi-scale sequential patterns, Hierarchical Periodic Memory Network is proposed by a hierarchical and periodical updating mechanism within a hierarchical memory network. Different from HPMN [14], we attempt to model the multi-scale sequential patterns explicitly and explain which scales are more important for the prediction task.

2.2 Topological Data Analysis and Persistent Homology

Topological data analysis (TDA) has been rapidly developed from theoretical aspects to applications in recent years [12]. Homology is an algebraic topological invariant that describes the holes in a space. The input to the persistent homology is given by a filtration in the topological space. Due to the variety of filtration diameters, persistent homology are widely used for capturing multi-scale topological features in data. Recent years have witnessed an increased interest in the application of persistent homology to machine learning problems. PCNN [10] investigates a way to use persistent homology in the framework of deep neural networks for dealing with audio signals. TSCTDA [18] generates a novel feature by extracting the structure of the attractor using topological data analysis to represent the transition rules of the time series.

3   Topological Interpretable Multi-scale Sequential Recommendation

To model the multi-scale user interests in an explainable and explicit way, we propose a topological data analysis based method for sequential recommendation, referred to as TIMe. Specifically, we first define the interpretable multi-scale sequential recommendation problem in terms of sequential pattern mining problem. Then, the whole network structure is depicted in detail. Finally, we discuss the relationship between TIMe and some other state-of-the-art sequential recommendation methods.

3.1 Formalization of Interpretable Multi-scale Sequential Recommendation

Let $U = \{u_1, u_2, \cdots, u_m\}$ be a set of users, and $I = \{i_1, i_2, \cdots, i_n\}$ be a set of items, where $|U| = m$, $|I| = n$. For each user u, a user behavior history before time t is represented by an ordered list: $L_u^{<t} = (I_u^1, I_u^2, \cdots, I_u^{t-1})$, where $I_u^{t-1} \subset I$ means a set of items interacted by user u at time $t - 1$.

Sequential pattern mining aims at finding the complete set of frequent subsequences from a set of sequences given a support threshold [13]. In sequential recommendation task, we want to identify such subsequence patterns to explain why a certain item is to be predicted next time. The length of a causal subsequence varies much, so only long and short-term patterns are not complete enough for next item prediction. Multi-grains of subsequences are needed for better explanation. Hence, we define an interpretable multi-scale sequential recommendation problem as below.

For a single scale sequential recommendation task, short term sequence models discover local sequential patterns like $I_u^i \to I_u^t$ and long term sequence models identify global sequential patterns like $\bigcup_{i=1}^{t-1} I_u^i \to I_u^t$. For a two-scale sequential recommendation task, long and short term sequence models attempt to learn such local and global patterns jointly. For multi-scale sequential recommendation task, the goal is to obtain sequential patterns with different time scales $P \to I_u^t, \forall P \in P_u = \{P_u^j | P_u^j \in I, j = 1, \ldots, d\}$ to predict the next items I_u^t.

3.2 Our Framework

To tackle this interpretable multi-scale sequential recommendation task, we propose the topological data analysis based method, namely TIMe. It maps the item sequence into multi-scale interest tree to explicitly model user interests at different time scales within persistent homology through the topological transformation layer. For multi-scale interest tree, node representations are derived by our proposed tree aggregation block. A novel tree transformer is proposed to learn the dependency of next items on this tree. The whole architecture is described in Fig. 2.

3.3 Topological Transformation Layer

Given a user historical sequence $L_u^{<t}$, we treat each item $i \in L_u^{<t}$ as a point and the sequence can be taken as a point cloud. To characterize the multi-scale topological properties of $L_u^{<t}$ in sequential recommendation, we newly define item i's ϵ-ball, item complex and item complex sequence with diameter ϵ [12].

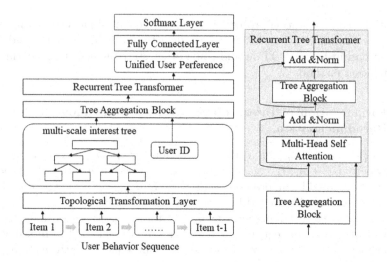

Fig. 2. Architecture of TIMe

Then Vietoris-Rips filtration method [25] is employed to construct the persistent homology. Specifically, we use item i's ϵ-ball based on Definition 1 to construct the topology for each item i.

Definition 1. Item i's ϵ-ball. $B_\epsilon(i) = \{y \in I : d_u(y, i) \leq \epsilon\}$ *is the ball with radius ϵ centered at i. $d_u(y, i) = \min_{t_y \in T_u(y), t_i \in T_u(i)} |t_y - t_i|$ is the time interval between items as the distance metric in topological space. $T_u(y)$ and $T_u(i)$ means the timestamp set of item y and i in the user u's historical sequence respectively.*

In Item i's ϵ-ball, edges between i and other items in this ball are constructed. An item simplex Δ is a topology with the item distance less than ϵ. Intuitively, we mainly focus on two kinds of simplexes: connected components and holes. Thus, the item complex can be built from item simplexes based on Definition 2.

Definition 2. Item Complex. *An item complex of diameter ϵ is a Vietoris-Rips complex $C(\epsilon) = \{\Delta | D(\Delta) \leq \epsilon\}$. $D(\Delta)$ is the diameter of the simplex Δ, which indicates the greatest item distance. The start time of item complex $C(\epsilon)$ is $t_s(C(\epsilon)) = \min_{\Delta \in C(\epsilon)} t_s(\Delta)$. The end time of item complex $C(\epsilon)$ is $t_e(C(\epsilon)) = \max_{\Delta \in C(\epsilon)} t_e(\Delta)$.*

Given a specific ϵ and a user sequence $L_u^{<t}$, a set of simplexes are generated from the sequence with the ϵ-ball method mentioned above. Each simplex Δ is tagged with its start time $t_s(\Delta)$ and end time $t_e(\Delta)$. For simplicity, we combine simplexes with intersected time span as an item complex, to make sure that there is no intersection between item complexes. Thus a sequence of item complexes are generated in temporal order, denoted as $L_u^{<t}(\epsilon)$ as Definition 3.

Algorithm 1. Multi-Scale Interest Tree Construction

Input: item complex sequences $L_u^{<t}$ with diameters $\epsilon_{1 \times d}$
Output: $\mathcal{T}_u^{<t}$
1: create an empty tree $\mathcal{T}_u^{<t} := \text{Tree}()$
2: get max diameter: $\epsilon_{\max} = \epsilon[-1]$
3: get item complexes: $C_u^0(\epsilon_{\max}) = L_u^{<t}(\epsilon_{\max})[0]$
4: $\mathcal{T}_u^{<t}.\text{create\_node}(\text{data}=C_u^0(\epsilon_{\max}),\text{children}=\text{null})$
5: $p = 1$
6: **for** $i = 2, \ldots, d$ **do**
7:     **for** Cnode $\in \mathcal{T}_u^{<t}.\text{leaves}()$ **do**
8:         **for** $C_u^j(\epsilon[d-i]) \in L_u^{<t}(\epsilon[d-i])$ **do**
9:             **if** $C_u^j(\epsilon[d-i]) \subset \text{Cnode}.data$ **then**
10:                $\mathcal{T}_u^{<t}.\text{create\_node}(\text{data}=C_u^j(\epsilon[d-i]),\text{children}=\text{null,pos}=\text{p++})$
11:                $\mathcal{T}_u^{<t}[\text{Cnode.pos}].\text{children.add}(\mathcal{T}_u^{<t}[p-1])$
12: **return** $\mathcal{T}_u^{<t}$

Definition 3. *User u's Item Complex Sequence with diameter ϵ.* $L_u^{<t}(\epsilon)$ *is an item complex sequence, i.e.* $C_u^1(\epsilon), \cdots, C_u^k(\epsilon)$. $\forall C_u^p(\epsilon), C_u^q(\epsilon) \in L_u^{<t}(\epsilon), 1 \leq p < q \leq k$, *satisfy* $t_e(C_u^p(\epsilon)) \leq t_s(C_u^q(\epsilon))$.

The increasing ϵ produces a filtration over this topological space of sequence $L_u^{<t}$. For example, $\epsilon_1 = 1 \min \leq \epsilon_2 = 60 \min \leq \cdots$ generate item complex sequences satisfying $L_u^{<t}(\epsilon_1) \subset L_u^{<t}(\epsilon_2) \subset \cdots$. This subset inclusion is further attributed to the element inclusion between two sequences with one diameter smaller than the other. There exist $C_u^q(\epsilon_2) \in L_u^{<t}(\epsilon_2)$ satisfying $C_u^p(\epsilon_1) \subset C_u^q(\epsilon_2)$ for each $C_u^p(\epsilon_1) \in L_u^{<t}(\epsilon_1)$. To precisely describe inclusion relations between complexes of different scales, we construct a multi-scale interest tree $\mathcal{T}_u^{<t}$ from complex sequences with multiple diameters in a top-down manner through Algorithm 1.

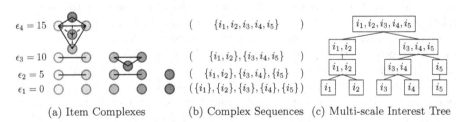

(a) Item Complexes   (b) Complex Sequences   (c) Multi-scale Interest Tree

Fig. 3. Illustration of an item sequence $(i_1, 0), (i_2, 1), (i_3, 12), (i_4, 13), (i_5, 15)$ through topological transformation layer as follows: (a)→(b)→(c).

Topological transformation layer transforms the item sequence into a multi-scale interest tree shown in Fig. 3. Through this layer, the basic unit of user interaction sequences are changed from item into item complex. When the diameter is very small ($\epsilon_1 = 0$), $L_u^{<t}(\epsilon_1)$'s topology are disconnected points as the last line in Fig. 3(a), and each point are an item complex, such as $C_u^0(0) = \{i_1\}$. When the diameter is very large ($\epsilon_4 = 15$), the topology of $L_u^{<t}(\epsilon_4)$ is a connected component, i.e., $C_u^0(15) = \{i_1, i_2, i_3, i_4, i_5\}$. Item complex sequence $L_u^{<t}(\epsilon)$ in Fig. 3(b) may exhibit appearance and disappearance of holes by changing the diameter ϵ. In a sense, the diameter parameter ϵ can be regarded as a resolution of the sequence $L_u^{<t}$. The inclusion relation between scales can be represented as multiscale interest tree through Algorithm 1 in Fig. 3(c). How to embed such a tree remains a question for the following section.

3.4 Tree Aggregation Block

Let $U \in \mathbb{R}^{m \times r}$ and $V \in \mathbb{R}^{n \times r}$ denote the embedding matrix of users and items respectively, where r is the latent dimension size. For each multi-scale interest tree $T_u^{<t}$, we propose to derive its representation from its leaves in a bottom-up manner based on item latent representations V, namely Tree Aggregation Block. For each leaf node $T_u^{<t}[p]$ at depth $d - 1$ (scale 1) of the tree ($T_u^{<t}[p]$.children = null), its representation $E_n[p]$ is equal to its item representation in V, and "itemid" attribute in Eq. (1) is to get item ids in the current item complex. For each non-leaf node at depth i (scale $d - i$, $0 \le i \le d - 2$) of $T_u^{<t}$, its representation $E_n[p]$ is derived from the aggregation of its children's representations. This process will proceed recurrently in a bottom-up manner until the root representation is derived. Here the AGG operation can be sum, mean, max and min in Eq. (1) and we use sum for experiments.

$$E_n[p] = \begin{cases} V[T_u^{<t}[p].data.itemid, :] & \text{otherwise} \\ \tanh(\text{AGG}(\{E_n[\text{Cnode.pos}]\}_{\text{Cnode} \in T_u^{<t}[p].\text{children}})) & !T_u^{<t}[p].\text{children} \end{cases} \quad (1)$$

3.5 Recurrent Tree Transformer

Traditional tree transformer is a transformer-based neural network to learn the relationship between elements in tree-structured data [4]. To identify which item complexes in multi-scale interest tree $T_u^{<t}$ are important for item prediction at time t, we propose to update tree node embedding by Tree Aggregation Block after self-attention, namely recurrent tree transformer in the right part of Fig. 2.

To encode the temporal order between item complexes in $T_u^{<t}$, we add positional embedding to the current node embedding E_n. For each item complex $C \in T_u^{<t}$, we compute its timestamp as the averaged timestamp of items in C as Eq. (2), where $t_C(k)$ means the timestamp that item k appears in C. Item complexes in the multiscale interest tree are sorted according to their timestamps. Each item complex's rank is taken as its position and encoded in the original way [19]. The corresponding position embedding matrix is denoted as E_p. The input embedding for tree nodes are computed as $E = E_n + E_p$.

$$t(C) = \frac{1}{|C.\text{itemid}|} \sum_{k \in C.\text{itemid}} t_C(k) \quad (2)$$

Self-attention only models the interaction between input nodes, but we focus on the node weight for the next item prediction. So we concatenate these tree nodes and an additional user node. With the recommendation timestamp as the user node's timestamp, its position is the number of tree nodes plus 1, which can be represented in the same way as [19], denoted as E_p^u. The initialized user node embedding is $E_n^u = U[u,:]$. We augment the tree node embedding matrix E with a row for user node as $X_u^0 = [E; E_n^u + E_p^u]$ for initialization.

Taking the concatenation of tree nodes and user node as input, our proposed recurrent tree transformer consists of two major building blocks: multi-head self-attention block and tree aggregation block. The input of the attention module consists of query, key, and value in Eq. (3) and Eq. (4). Here we only concern of the user node's effect on tree nodes by masking the unnecessary interactions from tree nodes to the user node. So we use a zero matrix $Mask$ and set $Mask[:,-1] = 1$ in Eq. (3), where τ is small enough, such as -10^9 in our experiments.

For the b-th recurrent tree transformer layer, the output of self-attention module is a weighted sum of the value in Eq. (4), where the weight matrix is determined by query and its corresponding key in Eq. (3).

$$A_u^b = \text{softmax}(\tau Mask + \frac{(X_u^{b-1}W_Q) \cdot (X_u^{b-1}W_K)^T}{\sqrt{l}}) \tag{3}$$

$$head_i^b = \text{Attention}(X_u^{b-1} \cdot W_Q, X_u^{b-1} \cdot W_K, X_u^{b-1} \cdot W_V) = A_u^b \cdot X_u^{b-1}W_V$$
$$S_u^b = \text{MH}(X_u^{b-1}) = \text{Concat}(head_1^b, head_2^b, \cdots, head_h^b)W_H \tag{4}$$

where $W_Q, W_K, W_V \in \mathbb{R}^{r \times r}$ are weight matrices for query, key and value respectively, and h is the number of heads. Then tree aggregation block is used to update tree node representations as $X_u^b = \text{TAB}(S_u^b)$, whose detailed implementations are described in the above subsection. Both dropout and layer normalization are utilized in two blocks.

We stack such B recurrent tree transformer layers. The output of the final layer is corresponding to the unified user interest representation $H_u^t = X_u^B[-1,:]$ derived from multi-scale interest tree $T_u^{<t}$. The last fully connected layer is used to model the interaction between items and users. Taking the multi-scale interest representation H_u^t as input, the weight matrix $W_I \in \mathbb{R}^{r \times n}$ for the fully connected layer can be used as the implicit item representation. It outputs scores over all the items for user u at time t by $O_u^t = H_u^t \cdot W_I$. The predicted probability that user u will buy or browse an item at time t is defined as $\hat{I}_u^t = softmax(O_u^t)$.

Next item prediction task at a certain time is reduced to a multi-class classification problem. Thus we use the cross entropy function as the optimization objective in Eq. (5) to learn network parameters.

$$L = -\sum I_u^t \cdot log(\hat{I}_u^t) + \lambda \|\Theta\| \tag{5}$$

I_u^t means the items that the user u is actually interacting with at time step t, λ is the regularization factor, Θ are network parameters.

3.6   Sequential Patterns Mined from TIMe

The interpretability of TIMe lies in the multi-scale interest tree. Take user u at time t for example, all candidate sequential patterns can be derived from multi-scale interest tree $\mathcal{T}_u^{<t}$ as follows

$$\{(\mathcal{T}_u^{<t}[p].data \rightarrow I_u^t, \rho[p]) | 0 \leq p \leq |\mathcal{T}.u^{<t}|, \rho = f(\{A_u^b\}_{b=1}^B)\}.$$

Support values ρ for all the candidate patterns are derived from the attention matrix A_u^b in Eq. (3). Intuitively, we adopt the attention matrix A_u^B from the final recurrent tree transformer layer as support values, $\rho = A_u^B[-1,:]$.

3.7   Discussion

Theoretically, we show that our proposed TIMe subsumes several existing methods when choosing proper $\{\epsilon_i\}_{i=1}^d$. The key lies in the Topological Transformation Layer (TTL). Here we consider the following three special cases of TTL. (1) $d = 1$ and ϵ_1 is small enough that each simplex contains only one item in the output sequence of TTL, such as SASRec [9] and BST [1]. (2) $d = 1$ and ϵ_1 is large enough that TTL output a sequence with only one item complex. All the items in this complex are taken as a set, such as matrix factorization based recommendation approaches [15]. (3) $d = 2$ and ϵ_1 is small enough while ϵ_2 is large enough. Some long and short sequence models [20,23] are of this kind.

4   Experiments

First, we compare TIMe with state-of-the-art baselines on public benchmark datasets. Then the role of each component is explored. Finally, parameter analysis and qualitative analysis are further discussed.

4.1   Experiment Setting

Datasets. Experiments are carried out on two public benchmark datasets, i.e. MovieLens-100k and Amazon Movies & TV. MovieLens-100k is a subset of the MovieLens dataset [5]. It contains $100,000$ anonymous ratings of approximately $1,682$ movies made by 943 MovieLens users. All selected users have rated at least 20 movies. Amazon Movies & TV is a subset of the Amazon Product dataset [6]. In this paper, the dataset is preprocessed to obtain a 50-core subset [21], i.e. every user rates in total at least 50 items and vice versa. The final dataset has $2,074$ users, $7,352$ movies and $268,586$ ratings. For both sets, we split the training and test dataset with the ratio $8 : 2$ according to the temporal order of user's historical behaviors. Given a minimum length M and a sequence L, we can construct $L[1 : M], L[1 : M + 1], \ldots, L[1 :]$ as training sequences.

Baseline Methods. Our baselines mainly include the following five groups. (1) General recommendation methods: ItemKNN [17] and BPRMF [15]. (2) Long term sequence models: SASRec [9], S^3Rec [24] and BST [1]. (3) Short term sequence models: STAMP [11] and GRU4REC [7]. (4) Long Short term sequence models: MARank [22]. (5) Multi-scale Sequence model: HPMN [14].

Implementation Details. We implement TIMe[1] and baselines based on NeuRec[2]. Generally, parameters with the best performance on the training set is chosen for the test set. For all the models, the batch size is 256 and the regularization coefficient is 0.001. For TIMe, the user and item embedding size are both 16. The minimum length of the training sequence is 16, the number of heads is 4 and the learning rate is 0.001. For MovieLens-100k, $\epsilon \in [60, 3.6 \times 10^3, 8.64 \times 10^4, 3.1536 \times 10^7]$ and $blocks = 1$. For Amazon Movies & TV, $\epsilon \in [8.64 \times 10^4, 6.048 \times 10^5, 3.1536 \times 10^7]$ and $blocks = 3$.

Evaluation Metrics. In this paper, we focus on the top-n recommendation and our evaluation measures are ranking based metrics, such as Precision [16], Recall [16], NDCG(Normalized Cumulative Gain) [8] and MRR(Mean Reciprocal Rank) [3]. All the metrics are at the cut-off 10.

4.2 Performance Analysis

Here we compare the performance of TIMe with five groups of baselines. Comparison results on both datasets are shown in Table 1. Under all evaluation metrics, performance differences between the bold type and other corresponding baselines are statistically significant with $p - value < 0.01$ in the two tailed paired t-tests. Compared with the other multi-scale sequential recommendation HPMN [14], the minimum and maximum relative improvement of TIMe occur under the evaluation of MRR@10 on MovieLens-100k and Recall@10 on Amazon Movies & TV respectively. Min.Imp.% and Max.Imp.% mean the relative performance improvement of TIMe compared with each baseline method under the corresponding evaluation condition respectively.

All sequential recommendation approaches outperform these two general recommendation methods on both datasets. The major reason lies in that sequential recommendation methods take the temporal order among items into consideration compared with general recommendation methods.

Most long term approaches (SASRec and S^3Rec) perform better than short term methods (STAMP and GRU4REC) on MovieLens 100k while they perform worse than short term methods on Amazon Movies & TV. This phenomenon coinsides with our intuition that the leading causal pattern for next item predictions is (global) long term user interests for some sequences, but (local) short term user interests for others. Long term model BST in Table 1 achieves better performances than short term methods on both sets because of explicitly modeling relations between causal patterns and the target item. Here comes the question: what about the combination of long and short term methods.

---

[1] https://github.com/hustyuantao/TIMe.
[2] https://github.com/wubinzzu/NeuRec.

Table 1. Performance comparison on two public benchmark datasets

Model	MovieLens-100k				Amazon Movies & TV				Min. Imp.%	Max. Imp.%
	Prec	Recall	NDCG	MRR	Prec	Recall	NDCG	MRR		
ItemKNN	.1601	.1192	.1931	.3676	.0411	.0202	.0435	.1063	14	81
BPRMF	.1616	.1196	.1894	.3507	.0376	.0189	.0385	.0967	19	92
SASRec	.1843	.1349	.2156	.3953	.0603	.0257	.0630	.1471	6.3	42
S^3Rec	.1887	.1357	.2247	.4065	.0649	.0269	.0675	.1498	3.3	36
BST	.1934	.1450	.2322	.4113	.0743	.0308	.0782	.1705	2.1	19
STAMP	.1763	.1262	.2064	.3698	.0703	.0301	.0736	.1591	14	22
GRU4REC	.1751	.1317	.2060	.3698	.0716	.0297	.0756	.1666	14	23
MARank	.1916	.1397	.2253	.3987	.0728	.0305	.0776	.1718	5.4	20
HPMN	.2055	.1448	.2419	.4185	.0792	.0331	.0839	.1884	.38	11
TIMe	**.2142**	**.1477**	**.2464**	**.4201**	**.0852**	**.0366**	**.0902**	**.1987**	–	–

Long short term methods, combination of long and short term methods, perform better than all single scale methods on both sets except BST. This comparison result tells us that long short term models with two scales are more flexible to be adapted to sequences with different leading causal patterns. The exception is mainly because of the explicitly modeling relations used in BST. Are two scale (long and short term) methods good enough?

It is natural to introduce additional scales, such as mid-term scales, to verify its effectiveness. The result is that multi-scale (more than two scales) models perform better than two scale (long and short term) methods consistently on both sets. For most cases, HPMN outperforms BST on both sets. The phenomenon empirically verifies the correctness of our motivation that mid-term scales play an essential role in the next item prediction task. Is HPMN good enough for multi-scale modeling of user interests?

Our proposed method TIMe outperforms HPMN by on both sets. On Movie-Lens 100k, the relative improvement of MRR@10 is at least 0.38%. On Amazon Movies & TV, the relative improvement of Recall@10 is 11%. The statistically significant improvement suggests the advantage of multi-scale modeling method in TIMe. One distinction is that TIMe explicitly models multi-scale causal patterns in the topological space with persistent homology for better explanations while HPMN only captures multi-scale implicit patterns in the latent space. The other is that both pattern relations and relations between the pattern and target item are embedded in recurrent tree transformer layer of TIMe, while only relations between the pattern and target item are computed in HPMN.

The absolute performance of each method is higher on MovieLens 100k than that on Amazon Movies & TV correspondingly. For Amazon Movies & TV, the number of users and items are more than that on MovieLens 100k. Due to the long tail effect, the data sparsity is much more serious on Amazon Movies & TV. However, our relative performance gain is higher on Amazon Movies & TV. The major reason lies in that TIMe generates multiple complex sequences from each item sequence through topological transformation layer. This can be treated as augmenting training sequences, which alleviates the data sparsity problem.

The relative MRR@10 improvement of TIMe is the smallest among four evaluation measures compared with other methods. It suggests that TIMe is better at identifying useful causal patterns for classifying which items will be the target rather than for the top ranked item (MRR@10).

4.3   Hyperparameter Sensitivity Analysis

We conduct comprehensive experiments on the effect of different settings of the number of time scales d and the number of recurrent tree transformer blocks B on datasets. Take the minimum and maximum diameters as 2 scales, and then gradually add intermediate diameters from small to large to obtain 3, 4 and 5 scales. Performance variations of TIMe with $B = 1$ and $d \in \{2, \ldots, 5\}$ on both sets are shown in Fig. 4(a) and (b) respectively. Performance variations of TIMe with $B \in \{1, 2, 3\}$ and $d = 4$ on MovieLens 100k, $d = 3$ on Amazon Movies & TV are shown in Fig. 4(c) and (d) respectively.

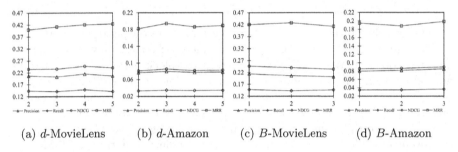

(a) d-MovieLens   (b) d-Amazon   (c) B-MovieLens   (d) B-Amazon

Fig. 4. Performance variations on MovieLens 100k and Amazon Movies & TV with change of the number of time scales d and blocks B.

With the increase of the scale number d from 2 to 5, the highest performance improvement of TIMe is 3–5% on MovieLens 100k and 4–7% on Amazon Movies & TV. TIMe achieves the highest point at $d = 4$ on MovieLens 100k and $d = 3$ on Amazon Movies & TV. After this point, its performance keeps steady on both sets. This suggests the scale number d can be determined by properly selection. The best choice $d > 2$ on both sets indicates that multi-scale modeling in TIMe plays a positive role in promoting the recommendation performance.

With the increase of the block number B from 1 to 3, the highest performance improvement of TIMe is 0.8–6% on MovieLens 100k and 2–6% on Amazon Movies & TV. The highest point is achieved by TIMe at $B = 1$ on MovieLens 100k and $B = 3$ on Amazon Movies & TV. With more training sequences, Amazon Movies & TV needs a more deeper model than MovieLens 100k.

4.4   Ablation Study

Tree **A**ggregation **B**lock (TAB) and **R**ecurrent **T**ree **T**ransformer (RTT) are two key components of TIMe. To explore their roles in the final recommendation performance, we do the following ablation studies on MovieLens 100k. To observe the accumulative effect of tree node embedding, we set the block size $B = 2$. (1) The recurrent tree transformer in TIMe is replaced with vanilla transformer, denoted as TIMe-$_{RTT}$. (2) The tree aggregation block is further removed from TIMe-$_{RTT}$, denoted as TIMe-$_{RTT-TAB}$. We compare the performance of the original method TIMe with TIMe-$_{RTT}$ and TIMe-$_{RTT-TAB}$ on MovieLens 100k shown in Table 2. Imp.% column corresponds to the relative performance improvement of each method compared with TIMe-$_{RTT-TAB}$.

Table 2. TIMe's different component effects on MovieLens-100k.

Model	Precision	Imp.%	Recall	Imp.%	NDCG	Imp.%	MRR	Imp.%
TIMe	.2057	3.68	.1426	1.78	.2397	2.79	.4275	2.86
TIMe-$_{RTT}$	.2016	1.61	.1418	1.21	.2350	0.77	.4197	0.99
TIMe-$_{RTT-TAB}$	.1984	–	.1401	–	.2332	–	.4156	–

The introduction of tree aggregation block improves the performance of TIMe by 1% or so under all evaluation measures. Different from the naive node embedding method in TIMe-$_{RTT-TAB}$, tree aggregation block in TIMe-$_{RTT}$ initializes node embedding which encodes parent-child relations in the multi-scale interest tree. The inclusion relation will be learned through the recurrent aggregation process. This is the major reason for TAB's performance improvement.

The addition of recurrent tree transformer brings about 2% performance improvement under all evaluation measures. Vanilla transformer in TIMe-$_{RTT}$ takes patterns from all scales in a temporal order as input, but ignore pattern relations from different scales, such as parent-child relations in the multi-scale interest tree. So we introduce tree aggregation block to vanilla transformer, and obtain recurrent tree transformer. This helps learn node/pattern embedding encoding both parent-child relations across scales and temporal order relations within a scale jointly. That's why the performance improvement of RTT is higher.

4.5   Qualitative Study

To figure out How TIMe works, we take one small example sequence with 32 movies from MovieLens-100k dataset as the input sequence, and attempt to predict the next movie. Through topological transformation layer, we obtain a multi-scale interest tree with four scales $d = 4$ in Fig. 5.

Intuitively each movie is represented as a IMDB category. At the minute scale, there are 23 nodes from a complex sequence at the bottom of this tree. At the day scale, we obtain a complex sequence with three patterns, denoted as ({drama, thriller, comedy, musical}, {romance, comedy, drama, action}, {romance, drama, comedy, sci-fi}). At the week scale, we obtain a complex sequence with only one node, denoted as ({drama, romance, comedy, action}). We obtain different compositions of movie genres at different scales. This combination ability of multi-scale interest tree makes the matching between user

Fig. 5. Example of multi-scale interest tree and attention visualization

interest and movies easy, because of the combinatorial nature of both user interest and movies.

To predict movie "Evita (1996)" for user 554, the learned attention map is depicted in the user's multi-scale interest tree with support values and colors in Fig. 5. Through this attention map, we identify many sequential patterns with the attention weight as the support value from TIMe. For movie "Evita (1996)" labeled with romance, drama and history, the most important pattern identified from TIMe is like "movies tagged with {romance, drama, comedy, sci-fi}" → "Evita (1996)" with support value 0.2. This is consistent with our intuition.

5 Conclusion

We propose a topological data analysis based framework for interpretable multi-scale sequential recommendation. Our proposed topological transformation layer maps the item sequence into complex sequences with different diameters in the form of multi-scale interest tree. The tree aggregation block is proposed to obtain node embedding from this tree in a recurrent aggregation process. Introducing this block to the vanilla transform, we obtain the recurrent tree transformer to derive the unified user interest representation. Sequential patterns for next item prediction will be obtained explicitly from the multi-scale interest tree and its corresponding attention matrix. Empirical studies verify the effectiveness and interpretability of our proposed method. In future, we will introduce the item content information into TIMe.

Acknowledgements. This research work was funded by the National Natural Science Foundation of China under Grant No. 62072447 and No. 11871145.

References

1. Chen, Q., Zhao, H., Li, W., Huang, P., Ou, W.: Behavior sequence transformer for e-commerce recommendation in Alibaba. In: DLP-KDD 2019 (2019)
2. Cho, K., van Merrienboer, B., Bahdanau, D., Bengio, Y.: On the properties of neural machine translation: encoder-decoder approaches. In: Proceedings of SSST@EMNLP 2014, pp. 103–111 (2014)
3. Craswell, N.: Mean reciprocal rank. In: Liu, L., Özsu, M.T. (eds.) Encyclopedia of Database Systems, p. 1703. Springer, US (2009). https://doi.org/10.1007/978-0-387-39940-9_488

4. Harer, J., Reale, C., Chin, P.: Tree-transformer: a transformer-based method for correction of tree-structured data (2019)
5. Harper, F.M., Konstan, J.A.: The MovieLens datasets: history and context. ACM Trans. Interact. Intell. Syst. 5(4), 1–19 (2015)
6. He, R., McAuley, J.: Ups and downs: modeling the visual evolution of fashion trends with one-class collaborative filtering. In: WWW 2016, pp. 507–517 (2016)
7. Hidasi, B., Karatzoglou, A., Baltrunas, L., Tikk, D.: Session-based recommendations with recurrent neural networks. In: International Conference on Learning Representations (2016)
8. Järvelin, K., Kekäläinen, J.: Cumulated gain-based evaluation of IR techniques. ACM Trans. Inf. Syst. 20(4), 422–446 (2002)
9. Kang, W., McAuley, J.: Self-attentive sequential recommendation. In: 2018 IEEE International Conference on Data Mining (ICDM), pp. 197–206 (2018)
10. Liu, J., Jeng, S., Yang, Y.: Applying topological persistence in convolutional neural network for music audio signals. CoRR abs/1608.07373 (2016)
11. Liu, Q., Zeng, Y., Mokhosi, R., Zhang, H.: Stamp: short-term attention/memory priority model for session-based recommendation. In: KDD 2018, pp. 1831–1839 (2018)
12. Obayashi, I., Hiraoka, Y., Kimura, M.: Persistence diagrams with linear machine learning models. J. Appl. Comput. Topol. 1(3), 421–449 (2018)
13. Quadrana, M., Cremonesi, P.: Sequence-aware recommendation. In: Proceedings of the 12th ACM Conference on Recommender Systems. In: RecSys 2018, pp. 539–540 (2018)
14. Ren, K., et al.: Lifelong sequential modeling with personalized memorization for user response prediction. In: SIGIR 2019, pp. 565–574 (2019)
15. Rendle, S., Freudenthaler, C., Gantner, Z., Schmidt-Thieme, L.: BPR: Bayesian personalized ranking from implicit feedback. In: UAI 2009, pp. 452–461 (2009)
16. Rijsbergen, C.J.V.: Information Retrieval, 2nd edn. Butterworth-Heinemann, Oxford (1979)
17. Sarwar, B., Karypis, G., Konstan, J., Riedl, J.: Item-based collaborative filtering recommendation algorithms. In: WWW 2001, pp. 285–295 (2001)
18. Umeda, Y.: Time series classification via topological data analysis. Trans. Jpn. Soc. Artif. Intell. 32(3), 228–239 (2017)
19. Vaswani, A., et al.: Attention is all you need. In: NIPS 2017, pp. 6000–6010 (2017)
20. Wang, P., Guo, J., Lan, Y., Xu, J., Wan, S., Cheng, X.: Learning hierarchical representation model for next basket recommendation. In: SIGIR 2015, pp. 403–412 (2015)
21. Yu, F., Liu, Q., Wu, S., Wang, L., Tan, T.: A dynamic recurrent model for next basket recommendation. In: SIGIR 2016, pp. 729–732 (2016)
22. Yu, L., Zhang, C., Liang, S., Zhang, X.: Multi-order attentive ranking model for sequential recommendation. In: The Thirty-Third AAAI Conference on Artificial Intelligence, AAAI, pp. 5709–5716 (2019)
23. Zhao, W., et al.: Leveraging long and short-term information in content-aware movie recommendation via adversarial training. IEEE Trans. Cybern. 50, 1–14 (2019)
24. Zhou, K., et al.: S3-rec: self-supervised learning for sequential recommendation with mutual information maximization. In: CIKM 2020, pp. 1893–1902 (2020)
25. Zhu, X.: Persistent homology: an introduction and a new text representation for natural language processing. In: IJCAI 2013, pp. 1953–1959 (2013)

SANS: Setwise Attentional Neural Similarity Method for Few-Shot Recommendation

Zhenghao Zhang[1,2,3], Tun Lu[1,2,3]($\boxtimes$), Dongsheng Li[4], Peng Zhang[1,2,3]($\boxtimes$),
Hansu Gu[5], and Ning Gu[1,2,3]

[1] School of Computer Science, Fudan University, Shanghai, China
{zhzhang18,lutun,zhangpeng_,ninggu}@fudan.edu.cn
[2] Shanghai Key Laboratory of Data Science, Fudan University, Shanghai, China
[3] Shanghai Institute of Intelligent Electronics and Systems, Shanghai, China
[4] Microsoft Research Asia, Shanghai, China
dongsli@microsoft.com
[5] Amazon.com, Seattle, USA

Abstract. Recommender systems generate personalized recommendations for users based on their historical data. However, if some users have few interactions in the training data, i.e., few-shot users, recommendations for them will be inaccurate. In this paper, we propose a setwise attentional neural similarity method (SANS) for the few-shot recommendation problem. Unlike general recommendation algorithms, we eliminate direct representations of few-shot users. First, a neural similarity method is proposed to effectively estimate the correlation between items. Then, we propose a setwise attention mechanism to obtain recommendation scores by aggregating the correlations between a candidate item and items in a candidate user's historical interactions. To facilitate model training in the few-shot scenario, training samples are generated by episode sampling, and each training sample is assigned with an adaptive weight to emphasize the importance of few-shot users. We simulate the few-shot recommendation problem on three real-world datasets and extensive results show that SANS can outperform the state-of-the-art recommendation algorithms in few-shot recommendation.

Keywords: Collaborative filtering · Few-shot learning · Neural networks · Top-N recommendation

1 Introduction

Recommender systems recommend items to users based on their historical interactions with other items. However, in practice, there are many newcomers or inactive users who have few interactions in online services, i.e., few-shot users, which makes it challenging to train accurate recommendation models for them. Due to the long-tail distribution of user activities in online services, these few-shot users are non-negligible and it is desirable to deliver high quality recommendations for these few-shot users.

© Springer Nature Switzerland AG 2021
C. S. Jensen et al. (Eds.): DASFAA 2021, LNCS 12683, pp. 69–84, 2021.
https://doi.org/10.1007/978-3-030-73200-4_5

Existing general-purpose recommendation algorithms cannot well address the few-shot recommendation problem. In many recent recommendation algorithms, especially deep learning-based ones [2,6], large number of training data are required in model learning, e.g., learn the embedding vectors of users and items. However, if a user has very few interactions, then the embedding vector of this user cannot be well learned, resulting in poor recommendation performance. For instance, many deep neural network-based methods adopt deep neural networks to learn the representations of users/items and capture the complex non-linear relationships among user/items, which may suffer from severe overfitting issue in the few-shot recommendation scenario. Item-based recommendation algorithms, such as FISM [8] and NAIS [5], can eliminate the overfitting issue on user modeling. However, how to effectively estimate the correlation between items (e.g., similarity) in the few-shot scenario is still an open question.

In this paper, we propose SANS, an item-based deep recommendation algorithm for few-shot recommendation. In SANS, we do not explicitly learn the representations of users like many existing works but represent each user by the set of items in his/her historical interactions. A neural similarity method is proposed to estimate the correlation between each pair of items, i.e., the probability that a user who likes one of the items will also like the other one. When recommending items for a user, we propose a setwise attention method which utilizes items in the user's historical interactions as a support set, estimates importance of each item in the support set via attention mechanism, and finally aggregates the correlations between the candidate item and the support set to generate the final prediction. To facilitate model training for few-shot users, we generate training samples by episode sampling and propose a new weighted loss function in which each training sample is assigned with an adaptive weight to emphasize the importance of few-shot users.

In summary, the main contributions of this paper are as follows:

- We propose SANS, a deep recommendation algorithm consisting of a neural similarity module and a setwise attention module to address the few-shot recommendation problem.
- We design a weighted loss function in which weights are adaptively assigned according to the number of user's interactions. Combined with episode sampling, the training of SANS is highly effective.
- We conduct extensive experiments on three real-world datasets, which demonstrate that SANS can substantially outperform state-of-the-art recommendation algorithms on few-shot users.

The rest of this paper is organized as following: Sect. 2 defines the few-shot recommendation problem and introduces item-based collaborative filtering. Section 3 proposes the network architecture of SANS, the weighted loss and its training procedure. Section 4 presents experimental results. Section 5 discusses the related work about collaborative filtering, cold-start recommendation and few-shot learning. Finally Sect. 6 concludes the paper.

2 Preliminaries

This section first defines the few-shot recommendation problem and then introduces the item-based collaborative filtering algorithm.

2.1 Few-Shot Recommendation

Let U and I donate the set of users and items, respectively. The training set S consists of the user-item tuples $S = \{(u,i) : u \in U, i \in I\}$. We define the set of items interacted by user u as $I_u^+ = \{i \in I : (u,i) \in S\}$. For each user-item tuple,

$$y_{ui} = \begin{cases} 1, & (u,i) \in S \\ 0, & (u,i) \notin S \end{cases}, \tag{1}$$

where $y_{ui} = 0$ means the interaction between user u and item i hasn't been observed. Then, we formulate the problem of few-shot recommendation as follows.

Definition 1. *(N-shot Recommendation) For a subset of users U^*, if $\forall u \in U^*$ satisfies $|I_u^+| = N$ (N > 0), the problem of recommending items to all users within U^* is N-shot recommendation.*

When N is small, e.g., 3 or 5, we can call it few-shot recommendation. There is a related problem in previous recommender system research called cold-start [14]. However, the difference between few-shot and cold-start is that few-shot users have no additional personal information other than few interactions. Few-shot users are special cold-start users.

2.2 Item-Based Collaborative Filtering

The item-based collaborative filtering method [13] uses similarities between candidate items and users' historical items to rank candidate items. For a user u with historical interactions I_u^+, the predicted score of user u on item i under the implicit feedback setting is:

$$\hat{y}_{ui} \propto \sum_{j \in I_u^+} a_{uj} s_{ij}, \tag{2}$$

where s_{ij} denotes the similarity between item i and item j. The similarity can be computed using different metrics such as cosine [12], Pearson [13], etc., or learned from data [5,8]. a_{uj} donates the preference of user u on item j when predicting u's preference on i. a_{uj} is usually set to 1 in existing methods, i.e., all historical interactions are equally important on predicting u's preference on i.

The item-based collaborative filtering is well suited for few-shot recommendation due to two reasons: 1) users are not explicitly modeled. Since there are no user-related model parameters, the difficulty in training user models in few-shot scenario doesn't exist; 2) new users with few interactions can also have recommendations, which is ideal for online services with newcomers everyday.

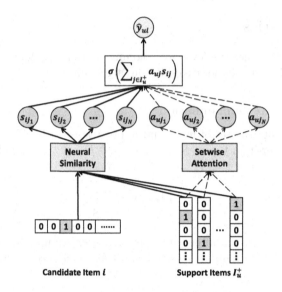

Fig. 1. The network architecture of SANS.

3   SANS: Setwise Attentional Neural Similarity Method

3.1   Algorithm Design

The Overall Architecture. As illustrated in Fig. 1, where $I_u^+ = \{j_1, j_2, \ldots, j_N\}$, SANS consists of two main components: 1) a neural similarity module, which estimates the correlation of two items using neural network, i.e., the similarity term in Eq. 2. More specifically, the neural similarity module outputs the possibility of users who interacted with one of the items will also interact the other item; 2) a setwise attention module, which estimates the preference of a user over each item in his/her historical interactions, i.e., the preference term in Eq. 2. More specifically, we assume that items from user history are not equally important in reflecting user preference, and we obtain the relative importance of different items via the proposed setwise attention module.

Neural Similarity. If each user has only one interaction, then the recommendation problem becomes estimating the probability that a user u who likes item i also likes another item j: $\Pr(y_{uj} = 1 | y_{ui} = 1)$. Let $\Pr(y_{uj} = 1, y_{ui} = 1)$ donates the joint probability that two items are favored by a user. The score of personalized item ranking for the user u who has only one interaction with item i is computed as follows:

$$\Pr(y_{uj} = 1 | y_{ui} = 1) = \frac{\Pr(y_{uj} = 1, y_{ui} = 1)}{\Pr(y_{ui} = 1)} = \Pr(y_{uj} = 1, y_{ui} = 1), \quad (3)$$

where $\Pr(y_{ui} = 1) = 1$ since the item i has already been in the historical interactions of user u. $\Pr(y_{uj} = 1, y_{ui} = 1)$ is the probability of the co-occurrence

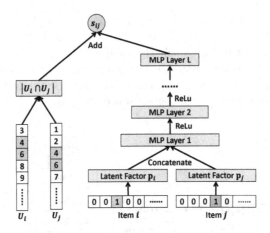

Fig. 2. The network architecture of the neural similarity module.

of item i and item j on user u, which can be approximated using the following equation:

$$\Pr(y_{uj} = 1, y_{ui} = 1) = \frac{|U_i^+ \cap U_j^+|}{|U|}, \tag{4}$$

where U_i/U_j is the set of users who interacted with item i/j. $|U|$ (the number of users) is a constant, which has no effect in ranking problems. Therefore, the probability that a user who likes one item also likes another item is proportional to the number of common users between two items:

$$\Pr(y_{uj} = 1 | y_{ui} = 1) \propto |U_i^+ \cap U_j^+|. \tag{5}$$

In addition, previous works [1,2] have shown that multi-layer perceptron (MLP) can help to capture the high-order relationships between entities. Therefore, we combine the above two ideas and propose our neural similarity method as shown in Fig. 2.

More formally, the neural similarity between item i and item j is as follows:

$$s_{ij} = |U_i^+ \cap U_j^+| + f_\theta(\begin{bmatrix} \mathbf{p}_i \\ \mathbf{p}_j \end{bmatrix}), \tag{6}$$

where $\mathbf{p}_i$ and $\mathbf{p}_j$ are embedding vectors for item i and item j, respectively. They are concatenated together and then passed to an MLP f_Θ:

$$\begin{aligned}
\mathbf{z}_1 &= \mathrm{ReLU}(\mathbf{W}_1^T \begin{bmatrix} \mathbf{p}_i \\ \mathbf{p}_j \end{bmatrix} + \mathbf{b}_1) \\
\mathbf{z}_2 &= \mathrm{ReLU}(\mathbf{W}_2^T \mathbf{z}_1 + \mathbf{b}_2) \\
&\cdots \\
f_\theta(\begin{bmatrix} \mathbf{p}_i \\ \mathbf{p}_j \end{bmatrix}) &= \mathbf{W}_L^T \mathbf{z}_{L-1} + b_L
\end{aligned} \tag{7}$$

where $\mathbf{W}_l$ and $\mathbf{b}_l$ are the weight matrix and bias of the l-th layer. They are represented as $\boldsymbol{\Theta} = (\mathbf{W}_1, \ldots, \mathbf{W}_L, \mathbf{b}_1, \ldots, b_L)$. The output of the MLP is finally added to $|U_i^+ \cap U_j^+|$. The MLP can benefit the similarity learning due to two reasons: 1) it can capture high-order non-linear relationships between items in addition to the number of common users; 2) it can estimate the similarity between two items without common users via representation learning.

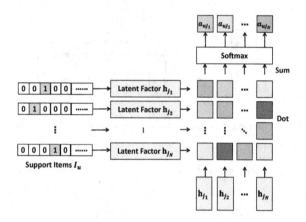

Fig. 3. The network architecture of the setwise attention module.

Setwise Attention. In the previous section, we have modeled the neural similarity between items, based on which the probability that a user who has more than one interactions will like another item can be obtained based on the weighted sum overs the similarities between a set of support items and the candidate item:

$$\hat{y}_{ui} = \sigma \left(\sum_{j \in I_u^+} a_{uj} s_{ij} \right). \tag{8}$$

σ is the sigmoid function, which converts the weighted sum to a value between 0 and 1. The preference a_{uj} should be based on the set of items interacted by user u. However, it is challenging to introduce parameters related to user u in the few-shot recommendation problem. To this end, we propose the setwise attention method as illustrated in Fig. 3, which is formally described as follows:

$$a_{uj} = \frac{\exp(\sum_{k \in I_u^+} \mathbf{h}_k^T \mathbf{h}_j)}{\sum_{i' \in I_u^+} \exp(\sum_{k \in I_u^+} \mathbf{h}_k^T \mathbf{h}_{i'})}, \tag{9}$$

where $\mathbf{h}_k$, $\mathbf{h}_j$ and $\mathbf{h}_{i'}$ are embedding vectors of items. The setwise attention module estimates a user's preferences over items in his or her historical interactions. If there are a few highly similar items in a user's historical data, we can infer that this user really likes this type of items and we should emphasize more on recommending items that are similar to these items. Otherwise, we know that the user has no preference differences among his/her historical items, so that it would be better to assign nearly equal weights to different items.

3.2 The Weighted Loss

For point-wise ranking problem, the binary cross-entropy loss function has been widely adopted. However, treating each training sample as equally important may not be optimal for model training [7]. Intuitively, users with many interactions contribute more to the total loss than users with few interactions. Thus, the model will converge when the users with many interactions are well trained but the few-shot users are usually underfitted due to fewer gradient updates. To remedy this, we propose a weighting mechanism to assign weights for each user based on the number of interacted items as follows:

$$c_u = c_0 \frac{|I_u^+|^\alpha}{\sum_{j\in U} |I_j^+|^\alpha}, \tag{10}$$

where c_0 sets the magnitude of weights and α controls the impact of the number of interactions on the weights. α is chosen in $[-1, 0]$. When $\alpha = 0$, all users have equal weights. When $\alpha = -1$, the weight is inversely proportional to the number of interactions.

The weights of users are then added into the binary cross-entropy loss forming the proposed weighted loss function for SANS as follows:

$$\mathcal{L} = \sum_{(u,i,I_u^S,y_{ui})\in\mathcal{D}} c_u(-y_{ui}\log\hat{y}_{uj} - (1 - y_{ui})\log(1 - \hat{y}_{uj}))$$
$$+ \lambda_\Theta\|\Theta\|^2 + \lambda_P\|\mathbf{P}\|^2 + \lambda_H\|\mathbf{H}\|^2, \tag{11}$$

where I_u^S is a support set sampled from I_u^+, which mimics few-shot recommendation in each training sample. λ_Θ, λ_P and λ_H are L2 regularization coefficients.

3.3 Model Training

In SANS, we have three sets of model parameters: Θ, $\mathbf{P}$ and $\mathbf{H}$. Θ and $\mathbf{P}$ are the parameters of the MLP and embedding matrix, respectively, in the neural similarity module. $\mathbf{H}$ is the embedding matrix for the setwise attention module.

Algorithm 1 presents the learning details. First, we calculate the weights for each user based on the number of interactions he/she has, then randomly initialize all parameters using Gaussian distribution. Training samples are dynamically generated in each iteration. For each user-item pair (u, i) in the dataset, a support set of items with size N is sampled from I_u^+, and a negative item is sampled from $I \setminus I_u^+$. Then, each quadruple—the user u, the positive item i or negative item j, the support set I_u^S, and the label y_{ui} is added to the collection of training samples. Finally, the SANS model is trained using training samples by the above episode sampling in each iteration, in which the model parameters can be updated by the stochastic gradient descent (SGD)-based methods.

Algorithm 1: LearnSANS

Data: Number of shots N, users U, items I, implicit feedback S, number of
epochs T, weight magnitude c_0, weight strength α and regularization
strength $(\lambda_\Theta, \lambda_P, \lambda_H)$

Result: SANS model weights (Θ, P, H)

1 **foreach** $u \in U$ **do**

2 $\quad$ $c_u \leftarrow c_0 \dfrac{|I_u^+|^\alpha}{\sum_{j \in U} |I_j^+|^\alpha}$;

3 Initialize (Θ, P, H) randomly;

4 **for** $i \leftarrow 1$ **to** T **do**

5 $\quad$ $\mathcal{D} = \emptyset$;

6 $\quad$ **foreach** $(u, i) \in S$ **do**

7 $\quad\quad$ Sample a support set I_u^S sized N from I_u^+;

8 $\quad\quad$ Sample an negative item j from $I \setminus I_u^+$;

9 $\quad\quad$ $\mathcal{D} \leftarrow \mathcal{D} \cap \{(u, i, I_u^S, 1)\}$;

10 $\quad\quad$ $\mathcal{D} \leftarrow \mathcal{D} \cap \{(u, j, I_u^S, 0)\}$;

11 $\quad$ Train SANS using $\mathcal{D}$ with weighted loss (Eq. 11);

---

4 Experiments

In this section, we compare SANS with state-of-the-art algorithms in few-shot
recommendation scenario aiming to answer the following research questions:

- **RQ1:** how does SANS perform compared with state-of-the-art recommendation algorithms in few-shot recommendation?
- **RQ2:** how does each component in SANS affect the performance of the overall model?
- **RQ3:** how does the performance of SANS change with different few-shot scenarios?

4.1 Experimental Settings

Dataset. We evaluate the proposed SANS method on three real-world datasets:
Last.FM[1], Steam[2] and Douban[3], which are publicly available. The statistics of
the three datasets are shown in Table 1.

- **Last.FM** was from HetRec 2011, which contains listening relationships between users and artists. The interactions from Last.FM are implicit.
- **Steam** was shared by Kaggle users, which contains purchase or play records of a set of Steam users. We convert all records into user-game tuples and remove duplicate ones.

---

[1] https://grouplens.org/datasets/hetrec-2011/.

[2] https://www.kaggle.com/tamber/steam-video-games.

[3] https://opendata.pku.edu.cn/dataset.xhtml?persistentId=doi:10.18170/DVN/LA9GRH.

- **Douban** was collect by Yin et al. [23], which contains ratings on books given by Douban users. The raw dataset is large, so we sample 5,000 users out of 383,033 users and treat all the ratings as implicit positive feedback.

Table 1. Statists of the three datasets.

Dataset	#Interaction	#User	#Item	Sparsity
Last.FM	82,155	1,885	6,953	99.37%
Steam	103,594	2,189	3,933	98.80%
Douban	199,053	5,000	34,604	99.88%

Evaluation Protocols. We use a user-based splitting method to split the dataset to simulate few-shot recommendation. First, we divide users into training users and test users by 8:2. All the interactions from training users are added to the training set. Then, for each user in the test user set, we sample N interactions from his or her interactions into the training set and put all the remaining interactions to the test set if this user has more than N interactions in his or her history. If the number of interactions from a user is less than N, we put this user back to the training set. We evaluate the recommendation performance by NDCG@10 [22] and Recall@10. For each user in the test set of Last.FM and Steam datasets, we rank all the items by predicated scores and exclude items that have already been interacted by the user in the training set to generate a top-10 list of recommended items. For the Douban dataset, we sampled 10,000 items and mix them with ground truth items as candidate items for each user. The NDCG@10 and Recall@10 values for that user can be calculated base on interactions in the test set. The performance on the entire dataset is reported by the average NDCG@10 and Recall@10 over all test users.

Compared Methods. We compare SANS with various types of methods including baseline method, item-based methods, matrix factorization-based methods, deep learning-based methods and metric learning-based method as follows:

- **ItemPop** recommends top-N popular items to users. It is a non-personalized baseline method.
- **ALS** [7] uses point-wise loss and treats all unknown feedback as negative to learn the matrix factorization model. By taking advantage of its mathematical property, their matrix factorization model can be trained using all negative feedback.
- **BPR** [12] uses pair-wise loss and collects negative samples by sampling to learn the matrix factorization model.
- **KNN** [13] first computes the similarity between each pair of items and then sorts items by the sum of their similarities to the user's interactions. We use cosine similarity as the similarity metric in the experiments.

- **NAIS** [5] is an item-based collaborative filtering method which models item similarity by the dot product of two embedding vectors. Besides, attention mechanism is used to generate weights for different items.
- **NeuMF** [6] is a deep learning-based collaborative filtering method, which uses deep neural networks instead of a linear function to model the interactions between users and items.
- **CFNet** [2] combines representation learning-based collaborative filtering approach and matching function-based collaborative filtering approach using neural networks to achieve higher performance.
- **LRML** [20] is a collaborative filtering method based on metric learning which learns embeddings of users and items in a unified hyperspace. Items are ranked by the Euclidean distance to a user by assuming that users will prefer items that are close to them in the hyperspace.

The experiments are implemented using Tensorflow. We use the released code from the authors to implement the following compared methods: NAIS[4], NeuMF[5], CFNet[6] and LRML[7].

Hyperparameter Settings. Hyperparameters of each method are tuned by the random search method. More specifically, all methods are tuned via a validation set and hyperparameters with the highest NDCG@10 are chosen as the optimal ones. We tuned the learning rates of all methods in $[0.001, 0.005, 0.01, 0.05]$ and the regularization coefficients in $[0.1, 0.01, 0.001]$. The optimal learning rates and regularization coefficients vary across datasets. For SANS, we fixed c_0 to 2,000 and tested the similarity embedding size of $[8, 16, 32, 64]$, the attention embedding size of $[4, 8, 12, 16]$ and the weight strength α of $[-0.5, -0.1, -0.05, -0.01]$. Finally, we set the dimension of the similarity embedding to 32, the dimension of the attention embedding to 4, and the weight strength α to -0.05. The architecture of MLP f_Θ is $64 \to 32 \to 16$. Deeper neural networks tend to achieve higher performance, but there is a law of diminishing marginal utility on the depth of networks. The SANS model is learned by the Adam optimizer [9].

4.2   Performance Comparison (RQ1)

We evaluate the performance of SANS and all the compared methods on three datasets with the number of shots increasing from 1 to 3 in Table 2. We have the following observations from the results:

- The performance of SANS is better than all other methods with N increasing from 1 to 3, which demonstrates the advantage of SANS on the few-shot recommendation.

---

[4] https://github.com/AaronHeee/Neural-Attentive-Item-Similarity-Model.
[5] https://github.com/hexiangnan/neural_collaborative_filtering.
[6] https://github.com/familyld/DeepCF.
[7] https://github.com/cheungdaven/DeepRec.

Table 2. Performance comparison between SANS and all compared methods in N-shot recommendation on three datasets. Relative improvements over the strongest baselines are also reported at the end of each table.

(a) Last.FM

Model	NDCG@10			Recall@10		
	N = 1	N = 2	N = 3	N = 1	N = 2	N = 3
ItemPop	0.2655	0.2600	0.2521	0.0548	0.0552	0.0548
BPR	0.2424	0.3016	0.3486	0.0510	0.0656	0.0762
ALS	0.3270	0.3659	0.3983	0.0698	0.0801	0.0890
KNN	0.2759	0.3361	0.3858	0.0631	0.0765	0.0885
NAIS	0.3392	0.3828	0.4045	0.0708	0.0815	0.0880
NeuMF	0.3940	0.4285	0.4558	0.0820	0.0922	0.1009
CFNet	0.2676	0.2610	0.2574	0.0554	0.0559	0.0565
LRML	0.3522	0.4061	0.4326	0.0738	0.0880	0.0963
SANS	**0.4205**	**0.4501**	**0.4839**	**0.0889**	**0.0978**	**0.1073**
Improvement	6.72%	5.05%	6.17%	8.36%	5.98%	6.32%

(b) Steam

Model	NDCG@10			Recall@10		
	N = 1	N = 2	N = 3	N = 1	N = 2	N = 3
ItemPop	0.3552	0.3432	0.3401	0.1210	0.1204	0.1226
BPR	0.3815	0.4380	0.4632	0.1567	0.1884	0.2100
ALS	0.3726	0.4402	0.4580	0.1442	0.1816	0.1976
KNN	0.3538	0.4551	0.4943	0.1411	0.1874	0.2127
NAIS	0.3891	0.4292	0.4461	0.1540	0.1769	0.1910
NeuMF	0.4361	0.4516	0.4551	0.1692	0.1874	0.2004
CFNet	0.3528	0.3390	0.3381	0.1175	0.1143	0.1184
LRML	0.4071	0.4377	0.4553	0.1460	0.1688	0.1976
SANS	**0.4806**	**0.5206**	**0.5223**	**0.1849**	**0.2102**	**0.2244**
Improvement	10.21%	14.30%	5.66%	9.29%	11.54%	5.50%

(c) Douban

Model	NDCG@10			Recall@10		
	N = 1	N = 2	N = 3	N = 1	N = 2	N = 3
ItemPop	0.0180	0.0169	0.0171	0.0055	0.0054	0.0057
BPR	0.0998	0.1187	0.1271	0.0276	0.0353	0.0383
ALS	0.0726	0.0838	0.0910	0.0199	0.0246	0.0272
KNN	0.0895	0.1245	0.1468	0.0245	0.0365	0.0436
NAIS	0.0462	0.0534	0.0602	0.0125	0.0150	0.0177
NeuMF	0.0803	0.0883	0.0939	0.0222	0.0262	0.0280
CFNet	0.0142	0.0138	0.0142	0.0041	0.0038	0.0043
LRML	0.0761	0.1017	0.1136	0.0213	0.0307	0.0350
SANS	**0.1161**	**0.1601**	**0.1841**	**0.0305**	**0.0450**	**0.0525**
Improvement	16.31%	28.52%	25.42%	10.28%	23.04%	20.43%

- When N = 1, most of the compared methods have very low NDCG@10 and Recall@10, and some of them (e.g., BPR in Last.FM and CFNet in Steam) even perform worse than ItemPop. This indicates that existing methods indeed cannot well address the few-shot recommendation problem.

- Complex methods based on advanced techniques, e.g., matrix factorization and neural networks do not always exhibit higher performance in few-shot recommendation. For instance, KNN outperforms almost all compared methods in the Douban dataset (only except BPR with N = 1). This indicates that complex models will easily overfit and be less desirable in the few-shot recommendation problems.
- SANS is more desirable due to: 1) overfitting will be less problematic because SANS does not learn user representations; 2) the proposed neural similarity method is more robust than conventional similarity methods due to the combination of simple and complex similarity modeling; 3) the proposed weighted loss function can emphasize few-shot users during model training, which can further alleviate inappropriate convergence on few-shot users.

4.3   Ablation Analysis (RQ2)

Here, we perform ablation analysis to investigate the impact of each component in SANS. The experimental results are shown in Table 3. SANS-SIM uses the linear part of neural similarity only, and SANS-MLP uses the MLP part of neural similarity, SANS-BCE replaces the weighted loss with binary cross-entropy loss, and SANS-AVG replaces setwise attention with the mean over neural similarities.

Table 3. Performance of SANS with each component removed in N-shot recommendation on the three datasets.

Dataset	Model	NDCG@10			Recall@10		
		N = 1	N = 2	N = 3	N = 1	N = 2	N = 3
Last.FM	SANS	**0.4205**	**0.4501**	**0.4839**	**0.0889**	**0.0978**	**0.1073**
	SANS-SIM	0.3703	0.4431	0.4732	0.0792	0.0967	0.1045
	SANS-MLP	0.3665	0.3809	0.3897	0.0763	0.0822	0.0862
	SANS-AVG	–	0.4443	0.4657	–	0.0964	0.1034
	SANS-BCE	0.4189	0.4477	0.4762	0.0886	0.0972	0.1053
Steam	SANS	**0.4806**	**0.5206**	**0.5223**	**0.1849**	**0.2102**	**0.2244**
	SANS-SIM	0.4592	0.4952	0.4943	0.1760	0.1957	0.2074
	SANS-MLP	0.4235	0.4283	0.4454	0.1613	0.1733	0.1895
	SANS-AVG	–	0.5085	0.5199	–	0.2080	0.2221
	SANS-BCE	0.4748	0.5114	0.5199	0.1818	0.2095	0.2209
Douban	SANS	**0.1161**	**0.1601**	**0.1841**	**0.0305**	0.0450	0.0525
	SANS-SIM	0.0826	0.1376	0.1684	0.0220	0.0390	0.0487
	SANS-MLP	0.0260	0.0346	0.0350	0.0073	0.0100	0.0099
	SANS-AVG	–	0.1592	0.1832	–	0.0449	0.0527
	SANS-BCE	0.1160	0.1596	0.1835	0.0304	**0.0451**	**0.0530**

Impact of Neural Similarity. SANS-SIM performs better than SANS-MLP and is closer to SANS, suggesting that it is reasonable to use the number of common users to measure the similarity between items. Adding MLP to the neural similarity yields the best-performing model because MLP can fit the nonlinear residual part of the neural similarity.

Impact of Setwise Attention. The attention mechanism is designed for recommendation that is more than one shot, so we only compare SANS-AVG with SANS in $N > 1$ scenarios. The attention mechanism has significant impacts on the Last.FM and Steam datasets, while the improvement is negligible on the Douban dataset. The reason may be due to that Douban is much more sparse and has much more items than the other datasets so that the interactions of the few-shot users are too random to provide any additional information.

Impact of Weighted Loss. The weighted loss can help to emphasize few-shot users when training the neural similarity module and setwise attention module in SANS. We can see from the results that the weighted loss can contribute to significant improvements on the Last.FM and Steam datasets but the improvement on the Douban dataset is negligible. Again, this should be due to the sparsity of the Douban dataset, so that most users are with very few ratings and giving higher weights to few-shot users does not make significant differences.

4.4 Analysis on Number of Shots (RQ3)

To find the best scenarios for SANS, we evaluate SANS with different numbers of shots. We split the three datasets with different numbers of shots from 2 to 16 by a step of 2. Due to space limitation, we only present the comparisons with the five best-performing methods on all datasets.

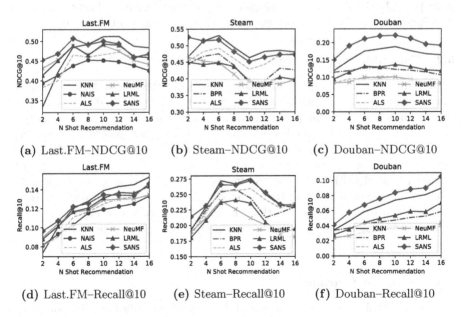

(a) Last.FM–NDCG@10   (b) Steam–NDCG@10   (c) Douban–NDCG@10

(d) Last.FM–Recall@10   (e) Steam–Recall@10   (f) Douban–Recall@10

Fig. 4. Performance comparison between SANS and five compared methods in N-shot recommendation with N varying from 2 to 16.

The experimental results are shown in Fig. 4. The relative improvements of SANS vary across different datasets. On the Last.FM dataset, SANS achieves the best performance when $N < 8$. On the Steam dataset, SANS achieves the best performance when $N \leq 4$. But on the Douban dataset, SANS consistently outperforms all the compared methods with significant margins. We have the following observations from the results: 1) SANS is desirable for few-shot recommendation on all datasets; 2) SANS is also more desirable on extremly sparse datasets, e.g., the Douban dataset, than the other methods even with lots of non-few-shot users; 3) KNN outperforms many recently proposed deep learning-based methods, which indicates that deep learning-based methods may not be appropriate for recommendations on few-shot users or on extremely sparse data due to higher chances of overfitting.

5   Related Work

5.1   Collaborative Filtering

Collaborative filtering-based recommendation algorithms achieve competitive performance in both rating prediction [11] and item ranking [2]. Classical collaborative filtering methods mainly includes matrix factorization-based methods which learn user/item feature vectors by various learning algorithms [7,12,17] and item-based methods which generate predictions based on item-item similarity matrix [5,8]. However, many classical methods are linear and thus cannot capture non-linear relationships between users and items, so that many deep learning-based collaborative filtering methods have been proposed in recent years [2,6,24]. Because of the representation power of deep neural networks, deep learning-based methods outperform classical methods in most scenarios.

5.2   Cold Start Recommendation

Cold-start recommendation methods can also solve the few-shot recommendation problem if there is additional information to be exploited. For example, the attribute-to-feature mapping method [4] learns the mapping from features of users or items to their embedding vectors. Then, they generate embedding vectors of new users or items based on their features. Social information is also useful to alleviate the cold start problem [15,16]. In addition, cross-domain recommendation algorithm [19] can use feedback from source domains to address cold-start recommendation in the target domain. However, additional information is not always available, so that the above methods may fail in practice. However, SANS does not require any additional information, i.e., more general than these methods.

5.3   Few-Shot Learning

Few-shot learning was first proposed for object classification in computer vision [3], which derives a new classifier for objects from new category using few training samples. The performance of few-shot learning has been significantly improved with the advances of deep learning and several neural network structures have been recently proposed [10, 18, 21]. Siamese neural networks [10] address one-shot learning by learning the similarity between samples using a network architecture composed of twin networks with shared weights. Matching networks [21] solve one-shot learning problems by conditional similarity. Prototypical networks [18] generate a prototypical vector for each few-shot category and predictions can be calculated based on distances to prototypes. To the best of our knowledge, there are very little prior works in few-shot recommendation.

6   Conclusion

This paper proposes SANS to address the few-shot recommendation problem. SANS consists of a neural similarity module to estimate the similarity of each pair of items and a setwise attention module to estimate user preferences. To facilitate training, we propose an adaptive weighted loss function with episodic sampling. Experimental studies on real datasets show that SANS can outperform state-of-the-art recommendation algorithms in the few-shot recommendation problem.

Acknowledgement. This work was supported by National Key Research and Development Project (No. 2018YFC0832303); National Natural Science Foundation of China (NSFC) under the Grants nos. 61932007 and 61902075.

References

1. Cheng, H.T., et al.: Wide & deep learning for recommender systems. In: Proceedings of the 1st Workshop on Deep Learning for Recommender Systems, pp. 7–10 (2016)
2. Deng, Z.H., Huang, L., Wang, C.D., Lai, J.H., Philip, S.Y.: DeepCF: a unified framework of representation learning and matching function learning in recommender system. In: Proceedings of the AAAI Conference on Artificial Intelligence, vol. 33, pp. 61–68 (2019)
3. Fei-Fei, L., Fergus, R., Perona, P.: One-shot learning of object categories. IEEE Trans. Pattern Anal. Mach. Intell. **28**(4), 594–611 (2006)
4. Gantner, Z., Drumond, L., Freudenthaler, C., Rendle, S., Schmidt-Thieme, L.: Learning attribute-to-feature mappings for cold-start recommendations. In: 2010 IEEE International Conference on Data Mining, pp. 176–185. IEEE (2010)
5. He, X., He, Z., Song, J., Liu, Z., Jiang, Y.G., Chua, T.S.: NAIS: neural attentive item similarity model for recommendation. IEEE Trans. Knowl. Data Eng. **30**(12), 2354–2366 (2018)
6. He, X., Liao, L., Zhang, H., Nie, L., Hu, X., Chua, T.S.: Neural collaborative filtering. In: Proceedings of the 26th International Conference on World Wide Web, pp. 173–182 (2017)
7. Hu, Y., Koren, Y., Volinsky, C.: Collaborative filtering for implicit feedback datasets. In: 2008 Eighth IEEE International Conference on Data Mining, pp. 263–272. IEEE (2008)

8. Kabbur, S., Ning, X., Karypis, G.: FISM: factored item similarity models for top-N recommender systems. In: Proceedings of the 19th ACM SIGKDD International Conference on Knowledge Discovery and Data Mining, pp. 659–667 (2013)
9. Kingma, D.P., Ba, J.: Adam: a method for stochastic optimization. arXiv preprint arXiv:1412.6980 (2014)
10. Koch, G., Zemel, R., Salakhutdinov, R.: Siamese neural networks for one-shot image recognition. In: ICML Deep Learning Workshop, vol. 2, Lille (2015)
11. Li, D., Chen, C., Lu, T., Chu, S., Gu, N.: Mixture matrix approximation for collaborative filtering. IEEE Trans. Knowl. Data Eng. (2019)
12. Rendle, S., Freudenthaler, C., Gantner, Z., Schmidt-Thieme, L.: BPR: Bayesian personalized ranking from implicit feedback. In: Proceedings of the Twenty-Fifth Conference on Uncertainty in Artificial Intelligence, pp. 452–461 (2009)
13. Sarwar, B., Karypis, G., Konstan, J., Riedl, J.: Item-based collaborative filtering recommendation algorithms. In: Proceedings of the 10th International Conference on World Wide Web, pp. 285–295 (2001)
14. Schein, A.I., Popescul, A., Ungar, L.H., Pennock, D.M.: Methods and metrics for cold-start recommendations. In: Proceedings of the 25th Annual International ACM SIGIR Conference on Research and Development in Information Retrieval, pp. 253–260 (2002)
15. Sedhain, S., Menon, A.K., Sanner, S., Xie, L., Braziunas, D.: Low-rank linear cold-start recommendation from social data. In: Thirty-first AAAI Conference on Artificial Intelligence (2017)
16. Sedhain, S., Sanner, S., Braziunas, D., Xie, L., Christensen, J.: Social collaborative filtering for cold-start recommendations. In: Proceedings of the 8th ACM Conference on Recommender systems, pp. 345–348 (2014)
17. Shi, Y., Karatzoglou, A., Baltrunas, L., Larson, M., Oliver, N., Hanjalic, A.: CLiMF: learning to maximize reciprocal rank with collaborative less-is-more filtering. In: Proceedings of the Sixth ACM Conference on Recommender Systems, pp. 139–146 (2012)
18. Snell, J., Swersky, K., Zemel, R.: Prototypical networks for few-shot learning. In: Advances in Neural Information Processing Systems, pp. 4077–4087 (2017)
19. Tang, J., Wu, S., Sun, J., Su, H.: Cross-domain collaboration recommendation. In: Proceedings of the 18th ACM SIGKDD International Conference on Knowledge Discovery and Data Mining, pp. 1285–1293 (2012)
20. Tay, Y., Anh Tuan, L., Hui, S.C.: Latent relational metric learning via memory-based attention for collaborative ranking. In: Proceedings of the 2018 World Wide Web Conference, pp. 729–739 (2018)
21. Vinyals, O., Blundell, C., Lillicrap, T., Wierstra, D., et al.: Matching networks for one shot learning. In: Advances in Neural Information Processing Systems, pp. 3630–3638 (2016)
22. Wang, Y., Wang, L., Li, Y., He, D., Chen, W., Liu, T.Y.: A theoretical analysis of NDCG ranking measures. In: Proceedings of the 26th Annual Conference on Learning Theory (COLT 2013), vol. 8, p. 6 (2013)
23. Yin, H., Cui, B., Li, J., Yao, J., Chen, C.: Challenging the long tail recommendation. arXiv preprint arXiv:1205.6700 (2012)
24. Zhang, S., Yao, L., Sun, A., Tay, Y.: Deep learning based recommender system: a survey and new perspectives. ACM Comput. Surv. (CSUR) **52**, 1–38 (2019)

Semi-supervised Factorization Machines for Review-Aware Recommendation

Junheng Huang, Fangyuan Luo, and Jun Wu$^{(\boxtimes)}$

School of Computer and Information Technology, Beijing Jiaotong University,
Beijing 100044, China
{19120362,20112027,wuj}@bjtu.edu.cn

Abstract. Textual reviews, as a useful supplementary of the interaction data, has been widely used to enhance the performance of recommender systems, especially when the interaction data is sparse. However, existing solutions to review-aware recommendation only focus on learning more informative features from reviews, yet ignore the insufficient number of training examples, resulting in limited performance improvements. To this end, we propose a co-training style semi-supervised review-aware recommendation model, called Collaborative Factorization Machines (CoFM), to augment the training dataset as well as increase its informativeness. Our CoFM employs two FMs as base predictors, each of which labels unlabeled examples for its peer predictor in the learning process. Specifically, a user-leaded FM and an item-leaded FM are separately built using different reviews to increase the diversity between two predictors. Furthermore, to exploit unlabeled data safely, the labeling confidence is estimated through validating the influence of the labeling of unlabeled examples on the labeled ones. The final prediction is made by linearly blending the outputs of two predictors. Extensive experiments on three real-world benchmarks demonstrate the superiority of CoFM over several state-of-the-art review-aware and semi-supervised recommendation schemes.

Keywords: Review-aware recommendation · Semi-supervised learning · Factorization Machines

1 Introduction

Nowadays, recommender systems [5] have been an indispensable tool in providing personalized Web services for different users in the situations of information overload. Collaborative Filtering (CF) [6], more specifically, Matrix Factorization (MF) [11], has been one of key techniques to build recommender systems. However, its performance is limited by the high sparsity and inferior expressiveness of the interaction data. One of promising solutions to improve prediction is exploiting the rich *side-information* concerning users and items [21] to complement interaction data, and a popularly used side-information is the textual

---

J. Huang and F. Luo—These authors contributed equally to this work.

C. S. Jensen et al. (Eds.): DASFAA 2021, LNCS 12683, pp. 85–99, 2021.
https://doi.org/10.1007/978-3-030-73200-4_6

reviews posed by users towards items (often existing alongside rating data), which gives birth to *review-aware recommendation*.

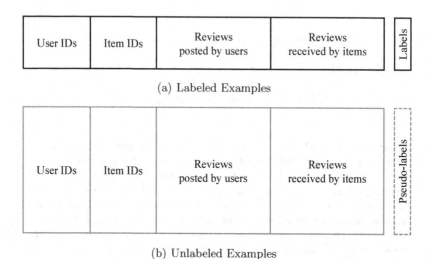

(a) Labeled Examples

(b) Unlabeled Examples

Fig. 1. An illustration of the motivation of this paper: Exploiting unlabeled examples to enhance review-aware recommendation.

Along this direction, some studies [2,13,16] combine MF [11] for predicting ratings with Latent Dirichlet Allocation (LDA) [3] for analyzing the content of reviews towards recommendation tasks, while others [7,20,26] try to learn better features from reviews for users and items through deep learning architectures like TextCNN [8]. Moreover, a recent empirical study [19] argues that the end-to-end learning style schemes for review-aware recommendation are not more effective than the shallow models integrating MF with LDA. In essence, review-aware recommendation is belonged to the family of hybrid recommendation [1] that leverages both interactions and rich side features of users and items for accurate predictions. Factorization Machines (FM) [18] is a prevalent and generic hybrid recommendation model, which can incorporate any side feature by concatenating them into a high-dimensional and sparse feature vector. Inspired by this, we illustrate the setup of a review-aware recommender system in Fig. 1(a), where the feature vectors of the transactions (user-item interactions) consist of user IDs (one-hot codes), item IDs (one-hot codes), and the textual features respectively extracted from the reviews posted by users and the reviews received by items[1], while the labels of such transactions are the ratings posed by users over items. The key advantage of FM is to learn low-dimensional embeddings for all the feature dimensions. Some recent end-to-end learning solutions to review-aware recommendation, such as [26] and [7], can be regarded as the extensions of FM.

---

[1] The textual features can be extracted by either LDA (e.g., [2,13,16]) or DNN (e.g., [7,20,26]).

Despite the success, the existing review-aware recommendation approaches are suboptimal, since they only dedicate to increasing the informativeness of training examples, but neglect of the insufficient amount of training data. From the view of machine learning, the observed ratings can be regarded as the labeled examples while the unobserved ones are the unlabeled examples. Due to the data sparsity, the number of labeled examples is much less than unlabeled ones, and thus it is challenging to train a reliable predictor purely based on the labeled examples. By taking reviews into consideration, the feature vectors become more informative, and thus the produced predictor often achieves better performance than a pure Collaborative Filtering (CF) algorithm (e.g., MF). Note that, given the sets of users and items, the number of training data used by review-aware recommender systems is as same as a pure CF technique (i.e., the sparsity of interaction data is not changed). We argue that, although labeled examples are expensive to obtain in recommender systems, unlabeled data is readily available and could be used as another data resource (beyond reviews) to aid the preference learning. Considering this, we intend to devise a semi-supervised review-aware recommendation model, illustrated by Fig. 1, which can exploit unlabeled examples (Fig. 1(b)) in addition to labeled ones (Fig. 1(a)) in order to better address the sparsity and inferior-informativeness of the interaction data. There have been a few studies carried out on designing the semi-supervised CF algorithms [24,25], but little effort has been dedicated to integrating unlabeled data into the learning process of review-aware recommendation. More importantly, such semi-supervised CF approaches are vulnerable in that they lack an effective measure of labeling confidence to choose appropriate unlabeled examples. As demonstrated by [12], if improper unlabeled examples were used, the performance of a semi-supervised learning algorithm might degrade sharply, and even perform worse than its supervised version.

To address the limitations of previous work, we propose a co-training style semi-supervised review-aware recommendation model, termed Collaborative Factorization Machines (CoFM). Our CoFM employs two FM models as the base predictors, each of which labels unlabeled examples for the other predictor in each round of co-training iterations. Specifically, we construct a user-leaded FM and an item-leaded FM by feeding different reviews to increase the diversity between predictors. To exploit the unlabeled examples safely, the labeling confidence is estimated through validating the influence of the labeling of unobserved examples on the observed ones. The final prediction is made by linear combination of the outputs from two base predictors. Our extensive empirical study shows encouraging results in comparison to state-of-the-art recommendation techniques, including both review-aware and semi-supervised solutions.

The rest of this paper is organized as follows. In Sect. 2, we review related work. Section 3 elaborates the proposed CoFM model, and Sect. 4 reports on the experimental results. Finally, Sect. 5 concludes this paper.

2 Related Work

In this section, we briefly review related work on review-aware recommendation and semi-supervised collaborative filtering.

2.1 Review-Aware Recommendation

A popular solution to improve recommendation accuracy in the situation of sparse data is to exploit textual reviews posted by users over items to complement user-item interactions. Roughly speaking, existing methods of review-aware recommendation can be divided into two categories.

The first manner focuses on topic modeling with textual reviews towards recommendation tasks. For example, Hidden Factors as Topics (HFT) [16] employed a LDA-like topic model on review text for users and items, and a MF model to fit the ratings. TopicMF [2], jointly modeled user ratings with MF and textual reviews with non-negative matrix factorization (NMF) to derive topics from the reviews. One main difference between HFT and TopicMF is that HFT learns the topics for each item, while TopicMF learns the topics for each review. Collaborative Topic Regression (CTR) [13] combined ideas of probabilistic matrix factorization (PMF) and LDA for recommendation tasks, which jointly optimizes the combined objective function of both PMF and LDA in an online learning fashion. Different from HFT and TopicMF, CTR is to pursuit the efficiency and scalability of the joint learning of MF and LDA. Another manner concentrates on learning specific features from textual reviews with deep neural networks towards recommendation tasks. For instances, DeepCoNN [26] learned embeddings for users and items from textual reviews by two parallel neural networks, and then coupled them by a FM in the last layer for rating prediction. D-Attn [20] modeled user preferences and item properties by convolutional neural networks with dual local and global attention. NARRE [7] proposed a novel attention mechanism to build recommendation systems and selected highly-useful reviews simultaneously. NRPA [15] learned a personalized attention recommendation model to select different words and reviews for different users and items.

Despite review-aware recommendation can improve the recommendation performance by enriching the informativeness of labeled training data, the data sparsity issue is not mitigated due to the unchanged labeled training size. This is the major problem we try to tackle in this paper.

2.2 Semi-supervised Collaborative Filtering

A canonical solution to data sparsity is data imputation which selects a set of user-item pairs whose values are unobserved, and then fills them with imputed values before making recommendations, such as RCF [24] and AutAI [17]. Such schemes are designed based on the idea of self-learning, a pioneer study of semi-supervised learning, which labels a few unlabel examples with high confidence to produce pseudo-labeled examples first, and then feed both label and pseudo-label examples to refine the model. Since the label examples are insufficient and the learned predictor is not reliable, mislabeling is unavoidable. Therefore, the expanded training set used for the next iteration will be noisy.

Disagreement-based semi-supervised learning employs 'multiple predictors' to smooth the labeling noisy. A prominent achievement in this area is the co-training paradigm [4]. The most related work to our work is CSEL [25] model which is a typical co-training style CF scheme. Concretely, CSEL generated two different SVD++ [9] models with different context, and then applied the predictions of each predictor on unlabeled examples to augment the training set of the other. Although effective, such a solution is weak in estimating the labeling confidence. Differently, we take labeling confidence validation into consideration to build a safer semi-supervised recommendation system.

3   The Proposed CoFM Approach

In this section, we elaborate our proposed CoFM model that aims at exploiting unlabeled examples in addition to labeled ones to alleviate the data sparsity problem by a co-training solution, which consists of three key layers: 1) Data Layer 2) Feature Layer 3) Learning Layer, shown in Fig. 2. We first give some necessary preliminaries and then describe our CoFM model in details.

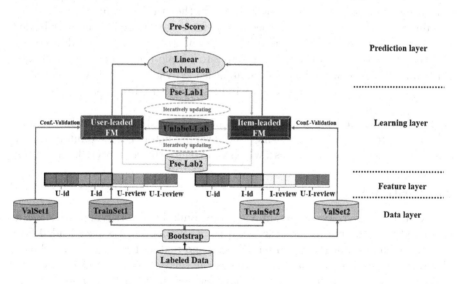

Fig. 2. The illustration of CoFM model, where the left and right parts are two different recommendation modules, bridged by a co-training mechanism.

3.1   Preliminaries

In a practical recommendation system, the interactions between m users and n items can be represented by a rating matrix $\mathbf{R} \in [0,1]^{m \times n}$, whose element r_{ui}

indicates the rating user u scored on item i, otherwise $r_{ui} = \varnothing$. Without loss of generality, we re-scale all ratings into the interval $[0,1]$.

From a machine learning view, such a matrix $\mathbf{R}$ can be regraded as a dataset $\mathcal{D} = \{(\mathbf{x}_t, y_t)\}_{t=1}^{m \times n}$, where $\mathbf{x}_t$ is a high-dimensional feature vector describing the transaction between a user and an item, concatenated by user ID and item ID (one-hot encoding), users' and items' review features, while $y_t = r_{ui}$ denotes the label of such a transaction. Furthermore, let $\mathcal{L} = \{(\mathbf{x}_t, y_t)|y_t \in [0,1]\}_{t=1}^{l}$ be the set of labeled examples (i.e., the observed user-item interactions), while $\mathcal{U} = \{(\mathbf{x}_t, y_t)|y_t = \varnothing\}_{t=l+1}^{m \times n}$ denotes the set of unlabeled examples (i.e., the unobserved user-item interactions), and $\mathcal{D} = \mathcal{L} \cup \mathcal{U}$. It is well known that $|\mathcal{L}| \ll |\mathcal{U}|$ in recommender systems, where $|\bullet|$ denotes the size of a set.

3.2 Data Layer

Generally speaking, in order to generate different base predictors for increasing model diversity, termed the user-leaded FM and the item-leaded FM respectively, one way is to train the models with different label examples by manipulating training set, while the other way is to build different views by manipulating the textual review information, which are both adopted in our paper. A straightforward way to manipulate training set is Bootstrap. It randomly samples a set of training examples $\mathcal{T}$ with replacement from the original training set, and the remain serve as the valid set $\mathcal{V}$. Specifically, for better modeling the user-leaded view, we collect user u's all reviews to represent its preference. In addition, we add the textual review of user u on item i to couple user u and item i closely. The same process can be done for the item-leaded view. All reviews received by item i and the user u's review on the item i can be used to represent the item i's review feature.

In this work, two base predictors are generated based on two different training subsets and textual reviews, in which such two predictors can be reinforced each other during a co-training process, which will be introduced in Sect. 3.4 in details.

3.3 Feature Layer

In the feature layer, two different views are built by diversifying the features extracted from textual reviews for two base predictors. For the user-leaded FM, its input includes user ID ($\mathbf{p}_u$), item ID ($\mathbf{q}_i$), and textual features extracted from all reviews posted by user u ($\mathbf{f}_u$) as well as the review posted by user u on item i ($\mathbf{f}_{ui}$). We refer $\mathbf{f}_{ui}$ as 'one-to-one review feature' in the below. Differently, for the item-leaded FM, we replace $\mathbf{f}_u$ with $\mathbf{f}_i$ (textual features extracted from all reviews received by item i), and remaining parts of the item-leaded FM's input are same with the user-leaded FM's. Here the textual features can be extracted by either traditional Latent Dirichlet Allocation (LDA) model [3]) or modern TextCNN [8] style techniques. A recent empirical study [19] reveals that using LDA can achieve competitive performance with TextCNN in the tasks of review-aware recommendation, and thus LDA is considered in this work due to its ease of use.

We can learn features from reviews for users and items through LDA or TextCNN [8], but this study [19] argues that review-aware recommendations using TextCNN are not more effective than the shallow models integrating MF with LDA, so LDA model is used for textual feature extraction in our model. In general, FM predicts the rating of user u on item i as follows:

$$\hat{y} = \omega_0 + \sum_{i=1}^{n} \omega_i x_i + \sum_{i=1}^{n} \sum_{j=i+1}^{n} < \mathbf{v}_i, \mathbf{v}_j > x_i x_j, \tag{1}$$

where ω_0 is the global bias and ω_i is the feature bias. $\mathbf{V} \in \mathbb{R}^{f \times n}$ is the latent matrix for all features, and every $< \mathbf{v}_i, \mathbf{v}_j >$ models the interaction between i-th and j-th feature dimensions. Therefore, $\mathbf{V}$ is the key reason why FM is an effective feature-based recommendation model, as it captures the rich information interaction. In our proposed model, $\mathbf{x} = [\mathbf{p}_u, \mathbf{q}_i, \mathbf{f}_u(\mathbf{f}_i), \mathbf{f}_{ui}]$ in Eq. (1) is concatenated by user u's and item i's one-hot encoded vector, user u's review feature $\mathbf{f}_u$ (item i's review feature $\mathbf{f}_i$) and one-to-one review feature $\mathbf{f}_{ui}$ for the user-leaded FM (the item-leaded FM).

3.4 Learning Layer

We randomly sample a set of unlabeled examples $\mathcal{U}_s$ from the set $\mathcal{U}$, and feed them into each base predictors to produce the pseudo-labeled examples $\hat{\mathcal{L}}$, then apply these pseudo-labeled examples to refine its peer (the other base predictor). It is noteworthy that the one-to-one review features are not available in these pseudo-labeled examples, so we take the average of user u's all one-to-one review features as its one-to-one review feature when modeling the user-leaded view, which is defined as follows:

$$\widetilde{\mathbf{f}}_u = \frac{1}{|\Omega_u|} \sum_{i \in \Omega_u} \mathbf{f}_{ui}, \tag{2}$$

where Ω_u is the observed rating set for user u and $|\bullet|$ is the size of a set. For the item-leaded view, we can obtain the one-to-one review feature of pseudo-labeled examples in the similar way,

$$\widetilde{\mathbf{f}}_i = \frac{1}{|\Omega_i|} \sum_{u \in \Omega_i} \mathbf{f}_{ui}, \tag{3}$$

where Ω_i is the observed rating set for item i.

One challenge here is how to determine the criteria for selecting unlabeled examples from $\mathcal{U}$ so as not to deteriorate the performance of base predictors. Inspired by semi-supervised regression [27], our mechanism for estimating the labeling confidence is designed based on an intuition that the error of a base predictor evaluated on the labeled example set should decrease if a confidently pseudo-labeled example is added into its training set. In other words, the confidently pseudo-labeled examples should be the ones that make the predictor consistent with the labeled example set. Benefited from bootstrap, we suggest measuring the confidence of pseudo-labeled examples on the valid set. Let $\Delta_{\hat{\mathcal{L}}}$ denotes the result of MSE evaluated on the valid set. Concretely, the confidence measure of pseudo-labeled examples is defined as

$$\Delta_{\hat{\mathcal{L}}} = \sum_{x \in \mathcal{V}} [(y - h(\mathbf{x}))^2 - (y - h^+(\mathbf{x}))^2], \tag{4}$$

where h denotes the original predictor generated from $\mathcal{T}$, while h^+ denotes the refined predictor learned from the enlarged training set $\mathcal{T} \bigcup \hat{\mathcal{L}}$. We claim that $\hat{\mathcal{L}}$ is confident and added into training set if $\Delta_{\hat{\mathcal{L}}} > 0$, otherwise it will be discarded.

3.5   Assembling Two Base Predictors

The final step of our proposed model is to assemble the results of two base predictors boosted by co-training. A promising way is the linear combination, the effectiveness of which has been verified in the Netflix contest [10]. We denote h_1 and h_2 as the two base predictors enhanced with unlabeled examples, the final prediction of our proposed model can be defined as:

$$h^*(\mathbf{x}) = \sum_{j=1}^{2} \alpha_j h_j(\mathbf{x}), \tag{5}$$

where α_j can be determined by a linear regression algorithm. To learn appropriate $\alpha = [\alpha_1, \alpha_2]^T$, a new training set $\bar{\mathcal{L}} = \{(\bar{\mathbf{x}}_t, y_t) | y_t \in [0, 1]\}_{t=1}^l$ is built based on $\mathcal{L} = \{(\mathbf{x}_t, y_t) | y_t \in [0, 1]\}_{t=1}^l$, where $\bar{\mathbf{x}}_t = [h_1(\mathbf{x}_t), h_2(\mathbf{x}_t)]^T$. A linear regression algorithm can learn a group of α which make a regressor fit $\bar{\mathcal{L}}$ best, i.e., such $\alpha = [\alpha_1, \alpha_2]^T$ achieve the best match between $h^*(\mathbf{x})$ and y_t. In this sense, such $\alpha = [\alpha_1, \alpha_2]^T$ can serve as the optimal weights for merging the two base predictors.

3.6   Algorithm Framework

Putting everything together, we summarize the proposed CoFM model in Algorithm 1, where $A_k(\mathcal{L}_k)$ returns a predictor generated from training set $\mathcal{L}_k$ by algorithm A_k (e.g., A_1, A_2 respectively denote the user-leaded FM and the item-leaded FM). The learning procedure stops when the maximum iterations T is reached, or there is no pseudo-labeled example that is able to reduce the MSE of any of base predictors on the labeled example set. Suggested by [4], a pool of unlabeled examples $\mathcal{U}_s$ ($|\mathcal{U}| \gg |\mathcal{U}_s|$) is utilized to reduce the computational load. At last, a linear combination of h_1 and h_2 is used to make the final prediction, i.e., $h^*(\mathbf{x}) = \sum_{j=1}^{2} \alpha_j h_j(\mathbf{x})$ where $\alpha = [\alpha_1, \alpha_2]^T$ is determined by a linear regression algorithm.

Algorithm 1. The CoFM Algorithm

Input: labeled example set $\mathcal{L}$, unlabeled example set $\mathcal{U}$, maximum iterations of co-training T, FM algorithm index A_1, A_2

1: **Initialization:**
2: $(\mathcal{T}_1, \mathcal{V}_1) \leftarrow Bootstrap(\mathcal{L}); (\mathcal{T}_2, \mathcal{V}_2) \leftarrow Bootstrap(\mathcal{L});$
3: Generate review features $\mathbf{f}_u, \mathbf{f}_i, \mathbf{f}_{ui}$ for each u, i and (u, i) pair by LDA;
4: $\mathcal{T}_1^* \leftarrow \{[\mathbf{p}_u, \mathbf{q}_i, \mathbf{f}_u, \mathbf{f}_{ui}] | (u, i) \in \mathcal{T}_1\}, \mathcal{V}_1^* \leftarrow \{[\mathbf{p}_u, \mathbf{q}_i, \mathbf{f}_u, \mathbf{f}_{ui}] | (u, i) \in \mathcal{V}_1\};$
5: $\mathcal{T}_2^* \leftarrow \{[\mathbf{p}_u, \mathbf{q}_i, \mathbf{f}_i, \mathbf{f}_{ui}] | (u, i) \in \mathcal{T}_2\}, \mathcal{V}_2^* \leftarrow \{[\mathbf{p}_u, \mathbf{q}_i, \mathbf{f}_i, \mathbf{f}_{ui}] | (u, i) \in \mathcal{V}_2\};$
6: $h_1 \leftarrow A_1(\mathcal{T}_1^*); h_2 \leftarrow A_2(\mathcal{T}_2^*)$
7: $round = 0;$
8: **while** $round \leq T$ **do**
9: $\quad round \leftarrow round + 1;$
10: $\quad$ **for** $k \in \{1, 2\}$ **do**
11: $\qquad$ Generate a mini-batch $\mathcal{U}_s$ by random sampling from $\mathcal{U}$;
12: $\qquad$ Generate $\widetilde{\mathbf{f}}_u, \widetilde{\mathbf{f}}_i$ according to Eq.(2) and Eq.(3) for (u, i) pair from $\mathcal{U}_s$
13: $\qquad$ **if** $k = 1$ **then**
14: $\qquad\quad \mathcal{U}_s^* \leftarrow \{[\mathbf{p}_u, \mathbf{q}_i, \mathbf{f}_i, \widetilde{\mathbf{f}}_i] | (u, i) \in \mathcal{U}_s\}$
15: $\qquad\quad$ % *each predictor labels unlabeled data for its peer*
16: $\qquad\quad \hat{\mathcal{L}} \leftarrow h_{3-k}(\mathcal{U}_s^*);$
17: $\qquad\quad \hat{\mathcal{L}}^* \leftarrow \{[\mathbf{p}_u, \mathbf{q}_i, \mathbf{f}_u, \widetilde{\mathbf{f}}_u] | (u, i) \in \hat{\mathcal{L}} \cup \mathcal{T}_k^*\};$
18: $\qquad$ **else**
19: $\qquad\quad \mathcal{U}_s^* \leftarrow \{[\mathbf{p}_u, \mathbf{q}_i, \mathbf{f}_u, \widetilde{\mathbf{f}}_u] | (u, i) \in \mathcal{U}_s\}$
20: $\qquad\quad \hat{\mathcal{L}} \leftarrow h_{3-k}(\mathcal{U}_s^*);$
21: $\qquad\quad \hat{\mathcal{L}}^* \leftarrow \{[\mathbf{p}_u, \mathbf{q}_i, \mathbf{f}_i, \widetilde{\mathbf{f}}_i] | (u, i) \in \hat{\mathcal{L}} \cup \mathcal{T}_k^*\};$
22: $\qquad$ **end if**
23: $\qquad h_k^+ \leftarrow A_k(\hat{\mathcal{L}}^*);$
24: $\qquad \Delta_{\hat{\mathcal{L}}^*} \leftarrow \sum_{\mathbf{x} \in \mathcal{V}_k}[(y - h_k(\mathbf{x}))^2 - (y - h_k^+(\mathbf{x}))^2];$ % *confidence validation*
25: $\qquad$ **if** $\Delta_{\hat{\mathcal{L}}^*} > 0$ **then**
26: $\qquad\quad \mathcal{T}_k^* \leftarrow \mathcal{T}_k^* \cup \hat{\mathcal{L}}; \mathcal{U} \leftarrow \mathcal{U} - \hat{\mathcal{L}};$
27: $\qquad$ **end if**
28: $\quad$ **end for**
29: $\quad$ **if** neither of $\mathcal{T}_1^*$ and $\mathcal{T}_2^*$ changes **then**
30: $\qquad$ **exit;**
31: $\quad$ **else**
32: $\qquad$ % *re-training with enlarged labeled sets*
33: $\qquad h_1 \leftarrow A_1(\mathcal{T}_1^*); h_2 \leftarrow A_2(\mathcal{T}_2^*);$
34: $\quad$ **end if**
35: **end while**
36: Create a new training set $\bar{\mathcal{L}}$ based on $\mathcal{L}$ and h_j $(j \in \{1, 2\});$
37: Learn α_js for h_js from $\bar{\mathcal{L}}$ by a linear regression algorithm;

Output: Merged predictor $h^*(\mathbf{x}) = \sum_{j=1}^2 \alpha_j h_j(\mathbf{x})$

---

4   Experiments

In this section, we conduct extensive experiments on three real-world datasets to demonstrate the effectiveness of our proposed CoFM model compared with several state-of-the-art review-aware recommendation algorithms and a state-of-the-art semi-supervised collaborative filtering technique. Table 1 summarizes the statistics of these datasets, whose rating densities are range from 0.089% to 0.798%, covering a broad range of data sparsity for rating prediction tasks.

Table 1. Statistics of three datasets.

Datasets	Users	Items	Ratings	Density
Music Instruments	1,429	900	10,261	0.789%
Office Products	4,905	2,420	53,228	0.448%
Video Games	24,303	10,672	231,577	0.089%

4.1   Experimental Setting

For each dataset, we randomly divide it into training set (80%) and testing set (20%). We randomly generate five splits and report the averaged performance.

We adopt Mean Square Error (MSE) to measure the recommendation quality, which is widely used for rating prediction in recommender systems. A lower MSE score indicates a better performance. Given a predicted rating $\hat{R}_{ui}$ and ground-truth rating R_{ui} from the user u for the item i, the MSE is calculated as:

$$MSE = \frac{\sum_{(u,i)}(R_{ui} - \hat{R}_{ui})^2}{|\mathcal{T}|}$$

where $|\mathcal{T}|$ denotes the size of testing set.

To validate the effectiveness of CoFM, we compare our CoFM approach with a conventional MF model, several state-of-the-art review-aware recommendation algorithms and a semi-supervised collaborative filtering technique, including

- **MF** [11] is a canonical collaborative filtering method which decomposes a user-item matrix into a shared low-dimensional latent space to recover the observed user-item interactions as well as predict the unobserved ones.
- **HFT** [16] is a pioneer study on review-aware recommendation, which extends MF with an additional regularizer that model the corpus likelihood using LDA.
- **DeepCoNN** [26] is the first deep learning model for review-aware recommendation, which utilizes the same CNN module to learn user and item embeddings based on their reviews.

- **D-Attn** [20] is an improved version of DeepCoNN, which adds a dual-attention layer at word-level before convolution, and preserves other components of DeepCoNN.
- **NARRE** [7] is another extended version of DeepCoNN, which adds an attention weight at review-level.
- **NRPA** [15] is a recently proposed method that introduces neural attention mechanism to build the recommender system using reviews. NRPA extends DeepCoNN by adding the word-level and review-level attention layer.
- **CSEL** [25] is a co-training style CF method which takes two data-discrepancy SVD++ models as the base predictors, and then such two predictors can improve each other during co-training iterations. Note that CSEL model takes context information into consideration. For fairness, we reconstruct CSEL model with review text information instead of contexts in this paper.

Moreover, to verify whether each component used by CoFM is useful or not, a series of degenerate variants of CoFM are also included in the comparisons:

- **U-FM (user-leaded FM) & I-FM (item-leaded FM)** are two base predictors used by CoFM. A major difference between them is that U-FM partially focuses on modeling user's reviews, while I-FM concentrates on item's reviews.
- **Ens (Ensemble of two predictors)** makes predictions by a linear blending of two base predictors(U-FM and I-FM), where the blending weights of base predictors are determined by a linear regression algorithm.
- **CoFM-wCV (CoFM without Confidence Validation)** is similar to CoFM, except that CoFM-wCV omits the confidence validation step in each co-training iteration.
- **CoFM-wBS (CoFM without Bootstrap)** is almost same with CoFM, but the former dose not use bootstrap to diversify the training sets for two base predictors.

The optimal experimental settings for all of above methods are determined by grid search. For our method, we set the number of topics $K = 50$ for LDA embedding vector, and feature dimensions of the factorization machine $f = 10$, learning rate $\eta = 0.001$, the maximum iterations $T = 50$, the pool size of unlabeled examples $\mathcal{U}_s = 500$ for Music Instrument dataset and Office Products dataset, and $\mathcal{U}_s = 2000$ for Video Games dataset.

4.2   Result Analysis

We first compare our proposed CoFM model with seven existing methods. The corresponding experimental results are summarized in Table 2, where the best performance is boldfaced and the percentages indicate the relative improvement of our approach over other compared methods.

By examining all methods, several observations can be drawn from the experimental results. First, we observe that five review-aware recommendation methods considerably outperform MF on all datasets, which indicates the effectiveness

Table 2. Performance of CoFM compared with the state of the arts

Dataset	Music Instruments		Office Products		Video Games	
Metric	MSE	Improve	MSE	Improve	MSE	Improve
MF	0.848	15.2%	0.79	9.37%	1.197	9.02%
HFT	0.766	6.54%	0.754	5.04%	1.126	3.4%
DeepCoNN	0.784	8.29%	0.773	7.37%	1.82	7.87%
D-Attn	0.762	5.64%	0.754	5.04%	1.145	4.89%
NARRE	0.775	7.23%	0.763	6.16%	1.136	4.14%
NRPA	0.751	4.26%	0.749	4.41%	1.122	2.94%
CSEL	0.748	4.03%	0.731	2.09%	1.120	2.85%
CoFM	**0.719**		**0.716**		**1.089**	

of complementing user-item interactions with textual reviews for rating prediction tasks. Second, among the four deep models (DeepConn, D-Attn, NARRE and NRPA), NRPA achieves the best performance due to its elaborate network architecture and dual attention mechanism. Third, we find that HFT achieves competitive performance with other deep review-aware recommendation models, which is consistent with [19]. This observation demonstrates that the benefits of deep learning based modeling techniques on textual review are overstated, so LDA model is used for textual feature extraction in our model. In addition, we observe that our proposed CoFM model and CSEL model achieves much better performance compared to HFT on three datasets, which indicates that the importance of exploiting unlabeled examples for rating prediction tasks. Finally, it is impressive that our proposed CoFM model consistently outperforms CSEL model all the time. The major difference between such two models is that CoFM model takes the safety of pseudo-labeled examples into consideration, while CSEL model exploits pseudo-labeled examples directly. This implies that confidence validation is very important to semi-supervised learning.

Next, we compare CoFM model with its two degenerate variants CoFM-wBS and CoFM-wCV, in order to verify the effectiveness of such two specific mechanisms designed for our CoFM model, i.e., bootstrap and confidence validation. Figure 3(a) illustrates the MSE of CoFM approach compared with CoFM-wBS, and we observe that CoFM outperforms CoFM-wBS on all datasets. It verifies that diversifying training data by bootstrap is useful to improve the effectiveness of co-training. As illustrated by Fig. 3(b), Fig. 3(c) and Fig. 3(d), there is a large gap between the performance curves of CoFM and CoFM-wCV. This observation again demonstrates that the confidence validation is beneficial to improve the performance of semi-supervised learning.

At last, we further compare CoFM model with its other three degenerate variants, including U-FM, I-FM and Ens. The corresponding quantitative results in terms of MSE are tabulated in Table 3. We observe that CoFM model and Ens

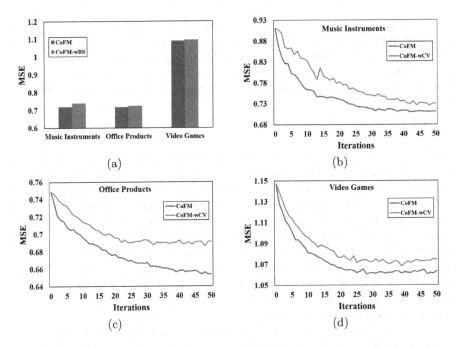

Fig. 3. Performance of CoFM compared with its degenerate variants.

Table 3. Performance of CoFM compared with its variants.

Dataset	Music Instruments		Office Products		Video Games	
Metric	MSE	Improve	MSE	Improve	MSE	Improve
U-FM	0.851	15.5%	0.738	2.98%	1.121	2.85%
I-FM	0.812	11.5%	0.748	4.28%	1.115	2.33%
Ens	0.768	6.38%	0.733	2.32%	1.107	1.63%
CoFM	**0.719**		**0.716**		**1.089**	

model outperform U-FM and I-FM on all datasets, which verifies the superiority of using multiple base predictors over single one. Comparing CoFM with Ens, we find that CoFM model always achieves better performance than Ens model. A major difference between such two schemes is that CoFM enhances the base predictors using unlabeled examples before assembling them, while Ens directly blends base predictors. It demonstrates that exploiting unlabeled data is helpful for improving recommendation performance. That is, one key motivation of this work is empirically verified. In addition, we also find that the improvement of CoFM model over its degenerate variants grows as the increasing of rating density. For example, CoFM model achieves the largest improvement on Movie Instrument whose rating density is much denser than Office Products and Video

Games. This phenomenon can be explained as follows. Given sparse datasets (e.g., Office Products and Video Games), it is so difficult to train a reliable base predictor in that the pseudo-labeled examples are not credible. In contrast, given denser datasets (e.g., Movie Instrument), the base predictors become stronger and can produce credible pseudo-labeled examples to further refine predictors.

5   Conclusions

Motivated by the fact that current review-aware recommendation is challenged by the problem of learning with insufficient training data, this paper presented a co-training style semi-supervised review-aware recommendation model, termed Collaborative Factorization Machines (CoFM). In sharp contrast to existing review-aware recommendation methods, CoFM learns embeddings for users and items by exploiting both labeled and unlabeled instances. Different from current semi-supervised CF techniques, we integrate the labeling confidence validation into our proposed framework to exploit unlabeled examples more safely. Through extensive experiments carried out on three benchmarks, we demonstrated that our CoFM consistently outperforms the state-of-the-art review-aware and semi-supervised recommendation approaches.

To our best knowledge, CoFM is the first semi-supervised framework for enhancing review-aware recommendation, so we believe that it has a great potential to advance real-world recommender systems since CoFM can effectively exploit the abounding unlabeled data. Inspired by the hashing-based recommendation [14,22,23], we will develop the discrete version of CoFM to boost its prediction efficiency.

Acknowledgments. The authors would like to thank the anonymous reviewers for their constructive suggestions. This work was supported in part by the National Natural Science Foundation of China under Grant 61671048 and the Fundamental Research Funds for the Central Universities under Grant 2019JBM316.

References

1. Adomavicius, G., Tuzhilin, A.: Toward the next generation of recommender systems: a survey of the state-of-the-art and possible extensions. IEEE T. Knowl. Data Eng. **17**(6), 734–749 (2005)
2. Bao, Y., Fang, H., Zhang, J.: TopicMF: simultaneously exploiting ratings and reviews for recommendation. In: AAAI, pp. 2–8 (2014)
3. Blei, D.M., Ng, A.Y., Jordan, M.I.: Latent Dirichlet allocation. J. Mach. Learn. Res. **3**, 993–1022 (2003)
4. Blum, A., Mitchell, T.: Combining labeled and unlabeled data with co-training. In: COLT, pp. 92–100 (1998)
5. Bobadilla, J., Ortega, F., Hernando, A., Gutierrez, A.: Recommender systems survey. Knowl.-Based Syst. **46**, 109–132 (2013)
6. Cacheda, F., Formoso, V.: Comparison of collaborative filtering algorithms: limitations of current techniques and proposals for scalable, high-performance recommender systems. ACM Trans. Web **5**(1), 1–33 (2011)

7. Chen, C., Zhang, M., Liu, Y., Ma, S.: Neural attentional rating regression with review-level explanations. In: WWW, pp. 1583–1592 (2018)
8. Kim, Y.: Convolutional neural networks for sentence classification. In: EMNLP, pp. 1746–1751 (2014)
9. Koren, Y.: Factorization meets the neighborhood: a multifaceted collaborative filtering model. In: SIGKDD, pp. 426–434 (2008)
10. Koren, Y.: The Bellkor solution to the Netflix grand prize. Technical report, Netflix prize documentation (2009)
11. Koren, Y., Bell, R., Volinsky, C.: Matrix factorization techniques for recommender systems. Computer **42**(8), 30–37 (2009)
12. Li, Y., Zhou, Z.: Towards making unlabeled data never hurt. IEEE Trans. Pattern Anal. **37**(1), 175–188 (2015)
13. Liu, C., Jin, T., Hoi, S.C.H., Zhao, P., Sun, J.: Collaborative topic regression for online recommender systems: an online and Bayesian approach. Mach. Learn. **106**(5), 651–670 (2017). https://doi.org/10.1007/s10994-016-5599-z
14. Liu, H., He, X., Feng, F., Nie, L., Liu, R., Zhang, H.: Discrete factorization machines for fast feature-based recommendation. In: IJCAI, pp. 3449–3455 (2018)
15. Liu, H., et al.: NRPA: neural recommendation with personalized attention. In: SIGIR, pp. 1233–1236 (2019)
16. McAuley, J.J., Leskovec, J.: Hidden factors and hidden topics: understanding rating dimensions with review text. In: RecSys, pp. 165–172 (2013)
17. Ren, Y., Li, G., Zhang, J., Zhou, W.: The efficient imputation method for neighborhood-based collaborative filtering. In: CIKM, pp. 684–693 (2012)
18. Rendle, S.: Factorization machines. In: ICDM, pp. 995–1000 (2010)
19. Sachdeva, N., McAuley, J.: How useful are reviews for recommendation? A critical review and potential improvements. In: SIGIR, pp. 1845–1848 (2020)
20. Seo, S., Huang, J., Yang, H., Liu, Y.: Interpretable convolutional neural networks with dual local and global attention for review rating prediction. In: RecSys, pp. 297–305 (2017)
21. Shi, Y., Larson, M., Hanjalic, A.: Collaborative filtering beyond the user-item matrix: a survey of the state of the art and future challenges. ACM Comput. Surv. **47**(1), 1–45 (2014)
22. Wu, J., Luo, F., Zhang, Y., Wang, H.: Semi-discrete matrix factorization. IEEE Intell. Syst. **35**(5), 73–83 (2020)
23. Zhang, H., Shen, F., Liu, W., He, X., Luan, H., Chua, T.S.: Discrete collaborative filtering. In: SIGIR, pp. 325–334 (2016)
24. Zhang, J., Pu, P.: A recursive prediction algorithm for collaborative filtering recommender systems. In: RecSys, pp. 57–64 (2007)
25. Zhang, M., Tang, J., Zhang, X., Xue, X.: Addressing cold start in recommender systems: a semi-supervised co-training algorithm. In: SIGKDD, pp. 73–82 (2014)
26. Zheng, L., Noroozi, V., Yu, P.S.: Joint deep modeling of users and items using reviews for recommendation. In: WSDM, pp. 425–434 (2017)
27. Zhou, Z.H., Li, M.: Semi-supervised regression with co-training. In: IJCAI, pp. 908–913 (2005)

DCAN: Deep Co-Attention Network by Modeling User Preference and News Lifecycle for News Recommendation

Lingkang Meng[1], Chongyang Shi[1(✉)], Shufeng Hao[2], and Xiangrui Su[1]

[1] School of Computer Science and Technology, Beijing Institute of Technology, Beijing, China
{lkmeng,cy_shi,suxiangrui}@bit.edu.cn
[2] College of Data Science, Taiyuan University of Technology, Shanxi, China
haoshufeng@tyut.edu.cn

Abstract. Personalized news recommendation systems aim to alleviate information overload and provide users with personalized reading suggestions. In general, each news has its own lifecycle that is depicted by a bell-shaped curve of clicks, which is highly likely to influence users' choices. However, existing methods typically depend on capturing user preference to make recommendations while ignoring the importance of news lifecycle. To fill this gap, we propose a Deep Co-Attention Network DCAN by modeling user preference and news lifecycle for news recommendation. The core of DCAN is a Co-Attention Net that fuses the user preference attention and news lifecycle attention together to model the dual influence of users' clicked news. In addition, in order to learn the comprehensive news representation, a Multi-Path CNN is proposed to extract multiple patterns from the news title, content and entities. Moreover, to better capture user preference and model news lifecycle, we present a User Preference LSTM and a News Lifecycle LSTM to extract sequential correlations from news representations and additional features. Extensive experimental results on two real-world news datasets demonstrate the significant superiority of our method and validate the effectiveness of our Co-Attention Net by means of visualization.

Keywords: News recommendation · Co-attention neural network · Recurrent neural network · Convolutional neural network

1 Introduction

In recent years, due to the rapid digital transformation of traditional newspaper companies and the increasing scale of news aggregation platforms, a massive number of news articles are published everyday. However, it is difficult to find interesting content for users from a large number of news. Therefore, personalized news recommendation systems are highly necessary to alleviate information overload and provide users with personalized reading suggestions [9,13].

© Springer Nature Switzerland AG 2021
C. S. Jensen et al. (Eds.): DASFAA 2021, LNCS 12683, pp. 100–114, 2021.
https://doi.org/10.1007/978-3-030-73200-4_7

Recently, a large number of deep learning-based methods have been applied to news recommendation because of their powerful ability to both learn news representation and capture user preference [21–24]. For example, Wang et al. [21] propose DKN to learn news representation via a knowledge-aware CNN from the news title and external knowledge, then capture user preference by applying an attention mechanism to user's clicked news and candidate news. Wu et al. [22] propose NAML to learn news representation using a news encoder from the news title, content and categories, then extract user preference with a user encoder featuring an attention mechanism focused on user's clicked news. Zhu et al. [24] propose DAN, which learns news representation via a parallel CNN from the news title and profile, then utilizes an attention-based neural network to learn user preference with respect to candidate news. Wu et al. [23] propose CPRS to learn news representation via a text encoder from the clicked news title and content, then utilize a multilevel attention network to obtain user preference by considering both click preference and reading satisfaction. However, these methods typically depend on capturing user preference to recommend news while ignoring news lifecycle, which may has the capacity to guide users' choices.

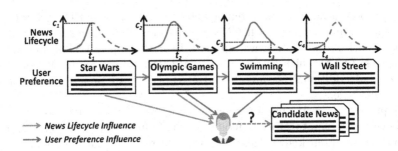

Fig. 1. An illustration of the influence of news lifecycle and user preference on users' choices.

To fill this gap, the following observations is made. First, news lifecycle may have a strong influence on users' choices. As shown in Fig. 1, given a user's clicked history, each news story has its own lifecycle that is depicted as a bell-shaped curve, where t represents user click's time and c represents the number of clicks. Intuitively, the first two news items have more clicks than other news items, which might result in their high influence on the user's choices. Second, fusing news lifecycle and user preference may provide a boosted capacity to influence users' choices. For example, in Fig. 1, if the user prefers to click news about "Olympic Games" and "Swimming", we can infer that the user is probably interested in sport news, which might result in their high influence on the user's choices. Thus, if we only consider unilateral influence, the recommendation results for the user will probably come to a certain deviation. And if we fuse news lifecycle and user preference, we may obtain the dual influence on the user's choices. For example, in Fig. 1, the second news article has both high clicks and user's interested content, probably resulting in its greater influence on the user's choices. On the contrary, the fourth news article is not outstanding in either aspect, which may mean that it is meaningless news in the user's click history.

In addition, news usually contains multiple information, such as the news title, content and entity. Since the news title, content and entities (e.g., person and location) can offer brief, detailed and key information of news respectively, it is necessary to conduct these information and learn the comprehensive news representation. Moreover, user preference usually changes rapidly and is easily affected by the context information (e.g., user click's time and location). For example, if the user live in Tokyo, he is more likely to click news about "Olympic Games". And when the user travel to New York, he will more probably click news about "Wall Street". Therefore, considering the sequential correlation and context information can better capture user preference.

Considering the above observations, in this paper, we propose a Deep Co-Attention Network DCAN by modeling user preference and news lifecycle for news recommendation. The main contributions of this paper are summarized below:

- We propose a novel news recommendation method named DCAN, which is equipped with a Multi-Path CNN which leans comprehensive news representation from multiple news inputs, a User Preference LSTM and a News Lifecycle LSTM which extract sequential correlations from news representations and additional features to better capture user preference and model news lifecycle.
- We propose a Co-Attention Net by fusing user preference attention and news lifecycle attention together to model the dual influence of users' clicked news. To the best of our knowledge, this is the first method that incorporates news lifecycle into news recommendation.
- Extensive experimental results on two real-world news datasets demonstrate the significant superiority of our DCAN and validate the effectiveness of our Co-Attention Net by means of visualization.

2   Related Work

Over the past decades, a large number of news recommendation methods have been proposed: these include collaborative filtering-based methods [2,15,19,20], content-based methods, [8,11,12] and hybrid methods [3,14,18]. For example, Rendle [20] proposes LibFM to make recommendations using featured-based matrix factorization. However, these methods often encounter limitations when faced with serious data sparsity or the cold-start problem in news recommendation.

Recently, deep learning-based methods [1,6,7] have attracted the attention of many researchers due to their strong ability to extract nonlinear relations and hidden features from complex data. For example, Huang et al. [7] propose DSSM to rank documents using word hashing and fully connected layers. Cheng et al.

[1] present Wide&Deep, which combines wide linear models and deep neural networks and make recommendations. Guo et al. [6] propose DeepFM, which merges factorization machines and deep neural networks for recommendation purposes. However, these methods are not specialized for news recommendation, resulting in their inability to improve experimental performance.

To address the above problem, some news recommendation methods employing deep neural networks have been proposed [21–24], such as DKN, NAML, DAN and CPRS. Compared to these prior works, our method not only features an improved capacity to learn news representation and capture user preference but also proposes a Co-Attention Net by fusing user preference and news lifecycle together, thereby achieving better performance than other methods.

3 Proposed Method

3.1 Problem Formulation and Architecture

In this paper, we consider multiple text of news N as text features $\{W^t, W^c, W^e\}$, where W^t, W^c and W^e are title, content and entities of news, respectively. Each W^t, W^c and W^e consist of a sequence of words, which are defined as $[w_1^t, w_2^t, ..., w_m^t]$, $[w_1^c, w_2^c, ..., w_n^c]$ and $[w_1^e, w_2^e, ..., w_l^e]$, where m, n and l are the number of words in title, content and entities of news, respectively. We consider context information of news N as context features, which contain user click's time, location, device and referrer. We also consider time feature t and click feature c of news N as lifecycle features, where t represents the time span from news publication to being clicked and c represents the click rate calculated by using current news clicks divided by total news clicks. As Fig. 2 shown, given a user's clicked news sequence $\{N_1, N_2, ..., N_t\}$ and a candidate news N_c, where N_i ($i = 1, ..., t$) is the i-th news clicked by the user, we aim to predict the click probability $\hat{p}$ of the user on candidate news N_c.

The complete architecture of our DCAN is illustrated in Fig. 2, DCAN takes a user's clicked news sequence $\{N_1, N_2, ..., N_t\}$ and a candidate news N_c as inputs. For each piece of news, we use a Multi-Path CNN to learn comprehensive news representation from multiple text inputs, i.e., news title, content and entities. Then, we apply a User Preference LSTM to extract preference-aware sequential correlations from news representations and context features, we also use a News Lifecycle LSTM to extract lifecycle-aware sequential correlations from news representations and lifecycle features. Afterwards, we adopt a Co-Attention Net fusing user preference attention and news lifecycle attention to model the dual influence to obtain predicted news representation. Finally, the predicted news representation and the candidate news representation are multiplied and fed into a dense layer to obtain the click probability $\hat{p}$.

3.2 Multi-Path CNN

Multi-Path CNN (MPCNN) is designed to learn comprehensive news representation from multiple text inputs. As shown in Fig. 2 (b), given a news article, we

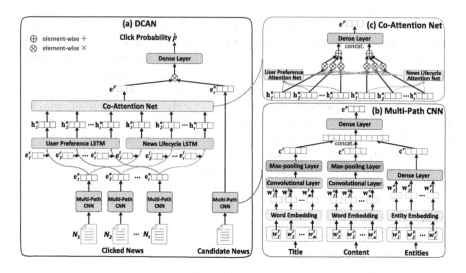

Fig. 2. The framework of our DCAN method.

use the news title, content and entities as inputs of MPCNN. To handle these multiple text inputs, MPCNN adopts two paths with convolutional layers [10] to respectively learn the title and content representation, as well as a path with a dense layer to learn the entity representation.

In the title path, given a news title $[w_1^t, w_2^t, ..., w_m^t]$, we generate m fixed-length word embeddings $[\mathbf{w}_1^t, \mathbf{w}_2^t, ..., \mathbf{w}_m^t]$ via the word embedding layer. Because of the existence of local contexts in the word sequence, we apply a convolutional layer to learn the title representation. Specifically, we sequentially concatenate all the word embeddings of the title into a title embedding matrix as $\mathbf{W}^t \in \mathcal{R}^{m \times d_1}$, where d_1 is the dimension of the word embeddings. We then adopt a convolution operation with filter $\mathbf{F}_t \in \mathcal{R}^{h_1 \times d_1}$ on $\mathbf{W}^t$, where h_1 denotes the window size of the filter. And a title feature $\hat{c}_i^t$ is extracted from a sub-matrix $\mathbf{W}_{i:i+h_1-1}^t \in \mathcal{R}^{h_1 \times d_1}$ by

$$\hat{c}_i^t = g(\mathbf{F}_t * \mathbf{W}_{i:i+h_1-1}^t + b_t), \tag{1}$$

where $g(\cdot)$ is a *ReLU* function, $*$ is the convolution operator and $b_t \in \mathcal{R}$ is a bias term. The filter $\mathbf{F}_t$ is employed sequentially at every available position in the title embedding matrix $\mathbf{W}^t$ to produce a title feature map $\hat{\mathbf{c}}^t = [\hat{c}_1^t, \hat{c}_2^t, ..., \hat{c}_{m-h_1+1}^t] \in \mathcal{R}^{m-h_1+1}$. Afterwards, we send the title feature map $\hat{\mathbf{c}}^t$ into the max-pooling layer to obtain the most valid title feature as

$$c^t = max(\hat{\mathbf{c}}^t) = max([\hat{c}_1^t, \hat{c}_2^t, ..., \hat{c}_{m-h_1+1}^t]). \tag{2}$$

We then utilize multiple filters to extract different title features and concatenate them sequentially, after which we obtain the title representation $\mathbf{c}^t = [c_1^t, c_2^t, ..., c_{k_t}^t] \in \mathcal{R}^{k_t}$, where k_t is the number of filters.

In the content path, we apply the same structure as the title path to learn the content representation. Given a news content $[w_1^c, w_2^c, ..., w_n^c]$, we generate

n fixed-length word embeddings $[\mathbf{w}_1^c, \mathbf{w}_2^c, ..., \mathbf{w}_n^c]$ via the word embedding layer, which is shared with the title path. Subsequently, we employ a convolutional layer with filter $\mathbf{F}_c \in \mathcal{R}^{h_2 \times d_1}$ to obtain the content feature map $\hat{c}^c \in \mathcal{R}^{n-h_2+1}$ from the content embedding matrix $\mathbf{W}^c \in \mathcal{R}^{n \times d_1}$, where h_2 represents the window size of the filter $\mathbf{F}_c$ and is different from the window size of the filter $\mathbf{F}_t$ in the title path. In the next step, a max-pooling layer is used to obtain the most valid content feature c^c from the content feature map $\hat{c}^c$. In a similar way to the process for the title path, we utilize multiple filters to extract different content features and concatenate them sequentially. As a result, we obtain the content representation $\mathbf{c}^c = [c_1^c, c_2^c, ..., c_{k_c}^c] \in \mathcal{R}^{k_c}$, where k_c is the number of filters.

In the entity path, because of the discrete nature of the news entity input, we adopt a dense layer to learn entity representation. Given the news entities $[w_1^e, w_2^e, ..., w_l^e]$, which include persons and locations extracted from the news articles, we use unique IDs to represent each of them individually and convert these IDs into low-dimensional embeddings $[\mathbf{w}_1^e, \mathbf{w}_2^e, ..., \mathbf{w}_l^e]$ via the entity embedding layer. We then concatenate all low-dimensional embeddings into an entity embedding vector $\mathbf{w}^e \in \mathcal{R}^{(l*d_2)}$, where d_2 is the dimension of low-dimensional embeddings. Moreover, we send $\mathbf{w}^e$ into the dense layer to learn the entity representation $\mathbf{c}^e \in \mathcal{R}^{k_e}$ by

$$\mathbf{c}^e = g(\mathbf{V}_e \mathbf{w}^e + b_e), \qquad (3)$$

where $g(\cdot)$ represents the $ReLU$ function, $\mathbf{V}_e \in \mathcal{R}^{k_e \times (l*d_2)}$ and $b_e \in \mathcal{R}^{k_e}$ are the weight matrix and bias vector.

Finally, we feed the concatenation of the multiple news representations $\mathbf{c}^t$, $\mathbf{c}^c$ and $\mathbf{c}^e$ into a dense layer and obtain the comprehensive news representation $\mathbf{e}^n \in \mathcal{R}^{k_n}$.

3.3 User Preference LSTM and News Lifecycle LSTM

User Preference LSTM (UPLSTM) is designed to extract preference-aware sequential correlations from news representations and context features. Since this context features can easily affect user preference, instead of simply applying an LSTM [4] to conduct news representations, UPLSTM incorporates context features to better capture user preference. This context features includes user click's time, location, device and referrer. We utilize cos and sin functions for continuous data to extract the periodicity of the feature. For example, the time features are calculated by $sin(2*\pi*i/24)$ and $cos(2*\pi*i/24)$, where i represents the i-th hour of the day. These features are then concatenated as time embedding. We then convert the discrete data (i.e., user click's location, device and referrer) into dense embeddings. Finally, we concatenate all embeddings as context representation $\mathbf{e}^c \in \mathcal{R}^{k_c}$.

As shown in Fig. 2 (a), given the news representations $[\mathbf{e}_1^n, \mathbf{e}_2^n, ..., \mathbf{e}_t^n]$ and context representations $[\mathbf{e}_1^c, \mathbf{e}_2^c, ..., \mathbf{e}_t^c]$ of the user's clicked news set $\{N_1, N_2, ..., N_t\}$, we employ UPLSTM to extract preference-aware sequential correlations. Specifically, in the t-step of the UPLSTM cell, which takes news representation $\mathbf{e}_t^n \in \mathcal{R}^{k_n}$, context representation $\mathbf{e}_t^c \in \mathcal{R}^{k_c}$, hidden state $\mathbf{h}_{t-1}^u \in \mathcal{R}^{k_s}$ and cell state $c_{t-1}^u \in \mathcal{R}^{k_s}$ as input, h_t^u and c_t^u are generated in the following steps:

$$f_t^u = \sigma(U_f^u e_t^n + V_f^u e_t^c + W_f^u h_{t-1}^u + b_f^u), \tag{4}$$

$$i_t^u = \sigma(U_i^u e_t^n + V_i^u e_t^c + W_i^u h_{t-1}^u + b_i^u), \tag{5}$$

$$\tilde{o}_t^u = \sigma(U_o^u e_t^n + V_o^u e_t^c + W_o^u h_{t-1}^u + b_o^u), \tag{6}$$

$$\tilde{c}_t^u = tanh(U_c^u e_t^n + V_c^u e_t^c + W_c^u h_{t-1}^u + b_c^u), \tag{7}$$

$$c_t^u = f_t^u \otimes c_{t-1}^u + i_t^u \otimes \tilde{c}_t^u, \tag{8}$$

$$h_t^u = \tilde{o}_t^u \otimes tanh(c_t^u), \tag{9}$$

where f_t^u, i_t^u and $\tilde{o}_t^u$ represent the forget, input and output operation of the UPLSTM cell, while $\sigma(\cdot)$ and $tanh(\cdot)$ represent sigmoid and hyperbolic tangent functions, respectively. $U^u \in \mathcal{R}^{k_s \times k_n}$, $V^u \in \mathcal{R}^{k_s \times k_c}$ and $W^u \in \mathcal{R}^{k_s \times k_s}$ are weight matrices, $b^u \in \mathcal{R}^{k_s}$ is bias vector and $\otimes$ is an element-wise multiplication operation.

News Lifecycle LSTM (NLLSTM) is designed to extract lifecycle-aware sequential correlations from news representations and lifecycle features. This lifecycle features contain time feature t and click feature c, where t represents the time span from news publication to being clicked, while c represents the click rate calculated by using current news clicks divided by total news clicks. Given the raw time feature t and click feature c, we generate the normalized features as freshness f and popularity p, which are calculated by

$$f = -N(log1p(t)), \tag{10}$$

$$p = N(c), \tag{11}$$

where $log1p(\cdot)$ is $log(x + 1)$, which can make input x more smooth. $N(\cdot)$ represents Z-score normalization and can make the input conform to Gaussian distribution. Then, we concatenate freshness f and popularity p to obtain the news lifecycle representation $e^l \in \mathcal{R}^2$. As shown in Fig. 2 (a), in a similar way to the process of the UPLSTM, given the news representations $[e_1^n, e_2^n, ..., e_t^n]$ and lifecycle representations $[e_1^l, e_2^l, ..., e_t^l]$ of the user's clicked news set $\{N_1, N_2, ..., N_t\}$, we employ NLLSTM to extract lifecycle-aware sequential correlations. Finally, we obtain the preference-aware sequential correlations $[h_1^u, h_2^u, ..., h_t^u]$ using the UPLSTM layer and obtain the lifecycle-aware sequential correlations $[h_1^n, h_2^n, ..., h_t^n]$ using the NLLSTM layer.

3.4   Co-Attention Net

Co-Attention Net is designed to model the dual influence (i.e., co-attention weight) of users' clicked news and learn predicted news representation by fusing together the advantages of news lifecycle attention net and user preference attention net, both of which are highly useful in identifying influential news.

As shown in Fig. 2 (c), to model user preference attention, we extract user preference from preference-aware sequential correlation h^u. We denote the user preference attention weight of the i-th news clicked by the user as α_i^u, which is calculated as follows:

$$a_i^u = \mathbf{v}_i^u tanh(\mathbf{W}_i^u \mathbf{h}_i^u + \mathbf{b}_i^u), \tag{12}$$

$$\alpha_i^u = \frac{exp(a_i^u)}{\sum_{j=1}^t exp(a_j^u)}, \tag{13}$$

where $\mathbf{v}_i^u \in \mathcal{R}^{k_u}$ is a query vector, $\mathbf{W}_i^u \in \mathcal{R}^{k_u \times k_s}$ and $\mathbf{b}_i^u \in \mathcal{R}^{k_u}$ are weight matrix and bias vector, where k_u is the number of user preference attention heads.

To model news lifecycle attention, we learn news lifecycle from lifecycle-aware sequential correlation $\mathbf{h}^n$. We denote the news lifecycle attention weight of the i-th news clicked by the user as α_i^n, which is calculated as follows:

$$a_i^n = \mathbf{v}_i^n tanh(\mathbf{W}_i^n \mathbf{h}_i^n + \mathbf{b}_i^n), \tag{14}$$

$$\alpha_i^n = \frac{exp(a_i^n)}{\sum_{j=1}^t exp(a_j^n)}, \tag{15}$$

where $\mathbf{v}_i^n \in \mathcal{R}^{k_n}$, $\mathbf{W}_i^n \in \mathcal{R}^{k_n \times 2}$ and $\mathbf{b}_i^n \in \mathcal{R}^{k_n}$, where k_n is the number of news lifecycle attention heads.

After we model the above two attentions, we apply a fusion function to model the co-attention weight. Here we define three types of fusion function f as

$$f(\alpha_i^n, \alpha_i^u) = vec(\alpha_i^n, \alpha_i^u) = \sigma(\mathbf{w}_{co}[\alpha_i^n, \alpha_i^u] + \mathbf{b}_{co}), \tag{16}$$

$$f(\alpha_i^n, \alpha_i^u) = sum(\alpha_i^n, \alpha_i^u) = \alpha_i^n + \alpha_i^u, \tag{17}$$

$$f(\alpha_i^n, \alpha_i^u) = mul(\alpha_i^n, \alpha_i^u) = \alpha_i^n * \alpha_i^u, \tag{18}$$

where $\sigma(\cdot)$ is a sigmoid function, $\mathbf{w}_{co} \in \mathcal{R}^2$ and $\mathbf{b}_{co} \in \mathcal{R}$ are weight vector and bias term. Finally, we calculate the co-attention weights with the preference-aware sequential correlations and lifecycle-aware sequential correlations to obtain the predicted news representation $\mathbf{e}^p$ as

$$\mathbf{e}^p = Dense([\sum_{i=1}^t f(\alpha_i^u, \alpha_i^n)\mathbf{h}_i^u, \sum_{i=1}^t f(\alpha_i^u, \alpha_i^n)\mathbf{h}_i^n]). \tag{19}$$

3.5   Loss Function

In Fig. 2 (a), given a predicted news representation $\mathbf{e}^p$ and a candidate news representation $\mathbf{e}_c^n$, we calculate the click probability of candidate news by an element-wise multiplication and a dense layer, that is, $\hat{p} = Dense(\mathbf{e}^p \otimes \mathbf{e}_c^n)$. Then, we minimize the negative log-likelihood function to train our method as

$$\mathcal{L} = -\sum_{x \in \Delta+} ylog(\hat{p}) - \sum_{x \in \Delta-} (1-y)log(1-\hat{p}), \tag{20}$$

where x represents candidate news, $\Delta+$ and $\Delta-$ are target news and negative samples. For target news, the label is $y = 1$, and $y = 0$ is the label for negative samples.

4 Experiments

4.1 Datasets

We use two real-world datasets in our experiments, namely, Globo[1] and Adressa[2] [5]. The Globo dataset contains about 3 million clicks, 314,000 users and more than 46,000 news articles over a period of 16 days from a Brazil news portal. The Adressa dataset contains about 113 million clicks, 398,545 users and 93,948 news articles over a period of 90 days from a Norwegian news portal. Moreover, both datasets contain news text information[3], user interaction information and context information. For both datasets, we select the first week to conduct experiments. Similar to [17], the datasets are ordered by time and grouped by hours. After each 6-hour training, all the compared models are evaluated on the news from the next hour and that hour's news is also used for training after the evaluation is done.

4.2 Baselines

We use the following state-of-the-art methods as baselines to evaluate the performance of our DCAN: (1) LibFM [20] is a remarkable feature-based matrix factorization method; (2) DSSM [7] is a deep structured semantic model for document ranking using word hashing and fully connected layers; (3) Wide&Deep [1] combines a wide linear channel and a deep neural network channel to make recommendations; (4) DeepFM [6] combines factorization machines and deep neural networks for make recommendations; (5) DKN [21] is a deep knowledge-aware network based on knowledge-aware CNNs and an attention mechanism; (6) NAML [22] is a neural attentive multi-view learning method based on news encoders and a user encoder; (7) DAN [24] is a deep attention network based on parallel CNNs and an attention-based RNN; (8) CPRS [23] is a multilevel attention network based on title encoders and content encoders, with a clicked predictor and a satisfaction predictor to make recommendations.

4.3 Experimental Settings

In our experiments, we utilize a large corpus pretrained by Word2vec [16] to represent the word embeddings of news titles, contents and entities, whose dimensions are set to 250. We adopt MPCNN, which offer 128 filters with window sizes of 5 in the title path and 10 in the content path, to learn news representations, whose dimensions are set to 250. We use UPLSTM and NLLSTM with 256 units and both attentions with 150 heads to model predicted news representations, whose dimensions are set to 250. The negative samples are taken from the popular news and the number of negative samples is set to 20. We employ Adam as our optimizer and the learning rate is set to 0.001. These key

---

[1] https://www.kaggle.com/gspmoreira/news-portal-user-interactions-by-globocom.
[2] http://reclab.idi.ntnu.no/dataset/.
[3] The Adressa dataset contains news title, content and entity information and the Globo dataset only contains news content information.

parameters are selected according to the validation set. The evaluation metrics in our experiments include Recall@5, Recall@10, MRR@5, MRR@10, NDCG@5 and NDCG@10. The key parameter settings for baselines are the same as configurations reported in DKN, and we adopt the authors' recommended configurations for NAML, DAN and CPRS.

4.4 Comparison Against Baselines

Table 1. Performance comparison on the Globo dataset.

Methods	Globo					
	Recall@5	Recall@10	MRR@5	MRR@10	NDCG@5	NDCG@10
LibFM	0.6172	0.8194	0.3775	0.4045	0.4368	0.5022
DSSM	0.6444	0.8498	0.4023	0.4259	0.4623	0.5259
Wide& Deep	0.6312	0.8267	0.3923	0.4187	0.4514	0.5148
DeepFM	0.6407	0.8468	0.3999	0.4150	0.4595	0.5202
DKN	0.6500	0.8365	0.4042	0.4227	0.4651	0.5203
NAML	0.6677	0.8636	0.4186	0.4460	0.4802	0.5447
DAN	0.6734	0.8701	0.4234	0.4505	0.4853	0.5496
CPRS	0.6760	0.8737	0.4228	0.4531	0.4855	0.5526
DCAN	**0.6903**	**0.8776**	**0.4444**	**0.4610**	**0.5053**	**0.5595**

Table 2. Performance comparison on the Adressa dataset.

Methods	Adressa					
	Recall@5	Recall@10	MRR@5	MRR@10	NDCG@5	NDCG@10
LibFM	0.5524	0.7718	0.3267	0.3565	0.3825	0.4538
DSSM	0.5861	0.8096	0.3493	0.3796	0.4079	0.4808
Wide& Deep	0.5748	0.7958	0.3435	0.3755	0.4007	0.4742
DeepFM	0.5835	0.8084	0.3466	0.3773	0.4051	0.4784
DKN	0.5920	0.8122	0.3563	0.3885	0.4146	0.4882
NAML	0.6158	0.8319	0.3696	0.3980	0.4305	0.5000
DAN	0.6208	0.8353	0.3733	0.4019	0.4345	0.5039
CPRS	0.6266	0.8409	0.3792	0.4086	0.4404	0.5104
DCAN	**0.6412**	**0.8466**	**0.3928**	**0.4178**	**0.4543**	**0.5189**

Table 1 and Table 2 show the overall performance of all the compared methods, with the best results highlighted in boldface. There are several noteworthy observations from Table 1 and Table 2:

- LibFM is worse than other models in experimental performance because of its traditional matrix factorization structure. It suggests that deep neural networks are effective in extracting nonlinear relations and hidden features from complicated data.

- DSSM, Wide&Deep and DeepFM have an excellent experimental performance, and they are all deep learning based methods. In particular, DSSM outperforms Wide&Deep and DeepFM, probably because DSSM utilizes the word hashing to model raw texts. However, these methods are not specialized for news recommendation, resulting in their inability to effectively learn news representation.
- DKN, NAML, DAN and CPRS outperform most of the algorithms, demonstrating their powerful capability to learn news representation and capture user preference. In particular, DAN considers the sequential correlations of users' clicked history, so DAN outperforms DKN and NAML in experimental performance. In addition, CPRS uses a multilevel attention network to capture user preference, it consequently outperforms other baselines in terms of experimental performance.
- Finally, these methods are inferior to our proposed method. We attribute the superiority of our DCAN to its three advantages: (1) DCAN proposes a Co-Attention Net by fusing user preference attention and news lifecycle attention together, and it can better model the dual influence of users' clicked news. (2) DCAN designs a MPCNN to better learn news representation from multiple news inputs, i.e., news title, content and entities. (3) DCAN presents UPLSTM and NLLSTM to extract sequential correlations from news representations and additional information to better capture user preference and model news lifecycle.

4.5   Comparison Among DCAN Variants

Table 3. Performance comparison among DCAN variants.

DCAN variants	Globo		Adressa	
	Recall@5	MRR@5	Recall@5	MRR@5
DCAN - news title	–	–	0.6303	0.3812
DCAN - news content	–	–	0.6207	0.3790
DCAN - news entities	–	–	0.6292	0.3835
DCAN with all news inputs	**0.6903**	**0.4444**	**0.6412**	**0.3928**
DCAN - context information	0.6759	0.4277	0.6214	0.3763
DCAN with context information	**0.6903**	**0.4444**	**0.6412**	**0.3928**
DCAN - both attentions	0.6605	0.4085	0.5899	0.3484
DCAN - user preference attention	0.6782	0.4291	0.6221	0.3735
DCAN - news lifecycle attention	0.6722	0.4258	0.6211	0.3749
DCAN with both attentions	**0.6903**	**0.4444**	**0.6412**	**0.3928**
DCAN with *mul* function	0.6737	0.4256	0.6201	0.3746
DCAN with *sum* function	0.6708	0.4267	0.6215	0.3754
DCAN with *vec* function	**0.6903**	**0.4444**	**0.6412**	**0.3928**

We compare DCAN variants to demonstrate the validity of the design of our DCAN based on the following four aspects: the usage of news title, content and entities; the usage of context information; the usage of news lifecycle attention and user preference attention; and the choice of fusion function. The results are shown in Table 3, and we can conclude that:

- The absence of either news input of DCAN can lead to worse performance, demonstrating the importance of various news inputs and the strong ability of our MPCNN to conduct multiple text information. In particular, DCAN without a news content has the lowest performance because the news content can provide richer information for representing news.
- DCAN with context information greatly improves performance compared with DCAN without context information, validating the importance of context information and the effectiveness of our UPLSTM in extracting preference-aware sequential correlations.
- DCAN with either attention is better than DCAN without both attentions, suggesting that attention mechanism is effective in modeling the influence of news. Further, DCAN with news lifecycle attention is a little higher than DCAN with user preference attention, probably because news lifecycle has the strong potential to find influential news. Moreover, DCAN with both attentions performs best in four DCAN variants, validating the effectiveness of our Co-Attention Net.
- The fusion function using *vec* has the best performance compared to other functions. The *sum* and *mul* functions have similar results. This is mainly because *vec* function with a deep structure is more effective to fuse both attentions than simple linear operation *sum* and *mul*.

4.6 Parameter Sensitivity of both Attention Heads

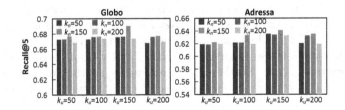

Fig. 3. Parameter sensitivity of both attention heads.

In this section, we analyze the effect of different numbers of user preference attention heads k_u and news lifecycle attention heads k_n on performance by using the term of Recall@5. Except for the parameters being analyzed, all other parameters are set as optimal configuration. The numbers of both attention heads are selected in set $\{50, 100, 150, 200\}$. According to Fig. 3, our DCAN achieves the best performance at $k_u = k_n = 150$, suggesting that such attention heads setting can better fuse the advantages of both attentions. Additionally, given the k_u, the performance is initially enhanced with the growth of k_n, and then it drops as k_n grows further. This is probably because that a smaller k_n has inadequate capacity to extract rich attention patterns, whereas a larger k_n can bring more noises. The case is similar for k_u when k_n is given.

4.7   Visualization of Attention Weights

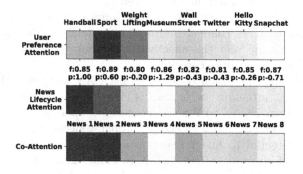

Fig. 4. Visualization of attention weights.

To intuitively demonstrate the effectiveness of the usage of both attentions, we randomly sample a user's clicked history from the Adressa dataset. As shown in Fig. 4, user preference attention can effectively capture user preference. For example, the first three news about sport are highlighted that is probably because the user is interested in sport. Additionally, the news lifecycle attention also shows the strong power of recognizing influential news. In Fig. 4, the first news has a relatively large freshness f and a large popularity p, resulting in its high influence. Moreover, the Co-Attention Net can fuse the advantages of the two attentions to model the dual influence of news. In Fig. 4, the first two news have high co-attention weights, which signifies that they have both high clicks and the user's interested content. Further, the fourth news has the lowest co-attention weight, proving that there are meaningless clicks in the user's click history.

5   Conclusion

In this paper, we proposed a Deep Co-Attention Network DCAN by modeling news lifecycle and user preference for new recommendation. DCAN has made remarkable contributions in the following three aspects: (1) To better model influence of users' clicked news, DCAN proposes a Co-Attention Net by fusing user preference attention and news lifecycle attention, which both have the strong ability to find influential news. (2) To better learn news representation, DCAN designs a MPCNN to learn various patterns from news title, content and entities, which are all useful for representing news. (3) To better capture user preference and model news lifecycle, DCAN presents UPLSTM and NLLSTM to extract sequential correlations from news representations and additional information. Extensive experimental results demonstrate the significant superiority of our DCAN, as well as the effectiveness of our Co-Attention Net.

Acknowledgment. This work is supported by the National Key R&D Program of China (2019YFB1406302, 2018YFB1003903), National Natural Science Foundation of China (No. 61502033, 61472034, 61772071, 61272361 and 61672098) and the Fundamental Research Funds for the Central Universities.

References

1. Cheng, H.T., et al.: Wide & deep learning for recommender systems. In: Proceedings of the 1st Workshop on Deep Learning for Recommender Systems, pp. 7–10. ACM (2016)
2. Das, A.S., Datar, M., Garg, A., Rajaram, S.: Google news personalization: scalable online collaborative filtering. In: Proceedings of the 16th International Conference on World Wide Web, pp. 271–280. ACM (2007)
3. De Francisci Morales, G., Gionis, A., Lucchese, C.: From chatter to headlines: harnessing the real-time web for personalized news recommendation. In: Proceedings of the fifth ACM International Conference on Web Search and Data Mining, pp. 153–162. ACM (2012)
4. Graves, A., Schmidhuber, J.: Framewise phoneme classification with bidirectional LSTM and other neural network architectures. Neural Netw. **18**(5–6), 602–610 (2005)
5. Gulla, J.A., Zhang, L., Liu, P., Özgöbek, Ö., Su, X.: The Adressa dataset for news recommendation. In: Proceedings of the International Conference on Web Intelligence, pp. 1042–1048. ACM (2017)
6. Guo, H., Tang, R., Ye, Y., Li, Z., He, X.: DeepFM: a factorization-machine based neural network for CTR prediction. In: Proceedings of the 26th International Joint Conference on Artificial Intelligence, pp. 1725–1731. AAAI Press (2017)
7. Huang, P.S., He, X., Gao, J., Deng, L., Acero, A., Heck, L.: Learning deep structured semantic models for web search using clickthrough data. In: Proceedings of the 22nd ACM International Conference on Information & Knowledge Management, pp. 2333–2338. ACM (2013)
8. IJntema, W., Goossen, F., Frasincar, F., Hogenboom, F.: Ontology-based news recommendation. In: Proceedings of the 2010 EDBT/ICDT Workshops, p. 16. ACM (2010)
9. Karimi, M., Jannach, D., Jugovac, M.: News recommender systems-survey and roads ahead. Inf. Process. Manag. **54**(6), 1203–1227 (2018)
10. Kim, Y.: Convolutional neural networks for sentence classification. In: Proceedings of the 2014 Conference on Empirical Methods in Natural Language Processing (EMNLP), pp. 1746–1751 (2014)
11. Kompan, M., Bieliková, M.: Content-based news recommendation. In: Buccafurri, F., Semeraro, G. (eds.) EC-Web 2010. LNBIP, vol. 61, pp. 61–72. Springer, Heidelberg (2010). https://doi.org/10.1007/978-3-642-15208-5_6

12. Li, L., Zheng, L., Li, T.: Logo: a long-short user interest integration in personalized news recommendation. In: Proceedings of the Fifth ACM Conference on Recommender Systems, pp. 317–320. ACM (2011)
13. Li, M., Wang, L.: A survey on personalized news recommendation technology. IEEE Access **7**, 145861–145879 (2019)
14. Liu, J., Dolan, P., Pedersen, E.R.: Personalized news recommendation based on click behavior. In: Proceedings of the 15th International Conference on Intelligent User Interfaces, pp. 31–40. ACM (2010)
15. Medo, M., Zhang, Y.C., Zhou, T.: Adaptive model for recommendation of news. EPL (Europhys. Lett.) **88**(3), 38005 (2009)
16. Mikolov, T., Chen, K., Corrado, G., Dean, J.: Efficient estimation of word representations in vector space. arXiv preprint arXiv:1301.3781 (2013)
17. de Souza Pereira Moreira, G., Ferreira, F., da Cunha, A.M.: News session-based recommendations using deep neural networks. In: Proceedings of the 3rd Workshop on Deep Learning for Recommender Systems, pp. 15–23. ACM (2018)
18. Phelan, O., McCarthy, K., Smyth, B.: Using Twitter to recommend real-time topical news. In: Proceedings of the Third ACM Conference on Recommender Systems, pp. 385–388. ACM (2009)
19. Rendle, S.: Factorization machines. In: 2010 IEEE International Conference on Data Mining, pp. 995–1000. IEEE (2010)
20. Rendle, S.: Factorization machines with libFM. ACM Trans. Intell. Syst. Technol. (TIST) **3**(3), 57 (2012)
21. Wang, H., Zhang, F., Xie, X., Guo, M.: DKN: deep knowledge-aware network for news recommendation. In: Proceedings of the 2018 World Wide Web Conference, pp. 1835–1844. International World Wide Web Conferences Steering Committee (2018)
22. Wu, C., Wu, F., An, M., Huang, J., Huang, Y., Xie, X.: Neural news recommendation with attentive multi-view learning. In: Proceedings of the 28th International Joint Conference on Artificial Intelligence, pp. 3863–3869. AAAI Press (2019)
23. Wu, C., Wu, F., Qi, T., Huang, Y.: User modeling with click preference and reading satisfaction for news recommendation. In: Proceedings of the 29th International Joint Conference on Artificial Intelligence, pp. 3023–3029. ijcai.org (2020)
24. Zhu, Q., Zhou, X., Song, Z., Tan, J., Guo, L.: DAN: deep attention neural network for news recommendation. In: Proceedings of the AAAI Conference on Artificial Intelligence, vol. 33, pp. 5973–5980 (2019)

Considering Interaction Sequence of Historical Items for Conversational Recommender System

Xintao Tian[1], Yongjing Hao[1], Pengpeng Zhao[1(✉)], Deqing Wang[2],
Yanchi Liu[3], and Victor S. Sheng[4]

[1] Institute of Artificial Intelligence, School of Computer Science and Technology,
Soochow University, Suzhou, China
{xttiansuda,yjhaozb}@stu.suda.edu.cn, ppzhao@suda.edu.cn
[2] School of Computer, Beihang University, Beijing, China
dqwang@buaa.edu.cn
[3] Rutgers University, New Jersey, USA
yanchi.liu@rutgers.edu
[4] University of Central Arkansas, Conway, USA
Victor.Sheng@ttu.edu

Abstract. Different from the traditional recommender systems with content-based and collaborative filtering, conversational recommender systems (CRS) can dynamically dialogue with users to capture fine-grained preferences. Although several efforts have been made for CRS, they neglect the importance of interaction sequences, which seek to capture the 'context' of users' activities based on actions they have performed recently. Therefore, we propose a framework that considers interaction **S**equence of historical items for **C**onversational **R**ecommendation (**SeqCR**). Specifically, SeqCR first scores candidate items through the sequence which users interact with. Then it can generate the recommendation list and attributes to be asked based on the scores. We restrict candidate attributes to the ones with high-scoring (high-relevance) items, which effectively reduces the search space of attributes and leads to user preferences that can be hit more quickly and accurately. Finally, SeqCR utilizes the policy network to decide whether to recommend or ask. We conduct extensive experiments on two datasets from MovieLens 10M and Yelp in multi-round conversational recommendation scenarios. Empirical results demonstrate our SeqCR significantly outperforms the state-of-the-art methods.

Keywords: Conversational recommendation · Interactive recommendation · Recommender system

1 Introduction

Recommender systems have been an effective way to seek a subset of items that satisfy user preferences from the item pool [8,11,17]. It can recommend long-tail items to the users or help them search for items they are interested in but

© Springer Nature Switzerland AG 2021
C. S. Jensen et al. (Eds.): DASFAA 2021, LNCS 12683, pp. 115–131, 2021.
https://doi.org/10.1007/978-3-030-73200-4_8

difficult to find. Traditional recommender systems conduct recommendations by inferring user preferences from user previous historical actions [8,15]. Although it has achieved remarkable success and has been standard fixtures in many scenarios, there are some inherent limitations. For example, since it needs to acquire the historical data of users, there is a cold start problem for users with less interactive data. Traditional recommender systems can only obtain user preferences passively rather than interact with users actively. These limitations prevent the system from conducting accurate recommendations.

Recently, the rapid development of the conversational recommender systems (CRS) brings revolutions to the aforementioned limitations. Different from the traditional recommendation system, it can capture and refine user preferences by proactively asking users specific questions. There are two components in the CRS – recommender component (RC, generating recommendation) and conversational component (CC, interacting with the user) [12]. The two components have been integrated to develop an effective CRS. [14] focuses on language understanding and item recommendation, [4,26] utilize knowledge graph to improve the performance of CRS. But they only recommend based on the context of the current conversation without the interaction history of users. Some works consider historical interactions and use reinforcement learning to determine the action of the dialogue. [21] uses the user's ratings and the user's query collected in the current conversational to generate recommendations. [12,13] utilize a factorization machine to estimate user's preferences, then refine user's current interests through dialogue. However, they neglect the importance of the interaction sequence of historical items (later called sequence) for recommendation. We argue that utilizing the sequences can not only better capture the user's 'context', but also better improve the quality of questions and recommendations raised by CRS.

To address the above problem, we present a conversational recommendation framework called SeqCR based on the sequence and the context of the conversation. Inspired by the recent success of CRS [13], this paper models CRS as the process of seeking items that satisfy user attribute preferences. We first utilize the interaction sequence to score candidate items, then utilize the policy network to determine the action. If this turn is a recommended action, SeqCR will regard the items with the high-scoring among the candidate items as the recommendation list. Otherwise, it will choose a suitable attribute of the high-scoring items to ask. Finally, it updates preferences and candidate items according to the feedback of users. In a conversation, SeqCR needs to alternate between recommendation and ask several times in order to minimize the number of turns. Our method recommends items that contain user-accepted attributes and don't contain user-rejected attributes.

In summary, the main contributions of this work are summarized as follows:

- To the best of our knowledge, it is the first time that the interaction sequence has been considered in the conversational recommender system.
- We propose a SeqCR framework to model conversational recommendation. It can make recommendations and ask questions based on history sequences.

Furthermore, we utilize high-scoring items to reduce the search space of candidate attributes.

- We examine the two datasets by simulating conversations, which demonstrate our method outperforms state-of-the-art CRS methods.

2 Related Work

The success of the recommendation system depends on whether it can accurately and timely provide relevant items to users. The traditional recommender systems have achieved significant success in business. For example, Amazon and Netflix will make personalized recommendations to users based on their historical behaviors (such as clicks, purchases, comments, etc.). At the initial stage, the recommendation system was mainly based on collaborative filtering (CF) such as matrix factorization [11,18] or content based [16] to infer user preferences. They depend on the interactive information between the user and the item, yet user-item interaction data is usually sparse. To tackle the data sparsity problem, [1] proposed to utilize the side information of items, such as reviews.

Sequential Recommendation. Early works on sequential recommendations are mainly based on Markov Chain (MC) assumption. It attempts to model an item-item transition matrix to capture sequential patterns among successive items. Rendle et al. [20] fused matrix factorization (MF) and an item-item transition for modeling global user preferences and short-term transitions, respectively. Tang et al. [22] used a CNN-based convolutional sequence embedding method. With the successful application of the self-attention network in NLP, [9] integrated self-attention with the sequential recommendation and achieved significant success. But sequential recommendation can only conduct recommended by utilizing the interaction sequence of historical items. Therefore it is difficult to capture current interest preferences.

Conversational Recommendation. The critical point of conversational recommendation systems is to understand the user's preferences fully, generate proper responses, furthermore make accurate recommendations based on the context. Early conversational recommender systems [6] mainly utilized predefined actions to interact with users. Li et al. [14] proposed a structure with four sub-components to understand user preferences in utterance, make recommendations, and use natural language to generate responses. Subsequently, [4,26] combined CRS with the knowledge graph. In term to develop an effective CRS, [21] integrates a belief tracker over semi-structured user queries and reinforcement learning (RL). [12] proposed a new solution named Estimation–Action–Reflection (EAR) to tackle the deep interaction between CC and RC. Recently, [13] proposed a general framework of conversational path reasoning, which models conversation as a path reasoning problem on a graph. However, none of them considers the impact of interaction sequences on the next recommendation. Therefore we design a novel approach to fuse sequence and CRS, which leads to better performance in generating recommendations and asking questions.

3 Problem Definition

Different from single-round of conversation, which only makes recommendations once and the conversation ends even if the user refuses [5], this paper follows a multi-round conversational scenario because it is the most realistic setting. CRS aims to understand the user intentions and recommends suitable items through multi-round of dialogue with the user. In CRS, the chat agent will analyze and learn user preferences based on the context of conversation, and generate questions or recommendations. When the system feels confident, it will recommend items to the user. Otherwise, if the information obtained is insufficient, the system will ask until the maximum turn. Successful recommendation is considered as the final goal.

Table 1. Main notations used in the paper.

Notation	Description
u, v, p	User, item, and attribute
$\mathcal{V}_p$	The item set that contain the attribute p
$\mathcal{V}_{cand}$	The candidate item set
$\mathcal{V}_u$	u's interaction sequence of historical items
$\mathcal{P}_v$	Attributes of item v
$\mathcal{P}_{u_a}$	The attribute set accepted by u in a conversation
$\mathcal{P}_{u_r}$	The attribute set rejected by u in a conversation
$\mathcal{P}_{cand}$	The candidate attribute set
$\mathcal{P}_{\mathcal{V}_{cand}}$	The attribute set of all candidate items
a	The action of SeqCR, either a_{ask} or a_{rec}

We now introduce the notations used to formalize our setting. As shown in Table 1, $u \in \mathcal{U}$ denotes a user from a user set $\mathcal{U}$, $v \in \mathcal{V}$ denotes an item from an item set $\mathcal{V}$. $\mathcal{P}_v$ represents the attribute set of item v. We define all attributes as $\mathcal{P}$, and $p \in \mathcal{P}$ is a specific attribute from $\mathcal{P}$. A conversation is started by the user with a specific attribute p_0 that the user likes [12,13,25]. In each turn t (until the recommended item is accepted or the maximum conversation turn is reached), the CRS will filter out the candidate set to retain items that contain user-accepted attributes and remove items that contain user-rejected attributes. Next, the CRS will choose an action to *recommend* or *ask*:

- *recommend*: If the action is *recommend*, we denote a recommended item list $\mathcal{V}_{rec} \subset \mathcal{V}_{cand}$ and the action as $a_{rec}(\mathcal{V}_{rec})$. Then users provide feedback to CRS according to their desired items. Assuming the feedback is positive, the conversation ends. Otherwise, the system removes the $\mathcal{V}_{rec}$ from the candidate item $\mathcal{V}_{cand}$ and moves to the next turn.

- ask: If the action is ask, we denote the asked attribute as $p_t \subset \mathcal{P}_{cand}$ and the action as $a_{ask}(p_t)$. If the feedback of user is positive, add p_t into $\mathcal{P}_{u_a}$ to indicate that the user likes this attribute. Otherwise, add it into $\mathcal{P}_{u_r}$ to denote that the user does not like this attribute in this conversation. Then system moves to the next turn.

4 Proposed Methods

We propose to utilize the interaction sequence of historical items for conversational recommendation (SeqCR). SeqCR uses the historical interaction sequence and the context of conversation to recommend the next item. Figure 1 shows the overall architecture, we can see that SeqCR mainly includes three modules. The first is a sequential module, which is mainly used to model the sequence of items that the user has interacted with. The second is a scoring module. It scores candidate items and attributes based on the knowledge learned in the sequence module. The last one is the policy network module, which is a reinforcement learning model for deciding whether to recommend or ask. The three modules are described in detail below.

4.1 Sequential Module

As shown in the bottom left corner of Fig. 1, given a interaction sequence of historical items $\mathcal{V}_u = (\mathcal{V}_u^1, \mathcal{V}_u^2, \mathcal{V}_u^3, ..., \mathcal{V}_u^{|\mathcal{V}_u|})$, then sequential module utilizes the embedding layer, self-attention block and prediction layer to predict the next item that user likes. We will introduce in detail below.

Our sequence module mainly refers to [9]. Since the length of the interaction sequence of each user is different, for convenience, we convert the training sequence $\mathcal{V}_u' = (\mathcal{V}_u^1, \mathcal{V}_u^2, \mathcal{V}_u^3, ..., \mathcal{V}_u^{|\mathcal{V}_u|-1})$ of each user to a fixed-length sequence $s = (s_1, s_2, s_3, ..., s_n)$, where n represents the maximum length of the new sequence s. If the interaction sequence is less than n, we add $n - (|\mathcal{V}_u| - 1)$ 'padding' items to the left of the sequence. If it is longer than n, only the sequence of the nearest n interaction items will be considered. In the embedding layer, we embed all items and get the embedding matrix $\mathbf{M} \in \mathbb{R}^{|\mathcal{V}| \times d}$. Subsequently, we apply a look-up operation to form the embedding matrix $\mathbf{E} \in \mathbb{R}^{n \times d}$ of sequence s, where $\mathbf{E}_i = \mathbf{M}_{s_i}$. Furthermore, we integrate $\mathbf{E}$ with a learnable positional embedding matrix $\mathbf{P} \in \mathbb{R}^{n \times d}$:

$$\mathbf{E}_S = \mathbf{E} + \mathbf{P}. \tag{1}$$

For self-attention, the three matrices $\mathbf{Q}$ (Query), $\mathbf{K}$ (Key), and $\mathbf{V}$ (Value) all come from the same input. Attention is defined by a scaled dot-product [23]:

$$\text{Attention}(\mathbf{Q}, \mathbf{K}, \mathbf{V}) = \text{softmax}(\frac{\mathbf{Q}\mathbf{K}^T}{\sqrt{d}})\mathbf{V}, \tag{2}$$

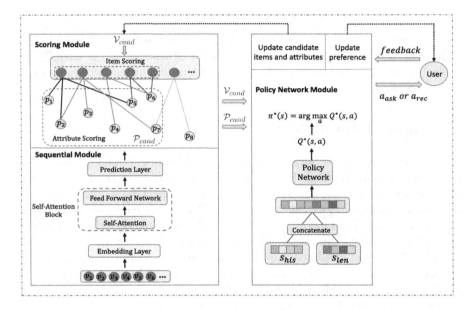

Fig. 1. Overview of the SeqCR framework. It scores the candidate items and attributes through the interaction sequence of historical items, and utilizes a policy network to decide whether to make recommendations or ask questions.

where d is the dimension of a query and a key vector. The self-attention block includes a self-attention layer and a point-wise feed-forward network. As mentioned above, when predicting the $(k + 1)$-th item, only the previous k items should be considered. However, the k-th output $(\mathbf{S}_k)$ in the self-attention layer also takes into account ones that after the k-th item. Therefore, we add a mask operation to prevent the connection between $\mathbf{Q}_i$ and $\mathbf{K}_j$ $(i < j)$. But self-attention is a linear model. In order to make the model nonlinear, we apply a point-wise two-layer feed-forward network to all $\mathbf{S}_k$. Based on the embedding layer, we learn more complex item transformations by stacking multiple self-attention blocks. Specifically, the b-th $(b > 1)$ self-attention block is defined as follows:

$$\mathbf{S}^{(b)} = \mathrm{SA}(\mathbf{F}^{(b-1)}) = \mathrm{Attention}(\mathbf{F}^{(b-1)}\mathbf{W}_Q, \mathbf{F}^{(b-1)}\mathbf{W}_K, \mathbf{F}^{(b-1)}\mathbf{W}_V),$$
$$\mathbf{F}_k^{(b)} = \mathrm{FFN}\left(\mathbf{S}_k^{(b)}\right) = \mathrm{ReLU}\left(\mathbf{S}_k^{(b)}\mathbf{W}_1 + \mathbf{b}_1\right)\mathbf{W}_2 + \mathbf{b}_2, \tag{3}$$

where the learnable matrices $\mathbf{W}_Q, \mathbf{W}_K, \mathbf{W}_V, \mathbf{W}_1, \mathbf{W}_2 \in \mathbb{R}^{d \times d}$ and $\mathbf{b}_1, \mathbf{b}_2$ are d-dimensional vectors. $\mathbf{S}_k^{(b)}$ denotes aggregating the first k $(k \in \{1, 2, 3, ..., n\})$ items' output of the self-attention layer in the b-th self-attention block. $\mathbf{S}^{(b)}$ and $\mathbf{F}^{(b)}$ represent the output of the self-attention layer and the feed-forward network in b-th block, respectively. The first block is defined as $\mathbf{S}^{(1)} = \mathrm{SA}(\mathbf{E}_S)$ and $\mathbf{F}^{(1)} = \mathrm{FFN}(\mathbf{S}^{(1)})$.

Multi-layer neural networks have a strong ability to learn features. However, simply adding more network layers will cause problems such as overfitting and consuming more training time. This is because when the network becomes deeper, the hidden danger of gradient disappearance will also increase, the model performance will decrease. But residual connections [7] can alleviate these problems. We first normalize the input x of both the self-attention layer and the feed forward network, then apply dropout to their output, next add the original input x as the output.

After b self-attention blocks, we predict the next item by utilizing $\mathbf{F}_k^{(b)}$. In other words, we predict the relevance of item i (in our method, the relevance is regarded as the score of item i) through an MF layer:

$$r_{i,k} = \mathbf{F}_k^{(b)} \mathbf{N}_i^T, \qquad (4)$$

where $r_{i,k}$ is the possibility of item i becoming the next item according to the first k items, and $\mathbf{N} \in \mathbb{R}^{|\mathcal{V}| \times d}$ is an item embedding matrix. In the network training phase, we define e_t as the expected output at time step t. If s_t is a padding item, we hope $e_t = $ <pad>, otherwise, we define the expected output as the next interaction item $e_t = s_{t+1}$ (if $t = n$, let $s_{t+1} = \mathcal{V}_u^{|\mathcal{V}^u|}$). We input the sequence and minimize the binary cross entropy loss for optimization.

$$L = - \sum_{\mathcal{V}^u \in \mathcal{S}} \sum_{k \in [1,2,...,n]} \left[\log \left(\sigma \left(r_{e_k,k} \right) \right) + \sum_{j \notin \mathcal{V}^u} \log \left(1 - \sigma \left(r_{j,k} \right) \right) \right]. \qquad (5)$$

4.2 Scoring Module

In the conversational recommender system, it will ask the user questions or make item recommendations. Before the conversation starts, the attribute set $\mathcal{P}_{u_a}$ accepted by the user and the attribute set $\mathcal{P}_{u_r}$ rejected by the user are empty, the candidate item set $\mathcal{V}_{cand}$ contains that the user has not interacted with, and the candidate attribute set $\mathcal{P}_{cand}$ is empty. The dialog starts with the user specifying a specific attribute p_0. The candidate item set is updated by $\mathcal{V}_{cand} = \mathcal{V}_{cand} \bigcap \mathcal{V}_{p_0}$. When the user starts a conversation with the system, the system needs to perform two scoring operations to determine which items to recommender or which attributes to ask. The two operations will be introduced in detail below.

Item Scoring. The first operation is the item scoring. In the top left corner of Fig. 1, green vertices indicate all candidate items $\mathcal{V}_{cand}$, and yellow vertices indicate attributes related to candidate items. We define the attribute set related to $\mathcal{V}_{cand}$ as follows:

$$\mathcal{P}_{\mathcal{V}_{cand}}^{|\mathcal{V}_{cand}|} = \left\{ \left(\mathcal{P}_{v_1} \bigcup \mathcal{P}_{v_2} \bigcup ... \bigcup \mathcal{P}_{v_{|\mathcal{V}_{cand}|}} \right) | v_i \in \mathcal{V}_{cand} \right\}. \qquad (6)$$

$\mathcal{P}_{\mathcal{V}_{cand}}^i$ represents the attribute set contained in the first i items in $\mathcal{V}_{cand}$. The link between the item and attribute vertice indicates that the item contains this attribute. After getting candidate items $\mathcal{V}_{cand}$, we take advantage of $\mathbf{F}_{|\mathcal{V}_u'|}^{(b)}$ ($\mathbf{F}_{|\mathcal{V}_u'|}^{(b)}$

obtained through feeding the historical sequence that the user u has interacted into the multi-layer self-attention block) to score $\mathcal{V}_{cand}$:

$$f_{item}(u, \mathcal{V}'_u, \mathcal{V}_{cand}) = \mathbf{F}^{(b)}_{|\mathcal{V}'_u|} \mathbf{N}^T_v \quad v \in \mathcal{V}_{cand},$$
$$\mathcal{V}_{cand} = \text{rank}(\mathcal{V}_{cand} \leftrightarrow f_{item}), \tag{7}$$

where $\mathcal{V}'_u$ is user u's training sequence of historical interaction items. The score of each item v in $\mathcal{V}_{cand}$ considers the information of the sequence $\mathcal{V}'_u$, reflecting the correlation between v and $\mathcal{V}'_u$. Then we rank $\mathcal{V}_{cand}$ from large to small according to the score. After ranking, the top items in the candidate items may become the next item to be recommended. Furthermore, the candidate attribute set is updated by

$$\mathcal{P}_{cand} = \mathcal{P}^L_{\mathcal{V}_{cand}} - \mathcal{P}_{u_a} - \mathcal{P}_{u_r}, \tag{8}$$

where $\mathcal{P}_{u_a}$ and $\mathcal{P}_{u_r}$ denote the attributes that the user accepts and rejects, respectively. In other words, $\mathcal{P}_{cand}$ is the attribute set that has not been interacted in the top L candidate items. If this turn is item recommendation, the system will recommend the top n items $\mathcal{V}_{rec}$ in the ranked $\mathcal{V}_{cand}$. Supposing the user accepts the item, the recommendation is successful and the conversation ends. Otherwise, the recommendation fails, system removes the recommended item from the $\mathcal{V}_{cand}$, i.e., it is updated by $\mathcal{V}_{cand} = \mathcal{V}_{cand} - \mathcal{V}_{rec}$. And the next turn is entered until the maximum turn is reached.

Attribute Scoring. The second operation is attribute scoring. SeqCR can decide which attribute to ask based on the current state. Therefore, it is important to find the appropriate attributes. Eliminating the uncertainty of the items is a better strategy. [24] has proven information entropy is an effective method. It is a measure to eliminate information uncertainty. The lower probability of an event, the greater information entropy that can be given when it happens. Therefore, we use information entropy to score attributes, which is defined as follows:

$$f_{att}(\mathcal{P}_{cand}, \mathcal{V}_{cand}) = -\text{prob}(p) \cdot \log_2(\text{prob}(p)) \quad p \in \mathcal{P}_{cand},$$
$$\text{prob}(p) = \frac{\sum_{v \in \mathcal{V}_{cand} \cap \mathcal{V}_p} \sigma(s_v)}{\sum_{v \in \mathcal{V}_{cand}} \sigma(s_v)}, \tag{9}$$

where σ is the sigmoid function, s_v is the score of item v. $\mathcal{V}_p$ denotes the items that include the attribute p. Instead of treating each item equally, we use the weighted entropy method to assign higher weights to important items. In short, if multiple items in the candidate items contain attribute p, then p is not a suitable attribute. Because our purpose is to ask some attributes that the user likes as much as possible, then filter $\mathcal{V}_{cand}$ to keep it containing the attributes accepted by the user as much as possible. When SeqCR asks the attribute p contained by multiple items, it will not be able to filter the candidate items effectively. If the system asks in this turn, it needs to choose a suitable attribute to ask. In our model, we consider the attribute $p_i \in \mathcal{P}_{cand}$ with the highest score as the attribute to be asked. Assuming that the user accepts p_i, we update the user-accepted attribute set $P_{u_a} = P_{u_a} \bigcup p_i$ and candidate items $\mathcal{V}_{cand} = \mathcal{V}_{cand} \bigcap \mathcal{V}_{p_i}$. Similarly, if a user rejects p_i, we update the user-rejected attribute set $P_{u_r} = P_{u_r} \bigcup p_i$ and candidate items $\mathcal{V}_{cand} = \mathcal{V}_{cand} - \mathcal{V}_{p_i}$. Therefore, the items in $\mathcal{V}_{cand}$ only contain P_{u_a}, don't contain P_{u_r}.

4.3 Policy Network Module

On the right side of Fig. 1, we utilize a policy network to learn whether to ask or recommend, and use the standard Deep Q-learning [19] for optimization. The system inputs state s in policy network, then get a $Q(s, a)$ value about two actions, indicates a reward for action a_{ask} or a_{rec}. We define s as follows:

$$s = s_{his} \oplus s_{len}. \tag{10}$$

s_{his} encodes the history of conversation. Its size is maximum turns T, where each dimension denotes user feedback at turn t [12]. The intuition is that if the user accepts the attribute multiple times, the next turn may be a good time to recommender. s_{len} encodes the length of the current candidate items. It means a system should recommend items when there are fewer candidate items. The reward contains five kinds of rewards [13], r_{rec_suc} and r_{rec_fail} indicate whether the recommendation is successful or not. r_{ask_suc} and r_{ask_fail} indicate whether the attributes asked are accepted by the user. It will give a r_{quit} if user quits or the conversation reaches the maximum number of turns. The intermediate reward r_t at turn t is the weighted sum of these five.

Although some of our components refer to Simple Conversational Path Reasoning(SCPR)[13], there are three significant differences between our model SeqCR and SCPR. First, SeqCR predicts the next item that user desires by scoring items based on the interaction sequence of historical items. Second, SeqCR takes the attributes of the items with the high-scoring as candidate attributes, which reduces the search space of attributes. Third, SeqCR believes that the attribute rejected by the user is also an important feature, and considers this feature when updating the candidate item set.

5   Experiments

In this section, we evaluate SeqCR on two real-world datasets. Our experiments are designed to answer the following research questions:

- **RQ1:** How does the performance of SeqCR compare to existing conversational recommendation methods?
- **RQ2:** Does restrict candidate attributes to the ones of the high-scoring candidate items really work?
- **RQ3:** Is it effective to consider the attributes rejected by the user?

5.1   Datasets

Due to the lack of conversation datasets, we adopted a template conversation method on the traditional recommendation datasets. SeqCR either recommends the items or asks a question about an attribute. The feedback of users is "yes" or "no". Our model SeqCR needs a timestamp to process the sequence. Therefore, we conduct experiments on *MovieLens 10M* (ml-10m)[1] for movies recommendation and *Yelp*[2] for businesses recommendation. Lei et al. [12] manually built a 2-layer taxonomy, which needs a lot of manpower and professional knowledge and is expensive for real usage. So we only use original attributes. Statistics about the datasets are shown in Table 2.

Table 2. Dataset statistics of MovieLens 10M and Yelp.

Dataset	MovieLens 10M	Yelp
#Users	69,861	25,000
#Items	6,673	152,975
#Attributes	14,162	2,083
#Interactions	8,065,720	1,437,418

5.2   Settings

5.2.1   Implementation Details

For all datasets, we treat reviews or ratings as historical items that users have interacted with and use timestamps to determine the sequence order of interaction. We discard users with fewer than ten interactions. We split the interaction sequence $\mathcal{V}_u$ of each user u into three parts: (1) the last interaction $\mathcal{V}_u^{|\mathcal{V}_u|}$ for testing, (2) the second most recent interaction $\mathcal{V}_u^{|\mathcal{V}_u|-1}$ for validation, and (3) all remaining interaction for training. Note that during the testing, the training sequence contains the training set and the validation set. Considering the patience of user and the requirement of SeqCR to capture user preferences, the maximum turn T is set as 15. Furthermore, we restrict the attributes of the top L = 10 high-scoring candidate items as candidate attributes. Following [12,13,21], the training process contains two parts: (1) An offline training for the scoring function in sequential module. We use two self-attention blocks ($b = 2$) and use *Adam* optimizer [10] to optimize. The learning rate and the batch size are 0.001 and 128, respectively. The maximum sequence length n is set to 200. The dropout rate is set 0.2. The goal is to score other items based on the interaction sequence of historical items. (2) An online training for conversational in a scoring module and a policy network module. We conduct conversations and train

---

[1] https://grouplens.org/datasets/movielens/.
[2] https://www.yelp.com/dataset/.

the policy network by using a user simulator. DQN parameters refer to [13]: $r_{rec\_suc}=1$, $r_{rec\_fail} = -0.1$, $r_{ask\_suc} = 0.01$, $r_{ask\_fail} = -0.1$, $r_{quit} = -0.3$, the sample batch size is 128, and discount factor γ is 0.999. For the policy network, we use the RMS optimizer and update the target network every 20 epochs.

5.2.2 User Simulator

It is unrealistic for CRS to chat with real humans during training, and therefore utilizing a user simulator is a common alternative [2]. We follow [12,13,21] to create a user simulator: given an observed user-item interaction (u, v), we regard v as the ground truth that user u wants to seek and the attributes $\mathcal{P}_v$ of item v are the preference attributes of u. In the beginning, we randomly selected an attribute $p_0 \in \mathcal{P}_v$ to start the dialogue. And (1) if the recommended item list contains v, the user accepts the recommendation, (2) user only accepts attribute $p_i \in \mathcal{P}_v$. This user simulator may have some flaws. For example, the user will reject attributes that they like but do not belong to $\mathcal{P}_v$. But it is a more practical method at the current stage.

5.2.3 Baseline

Due to the different settings of the CRS model and the distinction in the datasets, there are few suitable baselines. Therefore we use the following baselines to compare with:

- **AbsGreedy** [6]. The baseline can only make recommendations in each turn without asking, and then it will update the model based on user feedback until it is successfully recommended. Its performance is similar to Thompson Sampling [3].
- **Max Entropy.** This is a rule-based approach. The system either chooses an attribute with the maximum entropy to ask or recommend items according to a certain probability.
- **CRM** [21]. This is a CRS model that tracks user preferences through a belief tracker, then utilizes reinforcement learning (RL) to select an action. It is single-round recommendations. To achieve a fair comparison, we follow [12] to change it to multiple the multi-round dialogue.
- **EAR** [12]. Estimation–Action–Reflection (EAR) is a three stages solution, which integrates conversation component and recommender component. Its goal is to achieve accurate recommendations in fewer turns.
- **SCPR** [13]. This is a state-of-the-art method of CRS. It is the first to introduce a graph to conduct path reasoning. SCPR is a simple implementation of Conversational Path Reasoning (CPR), which reduces the attribute candidate space by using a graph structure. This inspires our SeqCR implementation hence being the most comparable model.

5.2.4 Evaluation Metrics

The evaluation metrics follow [13]. We use success rate(SR@t) [22] to measure the ratio of successful conversations, i.e., the ratio of successful recommendations in turn t. We also use average turns (AT) to represent the average number of turns for all conversations. If successfully recommend in the t turn, and the turn of this session is t. Assuming that the recommendation is not successful until the maximum turn T, the turn of the session is recorded as T. Therefore, the higher SR@t and lower AT indicate that model has better performance.

5.3   Performance Comparison with Existing Models (RQ1)

Table 3 shows the performance statistics of SeqCR and other baselines. We can see that our SeqCR model achieves higher SR and lower AT than state-of-the-art baselines, demonstrating the better performance of our model. In order to compare intuitively, we have drawn the SR@t $(1 \leqslant t \leqslant 15)$ of all models, where SCPR serves as the blue line of $y = 0$ in the figures and Success Rate *(SR*) [12] denotes the difference of SR between each method and SCPR.

Table 3. Performance comparison of different methods on the two real-world datasets. The best performance is highlighted in boldface. $(p < 0.01)$ (RQ1).

	MovieLens 10M				Yelp			
	SR@5	SR@10	SR@15	AT	SR@5	SR@10	SR@15	AT
Abs Greedy	0.630	0.703	0.728	5.72	**0.121**	0.187	0.242	12.73
Max Entropy	0.543	0.678	0.768	5.81	0.030	0.176	0.415	12.76
CRM	0.619	0.690	0.712	5.92	0.081	0.165	0.221	13.03
EAR	0.645	0.705	0.735	5.84	0.082	0.182	0.246	12.91
SCPR	0.621	0.782	0.951	4.89	0.047	0.303	0.552	11.77
SeqCR	**0.877**	**0.931**	**0.971**	**2.36**	0.055	**0.487**	**0.721**	**10.13**

As shown in Fig. 2, we can see that our model SeqCR is better than other baselines. Interestingly, we can find that SeqCR has a significant advantage in 2 to 6 turns of MovieLens 10M and 9 to 14 turns of Yelp, respectively. It demonstrates the effectiveness of training the sequence module by utilizing the sequence of historical items. Specifically, SeqCR scores candidate items based on historical sequences, then ranks the candidate items. After ranking, the more relevant items' rank is higher in the candidate items. The attributes we ask are selected from the top L candidate items. If ground truth ranks higher in the candidate items, it will be easier for SeqCR to ask about its attributes. In this way, our model can make accurate recommendations to the user in fewer turns. Simultaneously, we use a policy network to decide whether to recommend or ask, which significantly reduces the action space. Therefore, SeqCR will achieve a significant performance improvement in earlier turns. Compared with Yelp, the MovieLens 10M dataset has better sequence features (i.e., after ranking, the ground truth ranks higher in the candidate items.), so it has a more remarkable performance improvement in the first few turns.

Compared with the Yelp dataset, MoviesLens 10M has a higher SR* and a lower AT. Because there are more than 14000 attributes in the MovieLens 10M, some of which are characteristic (i.e., only a few items have these attributes). If SeqCR asks about these attributes and the user provides positive feedback, the length of candidate items can be greatly reduced so that the system can make accurate recommendations faster. This is why MovieLens 10M has higher SR* and lower AT. Simultaneously, we can clearly see that SR* of SeqCR shows a trend of the first upward and then downward. After ranking, assuming that the ground truth ranks higher, since SeqCR prioritizes the attributes of the top-ranked items, it is more likely to ask the attributes of ground truth according to the ranked candidate items, so SR* in the previous turns has an upward trend (i.e., it has a greater advantage than SCPR). When there are more dialogue turns, it indicates that ground truth ranks lower in the candidate items. It is difficult for SeqCR to ask about the attributes of the low-ranked items, which prevents the system from capturing users' preferences. Therefore, the SR* in the next few turns shows a downward trend. This denotes that considering the sequence of historical items can effectively reduce conversation turns.

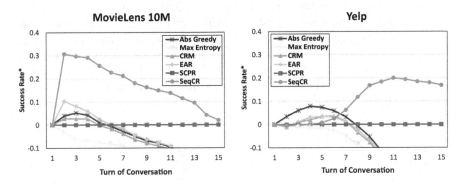

Fig. 2. The success Rate* of compared methods at different conversation turns on *MovieLens 10M* and *Yelp*. SCPR serves as the blue line of y = 0 (RQ1).

The two RL-based methods, CRM and EAR, show better performance at the beginning. But as the turns increase, their performance gradually decreases, even lower than Max Entropy. This is because their policy network not only decides whether to ask or recommend, but also decides which attribute is to ask. In other words, their action space is $|\mathcal{P}|+1$. For our datasets, they have 14162+1 and 2083+1 action spaces, respectively. It is very challenging for a policy network to handle such a large action space. Similarly, Abs Greedy demonstrates a great advantage in the first few turns. Because other models need to ask questions at the beginning of a conversation to capture user preferences, but Abs Greedy only makes recommendations until the recommendation is successful or the maximum turn is reached. Therefore, a higher success rate will be achieved in the early stage.

5.4   Evaluation of the Methods to Restrict Candidate Attribute (RQ2)

In order to prove the effectiveness of our method of restricting candidate attributes, we design a comparison experiment named SeqCR-L, where L means that we only consider the attributes of the first L high-scoring items as candidate attributes. SeqCR-all keeps the other parts unchanged, then considers the attributes of all candidate items as candidate attributes. As shown in Fig. 3, since MovieLens 10M has a better sequence feature, it has achieved better results in the first few turns. The Yelp has a poor sequence feature, the SR* is very low in the first few turns, but as the turns increase, SR* will get significantly improved. We choose the length L to be 10. When L is less than 10, it indicates that SeqCR restricts the candidate attributes too much, which prevents the model from finding suitable attributes to ask. When L is higher than 10, SeqCR has too few restrictions on candidate attributes, which increases the search space and reduces the hit rate of the ground-truth of attributes. Therefore, too many or too few restrictions on candidate attributes will reduce SR*. In the intermediate turns of MovieLens 10M, the $L = 10$ is not the best performance. The possible reason is that these poorly sequential data require more candidate attributes to determine the most suitable attributes.

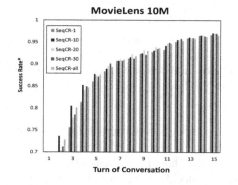
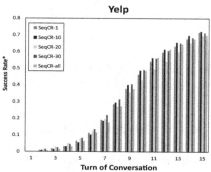

Fig. 3. Comparison of two methods to restrict candidate attributes. SeqCR and SeqCR+all use high-scoring candidate items and all candidate items, respectively. (RQ2).

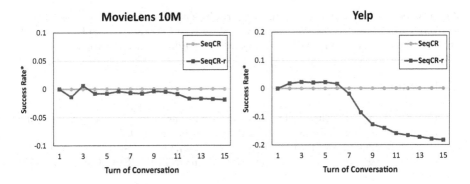

Fig. 4. Comparison of whether to consider attributes rejected by user (RQ3).

5.5 Evaluation of the Effectiveness to Consider Attributes Rejected by Users (RQ3)

At the same time, we believe that the rejected attributes are also an important feature. Different from SCPR, our model SeqCR removes items that contain attributes rejected by users. Because we think this negative feedback of attributes provides extra information. Based on such information, SeqCR can filter candidate items more accurately. We conducted a comparative experiment, SeqCR-r means that the rejected attributes are not considered in SeqCR. As shown in Fig. 4, SeqCR serves as $y = 0$. We can find that the performance of SeqCR-r is lower than SeqCR. It demonstrates that considering the attributes rejected by users is helpful to improve performance. On the Yelp dataset, one of its attributes belongs to multiple items, furthermore, it has a higher AT (i.e., there are more turns of conversations). Therefore, directly removing items in the first few turns will affect the choice of attributes. But as the turn grows, candidate items are getting more accurate, SeqCR will get a better performance improvement than SeqCR-r.

6 Conclusion

In this work, we are the first to introduce the interaction sequence of historical items in the conversational recommender system. Specifically, our model SeqCR has a sequential module, which can model user interaction sequences. Furthermore, it utilizes the scoring module to score candidate items and attributes. Finally, it can decide whether to ask or recommend through a strategy network module. SeqCR not only uses sequences to improve the quality of recommendations, but also restricts candidate attributes to the attributes of the first L high-scoring items, which greatly reduces the attribute candidate space. Extensive experiments show that our model outperforms state-of-the-art baselines. In the future, we will incorporate natural language and other information (such as comment content, knowledge graph) to make more accurate recommendations.

Acknowledgements. This research was partially supported by NSFC (No. 61876117, 61876217, 61872258, 61728205), ESP of the State Key Laboratory of Software Development Environment, and PAPD of Jiangsu Higher Education Institutions.

References

1. Bao, Y., Fang, H., Zhang, J.: TopicMF: simultaneously exploiting ratings and reviews for recommendation. In: AAAI, vol. 14, pp. 2–8 (2014)
2. Chandramohan, S., Geist, M., Lefevre, F., Pietquin, O.: User simulation in dialogue systems using inverse reinforcement learning. In: INTERSPEECH, pp. 1025–1028 (2011)
3. Chapelle, O., Li, L.: An empirical evaluation of Thompson sampling. In: NIPS, pp. 2249–2257 (2011)
4. Chen, Q., et al.: Towards knowledge-based recommender dialog system, pp. 1803–1813 (2019)
5. Christakopoulou, K., Beutel, A., Li, R., Jain, S., Chi, E.H.: Q&R: a two-stage approach toward interactive recommendation. In: KDD, pp. 139–148 (2018)
6. Christakopoulou, K., Radlinski, F., Hofmann, K.: Towards conversational recommender systems. In: KDD, pp. 815–824 (2016)
7. He, K., Zhang, X., Ren, S., Sun, J.: Deep residual learning for image recognition. In: CVPR, pp. 770–778
8. He, X., Liao, L., Zhang, H., Nie, L., Hu, X., Chua, T.S.: Neural collaborative filtering. In: WWW, pp. 173–182 (2017)
9. Kang, W.C., McAuley, J.: Self-attentive sequential recommendation. In: ICDM, pp. 197–206. IEEE (2018)
10. Kingma, D.P., Ba, J.: Adam: a method for stochastic optimization. In: ICLR (Poster) (2015)
11. Koren, Y., Bell, R., Volinsky, C.: Matrix factorization techniques for recommender systems. Computer **42**(8), 30–37 (2009)
12. Lei, W., et al.: Estimation-action-reflection: towards deep interaction between conversational and recommender systems. In: WSDM, pp. 304–312 (2020)
13. Lei, W., et al.: Interactive path reasoning on graph for conversational recommendation. In: KDD, pp. 2073–2083 (2020)
14. Li, R., Kahou, S.E., Schulz, H., Michalski, V., Charlin, L., Pal, C.: Towards deep conversational recommendations. In: NeurIPS, pp. 9725–9735 (2018)
15. Liu, J., et al.: Exploiting aesthetic preference in deep cross networks for cross-domain recommendation. In: WWW, pp. 2768–2774 (2020)
16. Lops, P., de Gemmis, M., Semeraro, G.: Content-based recommender systems: state of the art and trends. In: Ricci, F., Rokach, L., Shapira, B., Kantor, P.B. (eds.) Recommender Systems Handbook, pp. 73–105. Springer, Boston, MA (2011). https://doi.org/10.1007/978-0-387-85820-3_3
17. Luo, A., et al.: Collaborative self-attention network for session-based recommendation, pp. 2591–2597 (2020)
18. Mnih, A., Salakhutdinov, R.R.: Probabilistic matrix factorization. In: NIPS, pp. 1257–1264 (2008)
19. Mnih, V., et al.: Human-level control through deep reinforcement learning. Nature **518**(7540), 529–533 (2015)
20. Rendle, S., Freudenthaler, C., Schmidt-Thieme, L.: Factorizing personalized Markov chains for next-basket recommendation. In: WWW, pp. 811–820 (2010)

21. Sun, Y., Zhang, Y.: Conversational recommender system. In: SIGIR, pp. 235–244 (2018)
22. Tang, J., Wang, K.: Personalized top-n sequential recommendation via convolutional sequence embedding. In: WSDM, pp. 565–573 (2018)
23. Vaswani, A., et al.: Attention is all you need. In: NIPS, pp. 5998–6008 (2017)
24. Wu, J., Li, M., Lee, C.H.: A probabilistic framework for representing dialog systems and entropy-based dialog management through dynamic stochastic state evolution. IEEE/ACM Trans. Audio Speech Lang. Process. **23**(11), 2026–2035 (2015)
25. Zhang, Y., Chen, X., Ai, Q., Yang, L., Croft, W.B.: Towards conversational search and recommendation: system ask, user respond. In: CIKM, pp. 177–186 (2018)
26. Zhou, K., Zhao, W.X., Bian, S., Zhou, Y., Wen, J.R., Yu, J.: Improving conversational recommender systems via knowledge graph based semantic fusion. In: KDD, pp. 1006–1014 (2020)

Knowledge-Aware Hypergraph Neural Network for Recommender Systems

Binghao Liu[1], Pengpeng Zhao[1(✉)], Fuzhen Zhuang[2,3], Xuefeng Xian[4],
Yanchi Liu[5], and Victor S. Sheng[6]

[1] Institute of Artificial Intelligence, School of Computer Science and Technology,
Soochow University, Suzhou, China
bhliu1@stu.suda.edu.cn, ppzhao@suda.edu.cn
[2] Key Lab of Intelligent Information Processing of Chinese Academy of Sciences
(CAS), Institute of Computing Technology, CAS, Beijing 100190, China
[3] Xiamen Data Intelligence Academy of ICT, CAS, Beijing, China
zhuangfuzhen@ict.ac.cn
[4] Suzhou Vocational University, Suzhou, China
xianxuefeng@jssvc.edu.cn
[5] Rutgers University, New Brunswick, NJ, USA
yanchi.liu@rutgers.edu
[6] Department of Computer Science, Texas Tech University, Lubbock, USA
Victor.Sheng@ttu.edu

Abstract. Knowledge graph (KG) has been widely studied and
employed as auxiliary information to alleviate the cold start and sparsity problems of collaborative filtering in recommender systems. However, most of the existing KG-based recommendation models suffer from
the following drawbacks, i.e., insufficient modeling of high-order correlations among users, items, and entities, and simple aggregation strategies which fail to preserve the relational information in the neighborhood. In this paper, we propose a Knowledge-aware Hypergraph Neural Network (KHNN) framework to tackle the above issues. First, the
knowledge-aware hypergraph structure, which is composed of hyperedges, is employed for modeling users, items, and entities in the knowledge graph with explicit hybrid high-order correlations. Second, we propose a novel knowledge-aware hypergraph convolution method to aggregate different knowledge-based neighbors in hyperedge efficiently. Moreover, it can conduct the embedding propagation of high-order correlations explicitly and efficiently in knowledge-aware hypergraph. Finally,
we apply the proposed model on three real-world datasets, and the empirical results demonstrate that KHNN can achieve the best improvements
against other state-of-the-art methods.

Keywords: Recommender systems · Knowledge-aware hypergraph ·
Knowledge graph

© Springer Nature Switzerland AG 2021
C. S. Jensen et al. (Eds.): DASFAA 2021, LNCS 12683, pp. 132–147, 2021.
https://doi.org/10.1007/978-3-030-73200-4_9

1 Introduction

With the rapid development of the Internet, recommender systems (RS) [5,10, 19,23] have been widely deployed to alleviate the impact of information overloading [13]. A traditional recommendation method is collaborative filtering (CF), in which users and items are represented as ID-based vectors, and then the historical interactions of users and items are modeled by a operation such as inner product [15] or a network such as neural collaborative filtering [5]. However, the cold-start and sparsity problems generally exist in CF-based models. To address these issues, multiple types of side information have been explored for improving recommendation performance, such as item attributes [14], item reviews [26], and users' social networks [11].

Knowledge graph (KG), which is strong to model comprehensive side information, has attracted more and more attention in RS [16,19,20,22]. How to efficiently integrate the side information into latent representation vectors of users and items is important in the combination of knowledge graph and recommender systems. According to the methods dealing with the issue, existing KG-based recommender system models can be categorized into two types, path-based and graph neural network (GNN) based models. Path-based models [21] explore multiple meta-paths between target user and item in KG to infer user preference. Nevertheless, these models require domain knowledge. What's more, this type of models neglect abundant structural information stored in KG and cannot well explore and utilize the comprehensive correlations between the target user and item.

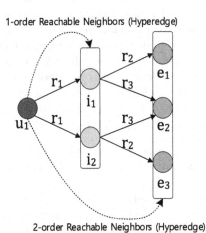

Fig. 1. An example of knowledge-aware hypergraph of u_1. u_1 is the target user whom we need to provide recommendation for. i_1 and i_2 are the items, and e_1, e_2, and e_3 are the entities in the knowledge graph. The link r_1 means i_1 and i_2 are interacted by u_1, r_2 and r_3 are defined relations in the knowledge graph.

As more and more graph neural networks emerge, several GNN-based models [20,22] have been proposed and indicate satisfactory improvements by model explicit high-order connectivity among entities in KG. However, these models still have two restrictions: (**L1**) High-order correlations between users, items, and entities in KG are essential for data modeling. These methods mainly apply GNN to enrich the representation of a target node by recursively aggregating their nearest neighbors in the KG, so have limitations on using high-order correlation between the target node and unoriginal neighbors. In addition, the graph structure, which is employed by existing methods, are insufficient to model and utilize high-order relation, as a typical graph can only store pairwise connections. (**L2**) During the neighborhood aggregation, simple aggregation operations such as element-wise mean/sum or max-pooling or weighted aggregation operations which only consider pair-wise connection are usually used as the aggregator to get the neighborhood embedding. This aggregation destroys rich relational information among neighborhood nodes. As a result, The neighborhood aggregation is forced to ignore the fine structure of neighborhood relations.

To address the above limitation of existing KG-based recommendation methods, we propose an end-to-end model named Knowledge-aware Hypergraph Neural Network (KHNN). Specifically, KHNN is equipped with two designs: (1) Knowledge-aware hypergraph construction, which is composed of initial hyperedge construction and knowledge hyperedge construction. For initial hyperedge convolution, we construct hyperedges based on the user-item interaction. For knowledge hyperedge construction, we construct hyperedges based on knowledge graph and initial hyperedge. Based on these generated hyperedges, we can construct two hypergraphs for the target user and item, respectively. As depicted in Fig. 1, item i_1 and item i_2 can be associated by target user u_1 without direct connections, and entity e_1, entity e_2, and entity e_3 in knowledge graph can be associated by item i_1 and item i_2 without direct connections. (**L1**) (2) Knowledge-aware hypergraph convolution, which is composed of neighborhood convolution and hyperedge convolution. For neighborhood convolution, we use a transform matrix to permute and weight entities in a hyperedge. For hyperedge convolution, we just concatenate the hyperedge features to get the final embeddings of users and items. (**L2**) To sum up, our contributions are as follows:

- We propose a knowledge-aware hypergraph which explicitly exploring and modeling the high-order correlations among users, items, and entities in the KG.
- We propose to apply a novel knowledge-aware hypergraph convolution method to model the rich relational information among neighborhood entities in a hyperedge and support the explicit and efficient embedding propagation of high-order correlations.
- We conduct sufficient experiments on all three recommendation scenarios. The results of the experiments show the superiority of KHNN over other state-of-the-art baselines.

2   Related Work

In this section, we will introduce the representative models in knowledge aware recommendation and hypergraph learning.

2.1   Knowledge-Aware Recommendation

Recently, Knowledge graph get more and more attention. Kg is widely applied in recommender systems for extending correlations between user historical interaction and candidate items. Some models apply KG-aware embeddings to improve the quality of item embeddings such as DKN [17] and CKE [25]. Nevertheless, these methods are insufficient to make use of high-order relations over KG. As a result, several models [21,24] have been applied to explore multiple semantic path (meta-path), which connect target users' historical interactions and candidate items over KG. Then the prediction function is learned through multiple path modeling and integrating. In spite of these models' effectiveness, the path-based models neglects comprehensive relational information stored in KG as each explored path is modeled independently. As a newly research domain, graph neural network has demonstrated its ability in learning good node embeddings in graph. RippleNet [16] diffuses users' underlying interest and explores their rich correlations in KG. KGCN-LS [18] and KGAT [20] conduct high-order information propagation by applying multiple KG-aware GNN layers. CKAN [22] utilizes the collaborative information by collaborative information propagation through users' historical interactions and it also apply a solution to combine collaborative information with knowledge information. Though GNN based models have obtained some performance improvement, they still suffer inferior performance, i.e., insufficient modeling of high-order correlations among users, items and entities in knowledge graph and simple aggregation strategies which fail to preserve the relational information in the neighborhood.

2.2   Hypergraph Learning

As a label propagation method, [27] first propose hypergraph learning to complex high-order relations. With the development of deep learning, hypergraph neural network have received much attention. Recent works focus on the learning of hyperedge weight. In these methods, the hyperedges or sub-hypergraphs with higher significance will be allocated larger weight [3]. Hypergraph neural network (HGNN) has been proposed as the first method to apply graph convolution on hypergraph structure [2]. Besides learning label propagation on hypergraph, dynamic hypergraph neural networks (DHGNN) [7] has been proposed to explore and model high-order relations among vertex in hypergraph. In addition, hypergraph has been applied in recommendation. In dual channel hypergraph collaborative filtering (DHCF) [6], hypergraph is used to learn the embeddings of users and items, as a result, these two types of information can be well connected while still maintaining their own features. Inspired by [6] and [7], we propose a way to apply the Knowledge-aware hypergraph for explicitly constructing knowledge-aware hyperedges to utilize the high-order correlations among users, items, and entities in KG.

3   Problem Definition

In this section, we introduce the formulation of the Knowledge-aware Hypergraph Neural Network. In the Knowledge-aware recommendation scenario, M users are represented as $\mathcal{U} = \{u_1, u_2, ..., u_M\}$, N items are represented as $\mathcal{V} = \{v_1, v_2, ..., v_N\}$ and a user-item interaction are represented as matrix $Y \in \mathbb{R}^{M \times N}$. The matrix Y is constructed according to users' implicit feedback, in which y_{uv} = 1 indicates that user u has interacted with item v, such as clicking, collecting, or purchasing; otherwise y_{uv} = 0. In addition, there is also a knowledge graph $\mathcal{G}_K = \{(h, r, t)|h, t \in \mathcal{E}, r \in \mathcal{R}\}$. In the knowledge graph, each knowledge triple (h, r, t) demonstrates that a relationship r exist between head entity h and tail entity t. And the sets of entities and relations in the knowledge graph are denoted as $\mathcal{E}$ and $\mathcal{R}$. For instance, the triple $(Back\ to\ the\ Future,\ directedby,\ Robert\ Zemeckis)$ represents the fact that $Robert\ Zemeckis$ is an director of the movie $Back\ to\ the\ Future$.

An edge can only connects two nodes in a typical graph. However, a hyperedge connects two or more nodes in a hypergraph [1]. $\mathcal{G}_H = (\mathcal{E}, \mathcal{H})$ denotes a knowledge-aware hypergraph, in which $\mathcal{E}$ denotes the entity set, and $\mathcal{H}$ represents the hyperedge set. A set $\mathcal{A} = \{(v, e)|v \in \mathcal{V}, e \in \mathcal{E}\}$ is used to conduct correlations between items and entities. In A, (v, e) indicates that item v can be correlated with entity e in the knowledge-aware hypergraph.

Given the knowledge-aware hypergraph $\mathcal{G}_H$ with historical interaction matrix Y, we aim to calculate the probability that user u would interact item v which he has not engaged with before. To be specific, our ultimate goal is to learn a prediction function $\hat{y}_{uv} = \mathcal{F}(u, v|\Theta, Y, \mathcal{G}_H)$, where $\hat{y}_{uv}$ denotes the predicted probability that user u will interact item v, and Θ indicates the model parameters of function $\mathcal{F}$.

4   The Proposed Method

In this section, the proposed KHNN framework is introduced. As represented in Fig. 2, the model contains three main components: 1) knowledge-aware hypergraph construction, which consists of initial hyperedge construction through user-item interactions and knowledge hyperedge construction in knowledge graph; 2) knowledge-aware hypergraph convolution, which consists of neighborhood convolution which aggregates entities to hyperedge and hyperedge convolution which aggregates vectors of hyperedge to user u and item v; and 3) prediction layer, which outputs the predicted click probability.

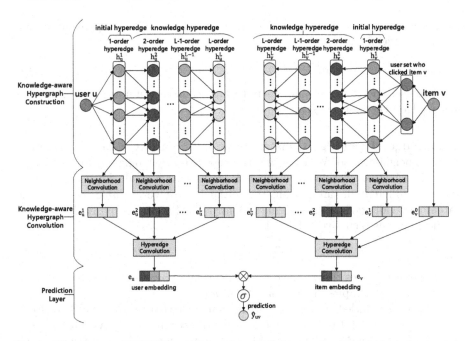

Fig. 2. The proposed KHNN model consists of three components: knowledge-aware hypergraph construction, knowledge-aware hypergraph convolution and prediction layer.

4.1 Knowledge-Aware Hypergraph Construction

As shown in Fig. 2, knowledge-aware hypergraph construction is composed of two major modules: initial hyperedge construction and knowledge hyperedge construction. The initial hyperedge is constructed by explicitly encoding user-item historical interaction into representations of both user and item. The knowledge hyperedge is constructed by entities which have high-order relation with source node in knowledge-aware hypergraph.

Initial Hyperedge Construction. It is obvious that users' historical interactions are able to representing the user's interest to some extent. As a result, we use user's interacted items to represent user u. The interacted item set of user u can be constructed as the initial hyperedge(entity set) in knowledge-aware hypergraph through the correlations between items and entities. Following, the initial hyperedge of user u is defined as:

$$h_u^1 = \{e|(v,e) \in \mathcal{A} \ and \ v \in \{v|y_{uv} = 1\}\} \tag{1}$$

Simplistically, users who have similar historical interactions can also be used to enrich the item's feature representation. We use items, which have been watched by the same user, to construct initial item set of item v as follows:

$$\mathcal{V}_v = \{v_u|u \in \{u|y_{uv} = 1\} \ and \ y_{uv_u} = 1\} \tag{2}$$

Integrating initial item set and correlation set, the initial hyperedge of item v is defined as follows:

$$h_v^1 = \{e|(v_u, e) \in \mathcal{A} \ and \ v_u \in \mathcal{V}_v\} \tag{3}$$

The user and item's initial hyperedges are constructed now for the following knowledge hyperedge construction.

Knowledge Hyperedge Construction. Entities in KG have close relations with its neighborhood. We construct the knowledge hyperedge by taking advantage of such close relations among entities in KG. As a result, knowledge hyperedges of different distances from the initial hyperedge are able to enrich the latent embedding of user and item efficiently. The knowledge hyperedge for user u and item v is recursively defined as:

$$h_o^l = \{t|(h, r, t) \in \mathcal{G}_K \ and \ h \in h_o^{l-1}\}, \quad l = 2, 3, ..., L \tag{4}$$

where l indicates the distance from the initial hyperedge, the subscript symbol o^1 is a uniform placeholder for symbol u or v.

Applying knowledge graph as auxiliary information to construct knowledge hyperedges is helpful to enrich the features of user and item. As illustrated in Fig. 2, the hyperedge is obtained through initial hyperedge construction and knowledge hyperedge construction. Based on these generated hyperedges, we can construct two hypergraphs for target user and target item.

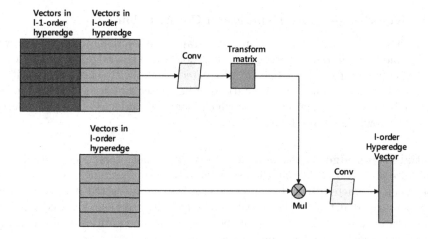

Fig. 3. Neighborhood convolution module. For k concat vectors, a k × k transform matrix is computed by convolution. We multiply transform matrix and input vector matrix to get permuted and weighted vector matrix. Then we apply a 1-dimension convolution to weighted vector matrix to get the 1-dimension hyperedge vector.

---

[1] Symbol o is used as a uniform placeholder for both user and item.

4.2 Knowledge-Aware Hypergraph Convolution

Knowledge-aware hypergraph convolution is composed of two sub-modules: neighborhood convolution sub-module and hyperedge convolution sub-module. Neighborhood convolution aggregates entity latent vectors to hyperedge and then hyperedge convolution aggregates hyperedge latent vectors to a single embedding for both user and item.

Neighborhood Convolution. Neighborhood convolution aggregates entity latent vectors to the hyperedge. A simple solution is pooling method like max pooling and average pooling. State-of-the-art methods like CKAN [22] employ a fixed, pre-computed transform matrix generated from triples in kg for vertex aggregation. However, such methods are unable to model comprehensive correlation among entities in neighborhood well. Inspired by DHGNN [7], we propose a novel neighborhood convolution method. We learn the transform matrix T from the entity vectors in both l-order and l-1-order hyperedges for vector permutation and weighting, as is shown in Fig. 3. The transform matrix obtained by convolution can make good use of the comprehensive correlation among entities in neighborhood. To be specific, we use a 1-d convolution to generate transform matrix T and use another 1-d convolution to aggregate the transformed vectors, as is described by Eq. 5 and Eq. 6.

$$T_o^l = conv_1(h_o^{l-1} || h_o^l) \quad l = 2, 3, ..., L \tag{5}$$
$$e_o^l = conv_2(T_o^l \cdot h_o^l) \quad l = 2, 3, ..., L \tag{6}$$

where the subscript o is a uniform placeholder for both user and item. $||$ is the concatenation operation. $conv_1$ and $conv_2$ are 1-dimension convolution but with different out channels.

Notice that, because the initial hyperedge has strong connections with original user and item. Consequently, the initial hyperedge are added for both user and item, as is described by Eq. 7 and Eq. 8.

$$T_o^1 = conv_1(h_o^1) \tag{7}$$
$$e_o^1 = conv_2(T_o^1 \cdot h_o^1) \tag{8}$$

It is noted that item v has its associated entities in kg to represent item v itself while user u does not. The entity, which have direct correlation with item v, contains most useful information to represent item v itself in kg. Consequently, the entity is added to the representation set of the item v and the entity is formulated as follows:

$$e_v^0 = e \quad (v, e) \in \mathcal{A} \tag{9}$$

After neighborhood convolution, we define the user representation set containing hyperedge embeddings and item representation set containing both hyperedge embeddings and additional embeddings for item v itself as follows:

$$\mathcal{T}_u = \left\{ e_u^1, e_u^2, ..., e_u^L \right\} \tag{10}$$
$$\mathcal{T}_v = \left\{ e_v^0, e_v^1, e_v^2, ..., e_v^L \right\} \tag{11}$$

Hyperedge Convolution. Hyperedge convolution aggregates hyperedge vector to user u and item v, as is illustrated in Fig. 2. We apply three types of aggregators to aggregate the multiple embeddings of hyperedge in Eq. 10 and Eq. 11 into a single embedding for both user and item.

Sum aggregator perform summation calculations of multiple embeddings in the set of representations:

$$agg_{sum}^o = \sum_{e_o \in \mathcal{T}_o} e_o \tag{12}$$

Pooling aggregator perform element-wise maximization calculations of multiple embeddings in the set of representations:

$$agg_{pool}^o = pool_{max}(\mathcal{T}_o) \tag{13}$$

Concat aggregator concatenates multiple embeddings in the set of representations:

$$agg_{concat}^o = e_o^{i_1} || e_o^{i_2} || ... || e_o^{i_n} \tag{14}$$

where $e_o^{i_k} \in \mathcal{T}_o$, and $||$ is the concatenation operation. Specifically, since the precondition of inner product operation is same dimensions of embeddings for both user and item, we abandon the initial hyperedge embedding in the item representation set to concatenate the rest representation vectors. The detail we will discuss in Sect. 5.5.

4.3   Optimization

Model Prediction. The single embedding of user's knowledge-aware hypergraph is defined as e_u. Analogously, as to item, e_v denote the item's knowledge-aware hypergraph. At last, we calculate the inner product and nonlinear transformation of embeddings to predict the user's score for the candidate item:

$$\hat{y}_{uv} = \sigma(e_u^T e_v) \tag{15}$$

Loss Function. For each user, we random select the same number of negative samples with positive samples to make sure the effectivity of model training. We will discuss the details of negative sampling in Sect. 5.1. Ultimately, we define the loss function of the model KHNN as follows:

$$\mathcal{L} = \sum_{u \in U} (\sum_{v \in \{v|(u,v) \in \mathcal{P}^+\}} \mathcal{J}(y_{uv}, \hat{y}_{uv}) - \sum_{v \in \{v|(u,v) \in \mathcal{P}^-\}} \mathcal{J}(y_{uv}, \hat{y}_{uv})) + \lambda ||\Theta||_2^2 \tag{16}$$

$\mathcal{J}$ is the cross-entropy loss, $\mathcal{P}^+$ is positive samples while $\mathcal{P}^-$ means the negative samples. Θ is the model parameters, and $||\Theta||_2^2$ is the L2-regularizer that parameterized by λ.

Table 1. Statistics of all the three datasets: Last.FM (Music), BookCrossing (Book) and Movie-Lens20M (Movie).

	Music	Book	Movie
# users	1,872	17,860	138,159
# items	3,846	14,967	16,954
# interactions	42,346	139,746	13,501,622
# inter-avg	23	8	98
# entities	9,366	77,903	102,569
# relations	60	25	32
# KG triples	15,518	15,1500	499,474
# link-avg	4	10	29

5   Experiments

In this section, we evaluate the KHNN model under three real-world scenarios to answer the following research questions:

- **Q1:** How does KHNN perform compared with state-of-the-art KG-based recommendation methods?
- **Q2:** How do different components (e.g., knowledge-aware neighborhood convolution and aggregator selection in hyperedge convolution) affect KHNN?
- **Q3:** How do different hyper-parameter settings (e.g. number of hyperedge, size of hyperedge) affect KHNN?

5.1   Dataset Description

We conduct experiments under three different scenarios: music, book and movie recommendations. The three datasets are different in size, sparsity and domain.

- **Last.FM**[2] contains 2 thousand users' listening information from Last.fm online music platform.
- **Book-Crossing**[3] contains 1 million ratings (ranging from 0 to 10) of books in the Book-Crossing community.
- **MovieLens-20M**[4] that contains 138 thousand users who have watched 27 thousand movies with 20 million ratings (ranging from 1 to 5) on the Movie-Lens platform.

The historical interactions in these three datasets are explicit feedback, as a result, these historical records need to be transformed into the implicit records in which 1 indicates the positive feedback. With regard to negative samples, we

---

[2] https://grouplens.org/datasets/hetrec-2011/.
[3] http://www2.informatik.uni-freiburg.de/cziegler/BX/.
[4] https://grouplens.org/datasets/movielens/.

randomly select same size with user's positive feedback from items which he has not interacted with.

The sub-KGs is constructed to guarantee the quality of entities in knowledge-aware hypergraph. We follow the work in [16,19] to construct sub-KGs from the Satori[5]. Each sub-KG, in which the confidence level of triple is higher than 0.9, is a subset of the whole KG. We abandon the items correlated with multiple entities and the items associated with no entity for accuracy. Table 1 is the summary of detailed statistics of the three datasets: Last.FM (music), Book-Crossing (book), MovieLens-20M (movie).

5.2   Baselines

We compare KHNN with following four types of recommendation models: CF-based method (BPRMF), embedding-based model (CKE), path-based model (PER) and GNN-based models(RippleNet, KGCN, KGNN-LS, KGAT, CKAN).

- **BPRMF** [12] is the Bayesian ranking model that uses Matrix Factorization (MF) as the prediction component for recommendation. It utilizes the user-item interaction to learn representations of users and items.
- **CKE** [25] is a typical embedding-based model, which use semantic embeddings derived from TransR [9] to enhance matrix factorization [12].
- **PER** [24] is a typical path-based model which use latent features derived from the meta-path in KG to denote the correlation between users and items in kg.
- **RippleNet** [16] is a GNN-based model which propagates user's interest to learn user and item embedding.
- **KGCN** [19] is a state-of-the-art GNN-based model which models the ultimate embedding of a candidate item v by aggregating the embedding of entities in the KG from neighbors of item v to item v itself.
- **KGNN-LS** [18] is another state-of-the-art GNN-based model which learn user-specific item embeddings by identifying important knowledge graph relationships for a given user
- **KGAT** [20] is another state-of-the-art GNN-based model, which considers user nodes as one type of entities in the knowledge graph and the interaction between users and items as one type of relation
- **CKAN** [22] is another state-of-the-art GNN-based model, which utilizes the collaborative information by collaborative information propagation through users' historical interactions and it also apply a solution to combine collaborative information with knowledge information.

5.3   Experimental Settings

In our experiments, these three datasets are divided into training, evaluation, and test sets with the proportion of 6:2:2. The following are the two recommendation

---

[5] https://searchengineland.com/library/bing/bing-satori.

Table 2. The result of AUC and F1 in CTR prediction.

Model	Last.FM		Book-Crossing		MovieLens-20M	
	AUC	F1	AUC	F1	AUC	F1
BPRMF	0.756	0.701	0.658	0.611	0.958	0.914
CKE	0.747	0.674	0.676	0.623	0.927	0.874
PER	0.641	0.603	0.605	0.572	0.838	0.792
RippleNet	0.776	0.702	0.721	0.647	0.976	0.927
KGCN	0.802	0.708	0.684	0.631	0.977	0.930
KGNN-LS	0.805	0.722	0.676	0.631	0.975	0.929
KGAT	0.829	0.742	0.731	0.654	0.976	0.928
CKAN	0.842	0.769	0.753	0.673	0.976	0.929
KHNN	**0.856**	**0.785**	**0.764**	**0.687**	**0.986**	**0.944**

scenarios we take into account. (1) As to click-through rate (CTR) prediction, we learn model parameters from the training set. Then we use the trained model to predict the interest of user's about the items in the test set. (2) As to top-K recommendation, the model, which learn model parameters from the training set, is used to choose K items with higher predicted score than other items in the test set. The metrics of AUC and F1 are used in CTR prediction and the metrics of Recall@K is selected in top-K recommendation. We use adam [8] to optimize all models in training. We set the batch size as 1024 in training. We apply the default Xavier initializer [4] to initialize the model's parameters.

We implement our KHNN model in PyTorch[6]. The best hyper-parameters are obtained by grid search. The learning rate is searched in $\{10^{-3}, 5 \times 10^{-3}, 10^{-2}, 5 \times 10^{-2}\}$. The embedding size is tuned among $\{8, 16, 32, 64, 128, 256\}$. The coefficient of L2 normalization is searched in $\{10^{-5}, 10^{-4}, 10^{-3}, 10^{-2}\}$. We search the set size of hyperedge in $\{16, 32, 64, 128\}$. It's note that we set same size of hyperedge for user and item.

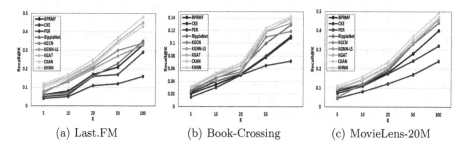

(a) Last.FM   (b) Book-Crossing   (c) MovieLens-20M

Fig. 4. The result of Recall@K in top-K recommendation.

---

[6] https://pytorch.org.

5.4   Performance Comparison (Q1)

The results of CTR prediction and top-K recommendation are presented in Table 2 and Fig. 4, respectively. Analyzing the experimental results, we can draw the following conclusions:

- KHNN shows overwhelming superiority over these state-of-the-art baselines on all three datasets. Specifically, KHNN achieve improvements over the state-of-the-art baselines w.r.t. AUC by 1.7% in Last.FM, 1.5% in Book-Crossing and 1% in MovieLens-20M.
- In CTR prediction and top-K recommendation, the KHNN's performance is excellent. Compared with CKAN, the performance of KHNN demonstrate the effectiveness of the knowledge-aware hypergraph convolution. The results' difference among KHNN, KGCN and KGNN-LS demonstrate the importance of explicitly encoding high-order correlation in knowledge graph.
- It is obvious that all models obtain the highest scores in the experiments on the MovieLens-20M dataset. One possible reason is that the number of average interactions and average links in kg are highest on the MovieLens-20M dataset. Consequently, there is not sufficient interactions and links in the poorer datasets for recommender models to learn the latent vectors specifically in kg-based models.
- According to the results of experiments, KG-based models achieve better performance than CF-based models in most cases. This experimental results illustrate that the usage of KG is helpful to capture underlying historical interaction between users and items.
- According to the results of KG-based models, the GNN-based models achieve better performance than the path-based models on all three datasets. This demonstrates the importance of modeling the high-order connectivity of neighbors in knowledge graph.

5.5   Study of KHNN (Q2)

Effect of Knowledge-Aware Neighborhood Convolution. To verify the impact of knowledge-aware neighborhood convolution, we study two variants of KHNN. For KHNN_{mean-1}, we apply mean aggregator to aggregate embeddings in 1-order neighborhood. For $\text{KHNN}_{mean-all}$, we apply mean aggregator to aggregate embeddings in all neighborhood. In the analyses of data from Table 3. From the results of experiments, we can observe that KHNN consistently perform better than KHNN_{mean-1} and $\text{KHNN}_{mean-all}$. We attribute the improvement to the neighborhood convolution, which is able to model comprehensive correlation among entities in neighborhood well.

Effect of Aggregator Selection in Hyperedge Convolution. To explore the impact of aggregator selection in hyperedge convolution, we apply Sum, Pool and Concat aggregator in KHNN to conduct hyperedge convolution. The results of experiment are shown in Table 4. We can draw the following conclusion:

agg_{concat} is superior to agg_{sum} and agg_{pool}. The results can be owed to that the Concat aggregator is able to retain more information stored in embeddings than Sum and Pool aggregators.

Table 3. The result of AUC w.r.t. effect of knowledge-aware neighborhood convolution.

Category	$KHNN_{mean-all}$	$KHNN_{mean-1}$	agg_{concat}
Music	0.811	0.840	**0.856**
Book	0.738	0.751	**0.764**
Movie	0.964	0.970	**0.986**

Table 4. The result of AUC w.r.t. different aggregators in hyperedge convolution.

Aggregator	agg_{sum}	agg_{pool}	agg_{concat}
Music	0.835	0.844	**0.856**
Book	0.740	0.752	**0.764**
Movie	0.970	0.972	**0.986**

5.6 Hyper-Parameter Study (Q3)

Effect of Number of Hyperedge. To explore how the number of hyperedge affects the performance, We vary the maximal number of hyperedge L of hyperedge construction. What's more, the concat aggregator should use the same number of embeddings used for calculation. However, the number of embeddings in T_u and T_v is different. As a result, we exclude e_v^1 in T_v as a compromise solution. For example, L = 2 means to aggregate e_v^0 and e_v^2 for item hyperedge convolution. Table 5 show the results of experiment with different number of hyperedge. It is obvious that the model achieve best performance when L is 3 in music, 2 in book and 2 in movie. The best results can be owed to the construction and convolution of high-order hyperedge provides sufficient auxiliary knowledge information. However, not only information but also noise are bring when the number of hyperedge is large.

Effect of Size of Hyperedge. We set the size of hyperedge form 16 to 128. The results is presented in Table 6. The best performance of music and book is obtained when the size is taken as 64. For movie recommendation, the best result is obtained when the size is taken as 128. One reasonable explanation is that not only knowledge information but also noise are bring by hyperedge.

Table 5. The result of AUC w.r.t. different number of hyperedge.

Number of Hyperedge	1	2	3	4
Music	0.835	0.842	**0.856**	0.850
Book	0.748	**0.764**	0.753	0.745
Movie	0.971	**0.986**	0.983	0.977

Table 6. The result of AUC w.r.t. different sizes of hyperedge.

Size	16	32	64	128
Music	0.827	0.842	**0.856**	0.848
Book	0.743	0.755	**0.764**	0.752
Movie	0.959	0.968	0.979	**0.986**

6   Conclusion

In this paper, we proposed a Knowledge-aware Hypergraph Neural Network (KHNN) framework. First, the knowledge-aware hypergraph structure is employed for modeling users items and entities in knowledge graph with explicit hybrid high-order correlations. Second, a novel knowledge-aware hypergraph convolution method is proposed to aggregate different knowledge-based neighbors in hyperedge efficiently and support the explicit embedding propagation of high-order correlations in knowledge-aware hypergraph. Our extensive experimental results on three real-world datasets demonstrated the superiority of KHNN over other models. As to future work, we will focus on how to design a better way to aggregate the multiple representations in hyperedge to get better performance.

Acknowledgements. This research was partially supported by NSFC (No. 61876117, 61876217, 61872258, 61728205), ESP of the State Key Laboratory of Software Development Environment, and PAPD of Jiangsu Higher Education Institutions.

References

1. Benson, A.R., Gleich, D.F., Leskovec, J.: Higher-order organization of complex networks. Science **353**(6295), 163–166 (2016)
2. Feng, Y., You, H., Zhang, Z., Ji, R., Gao, Y.: Hypergraph neural networks. In: AAAI, pp. 3558–3565. AAAI Press (2019)
3. Gao, Y., Wang, M., Tao, D., Ji, R., Dai, Q.: 3-D object retrieval and recognition with hypergraph analysis. IEEE Trans. Image Process. **21**(9), 4290–4303 (2012)
4. Glorot, X., Bengio, Y.: Understanding the difficulty of training deep feedforward neural networks. In: AISTATS. JMLR Proceedings, vol. 9, pp. 249–256. JMLR.org (2010)
5. He, X., Liao, L., Zhang, H., Nie, L., Hu, X., Chua, T.: Neural collaborative filtering. In: WWW, pp. 173–182. ACM (2017)

6. Ji, S., Feng, Y., Ji, R., Zhao, X., Tang, W., Gao, Y.: Dual channel hypergraph collaborative filtering. In: KDD, pp. 2020–2029. ACM (2020)
7. Jiang, J., Wei, Y., Feng, Y., Cao, J., Gao, Y.: Dynamic hypergraph neural networks. In: IJCAI, pp. 2635–2641. ijcai.org (2019)
8. Kingma, D.P., Ba, J.: Adam: a method for stochastic optimization. In: ICLR (Poster) (2015)
9. Lin, Y., Liu, Z., Sun, M., Liu, Y., Zhu, X.: Learning entity and relation embeddings for knowledge graph completion. In: AAAI, pp. 2181–2187. AAAI Press (2015)
10. Luo, A., et al.: Collaborative self-attention network for session-based recommendation. In: IJCAI, pp. 2591–2597. ijcai.org (2020)
11. Massa, P., Avesani, P.: Trust-aware recommender systems. In: RecSys, pp. 17–24. ACM (2007)
12. Rendle, S., Freudenthaler, C., Gantner, Z., Schmidt-Thieme, L.: BPR: Bayesian personalized ranking from implicit feedback. In: UAI, pp. 452–461. AUAI Press (2009)
13. Ricci, F., Rokach, L., Shapira, B.: Introduction to recommender systems handbook. In: Ricci, F., Rokach, L., Shapira, B., Kantor, P.B. (eds.) Recommender Systems Handbook, pp. 1–35. Springer, Boston, MA (2011). https://doi.org/10.1007/978-0-387-85820-3_1
14. Sen, S., Vig, J., Riedl, J.: Tagommenders: connecting users to items through tags. In: WWW, pp. 671–680. ACM (2009)
15. Wang, H., Wang, J., Zhao, M., Cao, J., Guo, M.: Joint topic-semantic-aware social recommendation for online voting. In: CIKM, pp. 347–356. ACM (2017)
16. Wang, H., et al.: RippleNet: propagating user preferences on the knowledge graph for recommender systems. In: CIKM, pp. 417–426. ACM (2018)
17. Wang, H., Zhang, F., Xie, X., Guo, M.: DKN: deep knowledge-aware network for news recommendation. In: WWW, pp. 1835–1844. ACM (2018)
18. Wang, H., et al.: Knowledge-aware graph neural networks with label smoothness regularization for recommender systems. In: KDD, pp. 968–977. ACM (2019)
19. Wang, H., Zhao, M., Xie, X., Li, W., Guo, M.: Knowledge graph convolutional networks for recommender systems. In: WWW, pp. 3307–3313. ACM (2019)
20. Wang, X., He, X., Cao, Y., Liu, M., Chua, T.: KGAT: knowledge graph attention network for recommendation. In: KDD, pp. 950–958. ACM (2019)
21. Wang, X., Wang, D., Xu, C., He, X., Cao, Y., Chua, T.: Explainable reasoning over knowledge graphs for recommendation. In: AAAI, pp. 5329–5336. AAAI Press (2019)
22. Wang, Z., Lin, G., Tan, H., Chen, Q., Liu, X.: CKAN: collaborative knowledge-aware attentive network for recommender systems. In: SIGIR, pp. 219–228. ACM (2020)
23. Xu, C., et al.: Long- and short-term self-attention network for sequential recommendation. Neurocomputing **423**, 580–589 (2021)
24. Yu, X., et al.: Personalized entity recommendation: a heterogeneous information network approach. In: WSDM, pp. 283–292. ACM (2014)
25. Zhang, F., Yuan, N.J., Lian, D., Xie, X., Ma, W.: Collaborative knowledge base embedding for recommender systems. In: KDD, pp. 353–362. ACM (2016)
26. Zheng, L., Noroozi, V., Yu, P.S.: Joint deep modeling of users and items using reviews for recommendation. In: WSDM, pp. 425–434. ACM (2017)
27. Zhou, D., Huang, J., Schölkopf, B.: Learning with hypergraphs: clustering, classification, and embedding. In: NIPS, pp. 1601–1608. MIT Press (2006)

Personalized Dynamic Knowledge-Aware Recommendation with Hybrid Explanations

Hao Sun[1], Zijian Wu[1], Yue Cui[1], Liwei Deng[1], Yan Zhao[2], and Kai Zheng[1(✉)]

[1] School of Computer Science and Engineering,
University of Electronic Science and Technology of China, Chengdu, China
{wuzijian,deng_liwei}@std.uestc.edu.cn, zhengkai@uestc.edu.cn
[2] Aalborg University, Aalborg, Denmark
yanz@cs.aau.dk

Abstract. Explainable recommendation is attracting more and more attention in both industry and research communities. While some existing models utilize reviews for improving the performance of recommender systems, most of them assume that user's preference is static and each review's importance is user-independent. However, it is intuitive that user's preference is always dynamically changing and reviews from similar users should be given more importance as they share similar tastes. Moreover, they achieve model explainability at either feature level that is too concise or review level that is too redundant. To deal with these problems, we propose a **P**ersonalized **D**ynamic **K**nowledge-aware **R**ecommender (**PDKR**) for dynamic user modeling and personalized item modeling. In particular, we model user's preference with defined entities and relations in sequential knowledge graphs and capture its dynamics with a novel interval-aware Gated Recurrent Unit (GRU). Furthermore, by leveraging self-attention mechanism, we can not only learn each review's user-specific importance, but also provide tailored explanations for each user at both feature level and review level. We conduct extensive experiments on three benchmark datasets from Amazon and Yelp and the results show that PDKR outperforms all the state-of-the-art recommendation approaches in rating prediction task while providing more effective explanations simultaneously.

Keywords: Recommender system · Knowledge graph · Gated Recurrent Unit · Attention mechanism · Rating prediction

1 Introduction

Explainable recommendation has been attracting increasing attention as previous studies show that it can not only improve user's acceptance of recommended items [8,16], but also improve the transparency, persuasiveness, effectiveness, trustworthiness and satisfaction of recommender systems [1,7,17,20]. To make

© Springer Nature Switzerland AG 2021
C. S. Jensen et al. (Eds.): DASFAA 2021, LNCS 12683, pp. 148–164, 2021.
https://doi.org/10.1007/978-3-030-73200-4_10

recommender systems more explainable, user reviews, which contain rich preference information, have been widely leveraged to improve the explainability of recommendation models. For example, HFT [11] is proposed to explain the recommendation results by linking each dimension of the latent vector with a hidden topic learned from reviews. Following this idea, RBLT [15] achieves recommendation explanations by showing the topic words that have the highest recommendability score learned from a unified semantic space. NARRE [3] utilizes an attention mechanism to select the most useful reviews as the explanation. DER [5] designs a "user-aware" attention network to select the most important sentence as the explanation.

Despite their improvement in explainability, these methods suffer from some inherent limitations. To begin with, while most of them take reviews as their model input, they fail to fully exploit side information, such as user's behavior patterns and item's attributes hidden in these reviews, and thus they would lead to poor performance when the data is sparse. More importantly, most of them represent each user as a static vector and do not take user's preference dynamics into consideration. It is worth mentioning that though some RNN-based methods including DER aim to capture user's dynamic preferences by modeling each item chronologically, we believe that modeling user's preference based on each single discrete item may not be robust since a single item usually cannot reflect user's preference and would introduce occasionality into the model. Besides, these models assume that each item has a global vector suitable for all users. However, considering different users may have very different tastes, and thus they may care about different attributes of the same item, we believe that each item's vector should contain more attribute information that the target user cares about. Therefore, it is intuitive to give higher weight to the reviews of the users who are similar to the target user and obtain a user-specific item representation.

To tackle the problems mentioned above, we design a novel recommender framework based on Knowledge Graph (KG), namely Personalized Dynamic Knowledge-aware Recommender (PDKR), for dynamic user modeling and personalized item modeling, while providing hybrid explanations at both feature level and review level. Firstly, we construct sequential knowledge graphs and extract triples that can reflect user's behavior patterns and item's attributes contained in each sequential knowledge graph. Then, we design a multi-layer graph attention network to capture user's short-term preference reflected in each sequential knowledge graph. Next, we capture user's dynamic preferences by adopting Gated Recurrent Units (GRUs) as the basic architecture to incorporate user's short-term preference and long-term preference. Though GRU is good at modeling sequential information, it does not consider the time interval information which is a very important signal since user's preference is relatively stable within a short period of time but can change dramatically when the time gap is large. To overcome this limitation, we incorporate time interval into the reset gate and the update gate to better distinguish user's historical and current preference. It is noteworthy that different from traditional RNN-based methods, we model user's dynamic preferences based on each sequence of items rather than each single discrete item. We further construct a self-attention layer, which could automatically learn the similarity between users, in order to make item representations more personalized. Finally, we improve the explainability of the recommendation results by showing selected reviews from the most similar users (review-level explanation) with featuring words highlighted (feature-level explanation).

The main contributions of this work are summarized as follows:

- We design a novel KG-based framework that leverages rich side information to achieve dynamic user modeling and user-specific item modeling simultaneously. To the best of our knowledge, it is the first attempt to combine knowledge graph and GRUs to capture user's dynamic preferences and address the data sparsity problem in rating prediction tasks.
- We modify the original GRU architecture by introducing time interval to make it aware of the user's preferences evolving with time. We also leverage a self-attention mechanism to learn the similarity between users and obtain user-specific item representation. By combining review-level and feature-level explanations, we can improve the effectiveness and satisfaction of the recommendation.
- We conduct extensive experiments on three benchmark datasets to evaluate the performance of our model. The favorable results verify our expectation that the proposed framework can reach a high prediction accuracy and provide more effective explanation at the same time.

2   Related Work

2.1   Explainable Recommendation

Explainable recommendation models have proved to be quite useful in ranges of applications. Many techniques have been proposed to improve the explainability of recommendation models. Early works [4,11,21] mainly focus on improving the explainability of matrix factorization models and collaborative filtering models while preserving their accuracy. For example, EFM [21] extends Matrix Factorization by aligning latent dimensions with explicit features and provides feature-level explanations. However, feature-level explanations can only provide very limited information and sometimes lead to ambiguity without sufficient contextual information. Recently, with the advance of deep learning, deep models [3,5,14] have been proposed to improve the accuracy and explainability of recommendation simultaneously. Basically, in these models, side information like reviews is utilized to enhance the explainability. NARRE [3], D-Attn [14] and DER [5] automatically learn the importance of different sentences and select the most important sentences as the review-level explanation of the recommendation. Although review-level explanation can provide more detailed description of items, sometimes they contain too much redundant information which is useless and distracting, thus affecting the quality of explanations.

To address the problems mentioned above, our model provides review-level explanations while highlighting the featuring words with their sentiment polarity suggested in corresponding colors. Therefore, users can choose to look at the featuring words only or refer to the review context when they cannot fully understand the featuring words.

2.2 Knowledge Graph Based Recommendation

Knowledge Graph (KG) based recommendation is attracting more and more attention since KG contains rich information about users and items, which can help enhance recommendation performance and address the problem of data sparsity. In general, existing KG-based recommendation studies can be divided into two categories. The first category is embedding-based methods. For example, Collaborative Knowledge based Embedding (CKE) [19] combines collaborative filtering with auxiliary knowledge embeddings such as text embedding and image embedding. The second category is path-based methods which leverage path information for recommendation. For example, Personalized Entity Recommendation (PER) [18] treats each knowledge graph as a heterogeneous information network and extracts meta-path-based latent features to represent the connectivity between users and items.

Although these models achieve promising results, most of them do not take user's preference dynamics into consideration. To address this issue, we aim to model user's dynamic preferences by chronologically organizing knowledge graphs as a sequence and utilizing a GRU architecture to incorporate user's historical and current preference.

3 Preliminaries

To formulate the problem, we first introduce the definitions of interaction sequence and sequential knowledge graph.

Definition 1 (Interaction Sequence). *Each interaction $o = (u, v, r, w, t)$ includes user ID u, item ID v, rating r, review w and time t, which denotes that user u interacts with item v at time t by rating r and review w. Given a historical interaction set $O = [o_1, o_2, .., o_{|O|}]$ of user u and sequence length λ, an interaction sequence $s_i = [o_{i,1}, o_{i,2}, ..., o_{i,\lambda}]$ is a chronologically-ordered continuous subset of O. The set of interaction sequences of user u is represented as $S_u = [s_1, s_2, ..., s_m]$, where m is the number of interaction sequences.*

Definition 2 (Sequential Knowledge Graph). *Given an interaction sequence s_i, a sequential knowledge graph denoted as $\mathcal{G}_i = (\mathcal{E}, \mathcal{R})$ can be constructed by extracting defined entities and relations from s_i, where $\mathcal{E}$ is the set of entities and $\mathcal{R}$ is the set of relations. All sequential knowledge graphs constructed from S_u is represented as $G_u = \{\mathcal{G}_1, \mathcal{G}_2, ..., \mathcal{G}_m\}$.*

Problem Statement. Given a set of users U and their historical interactions in the training set D_{tr}, we aim to accurately predict the rating r that a user u will assign to item v by learning the embeddings of users and items, and provide effective explanation at both review level and feature level.

Sequential Knowledge Graph Construction. This paper aims to leverage reviews for product recommendation, which usually provides three kinds of entities including user entity, item entity and feature entity. To capture user behavior patterns and item attributes, we convert the interaction between these entities into the following four groups of triplet facts, which are illustrated in Fig. 1.

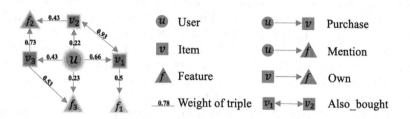

Fig. 1. One example of the sequential knowledge graph

(1) $(u, Purchase, v)$ represents the interaction that user u bought item v. The weight of the relation is defined as the rating that user u assigned to item v.

(2) (v, Own, f) represents the interaction that item v owns feature f. The weight of the relation is defined as the sentiment polarity that the user expressed on the feature, where 1 means the user likes the feature while 0.5 means the user dislikes the feature[1].

(3) $(u, Mention, f)$ represents the interaction that user u mentioned feature f more than once in her review history. The weight of the relation is defined as the frequency that user u mentioned feature f.

(4) $(v_1, AlsoBought, v_2)$ represents the interaction that item v_1 and item v_2 were bought together in one transaction more than once. The weight of the relation is defined as the frequency that the two items appeared in the same transaction.

An advantage of defining these relations is that user's behavior patterns and item attributes can be incorporated into a sequential knowledge graph in a unified manner. More specifically, by defining relation $Mention$ and $AlsoBought$, user's special preferences on some features and item's interdependence with each other can be captured. Besides, the strength of relations can be reflected by the weights of them. Note that other kinds of entities (such as categories and

---

[1] The sentiment polarity value should have been 1 (positive) or −1 (negative). We modify the negative value -1 to 0.5, which can be seen as how well the item performs on the feature.

brands) and their corresponding relations can also be added to capture specific user behavior patterns and item attributes.

All the weights are normalized within their relation types. With the entities and relations defined above, we can now construct sequential knowledge graphs by extracting triples (e_h, r, e_t) from each interaction sequence s_i, where e_h represents the head entity, e_t represents the tail entity and r represents the relation between them.

4   The Proposed Model

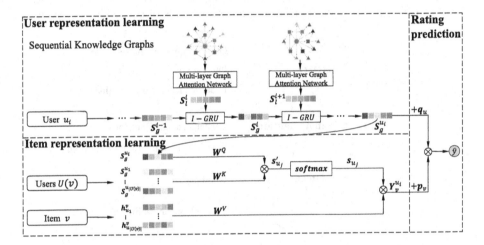

Fig. 2. Framework overview

As shown in Fig. 2, our proposed Personalized Dynamic Knowledge-aware Recommender (PDKR) consists of three parts: user representation learning, item representation learning and rating prediction. In the user representation learning part, we first design a multi-layer graph attention network to learn user's local sequence representation which represents user's short-term preference reflected in the current session. Then, we re-design the architecture of GRU to incorporate user's long-term preference reflected in previous sessions and user's short-term preference reflected in the current session, and obtain user's global sequence representation eventually. In the item representation learning part, we leverage a self-attention layer for user-specific item representation learning. Finally, we combine user representation and item representation to obtain the predicted rating in the rating prediction part.

4.1   Local Sequence Representation Learning

As illustrated in Fig. 3, given user u's sequential knowledge graphs $G_u = \{\mathcal{G}_1, \mathcal{G}_2, ..., \mathcal{G}_m\}$, we first update the embeddings of entities and relations, and then aggregate all the entity embeddings in the sequential knowledge graph $\mathcal{G}_i$ to obtain user u's local sequence representation S_l^i which represents user u's short-term preference reflected in the interaction sequence s_i.

Entity Embedding Update. To distinguish user entity from other kinds of entities, we first transform user entity embeddings and other kinds of entity embeddings into different latent space before the graph attention layer by applying the following equation:

$$h_i^{user/else} = W_1^{user/else} x_i, \tag{1}$$

where $x_i \in \mathbb{R}^d$ is the embedding of entity e_i, $W_1^{user}, W_1^{else} \in \mathbb{R}^{d \times d}$ are the transformation matrices of user entity and other kinds of entities respectively.

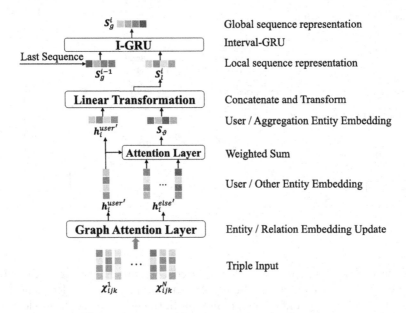

Fig. 3. Structure of user representation learning

In graph attention layer, we first calculate each triple's relative attention value, and then we concatenate the outputs of M independent attention layers where in each layer we calculate the weighted average of triple embedding separately:

$$\chi_{ijk} = W_2[h_i \| h_j \| g_k]$$
$$c_{ijk} = w_{ijk} LeakyReLU(W_3 \chi_{ijk})$$
$$\alpha_{ijk} = softmax(c_{ijk}) = \frac{exp(c_{ijk})}{\sum_{n \in N_i} \sum_{r \in R_{in}} exp(c_{inr})} \tag{2}$$
$$h_i' = \|_{m=1}^{M} \sigma(\sum_{j \in N_i} \alpha_{ijk}^m \chi_{ijk}^m),$$

where h_i and h_j denote embeddings of entities e_i and e_j respectively, χ_{ijk} denotes the triple (e_i, r_k, e_j), w_{ijk} denotes the weight of the relation in the triple (e_i, r_k, e_j), N_i denotes the neighbors of entity e_i and R_{in} denotes the set of relations between entity e_i and entity e_n. As a common practice of multi-head attention models, we employ averaging instead of concatenating in the final layer to get the final embedding of entities.

Relation Embedding Update. Similar to [13], linear transformation is then performed on the relation embedding matrix G to get the final relation embedding.

$$G^{'} = GW_R \qquad (3)$$

Generating Local Sequence Representation. Considering that different entities in the sequential knowledge graph should have different levels of priority and hence different weights, we further adopt the soft-attention mechanism to better model user's preference reflected in her interacted entities.

$$\beta_i = q^T \sigma(W_4 h^{user'} + W_5 h_i^{else'} + \omega)$$

$$S_\vartheta = \sum_{i=1}^{n} \beta_i h_i^{else'}, \qquad (4)$$

where $q \in \mathbb{R}^d$, $h_i^{else'}$ is the updated embedding of entities except for user entity, $h^{user'}$ is the updated embedding of user entity and n is the number of all entities involved in the sequential knowledge graph except for user entity.

Finally, we compute the local sequence representation of user u by taking linear transformation over the concatenation of user embedding and the aggregation of other entity embeddings $S_\vartheta{}^2$:

$$S_l = W_6[h^{user'} \| S_\vartheta] \qquad (5)$$

Sequential Knowledge Graph Learning. To better learn entity and relation embeddings, we employ TransE [2], which learns embeddings such that for a given valid triple (e_i, r_k, e_j), the condition $h_i + g_k \approx h_j$ holds. In particular, we try to learn entity and relation embeddings to minimize the L1-norm dissimilarity measure given by $g(i, k, j) = \|h_i + g_k - h_j\|_1$. The loss function of sequential knowledge graph learning is defined as follows:

$$\mathcal{L}_{KG} = \sum_{(i,k,j,j^{'}) \in \mathcal{T}} max\{g(i, k, j) - g(i, k, j^{'}) + \gamma, 0\}, \qquad (6)$$

where $\mathcal{T} = \{(i, k, j, j^{'}) | (i, k, j) \in \mathcal{G} \wedge (i, k, j^{'}) \notin \mathcal{G}\}$ and $(i, k, j^{'})$ is a negative triple (that does not exist in the sequential knowledge graph) constructed by replacing one entity in a valid triple randomly and $\gamma > 0$ is a margin hyper-parameter.

---

[2] For simplicity, the label for discriminating different users and sequences is omitted.

4.2   Global Sequence Representation Learning

Intuitively, a user's preference is relatively stable within a short period of time but could change dramatically after a long time. Therefore, it is necessary to take the time interval into consideration while modeling user's dynamic preferences. To prevent the model from being too complicated, we adopt GRU [6] as the basic architecture. However, it does not take the time interval information into consideration. So we re-design the architecture of GRU by simply incorporating time interval into the reset gate r_t and the update gate z_t. The newly designed GRU (called I-GRU) can learn when to infuse more historical information with its gate mechanism. The overall architecture of I-GRU is shown in Fig. 4.

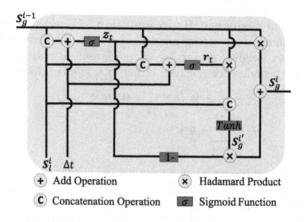

Fig. 4. The architecture of the new-designed I-GRU

Given user's previous global sequence representation S_g^{i-1} and current local sequence representation S_l^i, the current global sequence representation S_g^i can be calculated as follows:

$$
\begin{aligned}
z_t &= \sigma(W_z[S_l^i, S_g^{i-1}] - \Delta t w_t) \\
r_t &= \sigma(W_r[S_l^i, S_g^{i-1}] - \Delta t w_t) \\
S_g^{i'} &= Tanh(W_s[S_l^i, r_t \odot S_g^{i-1}]) \\
S_g^i &= z_t \odot S_g^{i-1} + (1 - z_t) \odot S_g^{i'},
\end{aligned}
\tag{7}
$$

where $w_t \in \mathbb{R}^d$, $\Delta t = t_i - t_{i-1}$ and t_i simply averages the interaction time of the sequence s_i. Therefore, when the time interval Δt is larger, which leads to smaller z_t and r_t, the influence of historical information S_g^{i-1} (long-term preference) is lowered by multiplying z_t and r_t, and the current input S_l^i (short-term preference) becomes the predominant factor, and vice versa.

4.3   Item Representation Learning

Traditional item modeling methods assume that each item has a global representation suitable for all users. However, considering different users may focus on different features for the same item, we believe that each item should have a user-specific representation for each user. Besides, it is intuitive that if user u_j is similar to target user u_i, then user u_j's review should be paid more attention when recommending items to target user u_i. To seamlessly integrate this observation into our model, we design a novel item representation learner adapted from self-attention, which could automatically identify the importance of different users' reviews.

To begin with, for each target user u_i, a query vector is defined as the linear transformation of the global sequence representation of u_i, which is denoted as $S_g^{u_i}$.

$$Q_{u_i} = W^Q S_g^{u_i} \tag{8}$$

Given an item v, we can get the set of users $U(v)$ who have reviews on item v. For each user $u_j \in U(v)$, we can calculate her key vector as follows:

$$K_{u_j} = W^K S_g^{u_j} \tag{9}$$

Similarly, for each user $u_j \in U(v)$, we can calculate her value vector by performing a linear transformation on her entity embedding of item v.

$$V_{u_j} = W^V h_{u_j}^v \tag{10}$$

Obtaining the query vector of the target user u_i and the key vector of user u_j, we can get the absolute attention value of u_j as follows:

$$s_{u_j}' = K_{u_j}^T Q_{u_i} \tag{11}$$

Softmax is applied on s_{u_j}' to get the relative attention value of each user u_j, which can be seen as the similarity between target user u_i and user u_j. It is worth mentioning that the review of the user who is the most similar to the target user u_i will be selected as the review-level explanation, which will be discussed later in the explainability study section.

$$s_{u_j} = softmax(s_{u_j}') = \frac{exp(s_{u_j}')}{\sum_{u_n \in U(v)} exp(s_{u_n}')} \tag{12}$$

To encapsulate more information contained in the item reviews, M independent attention layers are applied to calculate the item embedding, followed by an averaging layer to obtain the final user-specific item representation.

$$Y_v^{u_i} = \sigma\left(\frac{1}{M} \sum_{m=1}^{M} s_{u_j}^m V_{u_j}^m\right) \tag{13}$$

4.4 Prediction

Latent Factor Model (LFM) [9] aims to predict the rating by multiplying user's representation vector and item's representation vector based on matrix factorization. We extend it by introducing auxiliary embeddings q_u and p_v for more robust prediction. Specifically, the predicted rating is calculated as :

$$\widehat{R}_{u,v} = W_7^T(q_u + S_g^u) \odot (p_v + Y_v^u) + b_u + b_v + \mu, \tag{14}$$

where S_g^u and Y_v^u are user's global sequence representation vector and item's representation vector respectively, b_u, b_v and μ denote user bias, item bias and global bias respectively.

4.5 Training

Since the goal of this paper is rating prediction, which is actually a regression problem, we adopt Root Mean Square Error (RMSE) as the loss function:

$$\mathcal{L}_{rec} = \sum_{(u,v)\in\mathcal{D}_{tr}} (\widehat{R}_{u,v} - R_{u,v})^2, \tag{15}$$

where $\mathcal{D}_{tr}$ denotes the set of user-item pairs in the training set.

Finally, we have the objective function to optimize Eq. 6 and 15 jointly as follows:

$$\mathcal{L}_{PDKR} = \mathcal{L}_{rec} + \mathcal{L}_{KG} + \xi\|\Theta\|_2^2, \tag{16}$$

where Θ is the set of model parameters to be regularized.

5 Experiments

5.1 Datasets

In our experiments, we use three publicly available real-world datasets in different domains to evaluate our models and baselines. Two datasets are from Amazon[3]: **Movies_and_TV** and **Cell_Phones_and_Accessories**, which contain user reviews and ratings of products in movie entertainment category and electronic product category respectively. Another dataset is **Yelp**[4] which is a large-scale dataset consisting of restaurant reviews and ratings. Amongst them, Yelp is the largest dataset that contains more than 2.1 million reviews while Cell_Phones_and_Accessories is the smallest dataset which contains about 106 thousand reviews. These datasets are selected to cover different domains and scales, which can demonstrate the robustness of our model. Besides, we adopt the state-of-the-art approach described in [10,22] to extract features from reviews.

For each user interaction sequence, we divide them into 6:2:2 for training, validation and testing respectively. To ensure the KG quality, we filter out users and items with less than 20 ratings. The characteristics of our datasets are shown in Table 1.

---

[3] http://deepyeti.ucsd.edu/jianmo/amazon/.

[4] https://www.kaggle.com/yelp-dataset/yelp-dataset/data.

Table 1. Basic statistics of evaluation datasets

Datasets	#user	#items	#features	#reviews
Phones	4696	35011	697	106148
Movies	30161	41245	1987	1235945
Yelp	47944	58870	3660	2148994

5.2 Baselines and Implementation Details

To evaluate the performance of our proposed PDKR model, five state-of-the-art methods are selected as baselines:

- **PMF** [12] is a traditional matrix factorization method that uses Gaussian distribution to model the latent factors for users and items.
- **HFT** [11] is a traditional topic modeling based method that projects user's latent vector into the latent topic space with Latent Dirichlet Allocation (LDA).
- **DeepCoNN** [23] is a deep recommendation model that jointly models user behaviors and item properties using textual reviews. Authors have shown that it outperformed other topic modeling based methods.
- **NARRE** [3] is a state-of-the-art explainable recommendation method that first introduces a novel attention mechanism to explore the usefulness of reviews.
- **DER** [5] is a state-of-the-art explainable method that introduces time-aware GRU to model user's dynamic preferences and provide review-level explanation.

We implement our PDKR model using Pytorch. In our model, the batch size is fixed at 64, the learning rate is initialized as 5×10^{-5}, the margin hyper-parameter γ is initialized as 1 and the weight decay parameter ξ is initialized as 1×10^{-5}. The embedding size d is chosen in the range of $\{50, 100, 150, 200\}$ and the sequence length λ is chosen in the range of $\{3,5,7,9,11\}$. All the algorithms are implemented on an Intel(R) Xeon(R) CPU E5-2650 v4 @ 2.20 GHz and four GeForce GTX 2080 GPUs.

5.3 Evaluation Metric

To evaluate the performance of all methods, we select RMSE as the evaluation metric, which is widely used for rating prediction in recommender systems. Given a predicted rating $\widehat{R}_{u,v}$ and a ground-truth rating $R_{u,v}$ from user u to item v, the RMSE score is calculated as:

$$RMSE = \sqrt{\frac{1}{|\mathcal{D}_{ts}|} \sum_{(u,v) \in \mathcal{D}_{ts}} (\widehat{R}_{u,v} - R_{u,v})^2}, \qquad (17)$$

where $\mathcal{D}_{ts}$ is the set of user-item pairs in the testing set.

5.4   Evaluation on Rating Prediction

Table 2. Performances for all methods in terms of RMSE

Dataset	Phones	Movies	Yelp
PMF	1.2164	1.2980	1.3716
HFT	1.0525	1.1528	1.2401
DeepCoNN	0.9546	1.0486	1.1637
NARRE	0.9365	1.0323	1.1571
DER	0.9278	1.0331	1.1404
PDKR	**0.8371**	**0.9879**	**1.0492**
Improvement of PDKR	9.8%	4.4%	8.0%

Overall Performance Comparison. From the results shown in Table 2, we can see that: the simple PMF method performs the worst in all methods because it is a traditional collaborative filtering model which only takes the rating information into consideration and it fails to capture user's dynamic preferences. While HFT, DeepCoNN and NARRE all take review information as the model input, NARRE performs the best. This is because NARRE leverages a novel attention mechanism to identify more useful reviews, hence the impact of valuable reviews can be enhanced while the influence of noise from useless reviews can be reduced. DER performs better than NARRE because it not only takes the usefulness of reviews into account, but also leverages T-GRU to capture user's dynamic preferences. Our model PDKR achieves 9.8%, 4.4%, 8.0% relative improvement on Cell_Phones_and_Accessories, Movies_and_TV and Yelp respectively over DER, which demonstrates the superiority of our model. In our model, we not only fully exploit review information with the defined entities and relations in sequential knowledge graphs, but also propose a simple but powerful GRU variant to capture user's dynamic preferences. Moreover, by leveraging self-attention mechanism, we can obtain more user-specific item representation for each item. Therefore, the latent representations for both users and items are more accurate and eventually lead to higher accuracy in rating prediction.

Effect of Sequence Length. Sequence length λ is an important hyperparameter that directly influences how much information can be incorporated into a sequential knowledge graph and eventually influences RMSE. Due to space limit, we only report the results on Cell_Phones_and_Accessories dataset and Yelp dataset.

The sequence length is tuned in the range of {3,5,7,9,11}, and 5,7 is chosen to be the default value of λ on Cell_Phones_and_Accessories dataset and Yelp dataset respectively. From the results shown in Fig. 5, we can observe that both too high sequence length and too low sequence length would affect the model performance negatively. To explain, when the number of entities in a sequential knowledge graph is too large, it is difficult to find a suitable representation of

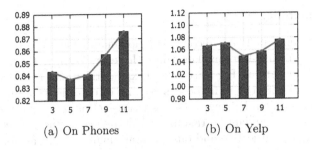

(a) On Phones   (b) On Yelp

Fig. 5. Effect of sequence length

the whole graph which contains too much information. On the other hand, when few entities are used, the representation of a sequential knowledge graph cannot reflect user's preference, which would lead to poor performance.

Effect of Data Sparsity. To simulate the effect of data sparsity, We randomly reduce the training data by a ratio of 5%, 10%, 15% and 20%.

Table 3. Effect of data sparsity on Cell_Phones_and_Accessories and Yelp

Mask	Model					
	Cell_Phones_and_Accesories					
	PMF	HFT	DeepConn	NARRE	DER	PDKR
0	1.2164	1.0525	0.9546	0.9365	0.9278	**0.8371**
5	1.2179	1.0748	0.9756	0.9464	0.9428	**0.8451**
10	1.2189	1.1037	0.9849	0.9541	0.9618	**0.8752**
15	1.2255	1.1102	1.0035	0.9827	0.9939	**0.8852**
20	1.2266	1.1583	1.0165	1.0105	1.0331	**0.9095**
Mask	Model					
	Yelp					
	PMF	HFT	DeepConn	NARRE	DER	PDKR
0	1.3716	1.2401	1.1637	1.1571	1.1404	**1.0492**
5	1.3809	1.2679	1.1715	1.1654	1.1491	**1.0548**
10	1.4121	1.2833	1.1817	1.1838	1.1664	**1.0627**
15	1.4247	1.3020	1.2081	1.1944	1.1792	**1.0802**
20	1.4592	1.3038	1.2313	1.2009	1.1799	**1.0928**

As shown in Table 3, when the data is sparser, all methods' accuracies decrease. However, PDKR still performs the best among all methods, which indicates the robustness of our model. One possible explanation is that, with the use of sequential knowledge graphs, we can explore more valuable information from user's reviews, which can remedy the problem of data sparsity to some extent.

5.5   Explainability Study

Previous works on explainable recommendation mainly focus on providing either feature-level or review-level explanations. However, both of them suffer from some inherent limitations. For feature-level explanations, simply providing a feature the user is interested in is too concise and often causes confusion without contextual information, leading to low user satisfaction. For review-level explanations, it is too redundant to take the whole review as the explanation since reviews are highly subjective and often contain many useless personal feelings, leading to low effectiveness. In this paper, we combine review-level explanations and feature-level explanations. In particular, we select the most important reviews of (dis)recommended item v for target user u_i, and highlight the featuring words and corresponding sentiment words in the reviews.

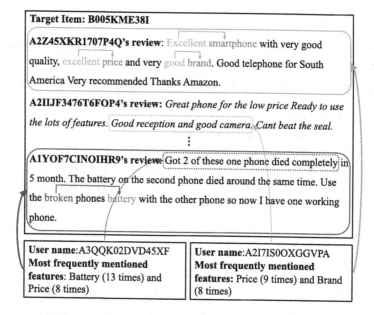

Fig. 6. The example of explanations generated for two different users on the same item. The most important review learned by NARRE is labeled by italic. The reviews from the most similar users selected by PDKR are presented in full line boxes while the sentences selected by DER are presented in dotted line boxes. Green and red suggest the explanation for why the target item is recommended and disrecommended respectively. The target user and corresponding explanation are linked by a directed arrow. (Color figure online)

To showcase the superiority of our model's explainability, we present an example in Fig. 6. Specifically, we compare the explanations of a target item generated by NARRE, DER and our model PDKR for different users. As we can see, NARRE highlights the same review for different users due to its user-independent attention mechanism while our model can provide personalized review information, which is more practical and reasonable in real scenarios. In the explanation generated by PDKR, featuring words are highlighted in blue while opinion words are highlighted in green when the sentiment is positive and in red otherwise. It should be noted that highlighting the featuring words gives users more options. For example, when the featuring word is clear enough as shown in user A2Z45XKR1707P4Q's review, user can choose to ignore the redundant body of the review. However, when the featuring word is confusing as shown in user A1YOF7CINOIHR9's review, user can refer to the context of the review to better understand why the battery is described by the word "broken", and thus both effectiveness and satisfaction of the recommendation can be improved. Though DER selects different sentences for different users, it does not exhibit such capability due to its lack of contextual information and would sometimes lead to confusion as shown in the explanation it generated for target user A3QQK02DVD45XF in the red dotted line box.

6 Conclusion

In this paper, we propose a novel knowledge-aware framework named PDKR, which models user's dynamic preferences and explores reviews' user-specific usefulness simultaneously in the context of explainable recommendation. Extensive experiments have been conducted on three large real-world datasets from Amazon and Yelp. The proposed PDKR outperforms all the state-of-the-art recommendation models in rating prediction task. Explainability study shows that we can provide personalized explanations for target user on each (dis)recommended item, which improves both effectiveness and satisfaction of the recommendation.

As for future work, we will investigate the potential of incorporating visual information into knowledge graphs for better recommendation performance and more comprehensive recommendation explanations.

Acknowledgements. This work is supported by NSFC (No. 61972069, 61836007, 61832017) and Sichuan Science and Technology Program under Grant 2020JDTD0007.

References

1. Bilgic, M., Mooney, R.J.: Explaining recommendations: satisfaction vs. promotion. In: Beyond Personalization Workshop, IUI, vol. 5, p. 153 (2005)
2. Bordes, A., Usunier, N., Garcia-Duran, A., Weston, J., Yakhnenko, O.: Translating embeddings for modeling multi-relational data. In: NIPS, pp. 2787–2795 (2013)
3. Chen, C., Zhang, M., Liu, Y., Ma, S.: Neural attentional rating regression with review-level explanations. In: WWW, pp. 1583–1592 (2018)
4. Chen, X., Qin, Z., Zhang, Y., Xu, T.: Learning to rank features for recommendation over multiple categories. In: SIGIR, pp. 305–314 (2016)
5. Chen, X., Zhang, Y., Qin, Z.: Dynamic explainable recommendation based on neural attentive models. In: AAAI, vol. 33, pp. 53–60 (2019)

6. Cho, K., et al.: Learning phrase representations using RNN encoder-decoder for statistical machine translation. arXiv preprint arXiv:1406.1078 (2014)
7. Cramer, H., et al.: The effects of transparency on trust in and acceptance of a content-based art recommender. User Model. User-Adap. Inter. **18**(5), 455 (2008). https://doi.org/10.1007/s11257-008-9051-3
8. Herlocker, J.L., Konstan, J.A., Riedl, J.: Explaining collaborative filtering recommendations. In: CSCW, pp. 241–250 (2000)
9. Koren, Y., Bell, R., Volinsky, C.: Matrix factorization techniques for recommender systems. Computer **42**(8), 30–37 (2009)
10. Lu, Y., Castellanos, M., Dayal, U., Zhai, C.: Automatic construction of a context-aware sentiment lexicon: an optimization approach. In: WWW, pp. 347–356 (2011)
11. McAuley, J., Leskovec, J.: Hidden factors and hidden topics: understanding rating dimensions with review text. In: RecSys, pp. 165–172 (2013)
12. Mnih, A., Salakhutdinov, R.R.: Probabilistic matrix factorization. In: NIPS, pp. 1257–1264 (2008)
13. Nathani, D., Chauhan, J., Sharma, C., Kaul, M.: Learning attention-based embeddings for relation prediction in knowledge graphs. arXiv preprint arXiv:1906.01195 (2019)
14. Seo, S., Huang, J., Yang, H., Liu, Y.: Interpretable convolutional neural networks with dual local and global attention for review rating prediction. In: RecSys, pp. 297–305 (2017)
15. Tan, Y., Zhang, M., Liu, Y., Ma, S.: Rating-boosted latent topics: understanding users and items with ratings and reviews. In: IJCAI, vol. 16, pp. 2640–2646 (2016)
16. Tintarev, N., Masthoff, J.: A survey of explanations in recommender systems. In: ICDE, pp. 801–810 (2007)
17. Tintarev, N., Masthoff, J.: Designing and evaluating explanations for recommender systems. In: Ricci, F., Rokach, L., Shapira, B., Kantor, P.B. (eds.) Recommender Systems Handbook, pp. 479–510. Springer, Boston, MA (2011). https://doi.org/10.1007/978-0-387-85820-3_15
18. Yu, X., et al.: Personalized entity recommendation: a heterogeneous information network approach. In: WSDM, pp. 283–292 (2014)
19. Zhang, F., Yuan, N.J., Lian, D., Xie, X., Ma, W.Y.: Collaborative knowledge base embedding for recommender systems. In: SIGKDD, pp. 353–362 (2016)
20. Zhang, Y., Chen, X.: Explainable recommendation: a survey and new perspectives. arXiv preprint arXiv:1804.11192 (2018)
21. Zhang, Y., Lai, G., Zhang, M., Zhang, Y., Liu, Y., Ma, S.: Explicit factor models for explainable recommendation based on phrase-level sentiment analysis. In: SIGIR, pp. 83–92 (2014)
22. Zhang, Y., Zhang, H., Zhang, M., Liu, Y., Ma, S.: Do users rate or review? Boost phrase-level sentiment labeling with review-level sentiment classification. In: SIGIR, pp. 1027–1030 (2014)
23. Zheng, L., Noroozi, V., Yu, P.S.: Joint deep modeling of users and items using reviews for recommendation. In: WSDM, pp. 425–434 (2017)

Graph Attention Collaborative Similarity Embedding for Recommender System

Jinbo Song[1](✉), Chao Chang[2], Fei Sun[3], Zhenyang Chen[2], Guoyong Hu[2], and Peng Jiang[2]

[1] The Institute of Computing Technology of the Chinese Academy of Sciences, Beijing, China
songjinbo18s@ict.ac.cn
[2] Beijing Kuaishou Technology Co., Ltd., Beijing, China
{changchao,chenzhenyang,huguoyong,jiangpeng}@kuaishou.com
[3] Alibaba Group, Hangzhou, China
ofey.sf@alibaba-inc.com

Abstract. We present Graph Attention Collaborative Similarity Embedding (GACSE), a new recommendation framework that exploits collaborative information in the user-item bipartite graph for representation learning. Our framework consists of two parts: the first part is to learn explicit graph collaborative filtering information such as user-item association through embedding propagation with attention mechanism, and the second part is to learn implicit graph collaborative information such as user-user similarities and item-item similarities through auxiliary loss. We design a new loss function that combines BPR loss with adaptive margin and similarity loss for the similarities learning. Extensive experiments on three benchmarks show that our model is consistently better than the latest state-of-the-art models.

Keywords: Recommendation systems · Collaborative Filtering · Graph Neural Networks

1 Introduction

Personalized recommendation plays a pivotal role in many internet scenarios, such as e-commerce, short video recommendations and advertising. Its core method is to analyze the user's potential preferences based on the user's historical behavior, to measure the possibility of the user to select a certain item and to tailor the recommendation results for the user.

One of the major topics to be investigated in the personalized recommendation is Collaborative Filtering (CF) which generates recommendations by taking advantage of the collective wisdom from all users. Matrix factorization (MF) [12] is one of the most popular CF model, which decomposes the interaction matrix between the user and item into discrete vectors and then calculates the inner product to predict the connected edges between the user and item. Neural Collaborative Filtering (NCF) [19] predict the future behavior of users by learning

© Springer Nature Switzerland AG 2021
C. S. Jensen et al. (Eds.): DASFAA 2021, LNCS 12683, pp. 165–178, 2021.
https://doi.org/10.1007/978-3-030-73200-4_11

the historical interactions between users and items. It employs neural network instead of traditional matrix factorization to enhance the non-linearity of the model. In general, there are two key components in learnable CF models—1) embeddings that represent users and items by vectors, and 2) interaction modeling, which formulates historical interactions upon the embeddings.

Despite their prevalence and effectiveness, we argue that these models are not sufficient to learn optimal embeddings. The major limitation is that the embeddings does not explicitly encode collaborative information propagated in user-item interaction graph. Following the idea of representation learning in graph embedding, Graph Neural Networks (GNN) are proposed to collect aggregate information from graph structure. Methods based on GraphSAGE [8] or GAT [23] have been applied to recommender systems. For example, NGCF [17] generates user and item embeddings based on the information propagation in the user-item bipartite graph; KGAT [24] adds a knowledge graph and entities attention based on the bipartite graph and uses entity information to more effectively model users and items. Inspired by the success of GNN in recommendation, we build a embedding propagation and aggregating architecture based on attention mechanism to learn the variable weight of each neighbor. The attention weight explicitly represents the relevance of interaction between user and item in bipartite graph.

Another limitation is that many existing model-based CF algorithms leverage only the user-item associations available in user-item bipartite graph. The effectiveness of these algorithms depends on the sparsity of the available user-item associations. Therefore, other types of collaborative relations, such as user-user similarity and item-item similarity, can also be considered for embedding learning. Some works [25,27] exploit higher-order proximity among users and items by taking random walks on the graph. A recent work [26] presents collaborative similarity embedding (CSE) to model direct and in-direct edges of user-item interactions. The effectiveness of these methods lies in sampling auxiliary information from graph to augment the data for representation learning.

Based on the above limitation and inspiration, in this paper, we propose a unified representation learning framework, called Graph Attention Collaborative Similarity Embedding (GACSE). In the framework, the embedding is learned from direct, user-item association through embedding propagation with attention mechanism, and indirect, user-user similarities and item-item similarities through auxiliary loss, user-item similarities in bipartite graph. Meanwhile, we combine adaptive margin in BPR loss [2] and similarity loss to optimize GACSE.

The contributions of this work are as follows:

- We propose GACSE, a graph based recommendation framework that combines both attention propagation & aggregation in graph and similarity embedding learning process.
- To optimize GACSE, we introduce a new loss function, which, to the best of our knowledge, is the first time to combine both BPR loss with adaptive margin and similarity loss for similarity embedding learning.

- We compare our model with state-of-the-art methods and demonstrate the effectiveness of our model through quantitative analysis on three benchmark datasets.
- We conduct a comprehensive ablation study to analyze the contributions of key components in our proposed model.

2 Related Work

In this section, we will briefly review several lines of works closely related to ours, including general recommendation and graph embedding-based recommendation.

2.1 General Recommendation

Recommender systems typically use Collaborative Filtering (CF) to model users' preferences based on their interaction histories [1,3]. Among the various CF methods, item-based neighborhood methods [5] estimate a user's preference on an item via measuring its similarities with the items in her/his interaction history using a item-to-item similarity matrix. User-based neighborhood methods find similar users to the current user using a user-to-user similarity matrix, following by recommending the items in her/his similar users' interaction history. Matrix Factorization (MF) [12,13] is another most popular one, which projects users and items into a shared vector space and estimate a user's preference on an item by the inner product between user's and items' vectors. BPR-MF [2] optimizes the matrix factorization with implicit feedback using a pairwise ranking loss. Recently, deep learning techniques has been revolutionizing the recommender systems dramatically. One line of deep learning based model seeks to take the place of conventional matrix factorization [7,11,19]. For example, Neural Collaborative Filtering (NCF) estimates user preferences via Multi-Layer Perceptions (MLP) instead of inner product.

2.2 Graph Based Recommendation

Another line is to integrate the distributed representations learning from user-item interaction graph. GC-MC [10] employs a graph convolution auto-encoder on user-item graph to solve the matrix completion task. HOP-Rec [15] employs label propagation and random walks on interaction graph to compute similarity scores for user-item pairs. NGCF [19] explicitly encodes the collaborative information of high-order relations by embedding propagation in user-item interaction graph. PinSage [16] utilizes efficient random walks and graph convolutions to generate embeddings which incorporate both graph structure as well as node feature information. Multi-GCCF [9] constructs two separate user-user and item-item graphs. It employs a multi-graph encoding layer to integrate the information provided by the user-item, user-user and item-item graphs.

3   Our Model

In this section, we introduce our proposed model called GACSE. The overall framework is illustrated in Fig. 1. There are three components in the model: (1) an embedding layer that can map users and items from one hot vector to initial embeddings; (2) an embedding propagation layer, which consists of two sub-layers: a warm-up layer that propagates and aggregates graph embeddings with equal weight, and an attention layer that uses attention mechanism to perform non-equal weight aggregation on the embedding of neighboring nodes; and (3) a prediction layer that concatenates embeddings from embedding layer and attention layer, then outputs affinity score between user and item. The following descriptions take user as central node, if there is no special instruction, it is also applicable to item as centre node.

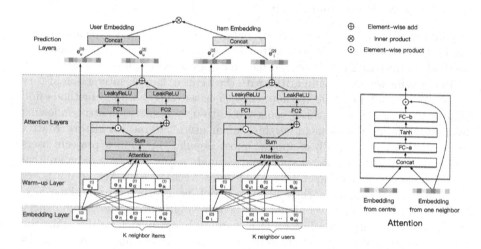

Fig. 1. An illustration of GACSE model architecture. The flow of embedding is presented by the arrowed lines. FC1 and FC2 are shared parameters on user embedding side and item embedding side. An illustration of attention aggregation is shown on the left. FC-a transforms the concatenated vector into a new vector. FC-b transforms vector into attention score.

3.1   Embedding Layer

Embedding layer aims at mapping the ids of user u and item i into embedding vectors $\mathbf{e}_u^{(0)} \in \mathbb{R}^d$ and $\mathbf{e}_i^{(0)} \in \mathbb{R}^d$, where d denotes the embedding dim. We use a trainable embedding lookup table to build our embedding layer for embedding propagation:

$$\mathbf{E} = [\,\overbrace{\mathbf{e}_{u_1}^{(0)}, \cdots, \mathbf{e}_{u_N}^{(0)}}^{\text{users embeddings}}\,,\ \overbrace{\mathbf{e}_{i_1}^{(0)}, \cdots, \mathbf{e}_{i_M}^{(0)}}^{\text{items embeddings}}\,] \tag{1}$$

where N is number of users and M is number of items.

3.2 Embedding Propagation Layer

In order to establish the embedding propagation architecture for collaborative information in graph, we define a embedding propagation layer. Our embedding propagation layer consists of two parts: (1) a warm-up layer and (2) an attention layer. Both layers have two steps: embedding propagation and aggregation.

Warm-Up Layer. To make the model expand the receptive field and grasp the structure of the user-item bipartite graph, we set up a warm-up layer based on the GNN [8,18] embedding propagation architecture.

Warm-Up Propagation. In the warm-up layer, we set all the embeddings to have equal weight. We define warm-up embedding propagation function as:

$$\begin{cases} \mathbf{m}_{i \to u}^{(0)} = \pi_{(u,i)}^{(0)} \mathbf{W}_0 \mathbf{e}_i^{(0)} \\ \mathbf{m}_{u \to u}^{(0)} = \mathbf{W}_0 \mathbf{e}_u^{(0)} \end{cases} \tag{2}$$

where $\mathbf{W}_0 \in \mathbb{R}^{d_1 \times d_0}$ is the trainable weight matrix to distill important information in embedding propagation. d_0 is the dimension of $\mathbf{e}^{(0)}$, and d_1 is the dimension of transformation. $\mathbf{m}_{i \to u}^{(0)}$ is the embedding from item i to user u. $\mathbf{m}_{u \to u}^{(0)}$ is self-connection of u. $\pi_{(u,i)}^{(0)}$ is the weight of the embedding that i passes to u.

In embedding propagation of warm-up layer, we set the propagation weight $\pi_{(u,i)}^0$ to be equal. Inspired by GCN [8], we define weight of each user's interacted item as:

$$\pi_{(u,i)}^{(0)} = \frac{1}{\sqrt{|\mathcal{N}_u||\mathcal{N}_i|}} \tag{3}$$

where $\mathcal{N}_u$ and $\mathcal{N}_i$ denote the first hop neighboring nodes of user u and item i.

Warm-Up Aggregation. After receiving the embeddings from neighbor nodes, we need to aggregate these embeddings. We define embeddings aggregation function in warm-up layer as:

$$\mathbf{e}_u^{(1)} = \sigma\big(\mathbf{m}_{u \to u}^{(0)} + \sum_{i \in \mathcal{N}_u} \mathbf{m}_{i \to u}^{(0)}\big) \tag{4}$$

where σ is nonlinear function such as LeakyReLU. Analogously, we can obtain item i's embedding $\mathbf{e}_i^{(1)}$.

Attention Layer. Next, in order to further encode the variable weight of neighbors, we build an embedding propagation and aggregation architecture based on the attention mechanism. The attention mechanism explicitly captures the relevance of interaction between user and item in bipartite graph.

Attention Propagation. Intuitively, the importance of each item that interacts with the user should be different. We introduce attention mechanism into embedding passing function:

$$\begin{cases} \mathbf{m}_{i \to u}^{(1)} = \pi_{(u,i)}^{(1)} \mathbf{e}_i^{(1)} \\ \mathbf{m}_{u \to u}^{(1)} = \mathbf{e}_u^{(1)} \end{cases} \tag{5}$$

where $\pi_{(u,i)}^{(1)}$ is attention weight.

Inspired by several kinds of attention score functions, we define score function of our model:

$$\text{score}(\mathbf{e}_u^{(1)}, \mathbf{e}_i^{(1)}) = \mathbf{V}^\top \tanh\left(\mathbf{P}\left[\mathbf{e}_u^{(1)} \| \mathbf{e}_i^{(1)}\right]\right) \tag{6}$$

where $\mathbf{V} \in \mathbb{R}^{d_2 \times 1}$ and $\mathbf{P} \in \mathbb{R}^{d_2 \times 2d_1}$ are trainable parameters. d_2 is the dimension of attention transformation. $\|$ denotes concatenate operation. After calculating attention score, we normalize it to get the attention weight via softmax function:

$$\pi_{(u,i)}^{(1)} = \frac{\exp(\text{score}(\mathbf{e}_u^{(1)}, \mathbf{e}_i^{(1)}))}{\sum_{j \in \mathcal{S}_u} \exp(\text{score}(\mathbf{e}_u^{(1)}, \mathbf{e}_j^{(1)}))} \tag{7}$$

where $\mathcal{S}_u$ is a set of user u's one hop neighboring items sampled in this minibatch.

Attention Aggregation. After attention massage passing, the attention aggregation function is defined as:

$$\begin{aligned} \mathbf{e}_u^{(2)} = &\sigma(\mathbf{W}_1(\mathbf{m}_{u \to u}^{(1)} + \sum_{i \in \mathcal{S}_u} \mathbf{m}_{i \to u}^{(1)})) + \\ &\sigma(\mathbf{W}_2(\mathbf{m}_{u \to u}^{(1)} \odot \sum_{i \in \mathcal{S}_u} \mathbf{m}_{i \to u}^{(1)})) \end{aligned} \tag{8}$$

where σ is LeakyReLU non-linear function. $\mathbf{W}_1, \mathbf{W}_2 \in \mathbb{R}^{d_3 \times d_1}$ are trainable parameters. d_3 is the dimension of attention aggregation. $\odot$ denotes element-wise product. Similar to NGCF, the aggregated embedding $\mathbf{e}_u^{(2)}$ does not only related to $\mathbf{e}_u^{(1)}$, but also encodes the interaction between $\mathbf{e}_u^{(1)}$ and $\mathbf{e}_i^{(1)}$. The interaction information can be represented by the element-wise product between $\mathbf{m}_{u \to u}$ and $\sum_{i \in \mathcal{S}_u} \mathbf{m}_{i \to u}$. Analogously, item i's attention layer embedding $\mathbf{e}_i^{(2)}$ can be obtained. We incorporate the attention mechanism to learn variable weight $\pi_{(u,i)}^{(1)}$ for each neighbor's propagated embedding $\mathbf{m}_{i \to u}^{(1)}$.

3.3   Model Prediction

After embedding passing and aggregation with attention mechanism, we obtained two different representations $\mathbf{e}_u^{(0)}$ and $\mathbf{e}_u^{(2)}$ of user node u; also analogous to item node i, we obtained $\mathbf{e}_i^{(0)}$ and $\mathbf{e}_i^{(2)}$. We choose to concatenate the two embeddings as follows:

$$\mathbf{e}_u^* = \mathbf{e}_u^{(0)} \| \mathbf{e}_u^{(2)}, \quad \mathbf{e}_i^* = \mathbf{e}_i^{(0)} \| \mathbf{e}_i^{(2)} \tag{9}$$

where $\|$ denotes the concatenate operation. In this way, we could predict the matching score between user and item by inner product:

$$y_{ui} = \mathbf{e}_u^{*\top} \mathbf{e}_i^* \tag{10}$$

More broadly, we can define the matching score between any two nodes a and b:

$$y_{ab} = \mathbf{e}_a^{*\top} \mathbf{e}_b^* \tag{11}$$

4 Optimization

To optimize the GACSE model, we carefully designed our loss function. Our loss function consists of two basic parts: BPR loss with adaptive margin and similarity loss.

4.1 BPR Loss with Adaptive Margin

We employ BRP loss for optimization, which considers the relative order between observed and unobserved interactions. In order to improve the model's discrimination of similar positive and negative samples, we define BPR loss with adaptive margin as:

$$\mathcal{L}_{\text{BPR}} = \frac{1}{|\mathcal{B}|} \sum_{(u,i,j) \in \mathcal{B}} -\sigma(y_{ui} - y_{uj} - \max(0, y_{ij})) \tag{12}$$

where $\mathcal{B} \subseteq \{(u,i,j)|(u,i) \in \mathcal{R}^+, (u,j) \in \mathcal{R}^-\}$ denotes the sampled data of mini-batch. $\mathcal{R}^+$ denotes observed interactions, and $\mathcal{R}^-$ is unobserved interactions. σ is softplus function. $\max(0, y_{ij})$ indicates that the more similar the positive and negative samples of a node are, the larger the margin of the loss function is.

4.2 Similarity Loss

Other types of collaborative relations, such as user-user similarity and item-item similarity in graph, can also be considered for embedding learning. The introduce of similarity loss for both user-user and item-item pair can reduce the sparsity problem by augmenting the data for representation learning. In this paper, the 2-order neighborhood proximity of a pair of users (or items) is defined as the similarity.

In order to avoid similarity loss affecting the embedding in the embedding propagation, we only calculate between $\mathbf{E}$ and context mapping embedding matrices $\mathbf{E}^{UC}$ and $\mathbf{E}^{IC}$ for users and items, respectively. Context mapping embedding matrices are defined:

$$\begin{aligned} \mathbf{E}^{UC} &= [\mathbf{e}_{u_1}^{UC}, \cdots, \mathbf{e}_{u_N}^{UC}] \\ \mathbf{E}^{IC} &= [\mathbf{e}_{i_1}^{IC}, \cdots, \mathbf{e}_{i_M}^{IC}] \end{aligned} \tag{13}$$

It should be noted that the dimensions of embeddings in $\mathbf{E}$, $\mathbf{E}^{UC}$ and $\mathbf{E}^{IC}$ are equal. The similarity loss for $\mathbf{e}^{(0)}$ with context embeddings $\mathbf{e}^{UC} \in \mathbf{E}^{UC}$ and $\mathbf{e}^{IC} \in \mathbf{E}^{IC}$ is defined as:

$$
\begin{aligned}
\mathcal{L}_{\text{similarity}} = -\sum \log(\sigma(\mathbf{e}_u^{(0)\mathrm{T}}\mathbf{e}_{\text{u-pos}}^{UC})) + \sum \log(\sigma(\mathbf{e}_u^{(0)\mathrm{T}}\mathbf{e}_{\text{u-neg}}^{UC})) \\
-\sum \log(\sigma(\mathbf{e}_i^{(0)\mathrm{T}}\mathbf{e}_{\text{i-pos}}^{IC})) + \sum \log(\sigma(\mathbf{e}_i^{(0)\mathrm{T}}\mathbf{e}_{\text{i-neg}}^{IC}))
\end{aligned}
\tag{14}
$$

where σ is sigmoid function. $\mathbf{e}_{\text{u-pos}}^{UC}$ and $\mathbf{e}_{\text{u-neg}}^{UC}$ are a positive and negative samples of user u, respectively. $\mathbf{e}_{\text{i-pos}}^{IC}$ and $\mathbf{e}_{\text{i-neg}}^{IC}$ are a positive and negative samples of item i, respectively. We employ random walk and negative sampling to construct positive and negative sample pairs for similarity loss.

4.3   Overall Loss Function

Finally, we get the overall loss function:

$$
\mathcal{L} = \mathcal{L}_{\text{BPR}} + \lambda_1 \mathcal{L}_{\text{similarity}} + \lambda_2 \|\Theta\|_2^2
\tag{15}
$$

where $\Theta = \{\mathbf{E}, \mathbf{E}^{UC}, \mathbf{E}^{IC}, \mathbf{W}_0, \mathbf{W}_1, \mathbf{W}_2, \mathbf{V}, \mathbf{P}\}$. λ_1 controls the strength of BPR loss and λ_2 controls the L_2 regularization strength to prevent overfitting. We use mini-batch Adam [28] to optimize the model and update the parameters of model.

5   Experiments

5.1   Datasets

We evaluate the proposed model on three real-world representative datasets: Gowalla[1], Yelp2018[2] and Amazon-book[3]. These datasets vary significantly in domains and sparsity. The statistics of the datasets are summarized in Table 1.

For each dataset, the training set is constructed by 80% of the historical interactions of each user, and the remaining as the test set. We randomly select 10% of interactions as a validation set from the training set to tune hyper-parameters. We employ negative sampling strategy to produce one negative item that the user did not act before and treat observed user-item interaction as a positive instance. To ensure the quality of the datasets, we use the 10-core setting, i.e., retaining users and items with at least ten interactions.

---

[1] https://snap.stanford.edu/data/loc-gowalla.htm.
[2] https://www.yelp.com/dataset/challeng.
[3] http://jmcauley.ucsd.edu/data/amazon/.

Table 1. Statistics of the datasets

	Gowalla	Yelp2018	Amazon-Book
#Users	29,858	45,919	52,643
#Items	40,981	45,538	91,599
#Interactions	1.027 m	1.185 m	2.984 m
Density	0.084%	0.056%	0.062%

5.2 Experimental Settings

To evaluate the effectiveness of top-K task in recommender system, we adopted Recall@K and NDCG@K, which has been widely used in [17,19]. In this paper, 1) we set K = 20; 2) all items that the user has not interacted with are the negative items; 3) all items is scored by each method in descend order except the positive ones used in the training set. Average metrics for all users in the test set is used for evaluation.

To verify the effectiveness of our approach, we compare it with the following baselines:

- **BPR-MF** [2] optimizes the matrix factorization with implicit feedback using a pairwise ranking loss.
- **NCF** [19] learns user's and item's embeddings from user-item interactions in a matrix factorization, which by a MLP instead of the inner product.
- **PinSage** [16] combines efficient random walks and graph convolutions to generate embeddings of nodes that incorporate both graph structure as well as node feature information.
- **GC-MC** [10] is a graph auto-encoder framework based on differentiable embedding passing on the bipartite interaction graph. The auto-encoder produces latent user and item representations, and they are used to reconstruct the rating links through a bilinear decoder.
- **NGCF** [17] explicitly encodes the collaborative signal of high-order relations by embedding propagation in user-item inter-action graph.
- **Multi-GCCF** [9] constructs two separate user-user and item-item graphs. It employs a multi-graph encoding layer to integrate the information provided by the user-item, user-user and item-item graphs.

We implement GACSE[4] with TensorFlow. The embedding size is fixed to 64 for all models. All models are optimized with the Adam optimizer, where the batch size is fixed at 1024. The learning rate of our model was set to 0.0001; λ_1 was set to 1×10^{-4}; λ_2 was set to 1×10^{-5}; Number of sampling neighbors was set to 64. Number of positive and negative samples for similarity loss was set to 5. All hyper-parameters of the above baselines are either followed the suggestion from the methods' author or turned on the validation sets. We report the results of each baseline under its optimal hyper-parameter settings.

---

[4] For reproducibility, we share the source code of GACSE online: https://github.com/ GACSE/GACSE.git.

Table 2. Overview performance comparison. Bold scores are the best in each column, while underlined scores are the second best. Improvements are statistically significant.

	Gowalla		Yelp2018		Amazon-Book	
	Recall	NDCG	Recall	NDCG	Recall	NDCG
BPR-MF	0.1291	0.1878	0.0494	0.0662	0.0250	0.0518
NCF	0.1326	0.1985	0.0513	0.0719	0.0253	0.0535
PinSage	0.1380	0.1947	0.0612	0.0750	0.0283	0.0545
GC-MC	0.1395	0.1960	0.0597	0.0741	0.0288	0.0551
NGCF	0.1547	<u>0.2237</u>	0.0581	0.0719	0.0344	0.0630
Multi-GCCF	<u>0.1595</u>	0.2126	<u>0.0667</u>	<u>0.0810</u>	<u>0.0363</u>	<u>0.0656</u>
GACSE	**0.1654**	**0.2328**	**0.0672**	**0.0836**	**0.0386**	**0.0703**
%Improv.	3.70%	4.06%	0.75%	3.21%	6.34%	7.16%

5.3   Performance Comparison

Overall Comparison. Table 2 summarized the best results of all models on three benchmark dataset. The last row is the improvements of GACSE relative to the best baseline.

BPR-MF method gives the worst performance on all datasets since the inner product cannot capture complex collaborative signals. NCF outperforms BPR-MF on all datasets consistently. Compared with BPR-MF, the main improvement of NCF is that MLP can model the nonlinear feature interactions between user and item embeddings.

Among all the baseline methods, graph based methods (e.g., PinStage, GC-MC, NGCF, Multi-GCCF) consistently outperform general methods (e.g., BPR-MF, NCF) on all datasets. The main improvement is that graph based model explicitly models the graph structure in embedding learning.

Multi-GCCF are the strongest baseline. It outperforms other baselines on all datasets except NDCG on Gowalla. NGCF gives the best performance of NDCG on Gowalla. They all employ embedding propagation to obtain neighbor's information and stack multiple embedding propagation layers to explore the high-order connectivity. This verifies the importance of capturing collaborative signal in the embedding function. Moreover, Multi-GCCF compared three different multi-grained representations fusion methods.

According to the results, GACSE preforms best among all baselines on three datasets in terms of all evaluation metrics. It improves over the best baseline method by 3.70%, 0.75%, 6.34% in terms of Recall on Gowalla, Yelp2018 and Amazon-book. It gains 4.06%, 3.21%, 7.16% NDCG improvements against the best baseline on Gowalla, Yelp2018 and Amazon-book respectively. Compared with Multi-GCCF and NGCF, GACSE builds an embedding propagation and aggregation architecture based on the attention mechanism. The attention mechanism enable GACSE to learn variable weights of embedding propagation for neighbors explicitly. Meanwhile it obtains high-order implicit collaborative information between user-user and item-item through similarity loss.

Table 3. Ablation studies of GACSE. GACSE-sl means GACSE without similarity loss. GACSE-am means GACSE without adaptive margin.

	Gowalla		Yelp2018	
	Recall	NDCG	Recall	NDCG
GACSE	0.1654	0.2328	0.0672	0.0836
GACSE-sl	0.1632	0.2334	0.0641	0.0805
	(−1.33%)	(+0.26%)	(−4.61%)	(−3.71%)
GACSE-am	0.1468	0.2149	0.0571	0.0728
	(−11.25%)	(−7.68%)	(−15.03%)	(−12.92%)

Ablation Analysis. Table 3 reports the influences of similarity loss and adaptive margin of GACSE on Gowalla and Yelp2018 datasets. As expected, the performance degrades greatly after removing adaptive margin and similarity loss. This confirms the importance of adaptive margin and similarity loss for embedding learning. Adaptive margin can improve model's discrimination for positive and negative sample with similar embeddings. Similarity loss for both user-user and item-item pair can reduce the sparsity problem by augmenting the data for representation learning. Similarity loss and adaptive margin can enhance the effectiveness of attention mechanism for embedding propagation and aggregation.

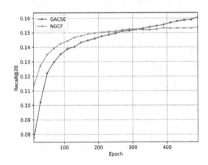
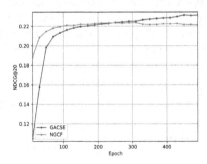

Fig. 2. Recall@20 on Gowalla   **Fig. 3.** NDCG@20 on Gowalla

Test Performance w.r.t. Epoch. Figures 2, 3, 4 and 5 show the test performance w.r.t. recall and NDCG of each epoch of MF and NGCF. We can see that, NGCF exhibits fast convergence than MF on three datasets. It is reasonable since

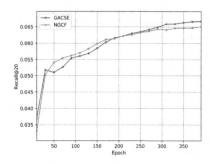
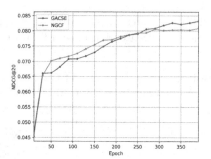

Fig. 4. Recall@20 on Yelp   **Fig. 5.** NDCG@20 on Yelp

indirectly connected users and items are involved when optimizing the interaction pairs in mini-batch. Such an observation demonstrates the better model capacity of NGCF and the effectiveness of performing embedding propagation in the embedding space.

6   Conclusion and Future Work

In this work, we explicitly incorporated collaborative signal and indirect similarities into the embedding function. We proposed a unified representation learning framework GACSE, in which the embedding is learned from direct user-item interaction through attention propagation, and indirect user-user similarities and item-item similarities through auxiliary loss, user-item similarities in bipartite graph. In addition, we combine adaptive margin in BPR loss and similarity loss to optimize GACSE. Extensive experimental results on three real-world datasets show that our model outperforms state-of-the-art baselines.

Several directions remain to be explored. A valuable direction is to incorporate rich side information into GACSE instead of just modeling user & item ids. Another interesting direction for the future work would be exploring multi-task & multi-object embedding learning on heterogeneous graph for recommender system.

References

1. Koren, Y., Bell, R.: Advances in collaborative filtering. In: Ricci, F., Rokach, L., Shapira, B. (eds.) Recommender Systems Handbook, pp. 77–118. Springer, Boston (2015). https://doi.org/10.1007/978-1-4899-7637-6_3
2. Rendle, S., et al.: BPR: Bayesian personalized ranking from implicit feedback. arXiv preprint arXiv:1205.2618 (2012)
3. Su, X., Khoshgoftaar, T.M.: A survey of collaborative filtering techniques. In: Advances in Artificial Intelligence 2009 (2009)

4. Koren, Y.: Factorization meets the neighborhood: a multifaceted collaborative filtering model. In: Proceedings of the 14th ACM SIGKDD International Conference on Knowledge Discovery and Data Mining (2008)
5. Linden, G., Smith, B., York, J.: Amazon.com recommendations: item-to-item collaborative filtering. IEEE Internet Comput. **7**(1), 76–80 (2003)
6. Kim, D., et al.: Convolutional matrix factorization for document context-aware recommendation. In: Proceedings of the 10th ACM Conference on Recommender Systems (2016)
7. Sedhain, S., et al.: AutoRec: autoencoders meet collaborative filtering. In: Proceedings of the 24th International Conference on World Wide Web (2015)
8. Kipf, T.N., Welling, M.: Semi-supervised classification with graph convolutional networks. arXiv preprint arXiv:1609.02907 (2016)
9. Sun, J., et al.: Multi-graph convolution collaborative filtering. In: 2019 IEEE International Conference on Data Mining (ICDM). IEEE (2019)
10. van den Berg, R., Kipf, T.N., Welling, M.: Graph convolutional matrix completion. arXiv preprint arXiv:1706.02263 (2017)
11. Wu, Y., et al.: Collaborative denoising auto-encoders for top-n recommender systems. In: Proceedings of the Ninth ACM International Conference on Web Search and Data Mining (2016)
12. Koren, Y., Bell, R., Volinsky, C.: Matrix factorization techniques for recommender systems. Computer **42**(8), 30–37 (2009)
13. Mnih, A., Salakhutdinov, R.R.: Probabilistic matrix factorization. In: Advances in Neural Information Processing Systems (2008)
14. Ebesu, T., Shen, B., Fang, Y.: Collaborative memory network for recommendation systems. In: The 41st International ACM SIGIR Conference on Research & Development in Information Retrieval (2018)
15. Yang, J.-H., et al.: HOP-rec: high-order proximity for implicit recommendation. In: Proceedings of the 12th ACM Conference on Recommender Systems (2018)
16. Ying, R., et al.: Graph convolutional neural networks for web-scale recommender systems. In: Proceedings of the 24th ACM SIGKDD International Conference on Knowledge Discovery & Data Mining (2018)
17. Wang, X., et al.: Neural graph collaborative filtering. In: Proceedings of the 42nd International ACM SIGIR Conference on Research and Development in Information Retrieval (2019)
18. Hamilton, W., Ying, Z., Leskovec, J.: Inductive representation learning on large graphs. In: Advances in Neural Information Processing Systems (2017)
19. He, X., et al.: Neural collaborative filtering. In: Proceedings of the 26th International Conference on World Wide Web (2017)
20. Christakopoulou, E., Karypis, G: Local item-item models for top-n recommendation. In: Proceedings of the 10th ACM Conference on Recommender Systems (2016)
21. Barkan, O., Koenigstein, N.: Item2vec: neural item embedding for collaborative filtering. In: 2016 IEEE 26th International Workshop on Machine Learning for Signal Processing (MLSP). IEEE (2016)
22. Hu, Y., Koren, Y., Volinsky, C.: Collaborative filtering for implicit feedback datasets. In: 2008 Eighth IEEE International Conference on Data Mining. IEEE (2008)
23. Veličković, P., et al.: Graph attention networks. arXiv preprint arXiv:1710.10903 (2017)

24. Wang, X., et al.: KGAT: knowledge graph attention network for recommendation. In: Proceedings of the 25th ACM SIGKDD International Conference on Knowledge Discovery & Data Mining (2019)
25. Grover, A., Leskovec, J.: node2vec: scalable feature learning for networks. In: Proceedings of the 22nd ACM SIGKDD International Conference on Knowledge Discovery and Data Mining (2016)
26. Chen, C.-M., et al.: Collaborative similarity embedding for recommender systems. In: The World Wide Web Conference (2019)
27. Liang, D., et al.: Factorization meets the item embedding: regularizing matrix factorization with item co-occurrence. In: Proceedings of the 10th ACM Conference on Recommender Systems (2016)
28. Kingma, D.P., Ba, J.: Adam: a method for stochastic optimization. arXiv preprint arXiv:1412.6980 (2014)

Learning Disentangled User Representation Based on Controllable VAE for Recommendation

Yunyi Li[1], Pengpeng Zhao[1(✉)], Deqing Wang[2], Xuefeng Xian[3], Yanchi Liu[4], and Victor S. Sheng[5]

[1] Institute of Artificial Intelligence, School of Computer Science and Technology, Soochow University, Suzhou, China
yylyyyl@stu.suda.edu.cn, ppzhao@suda.edu.cn
[2] Beihang University, Suzhou, China
dqwang@buaa.edu.cn
[3] Suzhou Vocational University, Suzhou, China
xianxuefeng@jssvc.edu.cn
[4] NEC Labs America, Princeton, USA
yanchi@nec-labs.com
[5] Department of Computer Science, Texas Tech University, Lubbock, USA
victor.sheng@ttu.edu

Abstract. User behaviour on purchasing is always driven by complex latent factors, which are highly disentangled in the real world. Learning latent factorized representation of users can uncover user intentions behind the observed data (i.e. user-item interaction) and improve the robustness and interpretability of the recommender system. However, existing collaborative filtering methods learning disentangled representation face problems of balancing the trade-off between reconstruction quality and disentanglement. In this paper, we propose a controllable variational autoencoder framework for collaborative filtering. Specifically, we adopt a modified Proportional-Integral-Derivative (PID) control to the β-VAE objective to automatically tune the hyperparameter β using the output of Kullback-Leibler divergence as feedback. We further introduce item embeddings to guide the system to learn representation related to the real-world concepts using a factorized Gaussian distribution. Experimental results show that our model can get a crucial improvement over state-of-the-art baselines. We further evaluate our model's effectiveness to control the trade-off between reconstruction error and disentanglement quality in the recommendation.

Keywords: Variational autoencoder · Disentangled representation learning · Recommender system

1 Introduction

The personalized recommendation is now popular, aiming to find user interests in different concepts. When users purchase on e-commerce platforms, many fac-

© Springer Nature Switzerland AG 2021
C. S. Jensen et al. (Eds.): DASFAA 2021, LNCS 12683, pp. 179–194, 2021.
https://doi.org/10.1007/978-3-030-73200-4_12

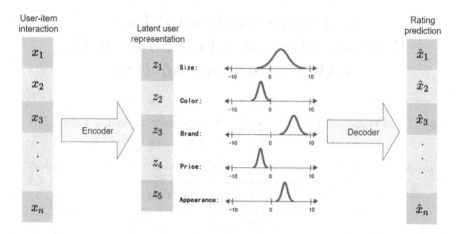

Fig. 1. The aim of learning a disentangled representation is to encode a latent user vector, each dimension of which stands for an independent concept of user intents. For example, the first dimension represents the user preference on size, the second dimension represents the preference on color, and so on.

tors are considered simultaneously (e.g. the price, size, or appearance), and these factors are always disentangled. Figure 1 shows that disentanglement learning is to find a factorized latent representation in which each unit is independent and corresponds to a single user intent in the real-world. Therefore, learning disentangled factors of users can uncover the real user preference hidden in the interaction data. These independent factors are related to real-world concepts and not sensitive to the misleading correlations of observed data, so they can enhance the robustness and interpretability of recommendation [8]. Early models like matrix factorization (MF) [25] project each user/item ID into a vectorized representation (i.e. embedding). Some following studies [5,16] introduce user-item interactions to extract the hidden feature of a user and combine embeddings of items to enrich the representation. However, these existing latent representation methods fail to disentangle latent factors, and are prone to preserve the mixed relationship of the factors mistakenly.

β-VAE [7], a modification of the Variational Autoencoder (VAE), is a common framework used to learn disentangled representation. β-VAE adds an extra hyperparameter β to the KL divergence term in the VAE objective. Multi-VAE [4] applies β-VAE to the recommender system first to learn the user representation with multinomial likelihood. The article introduces β to the VAE objective and sets it up to 0.2 to control the strength of regularization, weakening the constraint of the prior distribution. MacridVAE [8] is the first method to learn disentangled representation in recommendation based on user behaviour data. The author infers that user intentions are associated with several high-level concepts, and divides the latent representation into several groups regarding different concepts respectively. DGCF [21] devises a disentangled graph collaborative filtering model to disentangle user-item relationships. In these studies, they directly set $\beta < 1$ to achieve high reconstruction accuracy; the distribution of each dimension among the latent representation is still not independent. We claim that if we sacrifice the reconstruction ability a little, i.e., increasing the hyperparameter β, the model can learn a factorized representation that unearths real user intention, further improving the recommendation result. However, the β-VAE framework is sensitive to the choice of β. If we set β large (e.g. 50), the model can learn good disentangled representation, but the mutual information between the latent factor and the input data becomes too little to reconstruct the data. Therefore, we then face the challenge of how to achieve the trade-off between reconstruction accuracy and disentangling quality during training.

Moreover, research shows that unsupervised disentangled learning is theoretically impossible without inductive bias, and the increased disentanglement will not reduce the sample complexity of knowledge [5]. Based on this theory, it is worth integrating item information and user-item interaction to increase disentangled learning quality. [26] proposes a content-aware collaborative filtering framework that combines item image information and collaborative information together to learn item representations. However, the representation learned from item images may only contain appearance factors and ignore many other important features, such as price or material. We thus choose datasets with rich item information, such as user reviews and item categories, to learn the item embedding and integrate the embedding with our β-VAE model.

In this paper, we propose a controllable VAE framework based on β-VAE to learn a better user representation, each dimension of which is associated with the tailengleactual user preference in various concepts. In specific, we apply a Proportional-Integral-Derivative (PID) control to tune the value of β automatically. We modify the controller to fit the recommender system so that the system can achieve a trade-off between data reconstruction and disentangled factor representation. The input of the controller is the error between the true KL divergence in the past training period and the desired value set beforehand. Similar to the contribution of [2], the desired KL value is gradually increased from zero to an amount large enough to maintain enough mutual information from the input data. Moreover, we introduce a factorized Gaussian distribution into the encoder layer to uncover independent factors hidden in the item embedding and design a loss function based on the new distribution.

Our contributions can be concluded as follows:

- As far as we know, this is the first study to learn disentangled user representation in the recommender system by controlling the trade-off between data reconstruction accuracy bounds and Kullback-Leibler constraints.
- We adopt a non-linear PID control to the β-VAE framework to control how much information the model retains about each factor. We take item information into account using a factorized Gaussian distribution to increase the encoding capacity.
- We evaluate our model on three real-world datasets, and the results show Controllable VAE is able to achieve a good disentangled representation indicating user preference.

2   Related Work

2.1   Latent Representation Learning of CF

Existing Collaborative Filtering recommendations have contributed much to the latent representation learning of users and items. The most popular framework is Matrix Factorization (MF) [15,25], which leverages user/item ID into an embedding vector to learn the latent relationship between users and items. Further studies, such as NCF [22]. SVD++ [24] combines domain model and hidden factor model together, and proposes a new globally optimized neighborhood model. FISM [16] presents an item-based method that learns the item-item similarity matrix as the product of two low dimensional latent factor matrices. CoSAN [13] proposed a collaborative self-attention Network to learn the session representation. The autoencoder model is leveraged to learn user/item representation by the encoder-decoder framework, and bayesian inference theory is applied to the autoencoder to learn the distribution of user-item interactions, like CDAE [23], Mult-VAE [14], and AutoRec [18]. All the works above are committed to learning the representation, which indicates user intents and preference towards items. Our work pays attention to disentangled user representation and aims to learn independent distribution based on β-VAE model.

2.2   Disentangled Representation Learning

The goal of learning disentangled representation is not perfect inversion but finding independent factors. There are many methods proposed to improve the disentanglement of latent representation in generative models. InfoGAN [19] achieves unsupervised learning of disentangled representation by introducing mutual information to constrain the latent variable being associated with the characteristic output of the generated data. β-VAE adds a coefficient β to the KL divergence term in the VAE objective. Increasing the value of β can force the framework to learn a good disentangled representation, but resulting in increased reconstruction loss [2]. To counteract this trade-off between reconstruction and KL divergence, both β-TCVAE [3] and FactorVAE [11] decompose the KL term and put a heavy penalty on the total correlation to directly encourage factorized distribution in the latent vector. [12] uses moment matching in VAEs to penalize the co-variance between the latent dimensions.

These existing efforts have been majorly applied to the field of language modeling and image generation [7,19]. Few methods consider disentangled user representation in the recommender system. [8] divides user intentions into several high-level concepts and learns disentangled representation based on user behavior. [21] devises a disentangled graph model to learn user intents based on neural graph collaborative filtering. However, these works pay attention to distinguish the high-level concepts of user intents, and neglect the trade-off between the two terms in the β-VAE objective.

3 Preliminaries

3.1 β-VAE

The Variational Autoencoder (VAE) aims to learn the marginal likelihood of the data x from an unobserved continuous random variable z. The VAE framework consists of two parts: a recognition model and a generative model. The recognition model (i.e. the encoder) encodes the observed data x into a latent variable z with the posterior $p(z|x)$. And the generative model (i.e. the decoder) restitutes the latent representation to the productive output. Since the true distribution $p(z|x)$ is intractable, the model is trained with the aid of an approximate posterior distribution $q(z|x)$. The training objective of VAE is written as the tractable evidence lower bound (ELBO):

$$logp(x) \geq \mathcal{L}(\theta, \phi; x) = \text{ELBO} \\ = E_{q_\phi(z|x)}[logp_\theta(x|z)] - KL(q_\phi(z|x)\|p(z)) \tag{1}$$

The prior $p(z)$ and the posterior $q(z|x)$ are parameterized as standard Gaussian distributions. In order to estimate gradients of the lower bound, 'Reparameterization Trick' is used to transform the continuous distribution into a discrete relationship. The random variable z is parametrized as a differentiable transformation of a noise variable $\epsilon \sim \mathcal{N}(0,1)$:

$$z = \mu + \sigma \cdot \epsilon \tag{2}$$

where μ and σ represent the mean and variance of Gaussian distribution respectively. β-VAE is a common modification of VAE. It introduces an adjustable hyperparameter β to the original VAE objective:

$$\text{ELBO} = E_{q_\phi(z|x)}[logp_\theta(x|z)] - \beta(t)KL(q_\phi(z|x)\|p(z)) \tag{3}$$

Burgess et al. [2] discusses the effect of the hyperparameter from the perspective of information bottleneck, suggesting that β acts as a Lagrange multiplier, limiting the capacity of the bottleneck. Further researches [3,11] resolve the KL divergence and give an explanation of why putting a heavier constraint on the KL term leads to a more disentangled representation. When β is large, the framework tends to reduce the mutual information between the latent factor and the input data, resulting in a more independent latent representation.

From the perspective of the information bottleneck, the gap between the posterior distribution and the unit Gaussian prior is minimized by putting a large constraint on KL divergence. However, the mutual information between z and actual data x will become small and make it difficult to reconstruct from the latent representation.

3.2 PID Control

Proportional-Integral-Derivative (PID) control is a simple linear control framework that has been widely used in the industrial process. The general model is defined as:

$$y(t+1) = K_p e(t) + K_i \int_0^t e(\tau)d\tau + K_d \frac{de(t)}{dt}$$

$$e(t) = C - y(t)$$

(4)

where C is the desired value set beforehand. In specific, the control deviation is formed according to the desired value and the actual output value, and the deviation is formed by a linear combination of proportion, integral, and derivative to control the objective.

Each part of the controller has its own advantages and disadvantages. Proportional (P) control can respond to the change of errors quickly but cannot eliminate steady-state errors. Integral (I) control will continue to increase as long as there is an error in the system. That means if there is enough time, Integral control can completely eliminate errors. However, if the integral output changes too fast, it will cause over-integration and oscillation. Differential (D) control is a supplement to the P and I control, which reduces the overshoot and overcomes the oscillation, improving the stability of the system.

After understanding the effectiveness of each part of PID control, the method becomes relatively simple. In this model, I and D control are set to 0 at first, and the proportional gain is increased until the loop output starts to oscillate. When increasing the proportional gain, the system becomes faster, but it must ensure that the system does not become unstable. Once P control is set to get the desired quick response, the integral term will increase to stop the oscillation. The integral term will reduce the steady-state error but increase the overshoot. Reasonable overshoot is necessary for a fast system so that it can respond immediately to changes. Adjusting the integral term can achieve the minimum steady-state error. Increasing the derivative term will reduce the overshoot and produce a higher stability gain, but the system will become extremely sensitive to noise. In most cases, engineers need to weigh various characteristics of the control system when designing and then achieve trade-offs.

4   Methods

4.1   Controllable VAE Algorithm

Although the PID control has already been widely applied to the industrial field, such as the temperature control system [6], it is rarely used in recommender systems. Inspired by [17], we modify the PID control, illustrated in Fig. 2, to fit the variational autoencoder framework. First, we replace the proportional control with an unproportionate one, using the reciprocal of an exponential function. Second, we turn the integral and differential term from a successive process to a discrete one to fit the recommendation training model. Third, we add a constant value β_0 to determine the initial value range of $\beta(t)$. In general, our modified controllable model is defined as follows:

$$\beta(t) = \frac{K_p}{1 + exp(e(t))} - K_i \sum_{j=0}^{t} e(j) - K_d(e(t) - e(t-1)) + \beta_0$$

(5)

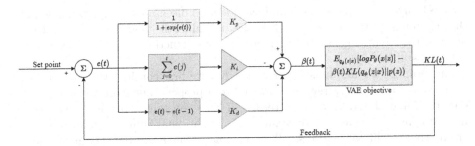

Fig. 2. Framework of our modified PID control. It combines a controller with the VAE framework to automatically tune the weight $\beta(t)$ via the output of the KL divergence in the objective.

where $e(t)$ stands for the error between the set point C and the KL feedback at training step t. With the help of the exponential function $\exp(.)$, the first term ranges between 0 and K_p, avoiding the instability of P control caused by the linear proportional term.

The second term sums the errors of all previous periods together. This integral term can create a progressively stronger correction. When the error of the controller remains negative (KL divergence above the set point), this term will continue to increase, leading to a larger $\beta(t)$ and encouraging KL divergence to shrink. In converse, when the error becomes positive (KL term below the set point), the second term will continue to decrease, leading to a lower $\beta(t)$ and forcing the KL divergence to grow. In both cases, the change of the integral control can force $\beta(t)$ to change in a direction that guides the KL divergence to approach the set point.

The third term is used to reflect the changing trend of the deviation signal. As soon as an error is found to change larger or smaller, a control signal is immediately yielded to resist its change and prevent the system from overshoot.

Finally, when the KL divergence is trained to an appropriate point, $\beta(t)$ will remain constant. Our model finds a balance between reconstruction and disentanglement. The user representation learned from our model can extract user intents over various concepts independently.

4.2 Parameter Choice for Controllable VAE

In this part, we discuss the choice of the parameters in the PID controller. Our goal is to ensure that the controller reacts to the error sufficiently and alters the output smoothly. We first consider the desired KL value C. As proposed in [2], C is used to evaluate how much information the model chooses to retain about each factor. The larger the desired value is, the more mutual information is retained by the latent representation z. On the other hand, if the KL divergence keeps very low continuously, the model will degenerate to an autoencoder model and cause KL-vanishing [1]. In our controllable VAE framework, we linearly increase C from a low value (e.g. 0) to a large one (e.g. 10). By doing so, the model will force KL divergence to be small and learn a good disentangled representation in the beginning. With the increase of the desired value, the model will progressively increase the KL divergence and decrease the reconstruction error until the system becomes stable.

In terms of the modulus K_p, K_i, and K_d, we first consider the exponential term. When the KL divergence is trained to a proper value close to the desired one, the integral and differential term will keep constant. Therefore, the exponential control acts as a determining role in the final value of β. In order to ensure that β will not be too small, we commonly set K_p to be one during training. The integral and differential term are used to make the controller sensitive to the change of errors, but if their sensitivity is too high, the system will be unstable and oscillate. Empirically, we find that a value between 0.5 and 1 stabilizes the training. Note that the parameters should not be too small to ensure the effectiveness of the controller.

4.3   Feature Disentanglement

It is difficult to learn a good disentangled representation under unsupervised training without inductive biases [5]. Therefore, we use additional information associated with the true label as supervising signals. Let $u \in \mathbb{R}$ denote the additional information, u_i is corresponded to the i-th concept in real world. We now define the posterior distribution $q_\phi(z|x)$ for latent z as:

$$q_\phi(z|x) = q_\phi(z|x, u)q(u) \tag{6}$$

where the latent substantive variable $q_\phi(z|x, u)$ is a factorized Gaussian distribution conditioning on the additional observation u:

$$q_\phi(z|x, u) \sim \mathcal{N}(\mu(x)\lambda(u), \sigma(x)\lambda^2(u)) \tag{7}$$

We then follow the variational inference paradigm and rewrite the evidence lower bound as follows:

$$\begin{aligned} \text{ELBO} = E_{q(u)}[E_{q_\phi(z|x,u)}[logp_\theta(x|z)] \\ - \beta(t)KL(q_\phi(z|x,u)||p(z))] + H[q(u)] \end{aligned} \tag{8}$$

The posterior distribution captures the aggregated structure of the latent variables based on the user feedback distribution. Therefore, the KL-term in Eq. 8 can be decomposed as:

$$\begin{aligned} KL(q_\phi(z|x, u)||p(z)) = I(z, x) + KL(q_\phi(z|u)|| \prod_j p(z_j)) \\ + \sum_j KL(q_\phi(z_j|u)||p(z_j)) \end{aligned} \tag{9}$$

The first term in Eq. 9 represents the Mutual Information (MI) between data x and the latent variable z. The second term is referred to as the Total Correlation (TC), which is a measure of the independence between the variables. A heavier penalty on the TC term forces the model to learn a factorized representation, each dimension of which is independent [3,11]. The third term is referred to as the dimension-wise KL. This term functions to prevent individual variables from deviating too far from their corresponding priors. Therefore, if we put a strong penalty on the KL divergence, i.e., on the total correlation, our model can find statistically independent factors from the observed data, inducing a more disentangled representation.

4.4   Details of Implementation

In this section, we provide details of the implementation: $p(z)$ (the prior), $q_\phi(z|x)$ (the encoder), $p_\theta(x|z)$ (the decoder). And we present the algorithm of the PID control. We optimize the parameter θ, ϕ to maximize the training objective following the implementation in [8].

Prior and Encoder. We set $p(z) \sim \mathcal{N}(0, I)$ to achieve factorized representation. The encoder $q_\phi(z|x)$ is to learn a latent representation z from the observed data x. We assume $q_\phi(z|x) = \prod_{i=1}^{N} q_\phi(z_i|x_i) = \prod_{i=1}^{N}(q_\phi(z_i|x_i, u_i)q(u_i))$. $q(u_i)$ denotes the item embedding encoded from the content vector, and $q_\phi(z_i|x_i, u_i)$ is a factorized Gaussian distribution $\sim \mathcal{N}(\mu_i(x_i)\lambda(u_i), \sigma_i(x_i)\lambda^2(u_i))$ to compute z_i with the combination of data vector x_i and item embedding u_i (see Eq. 7). The mean and the standard bias are calculated by a neural network f_{nn}. The item embedding u_i is encoded from the items' content information.

Decoder. The decoder aims to reconstruct data using the multinomial likelihood $p_\theta(x|z)$. We assume $p_\theta(x|z) = \prod_{i=1}^{N} g_\theta^i(z_i)$ and the function $g^i(.)$ is defined as a neural network parameterized by θ.

PID Control Algorithm. We summarize our PID control algorithm in Algorithm 1. The input of the controller is the error between the desired value C and the KL divergence at training step t. Line 4–6 calculates the P, I, D terms of the controller respectively. Line 8–9 limits $\beta(t)$ from not dropping too low and ensures VAE objective can operate normally. Note that the desired value C is gradually increased during training.

5   Experiment

5.1   Experimental Setup

In this section, we conduct experiments on three e-commerce datasets and answer the following research questions:

- **RQ1:** How does our controllable VAE model perform on recommendation compared with previous works?

Algorithm 1. PID Control Algorithm

Input: desired KL value C, the KL divergence feedback $KL(t)$, parameters $K_p, K_i, K_d, \beta_0, \beta_{min}$
Output: hyperparameter $\beta(t)$
1: Initialization: $I(0) = 0, \beta(0) = 0$
2: **for** $t = 1$ to N **do**
3:     $e(t) \leftarrow C - KL(t)$
4:     $P(t) \leftarrow \frac{K_p}{1+exp(e(t))}$
5:     $I(t) \leftarrow I(t-1) - K_i e(t)$
6:     $D(t) \leftarrow K_d(e(t) - D(t-1))$
7:     $\beta(t) \leftarrow P(t) + I(t) + D(t) + \beta_0$
8:     **if** $\beta(t) < \beta_{min}$ **then**
9:         $\beta(t) \leftarrow \beta_{min}$
10:    **return** $\beta(t)$

Table 1. Details for the three Amazon datasets.

Dataset	Users	Items	Interacts	Density
Beauty	8635	1483	18357	0.14%
Toys & Games	6231	11364	72063	0.10%
Clothing	10912	48059	373578	0.07%

- **RQ2:** How does each component in the VAE objective affect the results, and how does the controller enhance the robust of the recommender system?
- **RQ3:** How is the disentangled representation learned, and how does the representation enhance the interpretability of the recommender system?

Datasets. To evaluate the effectiveness of our method, we adopt three real-world publicly Amazon datasets [9,10]: Amazon-Beauty, Amazon-Toys& Games and Amazon-Clothing. Amazon datasets contain detailed item descriptions for us to learn the item embedding. We select users with more than 20 reviews in Clothing, 10 reviews in Toys & Games and 5 reviews in Beauty for different levels of item scale, as shown in Table 1. We extract UserID, ItemID, and the rating scores to indicate whether the user purchased the item. We randomly select 20% of the rated users as ground truth for testing and the remaining 70% and 10% data for training and validation.

Metrics. Following previous works [4,8], we use two ranking-based metrics: Recall@K and normalized discounted cumulative gain (NDCG@K). For each user, both metrics compare the predicted rank of the held-out items with their true rank. The Recall@K considers the proportion of cases ranked within the top-K predicted items. The NDCG@K accounts for the position of the hit by assigning higher scores to the top hits.

Table 2. Overall performance comparison.

Method	Beauty			Toy & Games			Clothing		
	N@20	N@50	R@50	N@20	N@50	R@50	N@20	N@50	R@50
Multi-DAE	0.00228	0.00249	0.00156	0.0431	0.0555	0.0905	0.469	0.474	0.448
Multi-VAE	0.00249	0.00256	0.00233	0.0458	0.0586	0.0970	0.472	0.477	0.448
CVAE	0.00267	0.00286	0.00248	0.0507	0.0656	0.1013	0.507	0.501	0.470
MacridVAE	0.00277	0.00282	0.00256	0.0615	0.0731	0.1148	0.519	0.524	0.477
Ours	**0.00303**	**0.00308**	**0.00267**	**0.0646**	**0.0780**	**0.1196**	**0.538**	**0.541**	**0.481**
Improv(%)	8.621	9.219	4.296	5.135	6.741	4.171	3.610	3.311	0.812

Baselines. We compare our method with the following competitive baselines, with particular emphasis on VAE-based methods for comparison. The same content information is used for all content-based methods.

- **Multi-DAE & Multi-VAE** [4]: This is a classic latent representation model based on denoising autoencoder and variational autoencoder. The implicit feedback is generated from a multinomial likelihood for the recommendation.
- **CVAE** [20]: This is a content-based model that learns the item information. It introduces a one-hot conditional label vector to the inference network to help distinguish cluster users.
- **Macrid-VAE** [8]: This is a state-of-the-art model using β-VAE to learn disentangled representation of user preference. In detail, it divides high-level concepts into several parts and infers a set of one-hot vectors to denote which concept each item respectively belongs to.

Parameter Setting. Each method's dimension of latent factors is empirically set to 100. The parameters of the PID control, i.e., K_p, K_i, K_d, are set to be $1, -1, -1$, respectively, as discussed in Sect. 4.2. The upper bound of the desired KL value is determined from the range $\{1, 5, 10, 20\}$. Other parameters are determined following the baseline [8]. Model parameters of the neural network are initialized randomly at first, and the normal distributions are initialized with mean 0 and standard deviation 0.001.

5.2 Recommendation Performance (RQ1)

We first compare the Top-K recommendation performance of our method with the baselines mentioned above. The results are listed in Table 2. We observe that our method outperforms the baselines across three datasets significantly, especially in small-scale datasets. In particular, its relative improvements over the strongest baselines w.r.t. NDCG@20 are 8.62%, 5.13%, and 3.61% in Amazon-Beauty, Toys & Games, and Clothing, respectively.

We attribute such improvements to the following aspects: 1) By introducing the modified PID control into the VAE objective, our model can achieve a good trade-off between the reconstruction loss and disentanglement representation quality, which has been neglected by the previous works. Our method is able to

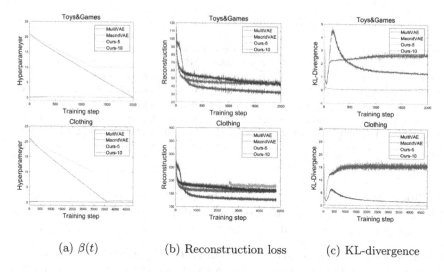

(a) $\beta(t)$ (b) Reconstruction loss   (c) KL-divergence

Fig. 3. Performance comparison for different methods on two datasets. The first row is the result of the experiment on Toys & Games (2000 training steps in total), and the second row is the result of the Clothing data (4800 training steps in total).

find more independent factorized user preference towards products and enhance the ability of prediction. 2) By taking item embedding into account, our model can extract more useful features from item content information and get better performance via semi-supervised training.

Comparing the results jointly across the three datasets, we find that our model achieves the best improvement on the Amazon-Beauty dataset, while the improvement on the Clothing dataset is much less than that on the Beauty. This might suggest that it is easy to find disentangled factors for clothes. For example, the characteristics of the dress and shoes are entirely different from each other, and the model can disentangle the categories easily with a set of one-hot parameter. But when it comes to the Beauty dataset, it is difficult to distinguish high-level concepts via previous methods. Hence, our model can learn a better user representation to disentangle latent factors that may be ignored by traditional VAE-based methods and improve the recommender results.

5.3 Trade-Off Evaluation (RQ2)

We evaluate how our model achieves a trade-off and improves the robustness of the recommender system in this section. Specifically, we train our model on the Amazon-Clothing and Toys & Games datasets with 400 training epochs. We record the value of β, reconstruction loss, and KL divergence at each training step and compare the results of different models in the same graph (shown in Fig. 3). Note that the total training step is 2000 for Toys & Games and 4800 for Clothing due to the different scales of the datasets. We have the following findings from the figure:

- As is shown in Fig. 3(a), the hyperparameter $\beta(t)$ of our model is gradually decreased from a large number (around 20) to a certain degree, while the previous models using β-VAE directly set it to 0.2. Specially, when the desired value C is set up to 10, $\beta(t)$ will drop to βmin. When C is set up to 5, $\beta(t)$ will drop to a certain degree (see around 10 in our experiments), indicating that the system finds a balanced trade-off.

- Figure 3(b) shows the comparison of the reconstruction loss for different models. We can see that the baseline, Multi-VAE, has the lowest reconstruction loss among the three methods. However, as is shown in Table 2, Multi-VAE does not make a great prediction. It indicates that lower reconstruction loss does not necessarily mean a better prediction effect. On the contrary, sacrificing reconstruction quality a little to learn a disentangled representation will lead to a better prediction result.

- When it comes to the comparison of KL divergence, shown in Fig. 3(c), we can see that the KL value of our model is much lower than the baselines, indicating that our model has learned an excellent disentangled representation. Compared with the oscillated curve of the baselines, the change in our model is smoother, and there is almost no difference between the values, no matter setting C up to 5 or 10. It indicates our training model is more stable and less sensitive to the choice of parameter. The controller can enhance the robustness of the system.

Above all, the training objective of VAE can be divided into two parts: the reconstruction function and the KL divergence. Reducing the reconstruction loss aims to increase the mutual information between the latent factor and the observed data, while lowering the KL divergence seeks to increase the independence between the dimensions of the user representation. The goal of recommendation is neither minimizing the reconstruction loss to regenerate user histories nor minimizing the KL divergence to produce random results. Therefore, reaching a balance between the reconstruction ability and disentanglement leads to a better recommender quality.

5.4 Disentanglement Learning (RQ3)

In this section, we visualize the effect of disentangled representation from the perspective of purchasing behaviour to see how does the disentangled representation enhance the interpretability of the recommender system. In specific, we randomly select five users and train our model on the Amazon-Clothing dataset to learn a user-item representation with 10 dimensions. Next, we select several user-item pairs with the largest value in each dimension and extract the side information (i.e. user reviews and item images), which possibly illustrates why users purchase the items. The details of our processes follow that in [8]. We show part of the side information in Table 3.

We analyze the contents of Table 3 one by one. Firstly, the interaction $(user_{2065}, item_{17130})$ and $(user_{2456}, item_{4527})$ has the highest value in dimension 4. We view the user reviews of these interactions and see that feature 'color'

Table 3. Samples of the disentangled representation learning on Clothing dataset.

z_dim	UserID	ItemID	Review	Image
z_4	2065	17130	Perfect! I have bras in a lot of colors and locally I can only find extenders in nude, white, and black. These fit my Victoria's Secret bras well. I've recommended these to friends.	
	2456	4527	awesome! Loved another color so I bought this one, too	
z_5	3431	10394	I got a size 14, they fit my waist but the thigh holes are big. They remind me of little boy shorts. I thought they would be more form fitting. I'll wear them just to be extra comfortable but that's it.	
	4229	19211	Oh gosh, these are wonderful, have bought more.	
z_8	2343	144	I bought this for my husband's birthday. He is tall and has trouble buying t-shirts that are long enough. This one is plenty long enough and is made with sturdy material. I think it will last a long time.	
	4229	25150	Comfy, good length, pretty	

is of high confidence being the user intents of the 4-th dimension. In terms of dimension 5, we find it difficult to exact common ground from the review information of interaction $(user_{3431}, item_{10394})$ and $(user_{4229}, item_{19211})$. We extract the images of the items and find they are both trousers. We thus assume that dimension 5 may represent the category of clothes. Similarly, the 8-th dimension is likely to represent the size of clothes, as the interaction $(user_{2343}, item_{144})$ and $(user_{4229}, item_{25150})$ both mention 'good length' in their reviews. In summary, we infer that the dimension 4,5,8 of the user latent representation is associated with 'color', 'species', and 'size' respectively.

In this section, we prove that our model can learn a disentangled representation by jointly analyzing user reviews and item images in the same latent dimension. Each dimension of the representation indicates users' different preferences on the products. The representation can be exposed to the users directly and enable the users to change the single value based on their preference, enhancing the interpretability and controllability of the recommender system. Note that some concepts may still be unknown by our existing knowledge, such as the material of the product or the reputation of the brand. This inspires us to explore a better way to combine item embedding with the user-item interaction.

6   Conclusion

In this paper, we proposed a general controllable variational autoencoder framework to learn a disentangled user representation in the recommendation. We introduced the PID controller to the recommender system. The modified nonlinear controller is used to automatically tune the hyperparameter in β-VAE objective using the feedback of KL divergence. We evaluated the effect of our model on three real-world datasets and achieved better prediction results over all the datasets, compared with the baselines. The results showed that learning a good user disentangled representation can improve recommendation quality. In future works, we will explore more effective ways to learn user intents and enhance the recommendation quality.

Acknowledgements. This research was partially supported by NSFC (No. 61876117, 61876217, 61872258, 61728205), ESP of the State Key Laboratory of Software Development Environment, and PAPD of Jiangsu Higher Education Institutions.

References

1. Bowman, S.R., Vilnis, L.: Generating sentences from a continuous space. arXiv preprint arXiv:1511.06349 (2015)
2. Burgess, C.P., Higgins, I.: Understanding disentangling in beta-VAE. arXiv preprint arXiv:1804.03599 (2018)
3. Chen, T.Q., Li, X., Grosse, R.B., Duvenaud, D.K.: Isolating sources of disentanglement in variational autoencoders. In: NeurIPS, pp. 2610–2620 (2018)
4. Dawen, L., Rahul, G.K., Matthew, D.H., Tony, J.: Variational autoencoders for collaborative filtering. In: WWW, pp. 689–698 (2018)
5. Francesco, L., Stefan, B., Mario, L., Sylvain, G., Bernhard, S., Olivier, B.: Challenging common assumptions in the unsupervised learning of disentangled representations. In: PMLR, pp. 4114–4124 (2019)
6. Graham, C.G., Stefan, F.G., Mario, E.S.: Classical PID control. Control System Design, Prentice Hall PTR (2005)
7. Irina, H.: Beta-VAE: learning basic visual concepts with a constrained variational framework. In: ICLR (2017)
8. Jianxin, M., Chang, Z., Peng, C., Hongxia, Y., Wenwu, Z.: Learning disentangled representations for recommendation. In: NeurIPS, pp. 5712–5723 (2019)
9. Jianmo, N., Jiacheng, L.: Justifying recommendations using distantly-labeled reviews and fined-grained aspects. In: EMNLP, pp. 188–197 (2019)
10. Julian, M., Christopher, T., Qinfeng, S., Anton, V.D.H.: Image-based recommendations on styles and substitutes. In: SIGIR, pp. 43–52 (2015)
11. Kim, H., Mnih, A.: Disentangling by factorising. In: ICML, pp. 2654–2663 (2018)

12. Kumar, A., Sattigeri, P., Balakrishnan, A.: Variational inference of disentangled latent concepts from unlabeled observations. In: ICLR (2018)
13. Luo, A., Zhao, P., Liu, Y.: Collaborative self-attention network for session-based recommendation. In: IJCAI, pp. 2591–2597 (2020)
14. Petar, V.: Graph attention networks. In: ICLR (2018)
15. Ruslan, S., Andriy, M., Geoffrey, E.H.: Restricted Boltzmann machines for collaborative filtering. In: ICML, pp. 791–798 (2007)
16. Santosh, K., Xia, N., George, K.: FISM: factored item similarity models for top-N recommender systems. In: KDD, pp. 659–667. ACM (2013)
17. Shao, H., Yao, S., Sun, D., Zhang, A.: ControlVAE: controllable variational autoencoder. In: ICML, pp. 8655–8664 (2020)
18. Suvash, S., Aditya, K.M., Scott, S., Lexing, X.: AutoRec: autoencoders meet collaborative filtering. In: WWW, pp. 111–112 (2015)
19. Xi, C., Yan, D., Rein, H., John, S., Ilya, S., Pieter, A.: InfoGAN: interpretable representation learning by information maximizing generative adversarial nets. In: NeurIPS, pp. 2172–2180 (2016)
20. Xiaopeng, L., James, S.: Collaborative variational autoencoder for recommender systems. In: SIGKDD, pp. 305–314 (2017)
21. Xiang, W., Hongye, J., An, Z., Xiangnan, H., Tat, S.C.: Disentangled graph collaborative filtering. In: SIGIR, pp. 1001–1010 (2020)
22. Xiangnan, H., Lizi, L., Hanwang, Z., Liqiang, N., Xia, H., Tat, S.C.: Neural collaborative filtering. In: WWW, pp. 173–182 (2017)
23. Yao, W., Christopher, D., Alice, X.Z., Martin, E.: Collaborative denoising autoencoders for top-N recommender systems. In: WSDM, pp. 153–162. ACM (2016)
24. Yehuda, K.: Factorization meets the neighborhood: a multifaceted collaborative filtering model. In: KDD, pp. 426–434. ACM (2018)
25. Yehuda, K., Robert, M.B., Chris, V.: Matrix factorization techniques for recommender systems. In: IEEE Computer, pp. 30–37 (2009)
26. Zhang, Y., Zhu, Z., He, Y.: Content-collaborative disentanglement representation learning for enhanced recommendation. In: RecSys, pp. 43–52. ACM (2020)

DFCN: An Effective Feature Interactions Learning Model for Recommender Systems

Wei Yang$^{(\boxtimes)}$ and Tianyu Hu

University of Chinese Academy of Sciences, Beijing, China
yangwei192@mails.ucas.ac.cn

Abstract. Data features in real industrial recommendation scenarios are diverse, high-dimensional and sparse. Effective feature crossing can improve the performance of recommendation, which is of great significance. Manual feature engineering is no longer applicable due to its high cost and low efficiency. Factorization machines introduce the second-order feature interactions to enhance learning ability. Deep neural networks (DNNs) have good nonlinear combination ability and can learn high-order feature interactions. However, DNNs implicitly learn feature interactions at the bit-wise level is not always effective. In this paper, we propose a novel factorization cross network (FCN), which is based on factorization to learn explicit feature crossing through neural network. FCN can learn low- and high-order feature interactions at the vector-wise level with linear time complexity. We introduce deep residual network (DRN) to learn implicit feature interactions. We further use learnable parameters to combine FCN and DRN, and name the new model as deep factorization cross network (DFCN). DFCN can automatically learn low- and high-order explicit and implicit feature interaction information. We have carried out comprehensive experiments on three real-world datasets. Experimental results demonstrate the effectiveness of DFCN, which performs best compared with other competitive models.

Keywords: Recommender systems · Feature interactions · Neural networks · Factorization machines

1 Introduction

With the development of the Internet, recommender systems have been widely used in scenarios such as e-commerce, music, advertising, and news [2,20]. As the scale of the Internet expands, the number of users and items increases exponentially. It has become an important research direction to mine effective feature interactions from massive data [28].

Large-scale recommendation systems have rich features [1,2,7], including attribute features, behavior features, context features, etc. These features are always diverse, high-dimensional and sparse [22], which makes it difficult to learn.

© Springer Nature Switzerland AG 2021
C. S. Jensen et al. (Eds.): DASFAA 2021, LNCS 12683, pp. 195–210, 2021.
https://doi.org/10.1007/978-3-030-73200-4_13

In order to effectively improve recommendation accuracy, it is necessary to learn feature interactions, which mainly include explicit interactions and implicit interactions [4,25]. Explicit feature interactions can be well understood. For example, we found in our research that women often buy skirts in summer to cool off. This is a typical interaction of gender, season, and product type. In addition to explicit feature interactions, many implicit feature interactions are also important [18,29]. A classic example is that men often choose to buy beer and diapers at the same time when shopping in the supermarket. Implicit feature interactions are often valuable but require effective models to mine [6].

Implicit feature interactions can learn hidden and invisible features [18], and explicit feature interactions can combine different features in a direct and effective way [13,19]. It is difficult to define whether explicit or implicit feature interactions are ultimately more effective. A better approach is to combine them and consider both explicit and implicit information [4,6,17,25,29]. A combination method is to use deep neural network (DNN) [16] to learn high-order feature interactions based on the explicit interaction features that have been constructed. Factorisation-machine supported neural network (FNN) [29] uses the pre-trained factorization machines for field embedding before applying DNN. Another combination method is to use two sub-networks to learn explicit and implicit feature interactions. DeepFM [6] uses factorization machine (FM) [19] to learn explicit feature interactions on the wide side, and uses DNN to learn implicit feature interactions on the deep side.

In addition, the expressive ability of low- and high-order combination features is also different. Low-order feature interactions can extract direct and effective information while high-order feature interactions can mine more complex feature information. The most efficient way is to combine explicit and implicit features of both low and high order [6,25]. Motivated by this, we propose a novel model, which is composed of a factorization cross network (FCN) and a deep residual network (DRN) [8]. FCN is a simple but effective network. It is based on factorization and performs second-order crossing of features at each layer. FCN transmits low-order information to the deep layer through residual connection, so the explicit feature interactions can be effectively learned. According to the fusion strategy proposed by us, we use learnable parameters to fuse low- and high-order explicit module FCN and implicit module DRN together, and name the new model as deep factorization cross network (DFCN). The main contributions of this paper can be summarized as follows:

- We propose a novel model DFCN, which can efficiently and automatically learn low- and high-order explicit and implicit feature interactions, avoiding manual feature engineering.
- We designed a simple but efficient model FCN. FCN can efficiently learn low- and high-order explicit feature interactions in a linear time complexity at the vector-wise level, and the order of feature interactions increases with the deepening of the network.
- We propose a new fusion strategy, which enables the model to automatically learn the weights between FCN and DRN for different datasets through

learnable parameters. We conducted comparison experiments on three real datasets, and our model DFCN and FCN perform best compared with other competitive networks.

2 Related Work

For the traditional small-scale recommendation scenarios with fewer users and items, the neighbor-based collaborative filtering algorithms [21,24] have been applied to the recommendation scenarios due to its interpretability. The collaborative filtering algorithms based on matrix factorization (MF) [15] introduce latent vectors to mine the deeper implicit information of users and items, and further improve the prediction accuracy. Although these collaborative filtering algorithms have some effect, they cannot be effectively applied to large-scale sparse datasets [19,28].

Logistic regression (LR) [11] has been widely used in recommender systems, which is easy to parallelize and process hundreds of millions of data. However, LR has limited learning ability and requires feature engineering [23] to increase the learning ability of the model. In practical application scenarios, different features often appear in combination, which requires feature interactions for expression. FMs [19] introduce the second-order cross terms of features by factorization, which can enhance the learning ability of the model. On the basis of FMs, field-aware factorization machines (FFMs) [13] improve model performance by introducing fields, taking into account the differences in interactions between different features.

With the continuous expansion of recommendation scenarios, data features show sparseness, continuous and discrete mixing, and high-dimensionality [22]. Traditional feature interaction methods can no longer meet the needs [28]. Many models combining FM and DNNs are beginning to be proposed, with the goal of learning low- and high-order explicit and implicit feature interactions. FNN [29] initializes model parameters based on FM pre-training, and then uses DNNs to learn high-order feature interactions. Product-based neural network (PNN) [18] designs a product layer to combine features with the inner product and outer product operations. Neural factorization machine (NFM) [9] refers to using DNN to learn high-order combination features based on second-order feature interactions in FM. Wide & Deep [4] incorporates feature engineering on the wide side and feature interactions of DNNs on the deep side. While DNNs is used to learn implicit high-order feature interactions on the deep side, DeepFM [6] directly uses FM on the wide side for second-order crossing of features, and Deep & Cross Network (DCN) [25] explicitly performs low- and high-order feature interactions by designing cross network on the wide side.

In addition, Deep matrix factorization (DMF) [27] is a deep neural network structure that maps users and projects into a potentially structured space. Neural collaborative filtering (NCF) [9] is proposed to replace the inner product of MF with any function through neural structure, so as to solve the expression limitation of MF. Moreover, deep interest network (DIN) [31] and deep interest evolution network (DIEN) [30] based on the sequence of user historical behavior are also proposed. By introducing attention mechanism to establish the relationship between different historical behaviors and candidate ad, DIN can effectively capture the diverse interest information of users.

3   Model

In this section we will elaborate on the architecture of Deep Factorization Cross Network (DFCN). A DFCN model starts with an embedding layer, followed by a factorization cross network and a deep residual network, and ended with the prediction layer which fuse the outputs of the two networks. The complete DFCN model is presented in Fig. 1.

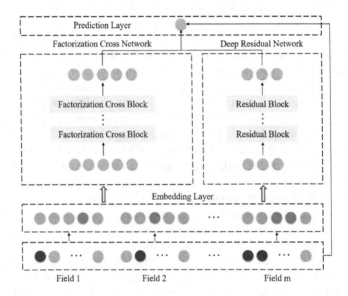

Fig. 1. The architecture of the Deep Factorization Cross Network (DFCN).

3.1   Embedding Layer

Data features in real industrial recommendation scenarios are diverse, high-dimensional and sparse [22]. For example, one input instance {user_id=u10, gender=female, season=summer, clicked_item_cates=[skirt, dance]} is normally encoded into high-dimensional sparse binary vector. Multi-hot encoding is usually used for multi-category features, while one-hot encoding is used for the others. In this way, the input instance can be encoded as:

$$\underbrace{[0,0,1,\cdots,0]}_{\text{user\_id}} \; \underbrace{[1,0]}_{\text{gender}} \; \underbrace{[0,1,0,0]}_{\text{season}} \; \underbrace{[1,0,0,1,\cdots,0]}_{\text{clicked\_item\_cates}} \qquad (1)$$

In order to solve the problem that high-dimensional sparse features are difficult to learn, we use embedding layer to map sparse vectors to low-dimensional dense vectors. Then all the embedding vectors are concatenated together to obtain the overall representation vector. The embedding layer is shown in Fig. 2, which is defined as follows:

$$c_i = w_i^e x_i \tag{2}$$

where $c_i \in R^k$ represents the embedding vector of the i-th field feature, x_i denotes the input vector, $w_i^e \in R^{k \times k}$ is the parameter matrix. By connecting the embedding vectors of each field, we obtain the dense representation vector as $c = [c_1^T, c_2^T, \cdots, c_m^T]$, where m denotes the number of fields.

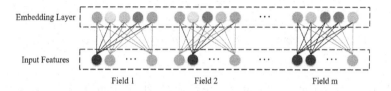

Fig. 2. The architecture of Embedding Layer.

3.2 Factorization Cross Network (FCN)

The core idea of FCN is to learn explicit feature interactions through neural network based on factorization. FCN can effectively learn low- and high-order explicit feature interactions. The factorization cross network is composed of factorization cross block. Figure 3 illustrates the architecture of FCN.

Through stacking factorization cross block, the low- and high-order interactive information can be continuously learned and transmitted to deep layers. As Fig. 3 shows, each factorization cross block has the following formula:

$$z_{l+1} = f_{FC}(z_l, w_l) + z_l \tag{3}$$

where $z_{l+1}, z_l \in R^k$ are vectors represent the outputs of the l-th and $l+1$-th factorization cross layer. The feature interaction function $f_{FC} : R^k \to R^k$ fits the residual of $z_{l+1} - z_l$. $w_l \in R^{k \times k}$ is the parameter matrix to be learned. k denotes the number of neurons in each layer.

The mapping function f_{FC} is the representation of the second-order feature interactions in factorization machines [19]. With the input $z_l = [e_1, e_2, \cdots, e_n]$ of i-th layer, the second-order interaction term of factorization machines is as follows:

$$f_{FM} = \sum_{i=1}^{n-1} \sum_{j=i+1}^{n} \langle v_i, v_j \rangle e_i e_j \tag{4}$$

where latent vector $v_j \in R^k$ represents the factorization information for each element $e_j \in R$, $k \in N_+$ is a hyper parameter that defines the dimension of

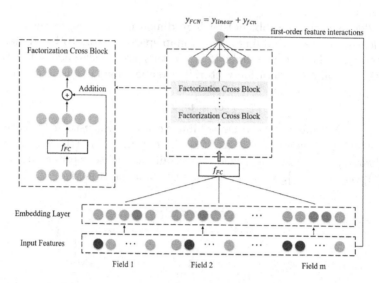

Fig. 3. The components and architecture of the Factorization Cross Network (FCN).

the latent vector, n represents the number of input features. By optimizing the operation, we can get the simplified formula as follows:

$$f_{FM} = \frac{1}{2} \sum_{t=1}^{k} ((\sum_{i=1}^{n} v_{i,t} e_i)^2 - \sum_{i=1}^{n} v_{i,t}^2 e_i^2) \tag{5}$$

Factorization machines take the sum of k computation results as the representation of interaction information. But from another point of view, k computation results can actually be viewed as a k-dimensional vector. Each element of this vector represents the second-order interaction information of the original features. We further take this k-dimensional vector as the k neurons of the neural network. Therefore, we can define the factorization cross function for vector-level feature interaction as:

$$f_{FC} = BN(\frac{1}{2}[(z_l w_l)^2 - z_l^2 w_l^2]) \tag{6}$$

where $w_l = [v_1, v_2, \cdots, v_j]^T \in R^{k \times k}$ is the parameter matrix of each layer, which is composed of latent vectors. $z_l = [e_1, e_2, \cdots, e_k] \in R^k$ is the input of each layer when we set n equal to k. BN refers to batch normalization [12], which is used to ensure the stability of data distribution. Through f_{FC} we can obtain the second-order feature interactions of each layer.

By stacking factorization cross blocks, the network can not only learn rich feature crossing, but also transmit low-order combination information to deep layer. Therefore, the output of the network contains both low- and high-order feature interactions. The order of feature interactions increases with the increase of the number of layers L.

Complexity Analysis. Let L denote the number of layers of FCN, and k denote the dimension of latent vector v_j. The time complexity of FCN is:

$$O(knL)$$

The time complexity is linear, and the calculation process of FCN can be easily achieved by matrix multiplication and addition. Compared with DNN, the computational complexity of FCN is very small. The space complexity is $O(k^2L)$, which is the same with DNN, will not take any more space than DNN.

The low computational complexity ensures the speed of the model, and the appropriate parameters ensure the expressiveness of the model. FCN can fully learn the combination information of low- and high-order features at a fast speed, and obtain the maximum expression ability.

3.3 Deep Residual Network (DRN)

FCN can learn explicit feature interactions by neural network, but it cannot learn implicit feature interactions. We use deep neural network to learn implicit feature interactions, which can be a supplement to FCN.

All the previous methods directly use DNN as the deep network. Although DNN can effectively learn implicit feature interactions, we hope that the network can not only learn high-order information, but also low-order information. Therefore, we use the deep residual network (DRN) [8] to learn the implicit combination of features, which is shown in Fig. 4.

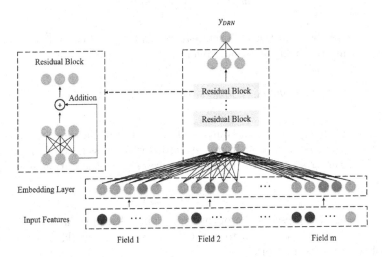

Fig. 4. The components and architecture of the Deep Residual Network (DRN).

The input of deep residual network is the vector generated by the embedding layer. FCN and DRN share the same embedding layer, which can be optimized by both subnetworks simultaneously to be more accurate. DRN is composed of residual block, which can implicitly learn nonlinear feature interactions. As Fig. 4 shows, each residual block has the following formula:

$$a_{l+1} = ReLU(w_l a_l + b_l) + a_l \tag{7}$$

where $a_{l+1} \in R^d$, $a_l \in R^d$ are the outputs of the $l+1$-th and l-th layer, $w_l \in R^{d \times d}$, $b_l \in R^d$ are model weight and bias, and $ReLU$ is an activation function. d denotes the number of neurons in each layer of the residual network.

By stacking residual blocks, the network can not only implicitly learn high-order feature interactions, but also transmit low-order feature interaction information to deep layers. The output of the last layer contains both low-order and high-order feature interactions.

3.4 Prediction Layer

The model consists of explicit feature interactions subnetwork and implicit feature interactions subnetwork. The accuracy and effectiveness of the model can be maximized by learning rich feature interactions.

FCN can learn explicit feature interactions of second-order and high-order through neural network, but first-order feature interactions are also important. In order to learn explicit information more effectively, we introduce a first-order combination of the original input features as follows:

$$y_{linear} = w_0 + \sum_{i=1}^{n} w_i x_i \tag{8}$$

where $w_i \in R$ is the coefficient of each feature, and $w_0 \in R$ is the bias. Then we get the learning result of the total explicit feature interactions as:

$$y_{FCN} = y_{linear} + y_{fcn} = w_0 + \sum_{i=1}^{n} w_i x_i + w_{L_1} z_{L_1} \tag{9}$$

where $y_{FCN} \in R$ represents the output of explicit learning, $w_{L_1} \in R^k$ is the weight vector, z_{L_1} is the output from the factorization cross network, L_1 denotes the depth of the cross network. In addition, we can get the output of DRN as follows:

$$y_{DRN} = w_{L_2} a_{L_2} \tag{10}$$

where $y_{DRN} \in R$ represents the output of implicit learning, $w_{L_2} \in R^d$ is the weight vector, a_{L_2} is the output from the deep residual network, L_2 denotes the depth of the residual network.

After obtaining the outputs of the two subnetworks, most previous work simply adds two outputs together, without taking into account the relationship between the two parts. We consider the output of the two parts to be of different importance. FCN can effectively learn explicit feature interactions, while DRN can learn implicit feature interactions. We cannot determine whether explicit or implicit information is more effective, and cannot simply assume that they are equally important.

Therefore, in order to combine the two parts more accurately and effectively, we introduce learnable parameters for the model to automatically learn the importance of different parts from the data. The final output of the model is:

$$p = sigmoid(\lambda_1 * y_{FCN} + \lambda_2 * y_{DRN}) \tag{11}$$

where $p \in (0, 1)$ denotes the output probability of each sample, sigmoid denotes the sigmoid function, $\lambda_1 \in [0, 1]$ and $\lambda_2 \in [0, 1]$ are weighted coefficients representing the importance of the two networks. λ_1 and λ_2 are designed as follows:

$$\lambda_1 = \frac{e^{\theta_1}}{e^{\theta_1} + e^{\theta_2}}$$
$$\lambda_2 = \frac{e^{\theta_2}}{e^{\theta_1} + e^{\theta_2}} \tag{12}$$

where $\theta_1 \in R$, $\theta_2 \in R$ are learnable parameters for weighing importance. By introducing learnable parameters, FCN and DRN can be effectively combined. In addition, the model also has good universality and can automatically learn personalized parameters for different datasets.

3.5 Network Learning

DFCN can be applied to a variety of tasks in recommended scenarios, including CTR, Ranking. For regression tasks, we can define mean square loss or mean absolute loss. For the ranking task, we can use pairwise ranking loss or triplet ranking loss. In this work, we mainly optimize Log loss for classification task, which is defined as:

$$Loss = -\frac{1}{N} \sum_{i=1}^{N} [y_i \log p_i + (1 - y_i) \log(1 - p_i)] \tag{13}$$

where $y_i \in \{0, 1\}$ is the true label of each sample, and p_i is the prediction probability of each sample, N is the total number of all input samples.

4   Experiment

In this section, we conduct extensive experiments to answer the following questions:

- **RQ1** How does our proposed FCN perform individually in learning feature interactions than other competitive methods?
- **RQ2** Is it necessary to combine explicit and implicit feature interactions as the final output? And how can we combine them?
- **RQ3** How does the settings of networks affect the performance of DFCN?

4.1 Experimental Settings

Datasets. We evaluate our proposed models on the following three datasets:

MovieLens[1] **Dataset** [7]. It is a widely adopted benchmark dataset in movie recommendation, which contains 6,040 users, 3,900 movies and 1,000,209 samples. We take the samples with rating greater than 3 as positive samples and the others as negative samples. The task is to predict whether a user would give a movie a positive rating based on historical behavior.

Frappe[2] **Dataset** [1]. Frappe is a context-aware mobile app recommender system. Researchers collect an app usage log as the frappe dataset, which con-tains user ID, item ID and context features such as weekday, weather, country and so on. It consists of 96203 entries by 957 users for 4082 apps used in various context. We randomly sample an unused app as a negative sample for each posi-tive sample.

Last.FM[3] **Dataset** [3]. It contains social networking, tagging, and music artist listening information from a set of 2000 users from Last.fm online music system. It consists of 92834 user-listened artist relations. Because of the sparsity of the dataset, we randomly sample an unused item as a negative sample for each positive sample.

The details of the datasets are presented in Table 1.

Table 1. Statistics of the evaluation datasets.

Datasets	#instances	#users	#items	#features
MovieLens	1,000,209	6,040	3,900	9,794
Frappe	192,406	957	4,082	5,382
Last.FM	185,668	1,892	17,632	19,524

Evaluation Metrics. We adopt the Area Under the ROC curve (AUC) to evaluate the predict performance of our model. AUC also considers the classifier's ability to classify positive and negative examples. In the case of unbalanced samples, AUC can still make a reasonable evaluation of the classifier. Besides, RelaImpr [10] metric is used to measure relative improvement over models, and RelaImpr is defined as:

$$RelaImpr = (\frac{AUC(measured\_model) - 0.5}{AUC(base\_model) - 0.5} - 1) \times 100\% \qquad (14)$$

Baselines. To evaluate the performance of our model DFCN, we compared with the following competitive models, which are specifically designed for sparse data prediction:

---

[1] https://grouplens.org/datasets/movielens/.

[2] https://www.baltrunas.info/research-menu/frappe.

[3] http://www.lastfm.com.

- FM [19]: Factorization machine is a model based on latent vector to solve the problem of feature interactions. FM and its variants have been widely used in real recommendation systems. We set FM as the base model.
- DNN [5]: Multi-layer fully connected neural network can effectively perform implicit feature interactions without manual feature engineering. It has been widely used in industrial scenarios as a basic model of deep learning.
- Wide & Deep [4]: Wide & Deep is an effective model for CTR prediction task, which consists of wide part and deep part. The wide part is constructed by feature engineering. The deep part is composed of MLP, which can automatically learn implicit feature interactions.
- PNN [18]: PNN designed a product layer to capture interactive patterns between features through inner product and outer product. Then high-order feature interactions are learned by multiple fully connection layers.
- NFM [9]: NFM performs a second-order combination of input features based on embedding vectors. On this basis, high-order feature interactions are learned through multiple fully connection layers.
- DeepFM [6]: DeepFM uses FM to learn explicit feature interactions on the wide side, and uses DNN to learn implicit feature interactions on the deep side. Moreover, wide part and deep part share the same embedding layer.
- DCN [25]: DCN is composed of cross network on the wide side and DNN on the deep side. The cross network can learn explicit feature interactions. DNN can learn implicit feature interactions and act as a complement to the cross network.

Parameter Settings. We randomly split each dataset into training set (90%) and testing set (10%), and we randomly take 10% from training set as validation set for tuning hyper-parameters. The embedding size was searched in [8, 16, 32, 64, 128, 256, 512], the number of hidden layers was searched in [1, 2, 3, 4, 5]. All methods were optimized with Adam [14], where the batch size was set to 512 with considering both training time and convergence rate. The learning rate was searched in [0.0005, 0.001, 0.005, 0.01, 0.05, 0.1]. To prevent overfitting, we used dropout for all neural network models. The dropout ratio was searched in [0, 0.1, 0.2, 0.3, 0.4, 0.5], respectively. Batch normalization was applied to neural networks for all methods. The stopping strategy was performed in training, where we stopped training process if the AUC score on validation set decreased for 5 successive epochs. Moreover, we empirically set the size of the hidden layer the same as the embedding size. For DFCN, we set the size of latent vector as embedding size for simplicity, and we applied batch normalization and dropout to avoid overfitting.

4.2   Performance Comparison with FCN (Q1)

We want to know how FCN performs on its own. It has been proved that it is necessary to construct high-order explicit feature interactions. For example, DCN introduces the cross layer to explicitly carry out high-order feature interactions.

Similarly, FCN learns low- and high-order explicit feature interactions through factorization cross block.

Table 2 shows the experimental results of all models on the three datasets. Surprisingly, our FCN performs better than other models on all datasets, including DeepFM with low- and high-order feature interactions, and DCN with cross network. For example, The AUC score of FCN on the MovieLens dataset is 0.8126, which is larger than 0.8121 of DeepFM and 0.8118 of DCN, and 1.76% higher than that of the base model FM. In fact, FCN can construct high-order feature interactions by combining features layer by layer, and transmit low-order feature information to the deep layer by adding residual connection. The experimental results strongly prove that our FCN model has a powerful ability in feature interactions, because it can learn both low-order and high-order feature interaction information. It is clear that FCN is a very simple but efficient feature interaction learning model.

Table 2. Performance of different models on MovieLens, Frappe and LastFM datasets.

Models	MovieLens		Frappe		LastFM	
	AUC	RelaImpr	AUC	RelaImpr	AUC	RelaImpr
FM	0.8072	0.000%	0.9795	0.00%	0.7906	0.00%
DNN	0.8082	0.33%	0.9814	0.40%	0.7920	0.48%
Wide&Deep	0.8105	1.07%	0.9838	0.90%	0.7995	3.06%
PNN	0.8113	1.34%	0.9845	1.04%	0.8002	3.30%
NFM	0.8116	1.43%	0.9844	1.02%	0.7989	2.86%
DeepFM	0.8121	1.60%	0.9847	1.08%	0.8011	3.61%
DCN	0.8118	1.49%	0.9849	1.13%	0.8006	3.44%
FCN	0.8126	1.76%	0.9851	1.17%	0.8013	3.68%
FCN & DRN	0.8126	1.76%	0.9848	1.11%	0.8007	3.48%
DFCN	**0.8134**	**2.02%**	**0.9865**	**1.50%**	**0.8022**	**3.99%**

4.3 Performance Comparison with DFCN (Q2)

It is well known that combining explicit and implicit information is effective. As shown in Table 2, DeepFM and DCN perform much better than FM and DNN. The proposed DFCN is the combination of explicit FCN and implicit DRN. However, how to effectively combine the two networks is a key problem. Due to the small parameter capacity of cross network, the expressive ability of DCN is limited. DNN of implicit high-order feature interactions should be combined as a supplement. The experiment in Q1 proves that FCN has a strong learning ability, and it is important to fuse FCN and DRN effectively. Therefore, in addition to the weighted fusion strategy proposed by us, we simply add the results of FCN and DRN for experimental comparison, just like most of the previous work.

Table 2 shows the experimental results of simple addition of FCN and DRN, as well as the performance of our comprehensive model DFCN. It can be seen that simply adding FCN and DRN does not lead to a meaningful improvement, indicating that we need to combine them in a more appropriate way. DFCN outperformed the base model by 2.02% on the MovieLens dataset, 1.50% on the Frappe dataset, and 3.99% on the LastFM dataset. DFCN achieves the best results on all datasets and makes a meaningful improvement over FCN. On the one hand, it proves that it is necessary to combine explicit and implicit features, on the other hand, it shows that our weighted fusion strategy is very effective. By weighted fusion of FCN and DRN, DFCN can fully learn feature interaction information, including low- and high-order explicit and implicit feature interactions. In addition, the model can automatically learn the weights of two subnetworks for different datasets.

4.4 Hyper-Parameter Study (Q3)

In this section, we study the impact of hyper-parameters on both FCN and DFCN.

Impact of the Number of Neurons per Layer. To simplify, we set the dimension of latent factor and hidden layer as embedding size. Embedding size increased gradually from 8 to 512, and the performance on the Frappe dataset is shown in Fig. 5 (a). The Frappe dataset contains rich contextual features that describe the data information in detail. Therefore, with the increase of the number of neurons, both DFCN and FCN can more fully learn the information contained in the features, so as to continuously improve the performance of the model.

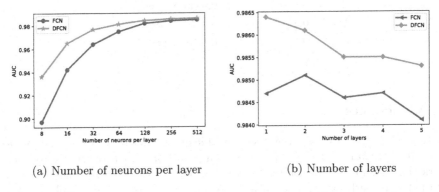

(a) Number of neurons per layer   (b) Number of layers

Fig. 5. Impact of number of neurons per layer and number of layers on performance.

Impact of the Depth of Network. We gradually increased the network depth from 1 to 5 to study the effect of network depth on the model. We can see the model performance on the Frappe dataset in Fig. 5 (b). With the increase of the number of network layers, the performance of DFCN decreases gradually. This is because the abundant features have fully described the characteristics of the dataset, and the data information can be well learned only by combining features of low order. High-order feature interactions will cause the model to learn too much noise from the training data, resulting in overfitting. The performance of FCN increases first and then decreases as the number of layers increases. This indicates that FCN is not sufficient to fully learn the data information by only performing low-order feature combination. But too much feature interactions will also lead to overfitting of the model.

Impact of the Dropout Ratio. Dropout enables the neural network to effectively avoid overfitting problems. We added dropout at each layer of FCN and DRN and kept their dropout ratios consistent. We gradually increased the dropout ratio from 0 to 0.5 at an interval of 0.1, and the performance on the Frappe dataset is shown in Table 3. The most suitable dropout ratio of DFCN is 0.3. The right dropout ratio can improve the generalization of the model. Too small dropout ratio leads to overfitting, while too large leads to underfitting. With the increase of dropout ratio, FCN performs better. It indicates that FCN has a strong learning ability and needs to use dropout to avoid overfitting.

Table 3. Impact of dropout ratio on performance.

Droupout Ratio	AUC	
	FCN	DFCN
0.0	0.9829	0.9840
0.1	0.9833	0.9859
0.2	0.9847	0.9860
0.3	0.9845	0.9864
0.4	0.9850	0.9862
0.5	0.9851	0.9862

5 Conclusion

In this paper, we propose a simple but efficient model named deep factorization cross network (DFCN), which is composed of a factorization cross network (FCN) and a deep residual network (DRN). FCN is a novel model, which can not only perform explicit feature crossing with linear time complexity, but also learn the low- and high-order feature interactions. According to the fusion strategy proposed by us, we use learnable parameters to fuse FCN and DRN together. Therefore, DFCN can effectively learn low- and high-order explicit and implicit combination information of features, avoiding manual feature engineering. Experiments show that our DFCN model performs best on the three real datasets compared with all other competitive models.

There are two directions for future study. First, we consider adding attention mechanisms [26] to the network to improve the learning ability of the model. Because the importance of the interaction between different features is not the same, it is necessary to use attention mechanism to capture different information. In addition, we also plan to explore the interaction of explicit and implicit feature vectors. By introducing the interaction matrix, the complex relationship between explicit and implicit feature vectors can be further explored.

References

1. Baltrunas, L., Church, K., Karatzoglou, A., Oliver, N.: Frappe: understanding the usage and perception of mobile app recommendations in-the-wild. arXiv preprint arXiv:1505.03014 (2015)
2. Bobadilla, J., Ortega, F., Hernando, A., Gutiérrez, A.: Recommender systems survey. Knowl.-Based Syst. **46**, 109–132 (2013)
3. Cantador, I., Brusilovsky, P., Kuflik, T.: Second workshop on information heterogeneity and fusion in recommender systems (hetrec2011). In: Proceedings of the Fifth ACM Conference on Recommender Systems, pp. 387–388 (2011)
4. Cheng, H.T., et al.: Wide & deep learning for recommender systems. In: Proceedings of the 1st Workshop on Deep Learning for Recommender Systems, pp. 7–10 (2016)
5. Covington, P., Adams, J., Sargin, E.: Deep neural networks for Youtube recommendations. In: Proceedings of the 10th ACM Conference on Recommender Systems, pp. 191–198 (2016)
6. Guo, H., Tang, R., Ye, Y., Li, Z., He, X.: DeepFM: a factorization-machine based neural network for CTR prediction. arXiv preprint arXiv:1703.04247 (2017)
7. Harper, F.M., Konstan, J.A.: The MovieLens datasets: history and context. ACM Trans. Interact. Intell. Syste. (TIIS) **5**(4), 1–19 (2015)
8. He, K., Zhang, X., Ren, S., Sun, J.: Deep residual learning for image recognition. In: Proceedings of the IEEE Conference on Computer Vision and Pattern Recognition, pp. 770–778 (2016)
9. He, X., Chua, T.S.: Neural factorization machines for sparse predictive analytics. In: Proceedings of the 40th International ACM SIGIR Conference on Research and Development in Information Retrieval, pp. 355–364 (2017)
10. He, X., et al.: Practical lessons from predicting clicks on ads at Facebook. In: Proceedings of the Eighth International Workshop on Data Mining for Online Advertising, pp. 1–9 (2014)
11. Hosmer Jr., D.W., Lemeshow, S., Sturdivant, R.X.: Applied Logistic Regression, vol. 398. Wiley, Hoboken (2013)
12. Ioffe, S., Szegedy, C.: Batch normalization: accelerating deep network training by reducing internal covariate shift. In: International Conference on Machine Learning, pp. 448–456. PMLR (2015)
13. Juan, Y., Zhuang, Y., Chin, W.S., Lin, C.J.: Field-aware factorization machines for CTR prediction. In: Proceedings of the 10th ACM Conference on Recommender Systems, pp. 43–50 (2016)
14. Kingma, D.P., Ba, J.: Adam: a method for stochastic optimization. arXiv preprint arXiv:1412.6980 (2014)
15. Koren, Y., Bell, R., Volinsky, C.: Matrix factorization techniques for recommender systems. Computer **42**(8), 30–37 (2009)

16. LeCun, Y., Bengio, Y., Hinton, G.: Deep learning. Nature **521**(7553), 436–444 (2015)
17. Lian, J., Zhou, X., Zhang, F., Chen, Z., Xie, X., Sun, G.: xDeepFM: combining explicit and implicit feature interactions for recommender systems. In: Proceedings of the 24th ACM SIGKDD International Conference on Knowledge Discovery & Data Mining, pp. 1754–1763 (2018)
18. Qu, Y., et al.: Product-based neural networks for user response prediction. In: 2016 IEEE 16th International Conference on Data Mining (ICDM), pp. 1149–1154. IEEE (2016)
19. Rendle, S.: Factorization machines. In: 2010 IEEE International Conference on Data Mining, pp. 995–1000. IEEE (2010)
20. Ricci, F., Rokach, L., Shapira, B.: Introduction to recommender systems handbook. In: Ricci, F., Rokach, L., Shapira, B., Kantor, P.B. (eds.) Recommender Systems Handbook, pp. 1–35. Springer, Boston (2011). https://doi.org/10.1007/978-0-387-85820-3_1
21. Sarwar, B., Karypis, G., Konstan, J., Riedl, J.: Item-based collaborative filtering recommendation algorithms. In: Proceedings of the 10th International Conference on World Wide Web, pp. 285–295 (2001)
22. Sarwar, B.M.: Sparsity, scalability, and distribution in recommender systems (2001)
23. Schifferer, B., Deotte, C., Oldridge, E.: Tutorial: feature engineering for recommender systems. In: Fourteenth ACM Conference on Recommender Systems, pp. 754–755 (2020)
24. Su, X., Khoshgoftaar, T.M.: A survey of collaborative filtering techniques. Adv. Artif. Intell. **2009** (2009)
25. Wang, R., Fu, B., Fu, G., Wang, M.: Deep & cross network for ad click predictions. In: Proceedings of the ADKDD 2017, pp. 1–7 (2017)
26. Xiao, J., Ye, H., He, X., Zhang, H., Wu, F., Chua, T.S.: Attentional factorization machines: learning the weight of feature interactions via attention networks. arXiv preprint arXiv:1708.04617 (2017)
27. Xue, H.J., Dai, X., Zhang, J., Huang, S., Chen, J.: Deep matrix factorization models for recommender systems. In: IJCAI, vol. 17, Melbourne, Australia, pp. 3203–3209 (2017)
28. Zhang, S., Yao, L., Sun, A., Tay, Y.: Deep learning based recommender system: a survey and new perspectives. ACM Comput. Surv. (CSUR) **52**(1), 1–38 (2019)
29. Zhang, W., Du, T., Wang, J.: Deep learning over multi-field categorical data. In: Ferro, N., et al. (eds.) ECIR 2016. LNCS, vol. 9626, pp. 45–57. Springer, Cham (2016). https://doi.org/10.1007/978-3-319-30671-1_4
30. Zhou, G., et al.: Deep interest evolution network for click-through rate prediction. In: Proceedings of the AAAI Conference on Artificial Intelligence, vol. 33, pp. 5941–5948 (2019)
31. Zhou, G., et al.: Deep interest network for click-through rate prediction. In: Proceedings of the 24th ACM SIGKDD International Conference on Knowledge Discovery & Data Mining, pp. 1059–1068 (2018)

Tell Me Where to Go Next: Improving POI Recommendation via Conversation

Changheng Li[1], Yongjing Hao[1], Pengpeng Zhao[1(✉)], Fuzhen Zhuang[2,3],
Yanchi Liu[4], and Victor S. Sheng[5]

[1] School of Computer Science and Technology,
Soochow University, Suzhou 215006, China
{chlijq,yjhaozb}@stu.suda.edu.cn,ppzhao@suda.edu.cn
[2] Key Lab of Intelligent Information Processing of Chinese Academy of Sciences
(CAS), Institute of Computing Technology, CAS, Beijing 100190, China
[3] Beijing Advanced Innovation Center for Imaging Theory and Technology,
Academy for Multidisciplinary Studies,
Capital Normal University, Beijing 100048, China
zhuangfuzhen@ict.ac.cn
[4] Management Science and Information Systems,
Rutgers University, Piscataway, NJ, USA
yanchi.liu@rutgers.edu
[5] Department of Computer Science,
University of Central Arkansas, Conway, USA
Victor.Sheng@ttu.edu

Abstract. Next Point-of-Interest (POI) recommendation estimates user preference on POIs according to past check-in history, suffering from the intrinsic limitation of obtaining dynamic user preferences. Conversational Recommendation System (CRS), which can collect dynamic user preferences through conversation, brings a solution to the above limitation. However, none of the existing CRS methods consider the spatio-temporal factors in the action selection phase, which are essential for POI conversational recommendation. In this paper, we propose a new Spatio-Temporal Conversational Recommendation System (STCRS) to fuse the spatio-temporal and dialogue information for next POI recommendation. Specifically, STCRS first learns the spatio-temporal information in the user's check-in history. Then reinforcement learning is used to decide which action (asking for an attribute or recommending POIs) to take at the next turn to achieve successful POI recommendation within as few turns as possible. Finally, our extensive experiments on two real-world datasets demonstrate significant improvements over the state-of-the-art methods.

Keywords: Conversational recommendation · Point-of-interest · Self-attention

© Springer Nature Switzerland AG 2021
C. S. Jensen et al. (Eds.): DASFAA 2021, LNCS 12683, pp. 211–227, 2021.
https://doi.org/10.1007/978-3-030-73200-4_14

1 Introduction

Next Point-of-Interest (POI) recommendation systems are emerging as an essential means of facilitating user's information seeking in many scenarios, like Restaurants and Food (e.g., Yelp) and travel (e.g., Trip Advisor). However, existing methods cannot communicate with users and can only obtain passive feedback from users in the process of POI recommendations since they solely infer user preference on POIs from the historical spatio-temporal check-ins. Users can not actively express his/her immediate preferences, which are often drifting with time. For instance, a user may not be interested in the Great Wall initially, but once he/she happens to watch a video about it, he/she may like it and then become interested in POIs nearby. Such limitation makes it hard to obtain dynamic user preferences, preventing the system from providing accurate POI recommendation.

The Conversational Recommendation System (CRS), which is a recently emerging research topic, brings a solution to the limitation mentioned above. It allows a recommendation system to dynamically obtain user preferences through dialogue and make recommendations appropriately. As the conversational recommendation system became a hot topic, the community began to make great efforts to explore its various settings. Li et al. [8] recommended movies by focusing on natural language understanding and generation. Liao et al. [9] built a multi-modal dialogue systems which can capture rich semantics in the visual modality such as product images. Sun and Zhang [13] get the user attribute preferences by analyzing user utterances and feed them into a policy network. But it only handles single-round recommendation and does not consider the interaction between Conversational Component and Recommend Component. The single-round recommendation ends the conversation after only one recommendation and will not recommend again if the recommendation fails. In contrast, Lei et al. [6] proposed a method on multi-round recommendation setting and considered the interaction between the Conversational Component and the Recommend Component. Multi-round conversational recommendation will continue to ask or recommend after the recommendation is rejected, until the maximum conversation round is reached. Lei et al. [7] proposed a graph-based CRS, which reduces the space of candidate attributes and items by introducing the graph structure.

However, none of the existing methods consider POI conversational recommendation, where spatio-temporal information is essential. Integrating spatio-temporal information can benefit POI conversational recommendation, significantly reducing candidate attribute and item space and hence shortening interaction turns.

To this end, in this paper, we propose a novel POI conversational recommendation framework called Spatio-Temporal Conversational Recommendation System (STCRS), in which an agent can assist users in finding POIs interactively. The agent contains two components, i.e., Spatio-Temporal POI Recommendation module and Spatio-Temporal Policy Network module. Specifically, the Spatio-Temporal POI Recommendation module performs POI prediction via modeling the user's spatio-temporal sequential check-ins and immediate preference confirmed by the user in conversation; the Spatio-Temporal Policy Network module decides which action to take based on state vector with spatio-temporal information. The action may be to ask the user if they like a certain attribute or recommend a ranked list of POIs. We train a policy network with reinforcement learning, maximizing the reward based on the conversation state which integrates spatio-temporal information. Inspired by Lei et al. [6], recommendations will be updated with users' online feedback.

To validate the effectiveness of STCRS, experiments are conducted on CA[1] and Ant Financial[2] datasets. Performances are compared with state-of-the-art CRS methods [6], which also use the information of user, POI, and attribute but do not use spatio-temporal information. We analyze each method's properties under different settings, including binary question and enumerated question. The experimental results show that STCRS outperforms the existing approaches.

In summary, our contributions are listed as follows:

- To the best of our knowledge, this is the first work to investigate POI conversational recommendation systems.
- We propose the STCRS framework to integrate spatio-temporal information into a conversational recommendation system for next POI recommendation.
- We conduct extensive experiments on two real-world datasets. Our experimental results show the superiority and effectiveness of STCRS, comparing with the state-of-the-art methods via comprehensive analysis.

2   Related Work

In this section, we give a brief review of next POI recommendation and conversational recommendation system.

2.1   Next POI Recommendation

The goal of the next POI Recommendation is to recommend a ranked list for user based on his/her historical check-ins. The next POI to be visited by the user should have a higher ranking. Cheng et al. [2] combines the localized region constraints with personalized Markov chains and predicts next POI through the transition probability. However, Markov chains cannot learn complex transitions between POIs. With the development of deep learning, researchers began to try to use neural networks to solve this problem. Feng et al. [4] captures the user transition patterns by using a metric embedding method to embed users and POIs into the same latent space. Xie et al. [15] embeds relationship among POI, Region, Time and Word into a shared low dimensional space. And it uses the linear combination of inner products to compute the score of POIs. Zhang et al. [17] leverages the temporal dependency in user's check-in sequence to

---

[1] https://github.com/WeiqiXu/FoursquareData.
[2] https://tianchi.aliyun.com/dataset/dataDetail?dataId=58.

model user's dynamic preferences. With the continuous development of research, methods of extending existing neural networks have also been proposed. Liu et al. [10] is the first method to model spatio-temporal information for the next POI recommendation, it models spatio-temporal information by replacing the simple transition matrices of RNN with spatio-temporal transition matrices. Zhao et al. [18] is proposed to incorporate spatio-temporal gates to learn the spatio-temporal information of check-in sequences. Xu et al. [16] and Luo et al. [11] use self-attention network for recommendation. J. Ni et al. [12] use a decay function and self-attention block to model time and distance intervals for next POI recommendation. Although significant progress has been made, the intrinsic limitation of obtaining dynamic user preference cannot be avoided.

2.2 Conversational Recommendation

Conversational recommendation system makes it possible to obtain user explicit feedback. Users can interact with CRS using natural language. There are different settings for various problems.

Li et al. [8] built a system which recommend movies through sentiment analysis and movies mentioned in the dialogue. But it only uses mentioned movies for recommender and recommender cannot help generate better dialogue. Chen et al. [1] solved the above two defects by introducing knowledge graph. [9] built a multi-modal dialogue systems which can understand the user's intention more clearly by using the visual information. Zhang [13] built a single-round CRS and used supervised learning and reinforcement learning to train a policy network. But it does not consider the interaction between Conversational Component and Recommend Component. Subsequently, Lei et al. [6] proposed a method on multi-round recommendation setting and consider the interaction between the Conversational Component and the Recommend Component. Recently, Lei et al. [7] proposed a graph-based CRS, the policy network of it has a smaller action space, so it does not require pre-training as adopted in [6,13]. Zhou et al. [19] improved the conversational recommendation by integrating both item-based preference sequence and attribute-based preference sequence. But they don't consider problem about the spatio-temporal factors.

We believe that obtaining a user's dynamic preference and recommend POIs upon attribute feedback is the key to POI conversational recommender system. However, none of the existing CRS works have considered the spatio-temporal information into the conversation. We believe that utilizing the spatio-temporal information will help decrease interaction turns and accurate next POI recommendations.

3   Preliminaries

In this section, we discuss how to integrate spatio-temporal and dialogue information to improve the effect of next POI recommendation. Our framework has

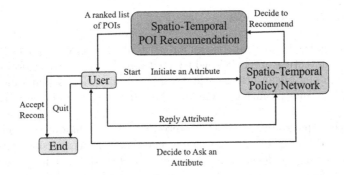

Fig. 1. The workflow of our Spatio-Temporal Conversational Recommendation System.

two components: the Spatio-Temporal POI Recommendation module and the Spatio-Temporal Policy Network module.

We now introduce the notation uesd to formalize our setting. Let u and U denote a user and the user set. v and V denote a POI and the POI set. Each POI v is associated with a set of attributes P_v, such as "School","Hotel", or "Theme Park" for businesses in CA. p and P denote a specific attribute and all attribute. The check-in records of each user are sorted into a sequence $L_u = (v_1^u, v_2^u, ..., v_{|L|}^u)$ by time. And each check-in record v_i^u is associated with its timestamp t_i^u and its geographic coordinates s_i^u.

STCRS aims to recommend POIs that users are interested in within as few turns as possible. The system asks questions based on L_u and the current time to determine the u's current preferences and makes personalized POI recommendations when appropriate to complete the conversation successfully.

Figure1 presents the workflow of our proposed STCRS framework. A CRS session is started with u's current preference attributes p^0, then the CRS removes POIs that do not contain attribute p^0 from the current POI candidate set V_{cand}. Then in each turn t, the STCRS needs to choose an action based on L_u: *recommend* or *ask*.

- If action is *recommend*, the Spatio-Temporal POI Recommendation module will rank the V_{cand}, and recommend a list. If the list contains the POI which user is interested in, user will accept the recommendation and this session ended successfully. Otherwise, user will reject the recommendation.
- If action is *ask*. User needs to clearly express whether he/she prefers the attribute selected by the Spatio-Temporal Policy Network (where the asked attribute is denoted as $p^t \in P$). If the feedback is positive, STCRS will keep the POIs which contain attribute p^t in the V_{cand} and add p^t into P_u (The user preferred attribute set determined through the dialogue). Otherwise, we only remove the POIs which contain attribute p^t.

A CRS session will continue as above until the POI is successfully recommended or the maximum number of turns is reached.

4   Proposed Methods

In this section, we will describe each module in our framework in detail. Figure 2 shows the architecture of STCRS, which contains the Spatio-Temporal POI Recommendation module and the Spatio-Temporal Policy Network module. STCRS executes a loop many times to complete the CRS session. This loop has the following steps:

First, the Spatio-Temporal POI Recommendation module models spatio-temporal information of L_u and scores candidate POIs. Second, the spatio-temporal information learned from L_u will be transformed into four Spatio-Temporal States (s_{st-ent}, s_{st-pre}, $s_{spatial}$ and $s_{temporal}$) and fed into the Spatio-Temporal Policy Network module. Third, the Spatio-Temporal Policy Network module decides which action (asking an attribute or recommending a POI list) to take based on the spatio-temporal and dialogue information. Then, the user simulator generates a reply to the action. Finally, using the information generated by the user simulator's reply to update the corresponding module. Specifically, if the user rejects the recommendation, we use the rejected POIs as negative sample to update the Spatio-Temporal POI Recommendation module. If the user gives feedback on attribute, we update the candidate set and dialogue state.

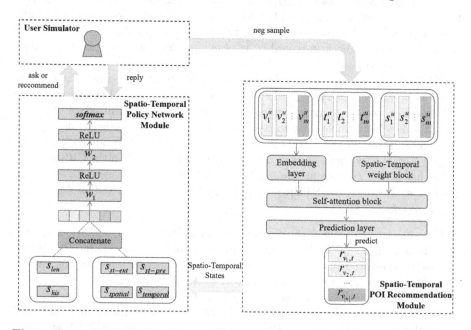

Fig. 2. The architecture of our proposed STCRS. The Spatio-Temporal POI Recommendation Module scores the candidate POIs and offers Spatio-Temporal States to the Spatio-Temporal Policy Network Module. The Spatio-Temporal Policy Network Module decides whether to ask or recommend at each turn. The user simulator responds to questions from the Spatio-Temporal Policy Network Module or recommendations from the Spatio-Temporal POI Recommendation Module. The reply of it is used to update the corresponding module.

4.1 Spatio-Temporal POI Recommendation Module

The goal of Spatio-Temporal POI Recommendation module is to calculate the scores of POIs that user may visit at next time step, given L_u.

To effectively merge spatio-temporal information in conversational recommendation, we use Spatio-Temporal Self-Attention Network (STSAN) to extract spatio-temporal information in L_u. STSAN consists of Embedding layer, Spatio-Temporal weight block, Self-Attention block and Prediction layer.

Embedding Layer: As the length of L_u is not fixed and the check-ins that are too early can not correctly reflect the u's current preferences, we only consider u's recent check-ins of a fixed length. Let m denote the fixed length. And we denote $\hat{L}_u = (v_1^u, v_2^u, ..., v_m^u)$ as the u's recent m check-ins. If the length of u's recent check-ins is less than m, we employ zero-padding to fill the left side of u's recent check-ins sequence until the sequence length is m. We create a POI embedding matrix $\mathbf{M} \in \mathbb{R}^{|V| \times d}$ to encode POI into a unique latent vector, where d is the latent dimension. And we create a positional matrix $\mathbf{P} \in \mathbb{R}^{m \times d}$ to encode m positional information in $\hat{L}_u$. As we mentioned above, each POI v in $\hat{L}_u$ has two embedding ($\mathbf{M}_v$ and $\mathbf{P}_i$, i is the position of v in $\hat{L}_u$). We add them to form the input matrix, the input matrix is defined as follows:

$$
\mathbf{E} = \begin{bmatrix} \mathbf{M}_{v_1} + \mathbf{P}_1 \\ \mathbf{M}_{v_2} + \mathbf{P}_2 \\ ... \\ \mathbf{M}_{v_m} + \mathbf{P}_m \end{bmatrix} \tag{1}
$$

Spatio-Temporal Weight Block: We calculate the temporal and spatial transition matrices $\mathbf{T}^u$ and $\mathbf{S}^u$ based on the temporal and spatial sequence associated with $\hat{L}_u$ (i.e., $(t_1^u, t_2^u, ..., t_m^u)$ and $(s_1^u, s_2^u, ..., s_m^u)$).

$$
\mathbf{T}_{ij}^u = \begin{cases} \Delta t_{ij}^u, & i \geqslant j, \\ 0, & i < j, \end{cases} \tag{2}
$$

$$
\mathbf{S}_{ij}^u = \begin{cases} \Delta d_{ij}^u, & i \geqslant j, \\ 0, & i < j, \end{cases} \tag{3}
$$

where Δt_{ij}^u and Δd_{ij}^u are the time and distance intervals between check-in v_i^u and check-in v_j^u. A decay function is used to convert Δt_{ij}^u and Δd_{ij}^u into a weight. Therefore the temporal and spatial weight matrix $\hat{\mathbf{T}}^u$ and $\hat{\mathbf{S}}^u$ can be calculated as follows:

$$
\hat{\mathbf{T}}_{ij}^u = \begin{cases} g(\Delta t_{ij}^u), & i \geqslant j, \\ 0, & i < j, \end{cases} \tag{4}
$$

$$
\hat{\mathbf{S}}_{ij}^u = \begin{cases} g(\Delta d_{ij}^u), & i \geqslant j, \\ 0, & i < j, \end{cases} \tag{5}
$$

where $g(x) = 1/log(e + x)$., we use $g(x)$ as the decay function. We utilize a weight factor ρ to balance the influence of the temporal and spatial information. We use the weight factor ρ as follows:

$$\mathbf{H} = \rho \cdot \hat{\mathbf{T}} + (1 - \rho) \cdot \hat{\mathbf{S}}, \tag{6}$$

where $0 < \rho < 1$. Finally we use linear transformation on $\mathbf{H}$:

$$\hat{\mathbf{H}} = \mathbf{W}\mathbf{H} + \mathbf{b}, \tag{7}$$

Self-attention Block: We convert the input matrix $\mathbf{E}$ obtained from the embedding layer as follow:

$$\mathbf{W}_{SA} = softmax(\frac{\mathbf{E}\mathbf{W}^Q(\mathbf{E}\mathbf{W}^K)^T}{\sqrt{d}}), \tag{8}$$

$$\mathbf{F} = \hat{\mathbf{H}}\mathbf{W}_{SA}(\mathbf{E}\mathbf{W}^V), \tag{9}$$

where $\mathbf{W}^Q, \mathbf{W}^K, \mathbf{W}^V \in \mathbb{R}^{d \times d}$ are used to project $\mathbf{E}$ into three matrices and $\hat{\mathbf{H}}$ is the output of the Spatio-Temporal weight block. We use layer normalization and residual connection on $\mathbf{F}$. Finally, we feed $\hat{\mathbf{F}}$ into a two-layer fully-connected layer.

$$\hat{\mathbf{F}} = \mathbf{E} + LayerNorm(\mathbf{F}), \tag{10}$$

$$\mathbf{O} = ReLU(\hat{\mathbf{F}}\mathbf{W}_1 + \mathbf{b}_1)\mathbf{W}_2 + \mathbf{b}_2. \tag{11}$$

Prediction Layer: We calculate the dot product of $\mathbf{M}_{v_i}$ and $\mathbf{O}_t$ to get a score $r_{v_i,t}$ of v_i. $\mathbf{O}_t$ is the t-th line of $\mathbf{O}$. The higher the score of v_i, the more likely v_i will be visited.

Network Training: We take the last POI of each user sequence L_u in the training set as a positive sample v_{pos}^u and perform negative sampling to form the positive and negative sample pairs (v_{pos}^u, v_{neg}^u). We optimize the network according to the following formula:

$$loss = -\sum_{u} \sum_{(v_{pos}^u, v_{neg}^u)} [log(\sigma(r_{v_{pos}^u,t})) + log(1 - \sigma(r_{v_{neg}^u,t}))]. \tag{12}$$

4.2   Spatio-Temporal Policy Network Module

The goal of Spatio-Temporal Policy Network module is to learn a policy which selects an action based on the dialogue state at each turn, in order to accomplish successful POI recommendation within as few conversation rounds as possible.

We use the policy network in deep reinforcement learning as our Spatio-Temporal Policy Network. For more introduction about reinforcement learning, please see [14]. The structure of the policy network is shown in the lower left part in Fig. 2. The basic component of reinforcement learning is state, action, reward and policy.

State: The state s_t is the description of the current conversation session. It is composed of s_{st-ent}, s_{st-pre}, s_{his}, s_{len}, $s_{spatial}$ and $s_{temporal}$. s_{st-ent}, s_{st-pre}, $s_{spatial}$ and $s_{temporal}$ come from the Spatio-Temporal POI Recommendation module and contain spatio-temporal information. We believe that such a design can effectively use spatio-temporal information.

- s_{st-ent}: This vector encodes the attribute entropy of the top-k POIs in V_{cand}. The intuition is that using u's historical spatio-temporal information to score V_{cand} and obtaining attribute entropy of the top-k POIs. The attribute with the larger entropy is asked, the more information we can get.
- s_{st-pre}: We treat the POI score as the POI's attribute score. And we use the score of top-k POIs in V_{cand} to calculate the score of attributes. For CA dataset, we calculate the average score of each attribute; For Ant Financial dataset, we calculate the score of each first-level attribute by dividing the total score of second-level attributes by the number of second-level attributes that are not repeated. We use tanh to transform the attribute score. The intuition is that the attributes with higher scores are more likely to be user preference attributes.
- s_{his}: This vector records the dialogue history. Its size is the number of maximum turns. Specifically, we use -1 to represent recommendation is rejected, 0 to present u dislikes the attribute we asking, 1 to present u gives positive feedback to the attribute we asking, 2 to present make a successful POI recommendation.
- s_{len}: This vector is the binary code of the length of V_{cand}. The shorter the length of V_{cand}, the greater the probability of successful recommendation.
- $s_{spatial}$: We denote the average geographic coordinates of $\hat{L}_u$ as $mean_{pos}^{his}$, the average geographic coordinates of the top-k POIs in V_{cand} as $mean_{pos}^{cand}$, the variance of the geographic coordinates of $\hat{L}_u$ as var_{pos}^{his}, and the variance of the geographic coordinates of the top-k POIs in V_{cand} as var_{pos}^{cand}. Using $mean_{pos}^{his} \oplus mean_{pos}^{cand} \oplus \tanh(var_{pos}^{his}/var_{pos}^{cand})$ as spatial information of the current dialogue round. $\oplus$ is used for vector concatenate. The intuition is that if the spatial information of V_{cand} and $\hat{L}_u$ is similar , a recommendation should be made.
- $s_{temporal}$: We assume that t_i^u is the time u seeked POI recommendations. According to the time period (morning or afternoon) when u visited v_i^u, count the attribute ratio of POIs during this time period in $\hat{L}_u$. We denote this ratio as f_{his}. And we count the attribute ratio of the top-k POIs in V_{cand}. We denote this ratio as f_{cand}. The cosine similarity of f_{his} and f_{cand} is denoted as cos_{his_cand}. Using $f_{his} \oplus f_{cand} \oplus cos_{his\_cand}$ as temporal information of the current dialogue round. The intuition is that if one attribute is visited multiple times by u in a period of time, u is more likely to prefer this attribute in the same period of time, and we should ask questions about it.

Action: The Spatio-Temporal Policy Network module needs to select an action a_t at time step t. Two kinds of actions can be selected. One is to make a POI recommendation. The other is to ask an attribute that the user may prefer in this CRS session. So the space of action is the number of attribute $|P| + 1$.

Reward: The reward follows [6], (1) r_{rec_suc}, we give a strongly positive reward when the POI recommendation is successful, (2) r_{rec_unsuc}, a slightly negative reward is given when the recommendation is rejected, (3) r_{ask_suc}, a strongly positive reward when the user accept the asked attribute, (4) r_{ask_unsuc}, a slightly negative reward, (5) r_{fail}, a strongly negative reward if the user quits the conversation, (6) r_{prev}, a slightly negative reward to avoid overly length conversations.

Policy: We denote the policy network as $\pi(a_t|s_t)$. It maps the current conversation state s_t into the action space. The Spatio-Temporal Policy Network module selects an action a_t according to the result of output layer and gets an immediate reward r_t at each turn. The goal of the Spatio-Temporal Policy Network module is to maximize the episodic expected reward of a CRS session. Policy Network will select high-value action after trial and error. The policy gradient method is used to optimize the network, formulated as follows:

$$\theta \leftarrow \theta - \alpha \nabla log \pi_\theta(a_t|s_t) R_t, \tag{13}$$

$$R_t = \sum_{t'=t}^{T} \gamma^{T-t'} r_{t'}, \tag{14}$$

where θ and α are the parameters and learning rate of policy network respectively, γ is the discount factor. Note that if θ is initialized randomly, the learning can converge slowly or fail. To address this issue, we follow [6] to conduct the rule-based pre-training.

5 Experiments

In this section, we conduct experiments to evaluate our proposed STCRS framework on two real-world datasets. Our experiment are guided by the following Research Questions(RQs).

- **RQ1.** How does STCRS perform compared to the state-of-the-art methods for conversational recommendation?
- **RQ2.** Is the design of the state vector effectively utilize the spatio-temporal information to complete the POI conversational recommendation?
- **RQ3.** How does the hyper-parameters affect the method performance (e.g., the discount factor γ and the learning rate α of the Spatio-Temporal Policy Network module)?

Table 1. Datasets statistics.

Dataset	CA	Ant Financial
#Users	2389	2481
#POIs	9144	1481
#Check-ins	93598	26808
#Attributes	34	142

5.1 Settings

5.1.1 Datasets
For better comparison and make POI conversational recommendation, we conduct experiments on two datasets: CA and Ant Financial. CA is a Foursquare dataset from users whose homes are in California, collected from January 2010 to February 2011 and used in [5]. Ant Financial is an Internet financial services company in China and the dataset provides shop information and Alipay user's payment log and users' browsing log from 07.01.2015 to 10.31.2016 (except 2015.12.12). Follow [6], we remove the duplicated user-POI check-ins in our datasets and only keep the first check-in. The statistics of the two datasets are summarized in Table 1. We sort the check-ins of each user by time and take the early 70% of user's check-ins as the training data, the last 10% as the testing data, the remaining 20% as the validation data.

For better comparison, we follow [6] to conduct experiments on CA for binary question scenario and Ant Financial for enumerated question scenario. In binary question scenario, the user answers yes or no when the user is asked a question about attribute. In enumerated question scenario, we build a 2-layer taxonomy which includes 15 first-layer categories and 142 second-layer categories. For example, The first-level category "city" includes 88 second-level categories, and each second-level category represents a specific city.

5.1.2 User Simulator
Conversational recommendation is a process of continuous interaction with users. CRS needs to interact with user to obtain the dynamic user preferences and make POI recommendations. But CRS is too expensive to be applied to real users to train from scratch. To solve these problems, we follow [6] to build a user simulator. When the user simulator simulates one conversation session for a user-POI (u, v) check-in, it restricts u to only prefer the attributes in P_v and only accepts the recommendation containing v.

5.1.3 Training Details
We set the length of recommendation list as 10, maximum turn as 15 on CA dataset, and maximum turn as 6 on Ant Financial dataset. Maximum turn of CA follows [6] and the standard for setting maximum turn of Ant Financial

is that the highest success rate of Max Entropy just exceeds 90%. Following [6,13], we perform two-stage training: (1) An offline training for Recommend Component. We use the training set to optimize Spatio-Temporal POI Recommendation module (Eq. (12)). The goal is to assign higher score to the check-in POI for each users. All hyper-parameters are tuned on the validation set: For CA dataset, the batch size is set as 64, the learning rate is 0.0001, the m is 10, the embedding size is 40, the dropout rate is 0.5, and the size of block and head is 1; For Ant Financial dataset, the learning rate is set to 0.001, the m is 9, the embedding size is 110, and the other are the same as the CA dataset. (2) An online training for Conversational Component. We use a user simulator to interact with STCRS to train the Spatio-Temporal Policy Network module using the validation set. The k is set as 100. The rewards are as follow: $r_{prev} = -0.01$, $r_{rec\_suc} = 1 + r_{prev}$, $r_{fail} = -0.3$, $r_{ask\_suc} = 0.1 + r_{prev}$, $r_{ask\_unsuc} = r_{prev}$. On CA dataset, $r_{rec\_unsuc} = r_{prev}$; On Ant Financial dataset, $r_{rec\_unsuc} = -0.1$. We use the AdamOptimizer to optimize the policy network.

5.1.4 Evaluation Metrics

To evaluation follows [6]. We use Success Rate (SR@t) [13] and Average Turns (AT) which is average conversation rounds for successful POI recommendations to measure the ratio of successful POI conversational recommendation and the effectiveness of conversation. Larger SR denotes better performance and shorter AT denotes more efficient conversation. In the offline training of Recommend Component, we use the NDCG@10 and HR@10 to find the best Recommend Component.

5.2 Baselines

To emphasize the importance of spatial-temporal information and the fairness of comparison, we compared our framework with the following CRS methods.

- **Max Entropy.** A ruled-based method. Generating random numbers based on the current number of candidate items to decide whether to ask or recommend. When asking a question, it chooses an attribute which has not been asked and has the maximum entropy in the candidate set. We use it for pre-training. Details can be found in [6].
- **Abs Greedy** [3]. This method only have a recommendation component. It only recommends items and updates itself when the recommendation is rejected, until it recommends the correct item or failed after reach the maximum number of turn.
- **CRM** [13]. This is a CRS method using reinforcement learning. It uses belief tracker to analyze the preferences expressed by user utterances. The output of belief tracker is fed into policy network for deciding which action should take at next step. We follow [6] to adapt it to the multi-round conversational recommendation scenario.
- **EAR** [6]. This method is based on multi-round conversational recommendation setting and emphasizes the interaction between conversation component

and recommendation component. Using BPR algorithm to update attribute-aware FM.

5.3   Performance Comparison (RQ1)

In this section, we compare our framework STCRS with four state-of-the-art baselines.

Table 2. Experimental results of STCRS and baselines (RQ1).

	CA				Ant Financial			
	SR@5	SR@10	SR@15	AT	SR@2	SR@4	SR@6	AT
Max Entropy	0.052	0.141	0.199	13.656	0.066	0.639	0.920	4.169
Abs Greedy	0.204	0.288	0.339	11.831	**0.381**	0.614	0.718	3.804
CRM	0.197	0.292	0.357	11.846	0.116	0.769	0.941	3.76
EAR	0.193	0.317	0.394	11.69	0.106	0.792	0.968	**3.673**
STCRS	**0.216**	**0.349**	**0.416**	**11.388**	0.058	**0.805**	**0.984**	3.879

Table 2 presents the statistics of method's performance. As can be clearly seen, our STCRS significantly outperforms the state-of-the-art baselines on various setting. This proves our hypothesis that considering spatio-temporal information in conversational recommendation can better make POI conversational recommendation. In order to better show the performance of STCRS and compare it with other baselines, we analyze the performance of STCRS in each round in Fig. 3.

Figure3 shows the Success Rate* (SR*) @t at different turns ($t = 1$ to 15 on CA and $t = 1$ to 6 on Ant Financial). SR* denotes the comparison of each method against the strongest baseline EAR, indicated as $y = 0$ in the figure.

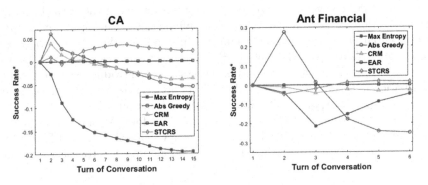

Fig. 3. Success Rate* of compared methods at different conversation turns on CA and Ant Financial (RQ1).

There is a common trend in two datasets. The performance of STCRS is weak at the beginning of a conversation, but it starts to grow and reach a stable state in the subsequent turns. The poor performance at first and the significant improvement afterward illustrate that it is difficult to successfully recommend the POI that the user is interested in by only using the Spatio-Temporal POI Recommendation module to extract the spatial-temporal information from the user's historical check-in records at the beginning of conversation, but the user's dynamic preferences obtained through the conversation can effectively improve the success rate of the POI recommendation. The subsequent excellent performance shows that introducing user's spatio-temporal information in the conversational recommendation can help the Spatio-Temporal Policy Network module choose the attributes that users are more likely to prefer to ask questions and improve the success rate of the POI conversational recommendation.

The performance of Max Entropy on Ant Financial gradually improve in turns 4–6. But the performance on CA continues to decline. The key reasons are that the POI in Ant Financial has more attribute information and the setting of Ant Financial is to ask enumerated question. User's response will sharply shrink the candidate POIs in this setting.

Comparing with EAR, STCRS has a greater advantage on CA. As we mentioned above, the POI in CA has fewer attributes and the setting of CA is to ask binary questions, so the performance of STCRS shows that the spatio-temporal information is helpful to choose the right attributes to ask questions and better complete POI conversational recommendations.

5.4   Ablation Studies on State Vector (RQ2)

In order to explore the effect of each part of the state vector, we remove or replace these parts one by one and check the change. Table 3 presents the statistics of our framework's performance on two conversation scenarios (binary questions and enumerated questions). s_{ent} and s_{pre} is the attribute entropy and attribute preference of all POIs in V_{cand}.

As can be clearly see, $s_{st\text{-}ent}$ is the most important part on two conversation scenarios. If we remove $s_{st\text{-}ent}$, although it obtains improvement at the beginning of conversation, SR@6 and SR@15 greatly suffers, due to the system makes POI recommendation before obtaining enough information. We replace $s_{st\text{-}ent}$ and $s_{st\text{-}pre}$ with s_{ent} and s_{pre}. Except for SR@2, the performance of other indicators have a decline. This shows that it is necessary to introduce spatio-temporal information in attribute entropy and attribute preference. Apart from $s_{st\text{-}ent}$ and $s_{st\text{-}pre}$, $s_{spatial}$ and $s_{temporal}$ also have a positive contribution to our framework. The spatio-temporal information is more important for CA (binary question). A reasonable explanation is that POI in CA has fewer attributes than POI in Ant Financial, so spatio-temporal information is more important to select attributes which users prefer at the current time.

Table 3. Performance of removing or replacing one component of state vector from STCRS (RQ2).

	CA				Ant Financial			
	SR@5	SR@10	SR@15	AT	SR@2	SR@4	SR@6	AT
$-s_{st\text{-}ent}$	0.218	0.296	0.35	11.664	**0.182**	0.629	0.836	4.024
$-s_{st\text{-}pre}$	0.229	0.344	0.404	11.352	0.088	0.779	0.976	3.911
$-s_{st\text{-}ent} + s_{ent}$	0.184	0.271	0.326	11.724	0.051	0.789	0.98	3.901
$-s_{st\text{-}pre} + s_{pre}$	**0.242**	0.342	0.4	**11.253**	0.069	0.778	0.973	3.921
$-s_{spatial}$	0.235	0.332	0.393	11.354	0.051	0.794	0.976	3.89
$-s_{temporal}$	0.201	0.295	0.352	11.839	0.074	0.794	0.977	**3.855**
STCRS	0.216	**0.349**	**0.416**	11.388	0.057	**0.813**	**0.982**	3.86

5.5 Sensitivity Analyses of Hyper-parameters (RQ3)

In this section, we explore the influences of the discount factor γ and the learning rate α in STCRS.

Influence of the Learning Rate α. We first fix optimizer of the policy network is Adam, γ is 0.6 on CA, γ is 0.8 on Ant Financial and vary α to explore the influence of α. We choose α in {0.0002, 0.0005, 0.001, 0.002}. As is shown in Table 4, when α is 0.0005, the performance of STCRS is best on two datasets. Although some indicators are better when α is set as other values, 0.0005 is the best value for the overall effect.

Influence of the Discount Factor γ. To explore the influence of γ, we fix the learning rate α as 0.001. And we search γ from {0.6, 0.7, 0.8, 0.9, 0.95, 0.99}.

Table 4. Influence of the learning rate α and the discount factor γ (RQ3).

		CA				Ant Financial			
		SR@5	SR@10	SR@15	AT	SR@2	SR@4	SR@6	AT
α	0.0002	**0.233**	0.333	0.404	11.351	0.051	0.794	0.978	3.883
	0.0005	0.228	**0.356**	**0.418**	**11.266**	0.051	**0.814**	**0.983**	3.923
	0.001	0.216	0.349	0.416	11.388	0.057	0.813	0.982	3.89
	0.002	0.198	0.29	0.338	11.885	**0.091**	0.779	0.97	**3.882**
γ	0.6	0.216	0.349	0.416	11.388	0.056	0.788	0.976	**3.867**
	0.7	0.224	0.331	0.395	11.45	**0.061**	0.801	0.982	3.917
	0.8	**0.23**	0.34	0.404	**11.372**	0.057	**0.813**	0.982	3.89
	0.9	0.164	0.319	0.385	11.847	0.058	0.805	0.984	3.879
	0.95	0.204	**0.352**	0.414	11.437	0.048	0.787	0.981	3.89
	0.99	0.169	0.337	**0.429**	11.645	0.036	0.801	**0.985**	4.038

From Table 4, we can see that the best γ for SR@2 is 0.7, the best γ for SR@5, SR@4 and AT on CA is 0.8, the best γ for SR@10 is 0.95, and the best γ for SR@15 and SR@6 is 0.99. From the perspective of successfully completing as many POI conversational recommendations as possible, the best γ is 0.99.

6 Conclusion

In this paper, we proposed a novel framework Spatio-Temporal Conversational Recommendation System (STCRS). We employed the Spatio-Temporal Self-Attention Network to extract the spatio-temporal information of user's check-in history, and used reinforcement learning to train a policy network to make decision at each turn. The state vector was designed carefully, which can build a bridge between Spatio-Temporal POI Recommendation module and Spatio-Temporal Policy Network module for communication. To the best of our knowledge, STCRS is the first method to use the spatio-temporal and dialogue information for next POI recommendation. We compared the Success Rate and the Average Turns of STCRS with CRS methods, and our experimental results show the improvement of our framework.

Acknowledgements. This research was partially supported by NSFC (No. 61876117, 61876217, 61872258, 61728205), ESP of the State Key Laboratory of Software Development Environment, and PAPD of Jiangsu Higher Education Institutions.

References

1. Chen, Q., et al.: Towards knowledge-based recommender dialog system. arXiv preprint arXiv:1908.05391 (2019)
2. Cheng, C., Yang, H., Lyu, M.R., King, I.: Where you like to go next: successive point-of-interest recommendation. In Proceedings of the 23rd International Joint Conference on Artificial Intelligence, pp. 2605–2611 (2013)
3. Christakopoulou, K., Radlinski, F., Hofmann, K.: Towards conversational recommender systems. In: Proceedings of the 22nd ACM SIGKDD International Conference on Knowledge Discovery and Data Mining, pp. 815–824 (2016)
4. Feng, S., Li, X., Zeng, Y., Cong, G., Chee, Y.M.: Personalized ranking metric embedding for next new poi recommendation. In: Proceedings of the 24th International Conference on Artificial Intelligence, pp. 2069–2075 (2015)
5. Gao, H., Tang, J., Liu, H.: gSCorr: modeling geo-social correlations for new check-ins on location-based social networks. In: Proceedings of the 21st ACM International Conference on Information and Knowledge Management, pp. 1582–1586 (2012)
6. Lei, W., et al.: Estimation-action-reflection: towards deep interaction between conversational and recommender systems. In: Proceedings of the 13th International Conference on Web Search and Data Mining, pp. 304–312 (2020)
7. Lei, W., et al.: Interactive path reasoning on graph for conversational recommendation. In: Proceedings of the 26th ACM SIGKDD International Conference on Knowledge Discovery & Data Mining, pp. 2073–2083 (2020)

8. Li, R., Kahou, S.E., Schulz, H., Michalski, V., Charlin, L., Pal, C.: Towards deep conversational recommendations. In: Advances in Neural Information Processing Systems, pp. 9725–9735 (2018)
9. Liao, L., Ma, Y., He, X., Hong, R., Chua, T.-S.: Knowledge-aware multimodal dialogue systems. In: Proceedings of the 26th ACM International Conference on Multimedia, pp. 801–809 (2018)
10. Liu, Q., Wu, S., Wang, L., Tan, T.: Predicting the next location: a recurrent model with spatial and temporal contexts. In: Thirtieth AAAI Conference on Artificial Intelligence (2016)
11. Luo, A., et al.: Collaborative self-attention network for session-based recommendation
12. Ni, J., et al.: Spatio-temporal self-attention network for next poi recommendation. In: Asia-Pacific Web (APWeb) and Web-Age Information Management (WAIM) Joint International Conference on Web and Big Data, pp. 409–423 (2020)
13. Sun, Y., Zhang, Y.: Conversational recommender system. In: The 41st International ACM SIGIR Conference on Research & Development in Information Retrieval, pp. 235–244 (2018)
14. Sutton, R.S., Barto, A.G.: Reinforcement Learning: An Introduction. MIT Press, Cambridge (2018)
15. Xie, M., Yin, H., Wang, H., Xu, F., Chen, W., Wang, S.: Learning graph-based poi embedding for location-based recommendation. In: Proceedings of the 25th ACM International on Conference on Information and Knowledge Management, pp. 15–24 (2016)
16. Xu, C., et al.: Long-and short-term self-attention network for sequential recommendation. Neurocomputing **423**, 580–589 (2021)
17. Zhang, Y., et al.: Sequential click prediction for sponsored search with recurrent neural networks. arXiv preprint arXiv:1404.5772 (2014)
18. Zhao, P., Zhu, H., Liu, Y., Li, Z., Xu, J., Sheng, V.S.: Where to go next: a spatio-temporal LSTM model for next poi recommendation. arXiv preprint arXiv:1806.06671 (2018)
19. Zhou, K., et al.: Leveraging historical interaction data for improving conversational recommender system. In: Proceedings of the 29th ACM International Conference on Information & Knowledge Management, pp. 2349–2352 (2020)

MISS: A Multi-user Identification Network for Shared-Account Session-Aware Recommendation

Xinyu Wen[1], Zhaohui Peng[1(✉)], Shanshan Huang[2], Senzhang Wang[3],
and Philip S. Yu[4]

[1] School of Computer Science and Technology, Shandong University, Qingdao, China
wenxinyu@mail.sdu.edu.cn, pzh@sdu.edu.cn
[2] Juhaokan Technology Co., Ltd., Qingdao, China
huangshanshan1@hisense.com
[3] School of Computer Science and Engineering, Central South University,
Changsha, China
szwang@csu.edu.cn
[4] Department of Computer Science, University of Illinois at Chicago, Chicago, USA
psyu@uic.edu

Abstract. The user's interactions with the system within a given time frame are organized into a session. The task of session-aware recommendation aims to predict the next interaction based on user's historical sessions and current session. Though existing methods have achieved promising results, they still have drawbacks in some aspects. First, most existing deep learning methods model a session as a sequence, but neglect the complex transition relationships between items. Second, a single account is usually regarded as a single user by default, where the scenario of multiple users sharing the same account is ignored. To this end, we propose a **M**ulti-user **I**dentification network named MISS for the **S**hared-account **S**ession-aware recommendation problem. MISS consists of two core components: one is the Dwell Graph Neural Network (DGNN), which incorporates item dwell time into the gated graph neural network to capture user interest drift across sessions. The other is a Multi-user Identification (MI) module, which draws on the attention mechanism to distinguish behaviors of different users under the same account. To verify the effectiveness of MISS, we construct two data sets with shared account characteristics from real-world smart TV watching logs. Extensive experiments conducted on the two data sets demonstrate that MISS evidently outperforms the state-of-the-art recommendation methods.

Keywords: Shared-account recommendation · Session-aware recommendation · Graph neural networks · Attention

1 Introduction

Recommender systems have a wide range of applications in many fields, such as e-commerce, news information websites, music and video websites [5,10,16].

© Springer Nature Switzerland AG 2021
C. S. Jensen et al. (Eds.): DASFAA 2021, LNCS 12683, pp. 228–243, 2021.
https://doi.org/10.1007/978-3-030-73200-4_15

User's interactions within a given time frame are usually organized in chronological order into a session. The goal of session-based recommendation aims to predict the next interaction for each independent session. Generally, the user's historical sessions reflect the user's long-term preferences, while the current session represents the main purpose at the moment. Therefore, some studies combine both the historical sessions and the current session to make recommendations, which is called session-aware recommendation.

Most session-aware recommendation studies are conducted on the session-based recommendation methods [31]. Recurrent neural network can reasonably model sequential characteristics in session [6,7,22]. Other neural networks, such as attention mechanism and graph neural networks, have also been successfully applied to session-based recommendation, capturing user interest in each session [26,28]. The aforementioned session-based recommendation methods only make recommendations for each anonymous session, but do not take into account the influence of long-term preferences in the user's historical sessions. Therefore, on this basis, many subsequent works have carried out research on session-aware recommendation methods. To capture user's long- and short-term preferences, the work of [17] proposes a hierarchical RNN framework to model user's interest drift across sessions. The work in [29] designs a hierarchical architecture based on GRU and Temporal Convolutional Network to capture the long-term interests and short-term interactions. In [31], the authors improve the session-graph in [26] and propose a personalized graph neural network with attention mechanism to model the effect of historical sessions on current session.

Despite the effectiveness, few existing studies have considered the shared-account problem, where multiple individual users share the same account in the system. Such shared-account scenarios are very common in many real-world application fields. Take Fig. 1 as an example, in the smart TV recommendation scenario, members of a family share the same account to watch videos. The watching logs of different members are mixed together and recorded as behaviors of a single account, which makes it harder to generate accurate personalized recommendations for each member. As shown in Fig. 1, three family members use the same account to watch their favorite video types. The man usually watches action films or war films while the woman prefers romantic love stories, and the little girl likes animated films. When the little girl is watching videos on the smart TV, she expects to be recommended for animated films that she will be interested in. However, since the watching logs of all members are mixed together, she will be recommended for not only animated films, but also action or romantic films that other members like. Identifying the individual users of the shared-account to recommend the items is the main challenge of shared-account recommendation. Prior recommendation methods for shared accounts usually capture user preferences by extracting latent features from high dimensional spaces that describe the relationships among users under the same account [23,25,30]. However, important sequential features are usually ignored in these studies, or they rely on explicit user ratings [15]. Therefore, these methods cannot be applied directly to shared-account session-aware recommendations.

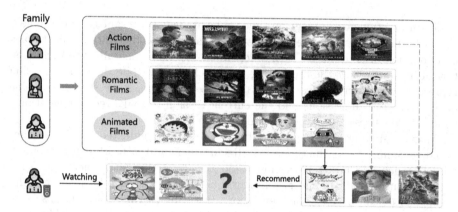

Fig. 1. An example of a shared-account scenario: multiple family members share the same account to watch videos on smart TV, and their watching logs are mixed together. When the little girl is watching videos on the family account, she will not only be recommended for her favorite animated films, but also for action movies or romantic movies that other family members prefer.

To address the above issues, we propose a novel multi-user identification framework MISS for shared-account session-aware recommendation problems. The MISS model consists of two core components: one is the Dwell Graph Neural Network (DGNN), which incorporates item dwell time into the gated graph neural network to capture complex items transition relationships across sessions. The other is a multi-user identification (MI) module, which utilizes attention mechanism to distinguish different user behaviors under the same account, thus to recommend the right item to the right user.

To verify the effectiveness of MISS, we construct two data sets with shared account characteristics, named FamTV-SA and FamTV-SAS. Both data sets are obtained from real-world smart TV watching logs. Extensive experiments are conducted on the two data sets, and the experimental results show that MISS significantly outperforms the state-of-the-art recommendation methods.

The main contributions of this work are summarized as follows:

– We introduce the task of shared-account session-aware recommendation, which has rarely been paid attention to in existing research. A novel MISS model is proposed to solve the problem.
– We propose a dwell graph neural network (DGNN) and a multi-user identification (MI) module, which can capture complex items transition relationships and distinguish different user behaviors under the same account.
– We conduct empirical studies on two real-world data sets. Extensive experiments demonstrate the performance of our proposed method and the contribution of each component.

2 Related Work

2.1 Session-Based Recommendation

Collaborative filtering algorithms have been widely used in traditional recommendation systems. The item-to-item approaches [14,19] recommended similar items for users, where the similarities were calculated based on the simultaneous occurrences in the same session. These methods are insufficient in considering items sequence order and give recommendation merely based on the last interaction, thus cannot well model the continuous preference information of items in the session. Subsequent researches applied the Markov chains to the recommendation method, predicting the next click based on the user's previous interaction. The work of [21] treated recommendation generation as sequential optimization problem and proposed a Markov Decision Process (MDPs) based method, calculating the next action through the state transition probabilities among items. In [3] the authors took playlists as Markov chains, and proposed Latent Markov Embedding (LME) to learn the representations of songs for playlists prediction. The main drawback of Markov-chains-based method is that the state space quickly becomes unmanageable when considering all possible sequences [11].

In recent years, deep neural networks have been successfully applied to recommendation [24]. The work of [6] first applied the recurrent neural networks (RNNs) to session-based recommendation, and introduced several modifications to classic RNNs making it more viable for this specific problem. The initial work has been extended in subsequent research. In [7] they proposed a number of parallel RNN architectures to model sessions based on rich items feature representations such as pictures and text description. The work of [22] proposed data augmentation technique and considered shifts in the input data distribution to enhance the performance of RNN-based models. In [8] a hybrid method that combines the RNN-based model with the kNN method was proposed, where the co-occurrence signals were used to predict sequential patterns. Recent work [26] modeled session sequences as graph-structured data, and utilized the Gated Graph Neural Networks to model complex transition relationships between items. However, these methods mentioned above can only use the current anonymous session or single sequence to make recommendations.

2.2 Session-Aware Recommendation

Session-aware recommendation aims to predict the next interaction based on the user's historical sessions and current session. In [17], the authors believed that historical interactions reflect user's interest, and thus proposed a hierarchical RNN model to capture user's long- and short-term preferences. The work [12] introduced a dual attentive neural network to exploit user's personalized preference and main purpose in current session. In [29] a hierarchical architecture that contains GRU and Temporal Convolutional Network was proposed, capturing both the long-term interests and the short-term interactions within sessions to output a dynamic user embedding and make recommendations. Recent work [31] proposed a personalized session-aware recommendation method, which utilized graph neural network to extract user personalized information and attention mechanism to model the effect of historical sessions on current sessions. Despite the effectiveness, none of the above studies take account of the shared account scenario.

2.3 Shared-Account Recommendation

There are relatively few recommendation studies focusing on shared accounts. The work of [30] used linear subspace clustering to study user identification, and recommended items that are most likely to be rated highly by each user. In [25], the authors supposed different users within a shared account get used to consuming services in different period. They decomposed users based on mining different preferences over different periods from consumption logs, and then utilized a standard User-KNN to make recommendation for each identified user. The work [1] used the Apriori algorithm to decompose users under the same account. By analyzing the similarity of the proportion of each type of items under a time period, the work [27] judged whether a session is generated by the same user and then made personalized recommendation to the identified users.

Although these studies have been proven to be effective in many applications, they are designed for static rating data, or ignore the important sequential characteristics of sessions. None of these methods can be directly applied in the shared-account session-aware recommendation scenario.

3 Method

3.1 Problem Formulation

Let $\mathcal{I}$ denote the set of all items and $\mathcal{K}$ denote the set of all accounts involved in the system. For each account $k \in \mathcal{K}$, the interactive items within a certain time frame are organized in chronological order as a session $s_i = \{v_{i,1}, v_{i,2}, ..., v_{i,m_i}\}$, where $v_{i,j} \in \mathcal{I}$ stands for an item the user has clicked within the session s_i, and m_i represents the total number of items in session s_i. All sessions of account k are represented as $\mathcal{S}^k = \{s_1, s_2, ..., s_{n_k}\}$, where n_k is the number of sessions of the account k. For convenience, we use n instead of n_k. The last session in S^k is the current session $\mathcal{S}_c^k = \{s_n\}$, and the remaining sessions are historical sessions $\mathcal{S}_h^k = \{s_1, s_2, ..., s_{n-1}\}$. The goal of session-aware recommendation is to predict the next interactive item v_{n,m_n+1} of current session S_c^k based on the historical sessions $\mathcal{S}_h^k$ and current session $\mathcal{S}_c^k$.

Different from traditional session-aware recommendation, our research focuses on the existence of shared accounts. In this scenario, a single account k may be shared by multiple users, which can be denoted as $\mathcal{U} = \{u_i\}_{i=1}^{|\mathcal{U}|}$, and u_i represents an individual user of the shared account. $\mathcal{S}^k$ is a mixture of behaviors from these users. Therefore, the key problem is to distinguish different user behaviors in historical sessions $\mathcal{S}_h^k$ and identify the operating user u_c of current session $\mathcal{S}_c^k$.

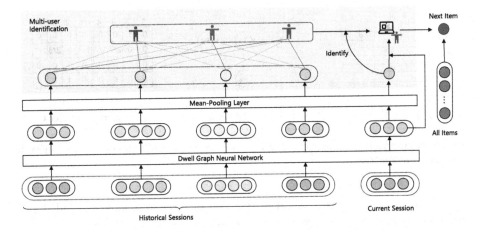

Fig. 2. An overview of the proposed MISS method.

3.2 Overview

Figure 2 provides an overview of the proposed MISS method. MISS consists of two main components: the dwell graph neural network (DGNN) and the multi-user identification (MI) module. For each account k, we model all sessions $\mathcal{S}^k$ as an account session graph $\mathcal{G}^k$. Then $\mathcal{G}^k$ is fed into the DGNN unit to capture complex transition relationships among items. We use the mean-pooling layer to generate session embedding. After that, the MI unit takes all sessions as input to distinguish the information of different latent co-users under the same account, and tries to filter the information belonging to the current operating user and generate the current user representation. Using the representation, we calculate the recommendation scores of all items, and the items with top-N scores will be the candidate items for recommendation.

3.3 Dwell Graph Neural Network (DGNN)

Session-Graph Construction. For each account $k \in \mathcal{K}$, all sessions $\mathcal{S}^k$ can be modeled as a directed graph $\mathcal{G}^k = (\mathcal{V}^k, \mathcal{E}^k)$. The node i in graph $\mathcal{G}^k$ represents an interactive item $v_i \in \mathcal{I}$ of the account k, and the edge $v_i \rightarrow v_j$ means an account interacts v_j after v_i within a session. Inspired by previous work [2,26], we make the following assumptions:

- In the edge $v_i \rightarrow v_j$, the effect of v_i on v_j is totally different from the effect of v_j on v_i.
- Longer dwell time on an item means greater interest in it.

To model these different transition relationships, we define two types of directed edges with different weights, the outgoing dwell edge with weights of w_{ij}^o and the incoming dwell edge with weights of w_{ij}^i. We set a time threshold t_d,

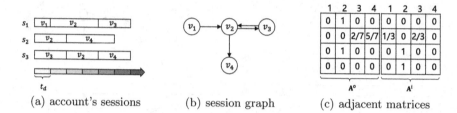

(a) account's sessions   (b) session graph   (c) adjacent matrices

Fig. 3. An example of all sessions of an account as well as corresponding session graph and adjacent matrices.

$d_{j,i\to j}$ means the time that the user stays on v_j after interacted v_i, and $d_{i,i\to j}$ represents time on v_i. The weights are calculated as:

$$w_{ij}^o = \frac{\lceil \frac{d_{j,i\to j}}{t_d} \rceil}{\sum_{v_i \to v_q} \lceil \frac{d_{q,i\to q}}{t_d} \rceil}, \tag{1}$$

$$w_{ij}^i = \frac{\lceil \frac{d_{i,i\to j}}{t_d} \rceil}{\sum_{v_i \to v_q} \lceil \frac{d_{i,i\to q}}{t_d} \rceil}. \tag{2}$$

The account session graph $\mathcal{G}^k$ can be represented by two adjacent matrices $\mathbf{A}_k^o$ and $\mathbf{A}_k^i$, which are written as:

$$\mathbf{A}_k^o[i][j] = w_{ij}^o, \tag{3}$$

$$\mathbf{A}_k^i[i][j] = w_{ij}^i. \tag{4}$$

Figure 3 shows an example of all sessions of an account as well as the corresponding session graph and adjacent matrices.

Graph Neural Network. Graph neural networks [20] are well-suited to solve the session-aware recommendation problem, because it can automatically extract features of session graphs with considerations of rich node connections [26]. The construction of the session graphs has been introduced above, and then we will demonstrate the learning process of node vectors in each session graph. At each time of node update, we first calculate the aggregated incoming and outgoing dwell information of node i as:

$$\mathbf{a}_i^t = \mathbf{A}_{k,i:}[\mathbf{v}_1^{t-1}, \mathbf{v}_2^{t-1}, .., \mathbf{v}_n^{t-1}]^\top \mathbf{W} + \mathbf{b}, \tag{5}$$

where $\mathbf{W} \in \mathbf{R}^{d \times 2d}$ controls the weight, $\mathbf{v}_i \in \mathbf{R}^d$ represents the latent vector of node i, and $\top$ denotes the transpose of the matrix. $\mathbf{A}_k \in \mathbf{R}^{n \times 2n}$ is the concatenation of the two adjacency matrices $\mathbf{A}_k^o$ and $\mathbf{A}_k^i$, and $\mathbf{A}_{k,i:} \in \mathbf{R}^{1 \times 2n}$ are the two columns of block in $\mathbf{A}_k$ corresponding to node i. Then we utilize the GRUs [4] to update each node's hidden state by incorporating the hidden state of other nodes at the previous timestep. The update functions are given as follows:

$$\mathbf{z}_i^t = \sigma(\mathbf{W}_z \mathbf{a}_i^t + \mathbf{U}_z \mathbf{v}_i^{t-1}), \tag{6}$$

$$\mathbf{r}_i^t = \sigma(\mathbf{W}_r \mathbf{a}_i^t + \mathbf{U}_r \mathbf{v}_i^{t-1}), \tag{7}$$

$$\widetilde{\mathbf{v}_i^t} = \tanh(\mathbf{W}_o \mathbf{a}_i^t + \mathbf{U}_o(\mathbf{r}_i^t \odot \mathbf{v}_i^{t-1})), \tag{8}$$

$$\mathbf{v}_i^t = (1 - \mathbf{z}_i^t) \odot \mathbf{v}_i^{t-1} + \mathbf{z}_i^t \odot \widetilde{\mathbf{v}_i^t}, \tag{9}$$

where $\mathbf{z}_i^t$ and $\mathbf{r}_i^t$ respectively represents the update and reset gate, $\sigma(\cdot)$ is the sigmoid function, and $\odot$ is the element-wise multiplication operator. $\mathbf{W}_z$, $\mathbf{U}_z$, $\mathbf{W}_r$, $\mathbf{U}_r$, $\mathbf{W}_o$, $\mathbf{U}_o$ are trainable weight parameters shared by all accounts.

After T propagation steps, we can obtain the final hidden state vector $\mathbf{v}_i^T$ of each node i in the session graph $\mathcal{G}^k$. For convenience, we use $\mathbf{v}_i$ instead. The final hidden state contains information from both the node itself and its T-order neighbors.

3.4   Multi-user Identification (MI)

In the shared account recommendation scenario, the behaviors of different latent co-users under the same account are mixed together. Most existing recommendation methods ignore this issue and generate a single embedding vector for each account. As a result, the recommendation for the currently operating user may be affected by the behavior of other users. Therefore, we generate multiple user vectors based on the historical sessions of the account, where each user vector represents a latent user under the account. Then the current user representation vector is obtained by comparing these user vectors with the information of the current session.

Latent Co-user Representations. Through the output of DGNN, we have obtained the embedding of account k's all sessions $\mathcal{S}^k$. For each session $s_i^k = \{v_{i,1}, v_{i,2}, ..., v_{i,m_i}\} \in \mathcal{S}^k$, we use mean-pooling to calculate its embedding representation $\mathbf{h}_i^k \in \mathbf{R}^d$ as:

$$\mathbf{h}_i^k = \text{mean}(\mathbf{v}_{i,1}, \mathbf{v}_{i,2}, ..., \mathbf{v}_{i,m_i}), \tag{10}$$

then the historical sessions $\mathcal{S}_h^k = \{s_1^k, s_2^k, ..., s_{n-1}^k\}$ can be represented as an matrix $\mathbf{H}_h^k = [\mathbf{h}_1^k, \mathbf{h}_2^k, ..., \mathbf{h}_{n-1}^k]$, and the current session $\mathcal{S}_c^k = \{s_n^k\}$ is denoted as $\mathbf{H}_c^k = [\mathbf{h}_n^k]$. In this work, we adopt the self-attentive method [13] to generate the multiple user representations. We assume that each account is shared by M latent co-users $\mathcal{U}_k = \{u_1, u_2, ..., u_M\}$. For each user u_i, we use the self-attention mechanism to obtain the weights vector $\mathbf{a} \in \mathbf{R}^{n-1}$:

$$\mathbf{a} = \text{softmax}(\mathbf{w}_2 \tanh(\mathbf{W}_1 \mathbf{H}_h^\top)), \tag{11}$$

where $\mathbf{w}_2 \in \mathbf{R}^{d_a}$ and $\mathbf{W}_1 \in \mathbf{R}^{d_a \times d}$ are trainable parameters shared by all accounts. The vector $\mathbf{a}$ represents the latent user u_i's attention weight of accounts historical behaviors. Then according to the attention weight, we sum up the embeddings of historical behaviors and obtain the vector representation of u_i, which can be calculated as:

$$\mathbf{u}_i = \mathbf{a}\mathbf{H}_h. \tag{12}$$

In order to obtain all latent users' representation vectors, we need to perform M times of attention. Thus we extend the vector $\mathbf{w}_2$ into a matrix $\mathbf{W}_2 \in \mathbf{R}^{M \times d_a}$, and the attention weights of the account can be represented as a matrix $\mathbf{A}$ as:

$$\mathbf{A} = \mathrm{softmax}(\mathbf{W}_2 \tanh(\mathbf{W}_1 \mathbf{H}_h^\top)), \tag{13}$$

and the final matrix of latent users can be written as:

$$\mathbf{U}_k = \mathbf{A}\mathbf{H}_h. \tag{14}$$

Current User Representation. After obtaining the latent co-user embeddings $\mathbf{U}_k$, for the current session embedding $\mathbf{H}_c^k$, we compute its weight vector $\mathbf{a}_c$ as:

$$\mathbf{a}_c = \mathrm{softmax}(\mathbf{H}_c \mathbf{U}_k^\top), \tag{15}$$

then the current user embedding $\mathbf{u}_c$ can be summed up by:

$$\mathbf{u}_c = \mathbf{a}_c \mathbf{U}_k. \tag{16}$$

At the same time, the user's local interest also plays a vital role in recommendation. We use the representation of the current session and the last clicked item to represent the user's local interest $\mathbf{z}_l \in \mathbf{R}^{2d}$. It can be written as:

$$\mathbf{z}_l = \mathbf{h}_n \parallel \mathbf{v}_{n,m}, \tag{17}$$

where $\parallel$ is the concatenation operation, $\mathbf{v}_{n,m}$ means the embedding of the last clicked item within current session, and $\mathbf{h}_n$ is the embedding of current session. The final user embedding $\mathbf{z}_u$ is computed as follows:

$$\mathbf{z}_u = [\mathbf{u}_c \parallel \mathbf{z}_l]\mathbf{W}_3, \tag{18}$$

where $\mathbf{W}_3 \in \mathbf{R}^{3d \times d}$ compresses the combined embedding vectors into the latent space $\mathbf{R}^d$.

3.5 Making Recommendation

After obtaining the representation vector $\mathbf{z}_u$ of the current user, for each item $v_i \in \mathcal{I}$ we calculate a recommendation score and obtain a score vector $\hat{\mathbf{z}}$ as:

$$\hat{\mathbf{z}} = \mathbf{z}_u^\top \mathbf{v}, \tag{19}$$

where $\mathbf{v}$ is the embedding of all items. Then a softmax function is applied to generate the probabilities that items will be interacted in the next time:

$$\hat{\mathbf{y}} = \mathrm{softmax}(\hat{\mathbf{z}}). \tag{20}$$

For each account session graph, we employee the cross-entropy loss between the predicted recommendation $\hat{\mathbf{y}}$ and the ground truth $\mathbf{y}$, which can be written as follows:

$$\mathcal{L}(\hat{\mathbf{y}}) = -\sum_{i=1}^{|\mathcal{I}|} (\mathbf{y}_i \log(\hat{\mathbf{y}}_i) + (1 - \mathbf{y}_i) \log(1 - \hat{\mathbf{y}}_i)). \tag{21}$$

The back-propagation through time (BPTT) algorithm is utilized to learn all parameters and the item embeddings in our proposed MISS.

4   Experiments

In this section, we will conduct experiments on two real-world data sets and demonstrate the efficacy of our proposed model MISS. We aim to answer the following research questions:

- **RQ1** Does our proposed MISS outperform the existing state-of-the-art methods in shared-account session-aware recommendation scenario? (Sect. 4.5)
- **RQ2** How well do the various components improve the performance of recommendation? (Sect. 4.6)
- **RQ3** Does incorporating dwell time help improve the performance of recommendations? (Sect. 4.6)
- **RQ4** How does the hyper-parameter M (the number of latent co-users) influence the recommendation performance? (Sect. 4.7)

4.1   Data Sets

Our model focuses on the shared account problem in session-aware recommendation, but there is no publicly available data set for the issue. Therefore, we collect 39k user watching logs from April 1st to June 30th 2020, and construct two data sets with shared account characteristics, named FamTV-SA and FamTV-SAS. The watching logs are obtained from a well-know smart TV platform, including which video is playing, when to start the playing and when to end. We analyze the watching logs of these accounts and find that the watching logs of some accounts contain children or educational videos, which indicates that this account may be shared by multiple family members. We mark such accounts as shared accounts. FamTV-SA only contains the watching logs of shared accounts, while FamTV-SAS is a mixed data set of both shared accounts and single-user accounts.

For both data sets, we first filter out users who have less than 10 watching logs, and merge the records of the same item watched by the same user within an adjacent time less than 10 min. Then we organize the same account interactive items with an interval of less than 30 min into a session in chronological order. After that, we filter poorly informative sessions by removing sessions which have less than 3 interactions, and ensure that each account has 5 sessions or more to have sufficient cross-session information.

For each account, we select 75% of sessions as the train set, 10% as the validation set and the remaining 15% as the test set. The statistics of the two data sets are shown in Table 1.

Table 1. Statistics of data sets after processing.

Data set	Accounts	Items	Sessions	Train sessions	Validation sessions	Test sessions
FamTV-SA	20,213	25,819	969,739	727,304	96,974	145,461
FamTV-SAS	22,347	25,921	1,082,119	811,589	108,212	162,318

4.2   Baseline Methods

We compare our proposed model with the following methods, including conventional, session-based and session-aware recommendation methods.

- **POP** recommends the most popular items to users.
- **Item-KNN** [19] recommends items similar to the previously clicked items in the session by calculating the cosine similarity between item vectors.
- **BPR-MF** [18] is a widely used matrix factorization method. Like [6], we represent the user feature vector with the average latent factors of items that appeared in the session so far.
- **GRU4Rec** [22] applies recurrent neural network to session-based recommendation.
- **SR-GNN** [26] uses the gated graph neural networks to model complex item transition relationships within sessions.
- **HGRU4Rec** [17] uses a hierarchical RNNs to model user's interest evolution, in which a session-based RNN to capture current session information while a user-level GRU depicting user history information.
- **HierTCN** [29] utilizes a hierarchical architecture that consists of RNN and Temporal Convolutional Network to capture the long-term interests and short-term interactions.

4.3   Evaluation Metrics

To evaluate the performance of different methods, we adopt Recall@N and MRR@N as the evaluation metrics, which are also widely used in other related works.

Recall@N is widely used as a measure for predictive accuracy, which represents the percentage of correct items among the top-N recommended items.

MRR@N (Mean Reciprocal Rank) is the average of reciprocal ranks of the correctly-recommended items. When the ground truth item is not in the recommendation list, the reciprocal rank is set to 0. MRR considers the rank of recommended items, and larger MRR values indicates that correct recommendations in the top of the ranking list.

In our experiment, N is set to 5, 10 and 20.

Table 2. Model Performance on data sets.

Methods	FamTV-SA						FamTV-SAS					
	Recall			MRR			Recall			MRR		
	@5	@10	@20	@5	@10	@20	@5	@10	@20	@5	@10	@20
POP	0.35	0.58	0.76	0.11	0.14	0.16	0.39	0.62	0.81	0.12	0.14	0.17
Item-KNN	8.40	11.23	14.66	5.97	6.12	6.37	8.64	11.40	14.89	6.05	6.34	6.71
BPR-MF	4.71	6.98	8.99	2.76	3.24	3.65	4.77	7.09	9.10	2.77	3.30	3.67
GRU4Rec	15.37	21.88	29.88	9.96	9.81	10.37	15.78	22.63	30.59	9.27	10.18	10.73
SR-GNN	22.29	30.28	38.69	13.65	14.87	15.62	22.32	30.40	38.77	13.71	14.98	15.70
HGRU4Rec	17.30	24.93	33.38	9.85	10.86	11.44	17.77	25.47	33.98	10.18	11.20	11.79
HierTCN	23.46	30.78	38.72	14.83	15.81	16.36	23.70	31.13	39.11	15.03	16.02	16.58
MISS	**24.61**	**33.01**	**42.00**	**15.07**	**16.19**	**16.81**	**24.81**	**33.17**	**42.13**	**15.22**	**16.33**	**16.87**

4.4 Parameter Setup

We set item embedding dimension $d = 100$ for both data sets. For the dell time threshold of the DGNN, we set $t_d = 300$. In the MI unit, we set a hyper-parameter M to indicate the number of users in the shared account. M is set to 4 for both data sets. As for the propagation step T, we set T to be 4. All parameters are initialized using a Gaussian distribution with a mean of 0 and a standard deviation of 0.1. The model is trained with Adam [9] optimizer, where the initial learning rate is set to 0.001. Moreover, the coefficient of L2 normalization is set to 0, and the batch size is 100.

4.5 Comparisons with Baselines (RQ1)

Results of MISS and other baselines on two data sets are shown in Table 2, with the best results highlighted in bold-face. Clearly, our proposed MISS model outperforms all baselines in terms of Recall and MRR on both data sets. We can make the following observations:

As we can see from Table 2, strong baseline HierTCN achieves best results among all baselines, which is also outperformed by MISS on both data sets. The performance improvement over HierTCN on shared-account-only data set FamTV-SA are 4.90%, 7.24% and 8.47% respectively in terms of Recall@5/10/20. This verifies the effectiveness of the two core components DGNN and MI in our model. The effect of DGNN and MI will be analyzed in Sect. 4.6.

It is obvious that RNN-based models perform better than traditional models, which demonstrates that RNN-based methods are good at modeling sequential information within sessions. The performance of MISS and HierTCN are better than pure session-based methods SR-GNN and GRU4Rec, which proves that taking historical sessions into account is effective in improving recommendations. Compared with GRU4Rec and HGRU4Rec, the performance of SR-GNN verifies the superiority of graph-based methods.

Table 3. Performance of MISS compared with four ablation models.

Methods	FamTV-SA						FamTV-SAS					
	Recall			MRR			Recall			MRR		
	@5	@10	@20	@5	@10	@20	@5	@10	@20	@5	@10	@20
MISS(-D)	24.54	32.93	41.91	14.97	16.10	16.72	24.73	33.08	42.05	15.14	16.22	16.78
MISS(-G)	22.97	29.24	35.90	13.72	14.67	15.11	23.04	29.35	36.07	13.87	14.87	15.48
MISS(-M)	23.51	30.90	38.67	13.97	14.87	15.76	23.74	31.26	38.87	14.16	15.09	15.93
MISS(-G-M)	21.34	27.51	34.02	11.74	12.58	13.07	21.37	27.64	34.11	11.81	12.70	13.28
MISS	**24.61**	**33.01**	**42.00**	**15.07**	**16.19**	**16.81**	**24.81**	**33.17**	**42.13**	**15.22**	**16.33**	**16.87**

4.6   Ablation Study (RQ2 & RQ3)

To verify the effectiveness of the proposed DGNN and MI modules, we compare our method with different variants and the results are shown in Table 3. MISS(-D): MISS without incorporating dwell time into GNN; MISS(-G): MISS has no DGNN component; MISS(-M): MISS without MI, that is, it makes recommendation without considering filtering information of different latent users; MISS(-G-M): MISS neither has the DGNN component nor the MI.

From Table 3, we can see that MISS consistently outperforms MISS(-G) and MISS(-M), which indicates the importance of modeling historical information and filtering latent users in shared accounts. MISS also performs better than MISS(-D). It proves that the assumption is reasonable that users' interests are related to their dwell time on items.

We can also observe that the performance of MISS(-M) is better than MISS(-G), which means that the DGNN module is more effective than the MI module in MISS. However, MISS(-G) still outperforms most of the baselines listed in Table 2. On FamTV-SAS, the gap between MISS(-M) and MISS is smaller than on FamTV-SA. One possible reason is that we assume the same number of latent users in all accounts, while there are a certain percentage of non-shared accounts in FamTV-SAS. Besides, in the real shared account scenario, different accounts are usually shared by a different number of latent users. The proposed MI module fails to identify the number of latent users; it remains as a topic for future work to design even better MI modules. Additionally, MISS(-G-M) gets the lowest performance amongst the four ablation methods. In summary, capturing users' historical interests and distinguishing information from different latent users are useful.

4.7   Influence of Hyper-parameters (RQ4)

In Sect. 3.3, we propose the multi-user identification module and introduce a hyper parameter M, which represents the number of latent co-users under the shared account. To study how the performance change with M, we compare the recommendation performance under the different M values while keeping other settings unchanged. Figure 4 illustrates the experimental results. We can find that for both data sets, the performance reaches the best values when M is 4.

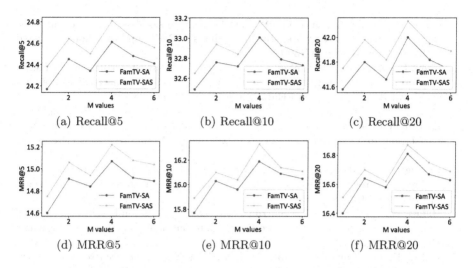

Fig. 4. Performance of MISS with different hyper-parameter M.

5   Conclusion

In this paper, we have proposed the new task of shared-account issue in the session-aware recommendation scenario. To address this task, we propose a novel framework named MISS. Through the Dwell Graph Neural Network and Multi-user Identification module, our framework is able to capture and distinguish the preference drift of different co-users under the same account. To evaluate the effectiveness of MISS, we construct two data sets with shared-account characteristics, while one contains only shared accounts and the other contains both shared and non-shared accounts. Experiments conducted on two data sets demonstrate that MISS evidently outperforms the state-of-the-art recommendation methods.

A limitation of MISS is that we assume the same number of latent users in all shared accounts. For future work, we will optimize the multi-user identification module to automatically detect the number of co-users, so as to achieve better performance.

Acknowledgements. This work is supported by National Natural Science Foundation of China (No. 62072282), Industrial Internet Innovation and Development Project in 2019 of China, Shandong Provincial Key Research and Development Program (Major Scientific and Technological Innovation Project) (No. 2019JZZY010105). This work is also supported in part by US NSF under Grants III-1763325, III-1909323, and SaTC-1930941.

References

1. Bajaj, P., Shekhar, S.: Experience individualization on online tv platforms through persona-based account decomposition. In: Proceedings of the 24th ACM international conference on Multimedia, pp. 252–256 (2016)

2. Bogina, V., Kuflik, T.: Incorporating dwell time in session-based recommendations with recurrent neural networks. In: Proceedings of RecTemp Workshop co-located with ACM RecSys, pp. 57–59 (2017)

3. Chen, S., Moore, J.L., Turnbull, D., Joachims, T.: Playlist prediction via metric embedding. In: Proceedings of the 18th ACM SIGKDD International Conference on Knowledge Discovery and Data Mining, pp. 714–722 (2012)

4. Cho, K., van Merrienboer, B., Gulcehre, C., Bougares, F., Schwenk, H., Bengio, Y.: Learning phrase representations using RNN encoder-decoder for statistical machine translation. In: Conference on Empirical Methods in Natural Language Processing (2014)

5. Covington, P., Adams, J., Sargin, E.: Deep neural networks for youtube recommendations. In: Proceedings of the 10th ACM Conference on Recommender Systems, pp. 191–198 (2016)

6. Hidasi, B., Karatzoglou, A., Baltrunas, L., Tikk, D.: Session-based recommendations with recurrent neural networks. In: International Conference on Learning Representations (2015)

7. Hidasi, B., Quadrana, M., Karatzoglou, A., Tikk, D.: Parallel recurrent neural network architectures for feature-rich session-based recommendations. In: Proceedings of the 10th ACM Conference on Recommender Systems, pp. 241–248 (2016)

8. Jannach, D., Ludewig, M.: When recurrent neural networks meet the neighborhood for session-based recommendation. In: Proceedings of the 11th ACM Conference on Recommender Systems, pp. 306–310 (2017)

9. Kingma, D., Ba, J.: Adam: a method for stochastic optimization. Comput. Sci. (2014). https://arxiv.org/abs/1412.6980

10. Li, C., et al.: Multi-interest network with dynamic routing for recommendation at Tmall. In: Proceedings of the 28th ACM International Conference on Information and Knowledge Management, pp. 2615–2623 (2019)

11. Li, J., Ren, P., Chen, Z., Ren, Z., Lian, T., Ma, J.: Neural attentive session-based recommendation. In: Proceedings of the 2017 ACM on Conference on Information and Knowledge Management, pp. 1419–1428 (2017)

12. Liang, T., Li, Y., Li, R., Gu, X., Habimana, O., Hu, Y.: Personalizing session-based recommendation with dual attentive neural network. In: International Joint Conference on Neural Networks, pp. 1–8 (2019)

13. Lin, Z., et al.: A structured self-attentive sentence embedding. In: International Conference on Learning Representations (2017)

14. Linden, G., Smith, B., York, J.: Amazon.com recommendations: item-to-item collaborative filtering. IEEE Internet Comput. 7(1), 76–80 (2003)

15. Ma, M., Ren, P., Lin, Y., Chen, Z., Rijke, M.D.: π-net: a parallel information-sharing network for shared-account cross-domain sequential recommendations. In: Proceedings of the 42nd International ACM SIGIR Conference, pp. 685–694 (2019)

16. Van den Oord, A., Dieleman, S., Schrauwen, B.: Deep content-based music recommendation. In: Proceedings of the 26th International Conference on Neural Information Processing Systems, pp. 2643–2651 (2013)

17. Quadrana, M., Karatzoglou, A., Hidasi, B., Cremonesi, P.: Personalizing session-based recommendations with hierarchical recurrent neural networks. In: Proceedings of the 11th ACM Conference on Recommender Systems, pp. 130–137 (2017)

18. Rendle, S., Freudenthaler, C., Gantner, Z., Schmidt-Thieme, L.: BPR: Bayesian personalized ranking from implicit feedback. In: Proceedings of the 25th Conference on Uncertainty in Artificial Intelligence, pp. 452–461 (2009)

19. Sarwar, B., Karypis, G., Konstan, J., Riedl, J.: Item-based collaborative filtering recommendation algorithms. In: Proceedings of the 10th International Conference on World Wide Web, pp. 285–295. Association for Computing Machinery (2001)

20. Scarselli, F., Gori, M., Tsoi, A.C., Hagenbuchner, M., Monfardini, G.: The graph neural network model. IEEE Trans. Neural Netw. **20**(1), 61–80 (2008)

21. Shani, G., Heckerman, D., Brafman, R.I.: An MDP-based recommender system. J. Mach. Learn. Res. **6**, 1265–1295 (2005)

22. Tan, Y.K., Xu, X., Liu, Y.: Improved recurrent neural networks for session-based recommendations. In: Proceedings of the 1st Workshop on Deep Learning for Recommender Systems, pp. 17–22 (2016)

23. Verstrepen, K., Goethals, B.: Top-n recommendation for shared accounts. In: Proceedings of the 9th ACM Conference on Recommender Systems, pp. 59–66 (2015)

24. Wang, S., Cao, J., Yu, P.: Deep learning for spatio-temporal data mining: a survey. IEEE Trans. Knowl. Data Eng. (2020). https://doi.org/10.1109/TKDE.2020.3025580

25. Wang, Z., Yang, Y., He, L., Gu, J.: User identification within a shared account: improving IP-TV recommender performance. In: Manolopoulos, Y., Trajcevski, G., Kon-Popovska, M. (eds.) ADBIS 2014. LNCS, vol. 8716, pp. 219–233. Springer, Cham (2014). https://doi.org/10.1007/978-3-319-10933-6_17

26. Wu, S., Tang, Y., Zhu, Y., Wang, L., Xie, X., Tan, T.: Session-based recommendation with graph neural networks. In: Proceedings of the 33rd AAAI Conference on Artificial Intelligence, vol. 33, pp. 346–353 (2019)

27. Yang, Y., Hu, Q., He, L., Ni, M., Wang, Z.: Adaptive temporal model for IPTV recommendation. In: Dong, X.L., Yu, X., Li, J., Sun, Y. (eds.) WAIM 2015. LNCS, vol. 9098, pp. 260–271. Springer, Cham (2015). https://doi.org/10.1007/978-3-319-21042-1_21

28. Ying, H., et al.: Sequential recommender system based on hierarchical attention network. In: Proceedings of the 27th International Joint Conference on Artificial Intelligence (2018)

29. You, J., Wang, Y., Pal, A., Eksombatchai, P., Rosenburg, C., Leskovec, J.: Hierarchical temporal convolutional networks for dynamic recommender systems. In: Proceedings of the 2019 World Wide Web Conference, pp. 2236–2246 (2019)

30. Zhang, A., Fawaz, N., Ioannidis, S., Montanari, A.: Guess who rated this movie: identifying users through subspace clustering. In: Proceedings of the 28th Conference on Uncertainty in Artificial Intelligence, pp. 944–953 (2012)

31. Zhang, M., Wu, S., Gao, M., Jiang, X., Xu, K., Wang, L.: Personalized graph neural networks with attention mechanism for session-aware recommendation. IEEE Trans. Knowl. Data Eng. (2020)

VizGRank: A Context-Aware Visualization Recommendation Method Based on Inherent Relations Between Visualizations

Qianfeng Gao[1,3], Zhenying He[2,3(✉)], Yinan Jing[2,3(✉)], Kai Zhang[2,3], and X. Sean Wang[1,2,3,4(✉)]

[1] School of Software, Fudan University, Shanghai, China
{qfgao18,xywangcs}@fudan.edu.cn
[2] School of Computer Science, Fudan University, Shanghai, China
{zhenying,jingyn,zhangk}@fudan.edu.cn
[3] Shanghai Key Laboratory of Data Science, Shanghai, China
[4] Shanghai Institute of Intelligent Electronics and Systems, Shanghai, China

Abstract. Visualization recommendation systems measure the importance of visualizations to make suggestions. While considering each visualization individually may be enough to gauge its importance in specific scenarios, it ignores the relations between visualizations under a visual analysis context. This paper is to study a strategy via a more general method called VizGRank which models the relations between visualizations as a graph, then calculates the importance of visualizations by adopting a graph-based algorithm. In this model, the relations derived from the visual encoding of the visualizations and the underlying data schema are used for recommendation. Due to the lack of public benchmarks, the effectiveness of the model is evaluated on the synthetic results from an existing public benchmark IDEBench as a workaround. However, since the existing benchmark is specific and synthetic and does not reflect the realistic scenarios of visualization recommendation completely, a new benchmark for visualization recommendation is designed and constructed by collecting real public datasets. Extensive experiments on both the public benchmark and the new benchmark demonstrate that the VizGRank can better capture the relative importance of visualization and outperforms the existing state-of-the-art method.

Keywords: Visualization recommendation · Benchmark · Visual analytics · Inherent relation

1 Introduction

During visual analysis, the analysts inspect and understand the datasets to gain interesting insights. Traditionally, the analysts iteratively browse visualizations

C. S. Jensen et al. (Eds.): DASFAA 2021, LNCS 12683, pp. 244–261, 2021.
https://doi.org/10.1007/978-3-030-73200-4_16

to search for satisfactory ones. Since this process is rather tedious and time-consuming, the analysts may end up having explored only a fraction of the visualization space and chosen an inferior visualization, possibly missing critical insights hidden in the data.

Visualization recommendation systems emerged to ease this effort by suggesting important visualizations. Below we use an example to show the principles of visualization recommendation. Consider a relational table *sales (Quantity, InvoiceDate, UnitPrice, Country, Sales, StockCode, CustomerID)* about sales data in an e-commerce application[1], as shown in Table 1.

Table 1. An excerpt of sales data from a UK retailer.

Quantity	InvoiceDate	UnitPrice	Country	Sales	StockCode	CustomerID
6	2010-01-12	2.55	EIRE	15.3	22727	12583
24	2010-04-12	3.75	France	90	10002	12662
8	2011-01-12	4.95	Germany	39.6	22326	14911
24	2011-03-02	1.79	Germany	42.96	21731	12472
...	...	...	...	...	...	...

According to different ranking metrics, different methods will recommend different visualizations. For example, Fig. 1 shows possible visualizations that may be recommended by different systems or methods. Figure 1(a) is a pie plot of *InvoiceDate* and *Quantity* recommended by Quickinsights [7] with high priority, since the first season of 2011 dominates by accounting for more than 50% of total sales. Figure 1(b) is a bar plot of *InvoiceDate*, *Sales*, and *Country* scored high in SeeDB [20] since the distributions of sales of these two countries demonstrate a large deviation. Figure 1(c) is a scatter plot of *Quantity*, *UnitPrice* and *Country* recommended by DeepEye [13], since the score is high according to their proposed partial order model.

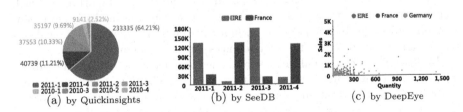

(a) by Quickinsights   (b) by SeeDB   (c) by DeepEye

Fig. 1. Possible recommended visualizations for the sales data in Table 1.

---

[1] https://www.kaggle.com/carrie1/ecommerce-data.

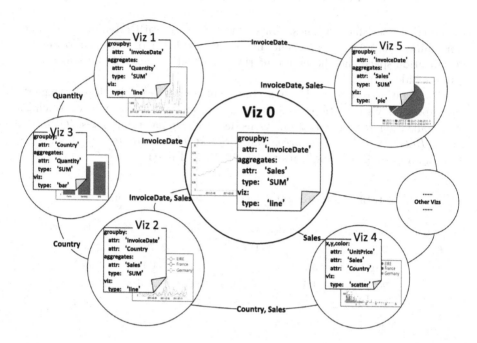

Fig. 2. The analysis space of the data in Table 1 in the form of a visualization graph.

In all the above methods, visualizations are ranked according to a specific *utility function* for *individual* visualizations. The utility function actually represents a user preference in the ranking. However, only using these utility functions has two limitations. On the one hand, it is subjective, since it depends on the interest and the knowledge of the analysts. On the other hand, it ignores the relation of visualizations. Adopting a certain utility function without knowing the visual analysis context, preferences may become biased and negatively impact recommendation effectiveness. Instead, we consider the importance of visualization based on the relation of visualizations. The relation inherently reflects the associations between visualizations and influences the relative importance of each visualization in the visual analysis context. In this way, the model is more general and objective since it is based on inherent relations between visualizations and it can also incorporate these preferences into the ranking.

Intuitively, various visualizations from one specific dataset are possibly linked [8]. During visual analysis, the analysts specify visualizations continuously and move from one visualization to another by creating new visualizations, modifying old ones, and revisiting previous ones. These visualizations constitute the analysis space as shown in Fig. 2, from which the analysts continuously pick visualizations for inspection.

As an example, suppose that the retailer wants to analyze the data in Table 1. To do so, she might start with the attribute *Sales*. She starts by examining a line plot of daily sales (node 'Viz0' in Fig. 2). Then she continues on overviews of *Sales* from different perspectives, for example, by adding another attribute *Country* to the current line plot (resulting in node 'Viz2' in Fig. 2) or by transforming *InvoiceDate* into seasonal granularity (getting a pie plot of sales as node 'Viz5' in Fig. 2). These plots are linked intrinsically since they present different aspects of the information related to *Sales*. As shown in Fig. 2, node 'Viz0' links with more visualizations. Thus, 'Viz0' is more likely to have higher importance than other visualizations. After inspecting visualizations of *Sales*, she turn to visualizations of *Quantity* shown as 'Viz1' in Fig. 2. Following this line, visualizations of *Quantity* are also intrinsically linked.

Based on the observation above, visualizations linking with more visualizations may represent the focuses of the analysts and thus should have higher relative importance. We believe that such an intrinsic relation will affect the importance of visualizations. Thus, we propose a model to capture in a graph the relations with a *relation function* and solve the ranking problem using a graph-based method. The relation function is defined mainly on the data schema and the visual encoding of visualizations. Besides, the utility function presents a user preference for a specific scenario on the ranking and should not be ignored in the recommendation. Thus, our model can also incorporate different utility functions into the ranking to implement a more comprehensive recommendation.

To verify the effectiveness of the proposed model, unlike previous methods [7,20,22] that usually conduct a user study which is difficult to reproduce, we adopt quantitative methods (e.g., precision and normalized discounted cumulative gain) to evaluate the performance of the recommendation. Due to the lack of benchmarks, evaluations are performed on the synthetic results from a public benchmark called IDEBench [8] as a workaround. However, due to its limitations, we also construct a new benchmark for evaluation ourselves.

IDEBench is a public benchmark of interaction logs for interactive data exploration. It contains workflows generated using a Markov chain. The query results generated from these workflows can be merged to generate an ordered list of visualizations (the more interactions and the more important) and be used to evaluate the performance of recommendation. However, IDEBench has a few obvious problems. IDEBench is designed for a specific scenario: (1) There is only one dataset available, possibly resulting in a bias. (2) Although IDEBench is designed for visual analysis, it focuses more on query performance than visualization results. (3) The workflows of IDEBench contain queries on binning and aggregating using *count* or *average*, mainly resulting in *histogram* and *binned scatter*. However, our scenario focuses on grouping and aggregations using *sum, average, and count*, resulting in various plots such as *bar, pie, scatter*, and *line*. IDEBench is synthetic and is not directly from real visualization results: (4) The workflows and query results are random since it is generated based on configured probability (e.g., choosing a specific attribute with a probability of 20%).

To provide an alternative to IDEBench and to replace or supplement user study, we construct a new benchmark using real datasets collected from Kaggle[2]. Kaggle is a popular data analysis website containing plenty of visualization

---

[2] https://www.kaggle.com/.

results in the notebooks. With these notebooks collected, we derive a visualization recommendation benchmark, called **KaggleBench**, with ordered visualizations by merging visualization results from different notebooks. KaggleBench has been published and openly available to the research community[3].

In conclusion, the main contributions of our work are as follow:

- We propose VizGRank, a graph-based method that takes advantage of the relations between visualizations to calculate the importance of visualizations.
- We construct a benchmark KaggleBench with 18 real datasets to enrich the public availability of benchmark datasets and to evaluate the proposed model. And KaggleBench (see Footnote 3) has been published for further research.
- Extensive experiments on both IDEBench [8] and KaggleBench demonstrate the effectiveness of our model VizGRank and we conclude that VizGRank achieves better results than the state-of-the-art method does.

The rest of the paper is organized as follows. We formulate the problem in Sect. 2 and explain the VizGRank model in Sect. 3. We elaborate on the construction of KaggleBench in Sect. 4, demonstrate the experiments and discuss the results in Sect. 5, and summarize related works in Sect. 6 before concluding with Sect. 7.

2   Problem Formulation

We first give a formal definition for visualization, relation function, and utility function. Then, we formulate the problem of the paper.

Visualization. Aggregation-based analysis dominates in the visual analysis [8, 12], thus we view a visualization as the composition of the group-by dimension, the aggregated measure, and the visualization type. For example, the dimension of the line plot (node 'Viz0') in Fig. 2 is *InvoiceDate*, the measure is *Sales*, the transformation is *raw* since it groups data by the raw InvoiceDate (it can be *season* of the pie plot in Fig. 1(a) and so on), the aggregation is *sum*, and the visualization type is *line*. Thus we define a visualization as:

Definition 1 (Visualization). *Given a relation $R(A)$, where A is the set of attributes, a visualization is determined by the group-by dimension (denoted d) and the aggregated measure (denoted m), the transformation of the group-by dimension d (denoted g), the aggregation of the measure m (denoted a) and the visualization type (denoted t). Hence, a visualization is defined as: $V_R(g(d), a(m), t)$.*

Relation Function. As the example in Fig. 2, the relation between 'Viz0' and 'Viz2' can be measured based on their common attributes *InvoiceDate* and *Sales*. Relation describes the connection between the two visualizations. We introduce a relation function to describe this intrinsic relation.

---

[3] https://github.com/vengeji/vizrec_bench.

Definition 2 (Relation Function). *For two visualizations V_i and V_j, $V_i \neq V_j$, the relation function F gives a score to measure relation between V_i and V_j. The relation function F is $F(V_i, V_j)$ that gives a relation score $\in [0, 1]$. We require the relation to be symmetric, i.e., $F(V_i, V_j) = F(V_j, V_i)$ for all V_i and V_j.*

Visualization Graph. Based on the given relation function, we can construct a visualization graph G. As in the graph shown in Fig. 2, 'Viz0' and 'Viz1' have common attributes *InvoiceDate* and 'Viz0' and 'Viz5' have common attributes *InvoiceDate* and *Sales*. Thus, they are connected via the relation function that considers the context (i.e., the similarity of the group-by and the measure attributes). Relation function can thus be used to define a visualization graph.

Definition 3 (Visualization Graph). *Given the relation function F, a set of visualizations V, a visualization graph G is an undirected edge labeled graph defined as $G = (V, E)$, where $E = \{((V_i, V_j), F(V_i, V_j))|V_i, V_j \in V$ and $V_i \neq V_j\}$, and $F(V_i, V_j)$ is used as the label of the edge (V_i, V_j).*

Utility Function. The utility function presents a user preference on the ranking of individual visualization and is also considered in our model. It is worth noting that the utility function is not a compulsory part, since there are cases where we may not have preferences on individual visualizations. Thus, by default, when no utility function is given, the utility score of each visualization is set to be 1.

Definition 4 (Utility Function). *The utility function U gives a score to describe the utility score of a visualization V_i. The higher the score, the greater the utility. The utility function U is a function that maps a visualization to a utility score, which is a number in $[0, 1]$.*

Importance. To measure the importance of each visualization, our model combines the relation function and the utility function to calculate the importance of each visualization. Thus, the importance of visualization is defined as:

Definition 5 (Importance). *Given relation function F, the visualization graph G, and the utility function U, the importance $I(V_i)$ of a visualization V_i is a score to measure the importance of each visualization V_i. We denote the importance of V_i as $I(V_i)$ when G, F, U are understood in the context.*

Definition 6 (Problem). *Given a relational dataset $R(A)$, a set V_R with n visualizations, a graph G constructed based on the relation function F, and a utility function U, find top k visualizations V_i that have the greatest importance value $I(V_i)$ in the graph G.*

3   VizGRank Model

In this section, we describe our model and show how the model works.

3.1   Relations Between Visualizations

Based on the example in Sect. 1 that the analysts specify the next visualizations based on previous ones, we design different relation functions to capture different relationships, namely *similarity-based and deviation-based*.

Similarity-Based Relation. As shown in the example in Sect. 1, the analyst focuses on certain visualizations and modifying repeatedly the attributes used in group-by and aggregation. The intuition is that another visualization contains relevant information (same attributes) as this one. Based on this, visualization having more neighbors with similar data schema and visual encoding is more important. So, first, we define the context of a visualization based on the attributes it contains, the transformations on the attributes, and the visualization type. Then, we propose a similarity-based relation based on the context defined.

Definition 7 (Context). *For a visualization $V_i(g(d), a(m), t)$, the context is defined as a set $C_{V_i} = \{g, d, a, m, t\}$.*

Then, we measure the similarity between two visualizations based on the context using the *Jaccard* similarity between two sets C_{V_i} and C_{V_i}:

$$sim(V_i, V_j) = \frac{|C_{V_i} \cap C_{V_j}|}{|C_{V_i} \cup C_{V_j}|} \tag{1}$$

Taking node 'Viz0' and 'Viz5' in Fig. 2 for example, the context of 'Viz0' is $\{raw, InvoiceDate, sum, sales, line\}$ and the context of 'Viz5' is $\{season, InvoiceDate, sum, sales, pie\}$. Thus, the similarity between 'Viz0' and 'Viz5' is $sim(Viz0, Viz5) = \frac{|\{InvoiceDate, sum, sales\}|}{|\{InvoiceDate, sum, sales, season, pie, raw, line\}|} = \frac{3}{7} = 0.429$.

Deviation-Based Relation. In the example, after analyzing the visualization 'Viz0' of the attribute *Sales*, the analyst turns to another visualization 'Viz1' of *Quantity* or 'Viz4' of *UnitPrice*. The context of visualizations with *Quantity* or *UnitPrice* is more deviated from *Sales* (i.e., with lower similarity). Thus, we define deviation-based relation to model this relation. Intuitively, deviation-based relation will promote visualizations with different contexts, improving coverage of different contexts. The deviation is defined as:

$$devi(V_i, V_j) = 1 - sim(V_i, V_j) \tag{2}$$

For example, the deviation of node 'Viz0' and 'Viz5' is $1 - \frac{3}{7} = 0.571$.

3.2   Calculating the Importance

We give an intuitive description that highly linked (by the relation) visualizations are more likely to be more important than visualizations with fewer linkages and a visualization has higher importance if the importance of its neighbors is high. Then the importance can be obtained by using algorithms like PageRank [3, 23]:

$$I(V_i) = \sum_{V_j \in V} \frac{F(V_i, V_j) \cdot I(V_j)}{\sum_{V_k \in V} F(V_j, V_k)} \tag{3}$$

Consider the graph in Fig. 2. The importance is calculated as shown in Fig. 3. First, we derive a relation matrix using the relation function, as shown in Fig. 3(a). Then we can obtain a stochastic matrix through row normalization (Fig. 3(b)). Finally, based on Markov Convergence Theorems [2], the importance of each visualization has a convergent state shown in Fig. 3(c).

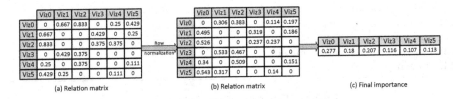

Fig. 3. The process of calculating the importance of each visualization in Fig. 2.

3.3 Incorporating Utility Function

Our model can incorporate a utility function to give a personalized ranking to a user preference. Then, the importance of visualizations is obtained by:

$$I(V_i) = (1 - \alpha) \cdot U(V_i) + \alpha \cdot \sum_{V_j \in V} \frac{F(V_i, V_j) \cdot I(V_j)}{\sum_{V_k \in V} F(V_j, V_k)} \tag{4}$$

where α is the damping factor that can be set between 0 and 1 accordingly.

Available Utility Functions. Previous methods [7, 13, 20] are feasible for the utility function of VizGRank. SeeDB [20] only considers the importance of bar plots in its recommendation and Quickinsights [7] considers complex filter conditions that do not fit well in our scenario. So, to enable comparison, we adopt the utility function in DeepEye [13]. DeepEye constructs a partial order model based on three factors including matching quality $M(v)$, the transformation quality $Q(v)$, and the importance of attributes $W(v)$ to recommend visualizations.

3.4 Different Schemes

There are different ways of using the VizGRank algorithm, with a choice of the relation and the utility function. Each choice is a scheme of VizGRank, as shown in Table 2. We can use relation without utility (e.g., **Sim** and **Devi**). And we can also incorporate a specific utility (e.g., **Sim-DE** and **Devi-DE**).

Algorithm 1. The VizGRank Algorithm

Input: The set of visualizations, V; The relation function, F; The utility function given, U

Output: an ordered list of Visualizations, V_{sorted}

1: $utility = [\,]$
2: **for** each $v_i \in V$ **do**
3:   $utility[v_i] = U(V_i)$ *or* 1 *if not given* U
4: **end for**
5: $k = 0;$ $importance = utility$
6: **while** $\delta > \epsilon$ **do**
7:   **for** each $v_i \in V$ **do**
8:     $importance_{k+1}(v_i) = (1 - \alpha) \cdot utility[v_i] + \alpha \cdot \sum_{v_j \in V} \frac{F(v_i,v_j) \cdot importance_k(v_j)}{\sum_{v_k \in V} F(v_j,v_k)}$
9:   **end for**
10:   $\delta = \|importance_{k+1} - importance_k\|_1$
11:   $k = k + 1$
12: **end while**
13: $V_{sorted} = Sort(V, importance_k)$
14: **return** V_{sorted};

Table 2. Different schemes of VizGRank.

Name	Relation	Utility
Sim	Similarity	–
Devi	Deviation	–
Sim-DE	Similarity	Partial order
Devi-DE	Deviation	Partial order

3.5   VizGRank Algorithm

The VizGRank algorithm is presented as pseudo-code in Algorithm 1. To enable comparison with DeepEye [13], the input visualization set V is generated based on their generation algorithm. First, we initialize the importance of each visualization based on the utility function given (line 1–5). Then, for each visualization V_i, we update the importance based on its neighbors' importance until the process convergences (line 6–13). Finally, we sort the visualization candidates and output an ordered list of visualizations (line 14) as the recommendation results.

4   Benchmark from Real Datasets

We construct a benchmark called KaggleBench to verify the VizGRank model. Next, we elaborate on the details of KaggleBench. A sample of the dataset in KaggleBench is shown in Fig. 4, see KaggleBench[4] for details.

---

[4] https://github.com/vengeji/vizrec_bench.

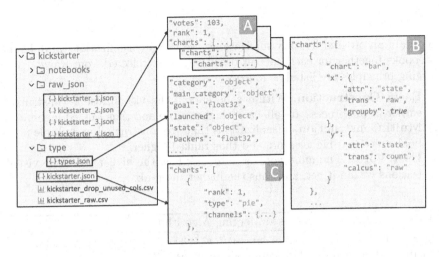

Fig. 4. A sample dataset in the KaggleBench. For more details see KaggleBench (see Footnote 4). For each dataset, we collect visualization results in JSON format recorded in files kickstarter_*.json (A, B). Then we merge results based on the votes to generate an ordered list with 25 visualizations in file kickstarter.json (C).

4.1 Criteria and Collection

Criteria. In Kaggle, each dataset has several notebooks containing analysis results (queries or visualizations). We have collected 18 datasets and corresponding notebooks as the source of our benchmark. To ensure the quality of the benchmark, datasets collected obey the following criteria:

(1) **Scenario Consistence.** Notebooks should contain visualization results.
(2) **Type Variety.** The dataset should contains various data types of temporal, categorical, and numerical. Notebooks should contain various visualization types of bar plot, line plot, pie plot, and scatter plot.
(3) **Number Limitation.** The number of attributes of the dataset used is within 5 to 30. Too few attributes make no sense for the recommendation, and too many attributes make it difficult to conduct a comprehensive analysis.
(4) **Coverage Constraint.** The notebooks should contain more than 15 different visualizations covering all the attributes in the dataset used.

Collection. We manually select notebooks that contain visualization outputs to guarantee **Scenario Consistence** and **Number Limitation**. To satisfy **Type Variety**, we only select datasets with multiple types of attributes and notebooks with multiple types of visualizations. We guarantee **Coverage Constraint** by dropping unused attributes and merging visualizations from other notebooks to reach more than 15 visualizations for each dataset.

4.2 Processing

To obtain an ordered list of visualizations, we merge visualizations in different notebooks together for each dataset based on votes collected from Kaggle. The merging principles are listed as follow:

(1) **Early Construction.** Within a notebook, the visualizations constructed earlier in the process are allocated higher score and thus ranked higher.
(2) **Multi Construction.** Visualizations that are constructed in multiple notebooks receive a higher score and thus ranked higher.
(3) **User Votes.** The more votes a notebook has, the higher score the visualizations within it get, and thus the higher the rank.

Algorithm 2. The Votes-based Merging Algorithm

Input: The list of notebooks of dataset R, nb^R; The list of votes of notebooks, $vote^R$
Output: an ordered list of visualizations for dataset R, V_R
1: sort nb^R according to $vote^R$
2: $merged[nb_i^R]$ = merge nb_i^R with other $nb_j^R \in nb^R$ in the order of $vote^R$
3: assign a rank score $|nb_i^R| - k + 1$ to each $v_k \in nb_i^R$, $|nb_i^R|$ is the length of nb_i^R
4: **for** each $nb_i^R \in merged$ **do**
5:    **for** each $v \in nb_i^R$ **do**
6:       V = find v in all other notebooks nb_j^R
7:       $v.score = \frac{vote_i^R * v.rank}{sum(vote^R)} + \sum_{v_j \in V} \frac{vote_j^R * v_j.rank}{sum(vote^R)}$
8:    **end for**
9:    $nb_i^R = Sort(nb_i^R, key = v.score)$
10: **end for**
11: **return** $merged$;

The algorithm is shown in Algorithm 2. We first sort all nb_i^R according to $vote_i^R$. For each visualization set V in nb_i^R, we merge visualizations of other notebooks to it according to the order of $vote_i^R$ (line 2). Next, we assign each visualization in it a rank score (line 3). Then for each visualization in each notebook, we find corresponding visualizations in other notebooks (line 6). Finally, we calculate the score of each visualization as the weighted sum of rank scores among different notebooks (line 7) and sort them (line 9) as the result.

Reviewing the principles above, we assign a rank score to each visualization according to the occurrence within each notebook (line 3) to follow **Early Construction** (e.g., the first plot of notebook receive a rank of 1 as shown in Fig. 4B.). Merging duplicate visualizations in different notebooks (line 6) make sure that visualizations created by multiple notebooks get a higher score, following **Multi Construction**. For example, there are 3 notebooks as shown in Fig. 4A, and we calculate the score of each visualization as the weighted sum of rank scores among different notebooks (line 7) based on **User Votes**.

The resulting sets of visualizations (Fig. 4C) of each dataset make up KaggleBench (see Footnote 4). To conclude, we have collected 18 datasets, each of which has a set of visualizations each of which has a rank score.

5   Experiments

5.1   Experiments Setup

Setup. All the following experiments are implemented in Python (3.7.4). They are conducted on a Ubuntu Linux 18.04.4 LTS machine with 32 Intel Xeon Silver 4208 CPU and 64GB RAM.

Datasets. Our experiments are performed on the following datasets:

- **Synthetic Dataset.** We merge all the default workflows of IDEBench [8] to generate a count-based ordered list of visualizations.
- **KaggleBench Dataset.** The benchmark described in Sect. 4.

Baselines. We compare our model VizGRank with the following baselines.

- **Random:** We shuffle the generated visualizations 10000 times and calculate the average performance as the baseline.
- **DeepEye [13]:** This is a state-of-the-art automatic visualization system.

Schemes. The schemes used in the experiments are summarized in Table 2. We first compare how different relations without utility (namely, **Sim** and **Devi**) function perform on KaggleBench and IDEBench. Then we compare our method in the given schemes with the state-of-the-art method DeepEye. We also incorporate DeepEye as a utility function in our model (namely, **Sim-DE** and **Devi-DE**) to demonstrate the flexibility of VizGRank.

Evaluation Metrics. We adopt commonly used metrics in recommendation systems[5] to measure the effectiveness of the top-K results of the recommendation as follows. Usually, the K is set to 10 or 20. However, we consider the relative importance of all visualizations, so we set the K to cover most results in the output. Thus, we set $K = \{5, 10, ..., 95, 100\}$ with a step of 5.

For each dataset in KaggleBench (18 in total) and IDEBench (only one), we run the algorithm and calculate the values of these metrics at different K. Then we average them across all the datasets as the performance indicator.

As a premise, for each dataset, some visualizations in the top-K output are the same as those in the benchmark, which we call *relevant visualizations*.

F1-score@K. F1-score@K is the evenly weighted harmonic mean of precision@K and recall@K. Precision@K refers to the proportion of relevant visualizations that appear in the top-K output. Recall@K refers to the fraction of relevant visualizations over the total amount of visualizations in the benchmark.

AP@K. AP@K (Average precision) takes into account both the number of relevant visualizations in the top-K and the positions of these visualizations. For each position i of relevant visualizations, we can get a precision@i, then average

---

[5] https://en.wikipedia.org/wiki/Evaluation_measures_(information_retrieval).

precision is the sum of each precision@i and normalized by the total number of visualizations in the corresponding dataset of the benchmark.

NDCG@K. NDCG@K (Normalized Discounted Cumulative Gain) considers the rank score and position of each relevant visualization in the output. Cumulative gain (CG) is the sum of rank scores of relevant visualizations. NDCG assumes highly relevant visualizations are more important and the greater the ranked position i of a relevant visualization the less important. Thus, the CG is discounted (DCG) by the position i using $1/log_2(i + 1)$. Since the output may vary in size of different datasets, to compare the performance the DCG is normalized by *ideal* DCG calculated by sorting visualizations according to rank scores.

5.2   Performance of Relations

We first experiment different relations without utility to see which relation captures the relation of visualizations during analysis better.

Evaluation on IDEBench. We compare our methods using IDEBench [8] first. The results are shown in Fig. 5.

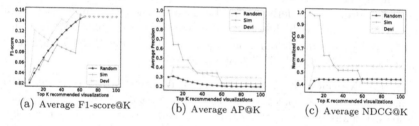

(a) Average F1-score@K   (b) Average AP@K   (c) Average NDCG@K

Fig. 5. The performance of different schemes without utility on IDEBench.

In terms of f1-score, the results of **Devi** are better than **Random**. **Sim** is worse than **Random**. However, for average precision and NDCG, the performance of **Sim** is high at the beginning and decreases as the number K increases. This is because in the result of **Sim** the first plot is relevant as in IDEBench, leading to a higher score at the first position. And the subsequent precipitous drop is due to the lack of relevant items in the following positions. Although the relevance of the first position is important, our work focuses more on the relative importance of visualizations. **Sim** is even worse than **Random** with large K, demonstrating the instability of the **Sim** method.

The limitations of IDEBench are also sources of this instability: as described in Sect. 1, there is only one dataset available in IDEBench, resulting in a bias, and the IDEBench is designed for interactive data exploration scenario, which does not fit well in our scenario. Thus, IDEBench is not completely applicable to our scenario.

Evaluation on KaggleBench. We then compare our methods using our benchmark. The results are shown in Fig. 6.

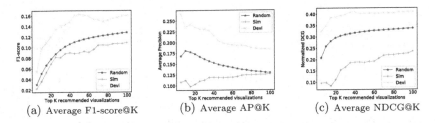

(a) Average F1-score@K   (b) Average AP@K   (c) Average NDCG@K

Fig. 6. The performance of different schemes without utility on KaggleBench.

The **Sim** relation has the worst performance in this experiment which is even worse than the Random. In fact, the similarity-based scheme results in most similar visualizations (e.g., most bar plots) clustered at the front of the recommendation list. Although as described in the example in Sect. 1, the analyst may focus on visualizations with similarity at certain periods, in a broader perspective, analyzing these visualizations is only a small part of the analysis. Therefore, the context similarity is not key to the relative importance of visualizations in this scenario.

The **Devi** scheme performs much better than **Sim**. This is intuitive: at the beginning, the analyst prefers to browse through as many different attributes as possible, then conduct further analysis. This is also known as the *Visual Information Seeking Mantra*: 'Overview first, zoom and filter, then details-on-demand' [8,17]. Besides, [15] also suggested that previous approaches recommend similar and redundant results, and the diversity of results should also be considered.

Results on our benchmark are more stable than on IDEBench since our benchmark contains more datasets and can cover more cases. In conclusion, the performance of the **Devi** is better on our benchmark since our scenario focuses on the initial phase of the visual analysis, where analysts are more likely to browse through different visualizations (i.e., with more deviation) to get familiar with the data. And on average, **Devi** is more able to capture the context and relation of visualizations during the analysis in this scenario.

5.3 Comparison with DeepEye

We compare our model without utility (i.e., **Sim** and **Devi**) with the state-of-the-art system DeepEye. Also, we experiment whether the utility function defined by DeepEye is useful on our evaluation dataset by incorporating DeepEye's utility function into our model (i.e., **Sim-DE** and **Devi-DE**).

Evaluation on KaggleBench. We first compare our model with DeepEye on KaggleBench. The results are shown in Fig. 7.

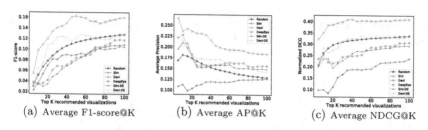

(a) Average F1-score@K   (b) Average AP@K   (c) Average NDCG@K

Fig. 7. Comparison of performance with DeepEye on KaggleBench.

The difference of performance is obvious: without incorporating the utility function, the **Devi** scheme outperforms **DeepEye** a lot in all metrics. And the result of DeepEye is even worse than the random shuffle-based result. It is easy to understand: DeepEye defines three factors including matching quality, transformation quality, and importance of attributes. The former two capture the perceptual quality of the visualization (i.e., how well the data matches the visualization). Only the third factor captures the importance of attributes. DeepEye lacks an awareness of the analysis context and is not general to most analysis scenarios. Thus, DeepEye's utility function is limited and can only be used as a preference upon the ranking. However, our model is based on the characteristics and relations of the underlying data itself and is more general in different scenarios with different data since the relation of visualizations is more inherent.

Besides, incorporating DeepEye's utility function into our model impedes the overall performance of our model, which results in the performance of our model being closer to that of the random method (as demonstrated by **Devi-DE**).

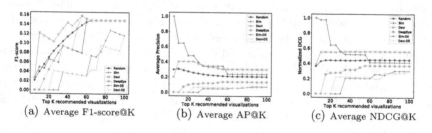

(a) Average F1-score@K   (b) Average AP@K   (c) Average NDCG@K

Fig. 8. Comparison of performance with DeepEye on IDEBench.

Evaluation on IDEBench. We then compare our model with DeepEye on IDEBench. The results are shown in Fig. 8.

Without incorporating DeepEyes's utility function, VizGRank performs better than DeepEye on IDEBench. However, due to the bias of IDEBench and the utility function, **Sim-DE** and **Devi-De** both perform worse than **Random** and **DeepEye**, suggesting that an inappropriate utility function can bring much bias to the result. **Sim** and **Devi** only use inherent features of the data to calculate the relative importance of the visualizations, making the model more general across different datasets.

6 Related Work

Most methods are based on utility functions capturing different information. We categorize them as *visual effectiveness*, *data content*, and *user goal/preferences*.

Visual Effectiveness. Methods in this category consider visual quality as the ranking metric. APT [14] introduces expressiveness rules to prune invalid visualizations and effectiveness rules to rank the remaining visualizations. Voyager [22] uses a set of expressiveness criteria based on perceptual effectiveness metrics. VizML [10] trains models to predict better visualization choices.

Data Content. Methods in this category try to measure the interestingness of data to make recommendation. Metircs for interestingness including: (1) simple descriptive statistics such as entropy, variance, skewness, kurtosis, and correlation. This approach is exemplified by DIVE [11] and DataSite [6]; (2) inferential statistics such as the chi-square test and the p-value test. Examples are AutoVis [21] and ETKI [19]; (3) specific score functions based on data content such as deviation in SeeDB [20] and similarity in zenvisage [18].

User Goal/Preference Methods in this category try to model the user. They investigate models that calculate how much a visualization satisfies the user's visualization or analysis goal/preference. Examples are HARVEST [9], VizAssist [1], and VizRec [16].

These methods regard visualizations as individuals and ignore the relations between visualizations. Inspired by the graph model for item recommendation such as [4,5], our method creatively considers the relations between visualizations for recommendation and can also incorporate previous methods as utility functions to the model, achieving a more comprehensive recommendation.

7 Conclusion

In this paper, we proposed VizGRank, a general graph-based model based on the relations between visualizations. To compare their performance fairly, we collected datasets to construct a benchmark called KaggleBench. Experimental results show that VizGRank performs well on both KaggleBench and IDEBench in the literature, which verifies the applicability and practicability of VizGRank. Discovering more generic relations (e.g., learning relations from examples) and relations based on more data features can be our future research direction.

Acknowledgement. This work is supported by the National Key R&D Program of China (No. 2018YFB1004404 and No. 2018YFB1402600), the NSFC (No. 61732004 and No. 61802066) and the Shanghai Sailing Program (No. 18YF1401300).

References

1. Bouali, F., Guettala, A., Venturini, G.: VizAssist: an interactive user assistant for visual data mining. Vis. Comput. **32**(11), 1447–1463 (2016). https://doi.org/10.1007/s00371-015-1132-9
2. Brémaud, P.: Markov Chains: Gibbs fields, Monte Carlo simulation, and Queues, vol. 31. Springer, Heidelberg (2013)
3. Brin, S., Page, L.: The anatomy of a large-scale hypertextual web search engine. Comput. Netw. **30**(1–7), 107–117 (1998)
4. Chen, J., Wang, C., Wang, J., Yu, P.S.: Recommendation for repeat consumption from user implicit feedback (extended abstract). In: ICDE, pp. 19–20 (2017)
5. Chen, J., Wang, X., Wang, C.: Understanding item consumption orders for right-order next-item recommendation. Knowl. Inf. Syst. **57**(1), 55–78 (2018). https://doi.org/10.1007/s10115-017-1122-5
6. Cui, Z., Badam, S.K., Yalçin, M.A., Elmqvist, N.: Datasite: proactive visual data exploration with computation of insight-based recommendations. Inf. Vis. **18**(2), 251–267 (2019)
7. Ding, R., Han, S., Xu, Y., Zhang, H., Zhang, D.: Quickinsights: quick and automatic discovery of insights from multi-dimensional data. In: SIGMOD, pp. 317–332 (2019)
8. Eichmann, P., Zgraggen, E., Binnig, C., Kraska, T., Idebench: A benchmark for interactive data exploration. In: SIGMOD, pp. 1555–1569 (2020)
9. Gotz, D., et al.: Harvest: an intelligent visual analytic tool for the masses. In: IUI, pp. 1–4 (2010)
10. Hu, K., Bakker, M.A., Li, S., Kraska, T., Hidalgo, C.: Vizml: a machine learning approach to visualization recommendation. In: CHI, pp. 1–12 (2019)
11. Hu, K., Orghian, D., Hidalgo, C.: Dive: a mixed-initiative system supporting integrated data exploration workflows. In: HILDA@SIGMOD, pp. 1–7 (2018)
12. Liu, Z., Jiang, B., Heer, J.: immens: real-time visual querying of big data, vol. 32, no. 3pt4, pp. 421–430 (2013)
13. Luo, Y., Qin, X., Tang, N., Li, G.: DeepEye: towards automatic data visualization. In: ICDE, pp. 101–112. IEEE (2018)
14. Mackinlay, J.: Automating the design of graphical presentations of relational information. ACM Trans. Graph. **5**(2), 110–141 (1986)
15. Mafrur, R., Sharaf, M.A., Khan, H.A.: Dive: diversifying view recommendation for visual data exploration. In: CIKM, pp. 1123–1132 (2018)
16. Mutlu, B., Veas, E., Trattner, C.: VizRec: recommending personalized visualizations. ACM TiiS **6**(4), 1–39 (2016)
17. Shneiderman, B.: The eyes have it: A task by data type taxonomy for information visualizations. In: The Craft of Information Visualization, pp. 364–371. Elsevier (2003)
18. Siddiqui, T., Kim, A., Lee, J., Karahalios, K., Parameswaran, A.G.: Effortless data exploration with zenvisage: an expressive and interactive visual analytics system. PVLDB **10**(4), 457–468 (2016)
19. Tang, B., Han, S., Yiu, M.L., Ding, R., Zhang, D.: Extracting top-k insights from multi-dimensional data. In: SIGMOD, pp. 1509–1524 (2017)
20. Vartak, M., Rahman, S., Madden, S., Parameswaran, A.G., Polyzotis, N.: SEEDB: efficient data-driven visualization recommendations to support visual analytics. PVLDB **8**(13), 2182–2193 (2015)

21. Wills, G., Wilkinson, L.: Autovis: automatic visualization. Inf. Vis. **9**(1), 47–69 (2010)
22. Wongsuphasawat, K., Moritz, D., Anand, A., Mackinlay, J.D., Howe, B., Heer, J.: Voyager: exploratory analysis via faceted browsing of visualization recommendations. IEEE Trans. Vis. Comput. Graph. **22**(1), 649–658 (2016)
23. Yoshida, A., et al.: New performance index "attractiveness factor" for evaluating websites via obtaining transition of users' interests. Data Sci. Eng. **5**(1), 48–64 (2020). https://doi.org/10.1007/s41019-019-00112-1

Deep User Representation Construction Model for Collaborative Filtering

Daomin Ji[1]([⊠]), Zhenglong Xiang[2,3], and Yuanxiang Li[1]

[1] School of Computer Science, Wuhan University, Wuhan 430072, China
{jidaomin,yxli}@whu.edu.cn
[2] Key Lab of Intelligent Optimization and Information Processing,
Minnan Normal University, Zhangzhou 363000, China
[3] School of Computer and Software,
Nanjing University of Information Science and Technology, Nanjing 210044, China
zlxiang@nuist.edu.cn

Abstract. Model-based collaborative filtering (CF) methods can be divided into user-item methods and item-item methods. In most cases, both of them can be seen as modeling the user-item interaction and the only difference between them is that they adopt different ways to build user representations. User-item methods obtain user representations by directly assigning each user a real-valued vector and do not consider users' historical item information. However, users' historical item information can reflect users' preferences to some extent and can alleviate the problem of data sparsity. Ignoring this information may lead to incomplete construction of user representations and vulnerability to data sparsity. Although existing item-item methods address this problem by using the users' historical items to build the user representations, they always use the same vector to represent the same historical item for different users, which may limit the expressiveness and further improvement of the models. In this paper, we propose Deep User Representation Construction Model (DURCM) to construct user presentations in a more effective and robust way. Specially, different from existing item-item methods that directly use historical item vectors to build user representations, we first adopt a conversion module to convert a user's historical item vectors into personalized item vectors, which enables that even the same item has different expressions for different users. Second, we design a special attention module to automatically assign weights to these personalized item vectors when constructing the users' final representations. We conduct comprehensive experiments on four real-world datasets and the results verify the effectiveness of our proposed methods.

Keywords: Recommendation systems · Collaborative filtering · User representation

1 Introduction

In the era of information explosion, recommender systems make great contribution to alleviating information overload. They help determine which information

© Springer Nature Switzerland AG 2021
C. S. Jensen et al. (Eds.): DASFAA 2021, LNCS 12683, pp. 262–278, 2021.
https://doi.org/10.1007/978-3-030-73200-4_17

should be offered to individual customers and allow users to quickly find the information that meet their need. Due to the efficacy and accuracy, recommender systems have been widely deployed in online services, such as e-commerce platforms [23], online video websites [3] and social media application [6].

Over the past decades, collaborative filtering (CF) methods have been widely employed to build recommendation systems. They predict users' preference for target items based on the users' past interactions [19]. The early research on collaborative filtering mainly focused on memory-based methods, which make recommendations by simply finding similar users or items from the user-item interaction matrix [13,17]. However, since their performance is so poor on large and sparse data, they are replaced by model-based collaborative filtering methods recently, which can provide better recommendation performance.

Model-based CF methods can be divided into two classes, user-item methods and item-item methods. User-item methods directly model the user-item interaction. For example, matrix factorization (MF), the most well-known user-item method, represents each user and each item with a low-dimensional vector and uses inner product to learn their interaction function. Item-item methods, such as FISM, capture the interaction between target items and users' historical items. Although they construct the models from a different perspective, by simple transformation, most of item-item methods can be seen as modeling user-item relations.

Although both user-item models and item-item models are well developed, they have corresponding disadvantages, which come from the ways they construct their user representations. In user-item methods, the user representations are directly obtained by associating each user with a real-valued vector. It do not consider information of users' historical items. However, users' historical items can reflect their preferences in some degree, ignoring these information may lead to incomplete construction of the user representations. Furthermore, using users' historical item information to construct user representations can alleviate the problem of data sparsity. While in item-item methods, user representations are exactly constructed based on users' historical item vectors, which makes them have better invulnerability to data sparsity. However, in terms of making recommendation on dense data, item-item methods are generally inferior to user-item methods. This is because existing item-item methods use the same vector to represent the same historical item for different users, which may limit the expressiveness and further improvement of the models.

In this paper, we mainly study how to construct the user representations in a more efficient and reasonable way. To address this problem, we propose *Deep User Representation Construction Model* (DURCM). In DURCM, there are two key steps to construct a user's final representations. First, different from existing item-item methods that directly use the historical item vectors to build the user representations, we use a conversion module to convert a user's historical item vectors to personalized item vectors. By the conversion, even the same historical item can have different representations for different users. Second, we use a special attention network to automatically assign different weights to these personalized item vectors according their importance to the users, and then integrate them into the users' final representations. Then we conduct extensive experiments on four real-world datasets to verify the effectiveness of our proposed method to construct the user representations.

2   Related Work

Model-based collaborative filtering methods have been widely investigated recently and they can be generally divided into two categories, user-item models and item-item models, according to the relations that they model.

User-item models usually directly capture the user-item relations. One classical example is matrix factorization (MF) [12,14,18]. In MF, each user and each item is associated with a low-dimensional vector. Then the users rating on the item is estimated by the inner product of the corresponding user vector and item vector. However, the inner product function limits the expressiveness of MF models are not suitable for capturing the complex user-item relations in real life. To solve the problem, He et al. [9] proposed NCF (Neural Collaborative Filtering) framework, which employs a multi-layer perceptron to learn the user-item interactions. NCF provides an effective framework to learn user-item relations based on deep neural network and several variants of NCF have been proposed, such as ONCF [7], NNCF [1] and NGCF [20]. Instead of using neural network to learn the user-item interaction function, DeepMF [22] uses DSSM, a deep learning model widely used in web search, to learn the user and item representations from the rating matrix, and then matches them by cosine similarity. To achieve a more robust model, DeepCF [4] unifies representation learning based CF methods and matching function learning based CF methods to combine their advantages and it demonstrates promising performance.

Instead of directly capture user-item relations, item-item models model the interaction between the target items and users' historical items. SLIM [15], one of the earliest model-based item-item models, tries to recover a sparse and non-negative matrix from the historical interaction data to learn the hidden item-item similarity in the user-item interaction matrix. Based on SLIM, Christakopoulou and Karypis proposed HOSLIM [2] to model high-order item-item relations. They first find a group of itemsets that are frequently co-rated by users, and then use an extended SLIM model to jointly learn the item-item similarity and item-itemset similarity. SLIM and its extensions are efficient to learn item-item similarity. However, if there exist two items that are not co-rated by any user in the rating matrix, it is impossible for these models to calculate their similarity. To solve this problem, inspired by matrix factorization methods, FISM [11] uses embedding-based strategy that represents each item as two low-dimensional vectors, one for being target item and the other for being historical item. Then a user's preference on a target item can be estimated by the summation of the inner product between the target item vector and each historical item vector. NAIS enhances FISM by introducing an attention network to discover important items among a user's historical items [8]. When interacted with the target items, different historical items will be assigned with different importance weights by the attention network. Similar to NCF model, DeepICF [21] uses a multi-layer perceptron to learn the complex high-order item-item interactions, which is not captured by previous work.

In most cases, both user-item methods and item-item methods can be seen as modeling the interaction between user vectors and item vectors, and the only difference between them is that they construct the user representations from two different ways, which leads to their corresponding disadvantages: user-item methods are more vulnerable to data sparsity and item-item methods can not provide highly accurate personalization recommendation results on dense data. This drives us to design a new model to construct user representations more effectively.

3 Preliminaries

3.1 Problem Statement

In this paper, we focus on implicit feedback since it is much more accessible and abundant than explicit feedback in real world [16]. Let M and N denote the number of users and items. Correspondingly, we have a partially observed user-item interaction matrix $Y \in \mathbb{R}^{M \times N}$. Each entry y_{ui} is a binary value and denotes whether there is an interaction between user u and item i. Then the recommendation problem with implicit feedback can be defined as estimating the values of these unobserved entries in user-item interaction matrix Y.

3.2 Analysis of User-Item Models and Item-Item Models

In this subsection, We take MF and FISM as examples to analyze the similarities and differences between user-item models and item-item models based on their forms.

MF is a classic example of user-item models. It represents each user and each item by a low-dimensional vector and uses inner product function to capture the user-item relations, which can be formulated as

$$\hat{y}_{ui} = p_u q_i, \tag{1}$$

where p_u and q_i denotes the corresponding user vector and item vector for user u and item i, and $\hat{y}_{ui}$ denotes the prediction score for y_{ui}.

FISM is an example of item-item models. It represents each item by two low-dimensional vectors, one for being the target item and the other for being a user's historical item. Then FISM also uses the inner product function to capture the item-item relations, which can be formulated as

$$\hat{y}_{ui} = (\frac{1}{|\mathcal{R}_u^+|^\alpha} \sum_{j \in \mathcal{R}_u^+} p_j) q_i, \tag{2}$$

where $\mathcal{R}_u^+$ denotes the index set of user u's historical items and α is a hyper-parameter that controls the normalization effect. p_j denotes the historical item vector and q_i denotes the target item vectors.

If we take the bracketed content in Eq. 2 as a whole, we can find that both MF and FISM can be regarded as modeling the interaction between user vectors and item vectors. The only difference is that the user vectors in the two models come from different sources. In MF, each user is directly assigned with a user vector and it does not consider users' historical item information, which leads to two major problems. First, users' historical item information can reflect their preferences in some degree, which may be important for user representation construction. Second, only using users' ID feature to construct user representations easily leads to invulnerability to data sparsity. While in FISM, user vectors are constructed based on historical item vectors, so FISM are more invulnerable to data sparsity than MF and construct the user representations more completely. However, it suffers from another problems. It uses the same vector to represent the same historical item of different users, which may limit the expressiveness and further improvement of the models. These conclusions can be easily generalized to most user-item methods and item-item methods and drive us to design a more complete and reasonable way to construct user representations.

4   Our Proposed Method

Figure 1 illustrates the architecture of our proposed model, Deep User Representation Construction Model. It mainly is composed of two parts. The lower part is used to construct the user representations, which is the focus of this model; the upper part is used to capture the interaction between the constructed user vectors and target item vectors. In the next, we will introduce them in details.

Input and Embedding Layer. DURCM takes the target item ID, the user's historical item IDs and the user ID as input. For target item i, we apply one-hot encoding on the ID feature. Then the item ID is projected into an embedding vector $q_i \in \mathbb{R}^k$, where k denotes the embedding size. For user u's historical items, we apply multi-hot encoding on their ID feature. Then for each historical item $j \in \mathcal{R}_u^+$, it will be projected into an embedding vector $s_j \in \mathbb{R}^k$. As for user ID, we also apply one-hot encoding to it and each user ID will be projected into two kinds of embedding vectors, conversion vector m_u and attention vector z_u. Note that the two vectors here are different from the user vectors used in user-item models. While the user vectors in user-item models represent the final representations of the users, the two kinds of vectors in our model are designed for participating in user representation construction and we will introduce their specific functions latter.

Thus, the outputs of embedding layer are an embedding vector q_i, a set of historical item vectors $\mathcal{S}_u = \{s_j | j \in \mathcal{R}_u^+\}$ and two special vectors m_u and z_u associated with user u.

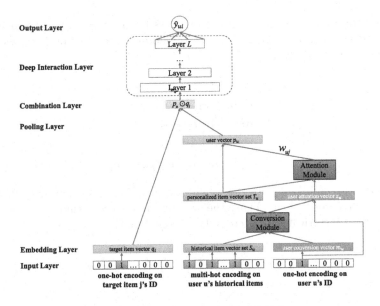

Fig. 1. The architecture of DURCM

Conversion Module. Different from most existing item-item models that construct the user representations directly based on the historical item vectors, DURCM first uses a special module named conversion module to convert the historical item vectors into personalized item vectors for different users. For user u's each historical item, its vector will be processed in conversion module as follows.

$$t_{uj} = f(s_j|\Theta_u), \tag{3}$$

where f denotes the conversion function, Θ_u denotes the parameters for conversion of user u's historical item vectors and t_{uj} denotes the corresponding personalized item vector for user u. For simplicity, we set the size of t_{uj} equal to that of s_j. An intuitive approach to implement Eq. 3 is linear transformation, which is defined as

$$t_{uj} = M_u s_j, \tag{4}$$

where M_u is a square matrix with trainable parameters.

However, in this case, we need to use Nk^2 parameters to learn the representations of all the M_u, which is easy to result in over-fitting problem when the user-item interaction matrix is highly sparse. Thus in this paper, we degenerate M_u to conversion vector m_u and use element-wise product to replace Eq. 4, which is formulated as

$$t_{uj} = m_u \odot s_j, \tag{5}$$

where $\odot$ denotes the element-wise product. Equation 5 only captures the linear interaction between the historical item vector and user conversion vector. To learn the non-linearity and complex pattern of the conversion process, we stack several multi-layer perceptron to process Eq. 5 further. Figure 2 illustrates the architecture of our designed conversion module. Then the personalized item vector can be obtained by

$$
\begin{aligned}
t_{uj} &= m_u \odot s_j, \\
t_1 &= ReLU(W_1 t_0 + b_1), \\
t_2 &= ReLU(W_2 t_1 + b_2), \\
&\cdots, \\
t_{uj} &= t_L = ReLU(W_L t_{L-1} + b_L),
\end{aligned}
\tag{6}
$$

where W_n, b_n and e_n denote the weight matrix, bias vector and output vector of the n-th hidden layer. Thus, the output of conversion module is a set of personalized historical item vectors $\mathcal{T}_u = \{s_j | j \in \mathcal{R}_u^+\}$.

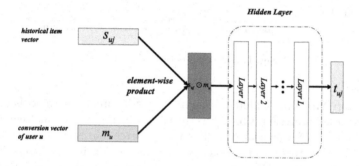

Fig. 2. The architecture of our proposed conversion module. Here softmax' layer is a special variant of standard softmax layer.

Pooling Layer. Now we can use these personalized item vectors to construct user u's representations. Since $\mathcal{T}_u$ may have different size for different users, we use a pooling layer to operate on $\mathcal{T}$ to produce a vector with fixed size. This process is defined as

$$
p_u = \sum_{j \in \mathcal{R}_u^+} w_{uj} t_{uj},
\tag{7}
$$

where p_u denotes the constructed user representation for u and w_{uj} denotes the importance of historical item j to user u. Notice we do not assign the item vectors with the same weights like FISM because this may lead to inferior performance.

To implement Eq. 7, we can regulate that w_{uj} is a trainable parameter and learn its value during the model training. But this simple strategy to assign the weights does not bring about a significant performance improvement to our model. To address this problem, we design a special attention network to learn the weights.

Inspired by the recent success of using neural networks to model the attention weights, we similarly use a multi-layer perception (MLP) to parameterize the attention function.

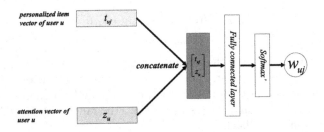

Fig. 3. The architecture of our proposed attention module

Figure 3 illustrate the architecture of our designed attention network. To learn the importance of item j to user u, we first associate each user u with an attention vector z_u, which is obtained from the embedding layer. Then for each personalized item vector t_j of user u, we concatenate it with the attention vector and feed the combined vector into a hidden layer to produce the corresponding attention weight. This process can be defined as

$$w_{uj} = a(t_{uj}) = softmax'(h^T ReLU(W\left[t_{uj} z_u\right] + b)), \tag{8}$$

where a denotes the attention function, $W \in \mathbb{R}^{k' \times k}$ and $b \in \mathbb{R}^{k'}$ denotes the weight matrix and bias vector of the hidden layer, and k' denotes the size of the hidden layer. $h \in \mathbb{R}^{k'}$ denotes the weights of the output layer of the attention network. softmax' is a variant of the softmax function, which is formulated as

$$softmax'(a(t_{uj})) = \frac{exp(a(t_{uj}))}{\left[\sum_{j \in \mathcal{R}_u^+} exp(a(t_{uj}))\right]^\beta}, \tag{9}$$

where β is a hyper-parameter that controls the normalization effect. Compared with the standard softmax function, softmax' is more suitable for CF datasets, since the variance of the number of historical items for different users is very large, as is argued in [8].

The Other Layers. We have obtained the user vector p_u and the target item vector q_i. Then we use a deep neural network following NCF (Neural Collaborative Filtering) framework to model their interactions, which can be formulated as

$$e_0 = p_u \odot q_i,$$
$$e_1 = ReLU(W_1 e_0 + b_1),$$
$$e_2 = ReLU(W_2 e_1 + b_2),$$
$$...,$$
$$e_L = ReLU(W_L e_{L-1} + b_L),$$
$$\hat{y}_{ui} = \sigma(h_{out} e_L),$$

(10)

where W_n, b_n and e_n denote the weight matrix, bias vector and output vector of the n-th deep interaction layer, and L denotes the depth of the deep interaction layer. h_{out} is the edge weights of the output layer and σ denotes the sigmoid function that maps the prediction score $\hat{y}_{ui}$ into the range of [0,1], which is more suitable for implicit feedback.

Objective Function. To learn the model parameters, we adopt the point-wise log loss as the loss function, which is defined as

$$\mathcal{L} = - \sum_{(u,i) \in \mathcal{R}^+ \mathcal{R}^-} [y_{ui} log(\hat{y}_{ui}) + (1 - y_{ui}) log(1 - \hat{y}_{ui})] + \lambda ||\Theta||^2, \quad (11)$$

where $\mathcal{R}^+$ denotes the set of positive training instances, which are equal to the whole observed user-item interactions. $\mathcal{R}^-$ denotes the set of negative training instances, which are sampled from the remaining unobserved interactions. For each positive instance in $\mathcal{R}^+$, we need to sample NS negative instances, where NS is a hyper-parameter that denotes the sampling ratio. λ is another hyper-parameter that controls the strength of L_2 regularization to avoid over-fitting. And Θ denotes the set of all the trainable parameters in the model.

Pre-training. Due the non-linearity of neural network model and the non-convexity of the objective function, optimization with gradient descent based methods can easily lead to local minimums [5]. Therefore, the initiation of model parameters plays an important role to the convergence and final performance of the model. We empirically find that random initiation of all the parameters has an negative impact on the performance of our proposed model. To address this problem, we pre-train DURCM with FISM since DURCM can be seen as an extension of FISM. Specifically, we first train FISM until convergence, then we use the trained item embedding vectors to initiate the corresponding parts of the parameters in DURCM. As for the other parameters including the conversion vectors and attention vectors, they are randomly initiated with a Gaussian distribution.

5   Experiments

To verify the effectiveness of our proposed methods, we conduct experiments to answer the following research questions:

- Does the proposed model DURCM outperform the state-of-the-art collaborative filtering methods?
- Are our proposed pre-training strategy really helpful for improving the performance of DURCM?
- How do the key modules in DURCM, attention module and conversion module, impact the model performance?

5.1 Experimental Setting

Datasets. We conduct experiments on four publicly accessible datasets: MovieLens 100K (ml-100k), MovieLens 1M (ml-1m), Amazon Music (AMusic) and Amazon Movies (AMovie)[1,2]. The former two are movie rating datasets that are widely used to test the performance of CF methods. Since the data in MovieLens is explicit feedback, we transform it into implicit data by retaining all the interaction signals but ignoring the rating values. Amazon datasets are comprised of users' reviews and ratings on different categories of products. We merge the review data and rating data and represent them as implicit data uniformly. Furthermore, since the original data of Amazon is highly sparse, we filter them such that only users with at least 20 interactions and items with at least 5 interactions are retained. The statistics of these four datasets are listed in the Table 1.

Table 1. Statistics of the datasets

Dataset	#user	#item	#rating	Sparsity
ml-100k	944	1,683	100,000	93.70%
ml-1m	6,040	3,706	1,000,209	95.53%
AMusic	1,776	12,929	46,087	99.67%
AMovie	16,136	70,649	895,240	99.92%

Evaluation Protocal. We adopted the widely used leave-one-out evaluation protocol to evaluate the performance of these models. For each user, we held-out the item of the last interaction to construct the test set and use the rest of data for training. Since it is too time-consuming to rank all items for each user during evaluation, following the common strategy, for each user we randomly sampled 100 items that are not interacted by the user, and then ranked the test item with the 100 sampling items according to their prediction scores. Furthermore, we adopted two widely used metric, HR (Hit Ratio) and NDCG (Normalized Discounted Cumulative Gain), to evaluation the performance of a ranked list. We truncated the ranked list at length of 10 for both metrics. As such, HR@10

---

[1] https://grouplens.org/datasets/movielens/.
[2] http://jmcauley.ucsd.edu/data/amazon/.

demonstrates whether the test item is present on the top-10 list, and NDCG@10 evaluates the ranking quality by assigning higher scores to the hits at top ranks. We calculate both metrics for each user and report the average scores.

Compared Methods. We compare DURCM with the following collaborative filtering method:

- **eALS** [10] is a well-known MF-based recommendation methods. It optimizes a special point-wise squared loss function by treating all missing data with a popularity-based weight.
- **JRL** [1] feeds the element-wise product of the user vector and item vector into a multi-layer perceptron to model user-item interaction.
- **NeuMF** [9] combines the hidden layers of two different deep learning based user-item methods to make final prediction. It can provide state-of-the-art performance. In this paper, we choose the pre-trained version.
- **FISM** [11] is a classic item-item model that uses inner product function to capture item-item relations.
- **DeepICF** [21] is a state-of-the-art item-item model that improves FISM by using a deep network to learn high-order item-item relations. In this paper, we choose the pre-trained version.

These methods are intentionally chosen to cover a diverse range of recommendation methods. Among these methods, the former three models are representatives of user-item models and the latter two are competitive item-item models. eALS and FISM are traditional linear models while the other three models are deep learning based CF methods. Furthermore, NeuMF and DeepICF can provide state-of-the-art performance.

Implementation Details. We implement DURCM based on Pytorch. To determine the hyper-parameters of DURCM, we randomly sample a positive item for each user in the training set to construct the validation set and tune the hyper-parameters on it. As for the regularization coefficient λ, we tune value in the range of $[10^{-6}, 10^{-5}, ..., 1]$. For normalization coefficient β, we test the value in the range of $[0.1, 1]$ at an interval of 0.1. Negative sampling ratio NS is searched in $[1, 2, 3, ..., 10]$. We employ three hidden layers for the deep interaction layer following the design of tower structure of [9]. Without additional mention, We pre-train DURCM as discussed in Sect. 4 and after feeding the model parameters, we tune the value of learning rate in the range of $[0.001, 0.05, 0.01, 0.1]$.

5.2 Performance Comparison (RQ1)

We first make a comparison between DURCM and the selected approaches. For a fair comparison, the embedding size is set to 64 for all these methods. The comparison is based on two aspects, recommendation accuracy and invulnerability to data sparsity.

Table 2. Comparison results of recommendation accuracy on the four datasets

Dataset	ml-100k		ml-1m		AMusic		AMovie	
Method	HR@10	NDCG@10	HR@10	NDCG@10	HR@10	NDCG@10	HR@10	NDCG@10
eALS	0.632	0.366	0.697	0.419	0.618	0.347	0.509	0.304
JRL	0.658	0.385	0.710	0.432	0.632	0.357	0.521	0.310
NeuMF	_0.672_	_0.398_	_0.728_	_0.446_	0.638	0.368	0.528	0.321
FISM	0.647	0.368	0.692	0.420	0.625	0.350	0.517	0.307
DeepICF	0.654	0.379	0.712	0.437	_0.643_	_0.372_	_0.533_	_0.326_
DURCM	**0.684**	**0.408**	**0.735**	**0.455**	**0.664**	**0.388**	**0.552**	**0.340**

Comparison of Recommendation Accuracy. Table 2 illustrates the performance of the six methods on the four datasets. The best and second best scores are shown bold and underlined, respectively. We can see that DURCM achieves the best performance on all the cases. Except DURCM, user-item model NeuMF provides the best performance on ml-100k and ml-1m. However, on Amazon datasets comprised of highly sparse data, the performance of NeuMF is surpassed by item-item model DeepICF. This is because user-item models perform well on dense data but are more vulnerable to data sparsity, as we discussed before. Compared with the second best scores achieved by the other methods, the average relative improvements of DURCM are 2.15% (on ml-100k), 1.49% (on ml-1m), 3.78% (on AMusic) and 3.93% (on AMovie). We attribute such improvements to that DURCM construct the user representations in a more robust and effective way. These results demonstrate the superiority of DURCM over state-of-the-art methods in terms of recommendation accuracy.

Invulnerability to Data Sparsity. To test the invulnerability to data sparsity of these methods, we first create datasets with different levels of data sparsity based on the four datasets. For each user, we randomly sample a subset of the user's whole interactions according to four different sample ratios, 20%, 40%, 60% and 80%, which leads to four derived datasets with different levels of data sparsity for each original dataset.

Figure 4 and Fig. 5 illustrate the performance and corresponding percentage decrease of performance of theses models on the datasets with different levels of sparsity in terms of HR@10. Since the results of NDCG@10 show almost the same trends, they are omitted for space limitation. At first, we can see that DURCM achieves the best performance on all the datasets with different sparsity levels. Compared with the best scores achieved by the other methods, the relative improvements of DURCM on different levels of data sparsity are 2.78% (Level 1), 3.02% (Level 2), 2.94% (Level 3) and 3.02% (Level 4). Furthermore, the performance of DURCM decreases at a slow rate with the increase of data sparsity. Besides DURCM, the item-item models (FISM and DeepICF) also achieve good performance on these datasets, and their percentage decrease of performance is close to that of DURCM. On the contrary, the user-item models (eALS, JRL and NeuMF) provide poor performance on the datasets of high level of sparsity. Their performance drops sharply as the data sparsity increases. In the worst case (AMovie of the highest level of sparsity), their performance has dropped by more than 55%, while the performance of DURCM and item-item models only drops by less than 45%. These results demonstrate DURCM have strong invulnerability to data sparsity and we attribute such ability to that DURCM encodes the signals of users' historical items into the user representation.

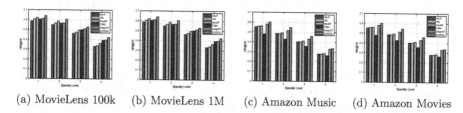

(a) MovieLens 100k   (b) MovieLens 1M   (c) Amazon Music   (d) Amazon Movies

Fig. 4. Performance of these methods on datasets with different levels of data sparsity

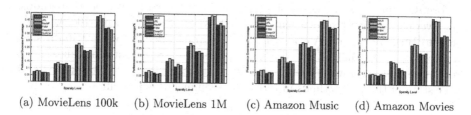

(a) MovieLens 100k   (b) MovieLens 1M   (c) Amazon Music   (d) Amazon Movies

Fig. 5. Percentage of performance decrease of these methods on datasets with different levels of data sparsity

5.3   Effectiveness of Pre-training (RQ2)

The initialization of parameters plays an important role in the convergence and final performance of neural network models. To verify the effectiveness of our proposed pre-training strategy, we also make a comparison between two versions of DURCM, with and without pre-training. For DURCM without pre-training, we use Adam to optimize it with random parameter initialization.

As is illustrated in Table 3, the DURCM with pre-training achieves the better performance in all the cases than the version without pre-training. On average, the relative improvements of pre-training strategy is 1.91% (on ml-100k), 1.37%(on ml-1m), 1.58% (on AMusic) and 1.54% (on AMovie), respectively. These results demonstrate the usefulness of our proposed pre-training method for DURCM initialization.

Table 3. Performance of DURCM with and without pre-training

Factors	ml-100k				ml-1m			
	Without pre-training		With pre-training		Without pre-training		With pre-training	
	HR@10	NDCG@10	HR@10	NDCG@10	HR@10	NDCG@10	HR@10	NDCG@10
k = 8	0.637	0.373	**0.650**	**0.381**	0.683	0.410	**0.691**	**0.416**
k = 16	0.652	0.381	**0.663**	**0.389**	0.704	0.430	**0.712**	**0.434**
k = 32	0.665	0.392	**0.674**	**0.399**	0.717	0.441	**0.726**	**0.448**
k = 64	0.673	0.398	**0.684**	**0.408**	0.723	0.447	**0.735**	**0.455**
Factors	AMusic				AMovie			
	Without pre-training		With pre-training		Without pre-training		With pre-training	
	HR@10	NDCG@10	HR@10	NDCG@10	HR@10	NDCG@10	HR@10	NDCG@10
k = 8	0.616	0.352	**0.622**	**0.357**	0.509	0.308	**0.516**	**0.313**
k = 16	0.631	0.366	**0.639**	**0.373**	0.524	0.319	**0.529**	**0.324**
k = 32	0.647	0.375	**0.658**	**0.382**	0.538	0.326	**0.542**	**0.332**
k = 64	0.653	0.381	**0.664**	**0.388**	0.542	0.332	**0.552**	**0.340**

5.4   Effectiveness of Key Modules (RQ3)

In DURCM, there are two key modules, conversion module and attention module, to construct the user representations. In this part, we perform comparison experiment to verify the effectiveness of the two modules. We create two additional kinds of DURCM, DURCM without conversion module (denoted as DURCM-woc) and DURCM without attention module (denoted as DURCM-woa). For DURCM without conversion module, we directly use the historical item vectors to construct the use vectors; for DURCM without attention module, we assign the personalized item vectors with the same normalized weights.

Table 4 illustrates the experimental results on the four datasets. We can see that the standard DURCM achieves the best performance on all the cases among the three versions. Compared with DURCM without conversion module, the average relative improvements of performance are 2.38% (on ml-100k), 1.92% (on ml-1m), 1.97% (on AMusic) and 2.37% (on AMovie). We attribute such improvements to that conversion module can personalize the historical item vectors for different users, which increases the expressiveness of our model. And compared with DURCM without attention module, the average relative improvements are 2.11% (on ml-100k), 1.89% (on ml-1m), 1.93% (on AMusic) and 2.47% (on AMovie). We believe these improvements come from attention module, which can automatically distinguish the important item for a user and then assign higher weights to the corresponding personalized item vectors when constructing the user representations.

Table 4. Comparison results of the three versions of DURCM on the four datasets

Factor	Dataset	DURCM		DURCM-woc		DURCM-woa	
		HR@10	NDCG@10	HR@10	NDCG@10	HR@10	NDCG@10
k = 8	ml-100k	0.650	**0.381**	0.638	0.372	0.640	0.373
	ml-1m	**0.691**	**0.416**	0.680	0.410	0.682	0.411
	Amusic	**0.622**	**0.357**	0.611	0.348	0.614	0.350
	Amovie	**0.516**	**0.313**	0.501	0.303	0.501	0.301
k = 16	ml-100k	**0.663**	**0.389**	0.647	0.380	0.646	0.381
	ml-1m	**0.712**	**0.434**	0.690	0.425	0.692	0.428
	Amusic	**0.639**	**0.373**	0.631	0.365	0.629	0.364
	Amovie	**0.529**	**0.324**	0.518	0.317	0.518	0.318
k = 32	ml-100k	**0.674**	**0.399**	0.654	0.391	0.657	0.390
	ml-1m	**0.726**	**0.448**	0.715	0.440	0.714	0.438
	Amusic	**0.658**	**0.382**	0.645	0.374	0.647	0.373
	Amovie	**0.542**	**0.332**	0.535	0.327	0.533	0.325
k = 64	ml-100k	**0.684**	**0.408**	0.669	0.398	0.672	0.401
	ml-1m	**0.735**	**0.455**	0.722	0.447	0.720	0.445
	Amusic	**0.664**	**0.388**	0.650	0.382	0.649	0.382
	Amovie	**0.552**	**0.340**	0.535	0.332	0.537	0.333

6 Conclusion and Future Work

In this paper, we first analyse the ways of user-item methods and item-item methods to construct their user representations and point out the corresponding drawbacks of them. Then we propose a novel model, DURCM, which utilizes users' historical items to recover the user representations in a more effective way. Before constructing user representations, DURCM uses conversion module to project the historical item vectors into different personalized item vectors for different users. Then the user representations are constructed based on these personalized item vectors with a special attention network to automatically assign corresponding weights according to their importance to the user. Finally, DURCM uses a multi-layer perceptron to model the interaction between the constructed user vectors and the target item vectors and make final prediction. Extensive experimental results demonstrate the effectiveness of our proposed method. As for future work, we plan to explore other kinds of conversion to investigate which conversion functions are efficient and can provide the best performance.

Acknowledgement. This work is supported by Key Lab of Intelligent Optimization and Information Processing, Minnan Normal University (NO. ZNYH202004) and the Starup Foundation for Talents of Nanjing University of Information Science and Technology.

References

1. Bai, T., Wen, J.R., Zhang, J., Zhao, W.X.: A neural collaborative filtering model with interaction-based neighborhood. In: Proceedings of the 2017 ACM on Conference on Information and Knowledge Management, pp. 1979–1982 (2017)
2. Christakopoulou, E., Karypis, G.: HOSLIM: higher-order sparse LInear method for top-N recommender systems. In: Tseng, V.S., Ho, T.B., Zhou, Z.-H., Chen, A.L.P., Kao, H.-Y. (eds.) PAKDD 2014. LNCS (LNAI), vol. 8444, pp. 38–49. Springer, Cham (2014). https://doi.org/10.1007/978-3-319-06605-9_4
3. Covington, P., Adams, J., Sargin, E.: Deep neural networks for youtube recommendations. In: Proceedings of 10th ACM Conference on Recommender Systems, pp. 191–198 (2016)
4. Deng, Z.H., Huang, L., Wang, C.D., Lai, J.H., Philip, S.Y.: DeepCF: a unified framework of representation learning and matching function learning in recommender system. In: In Proceedings of 33rd AAAI Conference on Artificial Intelligence, vol. 33, pp. 61–68 (2019)
5. Erhan, D., Bengio, Y., Courville, A., Manzagol, P.A., Vincent, P., Bengio, S.: Why does unsupervised pre-training help deep learning? J. Mach. Learn. **11**(Feb), 625–660 (2010)
6. Guy, I., Zwerdling, N., Ronen, I., Carmel, D., Uziel, E.: Social media recommendation based on people and tags. In: Proceedings of the 33rd International ACM SIGIR Conference on Research and Development in Information Retrieval, pp. 194–201 (2010)
7. He, X., Du, X., Wang, X., Tian, F., Tang, J., Chua, T.S.: Outer product-based neural collaborative filtering. In: In Proceedings of 27th International Joint Conference on Artificial Intelligence, pp. 2227–2233 (2018)
8. He, X., He, Z., Song, J., Liu, Z., Jiang, Y.G., Chua, T.S.: NAIS: neural attentive item similarity model for recommendation. IEEE Trans. Knowl. Data Eng. **30**(12), 2354–2366 (2018)
9. He, X., Liao, L., Zhang, H., Nie, L., Hu, X., Chua, T.S.: Neural collaborative filtering. In: Proceedings of 26th International Conference on World Wide Web, pp. 173–182 (2017)
10. He, X., Zhang, H., Kan, M.Y., Chua, T.S.: Fast matrix factorization for online recommendation with implicit feedback. In: Proceedings of 39th International ACM SIGIR Conference on Research on Development of Information Retrieval, pp. 549–558 (2016)
11. Kabbur, S., Ning, X., Karypis, G.: FISM: factored item similarity models for top-n recommender systems. In: Proceedings of the 19th ACM SIGKDD International Conference on Knowledge Discovery and Data Mining, pp. 659–667 (2013)
12. Koren, Y., Bell, R., Volinsky, C.: Matrix factorization techniques for recommender systems. Computer **42**(8), 30–37 (2009)
13. Linden, G., Smith, B., York, J.: Amazon.com recommendations: item-to-item collaborative filtering. IEEE Internet Comput. **7**(1), 76–80 (2003)
14. Mnih, A., Salakhutdinov, R.R.: Probabilistic matrix factorization. In: Proceedings of 21th Advance Neural Information Processing Systems, pp. 1257–1264 (2008)
15. Ning, X., Karypis, G.: Slim: sparse linear methods for top-n recommender systems. In: 2011 IEEE 11th International Conference on Data Mining, pp. 497–506. IEEE (2011)
16. Rendle, S., Freudenthaler, C., Gantner, Z., Schmidt-Thieme, L.: BPR: Bayesian personalized ranking from implicit feedback. In: Bilmes, J.A., Ng, A.Y. (eds.) Proceedings of 25th Conference on Uncertainty Artificial Intelligence

17. Sarwar, B., Karypis, G., Konstan, J., Riedl, J.: Item-based collaborative filtering recommendation algorithms. In: Proceedings of the 10th International Conference on World Wide Web, pp. 285–295 (2001)
18. Shi, Y., Larson, M., Hanjalic, A.: List-wise learning to rank with matrix factorization for collaborative filtering. In: Proceedings of the 4th ACM Conference on Recommender Systems, pp. 269–272 (2010)
19. Su, X., Khoshgoftaar, T.M.: A survey of collaborative filtering techniques. Adv. Artif. Intell. **2009** (2009)
20. Wang, X., He, X., Wang, M., Feng, F., Chua, T.S.: Neural graph collaborative filtering. In: Proceedings of the 42nd International ACM SIGIR Conference on Research and Development in Information Retrieval, pp. 165–174 (2019)
21. Xue, F., He, X., Wang, X., Xu, J., Liu, K., Hong, R.: Deep item-based collaborative filtering for top-n recommendation. ACM Trans. Inf. Syst. **37**(3), 1–25 (2019)
22. Xue, H.J., Dai, X., Zhang, J., Huang, S., Chen, J.: Deep matrix factorization models for recommender systems. In: Proceedings of 26th International Joint Conference on Artificial Intelligence, pp. 3203–3209 (2017)
23. Zhou, G., et al.: Deep interest network for click-through rate prediction. In: Proceedings of the 24th ACM SIGKDD International Conference on Knowledge Discovery & Data Mining, pp. 1059–1068 (2018)

DiCGAN: A Dilated Convolutional Generative Adversarial Network for Recommender Systems

Zhiqiang Guo[1], Chaoyang Wang[1], Jianjun Li[1(✉)], Guohui Li[2], and Peng Pan[1]

[1] School of Computer Science and Technology,
Huazhong University of Science and Technology, Wuhan, China
{zhiqiangguo,sunwardtree,jianjunli,panpeng}@hust.edu.cn
[2] School of Software Engineering, Huazhong University of Science and Technology,
Wuhan, China
guohuili@hust.edu.cn

Abstract. Generative Adversarial Network (GAN) has recently been introduced into the domain of recommendation due to its ability of learning the distribution of users' preferences. However, most existing GAN-based recommendation methods only exploit the user-item interactions, while ignoring to leverage the information between user's interacted items. On the other hand, Convolutional Neural Network (CNN) has shown its power in learning high-order correlations. In this paper, combining with the strengths of both GAN and CNN, we propose a Dilated Convolutional Generative Adversarial Network (DiCGAN) for recommendation, in which we first embed the interacted items of per user into an image in a latent space, and then use several dilated convolutional filters and a vertical convolutional filter to capture the high-order correlations among the interacted items. Moreover, an attention module is employed before convolution to generate attention maps for adaptive feature refinement. Experiments on several public datasets verify the superiority of DiCGAN over several baselines in terms of top-N recommendation. Further more, our experimental results show that when the dataset is more large and sparse, the performance gain of DiCGAN is also more significant, demonstrating the effectiveness of the CNN component in extracting high-order correlations from interacted data for better performance.

Keywords: Recommender systems · Generative Adversarial Network · Convolutional neural network · Attention module

1 Introduction

Since its debut in NeurIPS 2014, Generative Adversarial Networks (GAN) [6], which uses a discriminative model to guide the training of the generative model,

Supported by the National Natural Science Foundation of China under Grant No. 61672252, and the Fundamental Research Funds for the Central Universities under Grant No. 2019kfyXKJC021.

has enjoyed considerable success in tasks such as image generation [2] and natural language generation [28]. Recently, GAN has also gained increasing attention in recommender systems [1,4,16,18,20–22,24,27]. IRGAN [22] first utilizes a *theoretical minimax game* to iteratively optimize generative (G) and discriminative (D) models, which demonstrates GAN's potential in collaborative filtering based recommendation. But IRGAN suffers the inherent limitation of discrete item index generation approaches. That is, for the same item, the index is sometimes labeled as fake and other times as real, which makes D significantly confused. To address this problem, Chae *et al.* [1] proposed CFGAN, in which G tries to generate a plausible interact vector of a user composed of real-valued elements, rather than to sample a single item index that the user may be interested in. Though having solved the contradicting labels problem in IRGAN, CFGAN still has two limitations: 1) It requires much more space to store the interaction vectors for all users, and hence is more susceptible to data scale. Too large vector size (equals to the item size) makes CFGAN rather difficult to train and even fail to achieve significant ranking accuracy sometimes. 2) The same as IRGAN, CFGAN still only exploits the interaction vector of per user but ignores the high-order correlations among the user's interacted items, which limits its performance gains.

To address the above limitations, in this paper, we propose a Dilated Convolutional Generative Adversarial Network (DiCGAN) for recommender systems. The proposed DiCGAN first embeds the interacted items of per user into an *image* to compress the feature dimensions, and then leverages the convolutional neural network (CNN) to learn the high-order correlation features among user's interacted items, to represent user's preferences. In fact, CNN has shown its power in learning high-order correlations in recommender systems, such as Caser [17] and NextItNet [29]. The success of both GAN and CNN in recommendation motivates us to push forward to combine their strengths for further performance improvements.

In DiCGAN, we employ CNN to replace the multi-layer perceptron (MLP) part of the generator in CFGAN. Specifically, the CNN in our model contains several dilated convolutional filters [14,26], which are responsible for extracting high-order correlation features among a user's interacted items, and a vertical convolutional filter [17], which is responsible for extracting the high-order correlation features among the latent dimensions of a user's interacted items. Moreover, considering that the attention mechanism can help improve the representation of user's preference, we employ channel and spatial attention modules [23] to generate attention maps that can adaptively refine the feature maps in the channel and spatial axes, respectively.

Since GAN-based methods themselves can alleviate the data sparsity problem to some extent by augmenting user-item interaction information [5], with CNN's help on extracting high-order correlation features, DiCGAN can yield to remarkable performance improvement over existing GAN-based recommendation models, especially for high-sparsity datasets. In summary, the main contributions of this work are as follows:

- We combine GAN with CNN for recommendation by proposing a Dilated Convolutional Generative Adversarial Network (DiCGAN), in which a CNN that contains several dilated convolutional filters and a vertical convolutional filter is designed to serve as the generator of a GAN-based model, with the objective of extracting high-order correlation features for better performance.
- An attention module is employed in DiCGAN to generate attention maps that can be multiplied to the feature maps for adaptive refinement of features.
- Extensive experiments are conducted on three benchmark datasets and the results verify the superior performance of DiCGAN over state-of-the-art methods. Moreover, the experimental results also show that the proposed CNN component can yield to remarkable performance improvement, especially when dealing with large and sparse datasets.

2 Related Work

In recent years, GAN [6] has received increasing attention due to its ability to fit the complex data distributions. There are some good attempts on applying GAN for recommendation [1,22,24]. Motived by seqGAN [28], IRGAN [22] is first proposed to use policy gradient to solve the problem that the discrete data cannot be directly optimized by gradient descent as in the original GAN. However, due to the confusion of selection, discriminators in IRGAN cannot be properly trained to detect the difference between true and false items. To address this problem, by constructing multiple diverse item sets for a specific user, PD-GAN [24] is proposed to decrease the probability of assigning contradicting labels to the same items, but it ignores user preferences for different categories. CFGAN [1] addresses the contradicting labels problem of IRGAN by using a vector-wise adversarial training. But as mentioned in the introduction, there are still two limitations in CFGAN.

Due to its strong ability in learning complex high-level correlations, recently, there are also some research efforts on utilizing CNN to improve recommendation performance. For example, Convolution matrix factorization [12] alleviates the data sparsity problem by integrating CNN into probabilistic matrix factorization to capture contextual information of the documents. ONCF [9] learns high-order correlations among embedding dimensions by using CNN and the outer product. Caser [17] embeds the interacted items of per user into an image and employs CNN to learn user features for the sequential recommendation. But this method introduces max-pooling, which limits the model's expression ability in that it is impossible to fully utilize the location information of important data. To capture the relations between items, NextItNet [29] abandons the pooling layer and exploits dilated convolution to extracts feature from the image constructed by interacted items. The above-mentioned works indicate the potential of using CNNs for a better recommendation.

To address the shortcomings of the above mentioned GAN-based methods, DiCGAN improves the generative capacity of the generator by combining GAN with CNN, where CNN is used to extract higher-order correlation features among the interacted items. But different from previous CNN-based recommendation methods, our CNN component introduces dilated convolution and adopts the attention module to refine the convolution feature maps.

3 Preliminaries

We consider a recommender system with m users $U = \{u_1, u_2, \ldots, u_m\}$ and n items $I = \{i_1, i_2, \ldots, i_n\}$. Let $I_u = \{i_1^u, i_2^u, \ldots, i_t^u\}$ denote the set of t items that user u has interacted with. User's implicit feedback on items is represented by a sparse interaction matrix $Y \in \mathbb{R}^{m \times n}$, where an element y_{ui} is 1 if i is contained in I_u, and 0 otherwise. Moreover, we use $\mathbf{y}^u = Y_{u,*} \in \mathbb{R}^{n \times 1}$ to denote u's interaction across all items. The goal of collaborative filtering (CF) [11] is to predict u's preference score on i (denoted by $\hat{y}_{ui}$, where $i \in I \backslash I_u$), reflecting how much likely u will interact with i, and then recommend a subset of items with scores that are the highest to u.

Model-based CF methods [3,8,10,25] generally derive $\hat{y}_{ui}$ by assuming there is an underlying model, which can be abstracted as learning $\hat{y}_{ui} = f(u, i|\Theta)$, where Θ denotes the set of model parameters and f denotes the function that is parameterized by Θ and maps a given pair of user u and item i to a predicted score. Θ can be learned by optimizing an objective function with Y as training data. Existing model-based CF methods usually optimize the traditional point-wise or pair-wise objective functions to train the CF model.

GAN-based CF methods train the model in a different way. Instead of optimizing the point-wise or pairwise objective functions, GAN-based methods try to achieve more satisfactory accuracy in recommendation by borrowing the adversarial training approach through a competition process involving a generative model (G) and a discriminator model (D). In IRGAN, G tries to generate the indices of items relevant to a given user, and D tries to discriminate the user's ground truth items from those synthetically generated by G. In CFGAN, G tries to generate a plausible interact vector of a given user, while D tries to discriminate the user's ground truth interact vector from that generated by G. In this work, in view of the inherent limitation of IRGAN, we adopt vector-wise adversarial training similar to CFGAN to generate a plausible interaction vector of a user. But different from CFGAN, we employ CNN, rather than MLP, to extract the high-order correlations among user's interacted items.

4 Proposed Method

4.1 Framework Overview

As shown in Fig. 1, DiCGAN contains two parts: the generator (G) and the discriminator (D). G consists of an embedding layer, a convolutional layer and a fully connected layer. Firstly, the user-item interaction matrix is mapped into a latent space as an *image* with height L. Then, the convolutional layers apply several dilated filters and a vertical filter to get feature $\mathbf{z}_{dc}$ (high-order correlation feature among u's interacted items) and $\mathbf{z}_{vc}$ (high-order correlation feature

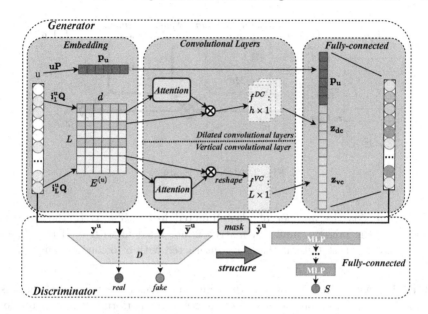

Fig. 1. Framework overview of DiCGAN

among the latent dimensions of u's interacted items) respectively. Moreover, an attention module is employed before convolution to generate attention maps for adaptive feature refinement. Finally, through a fully connected layer, the user feature ($\mathbf{p}_u$) combined with the item feature ($\mathbf{z}_{dc}$ and $\mathbf{z}_{vc}$) are transformed into a user-item interaction vector $\hat{\mathbf{y}}_u$, which reflects u's preference distribution on items. The D part tries to discriminate u's ground-truth interaction vector $\mathbf{y}^u$ from that generated by G. The training phase of DiCGAN is a *minimax* game, after training, G predicts the preference scores of user u for all the items, and the N items with the highest predicted scores are suggested to u.

4.2   Generator

Embedding Layer. Since we consider pure collaborative filtering, we use only the ID of a user and an item as the input feature. Specifically, we use $\mathbf{u} \in \mathbb{R}^{1 \times m}$ to denote the one-hot vector of u, and use $\mathbf{i}_k^u \in \mathbb{R}^{1 \times n}$ to denote the one-hot vector of item $i_k^u \in I_u$. Let $\mathbf{P} \in \mathbb{R}^{m \times d}$ and $\mathbf{Q} \in \mathbb{R}^{n \times d}$ denote the latent factor matrices for users and items, respectively, where d is the dimension of the latent space. We can get the latent representation of u by $\mathbf{p}_u = \mathbf{uP}$. For each item $i_k^u \in I_u$, we can get its latent representation by $\mathbf{q}_{i_k^u} = \mathbf{i}_k^u \mathbf{Q}$. Then, by retrieving the latent representations of L items in I_u and stacking them together, we can obtain an *image* represented by a matrix $E \in \mathbb{R}^{L \times d}$,

$$E = \begin{bmatrix} \mathbf{q}_{i_1^u} \\ \vdots \\ \mathbf{q}_{i_L^u} \end{bmatrix} \tag{1}$$

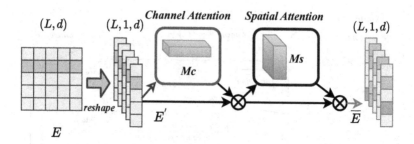

Fig. 2. Attention module

Recall that t is the number of u's interacted items. If $t < L$, the same as Caser [17], the missing values in matrix E are filled in with 0. If $t \geq L$, we only choose the first L items that have interacted with u to construct matrix E.

Attention Module. Considering that different items have different attentions, inspired by [23], we employ an attention model before performing the convolution operation. As shown in Fig. 2, we first conduct a simple reshape operation to convert E with shape (L, d) to E' with shape $(L, 1, d)$. Then, the input feature map is passed through a channel attention module to get the channel attention map,

$$M_c(E') = \sigma(\mathbf{W}_1(\mathbf{W}_0(E'_{cavg})) + \mathbf{W}_1(\mathbf{W}_0(E'_{cmax}))) \quad (2)$$

where $\mathbf{W}_0 \in \mathbb{R}^{d/\eta \times d}$ and $\mathbf{W}_1 \in \mathbb{R}^{d \times d/\eta}$ are weighted matrices, η represents the reduction ratio, E'_{cavg} and E'_{cmax} denote average-pooled feature and max-pooled feature to aggregate spatial information respectively, and $\sigma(\cdot)$ represents the Sigmoid function. Note that $\mathbf{W}_0$ and $\mathbf{W}_1$ are shared for both inputs and the ReLU activation function is followed by $\mathbf{W}_0$. Afterwards, E' and $M_c(E')$ are multiplied to get a feature map,

$$F = M_c(E') \otimes E' \quad (3)$$

where $\otimes$ represents the element-wise multiplication. Next, F is input into a spatial attention module to get a spatial attention map,

$$M_s(F) = \sigma(g(f^{AM}, [F_{savg} \oplus F_{smax}])) \quad (4)$$

where $g(\cdot)$ is the filter function, f^{AM} represents the convolution filter we used with size 7×1, F_{savg} and F_{smax} denote the average-pooled feature and max-pooled feature across the channels respectively, and $\oplus$ represents the concatenation operation. Finally,

$$\overline{E} = F \otimes M_s(F) \quad (5)$$

Dilated Convolutional Layers. This part is shown in the top half of the convolutional layer of G in Fig. 1. Due to its advantages of expanding the receptive

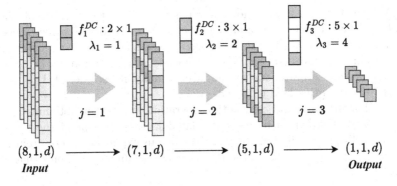

Fig. 3. An example of dilated convolution with three layers

field but without losing the accuracy and adding the number of layers, dilated convolution plays an important role in computer vision and natural language processing. Motivated by Wavenet [14], we utilize dilated convolution to construct our generator model. The basic idea of dilation is to extend the filter by using 0, so as to apply the convolution filter to the field with larger length than the original one.

For convenience, we use a standard CNN with filter $f^{s \times 1}$ to illustrate the dilated convolution process. The input of dilated convolution is the feature map after the attention module, i.e., $\overline{E}$. Suppose the number of the dilated convolution layer is M. We set the dilated factor of the j-th convolution layer to be $\lambda_j = s^{j-1}$. The height h_j of the dilated convolution filter $f_j^{DC} : h_j \times 1$ for the j-th dilated convolution layer can then be derived by $h_j = (s-1)\lambda_j + 1$. With stride of 1, the height of the feature map of the j-th dilated convolution layer is $l_j = l_{j-1} - h_j + 1$, and $l_0 = L$. It is not difficult to see that the receptive filed of this M-layer dilated convolution is s^M. The j-th dilated convolution operation $Dic_j(\overline{E}_{j-1})$, i.e., $g(f_j^{DC}, \overline{E}_{j-1})$ can be formally defined as,

$$\overline{E}_j = Dic_j(\overline{E}_{j-1}) = \begin{bmatrix} g_1(f_j^{DC}, \overline{E}_{j-1}^{[1:h_j]}) \\ g_2(f_j^{DC}, \overline{E}_{j-1}^{[2:h_j+1]}) \\ \vdots \\ g_{l_j}(f_j^{DC}, \overline{E}_{j-1}^{[l_j:l_{j-1}]}) \end{bmatrix} \tag{6}$$

where $g_{l_j}(\cdot)$ denotes one step operation of the filter function $g(\cdot)$, and $\overline{E}_{j-1}^{[x:y]}$ denotes the image from the x-th row of $\overline{E}_{j-1}$ to the y-th row of $\overline{E}_{j-1}$. Finally, through M-layer dilated convolution operation, we can get,

$$\mathcal{Z}_{dc} = Dic_M(Dic_{M-1}(\ldots(Dic_1(\overline{E}_0)))) \tag{7}$$

where $\overline{E}_0 = \overline{E}$. Since $\mathcal{Z}_{dc} \in \mathbb{R}^{1 \times 1 \times d}$, we reshape it to derive $\mathbf{z}_{dc} \in \mathbb{R}^{1 \times d}$.

Figure 3 gives an example for helping understand the dilated convolution process (assuming that the number of channels does not change during convolution). The standard convolution filter is set as $f^{2 \times 1}$, the number of dilated convolution layers is $M = 3$, and the dilated factor λ_j of the three layers are 1, 2 and 4 respectively. As can be observed, the shape of the input image is $(8, 1, d)$. Through the first dilated convolution layer, it becomes $(7, 1, d)$, and then becomes $(5, 1, d)$ after the second layer. Finally, it becomes $(1, 1, d)$. In this example, the receptive filed is $2^3 = 8$.

Note that in the actual convolution operation process for obtaining $\mathbf{z}_{dc}$, we use a variety of standard filters and dilated factors to expand the receptive filed. By utilizing the dilated convolution, we can explicitly encode item interdependencies to capture the high-order correlations among the items that u has interacted with.

Vertical Convolutional Layer. This part is shown in the bottom half of the convolutional layer of G in Fig. 1. We first perform a simple reshape operation to convert $\overline{E}$ with shape $(L, 1, d)$ to $\widetilde{E}$ with shape $(L, d, 1)$. In this layer, we use a vertical filter $f^{VC} \in \mathbb{R}^{L \times 1}$. Specifically, for $\widetilde{E}$, we slide d times from left to right, interact with each column of $\widetilde{E}$, to get the result of vertical convolution $\mathbf{z}_{vc} = [z_{vc}^1, z_{vc}^2, \ldots, z_{vc}^d] \in \mathbb{R}^{1 \times d}$. Obviously, this result is equal to the weighted sum over the L rows of $\widetilde{E}$ with f^{VC} as the weights. Therefore,

$$\mathbf{z}_{vc} = \sum_{b=1}^{L} f_b^{VC} \cdot \widetilde{E}^{[b]} \tag{8}$$

where f_b^{VC} is the b-th element of f^{VC}, and $\widetilde{E}^{[b]}$ denotes the b-th row of $\widetilde{E}$. By using vertical convolution, we aggregate the potential representations of u's interacted items, and capture the feature of each dimension of u's interacted items.

Fully Connected Layer. In order to capture u's general preference, we first concatenate u's latent feature $\mathbf{p}_u$, the feature $\mathbf{z}_{dc}$ from the dilated convolution layers, and the feature $\mathbf{z}_{vc}$ from the vertical convolution layer together. Then, we input them into a fully connected neural network, so as to project them to the output layer with n dimensions, and predict u's preference distribution $\hat{\mathbf{y}}^u \in \mathbb{R}^{n \times 1}$ by,

$$\hat{\mathbf{y}}^u = \mathbf{W}_{out}[\mathbf{p}_u \oplus \mathbf{z}_{dc} \oplus \mathbf{z}_{vc}]^T + \mathbf{b}_{out} \tag{9}$$

where $\mathbf{W}_{out} \in \mathbb{R}^{n \times 3d}$ and $\mathbf{b}_{out} \in \mathbb{R}^n$ are the weight matrix and bias vector of the output layer, respectively.

4.3   Discriminator

The discriminator part is shown in the bottom half of Fig. 1, which is a fully connected network with X layers. To alleviate the influence of sparsity on training and prevent G from finding an easy but useless solution to approximate the ground truth (e.g., generating the interaction vector having all of its elements as 1 but without considering u's relative preference at all), we employ a mask operation. Let $\mathbf{m}^u \in \mathbb{R}^{n \times 1}$ denote the selected negative items vector for u with a given negative sampling ratio γ. The mask operation is achieved by,

$$\overline{\mathbf{y}}^u = \hat{\mathbf{y}}^u \otimes (\mathbf{y}^u + \mathbf{m}^u) \tag{10}$$

Taking the real user preference distribution (interaction vector) $\mathbf{y}^u$ and the generated distribution $\overline{\mathbf{y}}^u$ as input $\mathbf{y}$, the discriminator outputs a single scalar value $S \in \mathbb{R}^1$, indicating the probability that its input comes from the real value rather than the generator. Formally,

$$\mathbf{h}_0 = \mathbf{W}_0^D \mathbf{y}$$
$$\mathbf{h}_1 = tanh(\mathbf{W}_1^D \mathbf{h}_0 + \mathbf{b}_1^D)$$
$$\cdots \cdots \tag{11}$$
$$\mathbf{h}_{X-1} = tanh(\mathbf{W}_{X-1}^D \mathbf{h}_{X-2} + \mathbf{b}_{X-1}^D)$$
$$S = \sigma(\mathbf{W}_X^D \mathbf{h}_{X-1} + \mathbf{b}_X^D),$$

where $tanh(\cdot)$ and $\sigma(\cdot)$ represent the Tanh and Sigmoid functions respectively, while $\mathbf{W}_x^D$ and $\mathbf{b}_x^D$ denote the weight matrix and bias of the x-th layer, respectively.

4.4   Optimization

We train a *minimax* game to unify G and D. Specifically, G attempts to generate u's preference distribution vector $(\hat{\mathbf{y}}_u)$ to cheat D, while D attempts to distinguish u's ground-truth preference distribution $(\mathbf{y}_u)$ from the one generated by G. The objective functions, denoted by J^G and J^D, are designed as follows,

---

Algorithm 1: Learning DiCGAN

Input: Interaction matrix Y, learning rate for G and D: μ_G and μ_D, minibatch
size of G and D: B_G and B_D.

Output: θ^G.

1 Initialize G's and D's parameters: θ^G and θ^D;
2 **while** *not converged* **do**
3   **for** *G-epoch* **do**
4   Sample minibatch of B_G users;
5   Generate fake preference distribution vectors $\{\hat{\mathbf{y}}^1, \hat{\mathbf{y}}^2, \ldots, \hat{\mathbf{y}}^{B_G}\}$;
6   Update G by $\theta^G \leftarrow \theta^G - \frac{\mu_G}{B_G} \cdot \nabla_{\theta^G} J^G$;
7   **end**
8   **for** *D-epoch* **do**
9   Sample minibatch of B_D users;
10   Get true preference distribution vectors $\{\mathbf{y}^1, \mathbf{y}^2, \ldots, \mathbf{y}^{B_D}\}$;
11   Get fake preference distribution vectors $\{\hat{\mathbf{y}}^1, \hat{\mathbf{y}}^2, \ldots, \hat{\mathbf{y}}^{B_D}\}$;
12   Update D by $\theta^D \leftarrow \theta^D - \frac{\mu_D}{B_D} \cdot \nabla_{\theta^G} J^D$;
13   **end**
14 **end**
15 **return** θ^G

Table 1. Statistics of the three datasets

Dataset	#User	#Item	#Interaction	Mean-Interaction	Sparsity
Ciao	1,081	8,444	24,755	22.90	0.9973
ABaby	6474	20358	100104	15.46	0.9992
AVideo	7,992	25,476	144,387	18.07	0.9993

$$J^G = \sum_u \log(1 - D(\hat{\mathbf{y}}^u | \mathbf{u}, \{\mathbf{i}_1^u, \mathbf{i}_2^u, \dots, \mathbf{i}_t^u\})) \tag{12}$$

$$J^D = -\sum_u (\log(D(\mathbf{y}^u)) + \log(1 - D(\hat{\mathbf{y}}^u | \mathbf{u}, \{\mathbf{i}_1^u, \mathbf{i}_2^u, \dots, \mathbf{i}_t^u\}))) \tag{13}$$

We use random gradient descent with minibatch and back propagation to train G and D, and alternately update the parameters θ^G and θ^D, i.e., fixing one when updating the other one. Algorithm 1 gives the detailed description of learning DiCGAN.

5   Experiments

To validate the effectiveness our proposed models, we conduct experiments to answer the following research questions:

- **RQ1.** Do our proposed DiCGAN models outperform the state-of-the-art top-N recommendation methods?
- **RQ2.** Does the convolutional neural network help improve the performance by learning from interactive data?
- **RQ3.** How do the key hyper-parameters affect the performance of DiCGAN?

5.1   Experimental Settings

Datasets. We conduct experiments on three publicly available datasets: Ciao DVD[1] [7] (**Ciao** for short), Amazon Baby (**ABaby** for short) and Amazon VideoGame (**AVideo** for short)[2] [13]. Note that all the three datasets provide users' explicit ratings on items, we convert them to 1 (indicating an interaction) to get implicit feedback data. Moreover, to ensure the quality of the datasets, we filter them to retain the users with at least 10 interactions and at most 100 interactions. The statistics of the datasets after preprocessing are summarized in Table 1.

---

[1] https://www.librec.net/datasets.html.
[2] http://jmcauley.ucsd.edu/data/amazon/.

Compared Methods. We compare our proposed method DiCGAN[3] with the following methods[4]:

- **BPR-MF** [15] optimizes the standard matrix factorization model with a pair-wise bayesian personalized ranking loss.
- **MLP** [10] takes advantage of multi-layer perceptron to replace inner product to learn the non-linear relationship between users and items.
- **IRGAN** [22] is the first GAN-based CF method, in which G tries to generate the indices of items relevant to a given user, while D tries to discriminates the user's ground truth items from those generated by G.
- **CFGAN** [1] is a GAN-based CF model in which G tries to generate a plausible interaction vector of a user composed of real-valued elements, rather than to sample a single item index that the user may be interested in.
- **Caser** [17] is the first method that utilizes convolutional filters to learn users' sequential patterns for sequential recommendation[5]. We have modified Caser to make it adaptable to our experiment settings.
- **DiC-Solo** is a degraded version of DiCGAN, which denotes the method that only contains the G component of DiCGAN using point-wise objective functions. In this way, DiC-Solo is a non-GAN-based but CNN-based method.

Evaluation Metrics and Methodology. To evaluate the performance and cover as many aspects of recommendation as possible, we employ three popular metrics for top-N recommendation: Precission@N, Recall@N and Normalized Discounted Cumulative Gain (NDCG@N). The first two metrics focus on how many correct items are recommended, while the third one accounts for the ranked quality of correct items in the recommendation set. We set N as 5 and 10. For each dataset, we hold the first 80% items in each user's interaction as the training set, and the remaining 20% items are used as the test set. Similar to [19], we use the test set as a positive sample and randomly select 9 times negative samples. Based on such a strategy, the recommendation methods can generate a ranked top-N list to evaluate the metrics as mentioned above.

Implementation Details. We first determine some hyper-parameters that empirically perform well, regardless of the combination of the other hyper-parameters or datasets used: height of the input *image* $L = 80$, learning rate $\mu^G = \mu^D = 0.001$, the number of dilated convolution layers $M = 7$, the standard convolution filter height $s = (3, 3, 5, 6, 5, 4, 3)$, dilated factor

---

[3] https://github.com/georgeguo-cn/dicgan.
[4] Since the source code of PD-GAN is not available and hard to reproduce, we do not include it as a competitor.
[5] Note that NextItNet [29] is also designed for sequential recommendation, but in NextItNet, each item corresponds to the probability of predicting the next item. Therefore, it cannot be modified to fit our experiment requirements, especially when $t < L$.

Table 2. Performance comparison in terms of Precision@N, Recall@N, NDCG@N. Best performance in existing methods is underlined and best performance in our models is in boldface. The percent of improvement is about the best results between ours and existing methods.

Datasets	Metric	Compare methods					Ours		Improv.
		BPR-MF	MLP	IRGAN	CFGAN	Caser	DiC-Solo	DiCGAN	
Ciao	Precision@5	0.2120	0.2118	0.2111	0.2120	0.2189	0.2209	**0.2229**	1.86%
	Recall@5	0.2515	0.2498	0.2375	0.2478	0.2581	0.2624	**0.2638**	2.19%
	NDCG@5	0.4048	0.4140	0.4157	0.4186	0.4132	0.4158	**0.4371**	4.41%
	Precision@10	0.1633	0.1647	0.1625	0.1626	0.1669	0.1680	**0.1685**	0.94%
	Recall@10	0.3732	0.3767	0.3654	0.3679	0.3728	0.3811	**0.3818**	1.34%
	NDCG@10	0.4387	0.4462	0.4470	0.4561	0.4471	0.4497	**0.4687**	2.76%
ABaby	Precision@5	0.1924	0.1928	0.1900	0.1930	0.1938	0.1974	**0.2014**	3.94%
	Recall@5	0.2849	0.2839	0.2808	0.2855	0.2861	0.2908	**0.2955**	3.27%
	NDCG@5	0.4601	0.4636	0.4586	0.4628	0.4645	0.4717	**0.4757**	2.40%
	Precision@10	0.1450	0.1469	0.1439	0.1447	0.1458	0.1488	**0.1509**	2.69%
	Recall@10	0.4235	0.4283	0.4207	0.4233	0.4257	0.4339	**0.4385**	2.38%
	NDCG@10	0.4957	0.5024	0.4961	0.5018	0.5025	0.5091	**0.5117**	1.81%
AVideo	Precision@5	0.2333	0.2359	0.2022	0.2032	0.2468	0.2719	**0.2721**	10.25%
	Recall@5	0.3127	0.3146	0.2701	0.2705	0.3312	0.3675	**0.3677**	11.01%
	NDCG@5	0.4908	0.5063	0.4645	0.4645	0.5236	0.5632	**0.5635**	7.61%
	Precision@10	0.1744	0.1792	0.1466	0.1462	0.1809	0.1953	**0.1953**	7.97%
	Recall@10	0.4552	0.4655	0.3794	0.3802	0.4727	0.5111	**0.5111**	8.13%
	NDCG@10	0.5230	0.5442	0.4942	0.4950	0.5517	0.5850	**0.5859**	6.21%

$\lambda = (1, 2, 4, 6, 4, 3, 1)$. For the key hyper-parameters, the negative sampling ratio $\gamma = 0.01$, the embedding size $d = 64$ and the number of hidden layers $X = 1$. In addition, we employ an early-stop strategy, in which the training process will stop if the precision do not increase after 50 steps. Note that different L corresponds to different data scales and different selection of the dilated factors. To see how the performance fare with different L, we have also conduct experiments with $L = 8, s = (2, 3, 3), \lambda = (1, 2, 1)$ and $L = 40, s = (3, 3, 4, 4, 3), \lambda = (1, 2, 4, 5, 3)$. The experimental results exhibit similar trends as that of $L = 80$, and thus are omitted due to space concern.

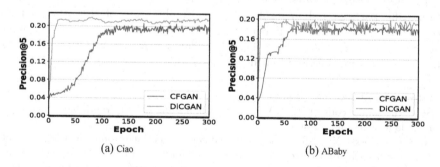

(a) Ciao   (b) ABaby

Fig. 4. Precision@5 during training on Ciao and ABaby.

5.2 Performance Comparison (RQ1)

Table 2 reports the summarized results for all the compared methods. We can see that DiCGAN in general achieves the best results among all the evaluated methods on the three datasets, which validates the effectiveness of our philosophy on combining CNN with GAN for better performance. Moreover, with the increase of the sparsity of the dataset, the performance improvement of DiCGAN is also more significant. Besides, we have the following key observations:

- Caser in general performs better than BPR-MF and MLP, while DiCGAN consistently outperforms CFGAN. This demonstrates the effectiveness of the convolutional structures on achieving better recommendation performance, due to their powerful ability in capturing high-order correlation features among user's interacted items. More concretely, the larger the item space is and the more sparse the dataset is, the smaller the proportion of 1 in the user's interaction vector, but the number of 1 does not vary much. For traditional neural network models that take the user's interaction item vectors as input, when the sparsity of the dataset increases, the effective data (with value 1) proportion will decrease, which may affect their ability to extract features. Different from them, DiCGAN depends on the stack of the interaction item vectors, which are only related to the number of interacted items but not its proportion. Moreover, as shown in Table 1, the mean number of interactions per user is independent with the scale and sparsity of the dataset. Therefore, our model can still extract abundant information, even for sparse datasets.
- DiCGAN consistently outperforms Caser and DiC-Solo, which are non-GAN-based but CNN-based methods. This demonstrates GAN's unique advantages in fitting approximate preference distribution, which is better than traditional CF methods that try to optimize point-wise objective functions. It is worth noting that the performance gap between DiCGAN and DiC-Solo is not significant, this is mainly because the structure of the discriminator in DiCGAN is relatively simple.
- DiCGAN and CFGAN achieve better results than IRGAN on the three datasets, mainly due to that they both use vector-wise adversarial training. Moreover, with the increase of item space, the improvement of DiCGAN over CFGAN is more significant. This result shows that CFGAN, which exploits user-item interaction vectors, is more susceptible to data sparsity, whereas DiCGAN, which only relies on user's interacted items, is less sensitive to data sparsity.

Figure 4 shows the Precision@5 trend of CFGAN and DiCGAN on Ciao and ABaby during training. As can be observed, in both datasets, DiCGAN exhibits faster learning speed than CFGAN. This is mainly due to that the generator of CFGAN is more sensitive to data scale. When the item space is large, CFGAN is more difficult to train and sometimes may even fail to achieve good performance. In contrast, DiCGAN utilizes the vectors of interacted items to perform convolution skillfully, which is less sensitive to the size of the datasets, and thus can achieve faster and better learning.

Table 3. Performance of CNN Components. Best performance is in boldface.

Datasets	Metric	DiG-Vc	DiG-Dic	DiG-WoA	DiCGAN
Ciao	Precision@5	0.2187	0.2178	0.2213	**0.2229**
	Recall@5	0.2555	0.2556	0.2599	**0.2638**
	NDCG@5	0.4234	0.4370	0.4280	**0.4371**
	Precision@10	0.1678	0.1651	**0.1686**	0.1685
	Recall@10	0.3826	0.3729	**0.3840**	0.3818
	NDCG@10	0.4618	0.4620	0.4594	**0.4687**
ABaby	Precision@5	0.1943	0.1928	0.1946	**0.2014**
	Recall@5	0.2859	0.2848	0.2863	**0.2955**
	NDCG@5	0.4677	0.4640	0.4678	**0.4757**
	Precision@10	0.1427	0.1414	0.1451	**0.1509**
	Recall@10	0.4158	0.4121	0.4241	**0.4385**
	NDCG@10	0.4995	0.4979	0.5029	**0.5117**
AVideo	Precision@5	0.2259	0.2243	0.2262	**0.2721**
	Recall@5	0.3018	0.2997	0.3021	**0.3677**
	NDCG@5	0.4938	0.4938	0.4944	**0.5635**
	Precision@10	0.1701	0.1693	0.1701	**0.1953**
	Recall@10	0.4435	0.4416	0.4442	**0.5111**
	NDCG@10	0.5248	0.5250	0.5275	**0.5859**

5.3   Effectiveness of CNN Layers (RQ2)

Performance of CNN Components. We use DiG-Vc (DiG-Vc is short for DiCGAN-Vc) and DiG-Dic to represent respectively the degraded versions of DiCGAN that only uses the vertical convolution in G (without attention), and that only uses the dilated convolution in G (without attention). Further, we use DiG-WoA to denote the variant of DiCGAN without the attention module in G.

As shown in Table 3, the performance of DiG-Dic is generally the same as that of DiG-Vc. But the model using both dilated convolution and vertical convolution (DiG-WoA) performs much better than the former two variants, demonstrating that a user's preference can be better captured by learning high-order correlation features from the perspective of user's interacted items and the latent dimensions of user's interacted items simultaneously. Moreover, the comparison between DiCGAN and DiG-WoA reveals that applying the attention model in general can achieve better results.

Impact of Pre-training. To show the effect of pre-training on DiCGAN, we conducted experiments to compare its performance with and without pre-training for generator. As shown in Table 4, the results of our model with pre-training are consistently better than those without pre-training. This validates the effectiveness and necessity of pre-training.

Table 4. Performance with/without pre-training. Best performance is in boldface.

Datasets	Metric	Without pre-training	With pre-training
Ciao	Precision@10	0.1681	**0.1685**
	Recall@10	0.3813	**0.3818**
	NDCG@10	0.4619	**0.4687**
ABaby	Precision@10	0.1453	**0.1509**
	Recall@10	0.4231	**0.4385**
	NDCG@10	0.5050	**0.5117**
AVideo	Precision@10	0.1802	**0.1953**
	Recall@10	0.4641	**0.5111**
	NDCG@10	0.5425	**0.5859**

5.4 Hyper-parameter Sensitivity (RQ3)

To evaluate the effect of the negative sampling ratio γ, embedding size d and hidden layer number X, we conduct experiments on Ciao by varying the value of them. Note that we have also conducted the same experiments on ABaby and AVedio, the results exhibit the similar trend and hence are omitted here due to space concern.

- **Negative Sampling Ratio.** As shown in Fig. 5 (a), DiCGAN achieves the best performance when γ is 0.01. With larger γ, G would pay too much attention to making outputs close to zero, rather than achieving its original goal of generating real preference distribution, and this may deteriorate the model performance.
- **Embedding Size.** Different embedding sizes carry different feature information. As shown in Fig. 5 (b), the performance of DiCGAN increases with the growth of d, and reaches the best when $d = 64$. This shows that the model can get better results by introducing more prediction factors to obtain a stronger representation ability. However, when embedding size is too large, it will affect the training of the model, resulting in the decline of the accuracy.

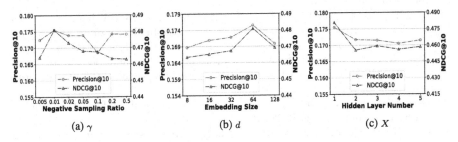

(a) γ (b) d (c) X

Fig. 5. Effect of Hyper-Parameter, i.e. γ, d, and X on Ciao

– **Hidden Layer Number.** As shown in Fig. 5 (c), the best performance in general is achieved when $X = 1$, and stacking more layers does not necessary improve the performance. The reason may be related to the training of the discriminator. The more hidden layers, the more parameters need to be trained, and the more difficult it is for the discriminator to reach a stable state.

6   Conclusion

In this paper, we proposed DiCGAN, a Dilated Convolutional Generative Adversarial Network for top-N recommendation. By expressing the interactions of per user into an *image*, we propose to exploit CNN's powerful high-order correlation learning ability in a GAN-based recommendation model. With the help of several dilated convolutional filters and a vertical convolutional filter, the high-order correlation features among user's interacted items are captured. Moreover, an attention module was employed to generate attention maps that can be multiplied to the feature map for adaptive refinement of features. Experimental results on three public datasets demonstrate that, compared with state-of-the-art GAN-based and CNN-based recommendation methods, DiCGAN can achieve remarkable performance improvement, and is less sensitive to data sparsity.

References

1. Chae, D.K., Kang, J.S., Kim, S.W., Lee, J.T.: CFGAN: a generic collaborative filtering framework based on generative adversarial networks. In: Proceedings of the 27th ACM International Conference on Information and Knowledge Management, pp. 137–146. ACM (2018)
2. Choi, Y., Choi, M., Kim, M., Ha, J.W., Kim, S., Choo, J.: Stargan: unified generative adversarial networks for multi-domain image-to-image translation. In: Proceedings of the IEEE Conference on Computer Vision and Pattern Recognition, pp. 8789–8797 (2018)
3. Deng, Z.H., Huang, L., Wang, C.D., Lai, J.H., Philip, S.Y.: DeepCF: a unified framework of representation learning and matching function learning in recommender system. In: Proceedings of the AAAI Conference on Artificial Intelligence, vol. 33, pp. 61–68 (2019)
4. Ding, R., Guo, G., Yan, X., Chen, B., Liu, Z., He, X.: BiGAN: collaborative filtering with bidirectional generative adversarial networks. In: Proceedings of the 2020 SIAM International Conference on Data Mining, pp. 82–90. SIAM (2020)
5. Gao, M., Zhang, J., Yu, J., Li, J., Wen, J., Xiong, Q.: Recommender systems based on generative adversarial networks: a problem-driven perspective. arXiv preprint arXiv:2003.02474 (2020)
6. Goodfellow, I., et al.: Generative adversarial nets. In: Neural Information Processing Systems, pp. 2672–2680 (2014)
7. Guo, G., Zhang, J., Yorke-Smith, N.: A novel Bayesian similarity measure for recommender systems. In: Proceedings of International Joint Conference on Artificial Intelligence (IJCAI), pp. 2619–2625 (2013)

8. Guo, H., Tang, R., Ye, Y., Li, Z., He, X.: DeepFM: a factorization-machine based neural network for CTR prediction. arXiv preprint arXiv:1703.04247 (2017)
9. He, X., Du, X., Wang, X., Tian, F., Tang, J., Chua, T.S.: Outer product-based neural collaborative filtering. arXiv preprint arXiv:1808.03912 (2018)
10. He, X., Liao, L., Zhang, H., Nie, L., Hu, X., Chua, T.S.: Neural collaborative filtering. In: Proceedings of the 26th International Conference on World Wide Web, pp. 173–182 (2017)
11. Hu, Y., Koren, Y., Volinsky, C.: Collaborative filtering for implicit feedback datasets. In: 2008 Eighth IEEE International Conference on Data Mining, pp. 263–272. IEEE (2008)
12. Kim, D., Park, C., Oh, J., Lee, S., Yu, H.: Convolutional matrix factorization for document context-aware recommendation. In: Proceedings of the 10th ACM Conference on Recommender Systems, pp. 233–240 (2016)
13. McAuley, J., Targett, C., Shi, Q., Van Den Hengel, A.: Image-based recommendations on styles and substitutes. In: Proceedings of the 38th International ACM SIGIR Conference on Research and Development in Information Retrieval, pp. 43–52. ACM (2015)
14. van den Oord, A., et al.: WaveNet: a generative model for raw audio. arXiv preprint arXiv:1609.03499 (2016)
15. Rendle, S., Freudenthaler, C., Gantner, Z., Schmidt-Thieme, L.: BPR: Bayesian personalized ranking from implicit feedback. In: Proceedings of the Twenty-Fifth Conference on Uncertainty in Artificial Intelligence, pp. 452–461 (2009)
16. Sun, C., Liu, H., Liu, M., Ren, Z., Gan, T., Nie, L.: LARA: attribute-to-feature adversarial learning for new-item recommendation. In: Proceedings of WSDM, pp. 582–590 (2020)
17. Tang, J., Wang, K.: Personalized top-n sequential recommendation via convolutional sequence embedding. In: Proceedings of the Eleventh ACM International Conference on Web Search and Data Mining, pp. 565–573. ACM (2018)
18. Tong, Y., Luo, Y., Zhang, Z., Sadiq, S., Cui, P.: Collaborative generative adversarial network for recommendation systems. In: 2019 IEEE 35th International Conference on Data Engineering Workshops (ICDEW), pp. 161–168. IEEE (2019)
19. Wang, C., Guo, Z., Li, J., Pan, P., Li, G.: A text-based deep reinforcement learning framework for interactive recommendation. In: Proceedings of the 24th European Conference on Artificial Intelligence, pp. 537–544 (2020)
20. Wang, H., Shao, N., Lian, D.: Adversarial binary collaborative filtering for implicit feedback. In: Proceedings of the AAAI Conference on Artificial Intelligence, vol. 33, pp. 5248–5255 (2019)
21. Wang, H., et al.: GraphGAN: graph representation learning with generative adversarial nets. arXiv preprint arXiv:1711.08267 (2017)
22. Wang, J., et al.: IRGAN: a minimax game for unifying generative and discriminative information retrieval models. In: Proceedings of the 40th International ACM SIGIR Conference on Research and Development in Information Retrieval, pp. 515–524. ACM (2017)
23. Woo, S., Park, J., Lee, J.-Y., Kweon, I.S.: CBAM: convolutional block attention module. In: Ferrari, V., Hebert, M., Sminchisescu, C., Weiss, Y. (eds.) ECCV 2018. LNCS, vol. 11211, pp. 3–19. Springer, Cham (2018). https://doi.org/10.1007/978-3-030-01234-2_1
24. Wu, Q., Liu, Y., Miao, C., Zhao, B., Zhao, Y., Guan, L.: PD-GAN: adversarial learning for personalized diversity-promoting recommendation. In: Proceedings of the 28th International Joint Conference on Artificial Intelligence, pp. 3870–3876 (2019)

25. Xue, H.J., Dai, X., Zhang, J., Huang, S., Chen, J.: Deep matrix factorization models for recommender systems. In: IJCAI, pp. 3203–3209 (2017)
26. Yu, F., Koltun, V.: Multi-scale context aggregation by dilated convolutions. In: International Conference on Learning Representations (ICLR) (2016)
27. Yu, H., Qian, T., Liang, Y., Liu, B.: AGTR: adversarial generation of target review for rating prediction. Data Sci. Eng. 5(4), 346–359 (2020). https://doi.org/10.1007/s41019-020-00141-1
28. Yu, L., Zhang, W., Wang, J., Yu, Y.: SeqGAN: sequence generative adversarial nets with policy gradient. In: Thirty-First AAAI Conference on Artificial Intelligence (2017)
29. Yuan, F., Karatzoglou, A., Arapakis, I., Jose, J.M., He, X.: A simple convolutional generative network for next item recommendation. In: Proceedings of the Twelfth ACM International Conference on Web Search and Data Mining, pp. 582–590. ACM (2019)

RE-KGR: Relation-Enhanced Knowledge Graph Reasoning for Recommendation

Ming He$^{(\boxtimes)}$, Hanyu Zhang, and Han Wen

Faculty of Information Technology, Beijing University of Technology, Beijing, China
heming@bjut.edu.cn, {zhanghanyu,wenhan}@emails.bjut.edu.cn

Abstract. A knowledge graph (KG) has been widely adopted to improve recommendation performance. The multi-hop user-item connections in a KG can provide reasons for recommending an item to a user. However, existing methods do not effectively leverage the relations of entities and interpretable paths in a KG. To address this limitation, in this paper, we propose a novel recommendation framework called relation-enhanced knowledge graph reasoning for recommendation (RE-KGR) that combines recommendation and explainability by reasoning user-item interaction paths (UIIPs). First, instead of applying an alignment algorithm for preprocessing, RE-KGR directly learns the semantic representation of entities from structured knowledge by stacking relation-based convolutional layers to take full advantage of the KG. Moreover, RE-KGR infers user preferences by calculating the sum of all UIIPs between users and items. Finally, RE-KGR selects several UIIPs with the highest probabilities as possible reasons for the recommendations. Extensive experiments on three real-world datasets demonstrate that our proposed method significantly outperforms several state-of-the-art baselines and achieves superior performance and explainability.

Keywords: Recommender systems · Graph neural networks · Knowledge graphs

1 Introduction

A knowledge graph (KG), which is a heterogeneous network composed of structured knowledge, has been widely applied in various fields, such as information retrieval and question answering. Inspired by the successful use of KG, researchers have attempted to link user-item interaction data to a KG, and constructed a collaborative knowledge graph (CKG) for recommendation tasks [12].

Existing KG-based methods can be classified into two categories. The first category is path-based (PB) methods [4], which extract various patterns of meta-paths carrying high-order information between entities and feed them into a prediction model to obtain recommendations. However, PB methods rely heavily on a manually designed meta-path extraction algorithm, which requires domain knowledge and is difficult to optimize in practice [11]. In addition, PB methods

© Springer Nature Switzerland AG 2021
C. S. Jensen et al. (Eds.): DASFAA 2021, LNCS 12683, pp. 297–305, 2021.
https://doi.org/10.1007/978-3-030-73200-4_19

also suffer from the path explosion phenomenon. The CKG has an extremely large path space as the number of path hops grows, which makes training more difficult. The second category is embedding-based (EB) methods [12,13], which align entity embeddings by KG embedding algorithms [7] and aggregate the entity embeddings from their neighbors to compute similarity. EB methods exhibit greater flexibility than PB methods in capturing high-order connections; however, existing EB methods do not fully use the relations between entities during high-order aggregation, which leads the model to inadequately learn the semantic representation of entities from structured knowledge.

To address these limitations, we propose relation-enhanced knowledge graph reasoning (RE-KGR), an end-to-end recommendation framework. Specifically, we employ a simple embedding method that is widely used in collaborative filtering (CF) based models [3,8] to encode entities and relations. Then, we project entities from the entity space to the corresponding relation space, and aggregate the entities and their neighbors. Moreover, we apply the attention mechanism [10,12] to help distinguish the importance of local neighbors. To alleviate the over-smoothing problem [6] and improve the information flow between convolutional layers, we adopt a dense connectivity pattern [5] that is widely used in convolutional neural networks. We then compute the similarity of adjacent entities for user-item interaction path (UIIP) reasoning and generate the recommendations and explanations. Experimental results on three real-world datasets show that our proposed approach significantly outperforms baseline methods.

2   Preliminaries

In a recommendation scenario, we typically have a set of users $\mathcal{U} = \{u_1, u_2, ..., u_M\}$ and set of items $\mathcal{I} = \{i_1, i_2, ..., i_N\}$, where M and N denote the number of users and items, respectively. We also have a KG $\mathcal{G}_{kg} = \{(h, r, t)|h, t \in \mathcal{E}, r \in \mathcal{R}\}$. Here h, r, and t denote the head, relation, and tail of a knowledge triplet, respectively, $\mathcal{E}$ and $\mathcal{R}$ denote the set of entities and relations in the KG. We map items into the KG via title matching to build the CKG $\mathcal{G} = \{(h, r, t)|h, t \in \mathcal{E} \cup \mathcal{U} \cup \mathcal{I}, r \in \mathcal{R} \cup \{Interact\}\}$. We formulate the recommendation task to be addressed in this paper as follows: Given a CKG $\mathcal{G}$, we aim to infer the probability $\hat{y}_{ui}$ that user u is interested in item i, and provide potential reasons $\mathcal{P}_{ui}^q$.

3   Methodology

The framework of RE-KGR is illustrated in Fig. 1, which consists of four main components: 1) an embedding layer, which parameterizes each entity and relation as a dense vector; 2) relation-based graph convolution (RGC) layers, which recursively aggregate entities by their neighbors' representations according to the relations between them; 3) a local similarity layer, which is applied to learn the similarity between connected entities; and 4) a prediction layer, which reasons UIIPs and infers possible interactions according to the local similarity scores.

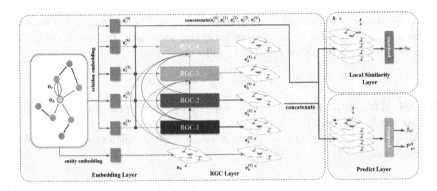

Fig. 1. The framework of relation-enhanced knowledge graph reasoning.

Furthermore, for RGC layers, we also adopt an attention mechanism [10, 12] to distinguish the importance of neighbors and dense connections [5] to reduce over-smoothing [6].

3.1 Embedding Layer

For a massive-scale CKG, the entities and relations are often trained on sparse binary features with one-hot encoding. We use the embedding table lookup operation to obtain the embeddings of entities and relations:

$$\mathbf{e}_i = \mathbf{E}_{entity}\mathbf{o}_i, \; \mathbf{e}_r^{(l)} = \mathbf{E}_{relation}^{(l)}\mathbf{o}_r, \qquad (1)$$

where $\mathbf{o}$ denotes one-hot vectors, $\mathbf{e}_i$ denotes the embedding of entity i, and $\mathbf{e}_r^{(l)}$ denotes the embedding of relation r used in the l-th RGC-layer.

3.2 RGC Layer

First-Order Aggregation. A CKG is a semantic network; thus, the entity h and t in a triplet $(h, r, t) \in \mathcal{G}$ usually contains various types and attributes. Therefore, we first project each entity t to a different semantic space conditioned to the relation r:

$$\mathbf{h}_t^r = \mathbf{W}_r \mathbf{e}_t. \qquad (2)$$

Here, $\mathbf{e}_t$ is the embedding of entity t, $\mathbf{W}_r \in \mathbb{R}^{k \times d}$ is the projection matrix.

To characterize the first-order proximity structure of entity h and aggregate its neighbors' representations, we compute the combination of h's closed neighborhood:

$$\mathbf{e}_h^{(1)} = \sigma \left(\sum_{(h,r,t) \in \mathcal{N}_h} \pi_{(h,r,t)} \mathbf{h}_t^r \right), \qquad (3)$$

where N_h denotes the closed neighborhood of h, $e_h^{(1)}$ is the first-order representation of h, $\sigma(\cdot)$ denotes an activation function, such as $ReLU(\cdot) = max(0, \cdot)$, and $\pi_{(h,r,t)}$ is the attention score defined as follows:

$$\pi_{(h,r,t)} = (M_{r^{-1}}\mathbf{h}_h^r + \mathbf{e}_r)^\top M_r \mathbf{h}_t^r. \tag{4}$$

Here, $M_{r^{-1}}, M_r$ are mapping matrices, and r and r^{-1} are a pair of inverse relations, such as *AuthorOf* and *WrittenBy*.

We employ the softmax function to normalize the coefficients across all closed neighborhoods of h:

$$\pi_{(h,r,t)} = \frac{\exp(\pi_{(h,r,t)})}{\sum_{(h,r',t') \in N_h} \exp(\pi_{(h,r',t')})}. \tag{5}$$

High-Order Aggregation. We further stack more aggregation layers to explore high-order connectivity. In addition, to reduce the over-smoothing problem and improve the information flow between RGC layers, we adopt dense connectivity, which is proposed in DenseNet [5]. More formally, in the l-th steps, we formulate the representation of an entity h as follows:

$$\mathbf{e}_h^{(l)} = \sigma\left(\sum_{(h,r,t) \in N_h} \pi_{(h,r,t)} \mathbf{h}_t^{r\,(l)}\right), \tag{6}$$

wherein the attention scores and projection of entities are formulated as follows:

$$\pi_{(h,r,t)} = (M_{r^{-1}}^{(l)} \mathbf{h}_h^{r\,(l)} + \mathbf{e}_r^{(l)})^\top M_r^{(l)} \mathbf{h}_t^{r\,(l)}, \tag{7}$$

$$\mathbf{h}_t^{r\,(l)} = \mathbf{W}^{(l)} \|_{k=0}^{l-1} \mathbf{e}_t^k. \tag{8}$$

Here, $\|$ is the concatenation operator, and $\mathbf{e}^{(0)}$ denotes initial embeddings.

3.3   Local Similarity Layer

We define the local similarity score between two connected entities h and t with relation r as follows:

$$s_{(h,r,t)} = (\mathbf{M}_{r^{-1}} \|_{k=0}^{L} \mathbf{e}_h^{(k)} + \|_{k=0}^{L} \mathbf{e}_r^{(k)})^\top \mathbf{M}_r \|_{k=0}^{L} \mathbf{e}_t^{(k)}, \tag{9}$$

where $\mathbf{M}_r$ and $\mathbf{M}_{r^{-1}}$ are mapping matrices.

3.4   Prediction Layer

RE-KGR uses the local similarity of entities to reason probable UIIPs and further infer user preferences. More formally, we use $\mathbf{P}_{UIIP} = \{(h,r,t)|(h,r,t) \in \mathcal{G}\}$ to describe an acyclic UIIP. The probability of the UIIP is calculated as follows:

$$p(\mathbf{P}_{UIIP}) = \prod_{(h,r,t) \in \mathbf{P}_{UIIP}} s_{(h,r,t)}, \tag{10}$$

wherein the local similarity of an entity h is normalized across its neighbors by adopting the softmax function.

To infer user preferences, we use $\mathcal{P}_{ui}$ to denote all acyclic UIIPs that start and end with user u and item i, respectively. Then, we combine the UIIPs' probabilities to compute the preference score of user u for item i:

$$\hat{y}_{ui} = \sum_{\mathbf{P}_{UIIP} \in \mathcal{P}_{(u,i)}} p(\mathbf{P}_{UIIP}). \tag{11}$$

3.5 Optimization

To increase the efficiency of training the recommendation model, we adopt BPR [9] loss:

$$\mathcal{L}_{CKG} = - \sum_{(h,r,t,t') \in \mathcal{T}} \ln sigmoid \left(s_{(h,r,t)} - s_{(h,r,t')} \right), \tag{12}$$

where $\mathcal{T} = \{(h,r,t,t')|(h,r,t) \in \mathcal{G}, (h,r,t') \notin \mathcal{G}\}$, (h,r,t') is a negative sample constructed by randomly replacing an entity in (h,r,t). In addition, we adopt L_2 normalization to prevent overfitting.

4 Experiments

4.1 Datasets and Evaluation Metrics

Three benchmark datasets were utilized for the evaluation of RE-KGR:

Amazon-Book[1]: Amazon-Book is a subclass of Amazon-Review, which is a widely used dataset for product recommendation.

Last-FM[2]: Last-FM is a music listening dataset collected from the Last.fm online music system, wherein tracks are viewed as items. We selected a subset of the dataset in which the timestamp was from January 2015 to June 2015.

Yelp2018[3]: Yelp2018 is a dataset from the 2018 edition of the Yelp challenge, which consists of user ratings of local businesses such as restaurants and bars.

We cleaned the data by applying the 10-core setting [12]. We randomly split the datasets into training (70%), validation (10%), and test (20%) sets. We used Freebase[4] to construct the KG for each dataset, and mapped items into the KG via title matching to build the CKG. For Yelp2018, we extracted exclusive KG data from the local business information network. To ensure the effectiveness of the extracted entities, we preprocess the three KG parts by removing sparsely connected entities (i.e., lower than 10 in both datasets) and retaining relations appearing in at least 50 triplets. We employed two widely used evaluation protocols to evaluate the performance of our model: recall@K and ndcg@K. We computed both protocols for each test user and reported the average score at K = 20.

---

[1] http://jmcauley.ucsd.edu/data/amazon.
[2] https://grouplens.org/datasets/hetrec-2011/.
[3] https://www.yelp.com/dataset/challenge.
[4] https://developers.google.com/freebase.

4.2   Baselines and Experimental Setup

To evaluate the effectiveness of RE-KGR, we compared our proposed model with CF methods (FM [8] and NFM [3]), EB methods (CKE [13], CFKG [1], and KGAT [12]), a PB method (RippleNet [11]), and a graph neural network-based method (GC-MC [2]). The embedding size of all models was fixed at 64 except for that of RippleNet, which was 16 due to its high computational cost. We applied dropout for NFM, GC-MC, and KGAT, where the ratio was tuned in $\{0, 0.1, \cdots, 0.8\}$. Furthermore, node dropout was employed for CG-MG and KGAT, where the ratio was searched in $\{0, 0.1, \cdots, 0.8\}$. For RippleNet, the number of hops and memory size was set to 2 and 8, respectively. For KGAT, we set the depth to 3 with hidden dimensions 64, 32, and 16. We selected 3 as the length of UIIPs inferred by our RE-KGR.

4.3   Performance Comparison

Table 1 presents an overall performance comparison of all methods for the top-K recommendation task on three datasets. RE-KGR significantly outperformed state-of-the-art methods on all datasets. Specifically, RE-KGR outperformed the state-of-the-art methods by $63.06\%, 40.23\%, 30.90\%$ in recall@K and $93.64\%, 76.91\%, 49.94\%$ in ndcg@K for the Amazon-Book, Last-FM, and Yelp2018 datasets, respectively.

Table 1. Performance comparison of all methods

Data	Metrics	FM	NFM	RippleNet	GC-MC	CKE	CFKG	KGAT	Ours	Improve
Amazon-Book	recall	0.1345	0.1366	0.1336	0.1316	0.1343	0.1142	0.1489*	**0.2428**	63.06%
	ndcg	0.0886	0.0913	0.0910	0.0874	0.0885	0.0770	0.1006*	**0.1948**	93.64%
Last-FM	recall	0.0778	0.0829	0.0791	0.0818	0.0736	0.0723	0.0870*	**0.1220**	40.23%
	ndcg	0.1181	0.1214	0.1238	0.1253	0.1184	0.1143	0.1325*	**0.2344**	76.91%
Yelp2018	recall	0.0627	0.0660	0.0664	0.0659	0.0657	0.0522	0.0712*	**0.0932**	30.90%
	ndcg	0.0768	0.0810	0.0822	0.0790	0.0805	0.0644	0.0867*	**0.1300**	49.94%

4.4   Study of RE-KGR

Effect of UIIP Reasoning and Relation-Based Graph Convolution. Table 2 shows the results of comparing RE-KGR with its variants. We replaced the prediction layer with direct calculation of the similarities between users and items as in [12], called w/o UIIP, and replaced the projection matrices (2) with a unique matrix, called w/o RGC.

Table 2. Effect of UIIP reasoning and RGC.

	Amazon-Book		Last-FM		Yelp2018	
	recall	ndcg	recall	ndcg	recall	ndcg
RE-KGR	**0.2428**	**0.1948**	**0.1220**	**0.2344**	**0.0932**	**0.1300**
w/o UIIP	0.1307	0.0984	0.0812	0.1332	0.0613	0.0810
w/o RGC	0.2279	0.1826	0.1101	0.2229	0.0878	0.1249

Table 3. Effect of model depth.

depth	Amazon-Book		Last-FM		Yelp2018	
	recall	ndcg	recall	ndcg	recall	ndcg
2	0.2179	0.1620	0.1165	0.2171	0.0665	0.0940
3	0.2421	**0.1956**	**0.1262**	**0.2420**	0.0919	0.1267
4	**0.2428**	0.1948	0.1220	0.2344	**0.0932**	**0.1300**

The experimental results in Table 2 can be summarized as follows:

1) UIIP reasoning was substantially superior to directly calculating similarities between users and items. This may be because the model optimized by maximizing the similarity between directly connected entities, which is measured via the inner product, is unavailable for indirectly connected entities. The lack of explicit constraints on different types of connections leads to the uncertain direction of vectors. The inner product reflects the linear correlation between vectors; however, it ignores the direction and therefore fails to compute the similarity between indirectly connected entities.

2) Removing RGC dramatically degraded the model's performance in each case. One possible reason is that disabled RGC weakens the model's ability to distinguish different types of connections, resulting in over-smoothing.

Effect of Model Depth. We varied the number of RGC layers to observe changes in the performance of RE-KGR. The results are presented in Table 3. Comparing the 2-layer and 3-layer model, it can be seen that increasing the depth of the model can effectively enhance performance. This suggests that deeper models can learn representations with more side information and make more accurate predictions. The further stacked 4-layer model led to marginal improvements and even reduced performance in some cases. We attribute this to local over-smoothing. More than 99% of items could be connected by UIIPs within three hops; therefore, too many layers of the model led to repeated aggregation of entities.

4.5 Case Study

To intuitively demonstrate the explainability of RE-KGR, we randomly selected one user, u_{55463}, from Amazon-Book and one recommended item, i_{940}. Figure 2 presents the visualized process of UIIP reasoning.

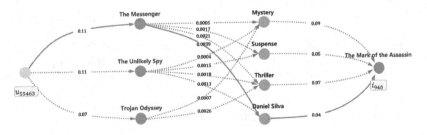

Fig. 2. Visualization of user-item interaction paths (UIIPs).

Benefiting from UIIP reasoning, the tracked UIIPs can be seen as reasons why an item is recommended for a user. For instance, the path ($u_{55463} \xrightarrow{0.11} The\,Messenger \xrightarrow{0.0839} Daniel\,Silva \xrightarrow{0.04} The\,Mark\,of\,the\,Assassin$) has the highest probability; therefore, we can generate the explanation "You might be interested in *The Mark of the Assassin* since you have read *The Messenger* written by the same author, Daniel Silva."

5 Conclusion

In this paper, we propose the RE-KGR framework that combines the advantages of both EB and PB recommendation methods. RE-KGR can learn entity representations with latent semantic information in a more flexible way than PB methods, and can provide more diverse explanations than EB methods. The proposed framework encodes structured knowledge and collaborative signal by recursively performing relation-based graph convolution, and provides explainable recommendations by reasoning UIIPs. Extensive experiments on three publicly available datasets demonstrate that our method can substantially improve the performance and explainability of knowledge-aware recommendation.

Acknowledgements. This work is supported in part by the Beijing Natural Science Foundation under grants 4192008.

References

1. Ai, Q., Azizi, V., Chen, X., Zhang, Y.: Learning heterogeneous knowledge base embeddings for explainable recommendation. Algorithms **11**(9), 137 (2018)
2. van den Berg, R., Kipf, T.N., Welling, M.: Graph convolutional matrix completion. arXiv preprint arXiv:1706.02263 (2017)
3. He, X., Chua, T.S.: Neural factorization machines for sparse predictive analytics. In: SIGIR, pp. 355–364 (2017)
4. Hu, B., Shi, C., Zhao, W.X., Yu, P.S.: Leveraging meta-path based context for top-n recommendation with a neural co-attention model. In: SIGKDD, pp. 1531–1540 (2018)
5. Huang, G., Liu, Z., Van Der Maaten, L., Weinberger, K.Q.: Densely connected convolutional networks. In: CVPR, pp. 4700–4708 (2017)
6. Li, G., Muller, M., Thabet, A., Ghanem, B.: DeepGCNs: Can GCNs go as deep as CNNs? In: ICCV, pp. 9267–9276 (2019)
7. Lin, Y., Liu, Z., Sun, M., Liu, Y., Zhu, X.: Learning entity and relation embeddings for knowledge graph completion. In: AAAI (2015)
8. Rendle, S.: Factorization machines. In: ICDM, pp. 995–1000. IEEE (2010)
9. Rendle, S., Freudenthaler, C., Gantner, Z., Lars, S.: BPR: Bayesian personalized ranking from implicit feedback. In: UAI 2009, Arlington, Virginia, United States, pp. 452–461 (2009)
10. Veličković, P., Cucurull, G., Casanova, A., Romero, A., Liò, P., Bengio, Y.: Graph attention networks. In: ICLR (2018)

11. Wang, H., et al.: RippleNet: propagating user preferences on the knowledge graph for recommender systems. In: ICKM, pp. 417–426 (2018)
12. Wang, X., He, X., Cao, Y., Liu, M., Chua, T.S.: KGAT: knowledge graph attention network for recommendation. In: SIGKDD, pp. 950–958 (2019)
13. Zhang, F., Yuan, N.J., Lian, D., Xie, X., Ma, W.Y.: Collaborative knowledge base embedding for recommender systems. In: SIGKDD, pp. 353–362 (2016)

LGCCF: A Linear Graph Convolutional Collaborative Filtering with Social Influence

Ming He$^{(\boxtimes)}$, Han Wen, and Hanyu Zhang

Faculty of Information Technology,
Beijing University of Technology, Beijing, China
heming@bjut.edu.cn, {wenhan,zhanghanyu}@emails.bjut.edu.cn

Abstract. Collaborative filtering (CF) is the dominant technique in personalized recommendation. It models user-item interactions to select the relevant items for a user, and it is widely applied in real recommender systems. Recently, graph convolutional network (GCN) has been incorporated into CF, and it achieves better performance in many recommendation scenarios. However, existing works usually suffer from limited performance due to data sparsity and high computational costs in large user-item graphs. In this paper, we propose a *linear graph convolutional CF* (LGCCF) framework that incorporates the social influence as side information to help improve recommendation and address the aforementioned issues. Specifically, LGCCF integrates the user-item interactions and the social influence into a unified GCN model to alleviate data sparsity. Furthermore, in the graph convolutional operations of LGCCF, we remove the nonlinear transformations and replace them with linear embedding propagations to overcome training difficulty and improve the recommendation performance. Finally, extensive experiments conducted on two real datasets show that LGCCF consistently outperforms the state-of-the-art recommendation methods.

Keywords: Recommender systems · Graph convolutional network · Collaborative filtering · Social network

1 Introduction

Collaborative filtering (CF) has seen great success due to its relatively high performance [9]. However, their recommendation performance is unsatisfactory due to the sparsity of user-item interaction data. Recently, the graph convolutional network (GCN) has been incorporated into CF achieving better performance in many recommendation scenarios, e.g., PinSage [10] and NGCF [7]. Despite the relative success of GCN-based recommendation, we argue that it still faces two challenges: (i) current GCN-based recommendation models usually only consider the user-item bipartite graph. However, the social influence enables users to build relationships and create different types of items. Hence, considering how

© Springer Nature Switzerland AG 2021
C. S. Jensen et al. (Eds.): DASFAA 2021, LNCS 12683, pp. 306–314, 2021.
https://doi.org/10.1007/978-3-030-73200-4_20

to incorporate social graphs into the GCN models will improve recommendation performance in the long run; (ii) we argue that the designs of these models are rather burdensome and difficult to train. Because GCN was originally designed for graph classification tasks, its many operations are unnecessary for the CF task. Therefore, it is very important to simplify the GCN model and make it efficient for recommendation.

In this paper, we propose a *linear graph convolutional CF* (LGCCF) with social influence to address these two challenges. First, we integrate the user-item bipartite graph and the user-user social graph into a unified GCN model through user nodes. Second, we carefully design a linear embedding propagation rule for graph convolutional operations, including the essential components of GCN neighborhood aggregation for CF. Empirically, we apply LGCCF to two real-world datasets, *Yelp* and *Flickr*, in extensive experiments. The results show that LGCCF achieves substantial gains over state-of-the-art GCN-based methods for recommendation.

2 Preliminaries

Problem Definition. In a recommendation scenario, we have a set of M users $U = \{u_1, u_2, \ldots, u_M\}$ and a set of N items $I = \{i_1, i_2, \ldots, i_N\}$. The historical user-item interactions can be represented as a bipartite graph $\mathcal{G}_1 = \langle U \cup I, \mathbf{R} \in \mathbb{R}^{M \times N} \rangle$, where U and I denote the user and item nodes, respectively, and $\mathbf{R}$ denotes the user-item interaction matrix. We use R_u and R_i to respectively denote the set of items interacted with by user u and the set of users who have interacted with item i. In addition to the user-item interactions, we use the social network as auxiliary information to enrich user-item interactions. The social network can be represented as a user-user social graph $\mathcal{G}_2 = \langle U, \mathbf{S} \in \mathbb{R}^{M \times M} \rangle$, where U is the user nodes and $\mathbf{S}$ represents the social matrix. We use S_u to represent the set of users that user u follows.

Given a user-item bipartite graph $\mathcal{G}_1$ and a user-user social graph $\mathcal{G}_2$, we aim to predict whether user u has potential interest in item i with which he has had no interaction before. Our goal is to learn a prediction function $\hat{r}_{ui} = \mathcal{F}(u, i)$, where $\hat{r}_{ui}$ denotes the probability that user u will click on item i.

Problems in GCN. Although the current GCN-based recommendation models are relatively successful, we argue that these models' propagation rule is not reasonable enough. Firstly, the role of nonlinear feature transformation in current GCN-based recommendation models is unclear. GCN was originally designed for node classification tasks. Therefore, it is not necessarily useful for the CF task. Secondly, nonlinear feature transformation incurs high computational costs and increases training difficulty. Recently, SGCN [8] demonstrated that nonlinear feature transformation has a negative effect on node classification tasks. Therefore, simplifying the GCN model to overcome training difficulty and high computational costs is a highly challenging problem.

3   Methodology

In this section, we describe the LGCCF whose structure is shown in Fig. 1.

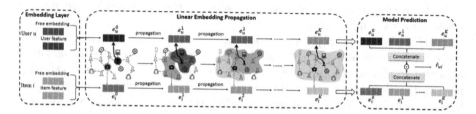

Fig. 1. The overall architecture of the proposed model, which contains three major components: embedding layer, linear embedding propagation, and model prediction

3.1   Architecture

Embedding Layer. Given a user-item bipartite graph $\mathcal{G}_1$ and a user-user social graph $\mathcal{G}_2$ as input, each user u (and item i) is encoded as a free embedding $e_u \in \mathbb{R}^d$ ($e_i \in \mathbb{R}^d$), where d is the free embedding size. And then, we additionally integrate user u's feature embedding $\mathbf{e}'_u \in \mathbb{R}^d$ (e.g., user profile) and item i's feature embedding $\mathbf{e}'_i \in \mathbb{R}^d$ (e.g., item text representation, item visual representation). Finally, by fusing the free embedding and feature embedding, we define the output of the embedding layer to user u and item i as:

$$\mathbf{e}_u^0 = \mathbf{e}_u + \mathbf{e}'_u, \quad \mathbf{e}_i^0 = \mathbf{e}_i + \mathbf{e}'_i, \tag{1}$$

where $+$ is the element-wise addition operation.

Linear Embedding Propagation. Next, we propose the linear embedding propagation layers, which adopt a simple weighted-sum aggregator and abandon the use of feature transformation and nonlinear activation, to recursively propagate each user u's and each item i's embeddings along higher-order connectivity.

User Embedding. For each user u, let $\mathbf{e}_u^k$ denote the k-th layer embedding. As users play a central role in both the user-item bipartite graph $\mathcal{G}_1$ and the user-user social graph $\mathcal{G}_2$, two aggregations are introduced to respectively process these two different graphs. The first is item aggregation, let $\mathbf{e}_{R_u}^{k+1}$ denote user u's aggregated embedding from the neighbor nodes R_u in the graph $\mathcal{G}_1$:

$$\mathbf{e}_{R_u}^{k+1} = \frac{1}{|R_u|} \sum_{a \in R_u} \mathbf{e}_a^k, \quad k = 0, 1, \cdots, K-1, \tag{2}$$

where $\frac{1}{|R_u|}$ means the item's neighbor nodes contribute the same weight to u. The other is social aggregation and it can help model users from the social perspective. Let $e_{S_u}^{k+1}$ represent the embedding of aggregation from the u's social neighbors at the $k+1$-th layer:

$$e_{S_u}^{k+1} = \frac{1}{|S_u|} \sum_{b \in S_u} e_b^k, \quad k = 0, 1, \cdots, K-1, \tag{3}$$

where $\frac{1}{|S_u|}$ implies the social neighbor nodes contribute the same weight to u.

Unlike the traditional GCN-based recommendation model, we use a simple linear embedding propagation rule to get user u's updated embedding e_u^{k+1} at the $k+1$-th layer after obtaining item aggregation and social aggregation:

$$e_u^{k+1} = e_u^k + (\gamma_1 e_{R_u}^{k+1} + \gamma_2 e_{S_u}^{k+1}), \quad k = 0, 1, \cdots, K-1, \tag{4}$$

where e_u^k ensures self-information from layer k can be retained at layer $k+1$. $e_{R_u}^{k+1}$ and $e_{S_u}^{k+1}$ are the item aggregation and social aggregation of user u, respectively. In order to distinguish between the influence of these two different neighbor types in the embedding propagation step, we set the item influence weight γ_1 and the social influence weight γ_2. In our experiments, we find that setting $\gamma_1 = \gamma_2 = 0.5$ generally leads to good performance.

Item Embedding. For each item i, we denote its k-th layer embedding as e_i^k. Item i's embedding is only influenced by its user neighbors R_i. We also use the linear embedding propagation rule to aggregate the item's embedding e_i^k with its user neighbors:

$$e_i^{k+1} = e_i^k + e_{R_i}^{k+1}, \quad e_{R_i}^{k+1} = \frac{1}{|R_i|} \sum_{c \in R_i} e_c^k, \quad k = 0, 1, \cdots, K-1, \tag{5}$$

where $\frac{1}{|R_i|}$ implies the user neighbor nodes contribute the same weight to i.

Model Prediction. After the linear embedding propagation processes with K times, we obtain multiple embeddings for user node u and item node i. Then, we concatenate them to constitute the final embeddings for user u and item i. The model prediction is defined as the inner product of user and item final embeddings:

$$e_u^* = e_u^0 \| \cdots \| e_u^K, \quad e_i^* = e_i^0 \| \cdots \| e_i^K, \quad \hat{r}_{ai} = e_u^{*\top} e_i^*. \tag{6}$$

3.2 Model Optimization

We adopt a ranking criterion, i.e. the Bayesian Personalized Ranking [5], to optimize the model parameters of LGCCF:

$$L_{BPR} = \sum_{(u,i,j) \in O} -\ln \sigma \left(\hat{r}_{ui} - \hat{r}_{uj} \right) + \lambda \|\Theta\|^2, \tag{7}$$

Table 1. Statistics of the datasets

Dataset	Yelp	Flickr
Number of Users	17237	8358
Number of Items	38342	82120
Number of Ratings	207945	327815
Rating Density	0.031%	0.048%
Number of Social Connections	143765	187273
Social Relations Density	0.048%	0.268%

where O denotes the positive and negative training data. $\sigma(x)$ is a sigmoid function. $\Theta = \{\mathbf{E}\}$ denotes the trainable model parameters; $\mathbf{E}$ is the embedding matrix for all user and item nodes. λ controls the L_2 regularization strength to prevent overfitting. We employ the mini-batch Adam [4] as the optimizer in our implementation to optimize the objective function. Moreover, the dimensions of the embedding in the linear embedding propagation layer are the same.

3.3 Time Complexity Analysis

$O\left(\sum_{k=1}^{K} |\mathcal{G}_1 + \mathcal{G}_2| d_l d_{l-1} + \sum_{k=1}^{K} |\mathcal{G}_1 + \mathcal{G}_2| d_l\right)$ is the overall time complexity of the classical GCN model, where $|\mathcal{G}_1 + \mathcal{G}_2|$ and d_l and d_{l-1} are the number of interactions in $\mathcal{G}_1$ and $\mathcal{G}_2$ and the current and previous transformation sizes, respectively. The time cost of classic GCN models usually result from the matrix multiplication $O\left(\sum_{k=1}^{K} |\mathcal{G}_1 + \mathcal{G}_2| d_l d_{l-1}\right)$. Here, we propose a computationally efficient method. Since it uses linear embedding propagation and does not have any hidden layers, we do not need the matrix multiplication for feature learning. Therefore, the overall time complexity of LGCCF is $O\left(\sum_{k=1}^{K} |\mathcal{G}_1 + \mathcal{G}_2| d_l\right)$, which is the time cost of the whole training epoch in the prediction layer.

4   Experiments

4.1   Experimental Settings

Datasets and Evaluation Metrics. We adopt two real-world datasets for empirical study, *Yelp* and *Flickr*, to evaluate the effectiveness of our model. Since *Yelp* is the explicit feedback data, we transform it into implicit feedback. We provide some statistics on these two datasets in Table 1. As our focus is recommending items to users, we use two commonly adopted ranking metrics for top-N recommendation evaluation: HR@N and NDCG@N.

Table 2. Overall performance comparison

	Yelp						Flickr					
	HR			NDCG			HR			NDCG		
	N=5	N=10	N=15	N=5	N=10	N=15	N=5	N=10	N=15	N=5	N=10	N=15
BPR	0.1695	0.2632	0.3252	0.1231	0.1554	0.1758	0.0651	0.0795	0.1037	0.0603	0.0628	0.0732
FM	0.1855	0.2825	0.3440	0.1341	0.1717	0.1876	0.0989	0.1233	0.1473	0.0866	0.0954	0.1062
SocialMF	0.1739	0.2785	0.3365	0.1324	0.1677	0.1841	0.0813	0.1174	0.1300	0.0723	0.0964	0.1061
TrustSVD	0.1882	0.2939	0.3688	0.1368	0.1749	0.1981	0.1089	0.1404	0.1738	0.0978	0.1083	0.1203
GraphRec	0.1915	0.2912	0.3623	0.1279	0.1812	0.1956	0.0931	0.1231	0.1482	0.0784	0.0930	0.0992
NGCF	0.1992	0.3042	0.3753	0.1450	0.1828	0.2041	0.0891	0.1189	0.1399	0.0819	0.0945	0.0998
SGCN	0.2036	0.3071	0.3823	0.1465	0.1872	0.2089	0.1053	0.1424	0.1612	0.0912	0.1107	0.1308
DiffNet++	0.2503	0.3694	0.4493	0.1841	0.2263	0.2497	0.1412	0.1832	0.2203	0.1269	0.1420	0.1544
LGCCF$_{w/o\ feature}$	0.2268	0.3364	0.4172	0.1692	0.2079	0.2309	0.1099	0.1395	0.1703	0.1001	0.1104	0.1214
LGCCF$_{w/o\ social}$	0.2403	0.3595	0.4394	0.1719	0.2160	0.2401	0.1606	0.2118	0.2506	0.1398	0.1575	0.1700
LGCCF	**0.2716**	**0.3887**	**0.4661**	**0.2013**	**0.2433**	**0.2667**	**0.2006**	**0.2507**	**0.2934**	**0.1782**	**0.1947**	**0.2091**
%Improv	8.51%	5.22%	3.94%	9.34%	7.51%	6.81%	42.06%	36.84%	33.18%	40.42%	37.11%	35.42%

Baselines and Parameter Settings. We compare our model with three groups of baselines, including traditional recommendation models (BPR [5] and FM [6]), social recommendation models (SocialMF [3] and TrustSVD [2]), and GNN-based recommendation models (GraphRec [1], NGCF [7], SGCN [8], and DiffNet++ [9]). Moreover, to further verify the efficacy of the social network and the feature embeddings, we designed two variants of the LGCCF: LGCCF$_{w/o\ social}$, where LGCCF's user-user social graph is removed; and LGCCF$_{w/o\ feature}$, where the feature embeddings are removed in Eq. 1.

The optimal parameter settings for all the comparison methods are achieved by either empirical study or adopting the original papers' settings. The embedding size is fixed at 64 for all models, and the batch size is fixed at 512. In our model, we try the regularization parameter λ in the range $\{0.0001, 0.001, 0.01, 0.1\}$ and the learning rate γ in the range $\{0.001, 0.003, 0.01, 0.03\}$. We find $\lambda = 0.001$ and $\gamma = 0.003$ achieve the best performance.

4.2   Performance Study

Performance Comparison. We first compare the recommendation performance of all methods. The performance comparison results are presented in Table 2. Thus, we have the following observations:

We note that our method achieves the best performance among all the methods on both datasets in terms of HR and NDCG. Our model provides a linear graph convolutional operation to integrate the user-item bipartite graph and the user-user social graph. We assume this is because the feature transformation operations and nonlinear operations may not benefit the recommendation performance. These results demonstrate the effectiveness of our model. In Table 2, LGCCF$_{w/o\ social}$ performs worse than LGCCF. It confirms that combining the user-user social graph with GCN can help in learning user or item embeddings and improve recommendation performance. We can also see that without user

Table 3. Performance comparison between LGCCF and LGCCF$_{nonlinear}$

Dataset		Yelp			Flickr		
Layer	Method	HR@10	NDCG@10	Time Cost	HR@10	NDCG@10	Time Cost
1 Layer	LGCCF$_{nonlinear}$	0.3618	0.2260	10 s	0.1832	0.1391	13 s
	LGCCF	**0.3735**	**0.2362**	**9 s**	**0.2090**	**0.1614**	**12 s**
2 Layer	LGCCF$_{nonlinear}$	0.3658	0.2263	15 s	0.1858	0.1427	20 s
	LGCCF	**0.3820**	**0.2407**	**14 s**	**0.2426**	**0.1859**	**16 s**
3 Layer	LGCCF$_{nonlinear}$	0.3684	0.2287	22 s	0.1945	0.1486	23 s
	LGCCF	**0.3887**	**0.2433**	**17 s**	**0.2507**	**0.1940**	**20 s**
4 Layer	LGCCF$_{nonlinear}$	0.3588	0.2210	27 s	0.1805	0.1369	29 s
	LGCCF	**0.3809**	**0.2353**	**24 s**	**0.2421**	**0.1829**	**26 s**

and item features, the performance of LGCCF$_{w/o\ feature}$ significantly deteriorates. This confirms that both user and item features are important for learning node latent factors in the graph and boosting the recommendation performance.

Effectiveness Analysis of Linear Embedding Propagation. We perform the analysis to demonstrate the validity of our proposed linear embedding propagation rule. We design a variant of the LGCCF: LGCCF$_{nonlinear}$. This variant of LGCCF retains feature transformation and nonlinear activation. A thorough comparison with LGCCF is carried out. Table 3 shows the outputs at various layers (1 to 4). The following are the main observations:

In all cases, LGCCF outperforms LGCCF$_{nonlinear}$ by a large margin, but the training time of LGCCF was always less than the that of LGCCF$_{nonlinear}$. These results illustrate that nonlinear feature transformations in traditional GCN models are unnecessary for the recommendation task and removing them can greatly improve the recommendation performance and reduce training difficulty. Table 3

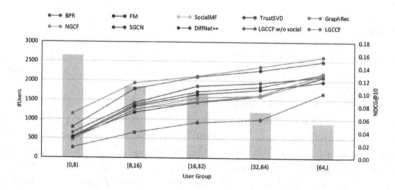

Fig. 2. Performance under different data sparsity on *Flickr*: The background histograms indicate the number of users involved in each group, and the lines demonstrate the performance with regard to NDCG@10

also shows the results of LGCCF at different depths. When the number of layers is 3, LGCCF achieves the best performance. This suggests that increasing the depths enables the efficient modeling of higher-order connectivity. However, when we further increase the layers to 4, the performance drops. This might be because adding more layers introduces unnecessary neighbors and noise to the representation learning.

Performance Under Different Data Sparsity. We show the performance of various models under different sparsity levels. Figure 2 illustrates the results with regard to NDCG@10 on different user groups in the *Flickr* dataset. Our proposed model, LGCCF, outperforms all other baselines on all user groups. $LGCCF_{w/o\ social}$ performs worse than LGCCF. This reconfirms the efficacy of our model, which uses the higher-order connectivity and social graphs to solve the data sparsity issue.

5 Conclusion

In this paper, we present LGCCF, a linear graph convolutional collaborative filtering with social influence. We integrate the user-item bipartite graph and the user-user social graph into one graph. We further design a new GCN model to capture the higher-order representations of the nodes in this graph. In this GCN model, we remove the nonlinear transformations and replace them with linear embedding propagations. We compare LGCCF with state-of-the-art models on two real-world datasets, and the superior results of LGCCF on top-N recommendation demonstrate its effectiveness.

Acknowledgement. This work is supported in part by the Beijing Natural Science Foundation under grants 4192008.

References

1. Fan, W., et al.: Graph neural networks for social recommendation. In: Proceedings of the 28th International Conference on World Wide Web, pp. 417–426 (2019)
2. Guo, G., Zhang, J., Yorke-Smith, N.: TrustSVD: collaborative filtering with both the explicit and implicit influence of user trust and of item ratings. In: Proceedings of the 29th AAAI Conference on Artificial Intelligence, pp. 123–129 (2015)
3. Jamali, M., Ester, M.: A matrix factorization technique with trust propagation for recommendation in social networks. In: Proceedings of the 4th ACM Conference on Recommender Systems, pp. 135–142 (2010)
4. Kingma, D.P., Ba, J.: Adam: a method for stochastic optimization. arXiv preprint arXiv:1412.6980 (2014)
5. Rendle, S., Freudenthaler, C., Gantner, Z., Schmidt-Thieme, L.: BPR: Bayesian personalized ranking from implicit feedback. In: Proceedings of the 25th Conference on Uncertainty in Artificial Intelligence, pp. 452–461 (2009)

6. Rendle, S., Gantner, Z., Freudenthaler, C., Schmidt-Thieme, L.: Fast context-aware recommendations with factorization machines. In: Proceedings of the 34th International ACM SIGIR Conference on Research and Development in Information Retrieval, pp. 635–644 (2011)
7. Wang, X., He, X., Wang, M., Feng, F., Chua, T.S.: Neural graph collaborative filtering. In: Proceedings of the 42nd International ACM SIGIR Conference on Research and Development in Information Retrieval, pp. 165–174 (2019)
8. Wu, F., Zhang, T., Holanda de Souza, A., Fifty, C., Yu, T., Weinberger, K.Q.: Simplifying graph convolutional networks. In: Proceedings of 36th International Conference on Machine Learning, pp. 6861–6871 (2019)
9. Wu, L., Li, J., Sun, P., Ge, Y., Wang, M.: DiffNet++: a neural influence and interest diffusion network for social recommendation. arXiv preprint arXiv:2002.00844 (2020)
10. Ying, R., He, R., Chen, K., Eksombatchai, P., Hamilton, W.L., Leskovec, J.: Graph convolutional neural networks for web-scale recommender systems. In: Proceedings of the 24th ACM SIGKDD International Conference on Knowledge Discovery and Data Mining, pp. 974–983 (2018)

Sirius: Sequential Recommendation with Feature Augmented Graph Neural Networks

Xinzhou Dong[1,2], Beihong Jin[1,2(⊠)], Wei Zhuo[3], Beibei Li[1,2], and Taofeng Xue[1,2]

[1] State Key Laboratory of Computer Science, Institute of Software, Chinese Academy of Sciences, Beijing, China
Beihong@iscas.ac.cn
[2] University of Chinese Academy of Sciences, Beijing, China
[3] MX Media Co., Ltd, Singapore, Singapore

Abstract. Many practical recommender systems recommend personalized items for different users by mining user-item interaction sequences. The interaction sequences, as a whole, imply the manifold collaborative relations among users and items. Further, from the view of users, the item orders and time intervals between interactions could expose the evolution of user interests, and from the view of items, attributes of the items on interaction sequences may reveal the variation of item popularity. However, most of the existing recommendation models ignore those valuable information, and cannot fully explore the intrinsic implication of interaction sequences. In the paper, we propose a method named Sirius, which develops GNNs (Graph Neural Networks) to model the collaborative relations and capture the dynamics of time and attribute features in sequences. We give the workflow of the Sirius method, and describe the implementations about graph construction, item embedding generation, sequence embedding generation and next-item prediction. Finally, we give an example of Sirius recommendations, which visually shows the impact of feature information on the recommendation results. At present, Sirius has been adopted by MX Player, one of India's largest streaming platforms, recommending movies for thousands of users.

Keywords: Recommender system · Deep learning · Graph neural network · Sequential recommendation

1 Introduction

In recent years, recommendation has been an effective way to solve the information overload and meet personalized requirements of different users. Among recommendation tasks, sequential recommendation refers to recommending items to users according to the user-item interaction sequences in the recent period. For example, on e-commercial platforms, products are recommended to users on the

© Springer Nature Switzerland AG 2021
C. S. Jensen et al. (Eds.): DASFAA 2021, LNCS 12683, pp. 315–320, 2021.
https://doi.org/10.1007/978-3-030-73200-4_21

basis of the recent user-product clicking records. Likewise, on video streaming platforms, recommending videos to users is largely based on historical watching records.

The early sequential recommendation methods often model an interaction sequence as a kth-order Markov chain, and predict the user's next action by previous k actions [5,13]. Obviously, under the assumption that a user's next action relates to previous k actions, these methods can capture short-term dependency in the sequence and perform well on recommendation. However, in reality, user's next action might have a certain relation to earlier actions instead of previous k actions.

With the boom of deep learning, a variety of different methods have been applied successively to the sequential recommendation. First, inspired by sequence modeling capability of RNNs (Recurrent Neural Networks), RNN-based methods [3,4] are proposed. They can capture the long-term dependency in the sequence, but they are prone to generate fake dependency and cannot explicitly model the complex transitions between items in the sequence. Next, in order to differentially treat the interactions or items, attention mechanisms, used alone or added in other neural networks as an extra component, are applied to sequential recommendation [6–8]. Recently, GNNs which combine the flexible expressiveness of graph data and the strong learning capability of neural networks, have emerged as a promising way to achieve sequential recommendation. The advantage of GNNs lies in the capability of capturing complex transitions of items, which helps generate effective embeddings for items, users, even interaction sequences [15,16].

We note that an interaction sequence is in essence a time series, in detail, two items in the sequence not only have a relative order, but also have a time interval, often differing from the other two. If an item in the sequence has a large time interval with the previous item, it means that its relation with the previous item is likely to be weak, and it may be more related to the user's long-term interests. That is, the time feature accompanied by an interaction sequence, depicted by the timestamps of items in the sequence, might reflect the evolution of user interests with a high probability. Recently, some work has payed attention to the temporal information in the sequences [7,9,17].

On the other hand, an interaction sequence is also associated with the attribute feature which is depicted by the attributes of items in the sequence. If the interaction sequences are observed from the view of item attributes, then the roles of different attributes in attracting users could be understood and further the variation of item popularity could be discovered. For example, if most of movies in a movie sequence are found to belong to the sci-fi genre, then this directly indicates that sci-fi movies are the user's favorite movies.

Unfortunately, from existing work, we discover none of sequential recommendation models or methods can simultaneously model complex transition relationships, time feature and attribute feature in sequences. Therefore, modeling the interaction sequences in a more accurate and comprehensive way to achieve more effective recommendations is still a challenge.

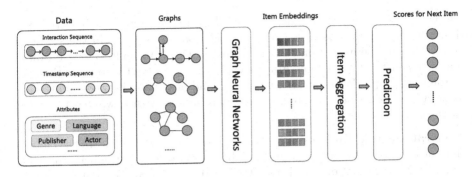

Fig. 1. Workflow of Sirius.

In the paper, we propose the Sirius method for sequential recommendation. Sirius can both model the collaborative relations and capture features in interaction sequences. In the following section, we give the overall workflow of the Sirius method, and outline our implementation of each step in the method as well as some possible implementations different from our choices.

2 Overview of Sirius

We denote the item set by $V = \{v_1, v_2, \ldots, v_{|V|}\}$, and the historical interaction sequence of user u over a period of time by $s_u = [v_{u,1}, v_{u,2}, \ldots, v_{u,n}]$, where $v_{u,i} \in V, i = 1, 2, \ldots, n$, and $\{v_{u,i}\}$ is sorted by the timestamp of user u interacting item v in ascending order. At the same time, we extract the timestamp for each item in sequence s_u to form timestamp sequence $t_{s_u} = [t_{u,1}, t_{u,2}, \ldots, t_{u,n}]$. Furthermore, we confine the item description to k attributes, and then define a group of attribute mapping functions, i.e., $p_j(v), v \in V, j = 1, 2, \ldots k$, each of which maps the item v to one or multiple values in A_j according to attribute j of item v, where A_j is the range of attribute j.

Our goal is to build a model to predict $v_{u,n+1}$ for user u, where the input of the model is the set of item sequences $\{s_u\}$, the set of corresponding timestamp sequences $\{t_{s_u}\}$ and attribute mapping functions $p_j(v), j = 1, 2, \ldots k$.

For the goal, in the Sirius method, we first construct graphs from an interaction sequence. Next, we build GNNs to learn embedding vectors of nodes on graphs. Then, we generate the representation vector of the sequence through the item aggregation. Finally, for each user, we score each item according to the sequence representation vector and the corresponding item embedding, generating the next-item recommendation. Figure 1 gives the workflow of the Sirius method. Sirius is mainly composed of four steps, each of which may be implemented in different ways in practice.

Building Graphs. Graph-based representation is more expressive, especially to represent the complex transition relationships of the items in the sequence, therefore we convert sequence data (including items and various features) into graph-based representations. Given a sequence, we build an item graph, using all distinct items in the sequence as nodes on the graph and constructing directed edges according to the adjacent relationship of items in the sequence [15], although there are some other strategies [11,12] to build the item graph. Then we construct k attribute graphs where the nodes are the same as the ones in the item graph. We treat attributes as a kind of relation between items, constructing the edges between item nodes according to whether two items have the same attribute value.

Generating Item Embeddings. We need to design GNNs to learn the embedding of nodes on the graph, using the above graphs as input. There are many different implementations, such as GGNN [10] used in SR-GNN [15] and GC-SAN [16], and GAT [14] with weights used in FGNN [12]. We borrow the GNN components in the message passing model [1] and design two kinds of GNNs (i.e., I-GNN and A-GNN) for two kinds of graphs to update node embeddings by three stages: message construction, message propagation and embedding update. In particular, we treat the time interval between items as a feature of corresponding edge on the item graph, regarding this time interval as a discrete feature and generating the corresponding embedding. During the message propagation of I-GNN, the embedding of the time interval between items is first transformed and then fused into the message vector. Although the time interval is also able to be modeled as a continuous feature, the experimental results do not support this practice. Further, since for an item, we have $1 + k$ item embeddings derived from $1 + k$ GNNs (i.e., one from I-GNN and the others from A-GNNs), we apply a gated mechanism to fuse multiple item embeddings, generating the final item embedding.

Generating Sequence Embedding. After getting the embedding vectors of items, we need to gather the vectors of all the items to generate the representation vector of the sequence. There are many kinds of aggregation methods, the simplest of which is the pooling method (e.g., mean pooling, max pooling, etc.). Some models use more complex aggregation. For example, in FGNN [12], GRU units are used to learn the aggregation order of nodes. In addition, some other features, such as the position of the item in the sequence, can also be considered to better determine the importance of different nodes. Our choice is to aggregate item embeddings from the view of the user's short-term and long-term interests. We take the last item in the sequence as the short-term interest and the previous items as the long-term interest, and obtain the representation vector of the sequence through the attention mechanism. Further, we consider the time distance of the item to the last interaction when calculating the attention coefficient of the item, integrating the time decay effect into the sequence embedding.

Predicting Next Item. The key to the prediction is how to calculate the item score. The most commonly used method is to calculate the inner product of the sequence representation vector and the item embedding vector, but such a calculation tends to score high on popular items [2], so a more appropriate method is to use cosine similarity. In our prediction step, we also consider the prediction interval (i.e., the interval between the prediction time and the time

Fig. 2. Recommendations of Sirius.

of the last item in the sequence), because the user's interest may change to varying degrees, depending on the length of the prediction interval. For fusing the prediction interval into the sequence representation vector, we first obtain the embedding vector of the prediction interval (via the similar method used for the time interval of items), and then fuse it into the sequence representation vector by element-wise addition, where the element-wise multiplication is also workable but is not adopted, due to the relatively poor experimental results.

3 Recommendation Case

In this section, we give a recommendation case, which visually shows the influence of time and attribute features in the sequence on the recommendation results.

As shown in Fig. 2, the left side shows a historical interaction sequence, containing six movies. The genres of each movie are listed on the top of its poster. All the six movies are in Hindi whose time intervals are roughly one day. From the sequence, we notice that the user mainly watches action movies but the latest watched movie has a different genre. Sirius gives two different recommendation results, depending on the time of the user's next visit. If the user visits the MX player App again within one day, the recommendation results from Sirius are in the top right corner, which are in line with the user's most recent viewing behavior, because Sirius tends to infer the recent interest of the user. If the user visits the App after 10 days, the recommendation results from Sirius are in the bottom right corner. The last two recommended movies reveal that Sirius considers the user's general interest. In addition, all the recommendation results are movies in Hindi, illustrating Sirius has known that the user's language preference is Hindi.

Acknowledgement. This work was supported by the National Natural Science Foundation of China under Grant No. 62072450 and the 2019 joint project with MX Media.

References

1. Gilmer, J., Schoenholz, S.S., Riley, P.F., Vinyals, O., Dahl, G.E.: Neural message passing for quantum chemistry. In: International Conference on Machine Learning, pp. 1263–1272. PMLR (2017)

2. Gupta, P., Garg, D., Malhotra, P., Vig, L., Shroff, G.: Niser: Normalized item and session representations with graph neural networks. arXiv preprint arXiv:1909.04276 (2019)
3. Hidasi, B., Karatzoglou, A.: Recurrent neural networks with top-k gains for session-based recommendations. In: Proceedings of the 27th ACM International Conference on Information and Knowledge Management, pp. 843–852 (2018)
4. Hidasi, B., Karatzoglou, A., Baltrunas, L., Tikk, D.: Session-based recommendations with recurrent neural networks. In: 4th International Conference on Learning Representations, ICLR 2016, San Juan, Puerto Rico, May 2–4, 2016, Conference Track Proceedings (2016)
5. Hosseinzadeh Aghdam, M., Hariri, N., Mobasher, B., Burke, R.: Adapting recommendations to contextual changes using hierarchical hidden Markov models. In: Proceedings of the 9th ACM Conference on Recommender Systems, pp. 241–244 (2015)
6. Kang, W.C., McAuley, J.: Self-attentive sequential recommendation. In: 2018 IEEE International Conference on Data Mining (ICDM), pp. 197–206. IEEE (2018)
7. Li, J., Wang, Y., McAuley, J.: Time interval aware self-attention for sequential recommendation. In: Proceedings of the 13th International Conference on Web Search and Data Mining, pp. 322–330 (2020)
8. Li, J., Ren, P., Chen, Z., Ren, Z., Lian, T., Ma, J.: Neural attentive session-based recommendation. In: Proceedings of the 2017 ACM on Conference on Information and Knowledge Management, pp. 1419–1428 (2017)
9. Li, R., Shen, Y., Zhu, Y.: Next point-of-interest recommendation with temporal and multi-level context attention. In: 2018 IEEE International Conference on Data Mining (ICDM), pp. 1110–1115. IEEE (2018)
10. Li, Y., Tarlow, D., Brockschmidt, M., Zemel, R.S.: Gated graph sequence neural networks. In: 4th International Conference on Learning Representations, ICLR 2016, San Juan, Puerto Rico, May 2–4, 2016, Conference Track Proceedings (2016)
11. Pan, Z., Cai, F., Chen, W., Chen, H., de Rijke, M.: Star graph neural networks for session-based recommendation. In: Proceedings of the 29th ACM International Conference on Information & Knowledge Management, pp. 1195–1204 (2020)
12. Qiu, R., Li, J., Huang, Z., Yin, H.: Rethinking the item order in session-based recommendation with graph neural networks. In: Proceedings of the 28th ACM International Conference on Information and Knowledge Management, pp. 579–588 (2019)
13. Rendle, S., Freudenthaler, C., Schmidt-Thieme, L.: Factorizing personalized markov chains for next-basket recommendation. In: Proceedings of the 19th international conference on World wide web. pp. 811–820 (2010)
14. Velickovic, P., Cucurull, G., Casanova, A., Romero, A., Liò, P., Bengio, Y.: Graph attention networks. CoRR (2017)
15. Wu, S., Tang, Y., Zhu, Y., Wang, L., Xie, X., Tan, T.: Session-based recommendation with graph neural networks. In: Proceedings of the AAAI Conference on Artificial Intelligence, vol. 33, pp. 346–353 (2019)
16. Xu, C., et al.: Graph contextualized self-attention network for session-based recommendation. In: IJCAI, pp. 3940–3946 (2019)
17. Ye, W., Wang, S., Chen, X., Wang, X., Qin, Z., Yin, D.: Time matters: sequential recommendation with complex temporal information. In: Proceedings of the 43rd International ACM SIGIR Conference on Research and Development in Information Retrieval, pp. 1459–1468 (2020)

Combining Meta-path Instances into Layer-Wise Graphs for Recommendation

Mingda Qian[1,2], Bo Li[1], Xiaoyan Gu[1(✉)], Zhuo Wang[3], Feifei Dai[1,2], and Weiping Wang[1]

[1] Institute of Information Engineering, Chinese Academy of Sciences, Beijing, China
{qianmingda,libo,guxiaoyan,daifeifei,wangweiping}@iie.ac.cn
[2] School of Cyber Security, University of Chinese Academy of Sciences, Beijing, China
[3] Sangfor Inc., Shenzhen, China
wangzhuo@hit-cs.com

Abstract. In the recommendation area, the concept of meta-path is famous for inferring explicit and effective relationships between nodes such as users and items. To extract useful information from the instances of meta-paths, existing methods embed meta-path instances separately. However, they ignore the complicated semantics presented by multiple instances. These complicated semantics not only provide additional information but also affect the semantics of single instances. Without considering the complicated semantics, the information extracted from the instances may be incomplete and less effective. To solve the problem, we propose to learn the complicated semantics by combining meta-path instances into layer-wise graphs (instance-graphs) for recommendation. Following the idea, we develop an Instance-Graph based Recommendation method (IGR). IGR combines meta-path instances into layer-wise instance-graphs. Then, the instance-graphs are investigated layer by layer to generate effective embeddings. Finally, these embeddings are discriminatively merged into user/item embeddings to make predictions. Extensive experimental results show that IGR outperforms various state-of-the-arts recommendation methods.

Keywords: Meta-paths · Recommender systems · Neural networks · Heterogeneous information networks

1 Introduction

In the recommendation area, meta-paths are widely used to capture structural features and extract semantics from heterogeneous information networks (HIN) [5,9] for recommendation. However, existing meta-path based recommendation methods embed each meta-path instance separately, which ignores the complicated semantics presented by multiple instances. For example, with the three meta-path instances shown in Fig. 1(a) as input, existing methods separately

© Springer Nature Switzerland AG 2021
C. S. Jensen et al. (Eds.): DASFAA 2021, LNCS 12683, pp. 321–329, 2021.
https://doi.org/10.1007/978-3-030-73200-4_22

embed the instances and predict the preference of Sam for *I Am Legend*. Nevertheless, in practice, these instances together present a complicated semantic that Sam and Ben have watched three same movies (i.e., *Titanic*, *Skyfall*, and *Casino Royale*) shown in Fig. 1(b). As existing methods may not capture the complicated semantics, the similarity between Sam and Ben may be underestimated, and the three meta-path instances may be regarded as less convincing. Hence, the movie *I Am Legend* that Ben has watched may not be recommended to Sam, which is actually a good recommendation.

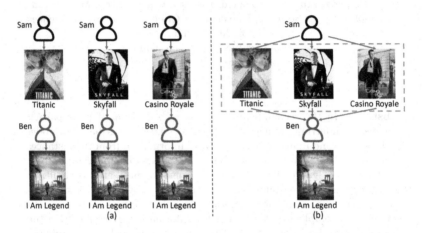

Fig. 1. In Figure (a), three meta-path instances which link user Sam and movie *I Am Legend* are presented. In Figure (b), the three instances are combined according to the shared components (i.e., Sam, and *Ben→I Am Legend*), and the instances together present a complicated semantic that Sam and Ben have watched three same movies.

To learn the complicated semantics presented by multiple instances, we propose a new framework to utilize meta-path. The framework consists of two steps. First, for each meta-path, its instances are combined into a layer-wise instance-graph according to their shared components. Second, the instance-graphs are embedded for the following recommendation. The framework has two major advantages: (1) With the shared components (i.e., nodes, edges, and sub-paths) of instances, the complicated semantics can be naturally revealed. Consider the example in Fig. 1, with the shared node Sam and sub-path *Ben→I Am Legend*, the three instances are properly combined, and the complicated semantic that Sam and Ben have watched three same movies is easily revealed. (2) the semantics of single instances can also be captured. The reason is that an instance-graph and its corresponding meta-path have similar structures, and single instances can be regarded as sub-graphs.

Following the framework, we propose an Instance-Graph based Recommendation method (IGR) which is illustrated in Fig. 2. IGR implements the framework with three steps. First, to reveal the complicated semantics, we propose a meta-path instance combining module which combines instances and generates layer-wise instance-graphs. Second, to investigate the instance-graphs, we propose a sequential instance-graph embedding module. In this module, as the instance-graphs are layer-wise and sequential, we propose a Layer-wise Graph Convolutional Network (LGCN) to embed them layer by layer in the order of the corresponding meta-paths. Third, to evaluate the importance of instance-graphs for users/items, we apply an attention guided merging module that discriminatively evaluates and merges the instance-graph embeddings.

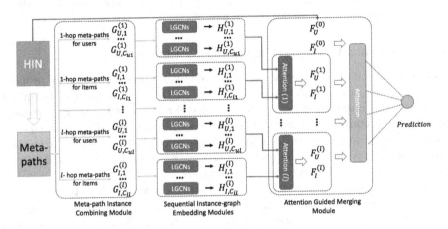

Fig. 2. The overall architecture of IGR.

2 Preliminaries

In this section, we introduce three definitions that are frequently used in this paper.

Definition 1. Heterogeneous Information Network. On a graph $\mathcal{G} = (\mathcal{V}, \mathcal{E})$, two functions $\phi : \mathcal{V} \rightarrow \mathcal{A}$ and $\varphi : \mathcal{E} \rightarrow \mathcal{R}$ can be defined. The functions map nodes and edges to node and edge types, respectively. A graph is a HIN if $|\mathcal{A}| + |\mathcal{R}| > 2$.

In this work, the entities such as the users, items, and auxiliary data are taken as nodes on HIN, and the relationships between entities are represented as edges.

Definition 2. Meta-path. A meta-path is a conceptual path that has limitations on the types of nodes and edges.

Meta-path can be represented in the form of $\mathcal{A}_0 \xrightarrow{\mathcal{R}_1} \mathcal{A}_1 \xrightarrow{\mathcal{R}_2} \cdots \xrightarrow{\mathcal{R}_l} \mathcal{A}_l$. For clarity, the meta-path can be abbreviated as $\mathcal{A}_0 \mathcal{A}_1 \cdots \mathcal{A}_l$. Such a meta-path contains l edges and $l + 1$ nodes, so it is called a l-**hop meta-path**. A meta-path instance belongs to this meta-path can be represented as $v_0 v_1 \cdots v_l$, where $v_0 \in \mathcal{A}_0, \cdots, v_l \in \mathcal{A}_l$.

Definition 3. Instance-Graph. In this work, we define instance-graphs as the layer-wise graphs which are generated by combining meta-path instances.

3 Proposed Method

Following the proposed framework, we propose an Instance-Graph based Recommendation method (IGR) which reveals and captures the complicated semantics presented by multiple instances. As shown in Fig. 2, IGR contains a meta-path instance combining module, several sequential instance-graph embedding modules, and an attention guided merging module.

3.1 Meta-path Instances Combining Module

To reveal the complicated semantics presented by multiple instances, we propose the meta-path instance combining module. This module combines the meta-path instances into layer-wise instance-graphs. For each meta-path, its instances are aligned according to their starting nodes, and the shared components are combined according to their positions. The generated instance-graphs are called layer-wise graphs because the nodes in the same position have the same type, and these nodes can be regarded as node layers.

Such a module has two advantages. First, it is appropriate to utilize the shared components to connect instances, since the shared components play similar roles in the same positions in different instances. Second, either a single instance or a combination of multiple instances is a subgraph of the corresponding instance-graph.

Besides, as the nodes are merged according to their positions, one node only appears once in a layer. With this premise, the instance-graphs can be built by generating node layers with nodes in the same positions and then linking edges between layers, which avoids the cost of listing all instances and merging them.

3.2 Sequential Instance-Graph Embedding Module

To extract information from the instance-graphs generated by the above module, we propose a sequential instance-graph embedding module. In this module, we treat every two adjacent layers in the instance-graphs as a directed graph. With the above premise, an instance-graph is regarded as a series of directed graphs that only contain two adjacent layers. To embed these two-layer directed graphs, we propose a Layer-wise Graph Convolutional Network (LGCN). Through stacking the LGCN layers, the directed graphs are sequentially embedded and form the embeddings of instance graphs.

Single LGCN Layer. For each directed graph, we apply one LGCN layer. Given a two-layer directed graph represented as $\mathcal{A}_0 \xrightarrow{\mathcal{R}_1} \mathcal{A}_1$ in an instance-graph, LGCN layer is defined as:

$$\mathcal{L}^{(1)} = (D^{(1)})^{-\frac{1}{2}} A^{(1)} (D^{(0)})^{-\frac{1}{2}}, \qquad (1)$$

$$H^{(1)} = Tanh(\mathcal{L}^{(1)} H^{(0)} W^{(1)} + b^{(1)}), \tag{2}$$

in which $A^{(1)} \in \mathbb{R}^{n_1 \times n_0}$ is the adjacent matrix between $\mathcal{A}_0$ and $\mathcal{A}_1$, $H^{(0)} \in \mathbb{R}^{n_0 \times d_0}$ and $H^{(1)} \in \mathbb{R}^{n_1 \times d_1}$ are the feature vectors of the start nodes and the end nodes, $W^{(1)} \in \mathbb{R}^{d_0 \times d_1}$ and $b^{(1)} \in \mathbb{R}^{1 \times d_1}$ are parameters of the dense layer. n represents the number of nodes on a layer, and d denotes the feature length. $D^{(0)}$ and $D^{(1)}$ are diagonal matrices, where $D_{ii}^{(0)} = \sum_{j=1}^{n_1} A_{ji}$ and $D_{jj}^{(1)} = \sum_{i=1}^{n_0} A_{ji}$.

The major difference between LGCN layers and general GCN layers is the definition of nodes. LGCN layers define nodes as start or end nodes and have a clear direction between two types of nodes. In GCN layers, the nodes simultaneously act as start and end nodes, making them hard to handle the node layers in the instance-graphs.

Stacking LGCN Layers. As a single LGCN layer transfers the information of the start nodes to the end nodes, LGCN layers can be stacked by taking the output of a layer as another layer's input. Through stacking LGCN layers along the instance-graphs, the instance-graph embeddings are generated. The instance-graph embeddings are in the form of collections of feature vectors.

As the instance-graph embeddings are generated for recommendation, we select meta-paths end with *User* or *Item*. Since meta-paths have different semantics, we apply different LGCNs for different instance-graphs. Following [1], we randomly block nodes with a probability p to avoid overfitting.

3.3 Attention Guided Merging Module

To evaluate the importance of instance-graphs for users/items, we propose an attention guided merging module. Given instance-graph embeddings, this module discriminatively evaluates and merges the instance-graph embeddings.

Suppose that there are C_{uk} k-layer instance-graph embeddings $H_{U,1}^{(k)}, \cdots, H_{U,C_{uk}}^{(k)} \in \mathbb{R}^{n \times d}$, they are rearranged into $\hat{H}_{U,1}^{(k)}, \cdots, \hat{H}_{U,n}^{(k)} \in \mathbb{R}^{C_{uk} \times d}$ which denote the instance-graph embeddings for users $u_1, \cdots, u_n$, respectively.

To evaluate the importance of different instance-graphs of the same length, we apply an attention layer on the rows of $\hat{H}_{U,i}^{(k)}$ since these rows are generated from instance-graphs of length k. The attention layer is inspired by [10] and defined as:

$$\alpha_{U,i}^{(k)} = Softmax(W_{U,2}^{(k)} tanh(W_{U,1}^{(k)} (\hat{H}_{U,i}^{(k)})^T)), \tag{3}$$

$$F_{U,i}^{(k)} = \alpha_{U,i}^{(k)} \hat{H}_{U,i}^{(k)}. \tag{4}$$

In this layer, the embeddings of instance-graphs of the same lengths are merged.

To further merge the embeddings $F^{(k)}$ into final user/item embeddings F, we apply another attention layer. Suppose the lengths of selected meta-paths are in $\{1, \cdots, l\}$, the attention layer is defined as follows:

$$\beta_{U,i}^{(k)} = Softmax(W_{U,4} tanh(W_{U,3} (F_{U,i}^{(k)})^T)), \tag{5}$$

Table 1. Statistics of three datasets and the meta-paths selected from them.

Dataset	User	Item	Interactions	Auxiliary data	Selected meta-paths
Frappe	957 Users	4,082 Apps	96,203	User−Nation, User−City	AU, NU, CU | UA UAU, UNU, UCU | AUA AUAU | UAUA, UNUA, UCUA
Last.fm	1,892 Users	17,632 Artists	92,834	User−User, Artist−Tag	AU, UU | UA, TA UAU, UUU | AUA, UUA AUAU, ATAU | UAUA
MovieLens 1M	6,040 Users	3,706 Movies	1,000,209	User−Occupation, Movie−Type	MU, OU | UM, TM UMU, TMU | MUM, OUM MUMU, MTMU | UMUM, UOUM

$$F_{U,i} = \sum_{k=0}^{l} \beta_{U,i}^{(k)} F_{U,i}^{(k)}. \tag{6}$$

In the above equations, $\alpha_{U,i}^{(k)} \in \mathbb{R}^{1 \times C_{uk}}, \beta_{U,i}^{(k)} \in \mathbb{R}^{1 \times l}$ are the weights generated by the attention layers. $W_{U,1}^{(k)} \in \mathbb{R}^{d' \times d}$, $W_{U,2}^{(k)} \in \mathbb{R}^{1 \times d'}$, $W_{U,3} \in \mathbb{R}^{d' \times d}$, $W_{U,4} \in \mathbb{R}^{1 \times d'}$ are attention parameters which are different for users and items. The embedding $F_i^{(0)}$ is the direct embedding of node i.

3.4 Training Details

To predict the preferences of users for items with the user/item embeddings, we apply bilinear operation on user and item embeddings $F_{U,i}, F_{I,j}$:

$$y = F_{U,i} Q(F_{I,j})^T, \tag{7}$$

in which $Q \in \mathbb{R}^{d \times d}$ is a diagonal parameter matrix. To train IGR, a pair-wise loss is applied, which is defined as:

$$\mathcal{L} = \sum -ln(sigmoid(y^+ - y^-)), \tag{8}$$

in which y^+ and y^- are the predicted scores of a pair of positive and negative samples. The user-item interactions are regarded as positive samples, and the negative samples are randomly sampled from items that have not been interacted by users.

3.5 Pre-training

As the instance-graphs have various semantics and are embedded through different numbers of LGCN layers, the LGCNs may disturb each other while training. To avoid this problem, IGR is pre-trained to stable the parameters in LGCNs. In detail, the parameters for the instance-graphs with different lengths are separately pre-trained, and then these parameters are taken as the initialization of the final IGR.

Table 2. Overall performance on three datasets.

Model	Frappe		Last.fm		MovieLens 1M	
	$HR@10$	$NDCG@10$	$HR@10$	$NDCG@10$	$HR@10$	$NDCG@10$
NIRec	0.708	0.561	0.776	0.631	0.706	0.430
CFM	0.698	0.567	0.835	0.682	0.716	0.440
GCMC	0.718	0.572	0.839	0.683	0.721	0.442
Ours	**0.741**	**0.587**	**0.842**	**0.687**	**0.726**	**0.446**

4 Experiment

In this section, we perform comprehensive experiments on three popular datasets: Frappe[1], Last.fm[2], and MovieLens 1M[3]. These datasets contain both auxiliary data and user history interactions with items. The statistics of these datasets and the meta-paths selected from the datasets are presented in Table 1.

For IGR, Adam [8] is applied as the optimizer. The keep probability of node dropout is set to 0.5, the learning rate is 10^{-3}, and the batch size for optimization is 1024 positive and negative sample pairs. In Frappe and Last.fm datasets, the direct embeddings of users/items $F^{(0)}$ take part in the recommendation, while these embeddings are not used in the MovieLens dataset. For a fair comparison, the length of embeddings of all methods is set to 64, and other parameters are tuned as the baselines proposed.

We select **NIRec** [7], **CFM** [11], and **GCMC** [1] as our baselines. For a fair comparison, all the baselines along with IGR are given same auxiliary data.

For evaluation, we adopt the widely used leave-one-out method [3,4]. The method holdouts the latest interactions of each user as the test set, the second latest interactions as the validation set, and the remaining interactions are the training set. For the Frappe and Last.fm dataset, since there is no timestamp information, we randomly select the interactions. For each positive sample in the test set, 99 items that do not interact with the user are randomly sampled to be negative samples. To evaluate the models, we apply two popular metrics HR and NDCG [6].

Following the discussion in [2], we get the number of training epochs from validation sets to get more convincing results. In the test sets, we further require both the positive samples and the negative samples to be unique, which is not required in [2,4].

4.1 Performance Comparison

The overall performance of IGR and all baselines are presented in Table 2. On three datasets, IGR consistently outperforms state-of-the-art methods on both

---

[1] http://baltrunas.info/research-menu/frappe.

[2] http://www.dtic.upf.edu/ocelma/MusicRecommendationDataset.

[3] https://grouplens.org/datasets/movielens/latest/.

Table 3. Ablation study on three datasets.

Model	Frappe		Last.fm		MovieLens 1M	
	$HR@10$	$NDCG@10$	$HR@10$	$NDCG@10$	$HR@10$	$NDCG@10$
Default	**0.741**	**0.587**	**0.842**	**0.687**	**0.726**	**0.446**
Remove attention	0.728	0.581	0.815	0.656	0.721	0.443
Remove 1-hop	0.725	0.578	0.790	0.644	0.689	0.412
Remove 2-hop	0.719	0.569	0.784	0.631	0.721	0.440
Remove 3-hop	0.725	0.576	0.795	0.640	0.722	0.441

$HR@10$ and $NDCG@10$ evaluation. By combining meta-path instances into instance-graphs and applying LGCNs to embed these graphs, IGR is capable of capturing the complicated semantics presented by multiple instances, so as to generate effective embeddings for recommendation.

4.2   Ablation Study

Since there are several components in IGR, we analyze their impacts via an ablation study. The performances of IGR and its four variants on three datasets are presented in Table 3. We introduce and analyze the variants respectively as follows:

- *Remove Attention.* Without the attention guided merging module, the instance-graph embeddings are directly rearranged and stacked to form the final user/item embeddings. This variant impairs the performance due to the imbalance between different embeddings.
- *Remove meta-paths of different lengths.* We find that different meta-paths are suitable for different datasets. Presumably, this is because the connections between nodes have different degrees of closeness on different datasets, which affects the power of meta-paths of different lengths.

5   Conclusion

In this work, we aim to take the complicated semantics presented by multiple instances into account for meta-path based recommendation. To achieve this goal, we propose a novel framework to utilize meta-paths. That is, guided by shared components, we combine the meta-path instances into layer-wise instance-graphs and embed these graphs for recommendation. Following the idea, we propose an Instance-Graph Based Recommendation method (IGR). In IGR, for each meta-path, its instances are combined into a layer-wise instance-graph. Then, we propose a sequential instance-graph embedding module to extract information from the instance-graphs. After that, we propose an attention guided merging module to evaluate the importance of the instance-graphs for users/items. Extensive experiments are conducted on three datasets, and the results show that IGR outperforms various state-of-the-art methods.

References

1. van den Berg, R., Kipf, T.N., Welling, M.: Graph convolutional matrix completion. In: KDD (2018)
2. Dacrema, M.F., Cremonesi, P., Jannach, D.: Are we really making much progress? A worrying analysis of recent neural recommendation approaches. In: RecSys, pp. 101–109 (2019)
3. He, X., Du, X., Wang, X., Tian, F., Tang, J., Chua, T.: Outer product-based neural collaborative filtering. In: IJCAI, pp. 2227–2233 (2019)
4. He, X., Liao, L., Zhang, H., Nie, L., Hu, X., Chua, T.S.: Neural collaborative filtering. In: WWW, pp. 173–182 (2017)
5. Hu, B., Shi, C., Zhao, W.X., Yu, P.S.: Leveraging meta-path based context for top-N recommendation with a neural co-attention model. In: SIGKDD, pp. 1531–1540 (2018)
6. Järvelin, K., Kekäläinen, J.: IR evaluation methods for retrieving highly relevant documents. In: SIGIR Forum, pp. 243–250 (2017)
7. Jin, J., et al.: An efficient neighborhood-based interaction model for recommendation on heterogeneous graph. In: SIGKDD, pp. 75–84 (2020)
8. Kingma, D.P., Ba, J.: Adam: a method for stochastic optimization. In: ICLR (2015)
9. Sun, Y., Han, J., Yan, X., Yu, P.S., Wu, T.: PathSim: meta path-based top-K similarity search in heterogeneous information networks. Proc. VLDB Endow. 4(11), 992–1003 (2011)
10. Vaswani, A., et al.: Attention is all you need. In: NIPS (2017)
11. Xin, X., Chen, B., He, X., Wang, D., Ding, Y., Jose, J.: CFM: convolutional factorization machines for context-aware recommendation. In: IJCAI 2019 (2019)

GCAN: A Group-Wise Collaborative Adversarial Networks for Item Recommendation

Xuehan Sun[1], Tianyao Shi[1], Xiaofeng Gao[1(✉)], Xiang Li[2], and Guihai Chen[1]

[1] Shanghai Key Laboratory of Scalable Computing and Systems,
Department of Computer Science and Engineering,
Shanghai Jiao Tong University, Shanghai, China
{Peter_suntain,sthowling}@sjtu.edu.cn,
{gao-xf,g-chen}@cs.sjtu.edu.cn
[2] Beijing University of Chemical Technology, Beijing, China
lixiang@mail.buct.edu.cn

Abstract. Recommendation System aims to provide personalized recommendation for different users. Recently, Generative Adversarial Networks based recommendation systems have attracted considerable attention. In previous research, GAN has shown potential and flexibility to learn latent features of users' preferences. However, GANs are hard to train to converge and waste many processes of fulfilling empty data, especially when meeting with the data sparsity problem.

In this paper, we propose a new group-wise framework, namely Group-wise Collaborative Adversarial Networks (GCAN) to solve the data sparsity problem and enable GAN to converge faster. We combine GAN with traditional collaborative filtering methods to generate recommendations (CAN), and then propose *binary masking* and *sample shifting* to achieve GCAN. Binary masking separates binary user-item interaction and abstracts group-wise relationship from these binary vectors, while sample shifting is designed to avoid incorrect learning process. A noise corruption parameter is then introduced with experiments to show the robustness of GCAN. We compare GCAN with other baseline methods on Yelp and SC dataset, where GCAN achieves the state-of-the-art performances for personalized item recommendation.

Keywords: Item recommendation · Adversarial networks · Group-wise recommendation

This work was supported by the National Key R&D Program of China [2020YFB 1707903]; the National Natural Science Foundation of China [61872238, 71722007, 71931001], the Huawei Cloud [TC20201127009], the CCF-Tencent Open Fund [RAGR 20200105], and the Tencent Marketing Solution Rhino-Bird Focused Research Program [FR202001].

© Springer Nature Switzerland AG 2021
C. S. Jensen et al. (Eds.): DASFAA 2021, LNCS 12683, pp. 330–338, 2021.
https://doi.org/10.1007/978-3-030-73200-4_23

1 Introduction

Recommendation system (RS) aims to provide accurate personalized recommendations for different users. Most of the recommenders learn the behavior of users through user-item interactions, among which *Collaborative Filtering* (CF) stands out as an effective method to abstract latent relationship [6]. In recent years, deep learning based RS has garnered considerable attention for its effectiveness to learn the latent features and achieve excellent performances [10]. Since the success of IRGAN [9], *Generative Adversarial Networks* (GAN) has shown its great potential to employ Collaborative Filtering on deep learning models while utilizing the advantage of both methods [1,8]. However, when applying GAN on some real-life purchase scenarios of recommendation system, this model exhibits drawbacks: When meeting with *data sparsity problem* in the user-item interaction matrix, GAN has been criticized for its long training process [1,4,7]. Since it is difficult to get users' specific preferences on every item, many of the user-item interactions are left empty [2]. When training GAN to make predictions on such items, generator needs to fulfill these empty data one by one repeatedly and the discriminator also needs to distinguish them in the similar way, significantly prolonging the training process.

In this paper, we propose a new scheme for solving these two problems in GAN based RSs, named *Group-wise Collaborative Adversarial Networks* (GCAN). With GAN as the basic building block, we initially apply collaborative filtering methods on GAN and construct *Collaborative Adversarial Networks* (CAN). Then we extend CAN to group-wise with two techniques, *binary masking* and *sample shifting*: i) Binary masking separates binary interactions and abstracts group-wise relationship from this binary vector. Then we mask the generated group-wise relation from binary values to ranking values to obtain the recommendation of the top-k items. ii) Sample shifting is used to avoid the generator learning incorrectly. Through each iteration, we collect items outside the group relation and assemble them as other groups. Besides, for items with empty data, we set their values close to but not exactly zero. Such methods prevent the group-wise vector from making the whole missing items as one group and learn the latent features of empty data effectively. To verify the robustness to noise corruption, we introduce a random noise parameter to the user-item interactions and compare the resistance of noise of CAN and GCAN in experiments.

Our main contributions are summarized as follows:

- To the best of our knowledge, this is the first group recommendation system that combines collaborative filtering with generative adversarial networks on group-level from the perspective of items.
- Group-level recommendations are further integrated to improve the performance of GAN-based methods. As a byproduct, the data sparsity problem in the training process and incompetence in predicting real purchase amount can also be alleviated.
- Extensive experiments are conducted on two large data sets. Quantitative and qualitative analyses justify the effectiveness and rationality of group-wise collaborative adversarial networks. GCAN also outperforms other deep

learning based RS methods and achieves state-of-the-art performances for personalized item recommendation.

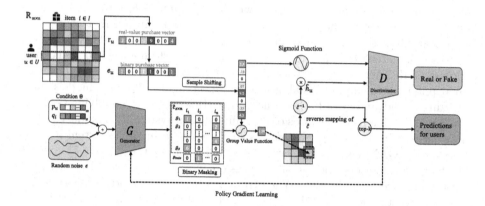

Fig. 1. Framework of GCAN.

2 Proposed Methodology

In this section, we propose *Collaborative Adversarial Networks* (CAN), introducing collaborative methods into Generative Adversarial Networks. To extend CAN to group-wise, we design two methods, *binary masking* and *sample shifting*, and propose *Group-wise Collaborative Adversarial Networks* (GCAN). Binary masking separate binary user-item interactions from the original matrix. Then, through group abstraction, we learn the group-wise relation from binary masking vectors. Sample shifting is proposed to promote the generator correctly learning the group-wise features. Then we introduce a noise parameter and further discuss the robustness of GCAN for noise corruption in Sect. 3.2 (Fig. 1).

2.1 Collaborative Adversarial Networks

Define r_{ui} as the implicit preference feedback by user u for item i, where $u \in$ user set U, and $i \in$ item set I. Each (u, i) pair represents the interactions between u and i. Hence, r_{ui} has a feedback value if there are interactions in (u, i), and is empty otherwise. Let $\mathbb{R} = (r_{ui})_{n \times m}$ form a sparse matrix. *Collaborative Adversarial Networks* (CAN) introduces traditional collaborative filtering methods into generative adversarial networks. Given the interaction matrix $\mathbb{R}_{m \times n}$, the target for CAN is to predict a accurate results $\hat{r}_{ui}$ for user u's preferences of item i. CAN then train the generator G and discriminator D through a minimax adversarial process:

$$Loss_{(G,D)}(r|\Theta) = \mathbb{E}_{r \sim P_{data}(r)} \left[\log D(r|\Theta) \right] + \mathbb{E}_{\hat{r} \sim P_{\phi}(\hat{r})} \left[\log(1 - D(\hat{r}|\Theta)) \right] \quad (1)$$

where $p_{data(r)}$ denotes the true distribution of r, $p_{\phi(\hat{r})}$ denotes the distribution of generating predictions, Θ denotes given condition parameters and $\Theta = \{p_u, q_i\}$. p_u is a n-dimensional vector for user u's interactions with items; q_i is a m-dimensional vector for item i's corresponding records with users. The corresponding loss function for discriminator D is defined in Eq. (2).

$$\theta^* \simeq \arg\max \frac{1}{K} \sum_{u,i} \Big(\log \sigma(f_\theta(r|\Theta)) + \log(1 - \sigma(f_\theta(\hat{r}|\Theta))) \Big) \tag{2}$$

where K is the size of the sampling set, $\sigma(\cdot)$ is the sigmoid function, $f_\phi(r|\Theta)$ represents the latent semantics learning function for generator G and is defined through the collaborative filtering forms by matrix factorization $f_\phi(r|\Theta) = b_{\phi,i} + b_{\phi,u} + p_u^t q_i$, where $b_{\phi,i} + b_{\phi,u}$ denotes the total bias for item i and user u respectively. And for generator G, we originally intend to optimize it by minimizing the difference between the generated distribution and ground-truth. Then we reformulate it through maximize optimization, which is implemented in Eq. (3).

$$\phi^* = \arg\max \frac{1}{K} \sum_{u,i} \log \Big((1 + \exp f_\theta(\hat{r}|\Theta)) \Big) \tag{3}$$

2.2 Extension to Group-Wise

To extend CAN to group-wise recommendation, we introduce a mapping vector $c_{s \times n}$ representing the mapping from group set $\mathbb{G} = \{g_1, g_2, \ldots, g_s\}$ to item set $\mathbb{I} = \{i_1, i_2, \ldots, i_n\}$. Each $g_l, 1 \leq l \leq s$ is an m-dimensional vector, inside which the corresponding value is a zero-one value: when item i belongs to a certain group g_l, corresponding value equals to 1 and when i does not belong, it equals to 0. Then we extend loss functions for D and G to group-wise through Eq. (4).

$$Loss_{(G,D)}(h|\Theta) = \mathbb{E}_h [\log \sigma(f_\theta(h|\Theta))] + \mathbb{E}_{\hat{h}} \Big[\log(1 - \sigma(f_\theta(\hat{h}|\Theta))) \Big] \tag{4}$$

where $\hat{h} = \hat{r}_u \hat{c}^T$. The maximum size of a certain group is controlled by Ω_g, which is $\max_l \|g_l\| \leq \Omega_g, 1 \leq l \leq s$ and $\|\cdot\|$ represents the L_2-norm. We use $binary\ masking$ method to obtain group-wise relation from binary purchase vector. Binary masking method extends original element binary vector to group-wise ranking matrix. To accelerate the generator and prevent it from producing trivial results, $sample\ shifting$ is used to guide G to correctly learning the latent features.

Binary Masking. To learn the relationship of group-wise, we propose $Binary$ $Masking$ (BM) methods. When the generator produce group mapping binary vector $c_{s \times n}$, we use the ranking number as the matrix value instead of just 0 or 1. For example, the purchase vector is $\{0.81, 45.7, 69.0, 2.15\}$ and the relevant binary vector is $\{0, 1, 1, 0\}$. Then inside the group-relation vector, we mask the original $\{0, 1, 1, 0\}$ as $\{0, 1, 2, 0\}$. Binary masking methods represent the relation inside a certain group. When use the reverse mapping of $\hat{c}$, we sorted items inside group g_i by their ranking value. The values for the group matrix is calculated by $c_{li} = \frac{\sum_{i \in g} r_{ui} \odot H(g_l)}{\sum_l H(g_l)}$, where $H(\cdot)$ is the Heaviside Step Function.

Sample Shifting. When training generator G through the adversarial process, *trivial mapping vectors* can be generated, such as mapping all items into the same group g_l or mapping them all not belonging to g_l. Denote the whole item set as I, we use I_{zero} to represent items with no interaction with user u. We reconstruct part of the I_{zero} as values close to 0 but not exactly 0. Such reconstruction will not effect the final recommendation of GCAN, since their group values are small (close to 0), but this methods will prevent the generator from producing 0-value vectors, and promote the convergence process. We reconstruct items' values with $e_{ui} = 1$ to rank them better. The sample rate for changing values is α_1. Another kind of Sample Shifting is used to ensure that in each iteration, there will be enough new groups. During each iteration, we collect items $i \in I$ at rate α_2 and make them as a new independent groups. Hence, G avoid producing useless predictions, and concentrate on mapping non-purchased items to accurate recommendation:

$$
\begin{aligned}
Loss_{(G,D)} =& \mathbb{E}_h \left[\log \sigma(f_\theta(h|\Theta)) \right] + \mathbb{E}_{\hat{h}} \left[\log(1 - \sigma(f_\theta(\hat{h}|\Theta))) \right] \\
&+ \beta \sum_j ||\hat{h}_{(u,j)} - h_{(u,j)}||^2 + \gamma ||\hat{g}_l - g_l||^2
\end{aligned}
\tag{5}
$$

2.3 Noise Tolerance

A robust recommendation system should be insensitive to small corrupting random noise. To evaluate the robustness of GCAN, we introduce another parameter ϵ representing for random noise and further denote $D(\hat{h}|\Theta + \epsilon)$ for the probability. Hence, the total loss function of GCAN is:

$$
\begin{aligned}
Loss_{(G,D)} =& \mathbb{E}_h \left[\log D(h \odot e|\Theta + \epsilon) \right] + \mathbb{E}_{\hat{h}} \left[\log(1 - D(\hat{h} \odot \hat{e}|\Theta + \epsilon)) \right] \\
&+ \beta \sum_j ||\hat{h}_{(u,j)} - h_{(u,j)}||_2^2 + \gamma ||\hat{e_u} - e_u||_2^2
\end{aligned}
\tag{6}
$$

where β is the coefficient for controlling the importance of constructing shifted samples for item j and γ is the coefficient for controlling binary masking vectors, $||\epsilon|| \leq \Omega_\epsilon$, Ω_ϵ denotes the upper bound to control the maximum noise. The influence of noise will be further discussed in Sect. 3.2 about the robustness of GCAN.

3 Experiments

3.1 Experiments Settings

Data Sets. We conduct our experiments with two large datasets: SC and Yelp, the characteristics of which are described in Table 1. SC is an implicit feedback commercial dataset constructed by Sihailvcang, containing over 1,100,000 users' purchasing records, especially in fruits and vegetables. Yelp is a dataset consisting of different users' ratings on commercials. We filter out those users and items with less than 10 interactions.

Metrics. We use two metrics for performance evaluating: *Hit Ration* and *Normalized Discounted Cumulative Gain*. $HR@N = \frac{Hits}{TestNums}, NDCG@N = \sum_{p=1}^{N} \frac{DCG}{IDCG}$ where $Hits$ represents the hitting results, $TestNums$ represents the overall number, $DCG, IDCG$ represents the *discounted cumulative gain* (DCG), normalized DCG at position p which are calculated by: $DCG = \sum_{i=1}^{p} \frac{2^{rel_i}-1}{\log_2(i+1)}, IDCG = \sum_{i=1}^{|REL|} \frac{2^{rel_i}-1}{\log_2(i+1)}$ where rel_i is the graded relevance of the result at position i, $|REL|$ is the list of relevant items in the corpus up to position p.

Table 1. Descriptions of the experimented datasets

Dataset	# Interaction	# User	# Item	Sparsity
SC	97,638	1,629	462	95.73%
Yelp	730,790	25,677	25,815	99.89%

Baselines. We compare the performances of the proposed methods with the following methods:

BPR [5]: *Bayesian Personalized Ranking (BPR)* is a pairwise recommender model for item ranking which takes use of users' preferences for purchased and non-purchased item pairs. BPR assumes that in implicit feedback, clear observations should be attached more importance than unobserved ones.

SVD [3]: SVD is the basic model for matrix factorization. It maps user-item interaction into a joint latent semantic space along with user and item relative bias. It also evolves into some other models, like SVD++ and Time-Aware SVD. Here we use the traditional formation of SVD as the baseline.

IRGAN [9]: This method trains two models through adversarial training, a generative one G and a discriminative D. We implement IRGAN by the code released by the authors on github[1]. Policy-gradient based reinforcement learning is applied to updating G.

RAGAN [1]: This method is a Rating Augmentation framework based on Generative Adversarial Networks. It adds negative items to avoid biases to high ratings through one-class collaborative filtering. We set the hyper-parameters the same as is mentioned in their papers [1].

---

[1] https://github.com/geek-ai/irgan.

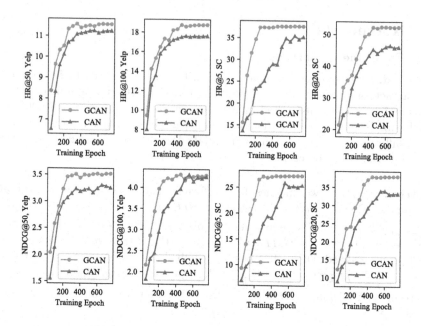

Fig. 2. Learning curves of CAN and GCAN

3.2 Results

Training Process. Figure 2 shows the learning curves generated by CAN and GCAN on Yelp. Compared to CAN methods, GCAN outperforms in two aspects: i) GCAN converges more quickly in around 300 epochs while CAN does not meet the trend of convergence until 500 epochs. ii) GCAN exceeds CAN in both HR@K and NDCG@K. Especially, GCAN achieves better performances on smaller data sets with the effect of group-wise training. There is an interesting observation that when N for top-K recommendation enlarges, the benefit of GCAN also becomes more obvious, which may arouse from users' specific tastes for groups of items.

Table 2. The impact of random noise on CAN and GCAN on HR.

	epsilon = 0.3		epsilon = 0.5		epsilon = 1.0	
	CAN	GCAN	CAN	GCAN	CAN	GCAN
Yelp	−6.7%	−3.3%	−8.3%	−5.1%	−14.9%	−7.2%
SC	−17.3%	−5.7%	−21.5%	−15.6%	−52.1%	−29.4%

Robustness to Noise Corruption. GCAN has a wider item range and a more accurate purchase prediction for group-wise vectors, thus enabling the training process more robust to noise. Table 2 reports the impact of random noise of GCAN and CAN. The results suggest that GCAN is less sensitive to random noise when compared with CAN. For example, GCAN exceeds CAN with 5.7% on SC compared with CAN'S 17.3%. This justifies that GCAN is more robust to random noise which is an important feature indicating the good generalization ability of our model.

3.3 Comparisons with Other Methods

We now compare the accuracy of GCAN with other baselines mentioned in Sect. 3.1. Table 3 illustrates the results for two datasets on HR@5, HR@20, NDCG@5, and NDCG@20. Our model GCAN achieves the best performances in most cases. For most recommendation, GCAN is able to recommend the user's preferred ones and rank them in correct order. Only on Yelp NDCG@20, GCAN achieves 4.29% compared to AMF's 4.37%. In all the other cases, GCAN outperforms the other baselines. Hence, we believe GCAN is a stat-of-the-art method for personalized item recommendation. The improvement of GCAN from other models is relatively remarkable. RAGAN is a newly proposed method, specially designed through rating augmentation to cope with the data sparsity problem in item recommendation. GCAN outperforms it by 6.25% on average, which is a remarkable increase allowing for the size of the user-item matrix. Furthermore, GCAN is modeled by only introducing group-wise purchase vectors, which is a rather easy implementation than what is proposed in RAGAN.

Table 3. Comparisons with baselines for recommendation performances under HR and NDCG. The best result of each column is highlighted in bold font formation. * indicates the best results of all previous methods. The last column IR represents the average improving rates of GCAN over other existing models.

	Yelp				SC				IR
	HR@5	HR@20	NDCG@5	NDCG@20	HR@5	HR@20	NDCG@5	NDCG@20	
BRP	0.0706	0.1359	0.0295	0.0384	0.2563	0.4132	0.1953	0.2763	35.38%
SVD	0.1074	0.1565	0.0244	0.0379	0.3193	0.4463	0.2451	0.3015	19.61%
IRGAN	0.1125*	0.1468	0.0326	0.0426	0.3528	0.4596	0.2542	0.332	10.14%
RAGAN	0.1119	0.1775*	0.0315	0.0396	0.3568	0.5015*	0.2638*	0.3514	6.25%
GCAN	**0.1156**	**0.1877**	**0.0351**	0.0429	**0.3767**	**0.5225**	**0.2727**	**0.3797**	–

4 Conclusions

In this paper, we proposed a new framework for item recommendation inspired by the group distribution feature of items. We identified the group distribution feature of user-item interaction and point out the hardness of embedding group-wise vector into adversarial networks. To cope with the problem, we proposed group-wise embedding with negative sampling and binary masking to avoid adversarial networks from over-fitting and converging too long. Our work achieves state-of-the-art methods on two large real-life datasets with metrics of Hit Ratio and Normalized Discounted Cumulative Gain.

For future work, we plan to employ the inspiration of group-wise to other recommendation systems, such as transfer learning and neural machines, which can cope with a wide range of common problems in recommender scenarios, such as cold-start, implicit feedback and so on. The challenge here is how to properly embedding the input data and correctly training them in a fast-convergence way. Lastly, it is still worth mentioning that our proposed methods group-wise embedding not only specifically used in item recommendation, and it is also widely used in other information retrieval areas, such as web search, text retrieval, question answering and so on. The potential of GCAN on these areas still needs exploring and we will further dig into these areas on extending GCAN to these fields.

References

1. Chae, D.K., Kang, J.S., Kim, S.W., Choi, J.: Rating augmentation with generative adversarial networks towards accurate collaborative filtering. In: The World Wide Web Conference (WWW), pp. 2616–2622 (2019)
2. Kim, J., Monteiro, R.D., Park, H.: Group sparsity in nonnegative matrix factorization. In: IEEE International Conference on Data Mining (ICDM), pp. 851–862 (2012)
3. Koren, Y., Bell, R., Volinsky, C.: Matrix factorization techniques for recommender systems. Computer **8**, 30–37 (2009)
4. Li, S., Kawale, J., Fu, Y.: Deep collaborative filtering via marginalized denoising auto-encoder. In: ACM International on Conference on Information and Knowledge Management (CIKM), pp. 811–820 (2015)
5. Rendle, S., Freudenthaler, C., Gantner, Z., Schmidt-Thieme, L.: BPR: Bayesian personalized ranking from implicit feedback. In: The Conference on Uncertainty in Artificial Intelligence (UAI), pp. 452–461 (2009)
6. Ricci, F., Rokach, L., Shapira, B., Kantor, P.B.: Recommender Systems Handbook, 2nd edn. (2015)
7. Tian, B., Zhang, Y., Chen, X., Xing, C., Li, C.: DRGAN: a GAN-based framework for doctor recommendation in Chinese on-line QA communities. In: Li, G., Yang, J., Gama, J., Natwichai, J., Tong, Y. (eds.) DASFAA 2019. LNCS, vol. 11448, pp. 444–447. Springer, Cham (2019). https://doi.org/10.1007/978-3-030-18590-9_63
8. Wang, H., et al.: GraphGAN: graph representation learning with generative adversarial nets. In: Association for the Advancement of Artificial Intelligence (AAAI) (2018)
9. Wang, J., et al.: IRGAN: a minimax game for unifying generative and discriminative information retrieval models. In: ACM International Conference on Research and Development in Information Retrieval (SIGIR), pp. 515–524 (2017)
10. Zhang, S., Yao, L., Sun, A., Tay, Y.: Deep learning based recommender system: a survey and new perspectives. ACM Comput. Surv. (CSUR) **52**(1), 5 (2019)

Emerging Applications

PEEP: A Parallel Execution Engine for Permissioned Blockchain Systems

Zhihao Chen[1], Xiaodong Qi[1], Xiaofan Du[1], Zhao Zhang[1,2], and Cheqing Jin[1(✉)]

[1] School of Data Science and Engineering, East China Normal University, Shanghai, China
{chenzh,xdqi,xfdu}@stu.ecnu.edu.cn, {zhzhang,cqjin}@dase.ecnu.edu.cn
[2] Guangxi Key Laboratory of Trusted Software, Guilin University of Electronic Technology, Guilin, China

Abstract. Unlike blockchain systems in public settings, the stricter trust model in permissioned blockchain opens an opportunity for pursuing higher throughput. Recently, as the consensus protocols are developed significantly, the existing serial execution manner of transactions becomes a key factor in limiting overall performance. However, it is not easy to extend the concurrency control protocols, widely used in database systems, to blockchain systems. In particular, there are two challenges to achieve parallel execution of transactions in blockchain as follows: (i) the final results of different replicas may diverge since most protocols just promise the effect of transactions equivalent to *some* serial order but this order may vary for every concurrent execution; and (ii) almost all state trees that are used to manage states of blockchain do not support fast concurrent updates. In the view of above challenges, we propose a parallel execution engine called **PEEP**, towards permissioned blockchain systems. Specifically, PEEP employs a deterministic concurrency mechanism to obtain a predetermined serial order for parallel execution, and offers parallel update operations on state tree, which can be implemented on any radix tree with Merkle property. Finally, the extensive experiments show that PEEP outperforms existing serial execution greatly.

Keywords: Blockchain · Permissioned · Execution optimization

1 Introduction

Blockchain technology provides data integrity, transparency and immutability to tackle trust problems among untrusted parties. For different goals, a number of blockchain systems have been implemented in both permissionless [12,17] and permissioned [1,4] environment respectively. Unlike blockchains in public settings, which are more concerned about the security and reliability at the cost of performance, permissioned blockchains, instead, seek higher throughput under stricter assumptions. For example, energy-efficient consensus protocols such as PBFT [6] or Raft [13] achieve much better performance among known and static

© Springer Nature Switzerland AG 2021
C. S. Jensen et al. (Eds.): DASFAA 2021, LNCS 12683, pp. 341–357, 2021.
https://doi.org/10.1007/978-3-030-73200-4_24

participants. With faster consensus, the execution of transactions takes dominion over the system performance beyond other factors. To corroborate our point, we have conducted experiments on PBFT consensus and transaction execution within four replicas in a private network. The results show that the performance of PBFT can reach over 15K TPS while the execution merely handles about 8K read/write operations per second. There is a big gap between them, which comes into our view.

Conventional blockchains commonly adopt sequential execution where transactions are executed one-by-one, to entail a consistent final state among all replicas. However, such sequential scheme cannot fully utilize the modern multi-core machines since most processors sit idle, yielding a low utilization of system resources and poor performance. Therefore, there are great potential benefits to empower concurrency over blockchains, but the core problem lies in that the effect of concurrent execution on different replicas may vary under most existing concurrency controls including both pessimistic and optimistic controls.

Some works try to incorporate concurrency in blockchain, such as Fabric [4] and its optimized version [15]. Fabric follows an *execute-order-validate* (EOV) architecture, where transactions are pre-executed on specified peers called *endorsers* and then proposed along with simulation results to ordering service for consensus. Next, peers commit each transaction after consensus if the validation for it passes. The simulations for all transactions are performed on endorsers concurrently. On the contrary, the validations for transactions still follow the serial manner, which couldn't fulfill the performance gap between ordering and execution. Therefore, the EOV architecture is not our focus.

In this paper, we present **PEEP**, a parallel execution engine exploiting concurrent execution of blockchain under *order-execute* architecture used in Tendermint [5]. More specifically, PEEP executes transactions following a two-phase design. Once a block is agreed by consensus, the first phase executes transactions concurrently with the coordination of a schedule layer, then the second phase applies the updates produced by the first phase to the underlying state trie. The state trie, i.e. Merkle Patricia trie (MPT) [17] in Ethereum, is a special Merkle tree used to manage states accessed by transactions. Surely, PEEP also parallelizes the updating on state trie to further improve the performance in contrast to existing sequential updating on Merkle trees. In addition, the commitment of state trie, including hash recalculation and node persistence, blocks the entire workflow, making higher throughput impossible. To eliminate this blockage, PEEP parallelizes this process to the consensus for next block and further increases the utilization of various system resources.

In summary, this paper makes the following contributions:

- We propose PEEP, a novel execution engine towards blockchain, which allows each replica to execute transactions independently and concurrently while promises deterministic serial order.
- We decouple the updating on state trie from execution and parallelize it separately. Specifically, we propose a lock-free parallel update algorithm and realize the non-blocked workflow by a *deferred commit* strategy to achieve better utilization of resources.

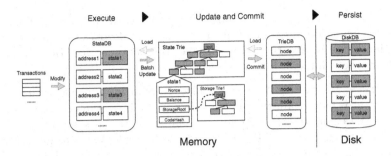

Fig. 1. Example of transaction execution flow.

– We implement a prototype for PEEP integrating above techniques and conduct comprehensive experiments. The results show that PEEP outperforms the serial and blocked execution significantly.

The rest of this paper is structured as follows. First, Sect. 2 introduces background and related works. Then, Sect. 3 overviews the design of PEEP, and Sect. 4 details the deterministic concurrent execution. Section 5 presents the parallel update algorithm and deferred commit strategy. Last, Sect. 6 reports thorough evaluations before Sect. 7 concludes this paper.

2 Background and Related Work

In this section, we provide necessary backgrounds and survey some related works.

Transaction Execution Flow. Figure 1 illustrates the detailed execution flow for transactions in the block agreed by a quorum of system. During the execution of every transaction, each replica reads/writes the latest states to a temporary memory cache called **StateDB**, which is built on top of the tries to reduce duplicate memory operations. This cache loads missed states from the underlying trie and will be reset per block. After all transactions are executed, the changes of states, denoted by green squares, are flushed to the trie in a batch. Subsequently, the hashes of dirty nodes are recalculated to generate a root hash that characterizes the global world state. Note that tries are logically organized in memory, so every dirty node on tries should be stored as a separate record to underlying key-value store for persistence.

Optimized Execution. Some works are proposed recently to enhance the performance of serial execution towards blockchain. A typical way is to exploit a two-phase execution framework [7,14] to parallelize the execution. In the first phase, the leader (or miner) executes transactions concurrently in an arbitrary order, and the other replicas repeat this concurrency based on the scheduling information forwarded by miner. ParBlockchain [2] adopts an order-execute paradigm and utilizes a dependency graph to keep determinism while leveraging transaction parallelism. Fabric and its variants follow a new paradigm, where transactions are first simulated concurrently and then each peer validates the results of simulation after consensus. Fabric++ [15] decreases the abort ratio in validation phase by aborting unserializable transaction earlier. However, in this paradigm, the validation is still performed serially.

State Trees. Merkle tree is the first choice to organize state data with authentication towards blockchain where fault tolerance [9] is guaranteed under a potentially hostile environment. Therefore, a number of variants are proposed, such as Merkle Bucket Tree [4], Merkle B+ tree and Merkle AVL tree [5]. Compared with them, trie with Merkle properties is more popular in blockchain systems for three reasons: (i) less space overhead, the keys of records are hidden in the path; (ii) structural invariance, the final structure of trie is just determined by inserted records but not the order; and (iii) multi-version store, all versions of every record are saved. The Merkle Patricia trie (MPT) [17] in Quorum and Libra Sparse Merkle trie (SMT) [3] are two typical implementations, which both persist each node of trie to a key-value store such as LevelDB or RocksDB. Additionally, to the best of our knowledge, few works explore concurrency on Merkle trees. Angela [10], as one of the few, proposed a highly concurrent Merkle tree through the finer grain conflict locking scheme.

3   System Overview

In this section, we overview the proposed execution engine PEEP, which achieves both parallelism and determined serializable order. We first present assumptions about our target system and then depict the architecture of PEEP.

3.1   Assumptions

The PEEP works in a blockchain system with three basic assumptions as follows.

Early Read/Write-Set Acquisition. The read/write keys of transactions handled by PEEP can be known in advance before execution by some techniques such as simulation or static analysis. This information of transfer transaction can be obtained easily, i.e., just the *from* and *to* address. However, it becomes complex if we want to gain the keys of states accessed by smart contract written in turing-completed language without any execution. We will discuss this case if this assumption doesn't hold in Sect. 4.3.

Order-Execute Workflow. Different from Fabric, PEEP handles blocks in an order-execute manner without a simulation phase. For the ordering service, PEEP needs the blockchain to employ a deterministic consensus protocol (i.e., no forks), such as the PBFT [6]. The transactions of each block are only executed after consensus on all replicas including the leader of PBFT. In some blockchains adopting PBFT, the leader executes transactions in every block before consensus to obtain the digest of world state and the other replicas execute transactions repeatedly to decides their votes during consensus. Instead, we move execution

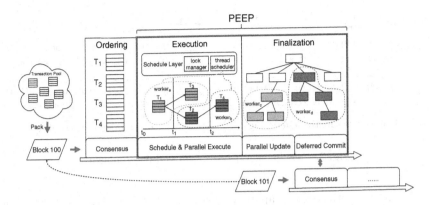

Fig. 2. System architecture.

to the end and allow some invalid transactions in the block to achieve better parallelism, which is elaborated in Sect. 5.2.

Adversary Model. The adversary model of PEEP is the same as PBFT consensus protocol, where at most f replicas can be malicious out of n replicas, such that $n \geq 3f + 1$. The faulty replicas can misbehave in arbitrary ways, including broadcasting conflict messages, keeping silent or even colluding.

3.2 Architecture

Figure 2 shows the workflow of our system, including three phases: ordering, execution and finalization. The latter two phases consist of our proposed PEEP. In the execution phase, the *schedule layer* coordinates the parallel execution of transactions, including the lock management for states and thread scheduling. Particularly, the schedule layer employs the ordered locking mechanism [16] to eliminate the non-determinism from parallel execution, resulting in identical results every time. In the finalization phase, the updates outputted by execution are applied to state trie in parallel to maximal the capacity of multi-core hardware. Note that the structural invariance of state trie is utilized to support parallel write operations. Besides, to make the entire workflow not be blocked by the commit of state tree, PEEP performs the commit of current block and consensus for next block in parallel, which is regarded as the deferred commit mechanism. More Specifically, PEEP distinguishes itself with the following three main techniques.

First, the schedule layer allows parallel execution with deterministic serial order but incurs no additional network communication among replicas for negotiation of the final result. Therefore, each replica can execute transactions in parallel independently. **Second**, PEEP accelerates the serial updates on state tree significantly by a parallel update algorithm. **Last**, PEEP prevents the commit of state tree from blocking the entire workflow by deferring this heavy process to consensus for next block. Therefore, various system resources of each

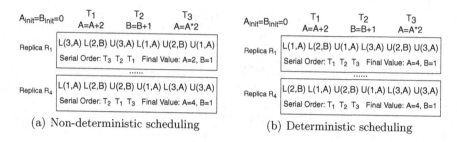

(a) Non-deterministic scheduling   (b) Deterministic scheduling

Fig. 3. Case for non-determinism and determinism.

replica are fully utilized, i.e., apart from the computational resources, the commit of state tree and consensus for block consume the disk I/O and network I/O respectively. To achieve this goal, PEEP needs to modify the structure of blocks slightly. However, we think this modification is acceptable and will detail it further in Sect. 5.2.

Summarized, the design of PEEP is a novel thus practical approach to improving overall performance towards permissioned blockchain systems, by concurrent execution of transactions, parallel update on state tree and deferred commit strategy.

4   Scheduling and Execution

4.1   The Case for Determinism

In blockchain, a crucial feature of parallel execution is deterministic, which assures consistency among multiple replicas. We further distinguish the non-deterministic and deterministic scheduling as depicted in Fig. 3. Three transactions T_1, T_2, T_3 are executed concurrently on four replicas $R_1 \sim R_4$, $L(i, s)$ indicates T_i acquires the lock on state s and $U(i, s)$ represents T_i releases the lock on state s. Figure 3(a) shows the results of an non-deterministic lock scheme, two-phase locking (2PL), while in Fig. 3(b), replicas leverage deterministic ordered locking to execute transactions. As can be seen, in Fig. 3(a), the execution result of R_1 is equivalent to serial order $\langle T_3, T_2, T_1 \rangle$ rather than $\langle T_1, T_2, T_3 \rangle$ defined by block, since T_3 acquires the lock on A before T_1 due to the unpredictable behaviour of thread scheduling. Furthermore, the result of R_1 diverges from that of R_4, yielding inconsistent states. On the contrary, Fig. 3(b) guarantees the determinism through an additional rule that the lock requested by multiple transactions should be strictly granted to them in the predefined order. Following this rule, T_1 always acquires lock on A before T_3 on all replicas, promising the determinism.

4.2   Parallel Execution

PEEP designs a schedule layer to take charge of scheduling transactions and producing final dirty states. The schedule layer contains two core components: (i)

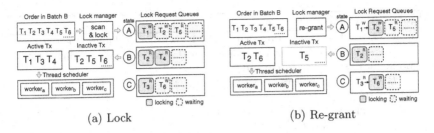

(a) Lock   (b) Re-grant

Fig. 4. Example of lock manager with ordered locking scheme.

lock manager, which grants locks to transactions following the ordered locking scheme; and (ii) thread scheduler, which coordinates the execution of transactions, such as worker thread assignment and management. The functions of both components are detailed as follows.

Lock Manager. The lock manager is implemented on a single thread, which grants locks to transactions according to the predefined order. At the beginning, all transactions stay inactive. Then, the lock manager scans inactive transactions and grants currently available locks to them based on their read/write-set. Once a transaction gathers all of its required locks, it becomes active and can be delivered to the buffer of the thread scheduler for execution. After a transaction finishes its execution, all its acquired locks are released for re-granting to other inactive transactions.

Thread Scheduler. The thread scheduler sets up multiple worker threads to execute transactions concurrently. The scheduler will assign every transaction in the buffer to an idle worker thread for execution. During the execution, the worker thread may abort a transaction if it encounters illegal access or error behaviors. This behavior is guaranteed to be identical among all replicas. After successful execution, the worker thread writes the updates of transactions to a state buffer, which will be flushed to state trie in a batch later.

Algorithm 1 describes pseudo-codes of the scheduler layer. When the scheduler layer receives a batch of transactions B, it forwards B to the lock manager for execution (lines 1–9). At the beginning, all transactions are put in the set S of inactive transactions and cannot be executed directly (line 2). In particular, the lock manager is single-threaded, which ensures the ordered lock granting. Then, the lock manager grants locks on transactions serially (lines 3–8). When dealing with a transaction t, the lock manager tries to grant locks on states of t's read-set and write-set by function $grant\_locks(\cdot)$ obeying the ordered locking scheme. The function $grant\_locks(\cdot)$ returns two variables cnt and $records$, namely the number of un-acquired locks and information of locks $info$ (line 5). Next, the lock manager updates the information by function $update\_locks\_information(\cdot)$, registering available locks to t (line 6). Last, if t obtains all locks on the states of its read/write-set (i.e., $cnt = 0$), t becomes active and is shifted from S to R for parallel execution (lines 7–9).

Algorithm 1: Schedule Layer

Input: B: batch of transactions with prior knowledge, R: buffer of active
transactions, W: collection of worker threads, D: buffer of dirty states

1 ▷ **Lock manager (single thread):**
2 $S \leftarrow B$ /*set of inactive transactions */
3 **for** $t \in S$ **do**
4   /*cnt: number of locks not acquired */
5      $cnt, info \leftarrow grant\_locks\,(t)$
6      $update\_locks\_information\,(t, info)$
7   **if** $cnt = 0$ **then**
8         $S \leftarrow delete(t)$
9         $R \leftarrow add\,(t, R)$

10 ▷ **Thread scheduler:**
11 **for** $w \in W$ **do**
12   /*run in parallel */
13      $t \leftarrow get\_transaction\,(R)$
14      $err, write\_set \leftarrow execute\,(t)$
15   **if** $err \neq null$ **then**
16         $abort\,(t)$
17         $release\_locks\,(t)$
18   **else**
19   **for** $\langle k, v \rangle \in write\_set$ **do**
20            $D \leftarrow write\_state\,(k, v, D)$
21         $release\_locks\,(t)$

All worker threads in the thread scheduler constantly try to fetch active transactions to execute (lines 10–21). For each worker thread w, if it fetches transaction t from buffer R successfully, w will execute t directly based on t's read-set. Afterward, PEEP writes the updates of transactions into a buffer D of dirty states first and then flush them to state trie for the parallel update. Therefore, during the execution, the worker thread may have to read values of states from both D and state trie. In this phase, since there are only read operations on state trie, all worker threads can access state trie concurrently with no contention. If t is aborted due to some reasons, such as transaction logic or mismatch between pre-acquired and real read/write-set, w will abort t and then release all relevant locks (lines 15–17). If the execution for t succeeds, the worker thread writes updates to buffer D, which is later flushed to state trie in a batch (lines 19–21).

Figure 4 gives an example to further elaborate Algorithm 1. The lock manager sets a queue for each state and serially push a transaction to queues respecting to states what it requests. When a lock is available, the lock manager grants it to the front transaction of the corresponding queue. More specifically, exclusive lock is granted to the write request while successive read requests can be granted shared locks. In Fig. 4(a), as for T_1, it has been granted the lock on state A. Based on this figure, T_1, T_3 and T_4 have acquired all of their locks and thus become active for scheduling. On the other hand, T_2, T_5 and T_6 remain inactive and wait for re-granted locks. After successful execution of T_1 and T_3, locks on them are released and re-granted to T_2 and T_6 by the lock manager, as shown in Fig. 4(b), which turns T_2 and T_6 into active.

4.3 Speculative Execution

If the assumption about early read/write acquisition doesn't hold, we still can use a speculative deterministic execution algorithm to execute transaction concurrently as proposed in Sparkle [11]. The basic principle obeys that the effects produced by the speculative deterministic execution must be always equivalent to the serial order defined by block. A transaction executed speculatively will be aborted and re-executed if its execution violates the predetermined order.

5 Parallel Update and Commit on State Trie

5.1 Parallel Update

The concurrent execution writes the latest states (also called dirty states) to the in-memory StateDB in the key-value form, and these dirty states are then flushed into state trie for digest calculation and persistence. However, the existing state tries apply updates in a completely serial manner, where the states are handled one after another. Instead, to enable parallel update on state trie, two main issues should be considered. (i) The nodes of state trie closer to the root face more conflicts, which is hard to deal with when handling updates concurrently. (ii) There's no requirement to optimize this processing because the bottleneck of system doesn't lie in the overhead for update operations on in-memory state trie but the cost of disk I/O for cache missing caused by node loading from underlying DiskDB as illustrated in Fig. 1.

For the first issue, a straightforward solution is to lock the entire state trie for each update. However, this approach is the same as the serial manner. Another practical idea is to design a locking mechanism based on the finer node granularity. Nevertheless, due to the existence of dependency between a node and its children, the nodes closer to the root will suffer from a higher race if multiple threads apply updates concurrently. Regarding the second issue, it is reasonable to assume that we maintain the entire latest state trie in memory since replicas in permissioned blockchain are more likely equipped with larger memory resources. For example, an MPT managing 1 million accounts with the latest states occupies merely about 200 MB memory space and a few Gigabyte space is enough for more than 10 million accounts. Note that we just need to cache the latest version of each state in memory and the historical versions are still persisted on DiskDB.

Based on the memory assumption, PEEP employs a lock-free parallel update algorithm on state trie to improve its performance. We decouple the read and

---

Algorithm 2: Parallel update on state trie

Input: ST: state trie, S: batch of dirty states, N_t: number of worker threads,
  W: collection of worker threads, M: map between node and list of tasks

1 ▷ **Stage1: task distribution**
2 $cnt \leftarrow 0$
3 **for** $\langle k, v \rangle \in S$ **do**
4     $idx \leftarrow cnt$ mod N_t, $cnt \leftarrow cnt + 1$
5     $task\_assign\,(W[idx], \langle k, v \rangle)$

6 ▷ **Stage2: parallel descending**
7 **for** $w \in W$ **do**
8   /*run in parallel */
9     $task\_set \leftarrow get\_tasks(w)$
10   **for** $\langle k, v \rangle \in task\_set$ **do**
11        $node_c \leftarrow find\_conflict\_node(ST, \langle k, v \rangle)$
12        $l_t \leftarrow M[node_c]$, $l_t \leftarrow append(l_t, \langle k, v \rangle)$, $M[node_c] \leftarrow l_t$

13 ▷ **Stage3: node distribution**
14 $cnt \leftarrow 0$
15 **for** $\langle node_c, l_t \rangle \in M$ **do**
16     $idx \leftarrow cnt$ mod N_t, $cnt \leftarrow cnt + 1$
17     $node\_assign(W[idx], \langle node_c, l_t \rangle)$

18 ▷ **Stage4: parallel node modification**
19 **for** $w \in W$ **do**
20   /*run in parallel */
21     $node\_set \leftarrow get\_nodes(w)$
22   **for** $node_c \in node\_set$ **do**
23        $l_t \leftarrow M[node_c]$
24   **for** $\langle k, v \rangle \in l_t$ **do**
25           $inplace\_update(ST, node_c, \langle k, v \rangle)$

---

write operation on it and consider the natural batch feature of blockchain. Specifically, the updates generated by transactions of the same block are merged into state trie in a batch. Therefore, we can analyze the conflicting relationship between nodes to be modified called *conflict nodes* of state trie and maximize the parallelism through a lock-free schedule.

For the parallel update on state trie, we regard each update $\langle k, v \rangle$, where k and v are the key and value of state respectively, outputted by transaction execution as a task. Note that if v is "$\perp$", it means the deletion of state with k. To achieve look-free, we first detect the potential contention among tasks and then put conflicting tasks, which modify the same node of trie, to the same worker thread. Roughly, the processing of a batch of tasks can be finished in four stages as presented in Algorithm 2. The input of the algorithm contains the state trie ST, a batch of dirty states S, a thread set W with N_t worker threads,

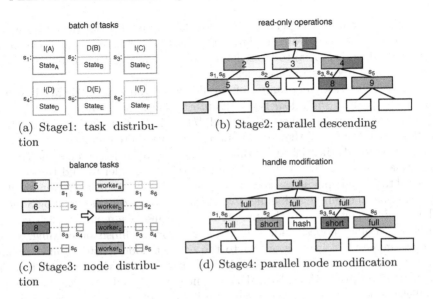

(a) Stage1: task distribution

(b) Stage2: parallel descending

(c) Stage3: node distribution

(d) Stage4: parallel node modification

Fig. 5. Example of parallel update.

and a map M between trie nodes and a list of tasks. We explain the map M in detail later. First, all tasks are dispatched to worker threads evenly through function $task\_assign(\cdot)$ by a modulo manner (lines 1–5).

Second, each worker thread handles tasks in parallel (lines 6–12). For each task $\langle k, v \rangle$, it descends to the *conflict node* $node_c$ by function $find\_conflict\_node(\cdot)$ (line 11). We regard $node_c$ as the conflict node if it should be modified when $\langle k, v \rangle$ is applied to ST. If multiple nodes should be modified, the node closest to root is the conflict node. For example, suppose state k is deleted from ST, since k stays in the leaf node, we should modify its parent to empty the pointer referring to it. Then the parent is the conflict node. If two tasks share the same conflict node, there is a race condition between them. We append them to a lock-free list l_t which accumulates tasks for the target conflict node (line 12). Note that map M records all conflict nodes and is also implemented as lock-free.

Third, the tasks accumulated on the same conflict node are handled by a single worker, which is similar to Stage1 (lines 13–17). Tasks are redistributed over all worker threads for lock-free parallel update. Last, each worker thread processes tasks serially (lines 21–25). In particular, for each $node_c$, worker thread w applies all updates in its list by function $inplace\_update(\cdot)$. It should be noted that a task $\langle k, v \rangle$ overwriting an existing state doesn't have conflict node, so we can apply the update directly in Stage2. Only the insertion and deletion would incur the modification on its parent.

Our algorithm can be applied to any state trie as well as storage trie to parallelize the batch updates. The key factor that our approach can work is the structural invariance feature of trie. The structural invariance means the final structure of a trie is independent of the order of updates but depends on the batch. Therefore, the parallel update still ensures the deterministic structure of state trie, which is crucial to blockchain. Besides, our algorithm can easily scale due to its lock-free design theoretically.

Figure 5 illustrates an example of parallel update on a state trie MPT, which compresses path by *short node* type. Figure 5(a) shows the six dirty states $s_1 \sim s_6$ represented by squares, where "$I(\cdot)$" and "$D(\cdot)$" indicate insertion and deletion of state respectively. All these tasks (states) are distributed uniformly over three worker threads distinguished by three colors. Figure 5(b) indicates the second stage that three worker threads descend on trie in parallel to find the conflict nodes of tasks and append tasks to corresponding lists. Take $worker_a$ represented by yellow as an example, it handles two tasks, task s_2 and s_6 assigned by Stage 1. Then it descends through the trie node $1, 3, 6$ and again node $1, 2, 5$ ending with two conflict nodes: node $5, 6$. After all workers finish descending, tasks are accumulated in lists according to the conflict nodes. Four conflict nodes, namely node $5, 6, 8, 9$, are dispatched with their tasks to three worker threads for final modification. Considering load balance, task s_2 and s_5 belong to different conflict nodes are assigned to the same worker thread, so that all threads obtain similar workloads as depicted in Fig. 5(c). Last, as shown in Fig. 5(d), each thread modifies conflict nodes based on tasks parallelly.

5.2   Deferred Commit Strategy

In some blockchains like Ethereum, the root hash of state trie, which reflects the effects of transactions in block B, is included in the block header. This requires the proposer (i.e., the leader in PBFT or miner in PoW) of B to execute transactions to generate the commitment and other replicas have to validate the execution of the proposer. As a result, the entire workflow is blocked by two-phase execution on the proposer and other replicas respectively, leading to poor performance. To increase the parallelism, we move the execution after consensus, so that both proposer and replicas can execute in parallel. This design will cause the invalid transactions may be included in block since a transaction may abort for some reasons by application logic, such as insufficient balance of transfer transaction. However, we consider this compromise is acceptable if we ensure invalid transaction won't take effect and mark it invalid explicitly like the Fabric. Through this approach, the state root cannot be fulfilled in current block header further, and we solve this issue next.

Another key factor to the performance is the commitment of state trie, including hash recalculation and node persistence. Specifically, our extension experiment shows that the overhead of commitment has exceeded the one of execution. Observe that hashing and persistence mainly consume the computation and disk I/O resources while the consensus consumes the network I/O. A natural choice is to perform them in parallel to utilize the system resources fully. To achieve this goal, we pipeline the commitment of state trie at block height h and the consensus at block height $h + 1$. In particular, when the execution finishes, each replica enters commit and consensus for block height h and $h + 1$ respectively, leading to a non-blocked workflow. Ideally, when the block B_{h+1} gets agreed, the commitment for block B_h is finished, then replicas can continue to process commitment for B_{h+1}. Therefore we use a parallel hash calculation and a write-oriented DiskDB LevelDB for persistence to accelerate the commitment.

To apply the deferred commit strategy, a left problem is state root padding. PEEP no longer requires each block to contain *its* state root. Instead, we misplace the state root and block header which means we can practically fill the state root for block B_h in the block header of block B_{h+2} at the cost of delayed authenticated query. Since the commitment for block height h can be finished before the consensus for B_{h+2}, we attach it to the header of B_{h+2} and then agree on it. Surely, we cannot deploy any authenticate query on state at block height h due to the lake of state root during this delay. On the other hand, since the faster consensus is available and PEEP optimizes the execution and commitment, we think this limitation is acceptable.

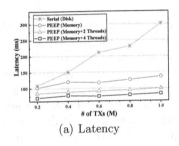

(a) Latency

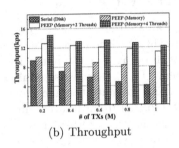

(b) Throughput

Fig. 6. Overall performance of PEEP against number of transactions.

6 Performance Evaluation

6.1 Implementation

PEEP is entirely written in Golang. We adopt go-Ethereum[1] as library and integrate an open-source PBFT realization in Hyperledger Fabric 0.6[2] into our system. In the schedule layer, we set up one thread to order transactions and multiple worker threads to execute. To support parallel execution of smart contracts, we set an EVM (Ethereum VM) pool with multiple instances to provide the execution environment. Besides, we reuse the MPT component in Ethereum and implement our parallel update algorithm upon it. Note that the number of workers that descend and process tasks is set the same as in the schedule layer. Meanwhile, we activate the parallel hash computation to accelerate hash calculation.

---

[1] https://github.com/ethereum/go-ethereum.
[2] https://github.com/hyperledger-archives/fabric.

6.2 Experimental Setup

All experiments are conducted on four qualified machines with an Intel Core i7-7700HQ CPU @ 2.80 GHz of 4 cores, 16 GB RAM and 1 TB disk space. These machines are connected via 10 GB Ethernet. We use Ubuntu 16.04 plus Go 1.14.2 and run every experiment three times to report the average.

Benchmark. We use benchmark proposed by BlockBench [8] to generate the test dataset consists of 1 million transactions. Specifically, a SmallBank contract as Macro-benchmark in BlockBench is deployed on one account and 1 million transactions are sent by 200 thousand accounts to invoke this smart contract under Zipfian distribution. A function called updateBalance is invoked by transactions, which leads to evenly one read and one write operations related to the invoker. Besides, a block is limited to contain at most 1 thousand transactions.

6.3 Overall Performance

We first evaluate the overall performance of PEEP in terms of throughput and latency per block without consensus. Figure 6(a) and Fig. 6(b) report the latency and throughput against the number of transactions respectively. In the former figure, Serial(Disk) represents serial execution under the case that there is no memory cache for state trie and storage trie, then all these tries' nodes should be directly loaded from disk, causing a serious amplification as the number of transactions increases. Other lines are implementations of PEEP varying from number of worker threads under our assumptions. Note that the amplification reduces sharply since most trie nodes can hit in memory due to a large memory cache. As shown in the latter figure, PEEP reaches over 14 thousand TPS which is a nearly 50% enhancement than the serial execution under the same assumption. Note that the overall transaction flow includes persistence at the end of execution which is hard to optimize. The improvement of PEEP is evident as can be seen in both figures. In conclusion, PEEP outperforms the serial ones significantly due to exploiting the multi-core processor's performance.

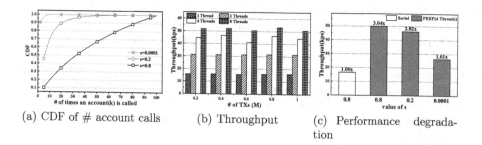

(a) CDF of # account calls   (b) Throughput   (c) Performance degradation

Fig. 7. Performance of parallel execution

6.4   Individual Analysis

Analysis of Schedule Layer. We calculate a CDF of the number of times accounts are called as Fig. 7(a). For example, $s = 0.2$ means 20% of the accounts are related to 90% of all transactions based on the figure. As can be seen, the lower the parameter s is, the higher contention will be. To evaluate the performance of parallel execution, we then conduct experiments based on $s = 0.8$ against number of worker threads shown in Fig. 7(b). As the number of worker threads increases, throughput keeps increasing, while the proportion improved reduces due to thread switching overhead and potentially single-point bottleneck of the lock thread since it grants locks sequentially. Furthermore, if a block contains a high-contention of transactions, throughput will be restrained due to the locking scheme. As depicted in Fig. 7(c), with the rise of contention caused by parameter s, the throughput decreases accordingly. Under extreme condition such as $s = 0.0001$, execution is closer to the serial manner and the performance lower by one order of magnitude consequently.

Performance of Parallel Trie. Two critical factors including the density of trie and the number of worker threads bear on the parallelism degree. We first measure the number of accessed trie nodes during updates per block to represent the density of trie. Since MPT is a compressed trie, it will expand as it becomes denser. In this experiment, each transaction directly modifies two different states. As reported in Fig. 8(a), the number of accessed nodes increases quickly then tends to stabilize as the number of block increases owing that trie becomes nearly completely expanded with no path compression, bringing higher parallelism at the cost of a slightly longer descending path. Figure 8(b) shows the empowerment of parallel operations compared to serial manner. Recall that all dirty nodes will be loaded during the execution phase, so update operations are fast due to no interaction with the disk. It reports that the parallel operations consists of parallel update and commit reduce the latency by 43%, which also contributes to overall performance enhancement.

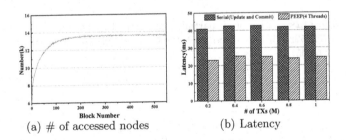

(a) # of accessed nodes   (b) Latency

Fig. 8. Performance of parallel trie.

Efficiency of Deferred Commit Strategy. We test the impact of the deferred commit strategy of PEEP. For the ordering phase, a block with 1K transactions consumes nearly 60 ms per round for consensus based on our measurement.

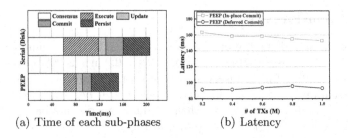

(a) Time of each sub-phases   (b) Latency

Fig. 9. Efficiency of deferred commit strategy

Figure 9(a) shows the proportion of each sub-phases under different assumptions. Serial(Disk) suffers read amplification and serial constraints that make the execution phase dominant while PEEP optimizes it. Besides, the commit and persist sub-phases share a nearly similar proportion with consensus so that it is practical to apply our deferred commit strategy. We implement the strategy with one round deferral, the average latency per block is reduced by 38% as reported in Fig. 9(b) since the asynchronous commitment allows the block to be confirmed early and does not obstruct the process for the next block.

7 Conclusion

This paper introduces PEEP, a parallel execution engine designed for permissioned blockchain systems for higher throughput. We apply a deterministic concurrent scheduling and a parallel update in particular to adding parallelism to execution. Besides, the deferred commit strategy is proposed for better utilization of system resources, yielding a non-blocked workflow. The experimental results support the practicability and effectiveness of our work. Future works are expected to exploit more efficient authenticated data structures and design complete pipelined execution of blockchain.

Acknowledgments. This work is partially supported by National Science Foundation of China (61972152, U1811264 and U1911203), Guangxi Key Laboratory of Trusted Software (kx202005).

References

1. CITA project (2020). https://github.com/citahub
2. Amiri, M.J., Agrawal, D., et al.: Parblockchain: leveraging transaction parallelism in permissioned blockchain systems. In: ICDCS, pp. 1337–1347. IEEE (2019)
3. Amsden, Z., Arora, R., et al.: The libra blockchain. Libra project, pp. 1–29 (2020). https://developers.libra.org/docs/assets/papers/the-libra-blockchain/2020-05-26.pdf
4. Androulaki, E., Barger, A., et al.: Hyperledger fabric: a distributed operating system for permissioned blockchains. In: EuroSys, pp. 30:1–30:15. ACM (2018)

5. Buchman, E.: Tendermint: byzantine fault tolerance in the age of blockchains. Ph.D. thesis (2016)
6. Castro, M., Liskov, B.: Practical byzantine fault tolerance. In: OSDI, pp. 173–186. USENIX Association (1999)
7. Dickerson, T., Gazzillo, P., Herlihy, M., Koskinen, E.: Adding concurrency to smart contracts. Distrib. Comput. 209–225 (2019). https://doi.org/10.1007/s00446-019-00357-z
8. Dinh, T.T.A., Wang, J., et al.: BLOCKBENCH: a framework for analyzing private blockchains. In: SIGMOD Conference, pp. 1085–1100. ACM (2017)
9. Ji, Y., Chai, Y., et al.: Smart intra-query fault tolerance for massive parallel processing databases. Data Sci. Eng. 5(1), 65–79 (2020)
10. Kalidhindi, J., Kazorian, A., et al.: Angela: a sparse, distributed, and highly concurrent Merkle tree (2018)
11. Li, Z., Romano, P., et al.: Sparkle: speculative deterministic concurrency control for partially replicated transactional stores. In: DSN, pp. 164–175. IEEE (2019)
12. Nakamoto, S., et al.: Bitcoin: a peer-to-peer electronic cash system (2008)
13. Ongaro, D., Ousterhout, J.K.: In search of an understandable consensus algorithm. In: USENIX Annual Technical Conference, pp. 305–319. USENIX Association (2014)
14. Pang, S., Qi, X., et al.: Concurrency protocol aiming at high performance of execution and replay for smart contracts. CoRR abs/1905.07169 (2019)
15. Sharma, A., Schuhknecht, F.M., et al.: Blurring the lines between blockchains and database systems: the case of hyperledger fabric. In: SIGMOD Conference, pp. 105–122. ACM (2019)
16. Thomson, A., Diamond, T., et al.: Calvin: fast distributed transactions for partitioned database systems. In: SIGMOD Conference, pp. 1–12. ACM (2012)
17. Wood, G.: Ethereum: a secure decentralised generalised transaction ledger. Ethereum Project Yellow Pap. 151, 1–32 (2014)

URIM: Utility-Oriented Role-Centric Incentive Mechanism Design for Blockchain-Based Crowdsensing

Zheng Xu[1,2,3], Chaofan Liu[1,2,3], Peng Zhang[1,2,3(✉)], Tun Lu[1,2,3(✉)], and Ning Gu[1,2,3]

[1] School of Computer Science, Fudan University, Shanghai, China
{zxu17,cfliu18,zhangpeng_,lutun,ninggu}@fudan.edu.cn
[2] Shanghai Key Laboratory of Data Science, Fudan University, Shanghai, China
[3] Shanghai Institute of Intelligent Electronics and Systems, Shanghai, China

Abstract. Crowdsensing is a prominent paradigm that collects data by outsourcing to individuals with sensing devices. However, most existing crowdsensing systems are based on centralized architecture which suffers from poor data quality, high service charge, single point of failure, etc. Some studies have explored decentralized architectures and implementations for crowdsensing based on blockchain, while incentive mechanisms for worker participation and miner participation, which serve as a crucial role in blockchain-based crowdsensing systems (BCSs), are ignored. To address this issue, we propose an incentive mechanism design named URIM to maximize participants' utilities, which consists of worker-centric and miner-centric incentive mechanisms for BCSs. For the worker-centric incentive mechanism, we model it as a reverse auction, in which dynamic programming is utilized to select workers, and payments are determined based on the Vickrey-Clarke-Groves scheme. We also prove this incentive mechanism is computationally efficient, individually rational and truthful. For the miner-centric incentive mechanism, we model interactions among the requester and miners as a Stackelberg game and adopt the backward induction to analyze its equilibrium at which the utilities of the requester and miners are optimized. Finally, we demonstrate the significant performance of URIM through extensive simulations.

Keywords: Crowdsensing · Blockchain · Incentive mechanism · Reverse auction · Game theory

1 Introduction

In recent years, sensing devices (such as smartphones, wearable devices and tablets) have been emerging in our daily life. The proliferation of devices capable of sensing and computing leads to the prosperity of a new sensing paradigm called crowdsensing. Crowdsensing leverages "humans-as-sensors" to enable traditional Internet of Things (IoT) application by combing perception capabilities

C. S. Jensen et al. (Eds.): DASFAA 2021, LNCS 12683, pp. 358–374, 2021.
https://doi.org/10.1007/978-3-030-73200-4_25

and crowdsourcing in many important fields, such as intelligent transportation, public safety, environmental monitoring and urban public management.

In the previous studies, most research on crowdsensing adopted centralized system architecture which generally consists of three roles: centralized platform, task requesters, and crowdsensing workers (also known as the data providers) [10]. However, there exists some drawbacks in the centralized crowdsensing systems, such as poor data quality, high service charge, single point of failure and privacy disclosure [13]. Therefore, some studies have explored decentralized techniques for crowdsensing systems, wherein a very popular solution is blockchain. Blockchain-based crowdsensing systems (BCSs) have following advantages. The decentralization, immutability and security of blockchain can facilitate the cooperation among mutually distrusted participants without service fees charged by the centralized platform and ensure the audibility of the crowdsensing data. Moreover, the decentralized blockchain architecture can avoid the single point of failure which may cause shutdown of traditional centralized systems. The anonymity of blockchain transactions can reduce risks of privacy disclosure.

Existing studies of BCSs mainly focus on architecture design and smart contracts implementation, while the incentive mechanisms for user engagement, which serve as a crucial role in crowdsensing systems, are ignored. Nowadays, most incentive mechanisms are designed for traditional centralized crowdsensing systems [15,19]. However, BCSs operate automatically via smart contracts without a centralized reliable intermediary. Thus, the incentive mechanism of BCSs is designed to optimize the utilities of participants when interacting with the blockchain, rather than the centralized platform. Although there are some studies on the incentive mechanisms for BCSs, they mainly focus on selection and reward allocation for workers [2,3,9]. They omit an exclusive and important role called miner in BCSs to handle and validate all operations. Existing incentive mechanisms generally involve the task requesters and workers, but ignore miners. Hence, existing incentive mechanisms are not fully compatible with BCSs. Due to the lack of appropriate incentive mechanisms, the utilities of participants cannot be maximized, and the efficiency of BCSs decreases.

There is an urgent need to design appropriate incentive mechanisms for BCSs, but it is a challenging task. First, BCSs allow participants to exchange data without a centralized truthful intermediary. It is crucial and challenging to build a system model of BCSs that can be compatible with holistic incentive mechanism design. Second, it is difficult to formalize how multiple roles interact with each other and how to optimize their utilities. Compared with traditional crowdsensing systems, a new role called miner is involved in BCSs. Miners directly determine the operating efficiency and security of BCSs, while how to select efficient miners in a safe and reliable way is not easy. Meanwhile, there are complex interactions between workers, miners and requesters, which aggravate the complexity of utility optimization in the incentive mechanism design.

Based on the above background, we focus on: *How to design holistic incentive mechanisms for BCSs to maximize utilities of participating roles?* To solve this problem, we propose a utility-oriented role-centric incentive mechanism design named URIM which aims to maximize utilities of participating roles in BCSs. According to the processes and participants of BCSs, workers and miners directly determines the performance of BCSs. Hence, URIM is designed to consist of worker-centric and miner-centric incentive mechanisms. In the worker-centric incentive mechanism (WCIM), the task requester publishes its task to BCSs through smart contracts. Then each worker submits its solution of the task and corresponding bidding price. Smart contracts on BCSs automatically selects workers by a dynamic programming algorithm and determines their payments by Vickrey-Clarke-Groves (VCG) scheme. The above interactions among the task requester, workers and smart contracts are modeled as a reverse auction. In the miner-centric incentive mechanism (MCIM), we first adopt cryptographic sortition to select eligible miners. For motivating miners to validate transactions and mine blocks, the task requester announces total transaction fees shared by all transactions related with its task. Then miners decide their mining strategies to validate different number of transactions and compete for the corresponding transaction fee. The above interactions among the task requester and miners are modeled as a Stackelberg game to optimize utilities of the task requester and miners. The main contributions of this paper are as follows:

- We propose a utility-oriented role-centric incentive mechanism design named URIM for BCSs. To the best of our knowledge, this is the first work on holistic incentive mechanism design for BCSs to ensure the utility maximization of all roles.
- We design a reverse auction based WCIM which adopts dynamic programming to select desirable workers and determine payments based on VCG scheme. We theoretically prove that WCIM is computationally efficient, individually rational and truthful.
- We design the MCIM that selects miners by cryptographic sortition and formulates mining competition by a two-stage Stackelberg game. Through backward induction, we analyze and validate the best response strategies of miners and the unique Stackelberg equilibrium where the utilities of the requester and miners are jointly maximized.
- We demonstrate the significant performance of URIM through extensive simulations.

The remainder of the paper is organized as follows. In Sect. 2, we review the related work of blockchain-based crowdsensing systems and incentive mechanisms for crowdsensing. In Sect. 3, we present the system model of BCSs with the design of URIM. We then present two compositions of URIM in Sect. 4 and 5. We present performance evaluations in Sect. 6. Finally, we conclude this paper in Sect. 7.

2 Related Work

In this section, we mainly review related research on the blockchain-based crowdsensing systems and incentive mechanisms for crowdsensing.

Blockchain-Based Crowdsensing Systems. Blockchain and automated execution of smart contracts greatly enhance the decentralized communication and cooperation without an intermediary in many fields [5]. In particular, the crowdsensing system can take the benefits of blockchain to achieve fair and trust-less collaboration. Crowdbc [13] proposes a decentralized crowdsensing framework based on blockchain and implements the main concepts in the framework through the usage of smart contracts. Zebralancer [14] shows how an anonymous decentralized crowdsensing system can be implemented on top of blockchain, which ensure the privacy of the crowdsensing data while preserves the transparency of blockchain systems. In [2], authors build a decentralized crowdsensing platform for data trading on blockchain. The research of homomorphic encryption for fair and secure BCSs is discussed in [21]. Zero-knowledge proof technique is proposed to enable data providers to submit data through a privacy preserving and secure way in BCSs [6]. The location privacy attack in the crowdsensing system is discussed in [20]. This work proposes a blockchain-based privacy preservation framework for protecting location of workers. These efforts are mainly aimed at how to implement crowdsensing on the blockchain.

Incentive Mechanism for Crowdsensing. The existing incentive mechanisms for crowdsensing systems can be divided into monetary and non-monetary incentive mechanisms. In monetary incentive mechanisms, auction is a common approach to pay workers with either real money or virtual tokens [7,11]. The radpvpc mechanism is proposed in [12], which aims to minimize the platform cost and maintain the participation level of workers. Yang *et al.* propose a reverse auction based incentive mechanism to determine winners and their payments [19]. Some studies have attempted to provide monetary incentive mechanisms based on BCSs. An *et al.* propose a blockchain-based crowdsensing data trading system with truthful and confidential incentive mechanisms in [2]. Truthful and cost-optimal incentives for mobile user participation are designed in [3]. Hu *et al.* design the workflow of BCSs with the help of automatic smart contracts and they leverage a three stage Stackelberg game to motivate participants in [9]. The above works [2,3,9] only consider the participation of workers, but they ignore the mining competition and consensus achievement. Non-monetary incentive mechanisms provide comprehensive and long-term incentives for participants. In [1], a reputation management framework is proposed to evaluate both the contributions quality and the trust level of participants. Crowdsensing data trustworthiness are quantified based on statistical and vote-based reputation scores in [17].

3   System Overview

In this section, we present the system model of BCSs with the holistic incentive mechanism design. As shown in Fig. 1, there generally exists three roles participating in BCSs: the task requester, workers and miners. The task requester and workers can also be regarded as miners when they join in the mining competition.

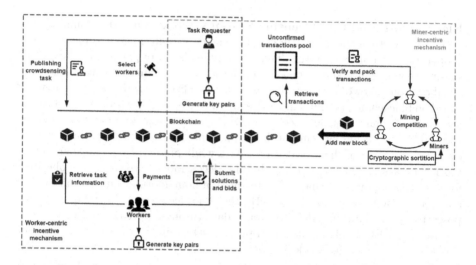

Fig. 1. System model of blockchain-based crowdsensing.

1. **Task requester.** Let R be the task requester (hereinafter referred to as requester), and R publishes crowdsensing task ST to the BCS. ST defines the sensing requirement including the task budget B, task duration and description of target data. To motivate workers to enhance the quality of sensing information, and miners to validate transactions related with ST, a certain amount of compensation will be paid to workers and miners respectively.
2. **Worker.** Let $W = \{w_1, w_2, \cdots, w_m\}$ be the set of all workers. Worker w_i should submit its solution SL_i and bidding price b_i before the task ending time if w_i is interested in task ST.
3. **Miner.** The responsibility of miners is to secure the blockchain network and to deal with every transaction in it. Each miner validate transactions in its block to compete for the mining reward which includes a specific block reward and fees sent with validated transactions. Miners are selected from those requesters and workers by a specific selection manner.

In Fig. 1, the workflow of BCSs with utility-oriented role-centric incentive mechanism contains five stages as follows:

1. **Participants register.** In the beginning, the requester and workers join in this system. Each registered participant has a unique public/private key pair to secure its transactions.
2. **Publish crowdsensing task.** In this stage, the requester publishes the crowdsensing task to BCSs via smart contracts.
3. **Submit crowdsensing information.** Workers can retrieve crowdsensing tasks they are interested in by interacting with blockchain. Workers submit crowdsensing solutions and bidding prices before the task deadline.
4. **Select workers and determine payments.** After receiving crowdsensing solutions, BCSs automatically select desirable workers and determine their payments via smart contracts.

5. **Select miners and generate blocks.** Based on participants' computing power, cryptographic sortition is adopted to randomly and unpredictably select miners. Selected miners pack bundled transactions into a block, then compete for mining rewards with proof of work (PoW) consensus.

The goal of incentive mechanism design for BCSs is to motivate participants and maximize their utilities. Other issues in the design and implementation of smart contracts in BCSs is out of the scope of our paper. People can refer to [2,13,21] for these issues. Both workers and miners play an important role in determining the performance of BCSs, so we need to design incentive mechanisms for them. In Stage 4, we propose a worker-centric incentive mechanism, in which eligible workers are paid a reasonable return according to their contribution and bidding price. In Stage 5, we propose a miner-centric incentive mechanism, in which miners compete with each other for a total transaction fee.

4   Worker-Centric Incentive Mechanism

In this section, we propose a reverse-auction-based worker-centric incentive mechanism named WCIM to select workers by dynamic programming, and determine their payments through VCG scheme.

After the requester R posts the task ST on BCSs, reverse-auction-based WCIM will output a subset of workers $S \in W$ as winners and determine the payment for each winner $w_i \in S$ by taking workers' contribution v_i and bidding price b_i as input. These processes are performed automatically on smart contracts. The sum of all winners' contributions is represented as $\sum_{w_i \in S} v_i$. The utility of R is the difference between winners' contributions and their social costs, which is represented as $\sum_{w_i \in S} (v_i - b_i)$. WCIM aims to maximize the utility of R while satisfying the budget control. Hence, WCIM can be formulated as follows:

$$Maximize \sum_{w_i \in S} (v_i - b_i), \quad Subject\ to \sum_{w_i \in S} b_i \leq B. \tag{1}$$

The WCIM is designed to satisfy properties as follows:

- **Computational Efficiency.** The incentive mechanism is computationally efficient if its computation runs in polynomial time.
- **Individual Rationality.** The incentive mechanism is individually rational if each worker has non-negative utility.
- **Truthfulness.** The incentive mechanism is truthful if no worker could obtain higher utility by reporting a false bid that deviates from its true cost no matter what others report.

4.1   Implementation of WCIM

From the above, WCIM consists of workers selection and payment determination.

Wokrers Selection. We reduce the workers selection to 0–1 Knapsack problem which is constructed as follows and use dynamic programming for the optimal solution in Algorithm 1.

Workers are denoted by set $W = \{w_1, \cdots, w_m\}$. Workers' bidding prices and contributions are denoted by $\{b_1, \cdots, b_m\}$ and $\{v_1, \cdots, v_m\}$. We map workers' bidding set $\{b_1, \cdots, b_m\}$ to a non-negative integer set $\{\beta b_1, \cdots, \beta b_m\}$ by multiplying each bidding price with amplification factor β. Meanwhile, we use $x_i \in \{0, 1\}$, $i \in \{1, \cdots, m\}$ to represent if worker w_i will be selected. The workers selection problem is constructed to determine $X = \{x_1, \cdots, x_m\}$ to

$$Maximize \sum_{i=1}^{m} (v_i - b_i) \cdot x_i, \quad Subject\ to \sum_{i=1}^{m} \beta b_i x_i \leq \beta B. \tag{2}$$

Payment Determination. We propose a VCG [18] based payment determination algorithm illustrated in Algorithm 2. In the generalized VCG auction, each winner is required to pay "harm" imposed on other workers, i.e., the difference between the optimal utility of the requester with and without this winner [18]. We define $V(S)_W^{-w_i}$ as the optimal utility of R excluding the contribution of worker w_i, which can be represented as

$$V(S)_W^{-w_i} = V(S)_W - (v_i - b_i). \tag{3}$$

Then, we define $V(S)_{W \setminus \{w_i\}}$ as the optimal utility of R excluding the participation of worker w_i. Thus, the payment p_i of worker w_i can be represented as

$$p_i = v_i - \left(V(S)_{W \setminus \{w_i\}} - V(S)_W^{-w_i} \right). \tag{4}$$

4.2   Theoretical Analysis of WCIM Properties

In this subsection, we prove that WCIM satisfies mentioned three properties: computational efficiency (Lemma 1), the individual rationality (Lemma 2) and the truthfulness (Lemma 3).

Lemma 1. *WCIM is computationally efficient.*

Proof. The winners selection has been reduced to 0–1 Knapsack problem, as illustrated in Algorithm 1 and it takes $O(n\beta B)$ time. The payment determination illustrated in Algorithm 2 takes $O(mn\beta B)$ time. WCIM is executed by Algorithm 1 and 2 sequentially and its running time is the sum of them. Hence, WCIM is a polynomial-time mechanism and computationally efficient.   □

Lemma 2. *WCIM is individually rational.*

Proof. Based on (4), we have

$$p_i = v_i - \left(V(S)_{W \setminus \{m_i\}} - V(S)_W^{-m_i} \right) = v_i - \left(V(S)_{W \setminus \{m_i\}} - V(S)_W + v_i - b_i \right)$$

$$= V(S)_W - V(S)_{W \setminus \{m_i\}} + b_i. \tag{5}$$

Algorithm 1. The Winners Selection Algorithm

Input: The worker set $\{w_1, w_2, \cdots, w_m\}$, their bid set $\{b_1, b_2, \cdots, b_m\}$, their contribution set $\{v_1, v_2, \cdots, v_m\}$ and the budget B, the amplification factor β;
Output: The selected worker set S;

```
 1: S ← ∅, H [i, j] ← 0;
 2: for i from 1 to m do
 3:     for j from 0 to βB do
 4:         if βb_i ≤ j && H [i − 1, j − βb_i] + (v_i − b_i) > H [i − 1, j] then
 5:             H [i, j] ← H [i − 1, j − βb_i] + (v_i − b_i);
 6:             X [i, j] ← 1;
 7:         else
 8:             H [i, j] ← H [i − 1, j];
 9:             X [i, j] ← 0;
10:         end if
11:     end for
12: end for
13: V ← H [m, βB];
14: B' ← 0;
15: for j from βB downto 0 do
16:     if H [m, j] == V then
17:         B' ← j/β;
18:     end if
19: end for
20: for i from m downto 0 do
21:     if X [i, βB'] == 1 then
22:         S ← S ∪ {w_i};
23:         B' = B' − b_i;
24:     end if
25: end for
26: return S;
```

Since S is the winner set, it is easy to find $V(S)_W \geq V(S)_{W \setminus \{m_i\}}$, and thus $p_i \geq b_i$. Hence, WCIM is individually rational. ☐

Lemma 3. *WCIM is truthful.*

Proof. If w_i reports a truthful bidding price, its utility can be represented as follows:

$$U(w_i) = p_i - b_i = V(S)_W - V(S)_{W \setminus \{w_i\}} + b_i - b_i$$
$$= V(S)_W - V(S)_{W \setminus \{w_i\}}.$$
(6)

In (6), w_i is unable to influence the value of $V(S)_{W \setminus \{w_i\}}$. After reporting the untruthful bidding price b'_i from the truthful bidding price b_i, the utility of w_i is changed as follows:

$$\Delta U(w_i) = U\left(w'_i\right) - U(w_i) = V(S)_{W'} - V(S)_{W' \setminus \{w_i\}} - \left(V(S)_W - V(S)_{W \setminus \{w_i\}}\right)$$
$$= V(S)_{W'} - V(S)_W$$
(7)

From Lemma 2, we can obtain $V(S)_{W'} \leq V(S)_W$, and then $\Delta U(w_i) \leq 0$. Thus, the worker w_i cannot get higher utility by reporting an untruthful bidding price. Hence, WCIM is truthful. ☐

Algorithm 2. The Payment Determination Algorithm

Input: Winner set $S = \{w_1, w_2, \cdots, w_m\}$, their bid set $\{b_1, b_2, \cdots, b_m\}$ and contribution set $\{v_1, v_2, \cdots, v_m\}$;

Output: The winner payment set P;

1: $P \leftarrow \oslash$;
2: **for** j from 1 to m **do**
3:    $V(S)_S^{-w_i} = V(S)_S - (v_i - b_i)$
4:    Calculate $V(S)_{S \setminus \{w_i\}}$ according to Algorithm 1
5:    $p_i = v_i - \left(V(S)_{S \setminus \{w_i\}} - V(S)_S^{-w_i} \right)$
6:    $P \leftarrow P \cup \{p_i\}$
7: **end for**
8: **return** P;

---

5 Miner-Centric Incentive Mechanism

In this section, we model the miner-centric incentive mechanism as a Stackelberg game to decide how to optimize the utility of the requester and miners.

5.1 Blockchain Mining with Crowdsensing

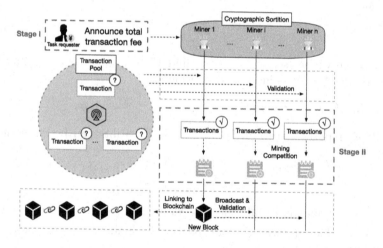

Fig. 2. Mining competition

As shown in Fig. 2, the block mining competition includes 5 steps that are, in order, miners selection, transaction fee announcement, transaction validation, block mining and block validation. In the beginning, eligible miners are selected by cryptographic sortition according to their computing power. Cryptographic sortition has sufficient randomness and unpredictability to eliminate manipulation of consensus led by the collusion among requesters, workers and miners [4]. We refer to and improve the work [16] so that participants with higher computing power could be selected with more chance.

Hence, a participant can be selected as the miner m_j at epoch τ if this condition is met: $\frac{HASH(\langle \tau \| rand(\tau) \rangle_{m_j})}{2^L} \cdot \frac{1-e^{-kI}}{1+e^{-kI}} \leq I_j^\tau$ where $rand(\tau)$ is a public randomness that can be extracted from the blockchain at epoch τ, $\langle \tau \| rand(\tau) \rangle_{m_j}$ is a signature of message $\tau \| rand(\tau)$ produced with private key of m_j, $HASH$ is a deterministic hash function, L is the bits of the output of $HASH$, I_j^τ is the fraction of m_j's computing power over all miners in BCSs at epoch τ.

Unlike traditional mining in which transaction fee is paid by each initiator, the requester R has to pay transaction fees to motivate miners to pack and validate transactions since all transactions are related to its task. After the selection of miners, R announces total transaction fee F that all miners compete for. If one miner successfully solves a crypto puzzle, it will broadcast its solution to BCSs. After the solution reaches Proof-of-Work (PoW) consensus, a new block is mined successfully and its miner obtains the mining reward which includes voluntary transaction fees of this block and a fixed block reward. The voluntary transaction fees depend on the number of transactions in the block, in other words, the miner can earn more voluntary transaction fees if it packs and validates more transactions.

Winning a mining reward depends on mining and propagating a block as quickly as possible. During the mining process, whether a miner can mine a new block depends on its relative computing power μ. However, the block may be orphaned by subsequent blocks and hence its miner will not be paid because of propagation time lag [8]. The occurrence of mining a block follows the Poisson distribution, and the probability of block being orphaned is $P_{orphan} = 1 - e^{-\lambda T(r)}$, where $\lambda = 1/600$ and $T(r)$ represents the block propagation time which depends linearly on r [8]. Therefore, the probability of winning the mining reward is denoted by $P = \mu(1 - P_{orphan}) = \mu e^{-\lambda \varepsilon r}$, where ε is a delay factor reflecting the impact of r on $T(r)$.

Given the set of selected miners, denoted by $M = \{m_1, \cdots, m_n\}$, each miner $m_j \in M$ decides to include r_j transactions in its block. The utility of m_j is determined by two parts: 1) the mining reward, and 2) the electricity and other costs associated with mining. Thus, the utility of m_j is presented by

$$U_m^j = \left(\frac{r_j F}{\sum_{m_n \in S} r_n} + D \right) P_j - c_j = \left(\frac{r_j F}{\sum_{m_n \in S} r_n} + D \right) \mu_j e^{-\lambda \varepsilon r_j} - c_j \quad (8)$$

where $\frac{r_j F}{\sum_{m_n \in S} r_n}$ means transaction fees obtained by m_j according to the ratio of its number of transactions, and D means the fixed block reward. The utility of the requester R is

$$U_R(F) = f(r_1, \cdots, r_n) - F \quad (9)$$

where $f(r_1, \cdots, r_n)$ is the satisfaction function with respect to the number of verified transactions from selected miners. We made a realistic and general assumption that $f(0, \cdots, 0) = 0$ and $f(r_1, \cdots, r_n)$ is a strictly concave function in variables $r_1, \cdots, r_n$ and monotonically increasing in each r_j [19].

5.2 Two-Stage Stackelberg Game Formulation

According to Sect. 5.1, we can formulate MCIM as a two-stage Stackelberg game. In the stage I of MCIM, R announces a total transaction fee F to motivate miners to pack transactions into their blocks. Since no rational miners will join in the mining competition with negative earnings, so we consider that $F > 0$. In the stage II of MCIM, each miner $m_j \in M$ decides to pack a different amount of transactions r_j in block mining competition to maximize its utility. Let $\Phi = \{r_1, \cdots, r_n\}$ denote the strategy profiles consisting of all miners' strategies and Φ_{-j} denotes the strategy profile excluding r_j. Thus in MCIM, the requester is the leader and miners are the followers. The objective of MCIM is to find the Stackelberg equilibrium where R can maximize its utility with the response strategies of miners, which is represented as follows:

– In stage I:

$$Maximize \ U_R^{\Phi}(F), \quad Subject \ to \ F > 0. \tag{10}$$

– In stage II:

$$Maximize \ U_m^j, \quad Subject \ to \ r_j \geq 0. \tag{11}$$

5.3 Equilibrium Analysis for MCIM

In this section, we analyze the optimal strategy of miners and the utility maximization of the requester. We apply the backward induction method to analyze the Stackelberg equilibrium of MCIM. In the stage II of MCIM, given the total transaction fee F, miners compete with each other to maximize their own utility by choosing their individual strategy, which can be considered as a block mining game (BMG) $G^M = \left\{M, \Phi, \{U_m^j\}_{m_j \in M}\right\}$, where M is the set of miners, Φ is miners' strategy set and U_m^j is the utility of miner m_j.

Definition 1. *A set of strategies $\Phi^* = \{r_1^*, \cdots, r_n^*\}$ is the Nash equilibrium of the BMG if $U_m^j\left(r_j^*, \Phi_{-j}^*\right) \geq U_m^j\left(r_j, \Phi_{-j}^*\right)$ for any $r_j \geq 0$.*

Theorem 1. *A Nash equilibrium in BMG $G^M = \left\{M, \Phi, \{U_m^j\}_{m_j \in M}\right\}$ exists.*

Proof. We compute the first order and second order derivatives of U_m^j defined in (8) with respect to r_j:

$$\frac{\partial U_m^j}{\partial r_j} = \left(\frac{F \sum_{m_{n \neq j} \in S} r_n}{\left(\sum_{m_n \in S} r_n\right)^2} - \lambda \varepsilon \left(\frac{F r_j}{\sum_{m_n \in S} r_n} + D\right)\right) \cdot \mu_j e^{-\lambda \varepsilon r_j} \tag{12}$$

and

$$
\begin{aligned}
\frac{\partial^2 U_m^j}{\partial r_j^2} = &-\lambda \epsilon \mu_j e^{-\lambda \varepsilon r_j} \left(\frac{F \sum_{m_{n \neq j} \in S} r_n}{\left(\sum_{m_n \in S} r_n\right)^2} - \lambda \varepsilon \left(\frac{F r_j}{\sum_{m_n \in S} r_n} + D\right)\right) - \\
&\mu_j e^{-\lambda \varepsilon r_j} \left(\frac{2F \sum_{m_{n \neq j} \in S} r_n}{\left(\sum_{m_n \in S} r_n\right)^3} + \lambda \varepsilon \left(\frac{F \sum_{m_n \in S} r_n - F r_j}{\left(\sum_{m_n \in S} r_n\right)^2}\right)\right) < 0.
\end{aligned}
\tag{13}
$$

Thus, U_m^j is strictly concave with respect to r_j. Hence, given any $F > 0$ and any strategy profile Φ_{-j} of the other miners, the best response strategy of m_j is unique when $r_j \geq 0$. Accordingly, the Nash equilibrium of noncooperative BMG G^M exists.

Further, by setting the first derivative of U_m^j to 0, we have

$$\left(\frac{F \sum_{m_n \neq j \in S} r_n}{\left(\sum_{m_n \in S} r_n \right)^2} - \lambda \varepsilon \left(\frac{F r_j}{\sum_{m_n \in S} r_n} + D \right) \right) \cdot \mu_j e^{-\lambda \varepsilon r_j} = 0, \qquad (14)$$

and we can obtain the best response strategy of m_j which is denoted as $\beta(r_j)$ in (15):

$$\beta(r_j) = \begin{cases} 0, & \text{otherwise.} \\ \dfrac{\sqrt{F \lambda \varepsilon \left(F \lambda \varepsilon \sum_{m_n \neq j \in S} r_n + 4D + 4F \right) \sum_{m_n \neq j \in S} r_n} - \lambda \varepsilon (2D + F) \sum_{m_n \neq j \in S} r_n}{2 \lambda \varepsilon (D + F)}, & F \geq 0. \end{cases} \qquad (15)$$

According to (14), we have

$$\lambda \varepsilon D \left(\sum_{m_n \in S} r_n \right)^2 + \lambda \varepsilon F r_j \sum_{m_n \in S} r_n = F \sum_{m_n \neq j \in S} r_n. \qquad (16)$$

Then, we can get $(\lambda \varepsilon |M| D + \lambda \varepsilon F) \sum_{m_n \in S} r_n = F(|M| - 1)$ by summing up (16) over all selected miners. Thus,

$$\sum_{m_n \in S} r_n = \frac{F(|M| - 1)}{\lambda \varepsilon |M| D + \lambda \varepsilon F}. \qquad (17)$$

By substituting (17) into (16), we have the unique Nash equilibrium for miner m_j in BMG, as shown in (18):

$$r_j^* = \frac{\left(F^2 + DF \right) (|M| - 1)}{\lambda \varepsilon |M| (|M| D + F) (F - D)}. \qquad (18)$$

□

According to the above analysis, the requester knows that there exists a unique Nash equilibrium for selected miners for any $F > 0$. Thus the requester can maximize its utility by choosing the optimal transaction fee F.

Theorem 2. *There exists the unique Stackelberg Equilibrium* $\left(F^*, r_j^{ne} \right)$ *in the MCIM game, where* F^* *is the unique maximizer of the requester's utility in (9) and* r_j^{ne} *is given by (18) with* F^*.

Proof. Since rational miners will not participate in mining when $F = 0$, $U_R(F) = 0$ for $F = 0$ and approaches to $-\infty$ when F goes to ∞. Note that $f(r_1, \cdots, r_n)$ is a strictly concave function in variables $\{r_1, \cdots, r_n\}$, hence $U_R(F)$ has a unique maximum value when $F = F^*$ that can be efficiently calculated by either bisection or Newton's method [19]. Therefore, there exists a unique Stackelberg Equilibrium $\left(F^*, r_j^{ne} \right)$ in MCIM. □

6   Performance Evaluation

In this section, we conduct extensive simulations to evaluate and investigate impacts of key parameters on performance of URIM. We present simulation settings, metrics and results as follows.

6.1   Simulation Setup

In the simulation, we consider multiple workers compete to complete the task, and submit their bidding prices and solutions. Then, miners are selected to pack transactions and compete for transaction fees. For evaluating the WCIM, the bidding price b and the contribution v of each worker is normally distributed over $[0.1, 1]$ and $[0.01, 0.5]$ respectively. The number of workers $|W|$ varies from 100 to 1000 with the increment of 100. For evaluating the MCIM, the utility of the requester is set as $U_R(F) = \theta \log \left(1 + \sum_{m_j \in M} r_j\right) - F$ that satisfies the assumptions in Sect. 5.1. We set θ to 10^4. The fixed block reward D is fixed at 100 and the delay factor is fixed at 10^{-4}. For each miner, its mining cost is randomly generated from $[10, 20]$ and μ is randomly generated from $[0.001, 0.1]$. All simulations are performed in Python 3.7.0 and Solidity 0.7.0 on a Windows machine with Intel Core i7-7700 CPU and 16 GB memory.

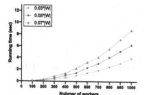

Fig. 3. Running time.

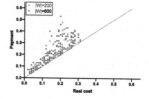

Fig. 4. Individual rationality.

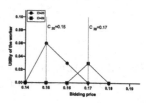

Fig. 5. Truthfulness.

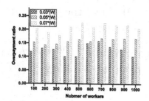

Fig. 6. Overpayment ratio at varied budgets.

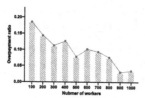

Fig. 7. Overpayment ratio at fixed budget.

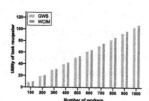

Fig. 8. Utility of requester.

6.2   Evaluation of the Worker-Centric Incentive Mechanism

To investigate the performance of WCIM, we present following metrics: running time, individual rationality, truthfulness, overpayment ratio and utility of requester. Furthermore, we compare WCIM with greedy winner selection (GWS) proposed in [2].

Running Time: We first demonstrate the running time of WCIM in Fig. 3. The budget is set as $|W|$ multiplied by $\{0.03, 0.05, 0.07\}$ respectively. We can find the running time increases slowly with increasing $|W|$. Additionally, the running time has slight changes when the budget increases. These results show WCIM can efficiently select workers and calculate their payments.

Individual Rationality and Truthfulness: Then we verify the individual rationality and truthfulness of WCIM. We demonstrate the individual rationality by comparing each payment and the related real cost (truthful bidding). We randomly set $|W|$ as 200 and 600 in Fig. 4, and we find each payment is greater than the related real cost. To verify the truthfulness, we randomly pick two winners (ID = 20, 29) and change their claimed bidding prices, then recalculate their utilities. We illustrate results in Fig. 5 and find two winners can only obtain their maximum utility if they bid the real cost $Cost_{20} = 0.15$, $Cost_{29} = 0.17$.

Overpayment Ratio: Figure 6 plots the overpayment ratio when $|W|$ changes from 100 to 1000 and the budget equals $|W|$ multiplied by $\{0.03, 0.05, 0.07\}$ respectively. Figure 7 shows the overpayment ratio decreases with the increase of $|W|$ when the budget is fixed at 1000. We find that the overpayment ratio is always less than 0.25, which means that the requester does not have to pay much extra money to induce truthfulness.

Utility of Requester: Figure 8 plots the utility of requester when $|W|$ change from 100 to 1000. $|W|$ multiplied by 0.05 is set as the budget. With the increase of workers and budget, more workers will be selected to complete the task and the utility of requester increase spontaneously. As seen from Fig. 8, WCIM outputs higher utility of requester than GWS because WCIM adopts dynamic programming which can always obtain the global optimal solution.

6.3   Evaluation of the Miner-Centric Incentive Mechanism

To evaluate MCIM, we reveal impacts of total transaction fee F and the number of miners $|M|$ on the number of total transactions TX and the utility of the requester $U_R(F)$.

Number of Total Transactions: Figure 9 depicts the impact of F on TX when $|M|$ is fixed at 1000. It is found that TX increases as F increases. This is because increased F incentivizes miners to pack more transactions into their blocks. Figure 10 depicts the impact of $|M|$ on TX when F is fixed at 20000. We can find that with the increase of $|M|$, TX increases with a slowdown, which is in line with (18). The reason is that more involved miners intensify the competition for the transaction fee, which incentivizes miners to validate more transactions.

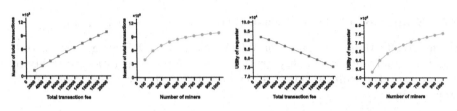

Fig. 9. Impact of F on TX.   **Fig. 10.** Impact of $|M|$ on TX.   **Fig. 11.** Impact of F on $U_R(F)$.   **Fig. 12.** Impact of $|M|$ on $U_R(F)$.

Fierce competition, however, reduces the probability of winning mining rewards. As a result, the growth trend of total transactions is slowing down.

Utility of Requester: For the utility of requester, we first evaluate the impact of F on it when $|M|$ is fixed at 1000 and present results in Fig. 11. We find that as F increases, $U_R(F)$ decreases gradually. The intuitive reason is that, the margin utility descends with more transaction fees. Although the requester announces more transaction fees, there is a diminishing marginal effect on the contributions of miners, which fails to cover the corresponding increase of F. As shown in Fig. 12, we evaluate the impact of $|M|$ on $U_R(F)$ when F is fixed at 20000. It is found that the requester can achieve greater utility when more miners join in the mining, which indeed demonstrates diminishing returns when $|M|$ increases and is in line with $U_R(F)$. Combining Fig. 11 and 12, we find the requester can optimize its utility with more miners and fewer F.

7   Conclusion

In this paper, we have proposed a utility-oriented role-centric incentive mechanism design named URIM for BCSs, which consists of worker-centric and miner-centric incentive mechanisms. Through both rigorous theoretical analyses and extensive simulations, we have demonstrated that the worker-centric incentive mechanism is computationally efficient, individually rational and truthful, and the miner-centric incentive mechanism can maximize the utility of the requester based on optimal strategies of miners. In the future work, we will further explore non-monetary incentive mechanisms for BCSs and evaluate our design in real-world applications.

Acknowledgements. This work was supported by the Scientific Research Program of Science and Technology Commission of Shanghai Municipality under Grant No. 19511102203.

References

1. Amintoosi, H., Kanhere, S.S.: A reputation framework for social participatory sensing systems. Mob. Netw. Appl. **19**(1), 88–100 (2014)
2. An, B., Xiao, M., Liu, A., Gao, G., Zhao, H.: Truthful crowdsensed data trading based on reverse auction and blockchain. In: Li, G., Yang, J., Gama, J., Natwichai, J., Tong, Y. (eds.) DASFAA 2019. LNCS, vol. 11446, pp. 292–309. Springer, Cham (2019). https://doi.org/10.1007/978-3-030-18576-3_18
3. Chatzopoulos, D., Gujar, S., Faltings, B., Hui, P.: Privacy preserving and cost optimal mobile crowdsensing using smart contracts on blockchain. In: 2018 IEEE 15th International Conference on Mobile Ad Hoc and Sensor Systems (MASS), pp. 442–450. IEEE (2018)
4. Conti, M., Gangwal, A., Todero, M.: Blockchain trilemma solver algorand has dilemma over undecidable messages. In: Proceedings of the 14th International Conference on Availability, Reliability and Security, pp. 1–8 (2019)
5. Crosby, M., Pattanayak, P., Verma, S., Kalyanaraman, V., et al.: Blockchain technology: beyond bitcoin. Appl. Innov. **2**(6–10), 71 (2016)
6. Duan, H., Zheng, Y., Du, Y., Zhou, A., Wang, C., Au, M.H.: Aggregating crowd wisdom via blockchain: a private, correct, and robust realization. In: 2019 IEEE International Conference on Pervasive Computing and Communications (PerCom2019), pp. 43–52. IEEE (2019)
7. Feng, Z., Zhu, Y., Zhang, Q., Ni, L.M., Vasilakos, A.V.: TRAC: truthful auction for location-aware collaborative sensing in mobile crowdsourcing. In: IEEE INFOCOM 2014-IEEE Conference on Computer Communications, pp. 1231–1239. IEEE (2014)
8. Houy, N.: The bitcoin mining game. Available at SSRN 2407834 (2014)
9. Hu, J., Yang, K., Wang, K., Zhang, K.: A blockchain-based reward mechanism for mobile crowdsensing. IEEE Trans. Comput. Soc. Syst. **7**(1), 178–191 (2020)
10. Huang, J., et al.: Blockchain-based mobile crowd sensing in industrial systems. IEEE Trans. Ind. Inf. **16**(10), 6553–6563 (2020)
11. Koutsopoulos, I.: Optimal incentive-driven design of participatory sensing systems. In: 2013 Proceedings IEEE INFOCOM, pp. 1402–1410. IEEE (2013)
12. Lakhani, K.: Innocentive.com (a) (harvard business school case no. 608–170). Harvard Business School, Cambridge (2008)
13. Li, M., et al.: CrowdBC: a blockchain-based decentralized framework for crowdsourcing. IEEE Trans. Parallel Distrib. Syst. **30**(6), 1251–1266 (2018)
14. Lu, Y., Tang, Q., Wang, G.: ZebraLancer: private and anonymous crowdsourcing system atop open blockchain. In: 2018 IEEE 38th International Conference on Distributed Computing Systems (ICDCS), pp. 853–865. IEEE (2018)
15. Ogie, R.I.: Adopting incentive mechanisms for large-scale participation in mobile crowdsensing: from literature review to a conceptual framework. Hum.-Cent. Comput. Inf. Sci. **6**(1), 24 (2016)
16. de Pedro, A.S., Levi, D., Cuende, L.I.: WitNet: a decentralized oracle network protocol. arXiv preprint arXiv:1711.09756 (2017)
17. Pouryazdan, M., Kantarci, B., Soyata, T., Foschini, L., Song, H.: Quantifying user reputation scores, data trustworthiness, and user incentives in mobile crowdsensing. IEEE Access **5**, 1382–1397 (2017)
18. Xu, J., Xiang, J., Yang, D.: Incentive mechanisms for time window dependent tasks in mobile crowdsensing. IEEE Trans. Wirel. Commun. **14**(11), 6353–6364 (2015)
19. Yang, D., Xue, G., Fang, X., Tang, J.: Incentive mechanisms for crowdsensing: crowdsourcing with smartphones. IEEE/ACM Trans. Netw. **24**(3), 1732–1744 (2015)

20. Yang, M., Zhu, T., Liang, K., Zhou, W., Deng, R.H.: A blockchain-based location privacy-preserving crowdsensing system. Futur. Gener. Comput. Syst. **94**, 408–418 (2019)
21. Zhang, J., Cui, W., Ma, J., Yang, C.: Blockchain-based secure and fair crowdsourcing scheme. Int. J. Distrib. Sens. Netw. **15**(7), 1550147719864890 (2019)

PAS: Enable Partial Consensus
in the Blockchain

Zihuan Xu, Siyuan Han, and Lei Chen[✉]

The Hong Kong University of Science and Technology, Hong Kong, China
{zxuav,shanaj,leichen}@cse.ust.hk

Abstract. Permissioned Blockchain enables distributed collaboration among organizations that may not trust each other. However, existing systems cannot efficiently support the ordering and execution of transactions in different workflows parallelly, which seriously affects system scalability and performances in terms of throughput and latency.

In this paper, we present a partial consensus mechanism named PAS to achieve fault tolerance and parallelism of transaction processing. In PAS, transactions in different workflows only need to be confirmed by the involved subset of nodes, which significantly enhances the system performance and scalability. Specifically, we introduce a novel data structure, called the hierarchical consensus tree (HCT). It is maintained in each node and used to coordinate the consensus process. HCT guarantees that the consistency reached in different sets of nodes is eventually agreed by all nodes without conflicts and rollbacks. Since there are many valid HCTs with different system improvements, we introduce an optimization problem, named OHCT, to obtain an HCT with respect to the optimal enhancement. We prove OHCT is NP-hard and propose a general framework with efficient algorithms to address it. Finally, we implement PAS on PBFT-based Hyperledger fabric and conduct extensive experiments to show the performance and scalability of PAS.

1 Introduction

The *permissioned Blockchain* (*e.g.*, Hyperledger [2], Multichain[1], and Tendermint [5]), where the node identities are controlled and known by each other, builds a dedicated environment to prompt accountable interactions among users. However, its performance and scalability issues caused by the underlying consensus mechanism arise many concerns [10].

Most existing permissioned chains require every node to maintain a single ledger and treat the system as a *replicated state machine* to reach global consistency by involving all nodes at any time [2], meanwhile, transactions are executed and ordered sequentially. Thus, it fails to parallelize the transactions that are not dependent on each other, which leads to low system performance and scalability. Consider a supply chain management example described in [24] where a role in the supply-chain workflows has multiple instances as shown in Fig. 1.

---

[1] https://www.multichain.com/.

© Springer Nature Switzerland AG 2021
C. S. Jensen et al. (Eds.): DASFAA 2021, LNCS 12683, pp. 375–392, 2021.
https://doi.org/10.1007/978-3-030-73200-4_26

Example 1. Suppose there are two under processing workflows: **Workflow 1:** A factory F_1 has produced some products P_1 and stored them in a warehouse W_2. Currently, a retailer R_3 places an order O_1 to purchase these products. **Workflow 2:** A retailer R_1 places an order to purchase product P_2 firstly. Now, factory F_1 confirms the order and places another order to P_2's material supplier S_1. Meanwhile, F_1 and S_1 agree to deliver the material M_2 by carrier C_1.

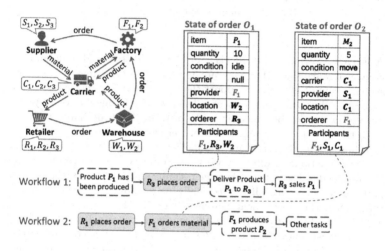

Fig. 1. Example in the supply-chain scenario

Here, companies collaborate to accomplish different supply-chain workflows which can be divided into several tasks with different service-level agreements (SLAs) agreed by related users to determine the data to read/write and the responsibility of each user. These SLAs can be represented in *smart contracts* [23]. Participants of each task modify the task data by transactions. Specifically, contracts O_1 and O_2 record data states of two tasks (R_3 orders product P_1 and F_1 orders material M_2). Since the data in O_1 and O_2 do not have overlap and dependency, participants of two tasks can order and execute task transactions internally and parallelly, because they are the only current valid modifiers of the task data. Notice that, each company can participate in multiple tasks simultaneously (*e.g.*, factory F_1 produces different products in both tasks) and join or leave a task at any time (*e.g.*, carrier C_1 only participates in O_2 for a while and after the delivery, later state of O_2 is irrelevant to C_1). Thus, to maintain a complete Blockchain modification history, the participation order of each user in each task needs to be preserved and globally agreed without conflicts.

In summary, to work in a more general scenario and obtain high performance and scalability, a Blockchain should satisfy at least the following three requirements: **(1)** Users can flexibly join or leave the modification processes of different data with separate SLAs. **(2)** Transactions modifying data without dependency can be ordered and executed in parallel. **(3)** The total order of transactions with dependencies and the users' participation order of the data modifications can be eventually agreed by the entire system under a Byzantine fault environment.

Most existing works satisfy *requirement (2)* by introducing sharding of nodes. For example, Hyperledger fabric v1.0+ uses channels and CAPER [1] uses applications to separate nodes. Such a mechanism enables processing transactions in different shards in parallel. However the settings of shards are static, it is inflexible for a user to join or leave a shard that fails to efficiently support *requirement (1)*. Moreover, to fulfill *requirement (3)*, fabric adopts a trusted channel to deal with cross-channel transactions [3] which breaks the decentralization principle of the Blockchain. While, to order the cross-application transaction in CAPER, all system nodes are involved which brings high communication cost in running the BFT consensus protocol. Especially, when the cross-application transactions occupy the majority, the system latency increase dramatically [1].

To overcome the shortcomings of existing approaches, in this paper, we propose a novel consensus mechanism called PAS. In particular, to satisfy *requirement (1)*, we separate the transactions into tasks. In a period, specific data can only be updated by users in one task. We use a special transaction to globally specify the task participants, such that a user can join or leave a task at any time. To satisfy *requirement (2)*, we order and execute transactions in different tasks in parallel. A scope of involved nodes reach strong consistency by the BFT protocol (*e.g.*, PBFT [7]). To satisfy *requirement (3)*, we propose a data structure named *hierarchical consensus tree* (HCT) to coordinate the consensus process to ensure the eventual agreement on every partial consensus without conflicts and rollbacks. Besides, as different HCT constructions can affect the system performance and scalability, to get the optimal HCT, we define an optimization problem. Though the problem is NP-hard, we managed to propose efficient solutions to it. We summarize our contributions as follows:

(i) We separate transaction types in the distributed collaboration scenario and identify the challenges to order them in parallel. Then we propose a partial consensus mechanism named PAS to address these challenges.

(ii) We propose a structure named hierarchical consensus tree (HCT) to support our consensus mechanism and introduce the OHCT problem to obtain an optimal HCT with the maximum system performance and scalability improvement.

(iii) We prove the NP-hardness and approximation hardness of the OHCT problem. Thus, we propose efficient algorithms to construct the HCT.

(iv) We implement PAS on PBFT-based Hyperledger Fabric as an example, and conduct extensive experiments to evaluate the system improvement as well as the effectiveness and efficiency of HCT construction algorithms.

The rest of the paper arranged as follows: Sect. 2 reviews related works. Section 3 overviews the PAS mechanism. Section 4 and Sect. 5 introduce the consensus in PAS and propose the HCT to realize the mechanism. Section 6 introduces the OHCT problem and the general framework with efficient algorithms to address it. Section 7 shows experiments and evaluations. We conclude in Sect. 8.

2   Related Work

Sharding Techniques. To achieve *requirements 1 and 2*, the sharding technique divides and maintains system states in several shards. The maintainers of shards consent in parallel on the execution order of transactions updating the states within each shard. However, to fulfill the *requirement 3*, the biggest challenge is to deal with the cross-shard transactions updating states in different shards simultaneously. Existing solutions (*i.e.*, RapidChain [27], OmniLedger [14] and Elastico [17]) are limited to the UTXOs transaction model. A recent work [9] applies sharding with SGX [18] under general workloads. It relies on a dedicated committee running BFT protocol to deal with cross-shard transactions. However, without a well-designed system state sharding schema, the majority of transactions can be cross-shard that is costly to deal with. Thus, in the worst case, the performance is merely the same as the system without sharding.

Hyperledger shards the users in different workflows by channels which are partitions of the network. However, channels are isolated from each other. It is inflexible for a node to join or leave a workflow at any time since the configuration of channels is fixed. Moreover, interactions between channels rely on a trusted channel [3] or an atomic commit protocol [2], which either breaks the decentralization or still treat the system as an entirety. Meanwhile, CAPER [1] adopts a similar idea to shard the users into applications based on their collaboration workflows and process transactions of each application internally. To solve the cross-application transactions, additional BFT protocols are designed. However, the protocols still need to involve all nodes. Moreover, if such transactions occupy the majority, the system latency increase dramatically [1].

Directed Acyclic Graph (DAG). By changing the chain-like structure to a directed acyclic graph, transactions can be appended to multiple branches in parallel which satisfy *requirements 1 and 2*. For example, IOTA [20] and Byteball [8] are two DAG-based permissioned chains. To satisfy the *requirement 3*, IOTA enforces each new transaction to pick two existing transactions as its predecessors and use the number of transactions' descendants and a PoW nonce as the proof to prevent conflicts. However, similar to PoW, the security of such protocol is nondeterministic and it is limited to cryptocurrency applications. While Byteball relies on a set of privileged users to order the transactions which breaks the decentralization principle and can easily become the performance bottleneck.

3   Overview of PAS

To satisfy all the requirements, PAS shards users into scopes. We use tasks and a special transaction to specify valid modifiers of a set of system states. We organize the scopes into a tree-like structure with each tree node accompanied by a Blockchain ledger. For each transaction, we determine which user scopes are compulsory to reach the consensus based on modified states and current valid state modifiers. Such that, the order of each transaction can be determined immediately after only a portion of all nodes reaching consensus. The result is eventually propagated to the system which significantly improves the parallelism.

Similar to prior works [9, 14, 19], we make two assumptions on the network: (i) Nodes are fully connected with each other. (ii) The message sent by an honest node can be eventually received by others within a maximum delay.

User and Validator. We denote *users* in PAS by $U = \{u_1, u_2, .., u_n\}$. Besides, like other permissioned chains, a set of nodes known as *validators* process transactions and ensure the consistency of states on behalf of the users. In PAS, we set the number of validators and users equally. Moreover, cryptography signatures are used to ensure the integrity and authenticity of a message sent by each node.

Consensus Scope and Ledger. With the similar idea of node partitions [12, 13, 25], in PAS, we partition validators to several disjoint sets named *consensus scope* with the cardinality in the range of $[k, 2k)$. Details will be discussed in Sect. 4. Validators in each scope maintain a ledger consisting of the transactions ordered within the scope. PAS works in epochs denoted by e. In each epoch, validators are shuffled to serve different scopes. Besides, each user is assigned to one scope to process transactions. Different from validators, the user-scope assignment is static. Therefore, the user-scope assignment also determines the structure of consensus scopes. We use $\mathcal{N} = \{\mathcal{N}_1, \ldots, \mathcal{N}_m\}$ to denote the consensus scopes where $\mathcal{N}_i$ also represents the users assigned to scope i ($\mathcal{N}_i \subset U$). The organization of the consensus scope is shown at the top of Fig. 2.

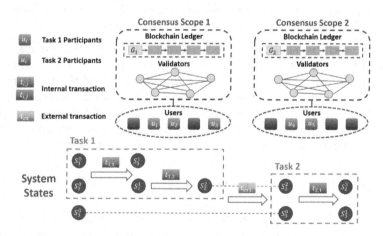

Fig. 2. Overview of the system

Data Model. Beyond UTXO-based model [19], we focus on the state-based model introduced in Ethereum [23]. Specifically, the system states denoted by $\mathcal{S} = \{S_1, S_2, \ldots, S_n\}$, can be created, updated or deleted by transactions.

Task and Participant. We define a *task* as successive modifications on a set of system states within a group of users (task *participants*). Besides, modifications may depend on other states within but not beyond the task (otherwise, those states should be involved in this task as well). Thus, the transaction execution order within a task is determined by its participants. In Fig. 2, there are two tasks with different users to update different states. For instance, once the system agrees on states (S_1, S_2) and participants (u_1, u_2, u_3) involved in *task 1*, until finishing the *task 1*, only its participants are authorized to generate transactions and determine their execution order to modify states in the task.

Transactions. We divide the transaction in PAS into two types based on the functionality. One is the ***internal transaction*** used to modify states in a task. In Fig. 2 *task 1*, u_1, u_2 and u_3 use transaction $t_{1,2}$ to modify S_2 from the original state S_2^1 to S_2^2. The other is the ***external transaction*** to change the task participants or create a new task. In Fig. 2, between *task 1* and *task 2*, an external transaction t_{ex1} changes the valid modifiers of S_2 with the confirmation of S_2's final state (S_2^2) after *task 1* which is also the initial state of S_2 in *task 2*. Importantly, the external transaction determines valid modifiers of each state.

Definition 1. *Transaction*. *A transaction is a tuple $t = (id, S, op, i, P, P', \Sigma)$, where id is the unique order of transactions modifying a set of states S. op is the operation with parameters. i is the initiator where $\forall i, u_i \in U$. P is the set of current valid modifiers of S where $P \subseteq U$ and P' is the new valid modifiers of S where $P' \subseteq U$. Σ is a set of signatures signed by validators who have ordered t.*

The id is to specify the order of transactions and the Σ is obtained during the consensus. For the validation of a transaction t, suppose a function $P(s)$ returns current valid modifiers of state s based on the records kept by each validator, if $\exists s \in t.S$, $P(s) \neq t.P$ or $t.i \notin P(s)$, t is treated as invalid.

Threat Model. In this paper, we assume the malicious nodes are less than $\frac{1}{3}$, since the BFT protocol is one of our security guarantee. The attackers are adaptive like described in [9,15,17,27] where the corruption of the validator takes time to achieve. Different from other sharding works [9,15,27], instead of assuming all the shards perform honestly, we do not assume all the consensus scopes perform honestly. Instead, nodes in a consensus scope can perform maliciously together to cheat others outside the scope for their profit.

4 Partial Consensus in PAS

Achieve Partial Consensus. As we assign users to scopes and rely on validators to process transactions, given a transaction t, after determining which users are relevant to t, validators in corresponding scopes run the BFT protocol to order and execute t. Besides, we need to identify in which consensus scope t has been consented. We denote $\mathcal{PC}(t) \subseteq \mathcal{N}$ as scopes where t has been ordered. Since validators in each scope are periodically shuffled, after current validators in $\mathcal{PC}(t)$ reach strong consistency, they will sign on t accompanied by the current epoch index e and commit t on the Blockchain ledger of the scope. Others can verify the authority of t through $t.\Sigma$. Notice that all existing BFT protocols are applicable for PAS. Examples are PBFT [7], Tendermint [5], Zyzzyva [16], Hot-Stuff [26] and MirBFT [21] where all of them can achieve $\lfloor \frac{n-1}{3} \rfloor$ fault-tolerant.

Validator Assignment. For safety, validators shuffling needs to be unbiased. Omniledger combines RandHound [22] with the verifiable random function (VRF) based leader election algorithm [11] to assign validators which can be used in PAS as well. Specifically, with a bounded message transmission delay δ, in each epoch, validators compute a hash value and gossip it to others for a time δ. Then, the one who gets the lowest value is selected as the leader to run the RandHound protocol to generate and broadcast a bias-resistant random number rnd_e with correctness proof. Finally, others can use rnd_e to get the validator-scope assignment. We refer the details and its security analysis to [15].

Consensus Scope Size. The size of a scope is the number of inside validators and we set it in the range of $[k, 2k)$ to seek a balance between security and performance. Fewer validators lead to lower latency and higher throughput [10]. However, it also reduces the safety, especially for the scope with the minimum size k. Since the validator assignment can be treated as random sampling without replacement, we consider the probability to form a fault scope directly. Suppose there are V validators with αV malicious, the random variable X denotes the number of malicious nodes in the scope with k validators. X should follow the hypergeometric distribution where $X \sim (k, V, \alpha V)$. Given a BFT protocol with f (e.g., for PBFT $f = \frac{n}{3}$) malicious tolerance, the probability to form a fault scope is $Pr[X \geq f] = \sum_{x=f}^{k} \frac{\binom{\alpha V}{x}\binom{(1-\alpha)V}{k-x}}{\binom{V}{k}}$. For example, with $V = 128, \alpha = 0.2, k = \frac{V}{8}$, by using PBFT, $Pr[X \geq \frac{k}{3}] = 6\%$. As stated in [9], when k is large enough (e.g., $k \geq 600$), the probability is considered negligible (e.g., $\leq 2^{-20}$).

5 Eventual Consistency

After reaching partial consensus, validators propagate the result to others. Thus, every node will eventually receive all consensus results. However, if we want to finalize a transaction t when only a subset of nodes reaches partial consensus on its execution order, it is vital to prevent the subsequential transactions conflicting with t from being adopted. Therefore, we first analyze possible conflicts.

Definition 2. Conflict Transactions. *For two valid transactions t_1 and t_2 within the same task $(t_1.P = t_2.P \bigwedge t_1.S \cap t_2.S \neq \emptyset))$ where t_1 is the first to reach partial consensus, if one of the three conditions holds, t_2 conflicts with t_1. 1. internal conflict: $t_1.op \neq t_2.op \bigwedge t_1.P' = t_2.P'$. 2. external conflict: $t_1.op = t_2.op \bigwedge t_1.P' \neq t_2.P'$. 3. dual-conflict: $t_1.op \neq t_2.op \bigwedge t_1.P' \neq t_2.P'$.*

We only focus on the conflict within a task, because, if t_1, t_2 are from two tasks, there must be one external transaction t_{ex} in between of the tasks. As long as we ensure that t_{ex} can be eventually agreed by the system without conflict (has been covered in condition 2 in Definition 2), t_1, t_2 do not have conflict anymore.

Internal Conflict. This happens when the participants of a task concurrently modify the same state with different operations by two transactions with the same id. Since task participants remain the same, validators need to reach partial consensus on their order are the same as well.

External Conflict. This happens when two external transactions t_1, t_2 (with the same id) change modifiers of the same state S with the final value (say S^*) to different users. To order t_1 or t_2 both current and new modifiers need to be involved. Suppose the current modifiers are in scope $\mathcal{N}_0$. When validators in $\mathcal{N}_0$ act maliciously, they can reach two conflict partial consensus (switching the modification authority to users in scopes $\mathcal{N}_1$ and $\mathcal{N}_2$) with the validators in $\mathcal{N}_1$ and $\mathcal{N}_2$ simultaneously. Meanwhile, users and validators in $\mathcal{N}_1$ and $\mathcal{N}_2$ cannot detect the deviation respectively which leads to the system inconsistency.

Dual-conflict. This happens when the modifier change and state update happen simultaneously. When shifting the state modifiers, an external transaction must specify it is based on which internal transaction to explicitly inform the final state values to new modifiers. For example, in Fig. 2, t_{ex1} is generated after $t_{1,2}$ ($t_{ex1}.id = t_{1,2}.id+1$) specifying the value of S_2 is S_2^2. For an internal transaction t_{in} ($t_{in}.id = t_{ex1}.id$) modifying S_2 based on the value S_2^2, if t_{ex1} reaches partial consensus first, t_{in} should not be accepted anymore, vice versa.

Hierarchical Consensus Tree. To prevent the conflicts, our idea is to control the consensus scopes where a transaction is ordered. We introduce a data structure named *hierarchical consensus tree* (HCT) kept by each node to organize the scopes and Blockchain ledgers and used to coordinate the consensus process. By leveraging a tree structure, any two scopes will share a common root. We restrict a transaction involving users in different scopes to be ordered by all validators under the common root of these scopes. Thus, the conflict can always be detected and we set rules to prevent the acceptance of conflicts.

For a transaction t, all possible combinations of $\mathcal{PC}(t)$ (scopes have reached consensus on t) form a join semi-lattice which is a partial order of the set include operation ($\subseteq$). Given a pair of scopes $\mathcal{N}_1$ and $\mathcal{N}_2$, a least upper bound (LUB) $\sqcup$ exists. $\bar{\mathcal{N}} = \mathcal{N}_1 \sqcup \mathcal{N}_2$ is a LUB of $\{\mathcal{N}_1, \mathcal{N}_2\}$ iff $\forall \mathcal{N}^*, \mathcal{N}_1 \subseteq \mathcal{N}^* \bigwedge \mathcal{N}_2 \subseteq \mathcal{N}^* \Rightarrow \mathcal{N}_1 \subseteq \bar{\mathcal{N}} \bigwedge \mathcal{N}_2 \subseteq \bar{\mathcal{N}} \bigwedge \bar{\mathcal{N}} \subseteq \mathcal{N}^*$. Based on LUB we can have the following definition.

Definition 3. *Monotonic Consensus Semi-lattice (MCSL). MCSL refers to $\mathcal{PC}(t)$ with the properties: (1) Forms a semi-lattice ordered by $\subseteq$. (2) Merging two scopes $\mathcal{N}_i$ and $\mathcal{N}_j$ involves consensus scopes included in the LUB of $\{\mathcal{N}_i, \mathcal{N}_j\}$. (3) Scope changing is non-decreasing ($\mathcal{PC}(t)$ can only accept new scopes).*

For a transaction t involving users in scopes $\mathcal{N}_1, \mathcal{N}_2$, it needs to be ordered at least by validators in $\mathcal{N}_1$ and $\mathcal{N}_2$. According to the MCSL, it equals to merge the partial consensus results in $\mathcal{N}_1$ and $\mathcal{N}_2$ and t should be ordered in all scopes included in the LUB of $\{\mathcal{N}_1, \mathcal{N}_2\}$. However, for another scope $\mathcal{N}_3$, it is still possible that $(\mathcal{N}_1 \sqcup \mathcal{N}_2) \cap (\mathcal{N}_1 \sqcup \mathcal{N}_3) = \mathcal{N}_1$ which means users in $\mathcal{N}_1$ can still generate conflict transactions without letting nodes in $\mathcal{N}_2$ and $\mathcal{N}_3$ be aware. Therefore, we further define the HCT to address this problem.

Definition 4. Hierarchical Consensus Tree (HCT). *HCT is a restricted MCSL with the constraint that each scope set can only have one ancestor. In an HCT, each leaf node is a single consensus scope. Each internal node is accompanied by a Blockchain ledger recording the transactions ordered by all validators covered by the consensus scope set of the internal node. Moreover, for two conflict transactions t_1 and t_2, if $\mathcal{PC}(t_2) \subset \mathcal{PC}(t_1)$, t_2 is treated as invalid.*

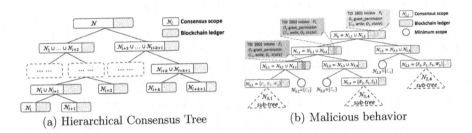

(a) Hierarchical Consensus Tree   (b) Malicious behavior

Fig. 3. Hierarchical Consensus Tree examples

Figure 3a shows an HCT example. The represented consensus scope of a tree node is the union of its two children's scopes. Precisely, each tree node ledger is maintained by validators in the scope of the tree node. A transaction involving users in different scopes should be ordered on the ledger of the tree node representing the LUB of these scopes. Now we prove HCT can prevent the conflict transactions being adopted during the eventual consistency process.

Theorem 1. *By following the hierarchical consensus tree to reach partial consensus, when a transaction t fulfills $(t.P \bigcup t.P') \subseteq \mathcal{PC}(t)$, its conflict transaction t^* cannot be accepted by any correct node.*

Proof. Suppose a transaction t has reached partial consensus and its later generated conflict transaction is represented as t^* (by Definition 2, $t.P = t^*.P$). We use S_1 to denote the set of consensus scopes where users of $t.P$ are assigned. Also, we use S_2, S_3 to denote the scopes where $t.P'$ and $t^*.P'$ are assigned respectively. Based on the HCT, $\mathcal{PC}(t) = S_1 \sqcup S_2$. For different conflicts:

Internal conflict $(t.op \neq t^*.op \bigwedge S_2 = S_3)$. To order t^*, it must have $\mathcal{PC}(t^*) = S_1 \sqcup S_3 = S_1 \sqcup S_2 = \mathcal{PC}(t)$. Since the order of t has reached partial consensus in $\mathcal{PC}(t)$ before t' is generated, honest nodes will treat t' as invalid.

External conflict $(t.op = t^*.op \bigwedge S_2 \neq S_3)$. To order t^*, it must have $\mathcal{PC}(t^*) = S_1 \sqcup S_3$. If $S_3 \subset (S_1 \sqcup S_2) \Rightarrow (S_1 \sqcup S_3) \subset (S_1 \sqcup S_2) \Rightarrow \mathcal{PC}(t^*) \subset \mathcal{PC}(t)$. According to Definition 4, t^* is invalid. If $S_3 \not\subset (S_1 \sqcup S_2)$, there must be at least one consensus scope $\mathcal{N}$ where $\mathcal{N} \notin (S_1 \sqcup S_2) \bigwedge \mathcal{PC}(t) \subset (S_1 \sqcup \{\mathcal{N}\})$. It means to order t^*, validators who have ordered t must be involved in the process. Thus, validators in $\mathcal{PC}(t)$ can prove that $t.\Sigma$ contains their signatures during the execution of the BFT protocol such that other honest validators in $S_1 \sqcup S_3$ can deny t^*.

Dual-conflict $(t.op \neq t'.op \wedge S_2 \neq S_3)$. If t is an external and t^* is an internal transaction, there must be $S_1 \neq S_2 \wedge S_1 = S_3$, meanwhile, $S_1 = (S_1 \sqcup S_3) \subset (S_1 \sqcup S_2) \Rightarrow \mathcal{PC}(t^*) \subset \mathcal{PC}(t)$. According to Definition 4, t^* is invalid and will be discarded. Else if t is an internal transaction and t^* is an external transaction, there must be $S_1 = S_2 \wedge S_1 \neq S_3$, we also consider two cases where $S_3 \subset (S_1 \sqcup S_2)$ or $S_3 \not\subset (S_1 \sqcup S_2)$. Thus, the rest proof is the same as external conflict.

Based on Example 1, we give an HCT example shown in Fig. 3b, to illustrate how it can prevent the generation of conflict transactions.

Example 2. Suppose a transaction t is sent by F_1 to appoint O_1's carrier as C_1 and allow C_1 to modify the state of $O_1.condition$ and t has been ordered by validators in $\mathcal{N}_{1,1}$. Meanwhile, F_1, C_2 and C_3 are malicious and try to generate an external conflict transaction t^* granting the same permission to another user:

Case 1: F_1 grants permission to C_2 which needs to be ordered by the validators in $\mathcal{N}_{2,1}$. Since $\mathcal{N}_{2,1} \subset \mathcal{N}_{1,1}$ and F_1, C_2 know the existence of t during their partial consensus, by Definition 4, t^* is invalid and will be discarded by the validators.

Case 2: F_1 grants permission to C_3 which needs to be ordered by the validators in $\mathcal{N}_0$. Since the honest validators in $\mathcal{N}_{1,1}$ has obtained the confirmation signatures on t, they can prove the existence of t to deny t^* during the running of the BFT protocol and other honest validators will discard t^* as well.

Notice that, for each node, in the path from its position to the tree root, the ledger of each internal node will always be up-to-date. Because whenever an update is made on those ledgers, the nodes will be involved in the consensus process to reach strong consistency. For instance, in Example 2, C_1 always knows the latest transactions committed on the ledgers of $\mathcal{N}_{2,2}, \mathcal{N}_{1,1}$ and $\mathcal{N}$.

6   HCT Optimization

The bottleneck to scale the system is the communication cost to consent the transactions [10], while, **transaction generation frequency (TGF)** of users (the interaction frequency between users) contributes to the cost.

Definition 5. *Interaction Frequency.* *A matrix $\mathcal{F}_{n \times n}$ where $n = |U|$ records the interaction frequency in users. $f_{i,j}$ in $\mathcal{F}$ is the generation frequency of transactions involving u_i and u_j where $u_i, u_j \in U$ and $f_{i,j} = f_{j,i}$.*

For instance, in Example 1, if retailer R_3 often places order O_1, the interaction frequency between F_1, W_2, F_1, R_3 and W_1, R_3 will be high. Notice that, the TGF is an estimation result in a period or determined by actual applications. It is used to establish and reconfigure the system based on the demands of users.

Optimal HCT Problem. The consensus message complexity denoted by $T(\mu)$ where μ is the number of involved validators, is determined by the used BFT protocol (*e.g.*, for PBFT $T(\mu) \in O(n^2)$). Observe that $T(\mu) \propto \mu$. Suppose users u_1 and u_2 are in scopes $\mathcal{N}_1$ and $\mathcal{N}_2$, the message complexity to consent transactions related to u_1 and u_2 is $f_{1,2}T(\mu_{1,2})$ where $\mu_{1,2}$ is the number of validators in the scopes under $\mathcal{N}_1 \sqcup \mathcal{N}_2$. Therefore, a well-structured HCT can further reduce the message complexity which leads to better system performance and scalability. We define the HCT optimization problem as below:

Definition 6. *Optimal Hierarchical Consensus Tree (OHCT) Problem.* *Given the interaction frequency matrix $\mathcal{F}_{n \times n}$, the complexity function $T(\cdot)$ of the BFT protocol and the scope validator cardinality constraint $[k, 2k]$. Our goal is:*

$$minimize \quad \sum_{u_i, u_j \in U} f_{i,j}\, T(\mu_{i,j}) \quad (\forall f_{i,j} : i \leq j)$$

$$subject\ to \quad \min_{u_i, u_j \in U} \{\mu_{i,j}\} \geq k, \ \max_{u_i, u_j \in U} \{\mu_{i,j}\} < 2k$$

Hardness Analysis. We prove the NP-hardness by studying a special case of OHCT and reducing the minimum bisection problem (MBP) [4] to it.

Theorem 2. *OHCT problem is NP-hard.*

Due to the space, we only show our proof sketch. Consider a special case of OHCT problem where $k = \frac{|U|}{2}$. In this case, we can only bisect users in two sets S_1, S_2 with all transactions completion cost as a constant $T(|U|)$ and minimize $T(|U|) \sum_{u_i \in S_1, u_j \in S_2} f_{i,j}$. Then, we can reduce the MBP to the case. Besides, we further analyze the hardness to get an approximation solution to OHCT.

Theorem 3. *There is no algorithm with constant approximation ratio for OHCT.*

The sketch of the proof is to use the conclusion in [6] that for a fixed $\epsilon > 0$, it is NP-hard to approximate the MBP problem with an additive term of $n^{2-\epsilon}$ [6].

Solution Framework. To solve the OHCT problem, a general framework is to **1. construct** an optimal HCT by regarding each user as a single consensus scope first and **2. pruning and merging** to fulfill the cardinality constraint.

Top-down Construction Algorithm. Intuitively, the greedy way to do the construction is to pair two users into a binary tree first. Then, each time we randomly pick one user from unassigned users. Starting from the tree root, we compare the normalized interaction frequency between the picked user and all users in the left and right sub-tree. Then we go into the root of sub-tree with higher normalized interaction frequency. We stop until we find the suitable leaf node position for all users. Details are shown in Algorithm 1.

Pruning And Merging Algorithm. After we obtain an HCT with each leaf to be one user, we perform DFS on the root. Each time, if the leaf number $|t|$ of an internal node is greater than $2k$, we continue. If $|t| \in [k, 2k)$, we group its leaves to one scope. If $|t| < k$, we reinsert each user in the leaf to the sibling sub-tree using the top-down construction algorithm. Because, since one sub-tree and its sibling are grouped under the same LUB by the construction algorithm, it means they have more frequent interactions. By merging the sub-tree into its sibling, it will bring less additional completion cost. Details are in Algorithm 2.

Complexity Analysis. Since $|U| = n$, for *Construction* with Algorithm 1, it recurrently decides the position for each user. In an average case, the algorithm takes $O(n^2 \log n)$. For *Pruning and Merging*, the worst case of its DFS takes $O(n)$ while the worst case cost of merge is $O(k \log n)$. In total, it takes $O(nk \log n)$. Since $k \ll n$, the total complexity is $O(n^2 \log n)$.

Bottom-up Construction Algorithm. Although the Algorithm 1 is efficient, its performance is affected by the input order of users. Thus, we can enhance the HCT construction by always considering every users. The idea is that each time we merge two sub-trees with the highest average transaction completion cost. Then, we recompute the cost between the new sub-tree and other sub-trees. We terminate until all users are rooted in the same tree. The intuition is to merge sub-trees with higher average transaction completion cost as earlier as possible to reduce the involved consensus scopes to the most. Details are in Algorithm 3.

Algorithm 1: Top-down HCT Construction

Input : Interaction frequency matrix $\mathcal{F}$ and Users U.
Output: tree root of a hierarchical consensus tree.
1 undetermined ← U;
2 find u_1 and u_2 with minimum interaction frequency;
3 HCTRoot.children ← $[u_1, u_2]$;
4 undetermined.remove(u_1 and u_2);
5 **foreach** user ∈ undetermined **do**
6 | tRoot ← HCTRoot;
7 | F_{left}, F_{right} ← 0, 0;
8 | **while** |tRoot.leaves| > 1 **do**
9 | | **foreach** leaf ∈ tRoot.leftChild.leaves **do**
10 | | | F_{left} += $\mathcal{F}[user][leaf]$;
11 | | **foreach** leaf ∈ tRoot.rightChild.leaves **do**
12 | | | F_{right} += $\mathcal{F}[user][leaf]$;
13 | | F_{left} /= tRoot.leftChild.leaves;
14 | | F_{right} /= tRoot.rightChild.leaves;
15 | | **if** F_{left} >= F_{right} **then**
16 | | | tRoot ← tRoot.leftChild
17 | | **else**
18 | | | tRoot ← tRoot.rightChild
19 | tRoot.children ← [tRoot, user];
20 **return** HCTRoot;

Algorithm 2: HCT Pruning And Merging

Input : root of a HCT and scope size constraint k.
Output: tree root of a hierarchical consensus tree.
1 push root in the search stack S_1;
2 mark all internal nodes as unchecked;
3 **while** S_1 is not empty **do**
4 | // |t|: number of leaves of tree t
5 | t ← S_1.getTop();
6 | **if** t.children are all checked **then**
7 | | S_1.pop() and continue while loop;
8 | t ← unchecked child node of t with fewer leaves;
9 | **if** |t| ≥ 2k **then**
10 | | S_1.push(t);
11 | **else if** |t| ≥ k **then**
12 | | merge all leaf nodes into one scope;
13 | | mark t as checked;
14 | **else**
15 | | Pruning branch t and save its leaves in L;
16 | | t.sibling.parent ← t.sibling.grandparent;
17 | | **foreach** l ∈ L **do**
18 | | | find position for l by searching t.sibling.
19 | S_1.pop();
20 **return** HCTRoot;

Complexity Analysis. In $n - 1$ rounds merging, the most time consuming part is to maintain the heap which provides the highest interaction frequency among sub-trees. It takes $O(max\{|t_i| \times |t_j|, (n - t)log(n^2)\}) \in O(nlogn^2)$ in average. Thus, using Algorithm 3 makes the time complexity of the framework be $O(n^2 logn^2)$.

Algorithm 3: Bottom-up HCT Construction

   Input : Interaction frequency matrix $\mathcal{F}$, Users U and cost function T.
   Output: tree root of a hierarchical consensus tree.

1 subTrees $\leftarrow$ U;
2 **foreach** $t_i \in subTrees$ **do**
3 | **foreach** $t_j \in subTrees$ **do**
4 | |_ $avgC(t_i, t_j) \leftarrow \mathcal{F}[i][j] \times T(2)$; //avgC: average completion cost

5 **while** $|subTrees| \, ! = 1$ **do**
6 | Merge t_i, t_j with the maximum $avgC(t_i, t_j)$ into t^* and remove t_i, t_j from subTrees;
7 | **foreach** $t \in subTrees$ **do**
8 | |_ $avgC(t^*, t) \leftarrow (\frac{avgC(t_i, t) \times |t_i|}{T(|t_i| + |t|)} + \frac{avgC(t_j, t) \times |t_j|}{T(|t_j| + |t|)}) \times \frac{T(|t_i| + |t_j| + |t|)}{(|t_i| + |t_j|)}$;

9 |_ subTrees.append(t^*);
10 **return** subTrees.first;

---

7 Experiment and Evaluation

In this section, we first evaluate the effectiveness and efficiency of our HCT construction algorithms on both real and synthetic datasets. Then, we choose *hyperledger fabric v0.6* as a permissioned Blockchain example to implement PAS and measure the performance and scalability enhancement brought by PAS.

7.1 HCT Construction Evaluation

Real Dataset. To obtain the real interaction frequency of Blockchain users, we use the dataset extracted from Ethereum blocks during the period from Dec 17, 2017, to Feb 23, 2018. It contains 14,393,250 unique addresses and 64,719,559 transactions. Specifically, we treat the token transform from one user to another as one task modifying the states of two account balances. We uniformly sample and group the unique addresses to form different sizes of user groups and obtain the interaction frequency distribution matrix among the groups.

Table 1. Synthetic datasets

Number of users $|U|$	4, 8, 16, **32**, 64, 128
σ of normal distribution	0.05, 0.1, **0.15**, 0.2, 0.25, 0.3
a of power-law distribution	1, 2, **3**, 4, 5, 6
Minimum scope size k	1, 4, **7**, 10, 13, 16

Synthetic Dataset. We generate synthetic interaction frequencies by following **uniform** (in the range of [0,1]), **normal** (with $\mu = 0.5$) and **power-law** (with $c = 1$) distributions. Table 1 shows the parameter settings we used in synthetic datasets and default values are in bold. Similarly, we construct transactions between users as the token transform. We first generate a $|U| \times |U|$ triangular contribution matrix (the sum of all elements is 1) by following the distributions mentioned above. This matrix denote the contribution of each pair of users to the transaction generation frequency (TGF mentioned in Sect. 6). Given the system

TGF, we can obtain the interaction frequency matrix by multiplying the TGF and the contribution matrix. Notice that, TGF is not selected as a parameter. Because the interaction frequency $f_{i,j}$ between two users p_i and p_j is computed by $TGF \times d_{i,j}$ where $d_{i,j}$ is the frequency contrition of user pair u_i and u_j. Thus, the overall message complexity is represented by $TGF \sum_{u_i, u_j \in U} d_{i,j} T(\mu_{i,j})$. For the same experiment settings, TGF is a constant which will not affect the result.

Implementation and Metrics. We implement our HCT construction algorithms in Python 3.7. The experiments are conducted on a server with Intel(R) Core(TM) i5 3.0 GHz CPU with 16 GB RAM. Each experiment is repeated 30 times and we report the average results. We choose PBFT as our baseline consensus protocol and set its $T(\mu) = \mu^2$. In each experiment case, we measure and compute the total message complexity result of each method and the **enhancement percentage** $(1 - \frac{algorithm}{baseline})$ compared with the baseline whose message complexity is $TGF|U|^2$. We compare our solution framework with the HCT construction algorithms of top-down, bottom-up and the random pair which randomly forms a valid HCT. The aim is to show the performance of the PAS even if in a random construction fashion.

Experimental Results. Since the real dataset tends to follow the normal distribution, due to the space, we only report the results on the real dataset and synthetic dataset following the power-law distribution.

Impact of Number of Users $|U|$. The first row of Fig. 4 shows the results of varying $|U|$ in both real and power-law synthetic dataset. The line chart shows the total consensus message complexity, while, the bar chart shows the enhancement percentage comparing with the baseline. For the top-down and bottom-up algorithms, the message complexity reduction is from 25% to 56%. The enhancement of the bottom-up is better than the top-down algorithm from 2% to 10% since the bottom-up always takes $T(\mu)$ into consideration, while, the time cost of bottom-up increases dramatically when $|U|$ increase. With increasing $|U|$ the enhancement of HCT also increases but the incremental speed becomes slower. Because HCT does not change the intrinsic complexity of the protocol itself. PBFT, as an example, with more validators, its $O(n^2)$ message complexity becomes obvious making the enhancement of HCT be a constant factor.

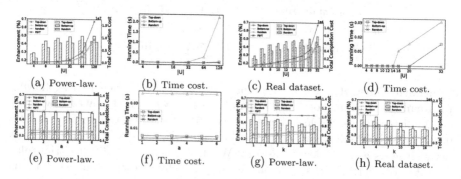

(a) Power-law.   (b) Time cost.   (c) Real dataset.   (d) Time cost.

(e) Power-law.   (f) Time cost.   (g) Power-law.   (h) Real dataset.

Fig. 4. Results of the comparison of HCT construction algorithms

Impact of a in Power-law Distribution. Figure 4e shows the impact of a in power-law distribution. With a larger a, the enhancement of HCTs built by three algorithms all decrease. Because a larger a means most of the interaction frequency is very small. Thus, the enhancement brought by the reduced validators in each consensus process becomes insignificant. Especially, the top-down algorithm is more sensitive to a, since it only considers a single node at each time which is more likely to reach local optimal.

Impact of Consensus Scope Cardinality k. Figure 4g and Fig. 4h show the results on varying k. With larger k, there are fewer consensus scopes in the HCT making the average number of validators need to be involved in each consensus process increase. Thus, the HCT enhancements all decrease. However, even if we only have 2 consensus scopes ($k = 16$), the performance enhancement can still reach 35% indicating to have a better balance between performance and security, it is not necessary to divide the consensus scope into extremely small ones.

Summary of the Results. From the above discussion, with HCT, the total consensus message complexity can always be reduced. Especially, bottom-up construction algorithm can achieve better performance than others with the reduction of PBFT message complexity by at least 30%. Moreover, the HCT enhancement is better when the interaction frequency distribution tends to follows a normal distribution. It means when all users frequently interact with each other, our mechanism can better improve the system. Besides, the cardinality constraint k will not influence the performance too much. The difference between $k = 1$ and $k = 16$ in a 32 nodes system is nearly 10%. Therefore, for better security, it is reasonable to set the k constraint higher.

7.2   PAS Evaluation

We evaluate the actual performance of PAS by implementing it on Hyperledger Fabric v0.6 and evaluate the systems with and without PAS.

Implementation and Metrics. The main aim of PAS is to make transactions be ordered in different consensus scopes based on involved users. Thus, the PAS system should support validators from the same Blockchain network in achieving consensus within different sub-networks. Therefore, we implement an external HCT module to make validators participate in the partial consensus of multiple subnets simultaneously. Specifically, the HCT module mainly does two things: **1. Compute and record the structure of HCT:** Since the HCT is constructed based on the estimated interaction frequency among users, after obtaining the interaction frequency from others (consensus may be required), nodes can build the HCT by running the construction algorithm by themselves. This process is only conducted when forming a new network, or the interaction frequency changes dramatically. Users (even only in a branch of the HCT) can decide to reconstruct the entire (or a sub) tree. **2. Routing transactions:** When a transaction is received by the validator, it will use the HCT module to determine in which scope to reach partial consensus to execute and order the transaction.

Simple key-value storage is implemented to record the valid modifiers of each system state obtained by the transaction history on the ledgers they maintain.

The experiments are conducted on the Azure cloud service cluster. We create validators with 16 GB RAM, 500 GB hard drive, running Ubuntu 18.04 LTS on each of them. They are connected to each other via a 1GB bandwidth network. The aim of our experiment is to measure the peak throughput of each system on varying the number of users/validators (from 4 to 16). We use the official chaincode transferring token between users as the task in the workflow. Then we use client nodes to simulate the transaction generation by following the interaction frequency distribution obtained from the real dataset. In each experiment, we steadily increase the overall system TGF to obtain the peak throughput which is measured by the completion rate from the time a transaction is generated until receiving the commit message of the block which contains the transaction.

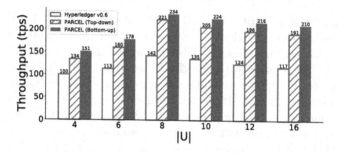

Fig. 5. Result of throughput on varying $|U|$

Experimental Results. Figure 5 shows the throughput comparison between PAS-based and original Hyperledger. The throughput of Hyperledger is between 100 to 142 tps, while, PAS can reach 134 to 234 tps. With the bottom-up constructed HCT, the performance enhancement is at least 50%. Meanwhile, despite the slightly lower enhancement of the top-down algorithm, as compared in Fig. 4 the lower time complexity of the top-down algorithm makes it more suitable for large scale systems. With more users, the enhancement ratio also increases which is similar to what we observed in the HCT construction experiments. Besides, with the nodes increasing, the throughput tends to decrease. In fact, in Hyperledger, to confirm connectivity between nodes, messages such as PeersMessage are sent periodically to check the connection status. When the number of validators in the network increases, such requests also increase, which affects the overall performance of the network to a certain extent. For example, the throughput of the two systems drops linearly from 8 to 16 nodes. Meanwhile, the dropping speed of HCT-based is slower than the original Hyperledger, which can also prove the better scalability of PAS.

8   Conclusions

In this paper, we introduce PAS, a consensus mechanism for permissioned Blockchain to satisfy the requirements in the general distributed collaboration scenario. Specifically, PAS enables a user to join or leave different tasks flexibly by using the transaction to specify the valid modifiers of each system state. We introduce the partial consensus to order transactions in different tasks in parallel. Moreover, to ensure the order of transactions determined by a set of nodes can be eventually agreed by all nodes with the BFT guarantee, we propose the hierarchical consensus tree (HCT) to coordinate the consensus process. When a transaction is ordered, the acceptance of its conflict transaction is strictly prevented. We also propose the OHCT problem to obtain an optimal HCT with the maximum system enhancement. We proved the NP-hardness and approximation hardness of the OHCT problem and propose a framework with efficient algorithms to solve it. Finally, we implement PAS on Hyperledger and conduct extensive experiments to evaluate it. The result shows that PAS can significantly improve system performance and scalability.

Acknowledgment. This work is partially supported by the Hong Kong RGC GRF Project 16213620, CRF Project C6030-18G, C1031-18G, C5026-18G, AOE Project AoE/E-603/18, China NSFC No. 61729201, Guangdong Basic and Applied Basic Research Foundation 2019B151530001, Hong Kong ITC ITF grants ITS/044/18FX and ITS/470/18FX, Microsoft Research Asia Collaborative Research Grant, Didi-HKUST joint research lab project, and Wechat and Webank Research Grants.

References

1. Amiri, M.J., Agrawal, D., Abbadi, A.E.: Caper: a cross-application permissioned blockchain. In: VLDB (2019)
2. Androulaki, E., et al.: Hyperledger fabric: a distributed operating system for permissioned blockchains. In: EuroSys (2018)
3. Androulaki, E., Cachin, C., De Caro, A., Kokoris-Kogias, E.: Channels: Horizontal scaling and confidentiality on permissioned blockchains. In: ESORICS (2018)
4. Arora, S., Karger, D., Karpinski, M.: Polynomial time approximation schemes for dense instances of np-hard problems. JCSS (1999)
5. Buchman, E., Kwon, J., Milosevic, Z.: The latest gossip on BFT consensus. arXiv preprint arXiv:1807.04938 (2018)
6. Bui, T.N., Jones, C.: Finding good approximate vertex and edge partitions is NP-hard. Inf. Process. Lett. (1992)
7. Castro, M., Liskov, B., et al.: Practical byzantine fault tolerance. In: OSDI (1999)
8. Churyumov, A.: Byteball: A decentralized system for storage and transfer of value (2016)
9. Dang, H., Dinh, T.T.A., Loghin, D., Chang, E.C., Lin, Q., Ooi, B.C.: Towards scaling blockchain systems via sharding. In: SIGMOD (2019)
10. Dinh, T.T.A., Wang, J., Chen, G., Liu, R., Ooi, B.C., Tan, K.L.: Blockbench: A framework for analyzing private blockchains. In: SIGMOD (2017)
11. Gilad, Y., Hemo, R., Micali, S., Vlachos, G., Zeldovich, N.: Algorand: scaling byzantine agreements for cryptocurrencies. In: SOSP (2017)

12. Han, S., Xu, Z., Chen, L.: Jupiter: a blockchain platform for mobile devices. In: 2018 IEEE 34th International Conference on Data Engineering (ICDE), pp. 1649–1652. IEEE (2018)
13. Han, S., Xu, Z., Zeng, Y., Chen, L.: Fluid: a blockchain based framework for crowdsourcing. In: Proceedings of the 2019 International Conference on Management of Data, pp. 1921–1924 (2019)
14. Kokoris-Kogias, E., Jovanovic, P., Gasser, L., Gailly, N., Ford, B.: Omniledger: a secure, scale-out, decentralized ledger. In: IEEE SP (2018)
15. Kokoris-Kogias, E., Jovanovic, P., Gasser, L., Gailly, N., Syta, E., Ford, B.: Omniledger: a secure, scale-out, decentralized ledger via sharding. In: SP (2018)
16. Kotla, R., Alvisi, L., Dahlin, M., Clement, A., Wong, E.: Zyzzyva: speculative byzantine fault tolerance. In: SIGOPS (2007)
17. Luu, L., Narayanan, V., Zheng, C., Baweja, K., Gilbert, S., Saxena, P.: A secure sharding protocol for open blockchains. In: SIGSAC (2016)
18. Mckeen, F., et al.: Innovative instructions and software model for isolated execution. In: Hasp@isca (2013)
19. Nakamoto, S.: Bitcoin: A peer-to-peer electronic cash system (2008)
20. Popov, S.: The tangle. cit. on (2016)
21. Stathakopoulou, C., David, T., Vukolić, M.: Mir-bft: High-throughput bft for blockchains. arXiv preprint arXiv:1906.05552 (2019)
22. Syta, E., et al.: Scalable bias-resistant distributed randomness. In: SP (2017)
23. Wood, G.: Ethereum: A secure decentralised generalised transaction ledger. Ethereum project yellow paper (2014)
24. Wüst, K., Gervais, A.: Do you need a blockchain? In: CVCBT (2018)
25. Xu, Z., Han, S., Chen, L.: Cub, a consensus unit-based storage scheme for blockchain system. In: 2018 IEEE 34th International Conference on Data Engineering (ICDE), pp. 173–184. IEEE (2018)
26. Yin, M., Malkhi, D., Reiter, M.K., Gueta, G.G., Abraham, I.: Hotstuff: Bft consensus with linearity and responsiveness. In: PODC (2019)
27. Zamani, M., Movahedi, M., Raykova, M.: Rapidchain: Scaling blockchain via full sharding. In: SIGSAC (2018)

Redesigning the Sorting Engine
for Persistent Memory

Yifan Hua[1], Kaixin Huang[1], Shengan Zheng[2], and Linpeng Huang[1(✉)]

[1] Shanghai Jiao Tong University, Shanghai, China
{huahuahuahua,Kaixinhuang,huang-lp}@sjtu.edu.cn
[2] Tsinghua University, Beijing, China
venero@tsinghua.edu.cn

Abstract. Emerging persistent memory (PM, also termed as non-volatile memory) technologies can promise large capacity, non-volatility, byte-addressability and DRAM-comparable access latency. Such amazing features have inspired a host of PM-based storage systems and applications that store and access data directly in PM. Sorting is an important function for many systems, but how to optimize sorting for PM-based systems has not been systematically studied yet. In this paper, we conduct extensive experiments for many existing sorting methods, including both conventional sorting algorithms adapted for PM and recently-proposed PM-friendly sorting techniques, on a real PM platform. The results indicate that these sorting methods all have drawbacks for various workloads. Some of the results are even counterintuitive compared to running on a DRAM-simulated platform in their papers. To the best of our knowledge, we are the first to perform a systematic study on the sorting issue for persistent memory. Based on our study, we propose an adaptive sorting engine, namely SmartSort, to optimize the sorting performance for different conditions. The experimental results demonstrate that SmartSort remarkably outperforms existing sorting methods in a variety of cases.

Keywords: Persistent memory · Sorting algorithm · Pointer-indirect · Wear-leveling

1 Introduction

Emerging persistent memory (PM) is a new type of non-volatile storage device. Unlike traditional SSD, HDD or Flash, PM technologies such as STT-RAM [1], PCM [2] and 3DXPoint [3] can provide byte-addressability, DRAM-comparable read and write latency. Applications can access data in PM with simple load/store instructions. PM has inspired a host of researches on redesigning persistent storage systems, such as memory management systems [9,28,29], file systems [7,8,30], KV-Stores [31,34] and DBMSs [10,35].

For many storage systems, sorting is one of the most commonly-used functions. For instance, the 'ORDER BY' SQL command will automatically call

© Springer Nature Switzerland AG 2021
C. S. Jensen et al. (Eds.): DASFAA 2021, LNCS 12683, pp. 393–412, 2021.
https://doi.org/10.1007/978-3-030-73200-4_27

the embedded sorting engine in a DBMS. The sorting component will sort the records in a table by a specified key[1]. While many PM-optimized storage systems have been proposed, only a few works discuss how to optimize sorting for PM. An intuitive thinking is simply to apply conventional sorting algorithms in PM-based systems.

However, such a naive migration has many limitations for traditional sorting methods. For internal sorting algorithms, first, since PM has the limited write endurance [5,6,32], simply migrating traditional sorting algorithms from DRAM to PM causes heavy write traffic to PM, which will reduce the lifespan of PM. This is mainly caused by the allocated PM space for large-size records and their swap during sorting. Second, long time consumption exists during record swap for large values. For instance, fixing the keys to be 8-byte in size and record number to be one million, sorting the records with 4 KB values will spend 26.1 s using quick sort while the records with 8-byte values only cost 224 ms. Large values also make it hard to exploit cache locality. Third, employing sorting methods directly in PM costs more time than with the assistance of DRAM in some cases (see more details in Sect. 3). For external sorting algorithms, data loaded from PM to DRAM for sorting as performed in DRAM-Disk architecture not only consumes large DRAM space, but also takes many runs in the merging phase. First, external sort is heavily dependent on DRAM resource. When the available DRAM space is scarce, it may suffer from frequent data migration between DRAM and PM, which leads to the read/write amplification problem. Second, since disk/SSD is block addressable, sorting records using external sort can only load block-size data from disk to DRAM, which induces heavy time overhead in both loading and sorting phases.

With a specific study for these commonly-used sorting algorithms, we find that no single sorting method can be the best-level fit (i.e., both time-efficient and memory space-efficient) for different workloads and situations. For instance, although quick sort in PM performs well for many cases, it is worse than external sort for large-size records when the DRAM space is sufficient in a DRAM-PM hybrid memory architecture[2]. External sort, on the contrary, has worse performance when the DRAM size is very small. We have verified these bottlenecks by conducting multiple experiments on the Intel Optane DC Persistent Memory (Optane) platform (see more details in Sect. 3).

Since it is reported that PM should have much higher write latency than read latency, and PM may suffer from the limited write endurance issue (e.g., PCM is reported to be worn out after 10^6–10^8 writes) [5,6,32], a few researchers have proposed PM-friendly sorting methods [12,13,36] to decrease PM writes. For example, segment sort [12] intends to trade off fewer writes for additional reads and allows a tunable combination of external sort and selection sort. The

---

[1] In this paper, we call the attribute for sorting in a record as key and the other attributes as value.

[2] In this paper, we assume that PM is always large enough to accommodate all records and unsorted records are initially stored in PM while DRAM is not always sufficient relative to PM.

reason is that although the read complexity of selection sort is $O(N^2)$, its write complexity is merely $O(N)$. Another work B*-sort [13] develops a binary tree-based structure for sorting records in PM, which has $O(N)$ complexity for writes and $O(NlogN)$ complexity for reads. Luo et al. [36] utilize a heap structure and observe that if a node is close to the heap root, it is more likely to be read and written frequently. Thus, in order to reduce the average writes to PM, nodes close to the root are placed into DRAM while those close to the leaves are placed into PM. All these methods are evaluated on a DRAM-simulated platform and show good experimental results. However, when we run them on the real PM hardware, their performance is far from the expected result (e.g., much worse than the simple quick sort, a conventional sorting algorithm; see more details in Sect. 3).

We believe that there are at least three reasons. First, these three PM-friendly sorting methods heavily rely on the assumption that the latency of PM read should be much better than PM write. However, this is not the case for Optane. A recently-published paper [14] shows that Optane's write latency is comparable to DRAM but its read latency is 2x–3x worse than DRAM. Second, they cannot exert the full potential of cache locality, which makes them worse than quick sort in actual execution. For instance, the selection sort used in segment sort has to scan the entire portion each turn. As for B*-sort, it links records with left-child and right-child pointers, and hence all reads and writes are made to be random. For NVMSort, the node swap in the heapify process has to search nodes in both PM and DRAM. Third, they introduce extra PM read and write overhead despite of the relatively lower time complexity. For example, B*-sort allocates PM for all the left-child and right-child pointers, extra tunnel lists and metadata, leading to heavy additional overhead.

The limitations that exist in both conventional sorting methods and PM-friendly sorting methods inspire us to rethink the design of sorting in persistent memory. We first notice that the byte-addressability feature of PM allows us to index records with simple pointers (i.e., data addresses), which is quite different from the DRAM-Disk storage architecture. We propose a pointer-indirect mechanism in this paper to speed up the sorting performance for large-size records. Based on the analysis that no single sorting method is the best fit for different conditions, we then design and implement an adaptive sorting engine, SmartSort. SmartSort can automatically pick the best-suited embedded sorting method in an ad-hoc style. Our contributions are summarized as follows.

- To the best of our knowledge, we are the first to make a systematic study for sorting methods in PM. We demonstrate that existing sorting methods have non-negligible limitations.
- Taking advantage of PM's byte-addressability, we propose a pointer-indirect sort mechanism, which not only reduces the PM read and write overhead, but also does good to PM wear-leveling.
- Combined with the advantages of various sorting methods, we develop an adaptive sorting engine, namely SmartSort, to minimize the sorting overhead in PM for different conditions.

- We conduct extensive experiments on the Optane platform and the results show that SmartSort remarkably outperforms existing sorting methods for various workloads and situations.

The rest of this paper is organized as follows. Section 2 and 3 introduce the background and motivation of our work, respectively. Section 4 presents our proposed pointer-indirect sort mechanism and adaptive sorting engine, namely SmartSort, in detail. We evaluate SmartSort in Sect. 5 and discuss related work in Sect. 6. In Sect. 7, we finally conclude this paper.

2   Background

2.1   Persistent Memory

Persistent Memory (PM) such as PCM [2], STT-RAM [1] and 3DXPoint [3] is a new type of memory technology that has large capacity, non-volatility, byte-addressability, and limited write endurance [5,6]. Attaching PM to the main memory bus provides a raw storage medium that can be orders of magnitude faster than modern persistent storage medium such as disk and SSD [30]. Intel Optane DC Persistent Memory (Optane for short) is the first commercially-available PM product [4]. The emergence of PM has inspired a lot of researches for building persistent storage systems and applications [7,9,28,29].

2.2   Review of Conventional Sorting Methods

Sorting is one of the most important components in many storage systems and indexing structures, such as DBMSs [10,35], KV-Stores [31,34], and B+-Trees [19,20]. Traditional sorting algorithms can be divided into two types: internal sort and external sort. Internal sort executes the sorting procedure for all records directly in memory space. By contrast, external sort is used for large data size that does not fit in memory and depends on a two-phase sorting procedure: 1) divide all the records into several chunks and each time load a single chunk into memory from disk/SSD to perform an internal sort (e.g., quick sort) on the chunk, then write out the sorted chunk to disk/SSD; 2) use merge sort in memory to combine multiple sorted chunk records into globally-sorted records. Table 1 provides the average time and space complexity for some representative internal sorting algorithms. Clearly, selection sort has the lowest write complexity while it suffers from high read complexity. Insertion sort has both high read complexity and write complexity. Other sorting algorithms, such as quick sort, merge sort and heap sort, achieve more balanced read and write complexity (i.e., both are $O(NlogN)$).

2.3   Sorting in Persistent Memory

Although there are a lot of researches on both PM-based system design and in-memory data sorting optimizations, few open discussions have been made

Table 1. Average time and space complexity of traditional sorting algorithms.

	Read time complexity	Write time complexity	Space complexity
Insertion sort	$O(N^2)$	$O(N^2)$	$O(1)$
Selection sort	$O(N^2)$	$O(N)$	$O(1)$
Quick sort	$O(NlogN)$	$O(NlogN)$	$O(logN)$
Merge sort	$O(NlogN)$	$O(NlogN)$	$O(N)$
Heap sort	$O(NlogN)$	$O(NlogN)$	$O(1)$

on combining these two points together and redesigning the sorting engine in persistent memory. An intuitive idea is to apply conventional sorting algorithms, such as quick sort [22], selection sort [23], merge sort [24], and external sort [25] for PM-based systems. However, such a naive migration for sorting methods in PM can lead to huge writes, which could reduce the lifespan of PM. In addition, sorting records directly in PM will lead to heavy time overhead with an increasing record size.

A few researchers have proposed PM-friendly sorting methods [12,13,36] by exploiting the unique features of persistent memory device. Due to the commonly-believed read/write asymmetry feature (i.e., write latency is much higher than read latency), they seek to trade off fewer writes for additional reads or minimize the write complexity assisted with special data structures. Segment sort [12] allows a tunable combination of external sort and selection sort. That is, α $(0 \leq \alpha \leq 1)$ of all records are sorted by external sort and the remained $(1 - \alpha)$ portion are sorted by selection sort. These two portions are then merged into the final sorted records. The reason is that although the read complexity of selection sort is $O(N^2)$, its write complexity is merely $O(N)$. Given that PM's read latency is much lower than write latency, segment sort is supposed to achieve better performance than simple external sort or selection sort with a proper α setting. B*-sort [13] develops a binary tree-based structure for sorting records in PM, which has $O(N)$ complexity for writes and $O(NlogN)$ complexity for reads. B*-sort also uses extra tunnel lists and register metadata to optimize the worst-case read complexity. Luo et al. [36] improve traditional heap sort by placing nodes near the heap root in DRAM and those near the leaves in PM to reduce writes in PM based on the observation that nodes close to the root are more likely to be accessed.

3   Motivation

Although the adapted conventional sorting algorithms and recently-proposed PM-friendly sorting techniques can be applied for persistent memory scenarios, we claim that both types of sorting methods have non-negligible limitations, such as high read/write complexity, severe write overhead due to large value size, and remarkable performance decrease caused by limited DRAM resource.

To observe the bottlenecks of existing sorting methods for PM, we conduct a series of experiments on randomly-generated records, each of which only contains a key (fixed 8-byte in size) and a value (varied-size). The detailed configuration of our experimental platform is provided in Sect. 5.1.

Table 2. Execution time (ms) of typical traditional sorting algorithms.

	10K	100K	1M	10M	100M	1B
Selection sort	123	12195	1232723	Too long	Too long	Too long
Insertion sort	186	18376	1911156	Too long	Too long	Too long
Quick sort	3.7	23	199.1	2581	24291	296782
Merge sort	4.6	28.4	343.2	4271	43871	569731
Heap sort	5.8	33.5	416.6	7975	78128	1526177

Table 3. Execution time (s) with different value size.

	In-PM				In-DRAM				PM-DRAM			
	8B	64B	512B	4 KB	8B	64B	512B	4 KB	8B	64B	512B	,KB
Selection sort	12.2	13.5	18.3	30.6	12.2	13.5	18.2	30.2	12.2	13.5	18.2	30.3
Insertion sort	18.4	27.9	130.2	Too long	18.4	27.6	127.9	Too long	18.4	27.6	127.9	Too long
Quick sort	0.02	0.04	0.14	0.95	0.02	0.03	0.07	0.41	0.02	0.03	0.09	0.53

3.1   Comparison of Typical Traditional Sorting Algorithms

Table 2 shows the time consumption of typical traditional sorting algorithms to sort records with 8-byte values from 10 thousand to 1 billion. We have two observations. First, selection sort has better performance among the algorithms with $O(N^2)$ time complexity. It has 1/3 less time consumption compared to insertion sort. Second, quick sort achieves the best time efficiency among all the compared sorting algorithms. Although merge sort and heap sort have the same read and write complexity as quick sort, they have lower performance in practical running. We infer that it is because quick sort can better utilize memory cache locality while merge sort always writes its temporary sorted results to new memory space and heap sort has more non-adjacent elements comparison. The drawback of quick sort, however, is that it has more writes than selection sort.

3.2   The Impact of Value Size

Table 3 shows the execution time of three traditional sorting algorithms (i.e. selection sort, insertion sort and quick sort) for sorting 100 thousand records with the value size growing. The In-PM mechanism means that the sorting procedure is directly executed in PM; the In-DRAM mechanism indicates that the records are loaded into and sorted in DRAM (but not stored back to PM); the PM-DRAM mechanism represents that the records are loaded into and sorted in DRAM, and finally stored back to PM.

From Table 3, we have three main observations. First, with the value size growing, the time consumption of sorting algorithms can increase remarkably. For instance, insertion sort and quick sort incur 7.1x and 4.7x time overhead in PM, respectively, when the value size increases from 8B to 512B. This is because more reads and writes are performed during the sorting procedure. Second, for both selection sort and insertion sort, the In-PM time consumption is similar to that of In-DRAM and PM-DRAM. This indicates that their time consumption for sorting is too heavy due to $O(N^2)$ time complexity. Third, for quick sort, the PM write overhead plays an important role in the final performance. It can be seen that In-DRAM and PM-DRAM quick sort outperform In-PM quick sort by 2.3x and 1.79x respectively, when the value size is 4 KB. Notice that compared to In-PM quick sort, PM-DRAM quick sort can also reduce writes to PM (i.e., merely N writes for storing sorted records), and hence alleviate the risk of PM wear out.

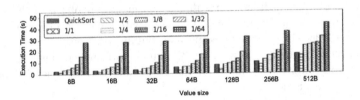

Fig. 1. The impact of available relative DRAM capacity on external sort.

3.3   The Impact of DRAM Capacity

In conventional storage architecture, data is first loaded from disk/SSD to DRAM, and the actual sorting procedure is executed in DRAM. However, compared to the durable storage device, DRAM space can be relatively more scarce, and hence it cannot store all the records in some cases. To address this problem, external sort is employed. Records in disk/SSD are divided into multiple chunks, each of which can be fitted in DRAM space and sorted. Each sorted chunk will be written back to disk/SSD and they will be merged as a final sorted file via properly using the limited DRAM resource. In a DRAM-PM hybrid architecture, external sort can still work in the similar way. Figure 1 shows the performance effect on external sort with different relative DRAM capacity. The number of records is ten million and the In-PM quick sort is considered as a baseline.

From Fig. 1, we can observe that the performance of external sort decreases with the relative DRAM capacity being smaller, but the sharp performance drop is mainly in the shift from full record capacity to 1/2 record capacity. For instance, with 8-byte value size, the time consumption climbs by 2.2x, when changing the DRAM capacity from full record space to 1/2 record space. However, only 18% extra time is incurred when shifting the DRAM capacity form 1/2 record space to 1/4 record space. Second, external sort is better than (In-PM) quick sort in performance with sufficient DRAM capacity and worse than (In-PM) quick sort if the DRAM capacity is smaller than the total record size. Notice that compared to In-PM quick sort, external sort has much better wear-leveling capability because it requires only $O(N)$ PM writes.

3.4   Performance Study of PM-Friendly Sorting Methods

Segment sort [12], B*-sort [13] and NVMSort [36] are three recently-proposed PM-friendly sorting methods that show good performance on a DRAM-simulated platform. Currently, since the real PM product is available, it should be interesting and useful to study their real performance.

Segment sort is a combination of external sort and selection sort, and its main idea is to trade off fewer writes for additional reads since PM is expected to have much higher write latency than read latency. Viglas et al. [12] believe that there will be an optimal ratio to make segment sort reach the best performance. However, we observe a different result on Optane-based platform. Figure 2(a) shows the time consumption of segment sort with different ratio to sort 100 thousand records. For simplicity while maintaining the spirit of segment sort, we replace the external sort with In-PM quick sort, which has higher write complexity but lower read complexity than selection sort. The merging phase is kept as the previous design.

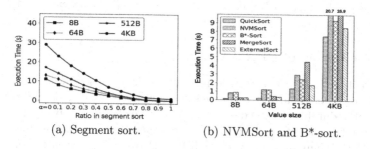

(a) Segment sort.   (b) NVMSort and B*-sort.

Fig. 2. Performance study for PM-friendly sorting methods.

In Fig. 2(a), $\alpha = 0$ means that segment sort only uses selection sort; by contrast, $\alpha = 1$ indicates that segment sort only utilizes quick sort. We can learn from Fig. 2(a) that as α decreases, the time overhead grows as well. In other words, the larger the ratio of quick sort is employed, the better the performance of segment sort is achieved. Segment sort can gain no time profits from selection sort by trading off fewer writes for additional reads. We believe that there are two reasons for it. First, the read latency is not better than write latency. A recent study on Optane's performance [14] shows that the random 8-byte read (i.e., load) latency can be 300 ns while the random 8-byte write (i.e., store) latency can be merely 100 ns. Optane's write can be faster than its read in terms of latency. While the read bandwidth of Optane is higher than write, we infer that the bottleneck for sorting is not bandwidth based on several experiments. For example, for 10 million 16-byte records, quick sort takes 304.99 s while random write takes merely 4.32 s. Thus, for Optane, segment sort just trades off faster writes for slower reads. Second, selection sort is not as efficient as quick sort in utilizing cache locality, and the portion of records using selection sort becomes the bottleneck in the entire sorting procedure. In conclusion, segment sort is worse than the simple quick sort algorithm.

B*-sort [13] adopts a binary search tree structure to reduce the complexity of PM writes to $O(N)$ and limit the average complexity of PM reads to $O(NlogN)$. To avoid the worst case for reads, it also utilizes additional tunnel lists and register metadata. NVMSort trades off part of PM writes for DRAM writes. While the theoretic complexity of B*-sort and NVMSort is much better than quick sort, the performance results on Optane-based platform are much worse. Figure 2(b) compares the time consumption among B*-sort, NVMSort, (In-PM) quick sort, (In-PM) merge sort and external sort for one million records when DRAM capacity is 1/2 the total record size. We can draw two takeaways from Fig. 2(b). First, for small-size values, B*-sort and NVMSort have much higher time overhead than quick sort. For instance, the time consumption of B*-sort is 6.1x and 5.5x higher than quick sort with the value size of 8B and 64B, respectively. There are two reasons for this: 1) B*-sort is a pointer-based data structure and hence incurs a lot of random reads and writes during sorting. NVMSort has a lot of non-adjacent data swap. They not only fail to utilize cache locality but also incur severe time overhead [3,4]; 2) the additional tunnel lists and register metadata in B*-sort add more PM-allocated overhead in the critical path. Second, for larger-size values, B*-sort can be comparable to quick sort. It is observed that with 4 KB value size, B*-sort is much better than NVMSort and merge sort while obtains merely less than 15% time consumption compared to quick sort. This is because each tree node access can benefit from sequential reads and writes. In conclusion, B*-sort is worse than the simple quick sort algorithm in performance but better for the wear-leveling goal.

4   SmartSort

As Sect. 3 demonstrates, both traditional sorting algorithms and recently-proposed PM-friendly sorting methods have limitations, which include 1) performance issue caused by large value size, limited DRAM capacity, and random PM read/write overhead; 2) wear-out concern caused by PM writes during sorting. Furthermore, no single sorting method can beat others in all cases. For instance, while quick sort is better in time efficiency than selection sort and B*-sort, it is not better in wear-leveling. While PM-DRAM quick sort can be better than In-PM quick sort, it heavily depends on the space consumption of DRAM and when the available DRAM capacity is limited, it will transform to external sort and the performance can drop sharply.

Based on these key observations and conclusions, we claim that it is necessary to redesign the sorting engine for persistent memory. In this paper, we propose an adaptive sorting engine, which is named SmartSort, to address the challenges we mentioned. In this section, we first present the overall structure of SmartSort, then introduce the PM-enabled pointer-indirect sort mechanism, and finally provide the details of how SmartSort works.

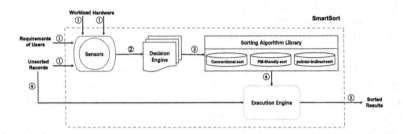

Fig. 3. Architecture of SmartSort.

4.1 Overview

SmartSort is an adaptive sorting engine that targets at providing the appropriate sorting technique for each workload according to the workload features and other significant related conditions. Figure 3 shows the architecture of Smart-Sort. SmartSort is composed of four core components: *Sensor (SS)*, *Decision Engine (DE)*, *Sorting Algorithm Library (SAL)* and *Execution Engine (EE)*. Among these components, *SS* is designed for extracting the useful information from unsorted records, requirements of users, workloads and the hardware in system (shown in ①), and conveying the information to *DE* (shown in ②). The information includes four aspects: 1) information of unsorted records, such as the size of total records, the number of total records, and the value size of records; 2) requirement information of users, such as the wear-leveling requirement of application and the persistence requirement of the sorted results; 3) workload information, such as the limited write size of PM and limited sorting time; 4) hardware information, such as the available DRAM capacity and the PM write endurance. Based on the above collected information, *DE* is responsible for selecting the appropriate sorting technique from *SAL* (shown in ③). *SAL* is a suite of sorting techniques, including adapted conventional sorting algorithms, existing PM-friendly sorting techniques and pointer-indirect-optimized sorting methods (see Sect. 4.2 for more details). After that, the selected sorting method will be transmitted to *EE*, and *EE* performs it on the unsorted records (shown in ④). Finally, the sorted results are generated as output (shown in ⑤). They may be either persisted in PM or simply copied to the user buffer.

4.2 PM-Enabled Pointer-Indirect Sort

For traditional DRAM-Disk storage architecture, records should be first loaded into DRAM for sorting, with internal or external sorting algorithms. The loading procedure is performed in a block-based style. That is, blocks containing the full record information (i.e., keys and values) are loaded into DRAM. The main bottleneck is the I/O overhead. Now, with PM, the records can be stored in PM directly for storage systems and applications, and the sorting procedure can be executed in PM as well. But as we point out in Sect. 3, the sorting performance drops remarkably with the growth of value size, which incurs more PM reads and writes. Some main memory database systems enable to use pointers for data indexing [37], which reduces the movement for the actual large-size value during scan or some other operations. Inspired by this, we devise the pointer-indirect sort mechanism to speed up the sorting performance for PM-resided records which have large-size values.

The key step of the the pointer-indirect sort mechanism is building the mapping <key, pointer> records, based on the original <key, value> records. Concretely, a new region (either in DRAM or PM) is created, and each <key, value> record is transformed to a much smaller <key, pointer> record, where *the pointer is an indirection (i.e., address) to the original <key, value> record*. Suppose that the key is fixed-size with 8B, then we can limit the <key, pointer> record to be merely 16B. Due to the space-efficiency of <key, pointer> records, with a given DRAM capacity, a much larger number of records may be loaded into DRAM for sorting when it is compared to traditional full-record loading mechanism.

Instead of directly conducting a sorting algorithm on large-size records, the pointer-indirect sort mechanism enables to sort the much smaller-size <key, pointer> records, and hence reduces a lot of PM reads and writes. It is also easy to read out sorted records via the sorted pointers. We have studied the result reading overhead in Sect. 5, which is very lightweight compared to the actual sorting overhead. In conclusion, the pointer-indirect sort mechanism is beneficial to both sorting performance and PM wear-leveling. Notice that the pointer-indirect mechanism is not limited to the use of a single sorting method. It can be combined with all existing sorting algorithms and techniques, such as

Table 4. Complicated sorting conditions and corresponding sorting methods.

Value size	Sufficient DRAM	Wear-leveling	Persistence	Best-suited sorting method
small (large)	Yes	Yes	Yes	(pointer-indirect) PM-DRAM quick sort
	Yes	Yes	No	(pointer-indirect) In-DRAM quick sort
	Yes	No	Yes	(pointer-indirect) In-PM quick sort
	Yes	No	No	(pointer-indirect) In-DRAM quick sort
	No	Yes	Yes	(pointer-indirect) external sort
	No	Yes	No	(pointer-indirect) In-PM B*-sort
	No	No	Yes	(pointer-indirect) In-PM quick sort
	No	No	No	

quick sort, external sort and B*-sort. As shown in Fig. 3, pointer-indirect sort is a core mechanism of the SAL component.

4.3 Adaptive Sorting

To achieve the best sorting performance with SmartSort, we should 1) clearly distinguish different workloads and situations as different conditions, and 2) select the most-suitable sorting method for the corresponding condition.

We study a variety of sorting conditions and provide the most appropriate sorting method for each condition in Table 4. We consider the value size to be small if it is smaller than 8 bytes, otherwise it is marked as large. The DRAM capacity is considered to be sufficient if DRAM can store all the <key, value> records or the corresponding <key, pointer> records using the pointer-indirect mechanism. The wear-leveling requirement indicates if the writes should be shorten to $O(N)$ times and the persistence requirement tells if the sorted results should be stored as a persistent object.

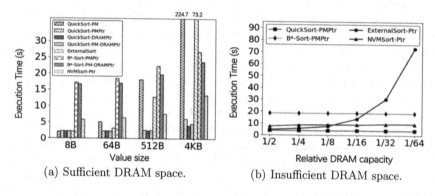

(a) Sufficient DRAM space.   (b) Insufficient DRAM space.

Fig. 4. Comparison of the execution time for different sorting methods.

The corresponding best-suited sorting method is carefully selected based on our experimental observations. As we discuss in Sect. 4.2, the pointer-indirect sort mechanism should gain many performance benefits when the value size is large. Figure 4(a) shows the time consumption for eight candidate sorting methods to sort ten million records when DRAM capacity is sufficient. We can observe that the performance of QuickSort-PMPtr (i.e., pointer-indirect In-PM quick sort) is much higher than QuickSort-PM (i.e., pure In-PM quick sort) when the value size varies from 64B to 4KB (e.g., 2.2x, 6.9x and 36.8x for 64B, 512B and 4KB values, respectively). However, when the value size is small (i.e., 8B), there is no need to employ the pointer-indirect mechanism because their performance is nearly equal.

From Fig. 4(a), we can also observe that using DRAM can bring benefits to the sorting performance. For instance, QuickSort-PM-DRAMPtr (i.e., pointer-indirect PM-DRAM quick sort) outperforms QuickSort-PMPtr (i.e., pointer-indirect In-PM quick sort) by up to 1.4x when the value size is 4KB. In addition, sorting in DRAM will reduce writes in PM a lot. Therefore, when wear-leveling is required, SmartSort should utilize DRAM for sorting. Concretely, when persistence for sorted results is required, it should utilize PM-DRAM quick sort; otherwise, it should use In-DRAM quick sort. By contrast, if wear-leveling is not a critical concern, SmartSort can simply choose In-PM quick sort and In-DRAM quick sort according to different persistence requirements.

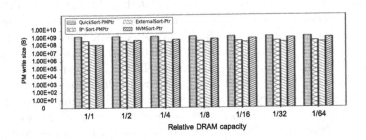

Fig. 5. Write size in PM for different pointer-indirect sorting algorithms.

When DRAM capacity is not sufficient, and if there is no restriction on wear-leveling, then it could be very straightforward to select the In-PM quick sort due to its efficiency. However, if wear-leveling is set as a target, then SmartSort should shift the sorting method. Notice that both external sort and B*-sort have the PM write complexity of $O(N)$, which should be the candidate sorting methods for this case. But which one is better? To answer this question, we conduct another experiment to compare the performance of B*-sort with external sort for different relative DRAM capacity. Figure 4(b) shows the results for sorting ten million records and QuickSort-PMPtr (i.e., pointer-indirect In-PM quick sort) is used as performance baseline. It can be observed that when the relative DRAM space is larger than or equal to 1/16, ExternalSort-Ptr (i.e., pointer-indirect external sort) is better than B*-Sort-PMPtr (i.e., pointer-indirect In-PM B*-sort). When the relative DRAM capacity gets even smaller, B*-Sort-PMPtr starts to outperform ExternalSort-Ptr. Although NVMSort-Ptr (i.e., pointer-indirect NVMSort) performs better than ExternalSort-Ptr and B*-Sort-PMPtr when the relative DRAM space is smaller than or equal to 1/16 in Fig. 4(b), its PM write size is comparable to that of QuickSort-PMPtr in that case as Fig. 5 shows. Thus, NVMSort-Ptr is not a suitable choice when wear-leveling is required. In our current implementation, we set the (relative DRAM capacity) switch boundary between external sort and B*-sort as 1/16. Compared with the baseline, SmartSort has a limited boundary of extra time overhead (i.e., 4.5x) to guarantee wear-leveling, which should be acceptable in practical use.

Based on the experiments and analysis above, we have demonstrated that each employed sorting technique in SmartSort is the best-suited one for the corresponding condition. Notice that the selection for a sorting method in SmartSort each time is not manually-configured. That is why we call SmartSort an adaptive sorting engine.

5   Experimental Evaluation

5.1   Experimental Setup

We implement SmartSort using C++ on a Linux server (CentOS 7.8) with 2.60 GHz Intel(R) Xeon(R) Gold 6240 CPU. This CPU has 36 physical cores, with a 24 MB L3 cache. We use 600 GB of overall Optane DIMM space in the maximum to ensure all records accommodated in this paper. DRAM size in Sect. 3 and 4 varies according to the record number, value size and relative ratio to PM as detailed in each picture. In Sect. 5, DRAM size is fixed as 4GB (large enough to store all records or pointers), 64 MB (nearly 1/3 of the total pointer size) and 4 MB (1/40 of the total pointer size). Throughout our experiments, the key size is fixed as 8B by default (i.e., keys are randomly-generated integers), and the value size is allowed to vary from 8B to 4KB. To guarantee data persistence and consistency in PM, similar to many prior works [7, 28], we properly utilize *clwb+sfence* instructions to force flushing out the records from caches. We use the standard benchmarks [27] to evaluate the sorting performance of SmartSort. Since the source code of B*-sort and NVMSort is not available in public, we implement them faithfully according to their papers. To demonstrate the benefit of SmartSort, we compare SmartSort with six traditional sorting algorithms (i.e., selection sort, insertion sort, external sort, quick sort, merge sort, heap sort) and three PM-friendly sorting techniques (i.e., segment sort, B*-sort, NVMSort).

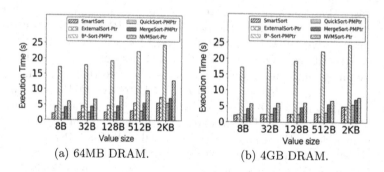

Fig. 6. The execution time for different sorting algorithms.

5.2   Sorting Performance for Different Workloads

Figure 6 compares the execution time for sorting ten million records between SmartSort and other sorting methods. The DRAM capacity in Fig. 6(a) and Fig. 6(b) is 64 MB (i.e., insufficient DRAM space) and 4 GB (sufficient DRAM space), respectively. We can observe that SmartSort is remarkably better than the other sorting methods and has good scalability with the value size growing.

For 64 MB DRAM capacity, SmartSort will adaptively select quick sort for 8-byte value size and pointer-indirect quick sort in PM for larger value size when wear-leveling is not a restricted factor. For 4 GB DRAM capacity, SmartSort will adaptively select PM-DRAM quick sort and pointer-indirect PM-DRAM quick sort (or In-DRAM quick sort and pointer-indirect In-DRAM quick sort if there is no need to persist the sorted results) for 8-byte and larger value size respectively.

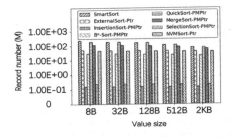

Fig. 7. Sorted record number in 1 min.

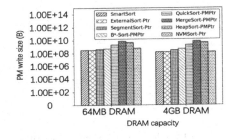

Fig. 8. Write size in PM.

Figure 7 shows the maximum number of records that can be sorted in one minute by different sorting methods. We set the DRAM capacity as 64 MB, which is insufficient to contain all records. In this case, SmartSort prefers In-PM quick sort for a small value size and pointer-indirect In-PM quick sort for a large value size. Insertion sort and selection sort complete the fewest records due to their $O(N^2)$ time complexity. Figure 8 shows the total PM write size (in bytes) of sorting ten million records for the compared methods when DRAM is 64 MB and 4 GB. For 64 MB DRAM capacity, SmartSort will adaptively select pointer-indirect external sort. For 4 GB DRAM capacity, SmartSort will adaptively select pointer-indirect PM-DRAM quick sort (or In-DRAM quick sort if there is no need to persist the sorted results).

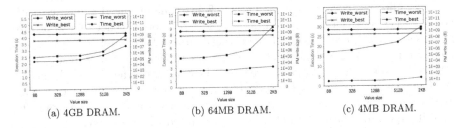

(a) 4GB DRAM.   (b) 64MB DRAM.   (c) 4MB DRAM.

Fig. 9. The best and worst performance to persist sorted result.

Given ten million records, Fig. 9 provides the upper bound and lower bound of SmartSort for both time consumption and PM writes with different DRAM

configurations when sorted results need to be persisted. With the decrease of
DRAM capacity, the upper bound of SmartSort's time consumption gradually
increases due to the use of (pointer-indirect) external sort and B*-sort. While the
gap between the worst-case time consumption and the best-case time consumption varies remarkably under different conditions, the gap between the worst-case
PM writes and the best-case PM writes is stable, which can be represented by
$O(NlogN)/O(N)$.

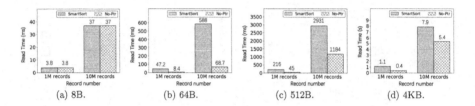

(a) 8B.   (b) 64B.   (c) 512B.   (d) 4KB.

Fig. 10. Time overhead of reading sorted records for different value size.

5.3 Comparison of Time Overhead on Reading Sorted Records

In addition to the sorting time cost, the overhead of reading sorted records (i.e.,
load all the sorted records to the user buffer) should be also studied since it
may be in the critical path of an application request (e.g., SELECT command
in DBMSs). Figure 10 compares the time overhead between SmartSort and the
non-pointer-indirect sorting algorithm.

It can be observed that when the value size is larger than 8 bytes, it is
slower to read out the records that are sorted by SmartSort. There are two
reasons. First, the pointer-indirect mechanism employed by SmartSort requires
an additional PM load operation for each record read. Second, the reads into the
actual records cannot exploit the cache locality since only the pointer records
are sorted. It can also be observed that with the value size getting larger, the
performance gap becomes smaller. Concretely, when the value size is 512B, the
pointer-indirect sort mechanism generates 2.48x time overhead compared with
normal quick sort, for reading ten million sorted records. When the value size
increases to 4 KB, the relative time overhead caused by pointer-indirect sort is
merely 1.46x. Although SmartSort requires more time to read sorted results from
PM to the user buffer for large-size records, the reading overhead is much smaller
than that of sorting (i.e., only 10.7% of quick sort for ten million records with
4KB values), and the sorting performance is improved by 12.3x. Therefore, the
overall request overhead can be remarkably reduced by SmartSort.

6 Related Work

Due to the interesting features of emerging persistent memory technologies, a
few researches have been proposed to optimize the sorting performance for PM.

For example, segment sort [12] assumes that a proper ratio between selection sort and external sort will lead to a better performance in PM, by trading off slower write operations for much faster read operations in PM. However, we have demonstrated that segment sort is consistently worse than quick sort on the Optane-based platform. B*-sort [13] utilizes the binary search tree structure to restrict the write complexity to $O(N)$ (i.e., write-once property) and maintains the average read complexity to $O(NlogN)$. It also develops a tunnel list structure and adds register metadata to optimize the worst-case read complexity to $O(NlogN)$ as well. Unfortunately, it is observed that although B*-sort has better wear-leveling effect, it is slower than the simple quick sort algorithm on the Optane-based platform. Based on a heap structure, Luo et al. [36] place nodes near the heap root in DRAM and those near the leaves in PM to reduce PM writes according to the observation that nodes near the root are more likely to be read or written. However, it runs on a DRAM-simulated platform and the write latency it sets is far longer than real PM. On the Optane-based platform, it performs worse than quick sort in our experiments. Compared to these PM-friendly sorting technique proposals, SmartSort provides a more comprehensive solution for the best-level sorting performance under different conditions.

Sorting is a significant function in many storage systems and index structures. The representative is B+-Tree, which internally sorts the records within one B+-Tree node for each insert operation. Some PM-optimized B+-Trees [18–20] have developed efficient techniques to minimize the sorting overhead. For instance, wB+-Tree [19] utilizes the indirection slot array, which is similar to our pointer-indirect mechanism in spirit, to avoid the actual sorting for records, and hence reduces a lot of PM write overhead. The limitation of the indirection array, however, is that the indirection number is limited (e.g., 8 or 16 in wB+-Tree). NV-Tree [18] only sorts records for In-DRAM inner nodes but leaves the records in In-PM leaf nodes out of order, thus totally avoiding the sorting overhead in PM. Each insert operation in NV-Tree just appends a new log to the last record (i.e., a log). The trade-off is the extra overhead of probing the entire node for each single read and the garbage collection overhead for invalid log records. Compared to the sorting techniques proposed in these B+-Trees, SmartSort is a more universal sorting engine, rather than being limited to sort only a small number of records (e.g., only in a B+-Tree node).

7   Conclusion

In this paper, we make a systematic study on sorting in PM and point out that existing sorting methods have limitations when using the real PM product. We propose an adaptive sorting engine, SmartSort, which can dynamically adjust its internal sorting technique to the corresponding condition to achieve the best-level performance. The experimental evaluation demonstrates the merit of SmartSort. We hope that SmartSort can inspire further researches in the area of PM sorting and we believe that more intelligent decisions on proper sorting techniques should be explored.

Acknowledgment. This work is supported by National Key Research & Development Program of China (No. 2018YFB1003302). Linpeng Huang is the corresponding author of this paper and also supported by SJTU-Huawei Innovation Research Lab Funding (No. FA2018091021-202004). Shengan Zheng is supported by China Postdoctoral Science Foundation (No. 2020M680570).

References

1. Kültürsay, E., Kandemir, M., Sivasubramaniam, A., Mutlu, O.: Evaluating STT-RAM as an energy-efficient main memory alternative. In: IEEE International Symposium on Performance Analysis of Systems and Software (ISPASS), Austin, TX, pp. 256–267 (2013)
2. Wong, H.-S.P., Raoux, S., et al.: Phase change memory. Proc. IEEE **98**(12), 2201–2227 (2010)
3. Hady, F.T., Foong, A., Veal, B., Williams, D.: Platform storage performance With 3D XPoint technology. Proc. IEEE **105**(9), 1822–1833 (2017)
4. Peng, I.B., Gokhale, M.B., Green, E.W.: System evaluation of the Intel optane byte-addressable NVM. In: Proceedings of the International Symposium on Memory Systems, pp. 304–315 (2019)
5. Qureshi, M.K., et al.: Enhancing lifetime and security of PCM-based main memory with start-gap wear leveling. In: 2009 42nd Annual IEEE/ACM International Symposium on Microarchitecture (MICRO) (2009)
6. Huang, K., Mei, Y., Huang, L.: Quail: using NVM write monitor to enable transparent wear-leveling. J. Syst. Archit. **102**, 101658 (2020)
7. Xu, J., Swanson, S.: NOVA: a log-structured file system for hybrid volatile/non-volatile main memories. In: Proceedings of the 14th USENIX Conference on File and Storage Technologies, pp. 323–338 (2016)
8. Zheng, S., Hoseinzadeh, M., Swanson, S.: Ziggurat: a tiered file system for non-volatile main memories and disks. In: Proceedings of the 17th USENIX Conference on File and Storage Technologies, pp. 207–219 (2019)
9. Coburn, J., Caulfield, A., Akel, A., et al.: NV-heaps: making persistent objects fast and safe with next-generation, non-volatile memories. In: Proceedings of the Sixteenth International Conference on Architectural Support for Programming Languages and Operating Systems, pp. 105–118 (2011)
10. Kaiyrakhmet, O., Lee, S., Nam, B., Noh, S.H., Choi, Y.: SLM-DB: single-level key-value store with persistent memory. In: Proceedings of the 17th USENIX Conference on File and Storage Technologies, pp. 191–205 (2019)
11. Seo, J., Kim, W.-H., Baek, W., Nam, B., Noh, S.H.: Failure-atomic slotted paging for persistent memory. SIGARCH Comput. Archit. News **45**(1), 91–104 (2017)
12. Viglas, S.D.: Write-limited sorts and joins for persistent memory. Proc. VLDB Endow. **7**(5), 413–424 (2014)
13. Liang, Y.-P., et al.: B*-sort: enabling Write-once Sorting for persistent memory. IEEE Trans. Comput.-Aided Design Integr. Circ. Syst. **PP**(99), 1 (2020)
14. Yang, J., Kim, J., Hoseinzadeh, M., et al.: An empirical guide to the behavior and use of scalable persistent memory. In: Proceedings of the 18th USENIX Conference on File and Storage Technologies, pp. 168–182 (2020)
15. Hwang, D., Kim, W., Won, Y., Nam, B.: Endurable transient inconsistency in byte-addressable persistent B+-tree. In: Proceedings of the 16th USENIX Conference on File and Storage Technologies, pp. 187–200 (2018)

16. Chen, Y., Lu, Y., Fang, K., Wang, Q., Shu, J.: uTree: a persistent B+-tree with low tail latency. Proc. VLDB Endow. **13**(12), 2634–2648 (2020)
17. Liu, M., Xing, J., Chen, K., Wu, Y.: Building scalable NVM-based B+tree with HTM. In: Proceedings of the 48th International Conference on Parallel Processing, pp. 1–10 (2019)
18. Yang, J., Wei, Q., Chen, C., Wang, C., Yong, K.L., He, B.: NV-Tree: reducing consistency cost for NVM-based single level systems. In: Proceedings of the 13th USENIX Conference on File and Storage Technologies, FAST 2015, pp. 167–181 (2015)
19. Chen, S., Jin, Q.: Persistent B+-trees in non-volatile main memory. PVLDB **8**(7), 786–797 (2015)
20. Oukid, I., Lasperas, J., Nica, A., Willhalm, T., Lehner, W.: FPTree: a hybrid SCM-dram persistent and concurrent b-tree for storage class memory. In: Proceedings of the 2016 International Conference on Management of Data, pp. 371–386 (2016)
21. Andrei, M., Lemke, C., et al.: Sorting with asymmetric read and write costs guy. In: Annual ACM Symposium on Parallelism in Algorithms and Architectures, vol. 2015, no. 6, pp. 1–12 (2015)
22. Woźniak, M., Marszałek, Z., Gabryel, M., Nowicki, R.K.: Preprocessing large data sets by the use of quick sort algorithm. In: Skulimowski, A.M.J., Kacprzyk, J. (eds.) Knowledge, Information and Creativity Support Systems: Recent Trends, Advances and Solutions. AISC, vol. 364, pp. 111–121. Springer, Cham (2016). https://doi.org/10.1007/978-3-319-19090-7_9
23. Hayfron-Acquah, J.B., Appiah, O., Riverson, K.: Improved selection sort algorithm. Int. J. Comput. Appl. (0975–8887) **110**(5), 29–33 (2015)
24. Marszałek, Z.: Parallelization of modified merge sort algorithm. Symmetry **9**(9), 176 (2017)
25. Khorasani, E., Paulovicks, B.D., Sheinin, V., Yeo, H.: Parallel implementation of external sort and join operations on a multi-core network-optimized system on a chip. In: Xiang, Y., Cuzzocrea, A., Hobbs, M., Zhou, W. (eds.) ICA3PP 2011. LNCS, vol. 7016, pp. 318–325. Springer, Heidelberg (2011). https://doi.org/10.1007/978-3-642-24650-0_27
26. Quartz. https://github.com/HewlettPackard/quartz. Accessed 27 Oct 2020
27. Sort Benchmark Home Page. http://sortbenchmark.org/. Accessed 27 Oct 2020
28. Huang, K., Li, S., Huang, L., Tan, K., Mei, H.: Lewat: a lightweight, efficient, and wear-aware transactional persistent memory system. IEEE Trans. Parallel Distrib. Syst. **32**(03), 649–664 (2021)
29. Volos, H., Tack, A.J., Swift, M.M.: Mnemosyne: lightweight persistent memory. ACM SIGARCH Comput. Archit. News **39**(1), 91–104 (2011)
30. Dulloor, S.R., et al.: System software for persistent memory. In: Proceedings of the Ninth European Conference on Computer Systems (2014)
31. Xia, F., et al.: HiKV: a hybrid index key-value store for DRAM-NVM memory systems. In: 2017 USENIX Annual Technical Conference (2017)
32. Psaropoulos, G., et al.: Bridging the latency gap between NVM and DRAM for latency-bound operations. In: Proceedings of the 15th International Workshop on Data Management on New Hardware (2019)
33. Wan, H., et al.: Empirical study of redo and undo logging in persistent memory. In: 2016 5th Non-Volatile Memory Systems and Applications Symposium (NVMSA) (2016)
34. Huang, Y., et al.: Closing the performance gap between volatile and persistent key-value stores using cross-referencing logs. In: 2018 USENIX Annual Technical Conference (2018)

35. DeBrabant, J., Arulraj, J., et al.: A prolegomenon on OLTP database systems for non-volatile memory. Proc. VLDB Endow. **7**(14), 57–63 (2014)
36. Luo, Y., Chu, Z., Jin, P., Wan, S.: Efficient sorting and join on NVM-based hybrid memory. In: Qiu, M. (ed.) ICA3PP 2020, Part I. LNCS, vol. 12452, pp. 15–30. Springer, Cham (2020). https://doi.org/10.1007/978-3-030-60245-1_2
37. Garcia-Molina, H., Salem, K.: Main memory database systems: an overview. IEEE Trans. Knowl. Data Eng. **4**(6), 509–516 (1992)

ImputeRNN: Imputing Missing Values in Electronic Medical Records

Jiawei Ouyang[1,3], Yuhao Zhang[2,3], Xiangrui Cai[2,3(✉)], Ying Zhang[1,3], and Xiaojie Yuan[1,2,3]

[1] College of Computer Science, Nankai University, Tianjin 300350, China
ouyangjiawei@mail.nankai.edu.cn,{yingzhang,yuanxj}@nankai.edu.cn
[2] College of Cyber Science, Nankai University, Tianjin 300350, China
zhangyuhao@mail.nankai.edu.cn,caixr@nankai.edu.cn
[3] Tianjin Key Laboratory of Network and Data Security Technology,
Tianjin 300350, China

Abstract. Electronic Medical Records (EMRs), which record visits of patients to the hospital, are the main resources for medical data analysis. However, plenty of missing values in EMRs limit the model capability for various researches in healthcare. Recently, many imputation methods have been proposed to address this challenging problem, but they fail to take medical bias into account. Medical bias is a ubiquitous phenomenon that the missingness of medical data is missing not at random because doctors prone to measure features related to the disease of patients. It reflects the physical conditions of patients, which helps impute missing data with accurate and practical values. In this paper, we propose a novel joint recurrent neural network (RNN) model called ImputeRNN, which considers medical bias for EMR imputation. We model the medical bias by an additional RNN based on a mask (missing or not) matrix, whose hidden vectors are incorporated into the model as contexts by a fusion layer. Extensive experiments on two real-world EMR datasets demonstrate that ImputeRNN outperforms state-of-the-art methods on imputation and downstream prediction tasks.

Keywords: Electronic Medical Records · Missing values · Imputation · Medical bias · Recurrent neural network

1 Introduction

Electronic Medical Records (EMRs) consist of substantial heterogeneous medical data, which provide abundant resources for carrying out extensive researches on human healthcare and medical diagnosis [15,22,30]. However, EMRs are always incomplete and contain plenty of missing values due to various reasons, such as collection fault, transmission errors, and so on [3,8,27]. The existence of missing values in EMR data leads to insufficient information and inaccurate analysis for medical researches [18,21,33]. Therefore, it is significant to address the problem of missing data in EMRs.

© Springer Nature Switzerland AG 2021
C. S. Jensen et al. (Eds.): DASFAA 2021, LNCS 12683, pp. 413–428, 2021.
https://doi.org/10.1007/978-3-030-73200-4_28

Imputation is useful for solving the problem of missing data [3, 25]. Compared with incomplete data, imputed data contains additional information that avoids inaccurate analysis [4, 28]. Massive work has shown that various analysis methods yield the best results based on imputed data [12, 14, 16, 17]. Thus it is necessary to impute missing values before data analysis. Traditional imputation methods are based on statistics and machine learning [5, 23]. However, the effectiveness of these methods is limited when there are a lot of missing values [10, 26]. Recently, some deep learning methods utilize finite observed values to impute incomplete data and achieve remarkable performance [14, 31]. State-of-the-art methods for EMR imputation are based on RNN, which captures temporal relations of data [2, 16, 17, 28]. Nevertheless, existing imputation methods typically concentrate on feature regularity of data. Due to the lack of consideration of medical bias, these methods fail to impute missing values appropriately in EMRs [21, 33].

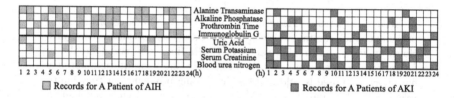

Fig. 1. An example of the medical bias in ICU EMRs. The abscissa is the time, and the ordinate is the medical feature. The upper four medical features are liver-related, and the lower four are kidney-related.

Medical bias is a ubiquitous phenomenon that the missingness of EMRs is missing not at random [1, 21, 29]. The measurements of patients are not recorded randomly, because doctors prone to measure patients for more related features [33]. Figure 1 describes an example of the medical bias in ICU EMRs. We find that the missingness of records for the patient with different diseases is distinct. The patient of AIH (Autoimmune Hepatitis) has more records on liver-related features, while the patient of AKI (Acute Kidney Injury) has more records on kidney-related features. It is a noteworthy medical bias that the missingness of features is related to the physical condition of patients. The related features have more records because their values are always abnormal and changeable that need to be recorded. Others have fewer records because their values are normal and stable that don't need to be recorded. Therefore, the missingness of EMRs reflects a medical bias that normal values have a higher missing probability than abnormal values [7, 20]. In conclusion, the medical bias helps determine the value change range of features, which is useful for EMR imputation.

In this paper, we propose a novel joint RNN model, called ImputeRNN, to utilize both values and the medical bias for EMR imputation. The model is made up of two RNN structures (an imputation GRU and a mask GRU). We exploit the imputation GRU to capture the regularity of values. Then we introduce a mask matrix to embody the medical bias and model it by the mask GRU. To employ medical bias, we incorporate the hidden vectors of the mask GRU into the imputation GRU by a fusion layer. Finally, we construct an imputation layer to impute missing values by previous information and current observed values. In this way, ImputeRNN achieves remarkable performance for EMR imputation. In summary, the contributions of this paper are as follows:

- We realize that medical bias is a universal phenomenon in EMRs. Reasonable use of medical bias is beneficial for EMR imputation.
- We propose ImputeRNN to incorporate medical bias for EMR imputation. We treat the missingness of data as an expression of medical bias, and then we design an additional RNN to model the medical bias.
- We conduct extensive experiments on two real-world EMR datasets. The results show that our model achieves state-of-the-art performance on both imputation and downstream prediction tasks.

2   Related Work

2.1   EMR Imputation

The existence of missing values in EMRs is an inevitable problem, and many traditional imputation methods are applied to address it. Fixed values filling [8] is a universal method that replaces missing values with statistic variables. MICE [27] imputes incomplete data by chained equations. KNN [6] utilizes the weighted average of k nearest similar neighbor samples to impute missing parts. Matrix Factorization [23] factorizes the original incomplete matrix into low-rank matrices and then fills missing values by recovering a complete matrix. GAIN [31] is a GAN-based imputation model that introduces a hint vector for training. These algorithms are suitable for most missing value scenarios, including EMR imputation. However, they do not consider the temporal information, which leads to a disappointing imputation performance for temporal medical data of EMRs.

Recently, some generative models based on RNN are proposed to impute time series, and they achieve noteworthy performance for EMRs imputation. M-RNN [32] interpolates within streams and imputes across streams. BRITS [2] is a bidirectional recurrent dynamical system for time series imputation. LGnet [28] designs a memory module that contains global information for imputation. Luo et al. propose a two-stage GAN model to impute multivariate time series [16]. To raise time efficiency and gain reasonable imputed values, they improve the previous GAN model with a compressing and reconstructing strategy [17]. This model got remarkable results in imputation and prediction tasks. However, these methods ignore the medical bias, which has a disadvantageous impact on EMR imputation. Zheng et al. [33] point to there is a strong bias in EMRs. They conduct a Hidden Markov Model (HMM) variant to capture temporal relation and medical bias, but this model has little ability to capture long-term temporal relations. Instead of the HMM, we construct a joint RNN model that has a well capacity for temporal relation mining. Meanwhile, our model pays attention to medical bias, which is significant for EMR imputation.

2.2 Medical Bias

Medical bias is a universal phenomenon in EMRs, and many medical papers have studied it. Phelan et al. [20] illustrate that interactions between patients and healthcare systems result in medical bias. Agniel et al. [1] design a retrospective observational study to evaluate the influence of bias on EMR researches. Vassy et al. [29] assess and address the medical bias with data visualization. MacNamee et al. [18] deal with medical bias in training data by stratified sampling and boosting. Pivovarov et al. [21] propose a method that leverages record frequency to identify and mitigate laboratory test bias in EMRs. In conclusion, medical bias is an inevitable problem in medical researches. Ignoring medical bias results in a misinterpretation of medical analysis. Therefore, we regard the medical bias as an auxiliary and incorporate it into our model for EMR imputation.

3 Preliminaries

For a patient, we take his medical record as a d-dimensional time series observed at $T = (t_1, \ldots, t_n)^\top$. We denote it by a value matrix $X = (x_1, \ldots, x_n)^\top \in \mathbb{R}^{n \times d}$. It includes n time steps and each step t_i contains d variables which represent d medical features of the patient. Then we express medical bias by missingness of data and we introduce a mask matrix $M = (m_1, \ldots, m_n)^\top \in \{0,1\}^{n \times d}$ to denote the missingness:

$$m_{ij} = \begin{cases} 1, \text{ if } x_{ij} \text{ is observed} \\ 0, \text{ otherwise} \end{cases},$$

where x_{ij} is the j-th variable of the i-th time step, m_{ij} is 1 if x_{ij} existed, otherwise 0.

Then we get two vital matrices that are concerned with time. An intuitive example is given as follows:

$$X = \begin{bmatrix} 2 & NA & 1 & \cdots & NA \\ NA & 9 & 7 & \cdots & NA \\ 14 & 26 & NA & \cdots & 32 \end{bmatrix}^\top, M = \begin{bmatrix} 1 & 0 & 1 & \cdots & 0 \\ 0 & 1 & 1 & \cdots & 0 \\ 1 & 1 & 0 & \cdots & 1 \end{bmatrix}^\top, T = \begin{bmatrix} 0 & 4 & 9 & \cdots & n \end{bmatrix}^\top.$$

Moreover, we deduce a time matrix $\delta \in \mathbb{R}^{n \times d}$ to represent the time interval from timestamp of last observed value to current timestamp. The formula and the example of time matrix δ are shown as follows:

$$\delta_{ij} = \begin{cases} 0, & \text{if } i = 0 \\ t_i - t_{i-1}, & \text{if } m_{(i-1)j} = 1, i > 0 \\ t_i - t_{i-1} + \delta_{(i-1)j}, & \text{if } m_{(i-1)j} = 0, i > 0 \end{cases}, \delta = \begin{bmatrix} 0 & 4 & 9 & \cdots & \delta_{n1} \\ 0 & 4 & 5 & \cdots & \delta_{n2} \\ 0 & 4 & 5 & \cdots & \delta_{n3} \end{bmatrix}^\top,$$

where $m_{(i-1)j}$ is missingness of the j-th variable at $(i-1)$-th time step, t_i and t_{i-1} are timestamps of the current step and the last step.

In this paper, we concentrate on filling missing values in incomplete EMRs. We propose a joint RNN model with two GRUs to learn regularities of values (X) and medical bias (M) simultaneously to recover missing values. Our purpose is to impute the vacancies of EMR data with accurate and suitable values.

4 Methodology

4.1 Architecture

Figure 2 describes the whole architecture of ImputeRNN. The model receives a value matrix and a mask matrix as inputs and generates imputed data as outputs. The backbone of ImputeRNN is a joint RNN model which is made up of two GRUs. We introduce a fusion layer to combine the hidden variables of these two GRUs. Then a new hidden variable, produced by the fusion layer, is filtered by the time decay gate. This variable is passed to the next cell of the imputation GRU and simultaneously is mapped to be an imputation candidate. The original value is mapped to be another candidate by feature regression. We design an imputation layer that receives these two candidates to generate final imputed data. The imputation GRU takes this imputed data as the input of cells. In the following subsections, we explicate modules of ImputeRNN in detail.

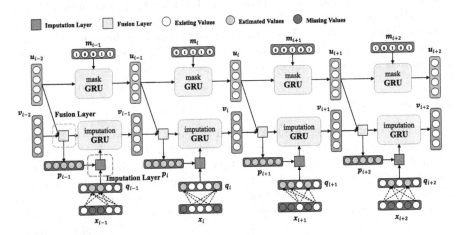

Fig. 2. The architecture of ImputeRNN. The mask GRU is modeled for the mask matrix M, and the imputation GRU is modeled for the value matrix X.

4.2 Joint RNN

The entire joint RNN model is assembled by two standard GRUs, which are typical RNN structures for capturing the temporal relation of data. The cell of GRU contains a reset gate and an update gate for saving and renewing stored memory information. The abbreviated formulas of these two GRUs are $\mathbf{GRU}_M$ and $\mathbf{GRU}_X$, which are modeled for the mask matrix and the value matrix. Besides, u_i and v_i are hidden variables of these two GRUs.

$$u_i = \mathbf{GRU}_M(m_i), \tag{1}$$
$$v_i = \mathbf{GRU}_X(x_i). \tag{2}$$

The upper GRU in Fig. 2 is the mask GRU trained to model the mask matrix M (medical bias). When a patient becomes healthier gradually, he visits hospitals less often. There are more missing values in data and more 0s in M. Considering the physical condition of patients, these missing values probably are normal or stable values in reality. It signifies that the regularities of the mask matrix M is useful for imputation. The i-th cell of the mask GRU takes m_i as input and produces a hidden variable u_i that is passed to the $(i + 1)$-th cells of two GRUs. The hidden variable u_i is an indicator that helps infer the physical condition of patients. It is crucial auxiliary information for EMR imputation.

The lower GRU in Fig. 2 is the imputation GRU trained to model the value matrix X (values). It is similar to the upper one, besides the hidden variables v_{i-1} and the input x_i are processed before they are transmitted to the i-th cell. The fusion layer merges v_{i-1} with u_{i-1} to produce a variable $\hat{c}_{i-1}$. This variable is filtered to be a new hidden variable c_{i-1} by the time decay gate γ_{i-1} which is calculated from time interval δ_{i-1}. Then the c_{i-1} is mapped to be an imputation candidate p_i. For the input x_i, a feature regression function maps it to be another candidate q_i. The imputation layer combines these two candidates to generate a variable o_i as the final result for EMR imputation. Meanwhile, this variable is the new input of the i-th imputation GRU cell.

4.3 Fusion Layer

In this subsection, we describe the internal design of the fusion layer. This layer combines information of missingness and value from the mask GRU and the imputation GRU. Then it introduces a time decay gate to filter the information according to the time interval. Although the missingness and the value have some inherent connections, the connotations of them are quite different. Moreover, the data types of them are also different: the missingness is boolean, while the value is numeric. Therefore it is reasonable to regard them as two kinds of information and merge them by different fusion strategies [11,19].

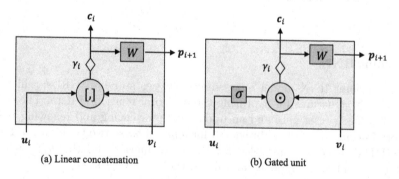

(a) Linear concatenation   (b) Gated unit

Fig. 3. The illustration of the fusion layer for combining information of missingness and values. (a) Linear concatenation strategy. (b) Gated unit strategy.

Figure 3 displays the structure of the fusion layer. There are two classic fusion strategies: linear concatenation and gated unit. They take hidden variables of two GRU as inputs u_i and v_i. Then they generate a new variable c_i for the next hidden cell of the imputation GRU. Meanwhile, they produce an imputation candidate variable p_i, which is transmitted to the imputation layer.

Linear concatenation is shown in Fig. 3 (a). This strategy combines two hidden variables by concatenation operation and endows equal attention weight to them. By this means, the missingness and the value contribute equivalent effectiveness to the model. The formula of Linear concatenation is:

$$\hat{c}_i = \tanh(W_c\,[u_i, v_i] + b_c), \tag{3}$$

where W_c is the weight parameter and b_c is the bias parameter.

Gated unit is shown in Fig. 3 (b). This strategy transforms the hidden variable u_i of the mask GRU to be a weighted gate through a sigmoid activation function [19]. By this means, the value plays a pivotal role in the model, and the missingness serves as a filter that not only preserves valid information but also discards what is futile. The formula of Gated unit is:

$$g_i = \sigma(W_g u_i + b_g), \tag{4}$$
$$\hat{c}_i = \tanh(W_c\,[g_i \odot v_i] + b_c), \tag{5}$$

where W_g is the weight parameter and b_g is the bias parameter.

Then we introduce a time decay gate γ_i to filter output hidden variable $\hat{c}_i$. This gate is derived from the time interval δ_i. Intuitively, the longer the time interval, the weaker the influence from the previous step. Thus we apply a monotonically decreasing function to calculate γ_i from δ_i. We restrict value range of γ_i from 0 to 1. The final hidden variable c_i is generated as follows:

$$\gamma_i = 1/e^{\max(0, W_\gamma \delta_i + b_\gamma)}, \tag{6}$$
$$c_i = \gamma_i \odot \hat{c}_i, \tag{7}$$

where W_γ is the weight parameter and b_γ is the bias parameter.

We think that the gated unit plays better than the linear concatenation for EMR imputation. Because the former takes unequal importance weight for the missingness and the value. The imputed values are more relevant to the value. Therefore, models should pay more attention to the value, and the missingness does duty for an assistant to improve imputation performance. Our experiments confirm that the gated unit is indeed superior to linear concatenation.

4.4 Imputation Layer

In this subsection, we describe the operating mechanism of the imputation layer. At each time step, this layer receives two variables p_i and q_i as imputation candidates. Then the mask vector m_i and the value vector x_i are combined into this layer as auxiliaries. In the end, the two candidates are integrated to generate the final precise imputed value o_i.

The first candidate is p_i which is transformed from the fusion hidden variable c_{i-1} of the fusion layer. The variable c_{i-1} is a representation that characterizes the patient's current latent physical condition by summarizing the previous regularities of the missingness and the value. Based on this representation, we obtain an imputation candidate by a fully connected layer:

$$p_i = W_p c_{i-1} + b_p, \qquad (8)$$

where W_p is the weight parameter and b_p is the bias parameter.

As for the other candidate q_i, we infer it from current observed values x_i by the feature regression. This method relies on feature correlations among all variables to estimate missing values. It generates the imputation candidate based on the patient's current observed physical condition. The unobserved values are zero so that they contribute little effect to q_i. Our feature regression function is:

$$q_i = W_q x_i + b_q, \qquad (9)$$

where W_q is the weight parameter and b_q is the bias parameter. The diagonal elements of W_q are 0s since we do not use x_{ij} to estimate q_{ij}.

Finally, we combine two candidates to generate the final imputed result o_i. We transform the missingness m_i into a gate parameter β_i. It decides the weight of two candidates. When most values exist, the candidate q_i is more important because observed values provide enough information for imputation. When there are plenty of missing values, the candidate p_i based on historic information plays a more significant role in imputation. We get the final result by:

$$\beta_i = \sigma(W_\beta m_i + b_\beta), \qquad (10)$$
$$\hat{o}_i = (1 - \beta_i) \odot p_i + \beta_i \odot q_i, \qquad (11)$$
$$o_i = (1 - m_i) \odot \hat{o}_i + m_i \odot x_i, \qquad (12)$$

where W_β is the weight parameter and b_β is the bias parameter.

4.5   Loss

To optimize ImputeRNN, we define a two-part loss. The first part is the *imputation loss* and the other is the *prediction loss*. The former is a squared error between estimated values and original observations on the non-missing part of data. It promotes imputed values to be as close to real values as possible. The latter is a loss function f on physiological labels such as mortality, ICD-9 codes. These labels manifest the physical condition of patients that determines the range of estimated values. As an example, we take mortality as target y and predict its result $\hat{y}$ by the last hidden variable of the imputation GRU:

$$\hat{y} = W_y p_n + b_y, \qquad (13)$$

where W_y is the weight parameter and b_y is the bias parameter.

We calculate the two-part loss by accumulating the *imputation loss* and the *prediction loss* with a hyper-parameter λ. This hyper-parameter decides the weight proportion of these two losses:

$$\mathcal{L} = \|(\boldsymbol{X} - \hat{\boldsymbol{O}}) \odot \boldsymbol{M}\|_2 + \lambda f(y, \hat{y}). \tag{14}$$

5 Experiments

In this section, we conduct various experiments on two real-world EMR datasets to compare our proposed model ImputeRNN with baselines.

5.1 Datasets

PhysioNet. It is a public medical dataset that comes from *PhysioNet Challenge 2012* [24]. It is made up of ICU records from 4000 patients where 554 of them died in the hospital. For the record of each patient, it is a multivariate time series collected in 48 h, and it consists of 41 features. It is worth noting that there are up to 80.67% missing values, which indeed influences the downstream analysis tasks. We divide the dataset into the training set, validation set, and testing set according to 80%, 10%, 10% to ensure the validity of our experiments.

MIMIC-III. It is an authoritative EMR database collected at Beth Israel Deaconess Medical Center [9]. It contains abundant medical features and is already a benchmark dataset for medical researches. Following the work of predecessors [22], we extract admission records from 11869 patients that 1031 of them died in the hospital. Then we filter features that may be insignificant (i.e. the missing rate of which is more than 98%), leaving 86 features. Moreover, we intercept the first 48-h record to align data. Then the missing rate of the dataset is 86.98%. At last, we take the same data division processing as Physionet.

5.2 Baselines and Implementation Details

To appraise the performance of ImputeRNN, we compare our model against ten imputation baselines of two categories.

- Non-RNN methods:
  Mean Imputation replaces the missing values by the corresponding global mean value of each variable. **KNN** [6] combines the weighted average of k nearest neighbor samples to impute. **Matrix Factorization** [23] factorizes the incomplete matrix into low-rank matrices and then fills vacancies by recovering. **MICE** [27] imputes the missing values by multiple imputations with chained equations. **GAIN** [31] is a GAN based imputation method that introduces a hint vector to reveal partial information about the missingness.

– RNN methods:
M-RNN [32] interpolates within data streams and imputes across streams simultaneously. **BRITS** [2] treats missing values as RNN variables and imputes them during backpropagation in a bidirectional recurrent system. **LGnet** [28] designs a memory module which contains global temporal information to impute missing values. **GAN-2-Stage** [16] is a two-stage GAN model for multivariate time series imputation, which including model training and "noise" vector optimization. **E^{2}GAN** [17] utilizes a compressing and reconstructing strategy to replace the "noise" vector optimization stage of GAN-2-Stage, and it achieves state-of-the-art imputation performance.

For evaluation, we implement two different fusion strategies on our model: ImputeRNN with linear concatenation (**ImputeRNN$_{lc}$**) and ImputeRNN with gate unit (**ImputeRNN$_{gu}$**). We compare them with baselines for imputation and prediction tasks. To evaluate the imputation performance of models, we calculate the distance between imputed values and original values by the root mean squared error (RMSE). Furthermore, we evaluate the mortality prediction performance on imputed data by the area under the ROC curve (AUC) score.

To ensure the comparability and validity of experiments, we guarantee the experiment settings of baselines are consistent with ours. We execute all models on the same hardware and process datasets identically. For non-deep learning baselines, we implement them by a versatile python package "fancyimpute[1]". For deep learning baselines, we implement them by TensorFlow 1.12 framework according to the corresponding papers. Moreover, we conduct the same training strategies for them, such as shuffle, early stopping, and drop out.

For our model, we construct ImputeRNN based on GRU and train them by the ADAM [13] optimizer. The learning rate of ADAM is a hyper-parameter that ranges from 0.001 to 0.01. The hidden unit number of GRU is 64 for PhysioNet, while 100 for MIMIC-III. The batch sizes are both 128. To avoid overfitting, we set the dropout rates to 0.5 for training. We normalize all input variables to be zero mean and unit standard deviation. Besides, we use early stopping on the validation set to find the best values for hyper-parameters, and then we report the results on the testing set from 10-fold cross validation.

5.3 Imputation Performance

To evaluate imputation performance, we randomly eliminate 10% of existed values and calculate RMSE between original values and estimated. Table 1 shows the imputation results on two datasets. We find that the methods based on RNN models achieve smaller results than Non-RNN models, which means better performance. We speculate that RNN helps the model take full advantage of the temporal information, which is a benefit for time series imputation. Moreover, all our models outperform the baselines for imputation performance on both datasets. The RMSE of ImputeRNN$_{gu}$ is 0.5312 on Physionet and 0.4722 on

---

[1] https://github.com/iskandr/fancyimpute.

Table 1. The Imputation performance in terms of the RMSE.

Categories	Methods	Physionet	MIMIC-III
Non-RNN	Mean	0.6132 ± 0.0001	0.5752 ± 0.0001
	KNN	0.5981 ± 0.0082	0.5369 ± 0.0077
	MF	0.6354 ± 0.0164	0.6423 ± 0.0204
	MICE	0.6314 ± 0.0113	0.6438 ± 0.0104
	GAIN	0.6241 ± 0.0074	0.5743 ± 0.0068
RNN	M-RNN	0.5874 ± 0.0043	0.5268 ± 0.0034
	BRITS	0.5726 ± 0.0030	0.5244 ± 0.0021
	LGnet	0.5646 ± 0.0040	0.5207 ± 0.0029
	GAN-2-Stage	0.5782 ± 0.0069	0.5217 ± 0.0046
	E^2GAN	0.5623 ± 0.0034	0.5131 ± 0.0020
Our models	ImputeRNN$_{lc}$	0.5423 ± 0.0027	0.4896 ± 0.0021
	ImputeRNN$_{gu}$	$\mathbf{0.5312 \pm 0.0025}$	$\mathbf{0.4722 \pm 0.0018}$

MIMIC-III, which are the best results on two datasets. For our models, we find that ImputeRNN$_{gu}$ performs better than ImputeRNN$_{lc}$. It demonstrates that the gate unit is a more suitable medical bias fusion strategy for EMR imputation than the linear concatenation. Additionally, we present the standard deviations of each RMSE result, and our models have a low-level variation of the standard deviation. It means that the results of our models are stable for imputation tasks.

5.4 Prediction Performance

In most cases, the missing data is imputed to assist downstream analysis models to be more efficient and powerful. Therefore, we conduct prediction experiments based on imputed complete datasets to validate the imputation performance indirectly. We take the in-hospital mortality as the prediction target, and we use the AUC score as the metric because of imbalanced datasets (i.e. the death is in the minority). We utilize different classifiers to predict the target. They are logistic regression (LR), random forest (RF), support vector machine with RBF kernel (SVM), and RNN.

Table 2 shows the mortality prediction performance on both datasets. All of our models outperform baselines, and the ImputeRNN$_{gu}$ achieves state-of-the-art results for every classifier. We discover the results of MIMIC-III are generally better than those of Physionet. We think that the more samples and features of MIMIC-III making a difference for the ability of models. For different classifiers, the RNN is superior to others, and the ImputeRNN$_{gu}$ achieves the best AUC scores on both datasets by the RNN classifier. They are 0.8867 for Physionet and 0.9057 for MIMIC-III. It is a remarkable improvement compared to the previous best model E^2GAN. Besides, the ImputeRNN$_{gu}$ performs better than the ImputeRNN$_{lc}$ on prediction tasks. This result is similar to the imputation

Table 2. The mortality prediction performance in terms of the AUC score.

Categories	Methods	Physionet				MIMIC-III			
		LR	RF	SVM	RNN	LR	RF	SVM	RNN
Non-RNN	Mean	.7013	.7391	.7943	.8273	.7646	.7842	.8275	.8410
	KNN	.7120	.7443	.7881	.8318	.7428	.7856	.8168	.8430
	MF	.6968	.7364	.7909	.8231	.7203	.7609	.8130	.8357
	MICE	.7057	.7291	.7762	.8193	.7183	.7782	.7913	.8208
	GAIN	.7054	.7371	.7936	.8267	.7601	.7868	.8253	.8397
RNN	M-RNN	.7149	.7727	.8060	.8473	.7791	.7973	.8326	.8671
	BRITS	.7321	.7870	.8149	.8502	.7752	.8172	.8411	.8719
	LGnet	.7458	.7816	.8117	.8719	.7948	.8073	.8391	.8804
	GAN-2-Stage	.7212	.7546	.8157	.8603	.7817	.8052	.8354	.8764
	E^2GAN	.7677	.7998	.8201	.8724	.8021	.8104	.8425	.8813
Our models	ImputeRNN$_{lc}$	.7784	.8053	.8412	.8819	.8184	.8225	.8614	.8982
	ImputeRNN$_{gu}$	**.7926**	**.8130**	**.8494**	**.8867**	**.8249**	**.8305**	**.8722**	**.9057**

performance. It proves that the appropriate use of medical bias is beneficial to impute EMRs with more realistic and logical values. These imputed values make sense for downstream applications. In conclusion, reasonable imputation is significant for downstream tasks, and the missing values imputed by our models truly enhance the prediction performance of downstream analysis methods.

5.5 Discussion

Time Efficiency. Firstly, we discuss the time efficiency of the training step to assess the computational complexity of the model. We compare our models with RNN-based baselines for how long the model needs to be trained. To ensure credible comparability, we set the same datasets, hardware, training strategies, and other parameters for all models. The result of time efficiency is shown in Table 3. We notice that all our models are more time-efficient compared to RNN-based baselines, especially GAN models. It demonstrates that ImputeRNN achieves the best performance with an acceptable complexity.

Missing Rate. We further consider the model imputation performance with different missing rates. We randomly select 10%, 20%,..., 90% data to discard from datasets. Then we impute the vacancies by ImputeRNN$_{gu}$ and baselines. Figure 4 presents the imputation results on two datasets. We find that RNN models perform better than Non-RNN models, especially for high missing rates, and our model performs the best. As the missing rate increases, the performance of all models becomes worse because insufficient data cannot provide enough information. However, ImputeRNN$_{gu}$ still achieves the best results for all missing rates. It indicates that our model is suitable for scenarios with high missing rates.

Loss Weight λ. We investigate the influence of loss weight λ in imputation and prediction tasks. It is a hyper-parameter that decides the proportion of

Table 3. Comparison of the time efficiency.

Categories	Methods	Physionet	MIMIC-III
RNN	M-RNN	$213 \pm 5s$	$669 \pm 3s$
	BRITS	$202 \pm 2s$	$566 \pm 4s$
	LGnet	$268 \pm 4s$	$693 \pm 4s$
	GAN-2-Stage	$2650 \pm 10s$	$5253 \pm 13s$
	E^2GAN	$274 \pm 5s$	$736 \pm 6s$
ImputeRNN	ImputeRNN$_{lc}$	$\mathbf{176 \pm 3s}$	$\mathbf{448 \pm 3s}$
	ImputeRNN$_{gu}$	$189 \pm 2s$	$453 \pm 4s$

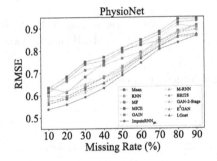
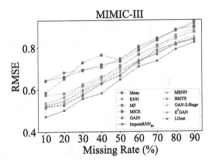

Fig. 4. The imputation performance comparison for different missing rate percentages on Physionet and MIMIC-III datasets.

imputation and prediction losses. When it becomes smaller, the imputation loss is stronger. Figure 5 shows the results of ImputeRNN$_{gu}$. We observe that the imputation task attains the minimum RMSE on two datasets when λ is 0.001 and 0.0001 rather than 0. It demonstrates that the prediction loss is useful for imputation. Similarly, the result of the prediction task is worse if λ becomes too big. The AUC scores on two datasets reach maximums when λ is 0.01 and 0.1. Comparing the optimal λ of two tasks, we find that λ of the prediction task is larger than the imputation task. It is reasonable because a slightly larger prediction loss makes imputed values more consistent with the prediction task. To summarize, both losses are essential for imputation and prediction.

5.6 Ablation Study

Finally, we research the effect of different modules in ImputeRNN. We remove feature regression (FR), prediction loss (PL), and mask GRU (MG) modules of ImputeRNN$_{gu}$ separately and conduct contrast tests. Table 4 shows the results of the ablation study. We find that the original complete model performs better than modified incomplete models. It indicates that all modules are necessary for imputation. We also observe that the model without mask GRU attains the worst

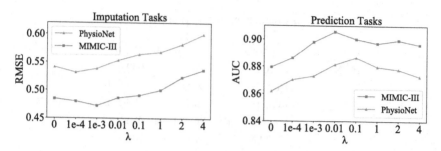

Fig. 5. The influence of loss weight λ in imputation and prediction tasks on Physionet and MIMIC-III datasets.

results. We conclude that mask GRU does play a vital role in ImputeRNN. It demonstrates that incorporating medical bias is significant for EMR imputation.

Table 4. The ablation study for imputation and prediction tasks.

Datasets	Methods	RMSE	AUC			
			LR	RF	SVM	RNN
Physionet	ImputeRNN$_{gu}$	**0.5312**	**0.7926**	**0.8130**	**0.8494**	**0.8867**
	Without FR	0.5491	0.7814	0.7943	0.8253	0.8759
	Without PL	0.5478	0.7797	0.7816	0.8124	0.8588
	Without MG	0.5907	0.7202	0.7460	0.8023	0.8344
MIMIC-III	ImputeRNN$_{gu}$	**0.4722**	**0.8249**	**0.8305**	**0.8722**	**0.9057**
	Without FR	0.4913	0.8103	0.8194	0.8520	0.8941
	Without PL	0.4846	0.8027	0.8135	0.8440	0.8795
	Without MG	0.5326	0.7699	0.7964	0.8304	0.8583

6   Conclusion

In this paper, we propose a novel joint RNN model called ImputeRNN to learn regularities of values and medical bias simultaneously to impute missing values in EMRs. We regard the missingness of data as an expression of medical bias and model it by a mask GRU. Then we apply a fusion layer to employ medical bias. Finally, we construct an imputation layer to impute missing values in EMRs. Empirical experiments on two real-world EMR datasets show that ImputeRNN achieves state-of-the-art performance for EMR imputation. For future work, we are interested in imputing EMR datasets that contain discrete variables.

Acknowledgements. This work is supported by Chinese Scientific and Technical Innovation Project 2030 (No. 2018AAA0102100), NSFC-General Technology Joint

Fund for Basic Research (No. U1936206, No. U1836109), National Natural Science Foundation of China (No. 61772289, No. U1903128, and No. 62002178), and Natural Science Foundation of Tianjin, China (No. 20JCQNJC01730).

References

1. Agniel, D., Kohane, I.S., Weber, G.M.: Biases in electronic health record data due to processes within the healthcare system: retrospective observational study. Br. Med. J. **361** (2018)
2. Cao, W., Wang, D., Li, J., Zhou, H., Li, L., Li, Y.: BRITS: bidirectional recurrent imputation for time series. In: Advances in Neural Information Processing Systems, NeurIPS, pp. 6776–6786 (2018)
3. Che, Z., Purushotham, S., Cho, K., Sontag, D.A., Liu, Y.: Recurrent neural networks for multivariate time series with missing values. Sci. Rep. **8**(1), 1–12 (2018)
4. Che, Z., Purushotham, S., Li, M.G., Jiang, B., Liu, Y.: Hierarchical deep generative models for multi-rate multivariate time series. In: International Conference on Machine Learning, ICML, vol. 80, pp. 783–792 (2018)
5. Fan, J., Zhang, Y., Udell, M.: Polynomial matrix completion for missing data imputation and transductive learning. In: Association for the Advancement of Artificial Intelligence, AAAI, pp. 3842–3849 (2020)
6. García-Laencina, P.J., Sancho-Gómez, J., Figueiras-Vidal, A.R., Verleysen, M.: K nearest neighbours with mutual information for simultaneous classification and missing data imputation. Neurocomputing **72**(7–9), 1483–1493 (2009)
7. Haneuse, S., Daniels, M.: A general framework for considering selection bias in EHR-based studies: what data are observed and why? Gener. Evid. Methods Improve Patient Outcomes **4**(1), 1203–1203 (2016)
8. Jerez, J.M., et al.: Missing data imputation using statistical and machine learning methods in a real breast cancer problem. Artif. Intell. Med. **50**(2), 105–115 (2010)
9. Johnson, A.E., et al.: MIMIC-III, a freely accessible critical care database. Sci. Data **3**(1), 1–9 (2016)
10. Khayati, M., Lerner, A., Tymchenko, Z., Cudré-Mauroux, P.: Mind the gap: an experimental evaluation of imputation of missing values techniques in time series. Proc. VLDB Endow. **13**(5), 768–782 (2020)
11. Kiela, D., Grave, E., Joulin, A., Mikolov, T.: Efficient large-scale multi-modal classification. In: Association for the Advancement of Artificial Intelligence, AAAI, pp. 5198–5204 (2018)
12. Kim, Y., Chi, M.: Temporal belief memory: imputing missing data during RNN training. In: International Joint Conference on Artificial Intelligence, IJCAI, pp. 2326–2332 (2018)
13. Kingma, D.P., Ba, J.: Adam: a method for stochastic optimization. In: International Conference on Learning Representations, ICLR (2015)
14. Li, S.C., Jiang, B., Marlin, B.M.: MisGAN: learning from incomplete data with generative adversarial networks. In: International Conference on Learning Representations, ICLR (2019)
15. Luo, J., Ye, M., Xiao, C., Ma, F.: HiTANet: hierarchical time-aware attention networks for risk prediction on electronic health records. In: Special Interest Group on Knowledge Discovery in Data, SIGKDD, pp. 647–656 (2020)
16. Luo, Y., Cai, X., Zhang, Y., Xu, J., Yuan, X.: Multivariate time series imputation with generative adversarial networks. In: Advances in Neural Information Processing Systems, NeurIPS, pp. 1603–1614 (2018)

17. Luo, Y., Zhang, Y., Cai, X., Yuan, X.: E^{2}GAN: end-to-end generative adversarial network for multivariate time series imputation. In: International Joint Conference on Artificial Intelligence, IJCAI, pp. 3094–3100 (2019)
18. MacNamee, B., Cunningham, P., Byrne, S., Corrigan, O.I.: The problem of bias in training data in regression problems in medical decision support. Artif. Intell. Med. 24(1), 51–70 (2002)
19. Ovalle, J.E.A., Solorio, T., Montes-y-Gómez, M., González, F.A.: Gated multi-modal units for information fusion. In: International Conference on Learning Representations, ICLR (2017)
20. Phelan, M., Bhavsar, N.A., Goldstein, B.A.: Illustrating informed presence bias in electronic health records data: how patient interactions with a health system can impact inference. Gener. Evid. Methods Improve Patient Outcomes 5(1), 22 (2017)
21. Pivovarov, R., Albers, D.J., Sepulveda, J.L., Elhadad, N.: Identifying and mitigating biases in EHR laboratory tests. Biomed. Inform. 51, 24–34 (2014)
22. Purushotham, S., Meng, C., Che, Z., Liu, Y.: Benchmarking deep learning models on large healthcare datasets. Biomed. Inform. 83, 112–134 (2018)
23. Salakhutdinov, R., Mnih, A.: Probabilistic matrix factorization. In: Advances in Neural Information Processing Systems, NeurIPS, pp. 1257–1264 (2007)
24. Silva, I., Moody, G., Scott, D.J., Celi, L.A., Mark, R.G.: Predicting in-hospital mortality of ICU patients: the PhysioNet/computing in cardiology challenge 2012. Comput. Cardiol. 39, 245–248 (2012)
25. Smieja, M., Struski, L., Tabor, J., Zielinski, B., Spurek, P.: Processing of missing data by neural networks. In: Advances in Neural Information Processing Systems, NeurIPS, pp. 2724–2734 (2018)
26. Sportisse, A., Boyer, C., Josse, J.: Estimation and imputation in probabilistic principal component analysis with missing not at random data. In: Advances in Neural Information Processing Systems, NeurIPS (2020)
27. Sterne, J.A., et al.: Multiple imputation for missing data in epidemiological and clinical research: potential and pitfalls. Br. Med. J. 338 (2009)
28. Tang, X., Yao, H., Sun, Y., Aggarwal, C.C., Mitra, P., Wang, S.: Joint modeling of local and global temporal dynamics for multivariate time series forecasting with missing values. In: Association for the Advancement of Artificial Intelligence, AAAI, pp. 5956–5963 (2020)
29. Vassy, J., et al.: Yield and bias in defining a cohort study baseline from electronic health record data. Biomed. Inform. 78, 54–59 (2018)
30. Yadav, P., Steinbach, M.S., Kumar, V., Simon, G.J.: Mining electronic health records (EHRs): a survey. ACM Comput. Surv. 50(6), 85:1–85:40 (2018)
31. Yoon, J., Jordon, J., van der Schaar, M.: GAIN: missing data imputation using generative adversarial nets. In: International Conference on Machine Learning, ICML, vol. 80, pp. 5675–5684 (2018)
32. Yoon, J., Zame, W.R., van der Schaar, M.: Estimating missing data in temporal data streams using multi-directional recurrent neural networks. IEEE Trans. Biomed. Eng. 66(5), 1477–1490 (2019)
33. Zheng, K., Gao, J., Ngiam, K.Y., Ooi, B.C., Yip, J.W.L.: Resolving the bias in electronic medical records. In: Special Interest Group on Knowledge Discovery in Data, SIGKDD, pp. 2171–2180 (2017)

Susceptible Temporal Patterns Discovery for Electronic Health Records via Adversarial Attack

Rui Zhang[1], Wei Zhang[2], Ning Liu[1], and Jianyong Wang[1(✉)]

[1] Tsinghua University, Beijing, China
{zhang-r19,jianyong}@tsinghua.eud.cn
[2] East China Normal University, Shanghai, China

Abstract. The recent advancements in deep neural networks (DNNs) are revolutionizing the healthcare domain. Although many studies try to build medical DNNs model based on historical Electronic Health Records (EHR) and have achieved promising performance in many clinical prediction tasks, recent studies show that DNNs are vulnerable to adversarial attacks. Much of the interest in adversarial examples has stemmed from their ability to shed light on possible limitations of DNNs. However, related research has been receiving sustained attention in computer vision community, how to design adversarial examples for EHR data remains a rarely investigated. To figure out this problem, we propose a novel approach for generating EHR adversarial examples, named as *TSAttack*, which explores temporal structure contained in EHR to achieve an effective and efficient attack. Based on the generated EHR adversarial examples, we further propose a procedure to discover susceptible temporal patterns (STP) in a patient's medical records, which provide clinical decision support for dynamic monitoring. Extensive experiments on the real-world longitudinal EHR database MIMIC-III have demonstrated the effectiveness of our approach is yielding better performance in adversarial settings.

Keywords: Adversarial attack · Medical data · Susceptible temporal patterns

1 Introduction

Electronic Health Records (EHR) contain rich historical clinical information about patients, how to create accurate predictive models from EHR data, which is one of the important research in clinical informatics. Recent years have witnessed the rapid development of deep learning (DL), a series of state-of-the-art medical predictive models based on DL have been designed for various clinical tasks [6,12,20]. In particular, many studies try to adapt powerful sequential neural models to analyze clinical time series data and achieve superior performance [9].

© Springer Nature Switzerland AG 2021
C. S. Jensen et al. (Eds.): DASFAA 2021, LNCS 12683, pp. 429–444, 2021.
https://doi.org/10.1007/978-3-030-73200-4_29

However, the "black box" nature of the DL leads to models that lack transparency. A well-trained DL model can be vulnerable to adversarial examples, which are maliciously crafted to mislead the model to output wrong predictions [17]. A primary reason is that adversarial samples are located in some areas in the high-dimensional feature space which are not learned during training. Due to the high-dimensionality characteristic of EHR data, the decision boundaries could be more vulnerable in the adversarial setting. Theoretically, for any given original example near the decision boundary, we can craft adversarial perturbations to generate a new sample that crosses the decision boundary of the model to the target class. Thus, adversarial examples expose regions of the input space where the model performs poorly, which can aid in understanding and improving the model. Additionally, EHR systems contain numerous data collecting and processing pipelines. There are many potential adversaries and are thus vulnerable at many different stages [7]. Most healthcare-associated applications are safety-critical, while eager to deploy DL technologies. Therefore, investigating the adversarial vulnerability on EHR data thus becomes a valuable research topic in both theory and practice.

Despite fruitful studies of the adversarial attack on computer vision community (e.g., image data) [5,11,13], how to craft adversarial examples for EHR data remains rarely explored. Sun et al. [16] are the first to consider adversarial example generation on deep predictive models for EHR data. Different from attacking image or textual data, their aim is to detect susceptible medical measurements by analyzing the adversarial perturbations, which provide guidance for clinicians/nurses. However, since the inherent characteristic of EHR data such as temporality, heterogeneity, and high-dimensionality, existing adversarial attack methods for the image or textual data cannot be directly applied to EHR data, how to develop an adversarial attacks approach on temporal EHR data still remains two major issues.

Challenges. (1) *static vs temporal*. EHR contains a sequence of multivariate clinical time series data. Compared with static image data, the most critical difference is the temporal structure contained in EHR data. Temporal information within the clinical time series data reflects changes in the patient's condition. In order to provide a more efficient method to attack temporal EHR data, we need to fully utilize and model the temporal information. (2) *semantic-less vs feature significance*. For image data, individual pixels value has no specific meaning. In contrast, each medical feature differs in clinical significance and corresponds to a specific clinical event, thus small changes in EHR data would easily change the clinical observations of a patient. Therefore, the ground truth label of an EHR adversarial example is more ambiguous.

Motivation. To address these issues, our idea is to design an effective and efficient EHR adversarial example generation method by incorporating temporal information into the attacking. The incorporated temporal information not only can lead to reducing the cost of attacking but also suggests the susceptible changes in the patient's condition. Specifically, adversarial perturbations in EHR data can inform which temporal patterns are susceptible. Therefore, beyond designing an EHR adversarial attack method, we are motivated by the question: "Which certain temporal patterns could be more susceptible to adversarial attack than others for a given patient?"

Overview. To summarize, in this paper, we propose a novel approach for generating EHR adversarial examples, *TSAttack* (**T**emporal **S**parse Adversarial **Attack**), which explores temporal structure contained in EHR to achieve sparsity. Inspired by the sparse attack method [4,5], *TSAttack* adopts an optimal adversarial attack strategy to generate EHR adversarial examples. To extract key temporal structures, we construct a novel temporal sliding window that applies on the adversarial perturbations. During the optimization procedure, *TSAttack* employs l_p-norm perturbation regularization and the proposed temporal sparsity regularization that encodes temporal structures simultaneously. Then, we extract the susceptible regions from adversarial examples according to the time-level susceptibility score, which is defined by the magnitude and structure of adversarial perturbation. To obtain Susceptible Temporal Patterns (STP), the raw time series data of susceptible regions are first abstracted into a series of higher-level, meaningful concepts (e.g., *'Decreasing Respiratory rate'*). Finally, we adopt temporal patterns mining algorithm to discover frequently temporal patterns within the susceptible regions.

To our knowledge, it is the first time that explores temporal structures when implementing adversarial attacks on EHR data. Furthermore, we provide a general procedure to discover susceptible temporal patterns in the patient's medical records based on adversarial perturbation. Extensive experiments on the real-world longitudinal EHR database MIMIC-III [10] have demonstrated the effectiveness of our approach.

2 Related Work

Adversarial Attack. Early works on adversarial attacks focus on computer vision community. The Fast Gradient Sign Method (FGSM) [8] is a single-step attack that uses the sign of the gradient of the loss function with respect to the input for crafting an adversarial image. Iterative Fast Gradient Method (IFGM) [11] is an iterative version of FGSM. DeepFool [13] first assumed the DNNs is linear, and then find the closest distance from the original input to the decision boundary of adversarial examples. Optimization-based attack methods C&W [4] and EAD [5] serve as another line of research, which found an adversarial example by jointly minimizing its l_2-norm distortion and a differentiable loss function based on the logit layer outputs of DNNs. The above studies are all based on images, [18] firstly presented the $l_{2,1}$-norm based optimization algorithm for adversarial videos based on the perturbation propagation.

A few prior works (e.g., [2,16]) propose to generate EHR adversarial examples. Sun et al. [16] adopted C&W method for attacking deep predictive models of EHR, which aims to detect susceptible locations in medical records. The work [2] proposed a saliency score based adversarial attack on longitudinal EHR data, named LAVA attack. In their method, medical features are binary coded so it is not suitable to directly apply this study to continuous value.

Temporal Patterns Mining. Since analysis on time-intervals, rather than on the raw time series data, may shed light on the interpretative and meaningful temporal patterns within the temporal dimension. Temporal patterns mining seeks for the most frequent patterns across a set of temporal relations amongst the symbolic intervals. A typical approach to represent the temporal relations among the time-intervals is Allen's temporal relations [1]. Moskovitch et al. [14] proposed KarmaLego framework that exploits the transitivity of temporal relations to generate candidates for finding frequent temporal patterns, which can be adapted as features to learn the clinical classifier.

3   Preliminaries

Assuming there are N patients' longitudinal EHR data. For each patient, the EHR data containing a sequence of multivariate observations. We use $\mathbf{x}^{(n)} = \{\boldsymbol{x}_1^{(n)}, \boldsymbol{x}_2^{(n)}, \boldsymbol{x}_3^{(n)}, ..., \boldsymbol{x}_{T^{(n)}}^{(n)}\}$ to represent an instance with $T^{(n)}$ visit records. To minimize clutter, in the following we drop the superscript (n) whenever it is unambiguous. Each visit $\boldsymbol{x}_t \in \mathbb{R}^D$ is a D-dimensional vector corresponding to clinical features. Given an input instance $\mathbf{x}$, medical prediction task aims to predict a future clinical event (e.g., mortality risk prediction, physiologic decompensation and phenotype classification). Formally, we denote the DNNs model by f is trained to learn the mapping from the input space to the label space : $\mathcal{X} \rightarrow \mathcal{Y}$. The network f is parameterized by θ and we denote it as f_θ.

Definition 1. *(EHR Adversarial Examples) Given an arbitrary $\mathbf{x} \in \mathcal{X}$, and a trained neural network model $f_\theta(\cdot)$, $\mathbf{x}^* = \{\boldsymbol{x}_1^*, \boldsymbol{x}_2^*, \boldsymbol{x}_3^*, ..., \boldsymbol{x}_T^*\}$ is an adversarial example of $\mathbf{x}$ when:*

$$\arg\min_{\mathbf{x}^*} \| \mathbf{x} - \mathbf{x}^* \|_p \qquad s.t. \ f_\theta(\mathbf{x}) \neq f_\theta(\mathbf{x}^*) \tag{1}$$

$\| \cdot \|_p$ *is the l_p norm distance, which is a metric to quantify the magnitude of the similarities between $\mathbf{x}$ and $\mathbf{x}^*$.*

To be explicit, we write the perturbation added to an original medical record $\mathbf{x}$ as: $\delta = \mathbf{x} - \mathbf{x}^*$. Here we focus on the setting of targeted attacks, the adversarial example targeting at label $\mathbf{y}_{c_*}$ can be generated through solving an optimization problem:

$$\arg\min_{\delta^*} \| \delta \|_p \qquad s.t. \ f_\theta(\mathbf{x} + \delta) = \mathbf{y}_{c_*} \tag{2}$$

Definition 2. *(Temporal Sliding Window) To better explore the temporal structure in EHR data, we introduce a novel temporal sliding window applies on the adversarial perturbations. A temporal sliding window $\mathcal{W}$ with stride s and size $D \times w$. To divide δ into a set of groups $\{\delta_{\mathcal{G}_i}\}$, $i \in \{1, 2, ..., |\mathcal{G}_i|\}$, where $|\mathcal{G}_i| = (T - w)/s + 1$.*

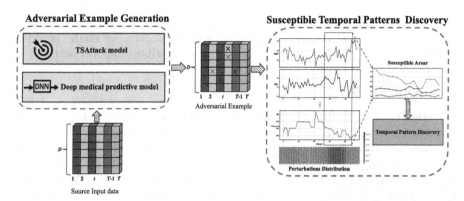

Fig. 1. Overview of the Susceptible Temporal Patterns (STP) discovery method. The proposed framework consists of two components: adversarial EHR examples are first generated by *TSAttack*, and then used to discover STP in patient's medical records, which provide clinical decision support for dynamic monitoring.

In the adversarial attacks step, we slide a window through time to capture temporal information hidden in the raw input space. When $s = 1$, the sliding window $\mathcal{W}$ moves one timestep at a time. Different splitting schemes can be selected by adjusting the value of stride s and size w.

Definition 3. *(Temporal Pattern) Temporal pattern is a higher-level, meaningful representation that aims to capture the dynamic behavior of the multivariate time series. We use $TP = (S, R)$ to denote a discovered temporal pattern, where $S = \{S_1, S_2, \ldots, S_k\}$ represents the set of abstract concepts and R is an upper triangular matrix that describes the temporal relations between each abstract concept.*

Here, we follow the widely adopted *Temporal Abstraction* (TA) [14,15] methods to represent time-stamped raw data as a set of time interval-based abstractions. In particular, we use the *Trend* abstraction, is denoted by the change directions (e.g., Decreasing, Increasing, Stable) of raw time series data. Then Allen's temporal relations [1] can be easily described all of the pairwise relations between time interval-based.

4   Methodology

In this section, our core idea is to discover susceptible temporal patterns in EHR data by adversarial attacking. We present the **T**emporal **S**parse Adversarial **Attack** approach to the EHR data, named *TSAttack*. Then we leverage adversarial examples to discover susceptible temporal patterns. The overall architecture for the proposed method is presented in Fig. 1. We start with introduce how to explore the temporal structure contained in EHR data to achieve a more effective attack. Based on the generated adversarial EHR data, we finally describe our novel perturbation strategy to discovering susceptible temporal patterns.

4.1   Attack Setting

Attack Strategy. In the attack setting, we adopt an optimal adversarial attack strategy to develop the components of adversarial example generation, since it has been shown to be effective in attacking image DNNs [4,5]. To obtain the adversarial sample $\mathbf{x}^*$, we aim to solve the optimization problem in (Eq. 2) to learn an effective perturbation strategy. However, it is hard to directly optimize. Existing works [4,5,16] turn to relax it into the following objective function:

$$\arg\min_{\delta^*} \| \delta \|_p - \ell(f_\theta(\mathbf{x}^*), \mathbf{y}_{c_*}) \tag{3}$$

Intuitively, the well-trained DNNs model corresponds to a decision boundary. When crafting adversarial examples based on $\mathbf{x}$, with iteratively optimizing the objective function, the sample will move towards the gradient of the loss function until it across the decision boundary.

Threat Models. For the medical prediction task, the Recurrent Neural Networks (RNNs) and its variants have achieved remarkable performance in various clinical time series data modeling [9]. Such as long short-term memory (LSTM) recurrent neural networks, which is designed to capture long term dependencies in sequence data. At time t, the LSTM updates the hidden vector of the state representation as: $h_t = \mathbf{LSTM}(\mathbf{x}_t, h_{t-1})$. Therefore, the temporal information hidden in the input sequence data can be effectively transmitted through the network. In particular, we assume adversarial perturbations share similar characteristics. Technically, adversarial perturbations can propagate between different visit steps. In other words, the added perturbations of the current $t-1$ visit can propagate to the next t visit, and then change the output of the predictive model.

4.2   Temporal Sparse Attack

Compare to statistic images, the main difference lies in the temporal structure contained in EHR data. The dimensionality of temporal EHR data is much higher as EHR data have both time and feature dimensions. On the one hand, there are potential interactions between medical features. On the other hand, there are temporal dependencies between the visit at different timesteps. Besides, the perturbation on EHR data would easily change the clinical observations of a patient. The dense adversarial attack in images domain is not suitable for temporal EHR data. Hence, we expect that the adversarial perturbations have the sparse property in both time and feature dimensions.

To address these issues, we introduce the proposed *TSAttack* for crafting EHR adversarial examples. Considering the temporal information hidden in the input data, *TSAttack* employs l_p-norm perturbation regularization and the proposed temporal sparsity regularization that encodes temporal structures simultaneously. Based on optimal adversarial attack strategy, we cast our problem as an optimization problem of the following form:

$$\arg\min_{\delta} \mathcal{F}(\mathbf{x} + \boldsymbol{\delta}, \mathbf{y}_{c_*}) + \lambda_1 \mathcal{L}(\boldsymbol{\delta}) + \lambda_2 \mathcal{D}(\boldsymbol{\delta}) \tag{4}$$

Where $\boldsymbol{\delta}$ is the adversarial perturbations that misclassify the generated adversarial example $\mathbf{x}^* = \mathbf{x} + \boldsymbol{\delta}$ to a target label $\mathbf{y}_{c_*}$. Here, $\mathcal{F}(\mathbf{x} + \boldsymbol{\delta}, \mathbf{y}_{c_*})$ denotes a loss function to measure the difference between the prediction and the target label. $\mathcal{L}(\boldsymbol{\delta})$ is a distortion function, which is used to control the magnitude of the adversarial perturbations. $\mathcal{D}(\boldsymbol{\delta})$ is the proposed temporal sparsity function can ensure the sparsity of generated adversarial perturbations. And the non-negative regularization parameters λ_1, λ_2 are constant to balance the two terms in the objective function. Next, we describe the details of the proposed method.

Adversarial Loss Function. Following C&W [4] and EAD [5] attack, we adopt the same loss function $\mathcal{F}(.)$ to enforces the DNNs model predicts its most likely class to be the target class $\mathbf{y}_{c_*}$. Given the original EHR data $\mathbf{x}$, and the threat model f_θ. To assign the target label $\mathbf{y}_{c_*}$ to the most probable label for $\mathbf{x}$, we would like to collect the outputs before the *Softmax* layer in the considered DNNs model: $\mathbf{Logit}(\mathbf{x}, \theta) = [[\mathbf{Logit}(\mathbf{x}, \theta)]_1, ..., [\mathbf{Logit}(\mathbf{x}, \theta)]_K] \in \mathbb{R}^K$. By minimizing the difference between the $[\mathbf{Logit}(\mathbf{x}, \theta)]_{c_*}$ and the maximum output in the logit layer, the prediction result closer to the target label $\mathbf{y}_{c_*}$. Formally, we optimize the following adversarial loss function:

$$\mathcal{F}(\mathbf{x} + \boldsymbol{\delta}, \mathbf{y}_{c_*}) = \max\{\max_{j \neq c_*}[\mathbf{Logit}(\mathbf{x} + \boldsymbol{\delta})]_j - [\mathbf{Logit}(\mathbf{x} + \boldsymbol{\delta})]_{c_*}, -\kappa\} \tag{5}$$

where $\kappa \geq 0$ is a confidence parameter that is usually set to zero.

Distortion Function. A key point for crafting adversarial examples is to evaluate the distance between the original examples and the adversarial ones. As mentioned before, we desire to generate adversarial samples that the perturbations are added on as few locations as possible. Following [16], we use l_1-norm as distortion function to evaluate the adversarial perturbations: $\mathcal{L}(\boldsymbol{\delta}) = \| \boldsymbol{\delta} \|_1 = \sum_n \sum_m |\delta_{nm}|$.

Temporal Sparsity Function. To perform the temporal sparse attack, we propose to adopt the sliding window strategy that injects time interval-wise sparsity into the adversarial attack. First, with the above defined temporal sliding window $\mathcal{W}$, we assume the adversarial perturbations $\boldsymbol{\delta}$ is split into a set of sub-regions $\{\boldsymbol{\delta}_{\mathcal{G}_i}\}$. Next, inspired by the setting of group Lasso [19] and $l_{2,1}$ norm [18], we apply temporal sparsity function across the $\{\boldsymbol{\delta}_{\mathcal{G}_i}\}$ as follows:

$$\mathcal{D}(\boldsymbol{\delta}) = \| \{\boldsymbol{\delta}_{\mathcal{G}_i}\} \|_{2,1} = \sum_i^{|\mathcal{G}_i|} \| \boldsymbol{\delta}_{\mathcal{G}_i} \|_2 \tag{6}$$

Algorithm 1: TSAttck

Input: A medical predictive DNNs model $f_\theta(\cdot)$; a clean EHR data $\mathbf{x}$, target
label $\mathbf{y}_{c_*}$; iterations number STEPS; step size α
Output: An adversarial EHR data $\mathbf{x}^*$
1 Initialize $\mathbf{x}_0^* = \mathbf{x}$
2 $i \leftarrow 0$
3 **while** $i \leq$ STEPS **do**
4 $\quad$ $\mathbf{x}_{i+1}^* \leftarrow \{\mathbf{x}_i^* - \alpha \cdot [\triangledown J(\mathbf{x}_i^*)]\}$
5 $\quad\quad$ where $J(.)$ is the objective function in Eq. 4
6 $\quad$ $\mathbf{x}_{i+1}^* \leftarrow Clip(\mathbf{x}_{i+1}^*, dom(\mathcal{X}))$
7 $\quad$ **if** $f_\theta(\mathbf{x}_{i+1}^*) = \mathbf{y}_{c_*}$ **then**
8 $\quad\quad$ $\mathbf{x}^* = \mathbf{x}_{i+1}^*$
9 $\quad\quad$ break
10 $\quad$ **end**
11 $\quad$ $i \leftarrow i + 1$
12 **end**
13 **return** $\mathbf{x}^*$

Different from directly optimize the l_p norm regularization, the temporal
sparsity function is expected to select a few key timesteps in EHR data to attack.
Thus, the adversarial perturbations added on a few timesteps can be propagated
along with the temporal structure to misleads DNNs. Besides, the distribution of
the sparse perturbation providing some insights into the vulnerability of different
time intervals of EHR on the adversarial attack. In our setting, by adjusting the
temporal sliding window size r, we can obtain different partition set $\{\delta_{\mathcal{G}_i}\}$ to
control the sparsity.

Training Algorithm. To generate EHR adversarial samples such that (Eq. 4)
is satisfied, we can implement iterative optimization algorithms to solve the
problem. Algorithm 1 describes the generation procedure via *TSAttack* method.
We can iteratively perturb $\mathbf{x}^*$ by minimizing objective function with first-order
optimization methods. Since the adversarial attack strategy cannot ensure per-
turbed features are still valid in clinical. For example, the value of *Respiratory
rate* cannot be negative. At each iteration, we clip the features of $\mathbf{x}_{i+1}^*$ into the
original input domain of $\mathcal{X}$: $\mathbf{x}_{i+1}^* \leftarrow Clip(\mathbf{x}_{i+1}^*, dom(\mathcal{X}))$. The $dom(\mathcal{X})$ includes
the statistical information for each medical feature, such as the maximum, mini-
mum, and variance. We wish to enhance perturb in locations of high variance to
have less recognizable modifications. In the implementation, we use the Adaptive
Moment Estimation (Adam) optimizer to train with a learning rate of 0.015.

4.3 Evaluation of Adversarial Perturbation

In this section, we introduce the perturbation evaluation metrics of EHR data
and then present how to compute the susceptibility score for the adversarial examples. Recall that the significant difference between attacking static

image data and temporal EHR data, we follow [16] and use three measures to quantify the degree of attacking. Let $\boldsymbol{\Delta} \in \mathbb{R}^{n \times D \times t}$ the optimal adversarial perturbation for all patients $\mathcal{X} = [\mathbf{x}^{(1)}, \mathbf{x}^{(2)}, ..., \mathbf{x}^{(n)}]$. The measure $M1(\boldsymbol{\Delta}, i, j) = \max_{1 \leq k \leq n}(|\boldsymbol{\Delta}_{k,i,j}|)$ indicates the global maximum perturbation for i timestep and medical feature j. The global average perturbation $M2(\boldsymbol{\Delta}, i, j) = \frac{1}{n}\sum_{k=1}^{n}(|\boldsymbol{\Delta}_{k,i,j}|)$ focuses on the magnitude of the adversarial perturbations at each time-feature grid. The measure $M3(\boldsymbol{\Delta}, i, j) = \frac{\|\boldsymbol{\Delta}_{.i,j}\|_0}{n}$ denotes the probability of being added perturbation.

Then, we can denote a *time-level susceptibility score* to indicate overall susceptibility or vulnerability at different timesteps via the following formula:

$$TS(\boldsymbol{\Delta}, i) = \sum_{j=1}^{D}\{M2(\boldsymbol{\Delta}, i, j) \odot M3(\boldsymbol{\Delta}, i, j)\} \tag{7}$$

4.4   Susceptible Temporal Patterns Mining

Here, we propose a new procedure to discover *Susceptible Temporal Patterns* (STP) in the patient's medical records, which provide clinical decision support for dynamic monitoring. Its core idea is based on temporal structure of the perturbation distribution, mining the frequent temporal patterns that exist in EHR adversarial examples. In specific, (1) We firstly extract the susceptible regions from EHR according to time-level susceptibility score. (2) Generate a collection of interval-based temporal patterns appearing in the susceptible region. (3) Finally, mining the frequent temporal patterns in the above collection.

Screening Susceptible Region. After generating adversarial sample x*, we take a further step and mining temporal patterns that are more possible to be attacked to fool the medical predictive model. Since different EHR data x requires a different subset of susceptible timesteps to attack, we first extract sub-regions that are highly vulnerable to the adversarial perturbations. It is worth noticing that thanks to the sparsity function and propagation of perturbations, *TSAttack* can select a few key sub-regions to apply spares attack. Thus, we assign the sub-regions near the maximum time-level susceptibility score as the susceptible region for analysis.

Modeling Temporal Patterns. To describe the multivariate time series data more intuitive, it is essential to transform the numeric variables into high-level semantic (usually with symbolic values) time intervals. Especially, without medical knowledge guidance that it will be hard to understand the raw EHR data. Additionally, time-intervals can reduce inherent random noise in the data (due to the irregularity of EHR). As defined in section, we construct temporal patterns by temporal abstraction and temporal relations methods. Figure 2 describes an example of temporal patterns of a series of raw time-point data.

In the temporal abstraction phase, numeric time-stamped data can be converted to time interval sequences using Knowledge-based TA. Following [14], we define a symbolic time interval E = <*start, end, sym*>. E.*start* and E.*end* denote

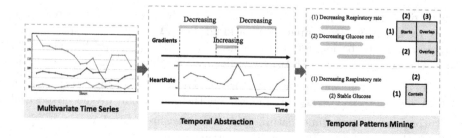

Fig. 2. Illustration of susceptible temporal patterns mining procedure.

the start timestep and end timestep of E, respectively. E.*sym* represents one of the abstract concepts set S, such as the Trend abstraction (e.g., Decreasing, Increasing, Stable). In the temporal relations phase, Allen's scheme [1] is widely adopted to describe the temporal relation between two time-intervals using 13 possible relations. Here, we choose the set R of 7 pairwise temporal relations in our task, include *Before, Meets, Overlaps, Contains, Finish-by, Equal, Starts.*

Temporal Patterns Mining. We adopt the KarmaLego algorithm [14] for mining frequent temporal patterns. Given a set of $\mathcal{T} = \{TP_1, TP_2, ..., TP_M\}$, where M is the total number of discovered temporal patterns in all susceptible regions. IF a TP_i has vertical support above the predefined minimal threshold *minSup*, it is referred to as *frequent temporal pattern* (FTP). The KarmaLego algorithm exploits the transitivity of temporal relations to significantly reduce the number of candidate temporal patterns.

5   Experiments and Analysis

In our experimental evaluation, we designed experiments to answer the following research questions:

- **RQ1:** Can the *TSAttack* effectively attack on EHR data? Whether the generated adversarial samples are meaningful in clinical?
- **RQ2:** How the proposed *TSAttack* method compare to other benchmark attack methods?
- **RQ3:** Cases for showing the susceptible temporal patterns discovery.

We begin by introducing the experiment settings include dataset and evaluation metric.

5.1   Experiment Settings

Dataset. To measure the performance of our proposed model, we used EHR data from MIMIC-III [10]. The MIMIC-III dataset is a publicly available EHR dataset, which is widely used in many medical deep predictive tasks [9,20].

MIMIC-III consists of medical records of 7,499 intensive care unit (ICU) patients between 2001 and 2012. Following [9], we introduce the problems of In-hospital mortality as our clinical prediction tasks. We use a subset of the MIMIC-III database containing more than 31 million clinical events that correspond to 17 clinical variables.

Evaluation Metrics. In our experiments, we adopt different metrics to evaluate predictive and attack performance.

Predictive Performance. For the medical predictive task, the positive and negative labels in the dataset are imbalanced. We adopt area under the receiver operating characteristic (**AUROC**) as the main metric. Besides, we also report the precision score (**Precision**) and F1-score (**F1**) to evaluate.

Attack Performance. We not only want to generate EHR adversarial samples successfully but also want to perturb as few locations as possible. Thus, attack success rate and distortion of adversarial samples are the main metrics of the corresponding task. We set $\mathcal{X}^*$ denotes the set of adversarial examples. Following the attack evaluation criterion [4,5,13], we use the attack success rate (**ASR**) to define the proportion of inputs that are generated as the adversarial examples:

$$\text{ASR} = \frac{1}{|\mathcal{X}^*|} \sum_{(\mathbf{x}^*, \mathbf{y}_t) \in \mathcal{X}^*} (f_\theta(\mathbf{x}^*) = \mathbf{y}_t) \tag{8}$$

As we introduced in Sect. 4.3, we aim to measure the adversarial perturbation in both magnitude and structure. The average M1, M2, and M3 distortion metrics of successful adversarial examples are also reported.

5.2 Empirical Analysis of TSAttack

In order to answer **RQ1**, we want to examine how the medical predictive DNNs model $f_\theta(\cdot)$ accuracy degrades on the adversarial examples compared with the performance on the clean data, i.e. datasets that are not attacked. Moreover, to verify whether the generated adversarial samples are meaningful in clinical, we conduct a novel experiment to test the EHR adversarial examples with the traditional scores methods.

Evaluation on Clean and Adversarial Examples. The experimental results are shown in Fig. 3. We report the attack performance of *TSAttack* by the extent to which it reduces the prediction performance of threat models. It is observed that our proposed method *TSAttack* achieves the high attack success rate on the In-hospital mortality task. While proposing the *TSAttack* method, we assume that exploiting perturbation propagation can enhance the effectiveness of adversarial attacks. To verify that, we also report the ASR of different iterations of *TSAttack* in Fig. 3. The GRU and LSTM model achieved a better performance drop than VanillaRNN. A possible reason is that GRU and LSTM were designed to capture long-term dependencies, and long memory is one of the important factors to the propagation of perturbations, which satisfy our expectations.

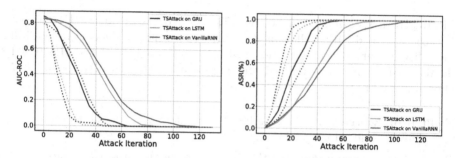

Fig. 3. The attack performance of *TSAttack* with different threat models on the MIMIC-III dataset. The solid line represents the prediction of mortality within a fixed time 48-h window, and dotted line represents within a fixed time 24-h window.

Table 1. Comparison of traditional scores system and DNNs model with *TSAttack*.

Methods	Clean			Attack		
	AUC-ROC	Precision	F1	AUC-ROC	Precision	F1
SAPS-II	0.701	0.611	0.158	0.689	0.601	0.148
APS-III	0.690	0.597	0.146	0.657	0.589	0.137
LSTM	**0.851**	0.731	0.662	**0.103**	0	0

Quality of the EHR Adversarial Samples. Since it is hard to directly evaluate whether an EHR adversarial example is meaningful in clinical or not, we design an experiment to analyze the quality of them. Before the extensive body of research on clinical predictions using deep learning, traditional statistics research modeling the risk of mortality via scoring systems based on knowledge predefined by experts. Such as SAPS-II (Simplified Acute Physiology Score) score and APACHE II (Acute Physiologic and Chronic Health Evaluation) score [3]. The ground truth label of adversarial examples unchanged only when attacking does not affect the scoring system's performance. The results are reported in Table 1. When attacking against LSTM, the predictive accuracy drops sharply. However, it has little impact on the scoring system. This shows that adversarial samples generated by *TSAttack*, are not only successfully adversarial attack the DNNs model but also still meaningful in clinical.

Attack Effects Under Different Parameter Setting. We further carefully examine the effect of two parameters in *TSAttack*. Overall, parameter λ_1 emphasis the magnitude of adversarial perturbations, we can see from Fig. 4(a) that increasing λ_1 leads to a larger adversarial perturbation. However, when $\lambda_1 < 0.001$, the attack rarely succeeds. For another parameter λ_2, our experiment results show that it controls the temporal structure and sparsity of adversarial perturbations. Figure 4(b) presents how the perturbation propagation impacts the distribution of adversarial perturbations added by *TSAttack*. In particu-

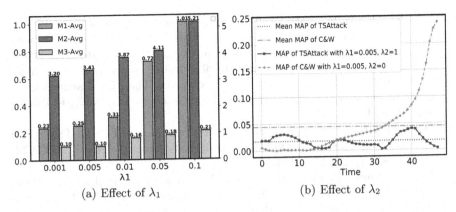

(a) Effect of λ_1 (b) Effect of λ_2

Fig. 4. Effect of parameters in *TSAttack*. (a) Attack performance of *TSAttack* by varying λ_1 (setting $\lambda_2 = 1$), lower numbers are better. (b) An example of distribution with adversarial perturbations added by *TSAttack* and C&W methods.

lar, the Mean Absolute Perturbation (MAP) of each timestep generated by the C&W method (when $\lambda_2 = 0$) focuses on the recent timesteps, which have a stronger influence on the prediction output but more likely to be detected. For *TSAttack* method, the adversarial perturbations present the characteristics of periodic propagation. While effectively reducing the cost of the attack, it can also reveal the vulnerable time-level regions. It provides the foundation for our subsequent analysis of STP discovery.

Table 2. Performance comparison of different adversarial attacks methods on MIMIC-III dataset.

Attack methods	Attack (0⇒1)				Attack (1⇒0)			
	ASR	M1-Avg	M2-Avg	M3-Avg	ASR	M1-Avg	M2-Avg	M3-Avg
FGM-l_1	11.5	8.244	8.244	0.160	86	3.407	3.407	0.070
FGM-l_2	100	0.419	16.534	0.350	100	0.112	7.143	0.140
FGM-l_∞	100	0.069	28.169	0.690	100	0.012	3.440	0.120
IFGM-l_1	98	0.185	9.962	0.190	100	0.033	2.188	0.100
IFGM-l_2	100	0.268	10.127	0.210	100	0.038	2.449	0.080
IFGM-l_∞	100	**0.047**	18.472	0.390	100	**0.015**	2.461	0.080
C&W	100	0.251	5.214	0.160	100	0.071	1.241	0.070
TSAttack	100	0.235	**3.198**	**0.100**	100	0.063	**0.723**	**0.050**

5.3   Attack Performance Comparison

To answer **RQ2**, we evaluated *TSAttack* against the several existing adversarial attack baselines. Since adversarial attack on temporal EHR is a novel task, there are very few baselines that we can compare with. We consider the following methods for comparison:

(1) **FGSM attack** [8]: Fast Gradient Sign Method is a one-step method to fast generate adversarial examples. It linearized the objective function with a first-order Taylor series approximation. Compared to the iterative attack strategy, FGSM addresses the demands of that need to generate a large number of adversarial examples with lowly expensive. The FGSM attacks using different perturbation regularizations are denoted by **FGSM**-l_1, **FGSM**-l_2 and **FGSM**-l_∞.

(2) **IFGSM attack** [11]: It is a basic iterative extension of FGSM. The FGSM attacks using different perturbation regularizations are denoted by **IFGSM**-l_1, **IFGSM**-l_2 and **IFGSM**-l_∞.

(3) **C&W attack** [4]: The proposed model C&W aimed to evaluate the defensive distillation strategy for mitigating the adversarial attacks. The work in [16] firstly adopted C&W method for attacking predictive models of medical records.

Attack Evaluation. Table 2 presents the results of all the comparison methods across different performance metrics. With the help of bringing in more temporal information, *TSAttack* method achieves the best performance with significantly reduces the M2 and M3 distortion metrics respectively. Since M2 and M3 distortion metrics effectively reflect that the magnitude and structure of the adversarial perturbation. For the attack success rate, both baselines and our method yield 100% ASR results. It is worth noting that M1 focuses on the maximum perturbation, which is consistent with l_∞ perturbation regularizations. Therefore, FGSM and IFGSM attack with l_∞ perturbation regularizations, leading to a drop in M1 metrics.

5.4   Detailed Analysis of STP Discovery

In this subsection, we conduct experiments to answer **RQ3**. We select correct classified samples to experiment. Following the same experimental setting in Sect. 5.1, we apply *TSAttack* on the test set to generate adversarial examples for STP discovery. In the experiment, we started by extracting the susceptible regions from adversarial examples according to the time-level susceptibility score. Then for susceptible time-level regions, we discover a total of 189,13 interval-based temporal patterns that appeared at least once within at least 20% of the adversarial examples. To discover STP sets that are informative, we select top-k STP according to the Information-Gain feature selection method. As shown in Table 3, we list the top-3 frequent STP that are most susceptible to In-hospital mortality prediction task in the test set of the ICU patient cohorts.

Table 3. The top-3 discovered STP and their support level.

Susceptible Temporal Patterns		Support
Pattern 1	(1) Increasing Respiratory rate (2) Decreasing Respiratory rate (3) Increasing DBP Time (1) → (2) Meets, (3) Before (2) Overlap	68.5%
Pattern 2	(1) Decreasing Respiratory rate (2) Decreasing Glucose rate (3) Increasing Respiratory rate Time (1) → (2) Starts, (3) Meets (2) Overlap	61%
Pattern 3	(1) Decreasing DBP (2) Increasing Respiratory rate (3) Stable DBP Time (1) → (2) Overlap, (3) Meets (2) Contains	58.5%

6 Conclusion

In this paper, we proposed a novel approach for generating EHR adversarial examples, *TSAttack*, which explores temporal structure contained in EHR to achieve an effective and efficient attack. By utilizing information hidden in the adversarial perturbations, we further developed a procedure to discover susceptible temporal patterns (STP) in a patient's medical records. To our knowledge, it is the first time that explores temporal structures when implementing adversarial attacks on EHR data. Extensive experiments conducted on MIMIC-III dataset demonstrated the effectiveness and interpretability of the proposed method.

Acknowledgement. This work was supported in part by National Key Research and Development Program of China under Grant No. 2020YFA0804503, National Natural Science Foundation of China under Grant No. 61532010 and 61521002, and Beijing Academy of Artificial Intelligence (BAAI).

References

1. Allen, J.F.: Maintaining knowledge about temporal intervals. Commun. ACM **26**(11), 832–843 (1983)
2. An, S., Xiao, C., Stewart, W.F., Sun, J.: Longitudinal adversarial attack on electronic health records data. In: WWW, pp. 2558–2564 (2019)

3. Bilgin, T.E., et al.: The comparison of the efficacy of scoring systems in organophosphate poisoning. Toxicol. Ind. Health **21**(7–8), 141 (2005)
4. Carlini, N., Wagner, D.: Towards evaluating the robustness of neural networks. In: IEEE Symposium on Security and Privacy, pp. 39–57 (2017)
5. Chen, P.Y., Sharma, Y., Zhang, H., Yi, J., Hsieh, C.J.: EAD: elastic-net attacks to deep neural networks via adversarial examples. In: AAAI, pp. 10–17 (2018)
6. Choi, E., Bahadori, M.T., Schuetz, A., Stewart, W.F., Sun, J.: Retain: interpretable predictive model in healthcare using reverse time attention mechanism. In: NIPS, pp. 3504–3512 (2016)
7. Finlayson, S.G., Chung, H.W., Kohane, I.S., et al.: Adversarial attacks against medical deep learning systems. arXiv preprint arXiv:1804.05296 (2018)
8. Goodfellow, I.J., Shlens, J., Szegedy, C.: Explaining and harnessing adversarial examples. In: ICLR (2015)
9. Harutyunyan, H., Khachatrian, H., Kale, D.C., Steeg, G.V., Galstyan, A.: Multitask learning and benchmarking with clinical time series data. Sci. Data **6**(1), 1–18 (2019)
10. Johnson, A.E.W., et al.: MIMIC-III, a freely accessible critical care database. Sci. Data **3**, 1–9 (2016)
11. Kurakin, A., Goodfellow, I., Bengio, S.: Adversarial machine learning at scale. In: ICLR (2017)
12. Ma, F., Chitta, R., Zhou, J., You, Q., Sun, T., Gao, J.: Dipole: diagnosis prediction in healthcare via attention-based bidirectional recurrent neural networks. In: SIGKDD, pp. 1903–1911 (2017)
13. Moosavi-Dezfooli, S. M., Fawzi, A., Frossard, P.: DeepFool: a simple and accurate method to fool deep neural networks. In: CVPR, pp. 2574–2582 (2016)
14. Moskovitch, R., Shahar, Y.: Fast time intervals mining using the transitivity of temporal relations. Knowl. Inf. Syst. **42**(1), 21–48 (2015)
15. Sheetrit, E., Nissim, N., Klimov, D., Shahar, Y: Temporal probabilistic profiles for sepsis prediction in the ICU. In: SIGKDD, pp. 2961–2969 (2019)
16. Sun, M., Tang, F., Yi, J., Wang, F., Zhou, J.: Identify susceptible locations in medical records via adversarial attacks on deep predictive models. In: SIGKDD, pp. 793–801 (2018)
17. Szegedy, C., et al.: Intriguing properties of neural networks. In: ICLR (2014)
18. Wei, X., Zhu, J., Su, H.: Sparse adversarial perturbations for videos. In: AAAI, pp. 8973–8980 (2019)
19. Yuan, M., Lin, Y.: Model selection and estimation in regression with grouped variables. J. Roy. Stat. Soc.: Ser. B (Stat. Methodol.) **68**(1), 49–67 (2006)
20. Zhang, X., Qian, B., Cao, S., Li, Y., Davidson, I.: INPREM: an interpretable and trustworthy predictive model for healthcare. In: SIGKDD, pp. 450–460 (2020)

A Decision Support System for Heart Failure Risk Prediction Based on Weighted Naive Bayes

Kehui Song[1,3], Shenglong Yu[1,3], Haiwei Zhang[2,3(✉)], Ying Zhang[1,3], Xiangrui Cai[2,3], and Xiaojie Yuan[1,2,3]

[1] College of Computer Science, Nankai University, Tianjin 300350, China
[2] College of Cyber Science, Nankai University, Tianjin 300350, China
[3] Tianjin Key Laboratory of Network and Data Security Technology, Tianjin 300350, China
{songkehui,yushenglong}@dbis.nankai.edu.cn,
{zhhaiwei,yingzhang,caixr,yuanxj}@nankai.edu.cn

Abstract. Heart failure (HF) affects the health of millions of people worldwide and the early detection of HF risk plays a vital role in prevention and prompt treatment. Various decision support systems based on machine learning have been presented recently to predict HF. However, the existing systems usually assumed that all features add equal weight to the prediction result, which could not properly simulate the diagnostic status. In this study, a decision support system is proposed for HF prediction using MSE Back Propagation Method (MSEBPM) and weighted naive Bayes. First, the feature selection method eliminates irrelevant features to improve accuracy and decrease computational times. Second, the proposed MSEBPM computes a weight vector for features based on their contributions, trying to minimize the MSE loss of the predicted class probabilities. Finally, the trained weight vector is applied to the weighted naive Bayes model for HF risk prediction. The proposed system is evaluated with a published dataset of 899 patients, and compared with conventional data mining techniques and other state-of-the-art systems. The results show that our proposed system leads to 82.96% accuracy in HF risk prediction, which suggests that it could be used to early detect HF in the clinic.

Keywords: Feature selection · Feature weighting · Weighted naive Bayes · Decision support system

1 Introduction

Heart failure (HF) is the main cause of morbidity and mortality nowadays. It has been proved that early detection of HF is crucial to properly treat patients before a heart attack, and thus this prompt treatment could prevent heart stroke to some degree. Although there have been large achievements in the medical

C. S. Jensen et al. (Eds.): DASFAA 2021, LNCS 12683, pp. 445–460, 2021.
https://doi.org/10.1007/978-3-030-73200-4_30

treatment for HF, the early detection of it still remains an enormous challenge [21]. Therefore, the development of an accurate and easy-to-use decision support system would be especially useful to aid physicians in the early detection of HF.

Recently, various decision support systems have already been proposed. A group of systems [12] directly utilize data mining models to predict HF risks. Meanwhile, another group is a set of hybrid systems [16,22], consisting of two stages. In the first stage, feature engineering techniques, including feature selection and feature weighting, are applied to select significant features or identify feature weights. In the second stage, the subset or weight vector is used as input to improve the HF prediction accuracy. According to our investigation, most of those hybrid systems put more emphasis on feature selection. They assumed that all features add equal weight to the prediction result, which could not properly simulate the diagnostic status.

In order to address this problem, we propose a decision support system for HF prediction using MSE Back Propagation Method (MSEBPM) and weighted naive Bayes. First, to eliminate the negative impact of irrelevant features, information gain is employed to select a set of significant features. In the medical field, a large number of features are stored in the database as the model input, including the one that is irrelevant to our current task. The random distributions of irrelevant features would add noise to the prediction results. Therefore, the reduced input features could improve the performance of the classification model in terms of accuracy as well as computational times. Second, the proposed MSEBPM is trying to calculate an optimal weight vector based on the contributions of the selected features. MSEBPM minimizes the difference between predicted results and the ground-truth labels under an MSE objective function. By utilizing backpropagation, the weight vector is converged to an optimal value. Third, the trained weight vector is applied to weighted naive Bayes, which is a variant of naive Bayes by adding additional weight parameters. This classification model could absorb feature weighting information, and therefore the accuracy could be further improved. The extensive experimental results demonstrate that our proposed system is more superior compared to other existing systems in HF risk prediction. The main contributions of this paper are as follows.

- First, a decision support system is proposed to predict HF risks by processing both physiological data and EMR data.
- Second, the information gain method is utilized to distinguish significant features from irrelevant ones, and MSEBPM is used to identify weight for each selected one. Both feature engineering techniques could improve the accuracy of HF prediction.
- Third, weighted naive Bayes is used as the classification model, adding weight information to the classification results to achieve a higher accuracy of nearly 83% in comparison with other state-of-the-art systems.

The rest of the paper is organized as follows. Section 2 briefly reviews the literature on existing HF prediction systems and relevant techniques. In Sect. 3, the overall structure of the decision support system is proposed. Section 4 presents the involved methods in detail, including feature selection, feature weighting, and the classification model. Section 5 presents and discusses the results of the experiments compared with other state-of-the-art methods. The paper is concluded in Sect. 6.

2   Related Work

This section briefly introduces works on existing heart diagnostic systems and relevant techniques, including the feature selection method and naive Bayes.

2.1   Heart Failure Detection Systems

In recent years, various data-driven decision support systems for diagnostic prediction have been intensively developed in the field of medical treatment, such as HF2HM [23], RESKO [14], SHMS [3].

More specifically, for HF detection, a few relevant works [7,13,18] are investigated by researchers. Recently, Machine Learning (ML) has been successfully used to improve prediction accuracy. For example, [2] applied the decision tree model to identify heart disease from patients. In [16], a novel kernel random forest ensemble is used to produce significantly better quality results. Furthermore, ANN-based methods have been widely adopted in medical diagnosis due to its powerful ability to address linear and non-linear problems. [10] uses ANN as the classification model for detecting the absence or presence of HF. However, those diagnostic systems suffer from a major limitation that each attribute is assumed to add equal weight to the diagnostic outcome, which is the key distinction between them and our proposed system. [20] utilizes Fuzzy_AHP to properly rank and compute a weight vector for HF attributes, and apply the weight vector to the ANN model for classification. This is the closest work to ours and will be implemented and compared in the experimental section.

Additionally, HF prediction systems always take sensor-based signals as their input, like ECG signals. Various signal analysis techniques are applied. [8] utilizes Discrete Wavelet Transform (DWT) to decompose heart rate signals into frequency sub-bands to extract features for HF risk prediction. [1] proposes a CNN model using two and five seconds durations of ECG signal segments for automated detection of heart disease. Due to the lack of sensor-based signals, our system solves the situation that signals have already been preprocessed by data managers, that is, signal analysis is not the scope of our study.

2.2   Feature Selection Methods

Feature selection methods are supposed to select a subset of features, which present most of the important information of classes. A survey [26] summarized two categories for feature selection. One is based on label information. According to the proportion of labeled samples, supervised [17], semi-supervised [6] and unsupervised [24] methods are used. Each feature is evaluated and ranked based on its correlation with class labels, and meanwhile trying to reduce the correlation to any other selected one.

The other category, consisting of filters, wrappers, and embedded methods, is based on different searching strategies. Filter methods rank all features based on statistical and intrinsic characters of the dataset, including information gain [19]. Wrapper methods utilize the learning algorithm or classifiers to evaluate the features, such as the Laplacian Score proposed in [9]. Embedded methods [5] incorporate the feature selection process as part of the training process. In our work, the information gain method is utilized to select a subset of features to improve the performance of our system.

2.3   Naive Bayes

Naive Bayes (NB) is one of the most widely used algorithms to classify samples. Given a sample $\mathbf{x} = (a_1, a_2, ..., a_N)$, NB uses Bayes formula (see Eq. 1) to estimate probability for each class, and the one with the largest value is chosen as the output label.

$$c(\mathbf{x}) = \arg\max_{c \in C} \frac{P(c) \prod_{i=1}^{N} P(a_i|c)}{\sum_{c' \in C} P(c') \prod_{i=1}^{N} P(a_i|c')}, \tag{1}$$

where C is the set of all class labels, a_i is the ith attribute of the test sample, and N is the total number of attributes. To simplify the algorithm, Eq. 1 is modified to:

$$c(\mathbf{x}) = \arg\max_{c \in C} P(c) \prod_{i=1}^{N} P(a_i|c). \tag{2}$$

When using Eq. 2 to classify $\mathbf{x}$, it is assumed that all attributes are fully independent given the class label. Only under this circumstance, $P(a_1, a_2, ..., a_N|c)$ is equal to $\prod_{i=1}^{N} P(a_i|c)$. However, this constraint condition could be rarely satisfied in the real world, which would definitely have a negative effect on classification performance. Therefore, this paper proposes an end-to-end pipeline to alleviate this negative influence by carefully selecting a subset of attributes and assigning them with different weights.

3   System Architecture

In this section, the structure of the proposed decision support system is discussed (see Fig. 1). This system is designed to provide physicians and patients with the predicted HF risk diagnosis by analyzing the input sensor data and EMR data.

The system has two main data sources. The first one is physiological data, which is collected by wearable sensors and transmitted to the medical database through wireless techniques (WIFI, Bluetooth, etc.). The physiological data consists of ECG, blood pressure, blood sugar, respiration rate, heart rate, cholesterol level, and EEG, which belong to the real-time monitoring data. The other

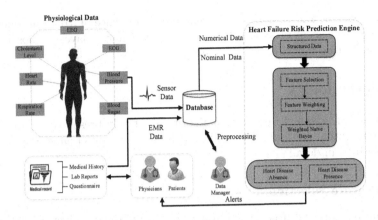

Fig. 1. The structure of the decision support system for HF risk prediction.

is EMR data, which includes medical history, lab reports and questionnaires provided by the collaboration between physicians and patients. The EMR data could offer basic information on patients, family history like diabetes history, and detailed clinical examinations. The two parts of data are both stored in the medical database. Afterward, data managers with specific domain expertise can preprocess original data and transform it into either numerical or nominal type for future prediction.

The HF risk prediction engine is the kernel part of our system. There are three steps to process and analyze the input structured data, and finally the prediction results of heart failure are sent back to physicians and patients for further medical treatment. The first step is feature selection. In the real world, medical records normally consist of a large number of features, including the one that is irrelevant to our current prediction task. Under this situation, a well-designed feature selection strategy would reduce noise and therefore improve the prediction performance afterward. Several feature selection methods [11,15,27,28], such as sequential forward selection, weighted least squares, rough sets, and univariate feature selection, are already implemented in healthcare systems. In our system, information gain is utilized for feature selection, and the detailed algorithm is presented in Sect. 4.1. The second step is feature weighting. After feature selection, the remaining features should share a weight vector $\mathbf{w} = (w_1, w_2, ..., w_N)$. The weight w_i for each feature (attribute) a_i reflects its relevance to the prediction task, that is, how much attention should be paid to each feature in the classification model. In our system, we propose MSEBPM to calculate the weight vector, which will be elaborated in Sect. 4.2. The third step is weighted naive Bayes, which is the classification model used in our system to predict the absence or presence of heart disease. It is a binary classification task. Weighted naive Bayes could introduce feature weights into the model and consider the different significance of each feature in the process of classification. The weight vector obtained in the second step works by increasing the influence of relevant features and by contrast decreasing the influence of irrelevant ones during the classification process. Therefore, weighted naive Bayes is chosen to be the classification model in our system.

4   Methodology

In this section, the feature selection algorithm is firstly presented, and then the feature weighting method MSEBPM as well as the weighted naive Bayes model are described in detail.

4.1   Feature Selection

Information gain [4] is used to determine which attribute in a given set of training feature vectors is most useful for discriminating between the classes to be learned, and indicate how important a given attribute is. The calculation of information gain is to find an attribute a_i that maximizes the difference between prior entropy and post entropy of the dataset. The information gain of the dataset D when using attribute a_i as the splitting one is defined as:

$$Gain_{a_i}(D) = Info(D) - Info_{a_i}(D), \tag{3}$$

where $Info(D)$ is the prior entropy and $Info_{a_i}(D)$ is the post entropy. They are defined in Eq. 4 and Eq. 5, respectively.

$$Info(D) = -\sum_{i=1}^{C} p_i \log_2(p_i), \tag{4}$$

where p_i is the prior probability of each class and C is the number of classes in dataset D.

$$Info_{a_i}(D) = \sum_{j=1}^{k} \frac{|D_j|}{|D|} \times Info(D_j), \tag{5}$$

where k is the number of values of attribute a_i, and dataset D can be split to k subset $(D_1, D_2, ..., D_k)$ according to the value of a_i. After measuring the information gain for each attribute a_i $(i = 1, 2, ..., N)$, the least important attributes are deleted to improve the effectiveness and efficiency of the classification model.

4.2   MSEBPM and Weighted Naive Bayes

Different from traditional naive Bayes, weighted naive Bayes adds an additional weight parameter w_i $(i = 1, 2, ..., N)$ to Bayes formula (see Eq. 1). The weighted naive Bayes is therefore defined as:

$$\hat{P}(c|\mathbf{x}) = \frac{\hat{P}(c) \prod_{i=1}^{N} \hat{P}(a_i|c)^{w_i}}{\sum_{c' \in C} \hat{P}(c') \prod_{i=1}^{N} \hat{P}(a_i|c')^{w_i}}. \tag{6}$$

$$\hat{c}(\mathbf{x}) = \arg \max_{c \in C} \hat{P}(c|\mathbf{x}) \tag{7}$$

The prior probability $\hat{P}(c)$ and the conditional probability $\hat{P}(a_i|c)$ are calculated by the m-estimation employed by [25], which are defined as:

$$\hat{P}(c) = \frac{\sum_{j=1}^{L} \delta(c_j, c) + \frac{1}{C}}{L + 1}, \tag{8}$$

$$\hat{P}(a_i|c) = \frac{\sum_{j=1}^{L} \delta(a_{ij}, a_i)\delta(c_j, c) + \frac{1}{L_i}}{\sum_{j=1}^{L} \delta(c_j, c) + 1}, \tag{9}$$

where the function $\delta(\alpha, \beta) = 1$ if $\alpha = \beta$ and otherwise $\delta(\alpha, \beta) = 0$. L is the number of training samples, C is the number of classes, c_j is the class label of jth training sample, a_{ij} is the ith feature value of the jth training sample, and L_i is the number of values for attribute a_i.

To obtain the weight vector $\mathbf{w} = (w_1, w_2, ..., w_N)$, we propose MSEBPM, which is MSE distance-based backpropagation method to learn the weight vector. The detailed learning procedure is depicted in Algorithm 1. It is observed that the weight vector is all initialized to 1, that is, we start the learning procedure from the original naive Bayes.

In our system, we use the mean squared error (MSE) as the objective function, and the goal is to minimize it. The objective function is defined as:

$$f(\mathbf{w}) = \frac{1}{2} \sum_{\mathbf{x} \in D} \sum_{c \in C} (P(c|\mathbf{x}) - \hat{P}(c|\mathbf{x}))^2, \tag{10}$$

where $P(c|\mathbf{x}) = 1$ if c is the true label of the sample $\mathbf{x}$, otherwise $P(c|\mathbf{x}) = 0$.

In order to execute the learning procedure presented in Algorithm 1, the gradient of the objective function $f(\mathbf{w})$ with respect to $\mathbf{w}$ should be derived. Under the situation of binary classification, Eq. 10 can be rewritten as:

$$f(\mathbf{w}) = \frac{1}{2} \sum_{\mathbf{x} \in D} (P(c_0|\mathbf{x}) - \hat{P}(c_0|\mathbf{x}))^2 + \frac{1}{2} \sum_{\mathbf{x} \in D} (P(c_1|\mathbf{x}) - \hat{P}(c_1|\mathbf{x}))^2. \tag{11}$$

Therefore, the derivative of objective function f is:

$$\frac{\partial f(\mathbf{w})}{\partial w_i} = - \sum_{\mathbf{x} \in D} \left((P(c_0|\mathbf{x}) - \hat{P}(c_0|\mathbf{x}))\frac{\partial \hat{P}(c_0|\mathbf{x})}{\partial w_i} + (P(c_1|\mathbf{x}) - \hat{P}(c_1|\mathbf{x}))\frac{\partial \hat{P}(c_1|\mathbf{x})}{\partial w_i} \right), \tag{12}$$

combining with Eq. 6, we have:

$$\frac{\partial \hat{P}(c_0|\mathbf{x})}{\partial w_i} = \frac{\partial \frac{\theta_{c_0}(\mathbf{w})}{\sum_{c \in C} \theta_c(\mathbf{w})}}{\partial w_i} = \frac{\frac{\partial \theta_{c_0}(\mathbf{w})}{\partial w_i}}{\sum_{c \in C} \theta_c(\mathbf{w})} - \frac{\theta_{c_0}(\mathbf{w})\frac{\partial \sum_{c \in C} \theta_c(\mathbf{w})}{\partial w_i}}{(\sum_{c \in C} \theta_c(\mathbf{w}))^2}$$

$$= \frac{1}{\sum_{c \in C} \theta_c(\mathbf{w})} \left[\frac{\partial \theta_{c_0}(\mathbf{w})}{\partial w_i} - \hat{P}(c|\mathbf{x})\frac{\partial \sum_{c \in C} \theta_c(\mathbf{w})}{\partial w_i} \right], \tag{13}$$

Algorithm 1. MSEBPM(D, $f(\mathbf{w})$, η)

Input: a training set D, an objective function $f(\mathbf{w})$, learning rate η
Output: the weight vector $\mathbf{w}$

1: initialize the weight vector $\mathbf{w} = (w_1, w_2, ..., w_N) = (1, 1, ..., 1)$
2: **while** not converged **do**
3:    **for** each instance $\mathbf{x}$ in D **do**
4:       estimate the prior probability $\hat{P}(c)$ of each class by Eq. 8
5:       estimate the conditional probability $\hat{P}(a_i|c)$ of each attribute a_i with respect
   to each class c by Eq. 9
6:       estimate the probability $\hat{P}(c|\mathbf{x})$ for each class by Eq. 6
7:       optimize the objective function $f(\mathbf{w})$ and update $\mathbf{w}$ for a batch of $\mathbf{x}$:
8:          $g_\mathbf{w} \leftarrow -\nabla_\mathbf{w} f(\mathbf{w})$
9:          $\mathbf{w} \leftarrow \mathbf{w} - \eta \times RMSProp(\mathbf{w}, g_\mathbf{w})$
10: **return** the trained weight vector $\mathbf{w}$

---

where $C = \{c_0, c_1\}$ in our binary classification task, and

$$\theta_c(\mathbf{w}) = \hat{P}(c) \prod_{i=1}^{N} \hat{P}(a_i|c)^{w_i}, \tag{14}$$

then we have:

$$\frac{\partial \theta_c(\mathbf{w})}{\partial w_i} = \theta_c(\mathbf{w}) \log \hat{P}(a_i|c). \tag{15}$$

Using Eq. 13, Eq. 14 and Eq. 15, the first part of derivative in Eq. 12 is calculated as:

$$\frac{\partial \hat{P}(c_0|\mathbf{x})}{\partial w_i} = \hat{P}(c_0|\mathbf{x}) \log \hat{P}(a_i|c_0) - \hat{P}(c_0|\mathbf{x}) \sum_{c \in C} \hat{P}(c|\mathbf{x}) \log \hat{P}(a_i|c). \tag{16}$$

Similarly, the second part of derivative in Eq. 12 is calculated as:

$$\frac{\partial \hat{P}(c_1|\mathbf{x})}{\partial w_i} = \hat{P}(c_1|\mathbf{x}) \log \hat{P}(a_i|c_1) - \hat{P}(c_1|\mathbf{x}) \sum_{c \in C} \hat{P}(c|\mathbf{x}) \log \hat{P}(a_i|c). \tag{17}$$

The derivative of the objective function is finally obtained by combining Eq. 12, Eq. 16 and Eq. 17. By repeatedly executing the backpropagation process on the training set, the weight vector $\mathbf{w}$ is converged to an optimal value. Then, as for a sample from the testing set, the class is estimated by the weighted naive Bayes formula (Eq. 6 and Eq. 7).

5   Experiments and Results

5.1   Dataset

The proposed method is evaluated with 4 heart disease datasets, which are collected from 4 medical institutions worldwide including Cleveland, Hungarian, Switzerland and Long Beach VA. These datasets are taken from the UCI

online ML and data mining repository (http://archive.ics.uci.edu/ml/datasets/
Heart+Disease). The dataset consists of 899 cases with 76 features (one predicted
feature included), and the statistical information is presented in Table 1.

Table 1. Statistical information of datasets.

Institution	Class 0	Class 1	Class 2	Class 3	Class 4	Total
Cleveland	157	50	31	32	12	282
Hungarian	188	37	26	28	15	294
Switzerland	8	48	32	30	5	123
Long Beach VA	51	56	41	42	10	200

There are 5 classes in the dataset. Class 0 represents the absence of heart
disease, while 1–4 indicate the presence of heart disease with different degrees.
In our system, only the absence or presence of heart disease will be predicted.
Therefore, class 1–4 are combined into one and the multi-label task is simplified
to a binary classification task.

5.2 Evaluation Metrics

Several metrics are used to illustrate the performance of our method compared
with others, as shown in the following:

1. **True positive (TP):** It is the number of abnormal cases that are correctly
   classified by the model.
2. **True negative (TN):** It is the number of normal cases that are correctly classified by the model.
3. **False positive (FP):** It is the number of normal cases that are wrongly classified as abnormal ones by the model.
4. **False negative (FN):** It is the number of abnormal cases that are wrongly
   classified as normal ones by the model.
5. **Acc:** It denotes the percentage of all cases that are correctly classified by the
   model, which is defined as:
   $Acc = \frac{TP+TN}{TP+TN+FP+FN}$
6. **Precision (P):** It measures the success of the classification model. We have:
   $Precision = \frac{TP}{TP+FP}$
7. **Recall (R):** It is the percentage of abnormal cases that are correctly classified
   by the model. We have:
   $Recall = \frac{TP}{TP+FN}$
8. **F1:** F1-score is the harmonic mean of precision and recall, which is defined
   as:
   $F_1 = 2 \cdot \frac{precision \cdot recall}{precision + recall}.$

5.3   Results and Discussion

This section presents the experimental results on different stages of our proposed decision support system, including results based on feature selection, results based on feature weighting and comparison results with other existing systems. All experiments are implemented using Python 3.7 and the dataset is randomly divided into 70% training set and 30% testing set.

Table 2. Comparison results of the classifiers before and after feature selection.

Method	Before feature selection				After feature selection			
	Acc (%)	P	R	F_1	Acc (%)	P	R	F_1
RF	78.89	0.81	0.84	0.82	80.74	0.84	0.84	0.84
KNN	71.48	0.75	0.77	0.76	71.48	0.77	0.74	0.75
NB	75.93	0.80	0.79	0.80	80.37	0.82	0.86	0.84
SVM	58.89	0.59	0.99	0.74	59.63	0.60	0.99	0.74
MLP	76.30	0.83	0.76	0.79	77.41	0.85	0.75	0.80
Ours	77.41	0.81	0.81	0.81	82.96	0.82	0.91	0.86

Results Based on Feature Selection. In order to prove that the proposed feature selection method works, various machine learning models are implemented, including Random Forest (RF), KNN, Naive Bayes (NB), Support Vector Machine (SVM) and Multi Layer Perceptron (MLP).

In this experiment, various numbers of features are selected to find the best subset for HF prediction. Figure 2 shows the changes in classification accuracy based on different numbers of selected features. It can be observed that all models perform best under the circumstances that 15 features are selected. Detailed information on the 15 selected features is listed in Table 4.

To further indicate the impact feature selection method has on the HF risk prediction task, Table 2 shows the comparison results of these machine learning models and our method before and after applying feature selection. It is observed that all models, except for KNN, have a higher accuracy after applying feature selection. In particular, the accuracy of naive Bayes is increased from 75.93% to 80.37%, which is a substantial performance increase. Different from other models, naive Bayes simply assumes that all input features are fully independent. However, this assumption is hard to be satisfied when a large number of features are used. Therefore, the feature selection method could eliminate noise and largely increase the accuracy by removing irrelevant and redundant features. The classification model used in our system is weighted naive Bayes, which is a variant of naive Bayes. The assumption that features should be independent is the same, and therefore the feature selection method also works in our proposed system. The accuracy of our method is increased from 77.41% to 82.96%, which indicate the efficiency of proposed feature selection method.

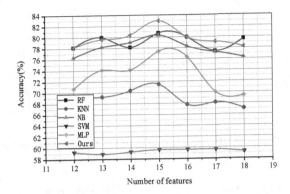

Fig. 2. The classification accuracy based on different numbers of selected features.

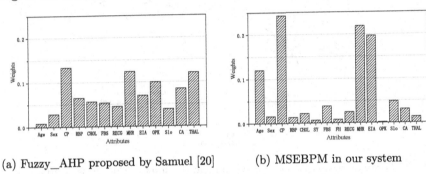

(a) Fuzzy_AHP proposed by Samuel [20]   (b) MSEBPM in our system

Fig. 3. The weights of each attribute in heart failure dataset.

Results Based on Feature Weighting. By investigating recent decision support systems, few of them apply feature weighting methods. Samuel et al. [20] propose Fuzzy_AHP to properly rank and compute the local weights of HF features. Therefore, we compare the weight vector calculated by our MSEBPM method with the one generated by Fuzzy_AHP.

Fuzzy_AHP calculates the weights of 13 selected attributes, which is illustrated in Fig. 3(a). It is observed that chest pain (CP), maximum heart rate (MHR) and heart status (THAL) are the top 3 most significant features. Meanwhile, our proposed MSEBPM obtains an optimal weight vector of the 15 selected features till the algorithm is converged. The weight distribution for each feature is shown in Fig. 3(b). The top 3 most significant features also include CP and MHR, but they have a higher difference compared with other values. This is quite close to the way of physicians' manual diagnosis, that is, pay much attention to 3–4 medial measurements to quickly evaluate the risk levels. Additionally, our method indicates the crucial significance of exercise induced angina (EIA), which is neglected by Fuzzy_AHP.

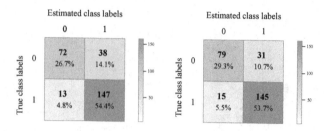

(a) Fuzzy_AHP combining (b) Our MSEBPM combining
with weighted NB   with weighted NB

Fig. 4. The confusion matrix of HF risk prediction.

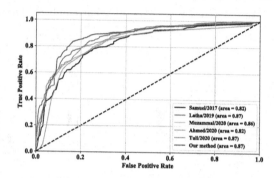

Fig. 5. Accuracy comparison of our system and the state-of-the-art using ROC curve.

Although Samuel et al. apply the weight vector to an ANN model for HF prediction, our experiment decides to apply it to weighted naive Bayes, which is the classification model in our system, to further demonstrate the effectiveness of the weight vector obtained by our MSEBPM. Weighted naive Bayes takes discrete features as input, and thus numeric features will be discretized through transformation rules presented in Table 4. The results are shown in Fig. 4. The accuracy of weighted naive Bayes using our MSEBPM is 82.96%, while the one using Fuzzy_AHP is 81.11%. The experimental results demonstrate that our MSEBPM is a more efficient and reasonable feature weighting method.

Comparison with Existing Systems. To further evaluate the performance of our proposed system, we compare it with other existing systems in terms of accuracy. The results are presented in Table 3.

Samuel et al. [20] use fuzzy_AHP for feature weighting and an ANN model for classification, achieving an accuracy of 75.19%. It is worth mentioning that the accuracy could be increased to 81.11% if we use our weighted naive Bayes in place of the ANN for classification. It suggests that the feature weights obtained by fuzzy_AHP are effective. However, the attributes are fed into the network by multiplying the value of each attribute with their corresponding weight, which is not the right way of fully utilizing the weight information. Latha et al. [12] directly use an ensemble classifier and achieve an accuracy of 78.52%. Muzammal et al. [16] utilize a correlation-based feature selection strategy and kernel random forest for classification, achieving an accuracy of 79.26%. Ahmed et al. [2] obtain an accuracy of 77.04% by using relief feature selection and decision tree. Tuli et al. [22] use PCA for feature selection and deep learning for classification, obtaining 79.26% accuracy. By contrast, the accuracy of our proposed system can achieve to nearly 83%, which suggests that it could be used for early HF detection in the clinic.

Table 3. Comparison results of recent works.

Authors/Year	Feature selection/Feature weighting	Classifiers	Overall accuracy (%)
Samuel et al. [20]/2017	–/Fuzzy analytic hierarchy process	Artificial neural network (ANN)	75.19
Latha et al. [12]/2019	–/–	Ensemble classifiers	78.52
Muzammal et al. [16]/2020	Correlation-based feature selection/–	Kernel random forest	79.26
Ahmed et al. [2]/2020	Relief feature selection/–	Decision tree	77.04
Tuli et al. [22]/2020	Principal component analysis/–	Deep learning	79.26
Our method	**Information gain/MSEBPM**	**Weighted naive Bayes**	**82.96**

In the medical field, ROC curve is always used to reflect the performance of a classification model at various thresholds settings. Figure 5 presents the ROC curve of the proposed method in comparison with others. It could be observed that the green line covers a larger area between the diagonal and the upper-left corner of the curves than others, which indicates that our method performs better than others in the prediction of HF risks.

Table 4. Detailed information of the 15 selected features from the heart disease dataset.

Label	Feature	Description	Discrete value	Range	Type
A_1	Age	Age of the patient	Young Medium Old Very old	<33 34–40 41–52 >52	Numeric
A_2	Sex	Gender of the patient	–	0, 1	Nominal
A_3	CP	4 types of chest pain: 1 = typical angina, 2 = atypical angina, 3 = non-anginal pain, and 4 = asymptomatic	–	1, 2, 3, 4	Nominal
A_4	RBP	The resting blood pressure in mm Hg on admission to hospital	Low Medium High Very high	<128 128–142 143–154 >154	Numeric
A_5	CHOL	The serum cholesterol in milligrams per deciliter (mg/dl)	Low Medium High Very high	<188 189–217 218–281 >281	Numeric
A_6	SY	The number of years as a smoker	Low Medium High Very high Seriously high	1–10 11–20 21–30 31–40 >40	Numeric
A_7	FBS	If the fasting blood sugar reaches 120 mg/dl (1 = true, 0 = false)	–	0, 1	Nominal
A_8	FH	If the family has history of coronary artery disease (1 = true, 0 = false)	–	0, 1	Nominal
A_9	RECG	The resting electrocardiographic results: 0 = normal, 1 = having ST-T wave abnormality, and 2= showing probable or definite left ventricular hypertrophy	–	0, 1, 2	Nominal
A_{10}	MHR	The maximum heart rate	Low Medium High	<112 112–152 >152	Numeric
A_{11}	EIA	If exercise induced angina happens (1 = true, 0 = false)	–	0, 1	Nominal
A_{12}	OPK	ST depression induced by exercise relative to rest	Low Risk Terrible	<1.5 1.5–2.55 >2.55	Numeric
A_{13}	Slo	The slope of the peak exercise ST segment: 1 = up slope, 2 = flat, and 3 = down slope	–	1, 2, 3	Nominal
A_{14}	CA	The number of major vessels colored by fluoroscopy	–	0, 1, 2	Nominal
A_{15}	THAL	Period of exercise test in minutes It demonstrates the heart status: 3 = normal, 6 = fixed defect, and 7 = reversible defect	–	3, 6, 7	Nominal

6   Conclusion

In this paper, we propose a decision support system for HF risk prediction. Three crucial issues are discussed, including feature selection using information gain, identification of the significance of each feature by employing MSEBPM, and heart disease prediction using a weighted naive Bayes model. We demonstrate that the weight information is beneficial for the weighted naive Bayes model to better predict HF. The results show that our system achieves remarkable improvements in HF detection compared with others.

Acknowledgements. This work is supported by the Chinese Scientific and Technical Innovation Project 2030 (No. 2018AAA0102100), NSFC-General Technology Joint Fund for Basic Research (No. U1936206, No. U1836109), and National Natural Science Foundation of China (No. 61772289, No. U1903128, and No. 62002178).

References

1. Acharya, U.R., Fujita, H., Lih, O.S., Adam, M., Tan, J.H., Chua, C.K.: Automated detection of coronary artery disease using different durations of ECG segments with convolutional neural network. Knowl.-Based Syst. **132**, 62–71 (2017)
2. Ahmed, H., Younis, E.M., Hendawi, A., Ali, A.A.: Heart disease identification from patients' social posts, machine learning solution on spark. Futur. Gener. Comput. Syst. **111**, 714–722 (2020)
3. Ali, F., El-Sappagh, S., Islam, S.R., Kwak, D., Ali, A., Imran, M., Kwak, K.S.: A smart healthcare monitoring system for heart disease prediction based on ensemble deep learning and feature fusion. Inf. Fusion **63**, 208–222 (2020)
4. Azhagusundari, B., Thanamani, A.S.: Feature selection based on information gain. Int. J. Innov. Technol. Explor. Eng. (IJITEE) **2**(2), 18–21 (2013)
5. Cai, D., Zhang, C., He, X.: Unsupervised feature selection for multi-cluster data. In: Proceedings of the 16th ACM SIGKDD International Conference on Knowledge Discovery and Data Mining, pp. 333–342 (2010)
6. Cheng, Q., Zhou, H., Cheng, J.: The fisher-Markov selector: fast selecting maximally separable feature subset for multiclass classification with applications to high-dimensional data. IEEE Trans. Pattern Anal. Mach. Intell. **33**(6), 1217–1233 (2010)
7. Dutta, A., Batabyal, T., Basu, M., Acton, S.T.: An efficient convolutional neural network for coronary heart disease prediction. Expert Syst. Appl. 113408 (2020)
8. Giri, D., et al.: Automated diagnosis of coronary artery disease affected patients using LDA, PCA, ICA and discrete wavelet transform. Knowl.-Based Syst. **37**, 274–282 (2013)
9. He, X., Cai, D., Niyogi, P.: Laplacian score for feature selection. In: Advances in Neural Information Processing Systems, pp. 507–514 (2006)
10. Jabeen, F., et al.: An IoT based efficient hybrid recommender system for cardiovascular disease. Peer-to-Peer Netw. Appl. **12**(5), 1263–1276 (2019)
11. Jiménez, F., Martínez, C., Marzano, E., Palma, J.T., Sánchez, G., Sciavicco, G.: Multiobjective evolutionary feature selection for fuzzy classification. IEEE Trans. Fuzzy Syst. **27**(5), 1085–1099 (2019)
12. Latha, C.B.C., Jeeva, S.C.: Improving the accuracy of prediction of heart disease risk based on ensemble classification techniques. Inf. Medi. Unlock. **16**, 100203 (2019)
13. Long, N.C., Meesad, P., Unger, H.: A highly accurate firefly based algorithm for heart disease prediction. Expert Syst. Appl. **42**(21), 8221–8231 (2015)
14. McGarry, K., Graham, Y., McDonald, S., Rashid, A.: RESKO: repositioning drugs by using side effects and knowledge from ontologies. Knowl.-Based Syst. **160**, 34–48 (2018)

15. Moran, M., Gordon, G.: Curious feature selection. Inf. Sci. **485**, 42–54 (2019)
16. Muzammal, M., Talat, R., Sodhro, A.H., Pirbhulal, S.: A multi-sensor data fusion enabled ensemble approach for medical data from body sensor networks. Inf. Fusion **53**, 155–164 (2020)
17. Nie, F., Huang, H., Cai, X., Ding, C.H.: Efficient and robust feature selection via joint 2, 1-norms minimization. In: Advances in Neural Information Processing Systems, pp. 1813–1821 (2010)
18. Pal, D., Mandana, K., Pal, S., Sarkar, D., Chakraborty, C.: Fuzzy expert system approach for coronary artery disease screening using clinical parameters. Knowl.-Based Syst. **36**, 162–174 (2012)
19. Raileanu, L.E., Stoffel, K.: Theoretical comparison between the gini index and information gain criteria. Ann. Math. Artif. Intell. **41**(1), 77–93 (2004)
20. Samuel, O.W., Asogbon, G.M., Sangaiah, A.K., Fang, P., Li, G.: An integrated decision support system based on ANN and fuzzy-AHP for heart failure risk prediction. Expert Syst. Appl. **68**, 163–172 (2017)
21. Tsai, C.H., et al.: Usefulness of heart rhythm complexity in heart failure detection and diagnosis. Sci. Rep. **10**(1), 1–8 (2020)
22. Tuli, S., et al.: HealthFog: an ensemble deep learning based smart healthcare system for automatic diagnosis of heart diseases in integrated IoT and fog computing environments. Futur. Gener. Comput. Syst. **104**, 187–200 (2020)
23. Wang, L., et al.: A hierarchical fusion framework to integrate homogeneous and heterogeneous classifiers for medical decision-making. Knowl.-Based Syst. **212**, 106517 (2020)
24. Yang, Y., Shen, H.T., Ma, Z., Huang, Z., Zhou, X.: 2, 1-norm regularized discriminative feature selection for unsupervised learning. In: IJCAI International Joint Conference on Artificial Intelligence (2011)
25. Zaidi, N.A., Cerquides, J., Carman, M.J., Webb, G.I.: Alleviating naive bayes attribute independence assumption by attribute weighting. J. Mach. Learn. Res. **14**(1), 1947–1988 (2013)
26. Zhang, R., Nie, F., Li, X., Wei, X.: Feature selection with multi-view data: a survey. Inf. Fusion **50**, 158–167 (2019)
27. Zhang, Y., Chen, C., Luo, M., Li, J., Yan, C., Zheng, Q.: Unsupervised hierarchical feature selection on networked data. In: Nah, Y., Cui, B., Lee, S.-W., Yu, J.X., Moon, Y.-S., Whang, S.E. (eds.) DASFAA 2020. LNCS, vol. 12114, pp. 137–153. Springer, Cham (2020). https://doi.org/10.1007/978-3-030-59419-0_9
28. Zhu, M., Zhu, H.: Learning a cost-effective strategy on incomplete medical data. In: Nah, Y., Cui, B., Lee, S.-W., Yu, J.X., Moon, Y.-S., Whang, S.E. (eds.) DASFAA 2020. LNCS, vol. 12113, pp. 175–191. Springer, Cham (2020). https://doi.org/10.1007/978-3-030-59416-9_11

Inheritance-Guided Hierarchical Assignment for Clinical Automatic Diagnosis

Yichao Du[1], Pengfei Luo[1], Xudong Hong[2], Tong Xu[1(✉)], Zhe Zhang[1], Chao Ren[1], Yi Zheng[3], and Enhong Chen[1]

[1] School of Computer Science and Technology,
University of Science and Technology of China, Hefei, China
{duyichao,pfluo,renchao,pb142090}@mail.ustc.edu.cn,
{tongxu,cheneh}@ustc.edu.cn
[2] Institution of Smart City Research (WuHu),
University of Science and Technology of China, Wuhu, China
xdhong@ahut.edu.cn
[3] HUAWEI Technologies, Hangzhou, China
zhengyi29@huawei.com

Abstract. Clinical diagnosis, which aims to assign diagnosis codes for a patient based on the clinical note, plays an essential role in clinical decision-making. Considering that manual diagnosis could be error-prone and time-consuming, many intelligent approaches based on clinical text mining have been proposed to perform automatic diagnosis. However, these methods may not achieve satisfactory results due to the following challenges. First, most of the diagnosis codes are rare, and the distribution is extremely unbalanced. Second, existing methods are challenging to capture the correlation between diagnosis codes. Third, the lengthy clinical note leads to the excessive dispersion of key information related to codes. To tackle these challenges, we propose a novel framework to combine the inheritance-guided hierarchical assignment and co-occurrence graph propagation for clinical automatic diagnosis. Specifically, we propose a hierarchical joint prediction strategy to address the challenge of unbalanced codes distribution. Then, we utilize graph convolutional neural networks to obtain the correlation and semantic representations of medical ontology. Furthermore, we introduce multi attention mechanisms to extract crucial information. Finally, extensive experiments on MIMIC-III dataset clearly validate the effectiveness of our method.

Keywords: Clinical automatic diagnosis · Hierarchical assignment · Co-occurrence graph · Graph Convolutional Network

1 Introduction

The clinical note is an essential part of Electronic Health Record (EHR), which contains lengthy and terminological text records about medical history, chief

© Springer Nature Switzerland AG 2021
C. S. Jensen et al. (Eds.): DASFAA 2021, LNCS 12683, pp. 461–477, 2021.
https://doi.org/10.1007/978-3-030-73200-4_31

complaint, current symptoms, and laboratory test results. To avoid the redundancy and ambiguity caused by the text, the World Health Organization recommends using the diagnosis codes in the International Classification of Diseases (ICD) for each disease, symptom, and sign to represent the patient's condition. The goal of clinical diagnosis is to assign the most likely diagnosis codes for the patient based on the clinical note. Traditionally, clinical diagnosis is completed by well-trained clinical coders, which is labor-intensive and error-prone because the diagnosis codes system is vast and growing. For example, in the United States, about 20% of patients are misdiagnosed at the primary care level, and one-third of the misdiagnosis will cause later severe injury to the patients [22].

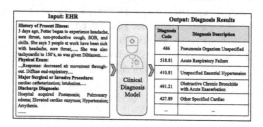

Fig. 1. Illustration of clinical automatic diagnosis task. The input and output of the model are EHR and diagnosis codes, respectively. The text related to the diagnosis code in the EHR is marked in colored font.

Consequently, the automatic clinical diagnosis based on EHR has aroused widespread attention in the industrial and academic circles [4]. Among the proposed methods, supervised machine learning methods were trained to learn shallow feature combinations for clinical note [7,19]. Recently, most deep learning models treated this task as a sequence learning problem, including used Convolutional Neural Networks [9,16] and Recurrent Neural Networks [3,21] to capture complex semantic information. On this basis, medical ontology was further introduced as auxiliary knowledge. Specifically, Bai et al. [1] incorporated the disease encyclopedia of Wikipedia into the model to enhance its predictive ability. Besides, the patient's history and demographic information could also be leveraged to enhance the prediction of future admissions [1,14,20]. Although these methods have made significant progress in automatic diagnosis, they may also fail due to the following challenges:

- **C1: The number of diagnosis codes is enormous, and the distribution is extremely unbalanced.** For example, the MIMIC-III [6] dataset, which is widely used for automatic diagnosis, contains 8,925 codes, but 4,344 appear less than five times in all data. The severe long-tail distribution makes it difficult to assign proper codes to rare diseases, which may cause irreparable damage to the patients.
- **C2: The correlations between diagnosis codes are greatly overlooked.** However, the medical relationship between diseases can help us identify diseases that are not clearly reflected by the clinical note. As shown in

Fig. 1, we can extract clues (colored fonts) from the text to assign diagnosis codes to the patient. For example, from the text "Hospital Acquired Pneumonia", we can easily infer the code "486 (Pneumonia Organism Unspecified)". Nevertheless, it is difficult to infer the code "410.81 (Acute Respiratory Failure)" only from the text. Fortunately, we can infer the code "410.81" from the relationship between it and the code "486", that is, "Pneumonia Organism Unspecified" will in all probability cause patients to have the symptom of "Acute Respiratory Failure".

– **C3: In clinical note, only a few key fragments can provide valuable information for automatic diagnosis.** For example, in the MIMIC-III dataset, clinical notes usually contain more than 1,500 tokens, but only a few tokens are related to specific diagnosis codes. Extracting crucial tokens for specific diagnosis codes is as tricky as finding a needle in a haystack.

To this end, we propose a model named Inheritance-guided Hierarchical Assignment with Co-occurrence-based Enhancement (IHCE) to address these challenges. First, for C1, we design a hierarchical assignment method based on the hierarchical inheritance structure of diagnosis codes defined by ICD, which makes assignment level by level. As shown in Fig. 2, "405.0 (Malignant renovascular hypertension)" and "405.1 (Benign secondary hypertension)" are mutually exclusive. Moreover, "405.01 (Malignant renovascular hypertension)" inherits the information of "405.0". Consequently, if we assign "405.0" at the high level, we will tend to further assign "405.01" instead of the children of "405.1". With the inheritance-guided hierarchical assignment, we can use the diagnostic results of a high level to guide the low level, which addresses the challenge of unbalanced distribution. Second, for C2, we construct a co-occurrence graph based on EHR data and use GCN to obtain the diagnosis codes' semantic representations. In this way, the representations of the diagnosis codes contain the correlation between diseases, which help us to assign codes to diseases for where it is challenging to find textual clues from the clinical note. Third, for C3, we enhance the ability to extract the tokens related to the diagnosis codes based on the attention mechanism which models the interaction between diagnosis codes' ontology representations and the clinical note. Finally, experiments on a real medical dataset show that IHCE is superior to the SOTA methods on all evaluation metrics.

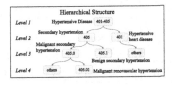

Fig. 2. An example of diagnosis codes' descriptors and their hierarchical inheritance structure based on ICD.

2   Related Work

2.1   Clinical Automatic Diagnosis

Clinical automatic diagnosis has become a research hot spot in medicine, aiming to solve manual diagnosis limitations. In recent years, deep learning technologies [9,16,21] have shown substantial advantages over traditional machine learning methods [7,19] and have been widely used for this task. Most researchers modeled this task as a multi-label text classification task based on the free text in EHR. Among them, Shi et al. [21] proposed a character-perceived LSTM network that generated written diagnosis descriptions and representations of diagnosis codes. Baumel et al. [3] proposed a hierarchical-GRU with a label-dependent attention layer to alleviate excessive text problem. Wang et al. [23] proposed a label-word joint embedding model and applied the cosine similarity to assign the codes. Moreover, some researchers incorporated external knowledge into the model [1,14,20]. For example, Knowledge Source Integration (KSI) [1] calculated the matching score between the clinical note and each knowledge document based on the intersection of clinical notes and external knowledge for this task. Our method is different from these methods, considering the hierarchy and co-occurrence relationship to achieve better performance in automatic diagnosis.

2.2   Graph Convolutional Network

In the past few years, Graph Convolutional Network (GCN) [8] has been widely used in various tasks to encode advanced graph structures, such as healthcare [11, 25], recommender systems [12], business analysis [10], machine translation [2], text classification [18,24]. Specifically, in order to promote the sharing of disease among patients, Liu et al. [11] applied GCN on text corpus to collect high-order neighbor information, and predicted for patients based on projection. Yao et al. [24] proposed Text-GCN, which was utilized to learn the representations of words and documents to improve text classification. Peng et al. [18] proposed a recursive regularized GCN to perform large-scale text classification on word co-occurrence graphs. Inspired by this, we apply GCN to obtain a good correlation between diagnosis codes and represent the medical ontology. Furthermore, we utilize the ontology representations as interactive information to improve the performance of automatic diagnosis.

3   Preliminaries

For a patient, the word sequence $S = \{w_1, w_2, ..., w_n\}$ of the patient's clinical note is included, where n is the length of S. Furthermore, a set of diagnosis codes $L = \{l_1, l_2, ..., l_{|L|}\} \in \{0,1\}^{|L|}$ are also contained to denote the diseases of the patient, where $|L|$ is the number of diagnosis codes. In addition, we also introduce hierarchical inheritance structure $\mathcal{L} = \{L^1, L^2, ..., L^T\}$ to expand L based on external knowledge (i.e., the hierarchical inheritance structure based on ICD in Fig. 2), where $L^t = \left\{ l_1^t, l_2^t, ..., l_{|L^t|}^t \right\}$ means all diagnosis codes of the level-t, and $\mathcal{T}$ is the total number of hierarchical levels. Note that, $L^{\mathcal{T}} = L$, which means that the last hierarchical level is the same as the patient's diagnosis codes. With above description, we can define the clinical automatic diagnosis task with inheritance guidance as follows:

Definition 1. *Given the patient's clinical note sequence S and the diagnosis codes hierarchical inheritance structure $\mathcal{L}$, our goal is to predict the patient's diagnosis codes set $\hat{L}^t = \left\{ \hat{l}_1^t, \hat{l}_2^t, ... \right\} \in \{0,1\}^{|\hat{L}^t|}$ level by level, and finally use the last level $\hat{L}^{\mathcal{T}}$ as the prediction of the patient's diagnosis.*

4   The Proposed Model IHCE

As shown in Fig. 3, IHCE mainly contains three components: (1) Document Encoding Layer (DEL), (2) Ontology Representation Layer (ORL), and (3) Hierarchical Prediction Layer (HPL). Specifically, we first utilize the DEL to obtain representations of the clinical note and diagnosis codes. Secondly, we apply the ORL to obtain the correlation and semantic representations of medical ontology. Finally, we design HPL to predict the patient's diagnosis codes based on hierarchical dependence and attention mechanism.

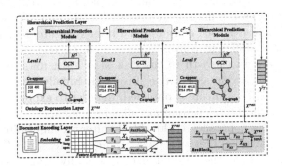

Fig. 3. The architecture of IHCE.

4.1   Document Encoding Layer

The goal of DEL is to generate unified representations for the clinical note and diagnosis codes. We first utilize the Embedding Module to encode the patient's clinical note and diagnosis codes. Then, we apply the Feature Extraction Module to enhance the semantic representation of the clinical note.

Embedding Module. First, given the word sequence $S = \{w_1, w_2, ..., w_n\}$, we use the word vector matrix $E = [e_1, e_2, ..., e_{|E|}] \in \mathbb{R}^{|E| \times d_e}$ to obtain the word embedding sequence $X = [x_1, x_2, ..., x_n] \in \mathbb{R}^{n \times d_e}$, where $|E|$ is the size of the vocabulary, and d_e is the dimension of the word vector. Similarly, we generate the diagnosis code ontology embedding for each code $l_i^t \in L^t$ via averaging the word embedding of its descriptor sequence:

$$v_i^t = \frac{1}{|N_i^t|} \sum_{j \in N_i^t} e_j, \quad i = 1, ..., |L^t|$$
$$V^t = \left[v_1^t, v_2^t, ..., v_{|L^t|}^t\right] \in \mathbb{R}^{|L^t| \times d_e} \quad , \tag{1}$$

where N_i^t is the text descriptor index set of l_i^t, and v_i^t denotes the word embedding of the l_i^t, and V^t indicates the representations of all codes of the level-t.

Feature Extraction Module. As shown in the lower part of the Fig. 3, we apply the multi-filter residual convolutional neural network [9] architecture for deep feature extraction on clinical note's embedding matrix X.

First, we utilize convolutional neural networks containing m filters to capture different length patterns of word sequence:

$$X_1 = F_1(X, W_1) = \tanh\left[..., W_1^T X^{j:j+s_1-1}, ...\right]$$
$$...$$
$$X_m = F_m(X, W_m) = \tanh\left[..., W_m^T X^{j:j+s_m-1}, ...\right], \text{where } j = 1, 2, ..., n, \tag{2}$$

Let us take the k-th operation as an example. $F_k(X, W_k)$ denotes the convolution operation on the matrix X, where $W_k \in \mathbb{R}^{(s_k \times d_e) \times d_c}$ is the parameter matrix, and d_c indicates each convolutional layer's feature mapping dimension. $s_1, s_2, ..., s_m$ denote different convolution kernel sizes, and $X^{j:j+s_k-1} \in \mathbb{R}^{s_k \times d_e}$ is the input matrix of the j-th to the $(j + s_k - 1)$-th rows in X. Note that, we set padding and stride as $floor(s_k/2)$ and 1. Finally, the feature matrices $X_k \in \mathbb{R}^{n \times d_c}, k = 1, 2, ..., m$ can be obtained. In order to express conciseness, the bias is ignored in all the calculation formulas in this paper.

Next, we connect m parallel residual blocks after the multi-filter convolutional layer, capturing longer text features by expanding the receptive field. Taking the k-th unit as an example, the residual block is formally defined as:

$$X_{k_1} = F_{k_1}(X_k, W_{k_1}) = \tanh\left[..., W_{k_1}^T X_k^{j:j+s_k-1}, ...\right],$$
$$X_{k_2} = F_{k_2}(X_{k_1}, W_{k_2}) = \left[..., W_{k_2}^T X_{k_1}^{j:j+s_k-1}, ...,\right],$$
$$X_{k_3} = F_{k_3}(X_k, W_{k_3}) = \left[..., W_{k_3}^T X_k^{j:j}, ...\right], \tag{3}$$
$$X_k^{res} = \tanh(X_{k_2} + X_{k_3}),$$

where $j = 1, 2, ..., n$, and W_{k_i} is the weight matrix of the k_i-th convolution layer in the residual block, specifically $W_{k_1} \in \mathbb{R}^{(s_k \times d_c) \times d_r}, W_{k_2} \in \mathbb{R}^{(s_k \times d_r) \times d_r}, W_{k_3} \in \mathbb{R}^{(1 \times d_c) \times d_r}$. The output of each residual block is $X_k^{res}, k = 1, 2, ..., m$, where d_r indicates the feature mapping dimension. Finally, we concatenate them together by rows to obtain an enhanced clinical note's representation:

$$X^{res} = \text{concat}(X_1^{res}, ..., X_m^{res}) \in \mathbb{R}^{n \times d_{res}}, \text{where } d_{res} = (m \times d_r). \tag{4}$$

4.2   Ontology Representation Layer

Comorbidities and complications manifest the correlation between the diagnosis codes ontology and play an auxiliary role for codes that are difficult to predict based on the clinical note alone. To this end, we first use co-occurrence features at each hierarchical level to construct a co-occurrence graph (co-graph) of diagnosis codes ontology. Then, we use GCN to capture the ontology's representations, which contain the correlation between the ontology. Here we take the level-t as an example to introduce the process.

Co-graph Construction. The co-graph is represented by $G^t = (L^t, E^t)$, where L^t and E^t indicate the diagnosis codes set and edge set of the level-t, respectively. For any diagnosis code l_i^t, if there is another code l_j^t in the EHR data that co-appears, there is an edge $e(l_i^t, l_j^t)$ between them. And the corresponding weight is calculated as follows:

$$e(l_i^t, l_j^t) = \frac{\text{count}(l_i^t, l_j^t)}{\sum_{l_k^t \in L^t} \text{count}(l_i^t, l_k^t)}, \tag{5}$$

where $\text{count}(\cdot, \cdot)$ indicates the number of times the two codes co-appear in the whole EHR dataset, which can represent prior knowledge. After that, the edge set E^t can be described as follows:

$$E^t = \{e(l_i^t, l_j^t) \mid l_i^t, l_j^t \in L^t\}. \tag{6}$$

Co-graph Propagation via GCN. Now we turn to represent the diagnosis codes. First, we can obtain the feature matrix $H^{t,(0)} = V^t \in \mathbb{R}^{|L^t| \times d_e}$ of the diagnosis codes ontology by Eq. (1). For the sake of simplicity, we omit the superscript t in the rest of this subsection. Then, we apply the GCN to propagate the representations of the diagnosis codes on the co-graph G, which takes the feature matrix $H^{(l)}$ and the matrix $\tilde{A}$ as input, and update the embedding of the codes by utilizing the information of adjacent codes:

$$H^{(l+1)} = \sigma\left(\tilde{D}^{-\frac{1}{2}} \tilde{A} \tilde{D}^{-\frac{1}{2}} H^{(l)} W^{(l)}\right), \tag{7}$$

where $\tilde{A} = A + I$, A is the adjacency matrix of G, I is the identity matrix, $\tilde{D}_{ii} = \sum_i \tilde{A}_{ij}$, and $W^{(l)}$ is a layer-specific trainable weight matrix. $\sigma(\cdot)$ denotes an activation function, such as the ReLU$(\cdot) = \max(0, \cdot)$. $H^{(l)} \in \mathbb{R}^{L \times d_g}$ is the matrix of activations in the l-th layer, where d_g indicates the hidden layer size of GCN. Then the last hidden layer is used to represent the diagnosis codes ontology, i.e., $H^t = H^{t,(l+1)} \in \mathbb{R}^{|L^t| \times d_g}$.

4.3   Hierarchical Prediction Layer

To simulate human diagnosis's gradual progress from shallow to deep, we propose an inheritance-guided hierarchical joint learning mechanism. To be specific,

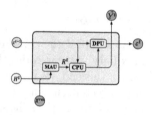

Fig. 4. Hierarchical prediction module.

according to the hierarchical structure of the codes, the patient is diagnosed progressively from coarse-grained to fine-grained.

Figure 4 shows the core module Hierarchical Prediction Module(HPM) of HPL. Specifically, HPM is mainly composed of three parts, namely Multi Attention Unit (MAU), Code Predicting Unit (CPU) and Dependency Passing Unit (DPU) respectively. For the level-t, the input of HPM includes three parts, i.e., the clinical note's representation X^{res}, the medical ontology representations H^t, and the dependency information c^{t-1} of the previous level:

$$
\begin{aligned}
R^t &= \mathrm{MAU}\left(X^{res}, H^t\right), \\
Y^t &= \mathrm{CPU}\left(c^{t-1}, R^t\right), \\
c^t &= \mathrm{DPU}\left(c^{t-1}, \tilde{Y}^t\right).
\end{aligned}
\tag{8}
$$

We first utilize the MAU part to obtain the correlation representation R^t between the clinical note and medical ontology. Next, the CPU part assigns the diagnosis codes $\tilde{Y}^t$ to the patient based on the R^t and c^{t-1}. Finally, the DPU part generates the level dependency information c^t for the next level based on the previous level's memory and the current level's assignment results. Note that we set c^0 to 0 since the current level is 0 and does not contain the previous level's information. Next, we introduce each unit of the HPM at level-t.

Multi Attention Unit. By the operations above, we can obtain the clinical note representation X^{res} and medical ontology representations H^t. Intuitively, the patient's clinical note is composed of a large number of lengthy text descriptions and different codes may focus on different aspects of the document. Therefore, for level-t, we need $|L^t|$ aspects to focus on different codes to represent the overall semantic of the whole clinical note. Next, we introduce the two attention mechanisms we use.

Ontology Guided Attention. For some diagnosis codes that are difficult to predict using only clinical text, we can improve it by interacting between the clinical note and medical ontology. First, we pass the document feature matrix X^{res} through a simple feed-forward neural network:

$$
O'_t = \tanh(W'_t \cdot (X^{res})^T),
\tag{9}
$$

where $W'_t \in \mathbb{R}^{d_g \times d_{res}}$ is the transform matrix, d_g is consistent with the dimension of the columns of H^t, and $O'_t \in \mathbb{R}^{d_g \times n}$ is the intermediate result. Then, for each code $l^t \in L^t$, we can generate the attention vector guided by the ontology:

$$\alpha_{l^t} = \text{softmax}(h_{l^t} \cdot O'_t), \tag{10}$$

where $h_{l^t} \in H^t$ is the feature vector of label l^t, and softmax$(\cdot)$ is the normalized exponential function for row operations. The attention $\alpha_{l^t} \in \mathbb{R}^{1 \times n}$ is then used to compute vector representation for each label:

$$x_i^{att'} = \alpha_{l^t} \cdot X^{res}, \tag{11}$$

Finally, we concatenate the $x_i^{att'}(i = 1, .., |L^t|)$ to obtain the ontology guided document representation, denoted as $X_t^{att'} = [x_1^{att'}, x_2^{att'}, ..., x_{|L^t|}^{att'}] \in \mathbb{R}^{|L^t| \times d_{res}}$.

Code Specific Attention. Similar to ontology guided attention, the code specific attention is formalized as:

$$\begin{aligned} O''_t &= \tanh(W''_t \cdot (X^{res})^T), \\ A''_t &= \text{softmax}(U''_t \cdot O''_t), \\ X_t^{att''} &= A''_t \cdot X^{res}, \end{aligned} \tag{12}$$

where $W''_t \in \mathbb{R}^{d_a \times d_{res}}$ is the intermediate parameter matrix. d_a is a hyperparameter, $O''_t \in \mathbb{R}^{d_a \times n}$ is the intermediate result matrix and $U''_t \in \mathbb{R}^{|L^t| \times d_a}$ is the code-specific attention parameter matrix. Finally, $X_t^{att''} \in \mathbb{R}^{|L^t| \times d_{res}}$ denotes code-specific document representation.

With the above description, we apply $R^t = \text{concat}(X_t^{att'}, X_t^{att''}) \in \mathbb{R}^{|L^t| \times 2d_{res}}$ as the output of the MAU.

Code Predicting Unit. For the level-t, we combine the result R^t of MAU with the inherited information c^{t-1} of the previous level to assign diagnosis codes to the patient. Specifically, the CPU uses a linear layer following a sigmoid transformation for each code:

$$\begin{aligned} X_t^{cls} &= \text{concat}(\text{broadcast}(c^{t-1}), R^t), \\ \tilde{Y}^t &= \sigma\left(X_t^{cls} \cdot W_y^t\right), \end{aligned} \tag{13}$$

where broadcast$(\cdot)$ is the process of making matrixes with different shapes have compatible shapes for arithmetic operations, $\sigma(\cdot)$ denotes an activation function, such as the sigmoid$(x) = \frac{1}{1+e^{-x}}$, $W_y^t \in \mathbb{R}^{(2d_{res}+d_c^{t-1}) \times 1}$ is the parameter of the CPU, and $\tilde{Y}^t \in \mathbb{R}^{|L^t| \times 1}$ is the prediction results of the level-t.

Dependency Passing Unit. We aim to preserve important information while reducing the harm caused by the previous level's error transmission. Therefore, we employ the combination of a linear layer and sigmoid function to imitate the gating mechanism to filter and integrate information as follows:

$$Z = \text{concat}((\tilde{Y}^t)^T, c^{t-1}),$$
$$c^t = \sigma(Z \cdot W_{dpu}^t), \tag{14}$$

where $Z \in \mathbb{R}^{1 \times (|L^t| + d_c^{t-1})}$ and $W_{dpu}^t \in \mathbb{R}^{(|L^t| + d_c^{t-1}) \times d_c^t}$ is the parameter matrix. Then, we can get the inter-level dependence $c^t \in \mathbb{R}^{1 \times d_c^t}$ based on the previous level's memory information and the prediction results of the current level.

4.4   Training

For training, we combine all levels of multi-label binary cross-entropy as the loss:

$$loss = \sum_t^{\mathcal{T}} loss^t = \sum_t^{\mathcal{T}} \sum_{i=1}^{L^t} [-y_i \log(\tilde{y}_i) - (1 - y_i) \log(1 - \tilde{y}_i)], \text{where } \tilde{y}_i \in \tilde{Y}^t,$$
$$\tag{15}$$

where $loss_t$ indicates the loss function of level-t.

5   Experiments

5.1   Dataset and Evaluation Metrics

In this paper, we conduct experiments on a real-world dataset: the MIMIC-III dataset, which is widely used in clinical automatic diagnosis. Following previous studies [9,16], we use the discharge summaries as the model's input and use the full codes and the top 50 most common codes for experiments. Specifically, for the MIMIC-III full setting, it includes the 8,925 codes, 47,719, 1,631, and 3,372 discharge summaries used for training, validation, and testing, respectively. For the MIMIC-III top-50 setting, it includes 8,067, 1,574, and 1,730 discharge summaries used for training, validation, and testing, respectively. In addition, we expand the codes from fine to coarse according to the hierarchical inheritance structure of ICD because EHR data only have the finest-grained codes (i.e. the level-4 in Table 1). The specific statistical results are shown in Table 1.

The evaluation metrics used in the experiments are Precision@K (K = 5, 8, and 15), Macro-F1, Micro-F1, Macro-AUC and Micro-AUC.

5.2   Implementation Details

We utilize PyTorch [17] to implement IHCE model and train it on a server with $4 \times$ V100 GPU. For the training setting, we use AdamW [13] for learning and set the learning rate and weight decay to 0.0001 and 0.00005, respectively. We set the dropout probability 0.4 and set the batch size to 16. We also apply an early stop mechanism, in which the training will stop if the Micro-F1 score on the validation set does not improve in 10 continuous epochs. Since our model has a number of hyperparameters, it is infeasible to search optimal values for

Table 1. The statistics of hierarchical levels.

Statistics	Full	Top-50
# codes in level-1	199	25
# codes in level-2	1,175	40
# codes in level-3	5,125	48
# codes in level-4	8,925	50
# avg codes per EHR in level-1	11.02	4.70
# avg codes per EHR in level-2	13.75	5.37
# avg codes per EHR in level-3	15.30	5.71
# avg codes per EHR in level-4	15.86	5.77

all hyperparameters. We keep the hyperparameters of the Feature Extraction Module consistent with Li [9]. Specifically, the word embedding dimension $d_e = 100$, the number of convolution kernels m in feature extraction is 6, and the size of the convolution kernels $s_1, s_2, ...s_m$ are set to "3, 5, 9, 15, 19, 25", $d_c = d_e$ and $d_r = 50$. Besides, we pre-train word embeddings on all the text in the training set using the word2vec [15] implemented by gensim[1]. The maximum length of a token sequence is 2,500, and the one that exceeds this length will be truncated. For the remaining parameters, we use the grid to search for the optimal hyperparameters. Specifically, we set the number of hidden layers to 1, and the hidden layer size $d_g = 300$ for GCN. In addition, we set $d_a=300$ for ORL's attention dimension, and $d_c^t = 500 (t = 1, 2, ..., T - 1)$ for all DPUs' parameters dimension.

5.3 Baselines

We compared IHCE with the following baselines, including machine learning and deep learning models:

- **LR:** which is a bag-of-words logistic regression model.
- **H-SVM** [19]: which designs a hierarchical SVM algorithm from root to leaf node by utilizing the hierarchical structure of diagnosis codes.
- **Bi-GRU** [16]: which employs bidirectional gated recurrent units to learn clinical note's representation for automatic diagnosis task.
- **C-MemNN** [20]: which combines the memory network with iterative compression memory representation to improve diagnosis accuracy.
- **C-LSTM-Att** [21]: which uses an LSTM-based language model to generate clinical note and diagnosis code representations as well as an attention mechanism to resolve the mismatch between notes and codes.
- **LEAM** [23]: which is proposed for text classification task by projecting labels and words in the same embedding space and using the cosine similarity to predict the label of text.

---

[1] https://radimrehurek.com/gensim/.

- **HARNNN** [5] which is initially used for multi-label text classification and considers the hierarchy of categories. We apply it to the automatic diagnosis.
- **CNN** [16]: which uses a single layer convolutional neural network and a max-pooling layer for automatic diagnosis task.
- **CAML and DR-CAML** [16]: which assign diagnosis codes based on clinical text by using CNN to aggregate information among the clinical note and attention mechanism to select the most relevant segment for each possible code. DR-CAML further uses text description as a regularization.
- **MultiResCNN** [9]: which utilizes multi-fliter convolutional neural networks and residual networks for automatic diagnosis and becomes the SOTA model on MIMIC-III.

5.4   Overall Performance

In this section, we compare the IHCE with existing works for clinical automatic diagnosis. Table 2 shows our overall performance on MIMIC-III full setting and MIMIC-III 50 setting. $T = 3$ means that our experiment is based on the last three levels (i.e., level-2 to level-4 in Table 1) in the hierarchy. Our model IHCE surpasses all baselines on both settings. The results indicate that IHCE is able to effectively perform clinical automatic diagnosis by exploiting the hierarchy and co-occurrence structure of the medical ontology and the attention mechanism. The specific analysis is as follows:

Table 2. Overall performance on MIMIC-III, where "–" means that the baseline did not report the result of the corresponding metric.

Models	MIMIC-III full						MIMIC-III top-50				
	AUC		F1-score		P@K		AUC		F1-score		P@K
	Macro	Micro	Macro	Micro	8	15	Macro	Micro	Macro	Micro	5
LR	56.1	93.7	1.1	27.2	54.2	41.1	82.9	86.4	47.7	53.3	54.6
H-SVM	–	–	–	44.1	–	–	–	–	–	–	–
C-MemNN	–	–	–	–	–	–	83.3	–	–	–	42.0
C-LSTM-Att	–	–	–	–	–	–	–	90.0	–	53.2	–
HARNN	–	–	–	40.5	–	–	–	–	–	–	–
BiGRU	82.2	97.1	3.8	41.7	58.5	44.5	82.8	86.8	48.4	54.9	59.1
LEAM	–	–	–	–	–	–	88.1	91.2	54.0	61.9	61.2
CNN	80.6	96.9	4.2	41.9	58.1	44.3	87.6	90.7	57.6	62.5	62.0
CAML	89.5	98.6	8.8	53.9	70.9	56.1	87.5	90.9	53.2	61.4	60.9
DR-CAML	89.7	98.5	8.6	52.9	69.0	54.8	88.4	91.6	57.6	63.3	61.8
MultiResCNN	91.0	98.6	8.5	55.2	73.4	58.4	89.9	92.8	60.6	67.0	64.1
IHCE ($T = 3$)	**92.9**	**98.9**	**10.4**	**57.3**	**73.5**	**58.7**	**91.0**	**93.6**	**64.7**	**69.6**	**65.2**

(1) In the MIMIC-III full setting, compared with the SOTA method Mul-tiResCNN, the IHCE improves Macro-AUC, Macro-F1 and Micro-F1 by 2.1%, 22.4% and 3.8%, respectively. It is worth noting that all models have low Macro-F1 scores on MIMIC-III full setting because the diagnosis codes space is too large, and the distribution is extremely unbalanced. Nevertheless, what is exciting is

that our model has **18.2%** and **22.4%** improvements in this metric compared to CAML and MultiReCNN, respectively. The reason is the IHCE considers hierarchical inheritance structure and dependencies. So the IHCE can assists the processing of low-frequency codes based on high-level prediction results. Similarly, we can observe that H-SVM with a hierarchical structure is better than BiGRU without a hierarchical structure in Micro-F1. However, the performance of H-SVM is lower than that of CAML and MultiReCNN because CAML and Multi-ReCNN utilize a primary attention mechanism to improve the ability to retrieve critical information. Furthermore, compared to CAML and MultiResCNN, our model has multiple attention mechanisms, so our model has more robust key information retrieval capabilities and surpasses them in all metrics.

(2) In the MIMIC-III top-50 setting, compared with the SOTA method MultiRescNN, the IHCE improves Macro-F1 and Micro-F1 by 6.8% and 3.9%, respectively. Although there are only 50 diagnosis codes in MIMIC-III top-50 setting, it still shows a slight long-tail effect. The IHCE has a significant improvement on the Macro-f1, indicating that our model can employ the hierarchical structure to alleviate this problem. It is worth noting that even though DR-CAML utilize codes description as regularization to assist in the allocation of diagnosis codes that are difficult to predict, the effect is still limited compared to CNN. However, the IHCE utilizes the co-occurrence structure between codes to solve this problem better.

5.5 Ablation Study

In this section, to verify each component's effectiveness in the IHCE, we perform ablation studies. The specific results are shown in Table 3. It is observed that removing each component will cause F1 to decrease, which illustrates the effectiveness of each component of our model. (1) **HPL's effectiveness:** After removing the HPL module, the macro-average metrics drop significantly, indicating that the inheritance-guided hierarchical assignment mechanism introduced by our IHCE has a significant effect on solving the long-tail effect. (2) **ORL's effectiveness:** After ORL is removed, the overall performance of IHCE declines because the method cannot model disease co-occurrence relationships. However, this ability is beneficial for assigning diseases for which it is not easy to find textual clues in the clinical note. (3) **Attention mechanism's effectiveness:** We only retain the *Code Specific Attention* module, which expands the attention mechanism in MultiResCNN and improves almost all metrics. It shows that our attention mechanism can better extract essential information to prevent the situation of finding a needle in a haystack.

5.6 Performance at Different Levels

In the clinical automatic diagnosis task, it is important to assign the diagnosis codes of the last level to the patient. It is also essential to evaluate the performance at different levels because, in some cases, a different granularity of codes may be required. Therefore, we compared the performance of IHCE and

Table 3. Ablation study results, where "w/o" indicates without.

Models	MIMIC-III full			MIMIC-III top-50		
	Macro-AUC	Macro-F1	Micro-F1	Macro-AUC	Macro-F1	Micro-F1
MultiResCNN (SOTA)	91.0	8.5	55.2	89.9	60.6	67.0
w/o ORL& HPL	91.0	8.7	55.9	89.9	61.2	66.9
w/o HPL	92.6	9.2	56.0	89.9	62.1	67.5
w/o ORL	**93.1**	10.0	56.7	90.6	63.6	68.5
IHCE ($\mathcal{T} = 3$)	92.9	**10.4**	**57.3**	**91.0**	**64.7**	**69.6**

IHCE-DPU at each hierarchical level. Note that this comparison is based on $\mathcal{T} = 3$. The IHCE-DPU ignores the dependency between the levels by removing the DPU in the HPM. In Fig. 5, we can see that the performance of IHCE at almost all levels is better than IHCE-DPU. Moreover, we can also notice that the performance on all metrics tend to decrease when the hierarchy deepens, and the trend on Macro-F1 in MIMIC-III full setting is the most obvious. The reason is that as the level deepens, the number of codes of this level will increase rapidly (e.g., the MIMIC-III full setting has 5,125, 8,925 unique codes in level-3 and level-4 respectively, as shown in Table 1). Moreover, we can notice that IHCE reduces this negative factor compared with IHCE-DPU by modeling the dependency among different hierarchical levels.

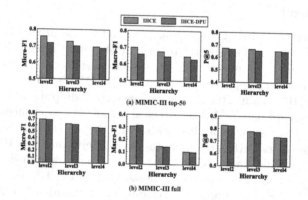

Fig. 5. Performance at different levels in hierarchy.

5.7   Effect of the Number of Hierarchical Levels

In this section, we turn to figure out the effect of the number of hierarchical levels, i.e., $\mathcal{T}$. To that end, a series of experiments are conducted to evaluate the effectiveness under different settings. Specifically, $\mathcal{T} = n$ means choosing the last n levels in Table 1. For example, $\mathcal{T} = 2$ means that we choose level-3 and level-4.

From Fig. 6, we can conclude that the models that consider hierarchical structure preform much better than models that do not. The performance rises when the number T of levels increases because high-level information has a guiding effect on the low level. However, the performance decreases when the T continuously increases. The reason is that when the number of codes between different levels is not an order of magnitude, errors caused by high-level results will still seriously affect low-level levels, although DPU has a mitigating effect. Specifically, for the MIMIC-III full setting, when $T = 4$, the model will extend level-1 with only 199 diagnosis codes, which is not in the same order of magnitude as other levels. For the MIMIC-III top-50 setting, each level's magnitude is not much different, and the impact of this error will also be reduced.

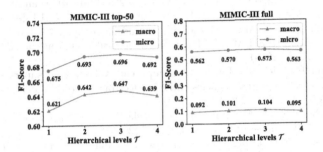

Fig. 6. Performance by varying the number of hierarchical levels.

6 Conclusion

In this paper, we proposed a novel Inheritance-guided Hierarchical Assignment with Co-occurrence-based Enhancement (IHCE) framework for clinical automatic diagnosis, which could jointly exploit code hierarchy and code co-occurrence. We utilized GCN to obtain the correlation between medical ontology. Moreover, we proposed a hierarchical joint prediction strategy based on the attention mechanism. Experimental results on real medical datasets show that our model has obtained state-of-the-art performance with substantial improvements in different evaluation metrics. We believe that our method can also be used for other tasks that require the application of hierarchical structure and label co-occurrence, such as hierarchical multi-label classification.

Acknowledgements. This research was partially supported by grants from the National Key Research and Development Program of China (Grant No. 2018YFB1402600), the National Natural Science Foundation of China (Grant No. 62072423), and the Key Research and Development Program of Anhui Province (No. 1804b06020377).

References

1. Bai, T., Vucetic, S.: Improving medical code prediction from clinical text via incorporating online knowledge sources. In: The World Wide Web Conference, pp. 72–82 (2019)
2. Bastings, J., Titov, I., Aziz, W., Marcheggiani, D., Sima'an, K.: Graph convolutional encoders for syntax-aware neural machine translation. arXiv preprint arXiv:1704.04675 (2017)
3. Baumel, T., Nassour-Kassis, J., Cohen, R., Elhadad, M., Elhadad, N.: Multi-label classification of patient notes a case study on ICD code assignment. arXiv preprint arXiv:1709.09587 (2017)
4. Esteva, A., et al.: A guide to deep learning in healthcare. Nat. Med. **25**(1), 24–29 (2019)
5. Huang, W., et al.: Hierarchical multi-label text classification: an attention-based recurrent network approach. In: Proceedings of the 28th ACM International Conference on Information and Knowledge Management, pp. 1051–1060 (2019)
6. Johnson, A.E., et al.: MIMIC-III, a freely accessible critical care database. Sci. Data **3**(1), 1–9 (2016)
7. Kavuluru, R., Rios, A., Lu, Y.: An empirical evaluation of supervised learning approaches in assigning diagnosis codes to electronic medical records. Artif. Intell. Med. **65**(2), 155–166 (2015)
8. Kipf, T.N., Welling, M.: Semi-supervised classification with graph convolutional networks. arXiv preprint arXiv:1609.02907 (2016)
9. Li, F., Yu, H.: ICD coding from clinical text using multi-filter residual convolutional neural network. In: AAAI, pp. 8180–8187 (2020)
10. Li, S., Zhou, J., Xu, T., Liu, H., Lu, X., Xiong, H.: Competitive analysis for points of interest. In: Proceedings of the 26th ACM SIGKDD International Conference on Knowledge Discovery & Data Mining, pp. 1265–1274 (2020)
11. Liu, N., Zhang, W., Li, X., Yuan, H., Wang, J.: Coupled graph convolutional neural networks for text-oriented clinical diagnosis inference. In: Nah, Y., Cui, B., Lee, S.-W., Yu, J.X., Moon, Y.-S., Whang, S.E. (eds.) DASFAA 2020. LNCS, vol. 12112, pp. 369–385. Springer, Cham (2020). https://doi.org/10.1007/978-3-030-59410-7_26
12. Liu, Y., Li, Z., Huang, W., Xu, T., Chen, E.H.: Exploiting structural and temporal influence for dynamic social-aware recommendation. J. Comput. Sci. Technol. **35**, 281–294 (2020)
13. Loshchilov, I., Hutter, F.: Decoupled weight decay regularization. arXiv preprint arXiv:1711.05101 (2017)
14. Ma, F., You, Q., Xiao, H., Chitta, R., Zhou, J., Gao, J.: KAME: knowledge-based attention model for diagnosis prediction in healthcare. In: Proceedings of the 27th ACM International Conference on Information and Knowledge Management, pp. 743–752 (2018)
15. Mikolov, T., Chen, K., Corrado, G., Dean, J.: Efficient estimation of word representations in vector space. arXiv preprint arXiv:1301.3781 (2013)
16. Mullenbach, J., Wiegreffe, S., Duke, J., Sun, J., Eisenstein, J.: Explainable prediction of medical codes from clinical text. In: Proceedings of the 2018 Conference of the North American Chapter of the Association for Computational Linguistics: Human Language Technologies, Volume 1 (Long Papers), pp. 1101–1111 (2018)
17. Paszke, A., et al.: PyTorch: an imperative style, high-performance deep learning library. In: Advances in Neural Information Processing Systems, pp. 8026–8037 (2019)

18. Peng, H., et al.: Large-scale hierarchical text classification with recursively regularized deep graph-CNN. In: Proceedings of the 2018 World Wide Web Conference, pp. 1063–1072 (2018)
19. Perotte, A., Pivovarov, R., Natarajan, K., Weiskopf, N., Wood, F., Elhadad, N.: Diagnosis code assignment: models and evaluation metrics. J. Am. Med. Inform. Assoc. **21**(2), 231–237 (2014)
20. Prakash, A., et al.: Condensed memory networks for clinical diagnostic inferencing. arXiv preprint arXiv:1612.01848 (2016)
21. Shi, H., Xie, P., Hu, Z., Zhang, M., Xing, E.P.: Towards automated ICD coding using deep learning. arXiv preprint arXiv:1711.04075 (2017)
22. Singh, H., Schiff, G.D., Graber, M.L., Onakpoya, I., Thompson, M.J.: The global burden of diagnostic errors in primary care. BMJ Qual. Saf. **26**(6), 484–494 (2017)
23. Wang, G., et al.: Joint embedding of words and labels for text classification. In: Proceedings of the 56th Annual Meeting of the Association for Computational Linguistics (Volume 1: Long Papers), pp. 2321–2331 (2018)
24. Yao, L., Mao, C., Luo, Y.: Graph convolutional networks for text classification. In: Proceedings of the AAAI Conference on Artificial Intelligence, vol. 33, pp. 7370–7377 (2019)
25. Yichao, D., Tong, X., Jianhui, M., Enhong, C., Yi Zheng, T.L., Guixian, T.: An automatic ICD coding method for clinical records based on deep neural network. Big Data Res. **6**(5)

BPTree: An Optimized Index with Batch Persistence on Optane DC PM

Chenchen Huang, Huiqi Hu$^{(\boxtimes)}$, and Aoying Zhou

School of Data Science and Engineering, East China Normal University,
Shanghai, China
cchuang@stu.ecnu.edu.cn, {hqhu,ayzhou}@dase.ecnu.edu.cn

Abstract. Intel Optane DC Persistent Memory (PM) is the first commercially available PM product. Although it meets many hypothesises about PM in previous studies, some other design considerations are observed in subsequent tests. For instance, 1) the internal data access granularity in Optane DC PM is 256B, accesses smaller than 256B will cause read/write amplification; 2) the locking overhead will be amplified when the PM operations are included in the critical area or the lock is added on PM. In this paper, we propose a novel persistent index called BPTree to fit with these new features. The core idea of BPTree is to buffer multiple writes in DRAM first, and later persist them in batches to PM to reduce the write amplification. We add a buffer layer in BPTree to enable the batch persistence, and design a GC-friendly log structure on PM to guarantee the buffer's durability. To improve the scalability, we also implement a hybrid concurrency control strategy to ensure most of the operations on PM are lock-free, and move the lock from PM to DRAM for the operations that must be locked. Our experiments on Optane DC PM show that BPTree reduces 256B PM writes by a factor of 1.95–2.48x compared to the state-of-the-art persistent indexes. Moreover, BPTree has better scalability in the concurrent environment.

1 Introduction

Technologies of persistent memory (PM) have been studied for over a decade. PM provides comparable performance and much higher capacity than DRAM. More important, it is persistent, which makes failure recovery faster. These properties make PM attractive for index structures (e.g., B+-Tree) in database: the index can be directly accessed and persisted in PM and recovered instantly, which saves a lot of rebuild time and eases the effort to manage a large index.

Intel unveiled Optane DC Persistent Memory (PM) in 2019, which is the first commercially available PM product. The previous studies about persistent B+-Tree [2,4,9,12] are all designed based on PM emulation or simulation. According to the existing available informations [5,10,11], we identified that Optane DC PM meets many hypothesises about PM in previous researches, such as higher write latency and power failure might cause inconsistent data. However, there are

© Springer Nature Switzerland AG 2021
C. S. Jensen et al. (Eds.): DASFAA 2021, LNCS 12683, pp. 478–486, 2021.
https://doi.org/10.1007/978-3-030-73200-4_32

some new design considerations when using Optane DC PM: read/write amplification and locking overhead amplification. These new characteristics make the previous studies unable to achieve the best performance on Optane DC PM [7], which provide new challenges for the design of persistent B+-Tree. LB+-Tree [8] is the latest work based on Optane DC PM, it designs a novel tree node structure to avoid a write operation acrossing multiple 256B. But its write amplification is still very serious, because the update area of a write operation is usually smaller than 256B. Moreover, it does not pay attention to the second problem, and still adds lock on PM, which limits the scalability.

In this paper, we present BPTree, an efficient persistent B+-Tree designed for Optane DC PM. To reduce the cost of consistency maintenance, BPTree adopts the *selective persistence scheme* like previous works [8,9], where non-leaf nodes are in DRAM and leaf nodes are in PM. In BPTree, we add a buffer leaf layer on top of the PM leaf layer to support batch persistence. Write operations are first buffered in DRAM, and then persisted in batches to PM to reduce the write amplification. To guarantee the buffer's durability, we also design a GC-friendly log structure on PM. Meanwhile, concurrency control is fundamental to high scalability. Existing persistent B+-Trees [2,8,9,12] most use locks to achieve concurrency control, the scalability is limited by the higher locking overhead on PM. In BPTree, we redesign a hybrid concurrency control strategy, which uses the concurrency protocol of B-Link tree [6] to handle the concurrency of in-memory structures, ensures most of the operations on PM are lock-free, and moves the lock from PM to DRAM for the operations that must be locked.

2   Background

Optane DC Persistent Memory. Intel Optane DC Persistent Memory (PM) is the first commercially available PM product, it has the common features offered by the previous PM assumptions: 1) byte-addressability, 2) non-volatility and 3) performance in the range of DRAM's. In addition to these, Optane DC PM has some other design considerations: 1) Read/Write amplification [10,11]. The internal data access granularity in Optane DC PM is 256B, accesses smaller than 256B will cause read/write amplification. 2) Locking overhead amplification. Since the write latency of PM is higher than DRAM, the locking time will be longer when PM operations are included in the critical area, resulting in lower concurrency of upper layer. Besides, if the lock is added on PM, frequent *lock* and *unlock* operations will make it becomes a hot spot, whose latency is much higher than that of random or sequential write [10].

Persistent B+-Trees. There are several considerations in designing persistent B+-Tree. To reduce persists, wB+-Tree [2] keeps leaf unsorted and relies on 8B atomic write for bitmap and logging to achieve consistency. Similarly to wB+-Tree, NV-Tree [12] employs unsorted leaf with bitmap and reduced persistence guarantee of internal nodes. Internal nodes are rebuilt from the persistent leaves on recovery. FPTree [9], a persistent and concurrent B+-Tree, employs such

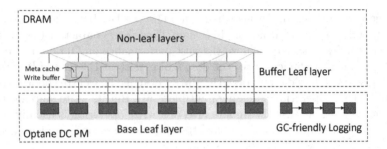

Fig. 1. Architecture of BPTree.

design. The internal nodes are placed in DRAM and leaves in PM. Bitmap, logging, and fingerprints are used for crash consistency and reducing PM reads. FASTFAIR [4], a persistent and concurrent B+-Tree, propose FAST and FAIR techniques to keep node sorted by exploiting properties of modern CPUs. The above works are all designed based on PM simulation or emulation, the main problem to be solved is how to reduce the number of PM writes under the premise of ensuring crash consistency. LB+-Tree [8] is a persistent B+-Tree designed on Optane DC PM, it uses entry moving technique to reduce the number of 256B writes in a write operation. However, its write amplification is still very serious, because the update area of a write operation is much smaller than 256B. Moreover, it doesn't pay attention to the locking overhead amplification, and still adds lock on PM, which limits the scalability.

3   Design

As shown in Fig. 1, BPTree adopts the *selective persistence scheme* [9] to reduce the overhead of consistency maintenance, where the non-leaf nodes are in DRAM and the leaf nodes are in Optane DC PM.

To reduce the write amplification on PM, we add a *buffer leaf layer* under the *non-leaf layers* to enable batch persistence, which is also DRAM resident. Write operations (Insert/Update/Delete) are first performed on the *buffer leaf layer*. When the buffer size reaches a threshold, we persist them in batches to the leaf nodes in PM (*base leaf layer*). We also design a GC-friendly logging scheme to guarantee the durability of *buffer leaf layer*.

The *buffer leaf layer* is not only beneficial for write, but also helps read. Point query first traverses the entries in buffer leaf node (*write buffer*), if the key is found, the query directly returns the latest value, which can avoid the access of base leaf node. Otherwise, the query accesses the base leaf node. Here, we cache the entry's metadata of base leaf node in buffer leaf, the query can first access the cached metadata (*meta cache*) to find the position of target value and then precisely get it from base leaf node, which further reduces the number of PM reads. For range queries, both leaf layers are scanned, and results are merged.

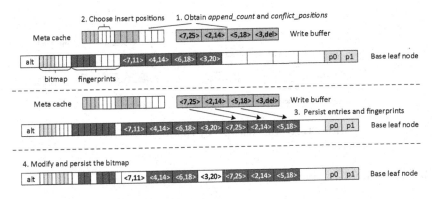

Fig. 2. Crash consistent batch persistence.

3.1   Buffer Leaf Layer

Buffer leaf layer is an in-memory structure and contains many memory blocks, each block serves one base leaf node at a time, which is called a buffer leaf node.

Buffer Leaf Node. As shown in Fig. 1, a buffer leaf node consists of two parts: *write buffer* and *meta cache*.

Write buffer is used to record the update of write operations, it's structure is the same as conventional sorted leaf nodes. Write operations are first recorded in write buffer, and then persisted into the base leaf node when the buffer is full.

Meta cache is used to cache the metadata (bitmap and fingerprints) of base leaf node. If there is no meta cache, the query requires first reading the metadata from base leaf and then accessing the target data. In this process, the metadata and target data may not in the same 256B, then the query needs multiple 256B PM reads. If the metadata is cached in buffer leaf node, we can get the target data with only one 256B PM read.

GC-Friendly Logging. Since *buffer leaf layer* is an in-DRAM structure, its durability needs to be guaranteed by a PM write-ahead log. For a write operation, we first persist the log through one PM write and then update the buffer leaf node. Here, the log content contains the key-value entry and a 4-bit checksum, where the checksum is used to check the invalid log caused by power failure.

In BPTree, we do not persist the entire *buffer leaf layer* into PM at intervals. Instead, a buffer leaf node is persisted as long as it is full, regardless of other buffer leaf nodes. After the persistence is completed, the corresponding logs will be cleared. If using the traditional log structure - all logs are stored in chronological order, the overhead of garbage collection (GC) in our design will be higher, because the logs of a buffer leaf node maybe not in a contiguous address space. To address this, we redesign a GC-friendly log structure using the advantage of PM byte addressing. Logs in BPTree are stored in the form of a linked list, and each linked list node records the logs of a buffer leaf node. After a buffer leaf node is persisted, the corresponding log node can be directly deleted from the linked list.

Fig. 3. Hybrid concurrency control in BPTree.

3.2 Base Leaf Layer

Base leaf layer is the complete leaf layer of BPTree, stored in PM, and all write operations will be batch persisted to it eventually.

Base Leaf Node. To reduce the number of PM writes, we adopt the unordered leaf node structure like FPtree [9]. As shown in Fig. 2, each base leaf node contains a metadata and an entries pool. The metadata consists of an alt bit, a bitmap, an array of fingerprints, and two next pointers. The alt bit specifies one of the two alternative next pointers. The bitmap tracks the allocation status of the key-value slots. The fingerprints array stores one-byte hashes of the keys stored at the corresponding positions of the entries pool, it serves to filter out unnecessary PM reads.

Crash Consistent Batch Persistence. Since *base leaf layer* is in PM, the batch persistence needs to be crash-consistent. For this, we exploit the append-only technique - never modify the old data before the persistence is completed.

When a buffer leaf is full, it will be persisted into the base leaf. The detailed persistence process is shown in Fig. 2: 1) Traverse the write buffer and meta cache to obtain two pieces of information: a. *append_count*: the number of entries to be appended in the base leaf. We only append insert and update, delete only need to modify the bitmap; b. *conflict_positions*: the positions of key conflict between write buffer and base leaf. Since the conflict keys will be rewritten, we require finding the positions of their old values to be released after the new entries are persisted; 2) Choose the suitable insert positions from bitmap according to the number of entries to be appended; 3) Persist the entries and corresponding fingerprints in the base leaf; 4) Modify and persist the bitmap atomically: the new inserted positions are set to 1, and the conflict positions are reset to 0. The base leaf node will split when it is also full. To reduce the logging overhead, we use the *logless node split* in LB+-Tree [8] to execute the split.

4   Concurrency Control

Most existing persistent B+-Trees [2,4,9,12] use locks to achieve concurrency control, but the overhead of locking will be amplified when it is added on PM or PM operations are included in the critical area. For this, we redesign a hybrid concurrency control strategy. Figure 3 gives the hybrid concurrency control flow of *insert* and *search*.

Concurrent insert: 1. Atomic find the leaf node using the method of OLFIT tree [1], which is lock-free. 2. Lock-free logging. We implement a lock-free log structure to avoid the higher locking overhead on PM. 3. Modify the buffer leaf by locking. 4. When a buffer leaf is full, it needs to be persisted into the base leaf. Here, before the base leaf split, we can achieve lock-free concurrency through asynchronous persistence (see Fig. 3(a)). When the base leaf split, since the asynchronous persistence will bring additional overhead, we execute synchronous persistence by locking (see Fig. 3(b)), where the lock is added on buffer leaf instead of base leaf. 5. After base leaf split, we continue to handle the concurrency write on *non-leaf layers* by locking.

Concurrent search: 1. Atomic read the non-leaf and buffer leaf without lock, the read atomicity is still ensured using the method of OLFIT tree [1]. 2. Lock-free access to the base leaf. Here, the base leaf read is atomic at any time, no other operations are required to determine the read state.

Next, we mainly discuss the concurrency of PM structures: lock-free logging, concurrent bath persistence and lock-free read on base leaf.

4.1 Lock-Free Logging

As we introduced in Subsect. 3.1, the log is stored as a linked list on the PM. In BPTree, the modifications on the log linked list include: Insert, Update and Delete. Among these operations, there is no conflict between Update and other operations. Then for the rest conflicts, we design a lock-free concurrency control method to solve them. To solve the conflict between insert, each thread corresponds to a log linked list, and the newly allocated log node only can be inserted into their own linked list. Next is the update operation, we can regard the log node as a part of the buffer leaf node, and then implement the concurrent update of the log node by locking the buffer leaf node. To solve the conflict between delete operations, the deleted log node is only marked at first, and then uniformly recycled by a background thread. At last, for the conflict between delete and insert, since the insert operation only occurs at the end of the linked list, we delay the GC of the last node to avoid the conflict.

4.2 Concurrent Batch Persistence

In a concurrent scenario, there are two ways of batch persistence: synchronous persistence and asynchronous persistence. 1) Synchronous persistence is to persist a full buffer leaf directly into the corresponding base leaf, and the new writes that on the same base leaf will be blocked. 2) In asynchronous persistence, a full buffer leaf will first become read-only and named as *static buffer leaf*, then a new buffer leaf is created to serve new writes, and the *static buffer leaf* is persisted into the corresponding base leaf in the background. Here, we named the buffer leaf serving write as *dynamic buffer leaf* to distinguish it from *static buffer leaf*.

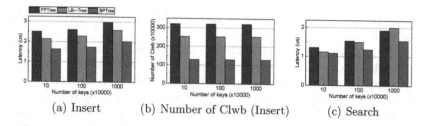

(a) Insert   (b) Number of Clwb (Insert)   (c) Search

Fig. 4. Single-threaded performance with different data sizes

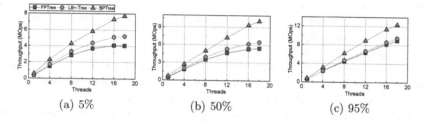

(a) 5%   (b) 50%   (c) 95%

Fig. 5. Multi-threaded performance under mixed workload with different read ratios

Asynchronous persistence can achieve higher write throughput by not blocking the later writes, but it brings additional load in some scenarios.

To obtain high throughput and avoid additional overhead, we adopt a selective persistence scheme. When the *dynamic buffer leaf* is full, we first determine whether the base leaf will split after inserting the new entries: if not split, the batch persistence can be executed asynchronously; if split, we execute the batch persistence synchronously.

4.3   Lock-Free Read on Base Leaf

In BPTree, the access of base leaf can ensure atomicity without any method. For the update operation on base leaf, we all use the append technique to keep all the old data not changed during the update, which is like maintaining two versions of data. Bitmap acts as a version number to ensure we read a snapshot of base leaf. Before modifying the bitmap, we can only read the old data, and the new data only can be read after bitmap is modified atomically. Note that, even after the bitmap is modified, the current read that getting the old bitmap can continue to read, because the old data will not change until the next update.

5   Experiment

We run experiments on a Linux (7.8) server equipped with two Inter(R) Xeon(R) Gold 6240M CPUs. The system is equipped with 186 GB DRAM and 1.5 TB Optane DC PM memory (6 × 256 GB DCPMMs) configured in the App Direct mode, each CPU has 3 DCPMMs. In our experiments, we avoid the NUMA effects by setting the CPU affinity of our programs to run on CPU 0, and making sure that the programs only access DRAM and PM local to the CPU. Yahoo! Cloud Serving Benchmark (YCSB) [3] is chose to generate the workloads. By default, we run the experiment under a uniform key distribution, and each run starts with a new tree pre-filled with 1 million records with 8-byte keys and 8-byte values. We evaluate the BPTree against other state of the art persistent B+-Tree structures including FPTree [9] and LB+-Tree [8]. Each persistent B+-Tree has the same node size, where 2 KB inner node and 1 KB leaf node.

5.1 Single-Threaded Performance

We first examine the single thread performance of each tree. Figure 4(a) shows the average latency of insert with varying data sizes. In general, BPTree has 1.27–1.31x lower latency than LB+-Tree and 1.46–1.53x than FPTree as the data size increases. We also test the amount of 256B PM writes generated by 1 million inserts, which can illustrate the reason for the latency gap. The experiment result is shown in Fig. 4(b), where BPTree has 1.95x lower amount of 256B PM writes than LB+-Tree and 2.48x than FPTree. Figure 4(c) gives the average latency of search with varying data sizes. In the figure, the latency of BPTree is always lower than that of LB+-Tree and FPTree. It is because the *buffer leaf layer* in BPTree contains a number of latest target values, which can avoid some reads of base leaf node. Besides, BPTree caches the bitmap and fingerprints of base leaf in the buffer leaf, further reduces the number of PM reads.

5.2 Multi-threaded Performance

Figure 5 shows the multi-threaded throughput of all trees under the mixed workload with different read ratios (5%, 50%, 95%). Firstly, with the increase of threads, the throughput of BPTree is 1.4–1.5x higher than that of LB+-Tree and 1.6–2x higher than that of FPTree. In terms of scalability, FPTree and LB+-Tree show the same trend of change, which reach the peak when the number of threads increases to 16. However, BPTree shows better scalability than FPTree and BPTree, its throughput increases almost linearly with the increase of threads. Secondly, with the increase of read ratio, the scalability of all trees are improved. When the read ratio increases from 5% to 50%, the max throughput of BPTree and FPTree are improved by 1.38x and 1.34x respectively, while LB+-Tree is only increased by 1.24x. This is because BPTree has better read performance than FPTree and LB+-Tree. In conclusion, BPTree has better scalability in the concurrent environment.

6 Conclusion

This paper presents a novel persistent B+-Tree called BPTree to meet the new challenges of Optane DC PM. In BPTree, we add a buffer layer to enable batch persistence to reduce the write amplification, and design a GC-friendly log structure on PM to guarantee the buffer's durability. We also implement a hybrid concurrency control strategy to ensure most of the operations on PM are lock-free, and move the lock from PM to DRAM for the operations that must be locked. Experimental results show that BPTree reduces 256B PM writes by a factor of 1.95–2.48x compare to the state-of-the-art persistent B+-trees. Moreover, BPTree has better scalability in the concurrent environment.

Acknowledgements. This work was supported by National Key R&D Program of China (2018YFB1003303), National Science Foundation of China under grant number 61772202, and Meituan Group. Thanks to the corresponding author Huiqi Hu.

References

1. Cha, S.K., Hwang, S., Kim, K.: Cache-conscious concurrency control of main-memory indexes on shared-memory multiprocessor systems. In: 27th International Conference on Very Large Data Bases, pp. 181–190 (2001)
2. Chen, S., Jin, Q.: Persistent B+-trees in non-volatile main memory. Proc. VLDB Endow. **8**(7), 786–797 (2015)
3. Cooper, B.F., Silberstein, A., Tam, E.: Benchmarking cloud serving systems with YCSB. In: 1st ACM Symposium on Cloud Computing, pp. 143–154 (2010)
4. Hwang, D., Kim, W., Won, Y.: Endurable transient inconsistency in byte-addressable persistent B+-tree. In: 16th USENIX Conference on File and Storage Technologies, pp. 187–200 (2018)
5. Izraelevitz, J., Yang, J., Zhang, L.: Basic performance measurements of the Intel optane DC persistent memory module. CoRR. abs/1903.05714 (2019)
6. Lehman, P.L., Yao, S.B.: Efficient locking for concurrent operations on B-trees. ACM Trans. Database Syst. **6**(4), 650–670 (1981)
7. Lersch, L., Hao, X., Oukid, I.: Evaluating persistent memory range indexes. Proc. VLDB Endow. **13**(4), 574–587 (2019)
8. Liu, J., Chen, S., Wang, L.: LB+-trees: optimizing persistent index performance on 3dxpoint memory. Proc. VLDB Endow. **13**(7), 1078–1090 (2020)
9. Oukid, I., Lasperas, J., Nica, A.: FPTree: a hybrid SCM-DRAM persistent and concurrent b-tree for storage class memory. In: 2016 International Conference on Management of Data, pp. 371–386 (2016)
10. Renen, A.V., Vogel, L., Leis, V.: Persistent memory I/O primitives. In: 15th International Workshop on Data Management on New Hardware, pp. 12:1–12:7 (2019)
11. Yang, J., Kim, J., Hoseinzadeh, M.: An empirical guide to the behavior and use of scalable persistent memory. In: 18th USENIX Conference on File and Storage Technologies, pp. 169–182 (2020)
12. Yang, J., Wei, Q., Chen, C.: NV-tree: reducing consistency cost for NVM-based single level systems. In: 13th USENIX Conference on File and Storage Technologies, pp. 167–181 (2015)

An Improved Dummy Generation Approach for Enhancing User Location Privacy

Shadaab Siddiqie[1(✉)], Anirban Mondal[2], and P. Krishna Reddy[1]

[1] IIIT Hyderabad, Gachibowli, Hyderabad 500032, Telangana, India
mashadaab.siddiqie@research.iiit.ac.in
[2] Ashoka University, Sonepat 131029, Haryana, India

Abstract. Location-based services (LBS), which provide personalized and timely information, entail privacy concerns such as unwanted leak of current user locations to potential stalkers. Existing works have proposed dummy generation techniques by creating a cloaking region (CR) such that the user's location is at a fixed distance from the center of CR. Hence, if the adversary somehow knows the location of the center of CR, the user's location would be vulnerable to attack. We propose an improved dummy generation approach for facilitating improved location privacy for mobile users. Our performance study demonstrates that our proposed approach is indeed effective in improving user location privacy.

Keywords: Location-based service · Privacy · Privacy preservation · Location privacy · Cloaking region · Dummy generation · Infeasible regions

1 Introduction

Advances in mobile communications technology coupled with the widespread use of GPS devices have resulted in the ever-increasing popularity of location-based services (LBS). Mobile users typically issue location-based queries. While such location-based queries constitute an essential class of mobile applications, the downside is that malicious entities (or even LBS servers themselves) may infer the location of any user by analyzing her location-based data [7]. This can lead to severe attacks e.g., illegal tracking of mobile users and leaking of mobile users' personal data to third parties [10]. The issue is to develop an approach to provide LBS to users while effectively reducing user location privacy violations.

Existing location-based privacy preserving techniques can be broadly categorized into three types, namely *anonymization, obfuscation* and *dummy generation*. In anonymization-based approaches, user locations are cloaked using anonymizers [4,6,11]. Anonymizers mix a given user's actual location with at least $k - 1$ other real users. Hence, the LBS provider cannot identify the actual user location with a probability of more than $1/k$. Such methods need to pool

© Springer Nature Switzerland AG 2021
C. S. Jensen et al. (Eds.): DASFAA 2021, LNCS 12683, pp. 487–495, 2021.
https://doi.org/10.1007/978-3-030-73200-4_33

user locations. They assume a trusted third-party server to mediate interactions between the users and the LBS server. However, it is often practically challenging to deploy a completely safe third-party server. Methods that use mobile peer-to-peer collaboration [4,6] suffer from the same location privacy problem since users share their location information with strangers. These methods also fail to anonymize a user's location if relatively few users are present nearby the user.

Obfuscation-based approaches [1,2,5] focus on substituting a given user's real location with a nearby intersection or landmark to obscure the user's real location. However, if the user is in a position with no appropriate intersections or landmarks in his/her vicinity, the substitute locations would be far from the user's actual location, thereby degrading the quality of LBS. Obfuscation-based approaches may also use the spatial transformation method [5], which distorts the actual user locations by adding random noise. However, as shown in [9], the amount of random noise necessary to prevent tracking attacks is enormous.

In dummy generation approaches [3,8,14,15], the user sends k-1 dummy locations along with the user's real location, thereby making k indistinguishable locations with $1/k$ probability for an adversary to find the real user. The LBS provider then provides a list of services from each location (i.e., the user's and the dummies' locations) in the query. The user can then filter out the list of services associated with the dummies' locations and choose only the information relevant to her actual location. The Circle-divided Dummy Generation (CDG) technique [15] generates dummies by considering an angle. Moreover, the Obstacle-based Dummy Generation (ODG) approach, which considers the surrounding environment, was proposed in [3]. Furthermore, the Efficient Dummy Generation (EDG) approach [14] created dummies not on the circumference of the circle, but rather on a thick strip of a circle (forming an annulus) to reduce the probability of exposing the user's location in areas with a high density of infeasible regions. A geographical region is defined as an *infeasible region* for an entity if the entity cannot possibly be physically present at that location. Examples include locations without any road infrastructure, restricted government facilities, military zones and forest areas for preserving endangered species.

Notably, *none* of the existing approaches works in regions with a higher number of infeasible regions, while keeping LBS users safe from an adversary, who somehow knows the center of the cloaking region (CR). Hence, there is an opportunity to improve the user location privacy from an adversary if we develop a mechanism to randomize the user's distance from the center of CR. In this regard, we propose the Annulus-based Gaussian Dummy Generation (AGDG) approach. AGDG randomizes the distance between the user and the center of CR by predetermining the user's placement using a virtual CR. Moreover, in case of areas with a higher number of infeasible regions, we randomize the placement of candidates based on the proposed annulus-based cloaking region with a Gaussian probability distribution. Hence, it becomes difficult to know the user's actual location even if an adversary somehow knows the center of CR. Furthermore, AGDG is more flexible with the construction of its CR w.r.t. existing approaches.

We conducted experiments to demonstrate that AGDG is indeed effective in providing improved location privacy w.r.t. existing approaches.

The paper is organized as follows. Section 2 presents our AGDG approach. Section 3 reports the performance study. Finally, we conclude in Sect. 4.

2   Proposed Approach

This section presents our proposed Annulus-based Gaussian Dummy Generation (AGDG) approach.

Basic Idea: The main idea behind our proposed AGDG approach is to randomize the distance between the center of our cloaking region and the user's location, thereby making it more difficult for an adversary to determine the actual location of the user. For this, we propose the notion of a virtual annulus-based cloaking region, where we predetermine the placement of the user as well as the dummies. Moreover, under our proposed AGDG approach, dummies are placed such that even in locations with a higher number of infeasible regions, location privacy is maintained. For this, we propose a notion of an annulus-based cloaking region with a Gaussian probability distribution for placing the candidates such that the distance between candidates is increased.

Proposed AGDG Approach: AGDG comprises the following steps: (1) Constructing the virtual cloaking region (VCR), (2) Determining the placement of the user in VCR, (3) Computing the real cloaking region, and (4) Determining the placement of dummies. Now we shall discuss each of these steps in detail.

Step 1 - Constructing the Virtual Cloaking Region (VCR): We construct a VCR to predetermine the user placement w.r.t. the VCR by randomizing the distance between the center of the VCR and the user. This makes our approach independent of the user's placement. To construct VCR, a virtual circle with center C is constructed using a user-defined cloaking area A_{min}. The radius R_{max} of the virtual circle should satisfy $\pi R_{max}^2 \geq A_{min}$. We simply choose $R_{max} = \sqrt{\frac{A_{min}}{\pi}}$. Another circle with radius R_{min} at the same center C is constructed, thereby forming an annulus (ring shape) with R_{max} and R_{min} as the outer radius and the inner radius respectively. Here, R_{min} is a user-specified constant. Since the virtual circle is used to create a VCR, we have more control over the range of R_{min}, which can range from 0 to R_{max}. To achieve k-anonymity, this annulus is divided into k equal sectors. These k sectors are denoted as $\langle S_1, S_2, \cdots, S_k \rangle$. Either the real user or one of the dummy users would later be placed in each sector. In the example in Fig. 1a, $k = 7$ and $\langle S_1, S_2, \cdots, S_7 \rangle$ represent the sectors of the annulus. Here, the angle projected by any sector is $\frac{2\pi}{7}$.

Step 2 - Determining the Placement of the User in VCR: Since the user's placement must be independent of the distance from the center of the cloaking region, a probability distribution at each point in the cloaking region is formed using a Gaussian distribution. This probability distribution ensures that the user in S_i is placed closer to ES_i, where S_i is a sector of the VCR and ES_i is

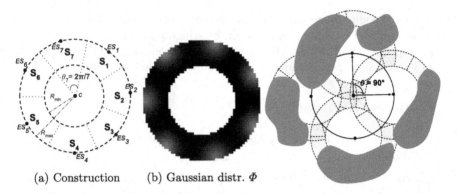

(a) Construction   (b) Gaussian distr. Φ

Fig. 1. Virtual cloaking region

Fig. 2. Multiple cloaking regions after rotation

at the edge of S_i, as depicted in Fig. 1a. This randomizes the distance between the center of the cloaking region and the user, while keeping maximum distance between any two candidates, thereby maximizing the cloaking region. Let (x, y) be a point in sector S_i. The probability distribution $\Phi_i(x,y)$ is given as follows, $\Phi_i(x,y) = \frac{1}{\sigma\sqrt{2\pi}}e^{-\frac{1}{2}\left(\frac{dist((x,y),ES_i)}{\sigma}\right)^2}$. Here, $dist((x,y), ES_i)$ is the distance between the point (x,y) and ES_i. Figure 1b shows the probability distribution Φ, where white represents a higher probability of placing the user, while black indicates a lower probability of placing the user. Then any random sector is selected, and the user's placement is determined in the virtual cloaking region using the Φ distribution. Then we superimpose our virtual cloaking region onto the real-world map with the determined users' placement coinciding with the actual geographical location of the real user on the map.

Step 3 - Computing the Final Cloaking Region: To place the annulus in the best possible region, we now rotate this annulus n times with an angle of $\theta = \frac{2\pi}{n}$ with the real user's position as the pivot (see Fig. 2). In Fig. 2, $n = 4$ and $\theta = 90°$. Here, the shaded regions are considered as infeasible regions, and n is constant. Then we deal with all infeasible regions using the equation $CR_f = (CR_{R_{max}} - CR_{R_{min}}) - CR_{ir}$ and select the final cloaking region with the least amount of infeasible regions (see Fig. 3a). CR_f is the area of the cloaking region after removing all infeasible regions. $CR_{R_{max}}$ is the area covered by the larger circle and equals πR_{max}^2. $CR_{R_{min}}$ is the area covered by the smaller circle and equals πR_{min}^2. CR_{ir} is the region occupied by infeasible regions.

After selecting the cloaking region with least infeasible regions as the final cloaking region, we readjust the sector sizes in the final cloaking region using the equation $\theta_f \times (i-1) \le S_i < \theta_f \times i$, where $\theta_f = \frac{2\pi - \theta_{ir}}{k}$. θ_{ir} is the total angle projected by the infeasible region at the center of annulus C (see Fig. 3b).

Step 4 - Placing Dummies in the Final Cloaking Region: Once the final cloaking region has been selected, we compute the appropriate placement of

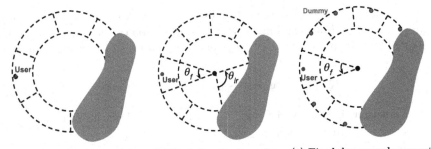

(a) Annulus with least θ_{ir} (b) Redefined sector size   (c) Final dummy placement

Fig. 3. Stages of annulus in AGDG

dummies in the final cloaking region. We normalize Φ such that sum of all probabilities of valid cells (locations with no infeasible regions) in a sector S_i equals 1. Any privacy-preserving scheme tries to maximize the similarity between the dummy and the real user's location. To achieve this, a new Gaussian distribution is required to increase the chance of dummies being placed at locations with query probability closer to that of the real user's. Query probability of a location is the probability that a user from that location has issued a query to the LBS provider in the past [12]; hence, it is based on querying history. Let (x, y) be a point in sector S_i and $P_{x,y}$ be the query probability at (x, y). Let (x_r, y_r) be the location of the real user and P_{x_r,y_r} be the query probability at (x_r, y_r). Then the probability distribution $\Psi_i(x, y)$ is given by the equation
$\Psi_i(x,y) = \frac{1}{\sigma\sqrt{2\pi}}e^{-\frac{1}{2}\left(\frac{|P_{x_r,y_r} - P_{x,y}|}{\sigma}\right)^2}$. To consider all infeasible regions and also to
ensure that the sum of the values of Ψ in a given sector equals 1, we normalize Ψ. Then we combine the two probability distributions as $\Omega = \frac{a\Psi + b\Phi}{a+b}$, where a and b are the weight coefficients for each distribution. Using probability distribution Ω, we now deploy dummies at each sector (see Fig. 3c).

Since we use virtual cloaking regions to determine the real user's placement, our approach is relatively safe from attackers with knowledge of the location of cloaking region's center. Additionally, since we use the annulus-based cloaking regions, our approach has better privacy-preserving performance even in locations with a higher number of infeasible regions. Since we try to maximize both CR and similarity between the user and the dummies' query probability, our approach can be reasonably expected to perform better than existing approaches.

3   Performance Evaluation

Our experiments consider a two-dimensional layout with 1000×1000 cells. Each cell has a dimension of 10×10 m^2. The basic CR requested by the user was assumed to be 1963.4 ($\pi \times 25 \times 25$) cells i.e., $R_{max} = 25$. Infeasible regions were randomly arranged in the layout, depending on the infeasible region ratio (IRR), which ranged from 0 to 0.9. Our performance study parameters were adopted from the existing work in [14]. Table 1 summarizes all the performance study parameters. Each experiment was conducted 100 times, hence the results presented here represent the average values over 100 runs of each experiment.

Table 1. Parameters used in our experiments

Parameter	Default	Variations	Parameter	Default	Variations
k	10	$[2, 3, \cdots, 29, 30]$	n	4	–
IRR	0.3	$[0, 0.1, \cdots, 0.8, 0.9]$	a, b	1	–
R_{min}	15	–	σ_Φ	0.001	–
R_{max}	25	–	σ_Ψ	2	–

To evaluate the performance of our approach, we use two metrics.

(a) Effective Cloaking Region (ECR): Effective Cloaking Region is a widely used metric [12] to compare the effectiveness of a privacy preservation algorithm. It is a measure of the maximum area covered by k location points ($k - 1$ dummy locations and a real user location). We have computed ECR by summing up the area of triangles formed by all the adjacent locations and the center of CR. $ECR = \sum_{i=1}^{k} Area(l_i, l_{(i+1)\%(k+1)}, C)$, where the $Area$ function returns the area of a triangle, given three vertices. Thus, ECR is equal to 0 for $k \leq 2$.

(b) Entropy (H): Entropy is widely used to measure the degree of anonymity in location-based services [13]. It indicates the uncertainty to determine the real location of an individual from all the candidates. Usually, query probability (p) of each possible location is used as accessory information to construct the entropy-based privacy metric. We thus assign each possible location a query-probability, denoted by p_i, and the sum of all probabilities p_i is 1. As a result, the entropy of identifying the real user from k candidate set can be computed by the equation $H = \sum_{i=1}^{k} -\frac{p_i}{\sum_{i=1}^{k} p_i} log_2 \left(\frac{p_i}{\sum_{i=1}^{k} p_i} \right)$. Thus, maximum entropy $H_{max} = log_2 k$ is achieved when all the k locations have the same probability of $1/k$.

We compare our proposed AGDG approach with the CDG [15], ODG [3] and EDG [14] approaches. We adapt these reference approaches with essentially the same setup as that of our approach to have a fair and meaningful comparison.

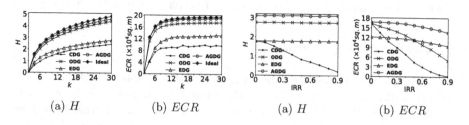

(a) H (b) ECR (a) H (b) ECR

Fig. 4. Effect of variation in k **Fig. 5.** Effect of variation in IRR

Effect of Varying the Number of Candidates: We first evaluate the relationship between k and the entropy (H) in Fig. 4a. Observe that in all the approaches, H increases with an increase in k. This is because the greater the number of dummies, the harder it will be for the adversary to find the real user. Both EDG and CDG have the worst performance because they do not consider the surrounding query probability. On the other hand, ODG and AGDG have higher H because, here candidates are placed by considering the query probability of the environment. It can be observed that AGDG has better performance as compared to that of ODG. Moreover, H in AGDG is close to the best possible value because we use Φ distribution to place dummies such that query probabilities of the locations are close to that of the real user.

Our experimental results at $IRR = 0.3$ are shown in Fig. 4b. Observe that CR increases with an increase in k because the greater the number of candidates, the more area they will cover in a given region. It can be observed that CDG has the worst performance because CDG does not consider the placement for infeasible regions. Hence, there is a chance of dummies being placed in the infeasible regions. On the other hand, AGDG has better performance than all the other approaches. This is because in AGDG, we place dummies using the Ψ distribution, thereby making candidates far apart from each other.

Effect of Varying the Ratio of Infeasible Regions: Privacy preservation schemes should ensure that dummy locations closely resemble the real user's location, even in locations with higher number of infeasible regions. We evaluate the entropy (H) with the probability of placing infeasible regions on the layout (IRR) in Fig. 5a. CDG has a drastic decrease in H as IRR increases. This is because CDG does not consider the placement for infeasible regions. All the other approaches have nearly constant H throughout the experiment since they consider feasible regions, thereby making them immune to any changes in the amount ratio of infeasible regions (IRR). EDG has less H than ODG and AGDG because EDG does not consider the query probability of its environment. On the other hand, AGDG has better performance than ODG because AGDG uses the Φ distribution to place dummies in locations such that their query probabilities are close to that of real user locations.

We evaluate the effective cloaking region with the probability of placing infeasible regions on the layout (IRR) in Fig. 5b. Observe that AGDG, ODG and CDG have similar ECR when there are no infeasible regions because here candidates are placed at the cloaking region's circumference when there are no infeasible regions. EDG has less ECR than AGDG, ODG and CDG since in EDG, dummies are placed in the annulus randomly. Observe that CDG has a drastic decrease in ECR as IRR increases. This is because CDG does not consider placement of infeasible regions. Similarly, even in ODG with an increase in IRR, ECR decreases. This is because in case of ODG, candidate placement is restricted only on the circumference of the circle. Hence, as IRR increases, candidates tend to form clusters along the circumference of the circle, thereby reducing ECR. On the other hand, both AGDG and EDG show a smaller decrease in CR as IRR increases, because here dummies are placed in annulus, unlike ODG and

CDG. Moreover, AGDG has better CR than EDG because in AGDG, we use Ψ probability distribution for the placement of candidates, whereas in EDG, candidates are placed randomly on the annulus.

4 Conclusion

The issue with LBS is that malicious entities may infer user locations by analyzing their location-based data. Incidentally, existing dummy generation and obfuscation-based approaches are not adequate to protect user location privacy for regions with a higher number of infeasible regions. Hence, we have proposed an improved dummy generation approach for facilitating improved location privacy for mobile users. Through experimental results, we have shown that our proposed approach is indeed effective in improving user location privacy. In the near future, we will extend our approach to handle more advanced adversaries.

References

1. Ardagna, C.A., Cremonini, M., di Vimercati, S.D.C., Samarati, P.: An obfuscation-based approach for protecting location privacy. IEEE Trans. Dependable Secure Comput. 8(1), 13–27 (2009)
2. Bordenabe, N.E., Chatzikokolakis, K., Palamidessi, C.: Optimal geo-indistinguishable mechanisms for location privacy. In: Conference on Computer and Communications Security, pp. 251–262. ACM (2014)
3. Cai, T.Y., Song, D.H., Youn, J.H., Lee, W.G., Kim, Y.K., Park, K.J.: Efficient dummy generation for protecting location privacy. J. Korea Inst. Inf. Electron. Commun. Technol. 9(6), 526–533 (2016)
4. Chow, C.Y., Mokbel, M.F., Liu, X.: Spatial cloaking for anonymous location-based services in mobile peer-to-peer environments. GeoInformatica 15(2), 351–380 (2011)
5. Duckham, M., Kulik, L.: A formal model of obfuscation and negotiation for location privacy. In: Gellersen, H.-W., Want, R., Schmidt, A. (eds.) Pervasive 2005. LNCS, vol. 3468, pp. 152–170. Springer, Heidelberg (2005). https://doi.org/10.1007/11428572_10
6. Ghinita, G., Kalnis, P., Skiadopoulos, S.: MobiHide: a mobilea peer-to-peer system for anonymous location-based queries. In: Papadias, D., Zhang, D., Kollios, G. (eds.) SSTD 2007. LNCS, vol. 4605, pp. 221–238. Springer, Heidelberg (2007). https://doi.org/10.1007/978-3-540-73540-3_13
7. Gruteser, M., Hoh, B.: On the anonymity of periodic location samples. In: Hutter, D., Ullmann, M. (eds.) SPC 2005. LNCS, vol. 3450, pp. 179–192. Springer, Heidelberg (2005). https://doi.org/10.1007/978-3-540-32004-3_19
8. Hara, T., Suzuki, A., Iwata, M., Arase, Y., Xie, X.: Dummy-based user location anonymization under real-world constraints. IEEE Access 4, 673–687 (2016)
9. Krumm, J.: Inference attacks on location tracks. In: LaMarca, A., Langheinrich, M., Truong, K.N. (eds.) Pervasive 2007. LNCS, vol. 4480, pp. 127–143. Springer, Heidelberg (2007). https://doi.org/10.1007/978-3-540-72037-9_8
10. Matsuo, Y., et al.: Inferring long-term user properties based on users' location history. In: International Joint Conferences on Artificial Intelligence, pp. 2159–2165 (2007)

11. Niu, B., Li, Q., Zhu, X., Cao, G., Li, H.: Achieving k-anonymity in privacy-aware location-based services. In: IEEE Conference on Computer Communications, pp. 754–762. IEEE INFOCOM (2014)
12. Niu, B., Zhang, Z., Li, X., Li, H.: Privacy-area aware dummy generation algorithms for location-based services. In: International Conference on Communications (ICC), pp. 957–962. IEEE (2014)
13. Serjantov, A., Danezis, G.: Towards an information theoretic metric for anonymity. In: Dingledine, R., Syverson, P. (eds.) PET 2002. LNCS, vol. 2482, pp. 41–53. Springer, Heidelberg (2003). https://doi.org/10.1007/3-540-36467-6_4
14. Song, D., Song, M., Shakhov, V., Park, K.: Efficient dummy generation for considering obstacles and protecting user location. Concurr. Comput.: Pract. Exp. **33**, e5146 (2019)
15. Zhao, H., Wan, J., Chen, Z.: A novel dummy-based kNN query anonymization method in mobile services. Int. J. Smart Home **10**(6), 137–154 (2016)

Industrial Papers

LinkLouvain: Link-Aware A/B Testing and Its Application on Online Marketing Campaign

Tianchi Cai(✉), Daxi Cheng(✉), Chen Liang(✉), Ziqi Liu, Lihong Gu,
Huizhi Xie, Zhiqiang Zhang, Xiaodong Zeng, and Jinjie Gu

Ant Financial Services Group, Hangzhou, China
{tianchi.ctc,daxi.cdx}@antfin.com, hi@liangchen.email

Abstract. A lot of online marketing campaigns aim to promote user interaction. The average treatment effect (ATE) of campaign strategies need to be monitored throughout the campaign. A/B testing is usually conducted for such needs, whereas the existence of user interaction can introduce interference to normal A/B testing. With the help of link prediction, we design a network A/B testing method LinkLouvain to minimize graph interference and it gives an accurate and sound estimate of the campaign's ATE. In this paper, we analyze the network A/B testing problem under a real-world online marketing campaign, describe our proposed LinkLouvain method, and evaluate it on real-world data. Our method achieves significant performance compared with others and is deployed in the online marketing campaign.

Keywords: Graph neural networks · Graph partitioning · Graph clustering · Network A/B testing

1 Introduction

Recently, Alipay launched an online marketing campaign that encourages users to invite others to join the campaign, so they can all receive discounts or cash rewards. Such user interaction-promoting services (IPS) are common to increase user engagement. Various strategies are developed for this campaign, and designing an A/B testing solution to quantify their average treatment effects is crucial. However, normal A/B testing solutions for IPS are improper because edges (user invitations) exist between different test groups and introduce bias to ATE; A/B testing addressing such interference is called network A/B testing. Under a thorough analysis of real-world graphs, we develop a graph clustering method LinkLouvain for network A/B testing and deploy it in the online marketing campaign. LinkLouvain has the following strengths:

1. Scalability. It conducts on graphs of billions of nodes and tens of billions of edges in 10 h.

T. Cai, D. Cheng and C. Liang— These authors contributed equally.

© Springer Nature Switzerland AG 2021
C. S. Jensen et al. (Eds.): DASFAA 2021, LNCS 12683, pp. 499–510, 2021.
https://doi.org/10.1007/978-3-030-73200-4_34

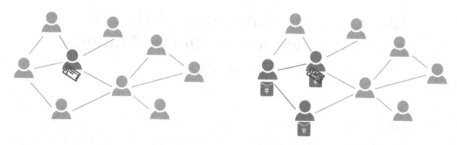

Fig. 1. Visualization of our online marketing campaign. Coupons are handed out to users (colored red in left figure), and users can invite their friends to join this marketing campaign, and they all receive a cash coupon (colored red in right figure). (Color figure online)

2. Simplicity. It is a static method that runs only once before the online marketing campaign. There is no need for additional streaming support (Fig. 1).
3. Effectiveness. It reduces network interference and reduces the heterogeneity of test groups throughout the campaign lifecycle (7 days). We develop two metrics *estimator bias* and *estimator variance* to measure the network interference and heterogeneity, respectively. Results show LinkLouvain outperforms others.

1.1   Interaction-Promoting Services (IPS)

For consumer-facing online products, encouraging user interactions is a common practice to increase user engagement. Some examples are 'People You May Know' on Facebook, 'Connections You May Know' on LinkedIn, and online marketing campaigns where coupons can be shared with others on Alipay. Such services, referred to as interaction-promoting services (IPS), are designed to encourage user interaction, and therefore benefit user engagement of the product.

All users and their interactions on the Alipay platform construct a real-world thorough social network with billions of nodes and tens of billions of edges. Users (nodes) and user invitations (edges) in our online marketing campaign form a subgraph of this thorough social network. Engaging nodes and edges increase throughout the campaign, and this time-evolving graph is always a subgraph of the social network.

In our paper, we analyze the growth properties of the network and the interference patterns from the campaign for a sound understanding of the following network A/B testing problem.

1.2   Network A/B Testing

For IPS, users who do not receive new services may still be affected through interactions with those who do receive new services. It introduces interference for user-level A/B testing; thus, a direct estimation of the ATE is no longer unbiased. Network A/B testing solutions are of great interest (Fig. 2).

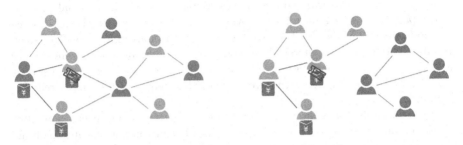

Fig. 2. Visualization of estimation bias in different A/B testing scheme. *Left*: In user-level randomization, users are randomly selected for the treatment (colored yellow) and the control group (colored cyan). However, the online marketing campaign in the treatment group may affect users in control group. In this case, both the treatment group and the control group has the same number of invited users, and the A/B testing misleadingly concludes that the treatment does not make any difference. *Right* Network A/B testing clusters users with interference together, and the cluster-level metrics show that the treatment group has more invited users. (Color figure online)

There are mainly two approaches to conduct an unbiased estimation of ATE under network interference. The first is afterward correction. For example, [4] assumes the interference is linear-additive, estimates the exposure probability, and weighs the estimation accordingly. The performance of this type of approach relies on making the right decision for the form of interference. We analyze the interference in a real social network, and in our case, however, the linear-additive model is over-simplified and a panacea solution is missing.

The other approach is to perform randomization at the cluster-level. That is, clusters of users, instead of users themselves, are used as randomization units. This approach assumes no/low interference between clusters. Our method falls into this category.

1.3   Graph Clustering

Many clustering algorithms have been studied to reduce the interference between their resulting clusters. [11] proposes a clustering algorithm r-net. Label propagation and modularity maximization algorithms are also studied in [4], and it suggests modularity maximization outperforms the other. However, these approaches usually assume their graphs are restricted-growth graphs (formally defined later) to perform better, which is hard to meet through our analysis of real social networks. Later, we'll introduce our LinkLouvain approach built on Louvain[1].

---

[1] A fast and parallel approximation for modularity maximization.

We also consider graph partitioning methods to generate balanced test groups with minimal edges between groups. Dynamic graphs at scale impose great challenges for graph partitioning. Most existing algorithms can not scale to billions of nodes. Graph theory based algorithms aiming to solve the optimal min-cut graph partitioning task have been proven NP-hard. Classical graph partitioning methods such as Metis [5] also have high computational complexity. To handle rapidly evolving graphs, classical methods are not favorable for efficiency issues and dynamic graph partitioning algorithms [7,8] are proposed by constantly updating labels and graph structure changes that require additional streaming support.

Our method also focuses on rapidly evolving graphs, however, in a more static manner. Unlike other static methods [9,10] that run periodically to obtain continuous partitioning results, we make a guess on the graph structure in the future (e.g. in a week) and partition the predicted graph for only once. In the beginning, we obtain an 'omniscient' view of all users and all possible interactions between them. Also, we have a campaign graph in the early stage campaign. Then we predict possible edges with graph neural networks (GNNs) to gain knowledge of a future snapshot of the campaign graph. The current snapshot and future snapshot are formed by invitations in the campaign, while the omniscient graph is irrelevant to specific applications. The predicted edges form a guess of the future snapshot, and it's then clustered by efficient graph clustering methods with linearithmic time such as Louvain. Finally, the clusters are randomly merged to p desired test groups for A/B testing.

2   Preliminaries

2.1   Problem Formulation

In this paper, we are interested in the task of network A/B testing. More specifically, we aim at estimating a precise and sound ATE. Estimating ATE when launching or updating IPS, however, is non-trivial. In the absence of interactions, user-level A/B testing is commonly used to estimate potential effects [6]. The estimation is unbiased if the *Stable Unit Treatment Value Assumption* (SUTVA) holds. This assumption requires the response of a unit (in this case, a user) to be invariant to treatments assigned to other units [1]. With this assumption, the *average treatment effect* (ATE) of a new service can be defined as

$$ATE = \frac{1}{N} \sum_i y_1(i) - y_0(i),$$

where $y_0(i)$ is the outcome for user i if not treated and $y_1(i)$ is the outcome for user i if treated. N represents the number of users.

However, the ground-truth ATE in real-world network A/B testing is impossible to obtain. Our work designs an estimator of ATE in the presence of network interference by splitting graph to clusters. The estimator is formulated as

$$\hat{ATE} = \frac{1}{M} \sum_i \sum_j y_1(q_j^i) - \frac{1}{N} \sum_i \sum_j y_0(c_j^i),$$

where q_j^i is i-th user in j-th cluster of the treatment group Q and c_j^i is i-th user in j-th cluster of the control group C. M and N represent numbers of users in Q and C, respectively.

Our goal is to design an ATE estimator that minimizes the estimation bias and variance. Therefore, the estimated ATE can guide business decisions.

2.2   Two Graphs of Interest

In our online marketing campaign, we have access to two graphs: a stable social graph $G = (V, E)$ and a time-evolving label graph $L = (V_L, E_L)$. We collect the social graph G containing all users of Alipay as nodes V and their historical interactions as edges E. It contains billions of nodes and tens of billions of edges and lays the foundation for predicting users' future interactions.

Additionally, as the new online marketing campaign goes on, we collect a label graph L, where users who participate in the online marketing campaign form node set V_L and user invitations form edge set E_L. L^0 and L^T represent the label graph in its early stage and by the end of the campaign of lifecycle T, respectively. It is called a label graph since the interaction data provides a strong hint for the form of interference. Previously, labeled data is less discussed because users already participated cannot join a new round of A/B testing. The novelty of LinkLouvain is that it uses link prediction to generalize the form of interference from this label graph to all users in the social graph G, and predicts an "estimator bias" for all edges V.

Various properties of the two collected graphs are analyzed in Sect. 4.1.

3   The Proposed Framework

To cluster a rapidly evolving graph, we train a GNN based link prediction model to predict possible edges in the evolving graph. Then we apply a traditional graph clustering algorithm such as Louvain to split the graph into small clusters. To use these clusters in A/B testing, we randomly combine them into desired p test groups. The procedure is shown in Fig. 3. *Label* comes from the edges (positive labels) in the current campaign graph and non-edges (negative labels) that exist only in the social graph G.

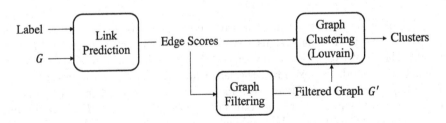

Fig. 3. Processing pipeline of the proposed LinkLouvain framework.

3.1 GNN Based Link Prediction Models

GNNs are a set of deep learning architectures that aggregate information from nodes' neighbors using neural networks. Deeper layers aggregate more distant neighbors, and the kth layer embedding of node v is

$$\mathbf{h}_v^k = \sigma(\mathbf{W}_k \cdot \mathrm{AGG}(\mathbf{h}_u^{k-1}, \forall u \in \mathcal{N}(v) \cup \{v\}))$$

where the initial embedding $\mathbf{h}_v^0 = \mathbf{x}_v$ is its node feature vector, σ is a non-linear function, and AGG is an aggregation function that differs in GNN architectures.

Figure 4 shows a naive GNN based link prediction algorithm with a twin-tower architecture. Each target node of an edge aggregates its own neighbors for K times. After aggregation of K-hop neighbors, the final embeddings $\mathbf{h}_A^K$ and $\mathbf{h}_B^K$ of two target nodes A and B are concatenated and fed to the final dense layer.

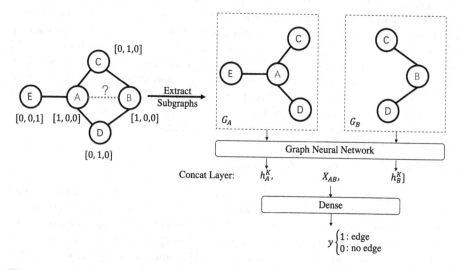

Fig. 4. Model architecture for link prediction. G: 1-hop neighborhood of a node; X: edge features; h: GNN embeddings; one-hot vectors: node labeling.

Moreover, we add structural features called node labeling [12] to naive link prediction. Node labeling assigns a one-hot vector to each node in the K-hop neighborhood of two target nodes A and B. It marks nodes' different roles in this neighborhood. For example, the left graph in Fig. 4 has 5 nodes in A and B's 1-hop neighborhood. There are three roles in the neighborhood: A and B are target nodes; C and D are nodes connecting both target nodes; E is a node that only connects to one target. The node labeling vector is appended to each node's original feature vector and tells GNN its relative location around the edge to be predicted. It helps GNN to have more accurate predictions on link existence.

Comparison of Link Prediction Models in Online Marketing Campaign. In the early stage of the online marketing campaign, we collect and sample user interactions as positive training samples and non-invitation relations as negative training samples. All training samples exist in our social graph G. There are 1.5 million positive edges and 1.5 million negative edges. Each edge has 128 features representing user interaction history. We compare the following models for the link prediction task:

- DNN: a dense neural network of five layers with layer size $[512, 256, 128, 64, 16]$.
- NG-LP: a naive GNN link prediction method with 2-hop neighbors ($K = 2$) and embedding size 64.
- NL-LP: a node labeling link prediction method with 2-hop neighbors ($K = 2$) and embedding size 64.

The main results are summarized in Table 1. F1[2], KS[3], and AUC [4] are widely used binary classification metrics. NL-LP performs the best by taking structural information into account.

Table 1. Link prediction task comparison.

	DNN	NG-LP	NL-LP
F1	0.88	0.89	**0.91**
KS	0.74	0.79	**0.84**
AUC	0.91	0.92	**0.96**

3.2   Graph Filtering

The output scores on edges represent possibilities of future online interactions. We filter out less possible edges and set the prediction score as edge weight. Graph filtering is crucial for a billion-node graph and the reasons are two-fold:

- Computation resources are limited for graphs of such size.
- Clustering algorithms like Louvain tend to generate unbalanced clusters when handling densely connected graphs. They undermine A/B testing performance heavily. Removing unnecessary edges help prevent long tails of resulting clusters.

However, if we set the threshold (γ) to abandon or keep an edge too high, we could drop too many possible edges. This introduces great bias on ATE estimates. We choose γ considering the trade-off between efficiency and effectiveness.

In the online marketing campaign, we set the threshold to be 0.5, and clustering the remaining graph costs 0.6 h.

---

[2] https://en.wikipedia.org/wiki/F1_score.
[3] https://en.wikipedia.org/wiki/Kolmogorov-Smirnov_test.
[4] https://en.wikipedia.org/wiki/Receiver_operating_characteristic.

3.3   Graph Clustering

To generate clusters of users as randomization units, we use Louvain to cluster the filtered graph G'. Clustering algorithms are well-discussed. In [4], researchers investigate several distributed clustering algorithms, such as label propagation and Louvain. Their result shows that Louvain performs better in preserving more intra-cluster edges and reducing network interference. Experiment results in the next section (Table 2) also support this conclusion.

The resulting clusters are finally randomly merged into partitions of the desired size p. These are the p test groups in A/B testing.

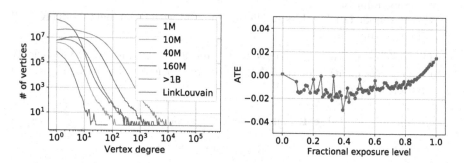

Fig. 5. *Left*: The vertex degree distribution of our real social network G at different growth levels, as well as G after graph filtering by LinkLouvain. *Right*: Since the treatment effect of a user depends on his/her neighborhood's treatment status, there exists interference. Moreover, this influence is non-linear to the fractional exposure level, and cannot be corrected afterward easily.

4   Application on Online Marketing Campaign

4.1   Patterns of Our Real-World Graphs

Though G is a large social network that does not change frequently, the size of L grows quickly as users joining our campaign. Therefore we can analyze the growth property of our social network. As the number of nodes in L reaches 1, 10, 40, 160 millions, we construct a subgraph of G with all nodes in L, and keep all edges between these nodes. Hence we can analyze the growth property of our social network retrospectively. We compare the graph properties of these four subgraphs of G, as well as the full graph G, which contains more than 1 billion nodes.

Maximum Degree Growth Is Unbounded. In Fig. 5 *Left*, we compare the degree distribution of G at different growth levels. We find that as the network grows larger, customers build more connections with each other, and the degree distribution shifts right. The long right tails of all five series suggest that the degree of this social network has a right-skewed distribution regardless of the network size. Moreover, diverged from bounded maximum degree assumptions [11], the maximum degree grows almost linearly to the number of nodes in the graph, and thus, unbounded.

In Fig. 5 *Left*, We also plot the degree distribution of the full graph G after graph filtering by LinkLouvain. It is clear that the degree distribution is less skewed compared to the original distribution of the full social graph G (labeled ">1B"). The intuition behind is that not all edges in G have the same influence on our online marketing campaign. Therefore we can eliminate many edges that are not likely to have interactions with LinkLouvain and hence reduce interference in our cluster-level randomization scheme.

Network Interference. We examine network interference patterns on our social network by estimating the ATE on different *neighborhood fractional exposure level* [4] (share of neighbors that are in the treatment group) to see if there is any pattern. We divide users into subgroups according to their different fractional exposure levels and plot estimate ATEs with respect to each group as a curve in Fig. 5 *Right*. We can draw two main conclusions. First, the treatment effect of a user depends on his/her neighborhood's treatment status, which means that the interference exists. Second, the interference does not follow a linear-additive pattern; in other words, the ATE is not linear with the fractional exposure level.

This explains the difficulty of using the afterward correction approach: there is no universal assumption for the form of interference suitable for all cases. The true form of interference might be complicated, and the linear-additive assumption might be over-simplified.

4.2   Metrics

Lower estimator bias and variance indicate more accurate and sound estimations. Here we introduce how to measure them.

Estimator Bias is measured by the degree of network interference between test groups. Clusters of users are randomly merged to p desired A/B test groups. Edges of graph L^T exist between test groups, and their interference is denoted as $I = \frac{|E^-|}{|E_L^T|}$, where E_L^T is the set of all edges (invitations in the campaign) in graph L^T, and E^- is the set of edges in graph L^T connecting nodes across test groups.

Estimator Variance represents the statistical power of designed estimators. To get higher statistical power, our estimator should generate clusters where the ATE metric of the clusters has a lower variance, which means online experiments are more sensitive. We use the following formula from [2] to calculate the variance of clusters in an estimator,

$$\text{Var}(\overline{Y}) \approx \frac{1}{K\mu_N^2} \left(\sigma_S^2 - 2\frac{\mu_S}{\mu_N}\sigma_{SN} + \frac{\mu_S^2}{\mu_N^2}\sigma_N^2 \right),$$

where $\overline{Y}$ is the total estimated conversion rate in this A/B testing group. K is the number of clusters in the group. S and N are the random variables of the sum of individual conversion and individual number respectively. μ and σ calculate the mean and variance/covariance of the corresponding random variables. We evaluate the metric variance with the same group size (1% of the total traffic).

4.3   Methods for Comparison

The methods in our comparative evaluation are as follows.

- Geo: the classical strategy to cluster users by their geographic locations.
- LinkLouvain: our proposed method with graph filtering threshold γ.
- Louvain: an ablation study that removes the link prediction stage and the graph filtering stage.
- HRLouvain: an ablation study that replaces the link prediction stage and the graph filtering stage by removing hotspots (nodes with more than θ neighbors).
- LinkLouvain-UW: an ablation study that replaces link prediction edge weight by 1 in our proposed method.
- LinkLabel: an ablation study with Louvain replaced by label propagation for graph clustering.

4.4   Evaluation

Table 2 summarizes the evaluation results of all the methods on our campaign. Metrics include estimator bias and estimator variance described in Sect. 4.2 as well as computation time. The number of clusters is also summarized for reference.

In general, LinkLouvain shows effectiveness in delivering precise and sound estimates and efficiency to run within 6 h.

Consistency. We compare three sets of threshold γ (0.2, 0.3, and 0.5) for LinkLouvain, and their key metrics are consistent. It leads to an easier tuning process during experiments.

Computational Performance. We run clustering methods with 40 workers on GRAPE [3]. Table 2 summarizes the computation time, and LinkLouvain with $\gamma = 0.5$ and HRLouvain with $\theta = 40$ are the most efficient in graph based methods.

Table 2. Evaluation summary. (Louvain runs for more than one day and drains computational resources. Its results are not available.)

Methods	# of clusters	I (%)	$Var(\bar{Y})$ (10^{-8})	Time (h)
Geo	346	**52**	47890	0.2
LinkLouvain, $\gamma = 0.2$	206M	**50**	1.17	12.7
LinkLouvain, $\gamma = 0.3$	248M	**49**	1.15	9.8
LinkLouvain, $\gamma = 0.5$	359M	**52**	**1.11**	**5.6**
Louvain	–	–	–	>24
HRLouvain, $\theta = 40$	367M	82	**1.10**	**5.4**
HRLouvain, $\theta = 100$	145M	67	32.45	12.1
HRLouvain, $\theta = 200$	98M	66	232.62	12.6
LinkLouvain-UW	442M	64	**1.07**	6.1
LinkLabel	351M	67	1.37	10.2

Comparison with Geo-Based Methods. A popular way to run A/B testing in online services is to use geographic regions as randomization units. It serves as a practical baseline for comparison. It is easy to use since it only requires locations of user queries. We compare our method with geo-based partitioning, and the results show we achieve much lower variance compared to this popular approach.

4.5   Online Results

The online campaign run for 7 days and our LinkLouvain ($\gamma = 0.5$) method was deployed to give estimates of ATE of different campaign strategies such as giving discount coupons or cash coupons. The ATE is the average payment made by users who receive coupons, and the ATE estimate of the best strategy is 1.05 times better than baseline (giving everyone a small amount of cash) without increasing the campaign budget. The A/B test was run on 2% users in the campaign, and after monitoring strategies for a day, the best coupon-distributing strategy was applied to 100% users. The performance of the campaign exceeds expectations with the help of LinkLouvain.

5   Conclusion

In this paper, we discuss network A/B testing motivated by interaction-promoting services. We analyze this problem in a real social graph and our label graph and develop LinkLouvain to address network A/B testing. The proposed approach is computationally efficient and achieves the preferable balance between estimator bias and estimator variance with the help of link prediction. It is deployed on a real marketing campaign and gives accurate and sound estimates of ATEs.

References

1. Cox, D.R., Cox, D.R.: Planning of Experiments, vol. 20. Wiley, New York (1958)
2. Deng, A., Knoblich, U., Lu, J.: Applying the delta method in metric analytics: a practical guide with novel ideas. In: Proceedings of the 24th ACM SIGKDD International Conference on Knowledge Discovery & Data Mining, pp. 233–242 (2018)
3. Fan, W., et al.: Parallelizing sequential graph computations. ACM Trans. Database Syst. (TODS) **43**(4), 1–39 (2018)
4. Gui, H., Xu, Y., Bhasin, A., Han, J.: Network A/B testing: from sampling to estimation. In: Proceedings of the 24th International Conference on World Wide Web, pp. 399–409 (2015)
5. Karypis, G., Kumar, V.: A fast and high quality multilevel scheme for partitioning irregular graphs. SIAM J. Sci. Comput. **20**(1), 359–392 (1998)
6. Kohavi, R., Longbotham, R., Sommerfield, D., Henne, R.M.: Controlled experiments on the web: survey and practical guide. Data Min. Knowl. Disc. **18**(1), 140–181 (2009)
7. Li, H., Yuan, H., Huang, J., Cui, J., Yoo, J.: Dynamic graph repartitioning: from single vertex to vertex group. In: Nah, Y., Cui, B., Lee, S.-W., Yu, J.X., Moon, Y.-S., Whang, S.E. (eds.) DASFAA 2020. LNCS, vol. 12113, pp. 482–497. Springer, Cham (2020). https://doi.org/10.1007/978-3-030-59416-9_29
8. Nicoara, D., Kamali, S., Daudjee, K., Chen, L.: Hermes: dynamic partitioning for distributed social network graph databases. In: EDBT, pp. 25–36 (2015)
9. Stanton, I., Kliot, G.: Streaming graph partitioning for large distributed graphs. In: Proceedings of the 18th ACM SIGKDD International Conference on Knowledge Discovery and Data Mining, pp. 1222–1230 (2012)
10. Tsourakakis, C., Gkantsidis, C., Radunovic, B., Vojnovic, M.: FENNEL: streaming graph partitioning for massive scale graphs. In: Proceedings of the 7th ACM International Conference on Web Search and Data Mining, pp. 333–342 (2014)
11. Ugander, J., Karrer, B., Backstrom, L., Kleinberg, J.: Graph cluster randomization: network exposure to multiple universes. In: Proceedings of the 19th ACM SIGKDD International Conference on Knowledge Discovery and Data Mining, pp. 329–337 (2013)
12. Zhang, M., Chen, Y.: Link prediction based on graph neural networks. In: Advances in Neural Information Processing Systems, pp. 5165–5175 (2018)

An Enhanced Convolutional Inference Model with Distillation for Retrieval-Based QA

Shuangyong Song[✉], Chao Wang, Xiao Pu, Zehui Wang, and Huan Chen

Alibaba Groups, Hangzhou 311121, China
{shuangyong.ssy,chaowang.wc,puxiao.px,zehui.wzh,
shiwan.ch}@alibaba-inc.com

Abstract. A common solution of automatic question-answering (QA) systems is retrieving the most similar question for a given user query from a QA knowledge base. Even though some models have got promising performance on this task, it may be hard for them to achieve a balance between accuracy and efficiency. In this paper, we propose an enhanced convolutional inference model with StructBert distillation, called StructBert-ECIM, to achieve such balance.

Keywords: Question answering · StructBert · Text matching

1 Introduction

Two kinds of techniques are commonly used in most QA systems: Information Retrieval (IR)-based ones and generation-based ones. We focus on building up an IR-based QA system, and question-knowledge matching is the most important technique. The enhanced sequential inference model (ESIM) [1] was proposed to measure the relationship between a pair of sentences by sequential encoding and attention-based alignment. Considering the good performance and decomposable implementation of ESIM, we refer to it and propose a modified version.

The emergence of transfer learning with large-scale language models (LM), such as Bert [2] (we call it GoogleBert, for differing from other per-train Bert model), has led to dramatic performance improvements across a broad range of tasks. However, GoogleBert does not make the most of underlying language structures. StructBert [4] was proposed which incorporates language structures into GoogleBert pre-training, getting better generalizability and adaptability.

The size and memory footprint of these large LMs make it difficult for them to be deployed in many scenarios. Recent research points to knowledge distillation as a potential solution, showing that when training data for a given task is abundant, it is possible to distill a large (teacher) LM into a small task-specific (student) model. In this paper, we propose an enhanced convolutional inference model with StructBert distillation, called StructBert-ECIM. We take the advantage of ESIM, and speed it up by integrating CNN-based component. In the

© Springer Nature Switzerland AG 2021
C. S. Jensen et al. (Eds.): DASFAA 2021, LNCS 12683, pp. 511–515, 2021.
https://doi.org/10.1007/978-3-030-73200-4_35

following of this paper, we will show the problem formalization and introduce
our proposed model, along with the experimental results.

2   The Proposed Model

The architecture of the proposed StructBert-ECIM is given in Fig. 1:

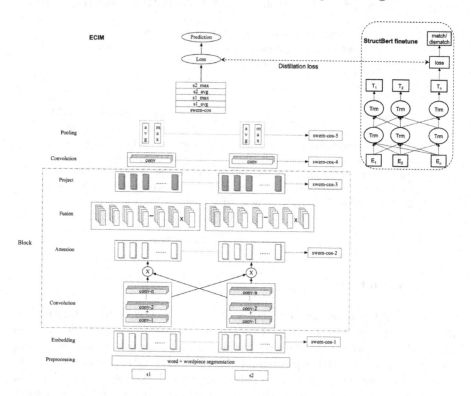

Fig. 1. Architecture of StructBert-ECIM.

1. **StructBert:** We utilize StructBert as the pre-training model. Dialog data
   from chatbots is a specific user generated content (UGC), riddled with typos
   and solecisms. StructBert can well fit UGC by leveraging the structural information.
2. **Enhanced Convolutional Inference Model:** ESIM consists of two information collection steps: one uses the sequence model to collect the context
   information of words, and the other uses the tree model to collect the clause
   information. With respect to the simple structure of ESIM, it can be used in
   online systems. However, the sequence model in ESIM is time-consuming, so
   we design an enhanced convolutional inference model (ECIM), a revision of
   ESIM, to improve the model speed by replacing the sequence model with a

convolution model. In Fig. 1, we can see the graphical presentation of a block, which consists of a convolution level, an attention level, a fusion level and a project level. These blocks are connected by an augmented version of residual connections.

3. **Distillation:** Fig. 2: 1) In **SepFinetune**, we first fine-tune the preliminary pre-trained StructBert by feeding the given in-domain labeled data, then get the predicted results for prepared unlabeled data. Afterwards we feed such an amount of predicted results to train ECIM, by treating the predicted value from structBert as soft labels. Finally we fine-tune the trained ECIM model with the help of the original labeled data, which aims at adapting the annotation domain better. 2) We propose the **JointFinetune**, in which we fine-tune the StructBert by feeding all public resources we have, and let the all-in-one fine-tuned structBert predict the unlabeled data from such different domains, and then collect all predicted data as training set for ECIM. Finally, we use the in-domain annotation set to fine-tune ECIM, which is identical to the SepFinetune process.

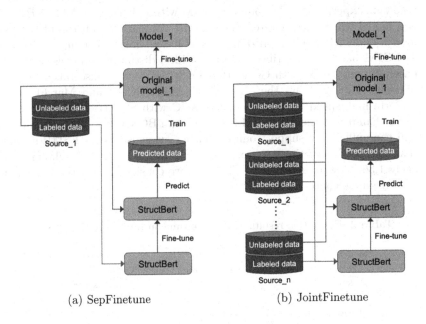

(a) SepFinetune   (b) JointFinetune

Fig. 2. Two pipelines for distillation mechanisms.

3 Experiments

1. **Data Description:** 1) **Gov:** we build chatbots for government affairs, such as medical insurance, tax and ID card. 2) **Timi:** a platform of chatbot solution

Table 1. Comparison of two distillation mechanisms.

Models	AUC			
	Gov	Timi	Youku	Quora
SepFinetune	0.753	0.854	0.832	0.864
JointFinetune	**0.768**	**0.864**	**0.841**	**0.889**

for merchants of *taobao* and *tmall*. 3) **Youku:** one of China's top online video platforms, which we build a chatbot for. 4) **Quora:** a dataset for the NLI task with two classes indicating whether one question is a paraphrase of the other, which contains about 400k question pairs. (https://www.kaggle.com/c/quora-question-pairs/data).

2. **Experimental Results:** AUC is employed as the evaluation metric, and Table 1 shows the comparison between two pipelines of distillation mechanisms. JointFinetune outperforms the other one on each subtask as LM can learn more semantic information. Therefore, we adopt JointFinetune in follow-up experiments. We compare ECIM with following models: 1) **BCNN** is a model which incorporates element-wise comparisons on top of the base model [5]; 2) **MatchPyramid** is a model based on sentence interactions [3]; 3) **ESIM** has been described before. Besides, distillation models of those 3 baselines and ECIM with GoogleBert or StructBert are also taken for comparisons. Table 2 shows the evaluation results. BCNN and MatchPyramid perform high speed, but their average AUC with StructBert distillation are lower than 0.8. Although ECIM is slower than BCNN and MatchPyramid, it still can sufficiently support online chatbot systems. As for ESIM, it shows that the speed is 27 times slower than ECIM, and the AUC results are averagely 1.36% lower than ECIM. Therefore we consider ECIM the best choice for our online chatbots.

Table 2. Evaluation results. (Inference time on Intel Core i7 CPUs.)

Models	AUC				Speed
	Gov	Timi	Youku	Quora	
BCNN	0.689	0.744	0.714	0.768	≈0.57ms
GoogleBert-BCNN	0.725	0.771	0.749	0.797	
StructBert-BCNN	0.744	0.800	0.782	0.833	
MatchPyramid	0.694	0.750	0.720	0.775	≈0.41ms
GoogleBert-MatchPyramid	0.727	0.778	0.765	0.805	
StructBert-MatchPyramid	0.757	0.803	0.799	0.839	
ESIM	0.696	0.788	0.779	0.825	≈221ms
GoogleBert-ESIM	0.728	0.829	0.790	0.840	
StructBert-ESIM	0.750	0.845	0.823	0.863	
ECIM	0.715	0.789	0.780	0.836	≈8.1 ms
GoogleBert-ECIM	0.739	0.831	0.806	0.861	
StructBert-ECIM	**0.768**	**0.864**	**0.841**	**0.889**	

4   Conclusion

In this paper, we presented the StructBert-ECIM model and demonstrated its effectiveness and efficiency on the QA task. Considering its high efficiency and strong performance, the model is quite suitable for a wide range of applications.

References

1. Chen, Q., Zhu, X., Ling, Z., Wei, S., Jiang, H., Inkpen, D.: Enhanced LSTM for natural language inference. In: ACL, no. 1, pp. 1657–1668 (2017)
2. Devlin, J., Chang, M.W., Lee, K., Toutanova, K.: BERT: pre-training of deep bidirectional transformers for language understanding. In: NAACL, pp. 4171–4186 (2019)
3. Pang, L., Lan, Y., Guo, J., Xu, J., Wan, S., Cheng, X.: Text matching as image recognition. In: AAAI, pp. 2793–2799 (2016)
4. Wang, W., et al.: StructBERT: incorporating language structures into pre-training for deep language understanding. In: ICLR 2020 (2020)
5. Yin, W., Schütze, H., Xiang, B., Zhou, B.: ABCNN: attention-based convolutional neural network for modeling sentence pairs. TACL 4, 259–272 (2016)

Familia: A Configurable Topic Modeling Framework for Industrial Text Engineering

Di Jiang[1], Yuanfeng Song[2](✉), Rongzhong Lian[1], Siqi Bao[1], Jinhua Peng[1], Huang He[1], Hua Wu[1], Chen Zhang[2], and Lei Chen[2]

[1] Baidu Inc., Beijing, China
{jiangdi,lianrongzhong,baosiqi,pengjinhua,hehuang,wu_hua}@baidu.com
[2] The Hong Kong University of Science and Technology, Hong Kong, China
{songyf,czhangad,leichen}@cse.ust.hk

Abstract. In this paper, we propose a configurable topic modeling framework named Familia. Familia supports an important line of topic models that are widely applicable in text engineering scenarios. In order to relieve burdens of software engineers without knowledge of Bayesian networks, Familia is able to conduct automatic parameter inference for a variety of topic models. Simply through changing the data organization of Familia, software engineers are able to easily explore a broad spectrum of existing topic models or even design their own topic models, and find the one that best suits the problem at hand. With its superior extendability, Familia has a novel sampling mechanism that strikes balance between effectiveness and efficiency of parameter inference. Furthermore, Familia is essentially a big topic modeling framework that supports parallel parameter inference and distributed parameter storage. The utilities and necessity of Familia are demonstrated in real-life industrial applications. Familia would significantly enlarge software engineers' arsenal of topic models and pave the way for utilizing highly customized topic models in real-life problems. Source code of Familia have been released at Github via https://github.com/baidu/Familia/.

Keywords: Topic modeling · Familia · Text engineering

1 Introduction

Topic models have become one kind of important tools for text engineering. In the last decade, a wide spectrum of topic models has been proposed in academia

---

The first two authors contributed equally and share the "co-first author" status. This work was partially conducted while Yuanfeng Song was with Baidu. This work is supported in part by the National Natural Science Foundation of China under Grant No. 61802230, Fostering Project of Dominant Discipline and Talent Team of Shandong Province Higher Education Institutions.

C. S. Jensen et al. (Eds.): DASFAA 2021, LNCS 12683, pp. 516–528, 2021.
https://doi.org/10.1007/978-3-030-73200-4_36

and demonstrates promising performance. However, for industrial topic modeling, Probabilistic Latent Semantic Analysis (PLSA) [7] and Latent Dirichlet Allocation (LDA) [1] are the working horses so far [2,16]. With the richness of the other topic models, we rarely witness employment of them in industrial applications. The huge gap between the abundance of topic models proposed in academia and their rare appearance in industry is mainly caused by the following reasons:

1. Most of existing topic models do not have industrial implementation for convenient usage. Implementing these topic models from scratch is both time-consuming and error-prone;
2. Although many tasks cannot be suitably supported by existing topic models, designing a proper topic model and the corresponding parameter inference algorithms are daunting for engineers;
3. Most advanced techniques for efficient parameter inference are exclusively designed for PLSA/LDA. Lacking highly efficient parameter inference algorithm impedes most topic models' applicability in industry.

Due to the above reasons, engineers' topic modeling choice is usually limited to PLSA or LDA, which, however, may not fit well their task at hand. Such improper practice heavily undermines the effectiveness of topic models in real-life applications.

In this paper, we propose a novel topic modeling framework, Familia, which is easily configurable and can be utilized as off-the-shelf tool for software engineers without much knowledge of Bayesian networks. Familia supports a broad line of topic models, which are of significant presence in the literature as well as heavily demanded in industry. Software engineers can investigate many topic models for their tasks at hand by simply changing the training data organization. Moreover, Familia takes over all the burdens of parameter inference, parallel computing and post-modeling utilities. Specifically, Familia contains two parameter inference methods: Gibbs sampling (GS) and Metropolis Hastings (MH) with alias table [10], based on which a hybrid sampling mechanism is designed to strike a balance between effectiveness and efficiency. To meet the requirements of topic modeling for massive data, Familia is inherently built upon the Parameter Server (PS) architecture [4,11,21]. Multiple computing nodes can be harnessed for parallel parameter inference and distributed storage. Furthermore, Familia contains multiple built-in post-modeling utilities such as dimensionality reduction and semantic matching, which can be readily applied in many downstream applications. The contributions of this paper are summarized as follows:

1. We opensource a novel framework, named Familia, to bridges the huge gap between topic model research in academia and industrial practice of topic modeling. To the best of our knowledge, it is the first framework that supports multiple topic models in a user-friendly manner.
2. We systematically investigate the performance of GS and MH with different topic models. We further propose a hybrid sampling mechanism to balance the effectiveness and efficiency of parameter inference. Based on the earned

insights, we provide practical suggestions on choosing sampling method for different topic models.

The rest of this paper is organized as follows. In Sect. 2, we review the related work. In Sect. 3, we discuss the mathematical foundations of Familia. In Sect. 4, we detail the parameter inference algorithms. In Sect. 5, we present the experimental results. Finally, we conclude this paper in Sect. 6.

2 Related Work

The present work is related to a wide range of topic models in the literature and the recent advances of sampling-based parameter inference algorithms.

2.1 Topic Models

LDA [1] plays an important role in the field of topic modeling. In the last decade, many extensions of LDA have been proposed to meet specific needs of many applications. For example, Topic-over-Time (TOT) [20] is presented to capture latent topics and their changes over time. Supervised LDA [13] captures the regularity of labelled documents by introducing response signals. GeoFolk [17] focuses on discovering latent topics from social media by using text features as well as spatial information. Sentence LDA [9] is proposed and it assumes all words in a single sentence are generated from one topic. A bilingual topic model is proposed in [5] as a language modeling framework, in which the topics are learned from query-title pairs. Pair Model [8] derives word-pairs from click-through data, and maps queries and documents into a topic space spanned by word-pairs. Multi-Faceted Topic Model (MfTM) [18] is proposed to capture the temporal characteristics of each topic by jointly modeling latent semantics among terms and entities. Although the aforementioned ones take only a small portion of topic models in the literature, they are suitable choices in many industrial scenarios. There are also many existing industrial topic model systems such as PLDA+ [12], LightLDA [23] and Gensim [15]. However, compared to Familia, these existing systems only support limited kinds of topic models.

2.2 Efficient Sampling Algorithms

Another major research trend of LDA parameter inference is to design efficient sampling algorithms. Conventional Gibbs sampling [6] has a complexity of K per sample when the topic amount is set to K. FastLDA [14] has a complexity of significantly less than K per sample. SparseLDA [22] contains both an algorithm and data structures for efficiently conducting Gibbs sampling. A much faster algorithm is proposed in [10] which scales linearly with the number of instantiated topics in the document. As one step further, LightLDA [23] achieves an $O(1)$ Metropolis-Hastings sampling algorithm, whose running cost is agnostic of model size. Recently, [3] proposed WarpLDA, which achieves both $O(1)$ time complexity per token and $O(K)$ scope of random access. With their effectiveness, these techniques are exclusively designed for LDA. Their applicability has never been explored in a broader scope of topic models, on which how to conduct efficient sampling is still an open question.

3 Generative Assumption

In this section, we discuss the details of Familia by describing its mathematical foundations. We describe Familia in conventional topic model terminologies, and three newly introduced terminologies are formally defined as follows:

1. *item: the basic unit of an observed variable, such as a word, a timestamp, etc.*
2. *factor: the basic unit in which all the item are generated by the same distribution and the same topic. In terms of the distribution being continuous or discrete, the factors can be further categorized as continuous factor and discrete factor*
3. *blob: the basic unit in which all the items are generated by the same topic*

The generative process is depicted in Algorithm 1. In order to generate a document d, we first draw θ_d, which is a Multinomial distribution over topics. Then, for each blob, we draw a topic z. Based upon z, for each discrete factor i, we generate discrete items u according to the corresponding discrete distribution ϕ_{iz}. The continuous items are generated in an analogous approach. Finally, if the topic model needs to capture the supervised signal (e.g., the category of the document or the rating of the quality of the document), the signal is further drawn from a Gaussian distribution that uses $\bar{z}_d$ as parameters. Specifically, $\bar{z}_d = \frac{1}{N}\sum_{n=1}^{N} z_n$, where N is the amount of tokens in the document.

It is easy to see that Algorithm 1 is a generic process for many topic models. A variety of topic models can be modeled by Algorithm 1, to name a few, LDA [1], Supervised LDA [13], Sentence LDA [9], TOT [20], Bilingual Topic Model [5], Pair Model [8], GeoFolk [17], LATM [19] and Multifaceted Topic Model [18]. Besides these existing models, users can design their own topic models for specific tasks as long as they follow the generative process in Algorithm 1.

4 Parameter Inference

We proceed to discuss how to conduct parameter inference for Familia. We detail the mathematical derivation for the most complicated scenario where there simultaneously exist discrete factors, continuous factors and the supervised response. The other simpler scenarios can be trivially derived based upon the following discussion. By translating the generative process of Algorithm 1 into joint distribution, we aim to maximize the likelihood

---

Algorithm 1: Generative Process of Familia

---

1 **for** *each topic $k \in 1, ..., K$* **do**
2   **for** *each discrete factor $i \in 1, ..., M$* **do**
3 | draw a discrete factor distribution $\phi_{ik} \sim$ Dirichlet(β_i)
4   **end**
5   **for** *each continuous factor $j \in 1, ..., N$* **do**
6 | generate a continuous factor distribution ψ_{jk}
7   **end**
8 **end**
9 **for** *each document $d \in 1, ..., D$* **do**
10   generate topic distribution $\theta_d \sim$ Dirichlet(α)
11   **for** *each blob b in d* **do**
12   generate a topic $z \sim \theta_d$
13   **for** *each discrete factor $i \in 1, ..., M$* **do**
14 | generate items $u \sim \phi_{iz}$
15   **end**
16   **if** *there exists continuous factor* **then**
17   **for** *each continuous factor $j \in 1, ..., N$* **do**
18 | generate items $v \sim \psi_{jz}$
19   **end**
20   **end**
21   **end**
22   **if** *there exists supervised signal* **then**
23 | draw signal $y_d \sim P(y_d|\bar{z}_d, \eta, \sigma^2)$, where $P(y_d|\bar{z}_d, \eta, \sigma^2) = N(y_d|\eta^T\bar{z}_d, \sigma^2)$
24   **end**
25 **end**

---

of the observed items and the supervised signals. The complete likelihood $P(\mathbf{u}_{1...M}, \mathbf{v}_{1...N}, \mathbf{y}_{1...D}, \mathbf{z}|\alpha, \beta, \Psi, \eta, \sigma)$ is presented as follows:

$$P(\mathbf{u}_{1...M}, \mathbf{v}_{1...N}, \mathbf{y}_{1...D}, \mathbf{z}|\alpha, \beta, \Psi, \eta, \sigma) =$$

$$P(\mathbf{z}|\alpha) \underbrace{\prod_{i=1}^{M} P(\mathbf{u}_i|\mathbf{z}, \beta)}_{discrete\ factors} \underbrace{\prod_{j=1}^{N} P(\mathbf{v}_j|\mathbf{z}, \Psi)}_{continuous\ factors} \underbrace{\prod_{d=1}^{D} P(y_d|\bar{z}_d, \eta, \sigma^2)}_{supervised\ signal}$$

$$= \left(\frac{\Gamma(\sum_{z=1}^{T}\alpha_z)}{\prod_{z=1}^{T}\Gamma(\alpha_z)}\right)^D \prod_{d=1}^{D} \frac{\prod_{z=1}^{T}\Gamma(m_{dz}+\alpha_z)}{\Gamma(\sum_{z=1}^{T}(m_{dz}+\alpha_z))} \qquad (1)$$

$$\prod_{i=i}^{M} \left(\frac{\Gamma(\sum_{u_i=1}^{U_i}\beta_{u_i})}{\prod_{u_i=1}^{U_i}\Gamma(\beta_{u_i})}\right)^T \prod_{z=1}^{T} \frac{\prod_{u_i=1}^{U_i}\Gamma(n_{zu_i}+\beta_{u_i})}{\Gamma(\sum_{u=1}^{U_i}(n_{zu_i}+\beta_{u_i}))}$$

$$\prod_{j=1}^{N}\prod_{d=1}^{D}\prod_{l=1}^{L_d} P(v_{jdl}|\psi_{z_{jdl}}) \prod_{d=1}^{D} P(y_d|\bar{z}_d, \eta, \sigma^2)$$

where m_{dz} is the number of sentences that are assigned to topic z in document d. n_{zv} is the number of times that v is assigned to topic z through Multinomial distribution and $\Gamma(\cdot)$ indicates Gamma function. The goal of parameter inference is to estimate Θ, Φ, Ψ, η and σ in Algorithm 1 through sampling the latent topic z of each blob. In this following two subsections, we describe how to sample $\mathbf{z}$ through Gibbs sampling and Metropolis Hastings based on Eq. (1). Utilizing the sampled z to estimate the parameters of discrete distributions is well-documented in literature [9]. We will detail how to utilize $\mathbf{z}$ to estimate the parameters of continuous distribution in Sect. 4.3.

4.1 Gibbs Sampling

By applying Bayes rule to Eq. (1), the full conditional of assigning topic k to blob b in document d is as follows:

$$P(z_b = k|\mathbf{u}_{1...M}, \mathbf{v}_{1...N}, \mathbf{y}_{1...D}, \mathbf{z}_{-b}, \alpha, \beta, \Psi, \eta, \sigma) =$$

$$\frac{m_{dk} + \alpha_k}{\sum_{k'=1}^{K}(m_{dk'} + \alpha_{k'})} \prod_{i=i}^{M} \frac{\Gamma(\sum_{u=1}^{U_i}(n_{ku} + \beta_u))}{\Gamma(\sum_{u=1}^{U_i}(n_{ku} + \beta_u + N_{bu}))} \tag{2}$$

$$\prod_{u_i \in b} \frac{\Gamma(n_{ku_i} + \beta_{u_i} + N_{bu_i})}{\Gamma(n_{ku_i} + \beta_{u_i})} \prod_{j=1}^{N} \prod_{l=1}^{L_b} P(v_{jl}|z_k) \times P(y_d|\bar{z}_d, \eta, \sigma^2)$$

In order to sample a new topic for the blob b, we need to calculate the above conditional probability for all topics and conduct normalization. Hence, the time complexity for sampling a topic for a blob is $O(K)$, where K is the topic amount.

4.2 Metropolis Hastings

If the topic amount K is large, the $O(K)$ per blob complexity is time-consuming. We now propose an efficient alternative based upon Metropolis Hastings (MH), which has been successfully applied in [10,23] for LDA. For the convenience of designing proper proposals for MH, we first conduct approximation to Eq. (2) as follows:

$$P(z_b = k|\mathbf{u}_{1...M}, \mathbf{v}_{1...N}, \mathbf{y}_{1...D}, \mathbf{z}_{-b}, \alpha, \beta, \Psi, \eta, \sigma) \approx$$

$$\frac{m_{dk} + \alpha_k}{\sum_{k'=1}^{K}(m_{dk'} + \alpha_{k'})} \prod_{i=i}^{M} \prod_{u_i \in b} \frac{n_{ku_i} + \beta_{u_i}}{\sum_{u=1}^{U_i}(n_{ku} + \beta_u)} \tag{3}$$

$$\prod_{j=1}^{N} \prod_{l=1}^{L_b} P(v_{jl}|z_k) \times P(y_d|\bar{z}_d, \eta, \sigma^2)$$

The above equation approximates Eq. (2) when an item appears multiple times with in a blob. Based on this approximation, we discuss the MH sampling method for Familia. MH needs proper proposals to work. We now discuss four proposals which fall into two major categories: document-based proposals and item-based proposals.

Document-Based Proposals. The first document-specific proposal is the *document-topic proposal*:

$$\Xi_d(k) \propto \frac{m_{dk} + \alpha_k}{\sum_{k'=1}^{K}(m_{dk'} + \alpha_{k'})} \tag{4}$$

The second document-specific proposal is the *supervised signal proposal*:

$$\Xi_{y_d}(k) \propto P(y_d|\bar{z}_d, \eta, \sigma^2) \tag{5}$$

Item-Based Proposals. The first item-based proposal is the *discrete item-topic proposal*, which is denoted as $\Xi_u(k)$:

$$\Xi_u(k) \propto \frac{n_{ku_i} + \beta_{u_i}}{\sum_{u=1}^{U_i}(n_{ku} + \beta_u)} \tag{6}$$

The second item-based proposal is the *continuous Item-topic proposal*, which is denoted as $\Xi_v(k)$:

$$\Xi_v(k) \propto P(v_{jl}|z_k) \tag{7}$$

It is easy to see that each proposal encourages the sparsity of their corresponding component in Eq. (3). For fast sampling from each proposal, a data structure named alias table [10,23] is utilized to reduce the sampling complexity. Theoretically, an alias table need to be created for each document and each item. A caveat is that document-topic proposal does not need to explicitly establish alias table [23], because sampling from the document-topic proposal can be cheaply simulated through returning the topic assignment of a randomly sampled blob in the document. The MH method is formally presented in Algorithm 2. When sampling a new topic for a blob, the algorithm sequentially utilizes document-based proposals and item-based proposals to update the topic candidates. For each topic candidate, the algorithm chooses whether to accept it according to the

---

Algorithm 2: Metropolis Hastings of Familia

1 **for** *each document $d \in 1, ..., D$* **do**
2   **for** *each blob b in d* **do**
3   **for** *a predefined number of MH steps* **do**
4   propose a topic z_1 based on document-topic proposal
5   update the topic to z_1' according to acceptance ratio
6   propose a topic z_2 based on alias table of y_d
7   update the topic to z_2' according to acceptance ratio **for** *each discrete factor $i \in 1, ..., M$* **do**
8   **for** *item u in this factor* **do**
9   propose a topic z_{u3} based on alias table of u
10   update the topic to z_{u3}' according to acceptance ratio
11   **end**
12   **end**
13   **for** *each continuous factor $j \in 1, ..., N$* **do**
14   **for** *item v in this factor* **do**
15   propose a topic z_{v4} based on alias table of v
16   update the topic to z_{v4}' according to acceptance ratio
17   **end**
18   **end**
19   **end**
20   **end**
21 **end**

---

acceptance ratio, just like the standard Metropolis Hastings. Note that this process can be repeated for several iterations and each iteration is formally defined as an *MH step*. In the experiment section, we will show the extent to which the number of MH steps can affect the performance of the MH method.

4.3   Issues of Continuous Distributions

For topic models with continuous factors, updating the parameters of continuous distributions is computationally expensive, especially for distributed environment in which the synchronization of these parameters is needed. In Familia, we choose a well-adopted practice [17,20]: we update the parameters of the continuous distributions after each major iteration (i.e., scanning through the whole corpus of training data). For Gaussian distributions, we straightforwardly update the parameters by the sample mean and sample variance. If the continuous distribution is Beta distribution $Beta(\psi_{k1}, \psi_{k2})$, we update the parameters ψ_{k1} and ψ_{k2} for the kth topic as follows:

$$\psi_{k1} = \bar{s}_k \left(\frac{\bar{s}_k(1 - \bar{s}_k)}{v_k^2} - 1 \right), \tag{8}$$

$$\psi_{k2} = (1 - \bar{s}_k) \left(\frac{\bar{s}_k(1 - \bar{s}_k)}{v_k^2} - 1 \right) \tag{9}$$

where $\bar{s}_k$ and v_k^2 denote the sample mean and biased sample variance of topic k's items. As for supervised signals [13], we denote the $(D \times K)$ matrix whose dth row is $\bar{z}_d^T$ as A, and the $D \times 1$ vector of supervised signals as Y, the η and σ are updated as follows:

$$\eta = (A^T A)^{-1} A^T Y, \tag{10}$$

$$\sigma = \frac{1}{D}(Y^T Y - Y^T A (A^T A)^{-1} A^T Y) \tag{11}$$

In practice, the above matrix manipulation can be approximated by techniques to reduce the computational cost and scale to large data set.

4.4   Hybrid Sampling Mechanism

So far, we have discussed two basic sampling techniques that are supported in Familia. The major advantage of MH is efficiency: the per blob sampling complexity can be reduced as low as $O(1)$ for some topic models through utilizing alias tables. However, as we will show later in Sect. 5, the models trained by GS are usually better than MH. Hence, it is desirable to design a mechanism that tradeoffs the efficiency of MH and effectiveness of GS. To meet this requirement, Familia enables the users to choose the sampling method for each iteration. Such flexibility makes it possible to investigate many hybrid sampling mechanisms which collectively apply GS and MH.

4.5   Data Organization

In Familia, data organization is critical for automating parameter inference because it provides basic information of Bayesian network structure of the topic model. Based on data organization, Familia can deduce each component in Eq. (2) and Eq. (3) and then conduct parameter inferences without imposing any burden on human to derive the mathematical equations. Documents are grouped as blocks to facilitate distributed computing. In each document, the blob is utilized as the basic unit whose content shares the same topic. A blob contains multiple factors and a factor can contain any amount of items. The data organization of Familia is presented as follows:

$$
\text{block} \begin{cases} \text{document}_1 \begin{cases} \text{blob}_1 \begin{cases} \text{supervised signal} \\ \text{factor}_1 \begin{cases} \text{item}_1 \\ \text{item}_2 \\ \dots \end{cases} \\ \text{factor}_2 \\ \dots\dots \end{cases} \\ \text{blob}_2 \\ \dots\dots \end{cases} \\ \text{document}_2 \\ \dots\dots \end{cases}
$$

Since the "item" in Familia data organization can be specialized into words(discrete), tags(discrete) or even timestamps(continuous), the above data organization is generic. It supports a variety of existing topic models such as LDA, Supervised LDA, Sentence LDA, TOT, Bilingual Topic Model, Pair Model, GeoFolk, LATM and Multifaceted Topic Model as well as user-designed models for special tasks as long as they follows the generative process in Algorithm 1.

5   Experiments

In this section, we report the experimental results. We first investigate the performance of a series of sampling methods across topic models and data sets in Sect. 5.1. Then we demonstrate the scalability of Familia in Sect. 5.2.

5.1   Performance of Sampling Methods

We systematically investigate the performance of different sampling methods in terms of LDA, Sentence LDA and TOT. In order to minimize effect beyond algorithmic performance, experiments in this subsection are conducted on a single computing node. NIPS dataset from UCI Bag of Words Data Set[1] is utilized for experiments of LDA. Amazon data[2] is utilized for Sentence LDA. As for TOT,

---

[1] https://archive.ics.uci.edu/ml/datasets/Bag+of+Words.
[2] http://uilab.kaist.ac.kr/research/WSDM11.

we utilize the 21 decades of U.S. Presidential State-of-the-Union Addresses[3] for the experiments. Due to space limitation, we present the results when the topic amount is set to 50. Similar insights can be obtained when the topic amount is set to other values.

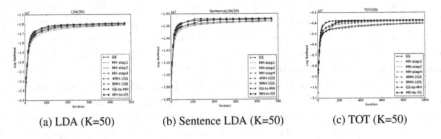

(a) LDA (K=50)   (b) Sentence LDA (K=50)   (c) TOT (K=50)

Fig. 1. Comparison of sampling methods (iteration) (Best Viewed in Color)

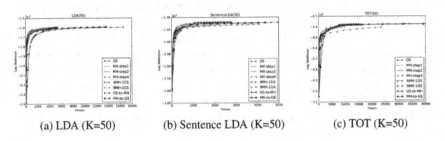

(a) LDA (K=50)   (b) Sentence LDA (K=50)   (c) TOT (K=50)

Fig. 2. Comparison of sampling methods (time) (Best Viewed in Color)

The log likelihood of each sampling methods is plotted against iteration in Fig. 1 and against time in Fig. 2. Therein, MH-step1 is MH with only one MH step. Analogously, MH-step2 and MH-step4 are MH with 2 and 4 MH steps respectively. 4MH-1GS is the sampling method that performs one iteration of GS after every 4 iterations of MH. 9MH-1GS performs one iteration of GS after every 9 iterations of MH. GS-to-MH starts with GS for the first 100 iterations and then switches to MH for the remaining iterations. MH-to-GS starts with MH for the first 100 iterations and then switches to MH for the remaining iterations. Note that space limitation prevents us from presenting more parameter settings for each sampling method. However, the results shown in Figs. 1 and 2 are sufficient to showcase our insights obtained from this study.

We first investigate the performance of MH in terms of the number of steps. From Fig. 1, we observe that larger MH steps usually result in better performance. In most cases, 4-step is better than 2-step and further better than 1-step.

---

[3] http://www.gutenberg.org/dirs/etext04/suall11.txt.

The experimental results of all the three topic models verify the above argument. We proceed to compare the performance of different sampling methods. A possible explanation is that more MH steps help the sampling method to explore more states if the Markov chain has low conductance. 4MH-1GS usually achieves the best performance while MH (including all the three MH methods with different steps) usually demonstrates the worst performance. The performance of GS and the other hybrid methods is between 4MH-1GS and MH. 4MH-1GS is better than 9MH-1GS, showing that fairly high frequency of switching MH and GS is effective to improve the model quality. The hybrid methods have higher chance to prevent the sampling algorithm getting "stuck" in a subset of Markov chain states. Hence, MH is not a good choice if the quality of the resultant model is highly valued. From Fig. 2, we observe that MH-step1 achieves a fairly good model within the least time while GS takes the longest time. The time consumed by hybrid sampling methods is between MH-step1 and GS. The superiority of MH in efficiency is achieved by reusing the alias tables, which can reduce the amortized time complexity to as low as $O(1)$ per blob. In contrast, the complexity of GS is $O(K)$ per blob, since it needs to calculate the probability for each topic.

Based upon above observations, we obtain the following important insights, which are valid across the three different topic models: 1. GS achieves higher likelihood than MH while MH consumes less time to achieve a fairly good result; 2. Some hybrid sampling methods can achieve even better result than GS while consumes less time than GS. 3. If the quality of the model is the emphasis, hybrid sampling methods like 4MH-1GS should be chosen because it achieves the best model quality with fairly good efficiency. If efficiency is the focus, MH may be chosen since it consumes the least time to generate a reasonably good model.

5.2   Scalability of Familia

We proceed to demonstrate the scalability of Familia with 1, 5, 10 and 20 computing nodes. The log likelihood of LDA trained on 10 nodes is presented in Fig. 3a, from which we observe that the results obtained from 10 nodes is aligned with those from a single node, showing that the quality of these three models trained by different sampling methods is not heavily affected by the distributed environment. Although all sampling methods achieve slightly lower log likelihood than those of single node, such degradation is modest in practice. In distributed environment, 4MH-1GS is the method with the best performance and MH methods usually achieve the lowest likelihood. The insights discussed in Sect. 5.1 still hold for training topic models in distributed environment. Similar phenomenon is observed for Sentence LDA and TOT or when the number of nodes is set to 5 or 20 and their results are skipped.

Another important question is how much speedup we obtain when multiple computing nodes are involved. The speedup analysis of LDA is presented in Fig. 3b. High speedup ratio is an indicator of low communication and synchronization cost. With training topic models with PS, low communication cost is primarily achieved by the sparsity of the model under training. The sparser the

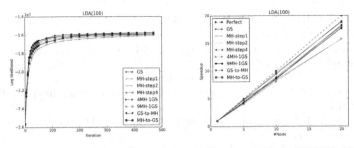

(a) Comparison of Sampling Methods (Parallel-Iteration): LDA (node=10, K=100)

(b) Speedup Analysis: LDA (K=100)

Fig. 3. Scalability analysis (Best Viewed in Color)

model is, the less the parameters that each worker needs to pull from servers. When sorted by speedup ratio, the ranking of these sampling methods varies from model to model, showing a specific sampling method has different capability of promoting the sparsity of a topic model. However, GS always has the best speedup ratio, indicating that GS is quite effective in promoting the sparsity of the model. Similar phenomenon is observed for Sentence LDA and TOT and their results are skipped due to space limitation.

6   Conclusion

In this paper, we propose a configurable topic modeling framework named Familia for industrial text engineering. The framework provides novel functionalities such as topic model customization, automatic parameter inference and post-modeling utilities. Based on the hybrid sampling mechanism of Familia, we further provide practical suggestions of choosing proper sampling methods for different topic models. Equipped with Familia, software engineers can easily test different assumptions of the latent structure of their data without tediously deriving mathematical equations and implementing sampling algorithms from scratch. Familia would help the technique of topic modeling to be utilized in more proper and convenient manner in industrial scenarios.

References

1. Blei, D.M., Ng, A.Y., Jordan, M.I.: Latent Dirichlet allocation. J. Mach. Learn. Res. **3**, 993–1022 (2003)
2. Borisov, A., Serdyukov, P., de Rijke, M.: Using metafeatures to increase the effectiveness of latent semantic models in web search. In: WWW (2016)
3. Chen, J., Li, K., Zhu, J., Chen, W.: WarpLDA: a cache efficient O(1) algorithm for latent Dirichlet allocation. stat 1050, 2 (2016)
4. Dean, J., et al.: Large scale distributed deep networks. In: NIPS (2012)

5. Gao, J., Toutanova, K., Yih, W.T.: Clickthrough-based latent semantic models for web search. In: SIGIR (2011)
6. Griffiths, T.L., Steyvers, M.: Finding scientific topics. Proc. Natl. Acad. Sci. **101**(suppl 1), 5228–5235 (2004)
7. Hofmann, T.: Probabilistic latent semantic indexing. In: SIGIR (1999)
8. Jagarlamudi, J., Gao, J.: Modeling click-through based word-pairs for web search. In: SIGIR (2013)
9. Jo, Y., Oh, A.H.: Aspect and sentiment unification model for online review analysis. In: WSDM (2011)
10. Li, A.Q., Ahmed, A., Ravi, S., Smola, A.J.: Reducing the sampling complexity of topic models. In: SIGKDD (2014)
11. Li, M., et al.: Parameter server for distributed machine learning. In: Big Learning NIPS Workshop, vol. 6, p. 2 (2013)
12. Liu, Z., Zhang, Y., Chang, E.Y., Sun, M.: PLDA+: parallel latent Dirichlet allocation with data placement and pipeline processing. TIST **2**(3), 26 (2011)
13. Mcauliffe, J.D., Blei, D.M.: Supervised topic models. In: Advances in Neural Information Processing Systems, pp. 121–128 (2008)
14. Porteous, I., Newman, D., Ihler, A., Asuncion, A., Smyth, P., Welling, M.: Fast collapsed Gibbs sampling for latent Dirichlet allocation. In: SIGKDD (2008)
15. Řehůřek, R., Sojka, P.: Software framework for topic modelling with large corpora. In: Proceedings of the LREC 2010 Workshop on New Challenges for NLP Frameworks, pp. 45–50. ELRA, Valletta, May 2010
16. Si, X., Chang, E.Y., Gyöngyi, Z., Sun, M.: Confucius and its intelligent disciples: integrating social with search. Proc. VLDB Endow. **3**(1–2), 1505–1516 (2010)
17. Sizov, S.: GeoFolk: latent spatial semantics in web 2.0 social media. In: WSDM (2010)
18. Vosecky, J., Jiang, D., Leung, K.W.T., Xing, K., Ng, W.: Integrating social and auxiliary semantics for multifaceted topic modeling in Twitter. TOIT **14**(4), 27 (2014)
19. Wang, C., Wang, J., Xie, X., Ma, W.Y.: Mining geographic knowledge using location aware topic model. In: Proceedings of the 4th ACM Workshop on Geographical Information Retrieval, pp. 65–70. ACM (2007)
20. Wang, X., McCallum, A.: Topics over time: a non-Markov continuous-time model of topical trends. In: SIGKDD (2006)
21. Xing, E.P., et al.: Petuum: a new platform for distributed machine learning on big data. IEEE Trans. Big Data **1**(2), 49–67 (2015)
22. Yao, L., Mimno, D., McCallum, A.: Efficient methods for topic model inference on streaming document collections. In: SIGKDD (2009)
23. Yuan, J., et al.: LightLDA: big topic models on modest computer clusters. In: WWW (2015)

Generating Personalized Titles Incorporating Advertisement Profile

Jingbing Wang[1], Zhuolin Hao[2], Minping Zhou[2], Jiaze Chen[2], Hao Zhou[2], Zhenqiao Song[2], Jinghao Wang[2], Jiandong Yang[2], and Shiguang Ni[3(✉)]

[1] Tsinghua-Berkeley Shenzhen Institute,
Tsinghua University, Shenzhen, China
[2] Beijing Youzhuju Network Technology Co. Ltd., Beijing, China
[3] Shenzhen International Graduate School,
Tsinghua University, Shenzhen, China
ni.shiguang@sz.tsinghua.edu.cn

Abstract. Advertisement (Ad) title plays a significant role in the effectiveness of online commercial advertising. However, it's difficult for most advertisers to think of attractive titles for their products. By mining keywords from current ad material, traditional retrieval methods and neural text generation models have been applied to solve this problem. However, few of them focus on personalized ad titles generation. Ad titles from different advertisers can be very diversified, and there is massive previous advertising data available, which can tell the style, content, and vocabulary of specific advertisers. Based on massive previous advertising data and current ad material, we propose an Ad-Profile-based Title Generation Network (APTGN) to automatically generate personalized titles for ads. The model utilizes massive advertising data and current ad material to construct a profile for each ad, which is further integrated into the generation model to help recognize the preferences of specific ads. Automatic evaluation metrics and online A/B testing both show that our model significantly outperforms all the baselines, increasing the adoption rate of recommendation titles by 27.22%. Through our deployed model, once an advertiser needs to customize an ad title for their products, satisfactory titles can be recommended automatically without bothering to write any words.

Keywords: Personalized advertisement title generation ·
Advertisement profile · Feature extraction

1 Introduction

Advertisement titles are often the first impression to attract potential consumers and thus trigger conversion behavior in online commercial advertising. When making ads, advertisers are usually inclined to use informative and intriguing titles to promoting their products. Those ads are exposed to potential consumers on the media. Consumers usually take a glance at the ad titles, and only when

© Springer Nature Switzerland AG 2021
C. S. Jensen et al. (Eds.): DASFAA 2021, LNCS 12683, pp. 529–540, 2021.
https://doi.org/10.1007/978-3-030-73200-4_37

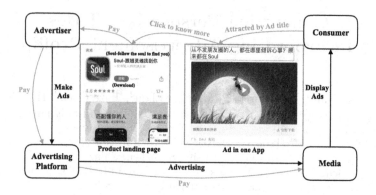

Fig. 1. Simple illustration of advertising process. Inside the red box is title for this ad. The translation is "Where do people who never post in WeChat Moments share their thoughts? It's in Soul.". (Color figure online)

attracted by the titles, will they continue to click on the ad to know more about the products. Thus, the ad titles serve as a bridge between customers and products, which is the key for advertisers and the related parties to make higher revenue through online advertising (Fig. 1).

However, how to generate correct and attractive ad titles that meet the advertisers' needs effectively remains a troublesome problem. First, making ad titles manually is a big challenge. In industrial applications, ad titles are commonly either written by the advertisers or generated based on designed templates. Larger advertisers may hire professional copywriters to write titles for their ads. Those titles are often more relevant, specified, and attractive but cost a lot of time and money. It's difficult for medium and smaller advertisers to generate titles in this way. The template-based method is another popular way to make titles [2]. But titles generated by pre-defined templates are stereotyped, even incorrect. Second, different advertisers have different preferences on the ad titles. It's difficult to generate personalized titles for each ad.

To free the advertisers of the ad title making process, some work has been done to utilize text generation algorithms to generate ad titles automatically. But only limited information (customer keywords [1] or texts on the landing pages [5]) is used to help generation. Thus, generated titles may fail to meet the different preferences of different advertisers. In personalized advertisement recommendation systems, user modeling has been widely used and proved to receive great success [4,10]. But few of those techniques have been explored in personalized advertisement generation.

In this paper, we propose an Ad-Profile-based Title Generation Network (APTGN) to automatically generate personalized titles for each ad. Ads from different advertisers usually differ a lot in their content and style while ads from the same advertiser share many similarities. Those features can be fully reflected in their previous ads, which are not utilized in current title generation methods. Therefore, to generate personalized titles, we first conduct an in-depth exploration of the recorded basic information of the advertiser and their previous ads. Twenty features are extracted to make a full depiction of the ad from three aspects, including basic information, stylistic information, and current information. Second, we use a feature extraction module to vectorize the constructed features. Given all the feature vectors, a transformer-based generation model is used to generate titles for each ad.

The main contributions of this paper can be summarized as follow. We design an Ad-Profile based Title Generation Network to automatically generate personalized and intriguing titles for advertisers. Leveraging massive related previous advertisement, this model can get the advertiser rid of making ad titles completely. This model has already been deployed to the online advertising platform in this company and the adoption rate of recommended titles has shown a continuous increase. Once the advertiser comes to the platform and starts making ads, satisfactory titles will be recommended immediately without requiring the advertiser to write any words, which speeds up the ad making process. Higher quality ad titles also gain more revenue for both advertisers and advertising platforms.

2   Ad-Profile-Based Title Generation Network

In this section, we describe in detail the proposed APTGN model. The overall model framework is shown in Fig. 2.

The advertising platform contains a large amount of previous ad data, including basic target information, query (keywords provided by advertisers to make ad titles, optionally), title, landing page, and so on. When an advertiser creating an ad on the advertising platform, all the related information can be retrieved on this platform. First, we use a feature construction module to build a profile for each ad, namely, twenty features are designed to depict the ad from three aspects, basic, stylistic, and current information. Those features are usually long and massive texts. Hence, a feature extraction module is designed to give a representation vector of each feature. Given all the features, a transformer-based generation model is used to generate titles for each ad automatically.

The problem can be formulated as follows: given an ad, which includes its landing page, query (optional), the model aims to generate a title making full use of all the available information. Namely, given all the available information $X = (x_1, \cdots, x_M)$, where each x_i represents one kind of related information, the title generation model aims to take X as input, and generates an ad title $Y = (y_1, y_2, \cdots, y_N)$ word by word. The log conditional probability can be formalized as:

$$logPr(Y|X;\theta) = \sum_{j=1}^{N} logPr(y_j|X, y_{\leq j};\theta) \tag{1}$$

where $y_{\leq j} = (y_1, y_2, \cdots, y_{j-1})$ and θ indicates model parameters.

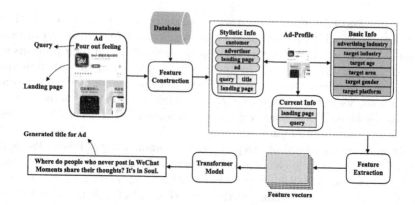

Fig. 2. Overview of Ad-Profile-based Title Generation Model.

2.1 Ad Profile Construction

When an advertiser comes to the advertising platform preparing ad titles, all the related information collected by the advertising platform can be used to assist in the auto-generation of titles. To help understand, a brief introduction of the hierarchical advertising system will be given first. In a general advertising system, a customer is a virtual unit, which can register multiple advertiser accounts. Each advertiser can release multiple ad groups. Each ad group contains multiple ads and each ad mainly includes one title, one image, or one video promoting the product [7]. Ad groups and ads can be updated over time.

The data collection platform in this company records the relationship among customers, advertisers, ad groups, and ads. The detailed information of each ad is also collected, including query, selected title, and the landing page of the ad. Based on this information, we construct a profile for each ad as follows.

Basic Info: The basic information of each ad mainly depicts its static property, which not only gives useful direction but also restriction of title generation.

- Advertiser Industry. The advertiser industry is important for title generation. Given query "苹果 (apple)" without knowing the industry information, the model may be confused at generating titles like "这种苹果非常好吃 (this kind of apple tastes good)" or "苹果电脑打折啦 (Apple computers are on sale)".
- Target Information. Target information covers the basic information of targeting audiences, including target industry, target area, target age, target gender, and target platform. With this information, the model can generate more personalized and correct titles. For example, given target gender "女 (female)", the model will prefer to generate "漂亮 (beautiful)" instead of "帅气 (handsome)".

Current Info: When generating titles for a given ad, the textual information we can make use of mainly comes from query and texts on the landing page, both are quite relative to the ad title.

- Landing Page. The landing page displays the details of the ad, which can be regarded as the embodiment of the condensed ad title. Thus it can provide much information for title generation.
- Query. Query reflects the current interest of the advertiser, which is most related to the title to be generated.

Stylistic Info: Similar ads can provide great insight for title generation of the current ad. The advertiser's preference (including style, content, vocab, and so on) for the current ad can be reflected in the related previous ads. Considering the hierarchy advertising system, we take the following four kinds of similarity into account.

- Customer. Ads from the same customer would share some similarities. As mentioned in Current Info, the landing page and query are the most direct references for title generation. So we retrace previous landing pages, queries, and titles from the same customer as the current ad to help generate titles.
- Advertiser. Same as Customer, previous landing pages, queries, and titles from the same advertiser are used as supplemental features to help title generation.
- Landing Page. Ads on the same landing page are usually similar. Therefore, previous queries and titles on the current landing page are integrated into the title generation model.
- Ad Group. Previous information from the same ad group can provide much useful help to current title generation. Same as the above, previous queries and titles from the same ad group as the current ad are used.

2.2   Feature Extraction

In Sect. 2.1, We have constructed basic, current and stylistic features for each ad. Basic information are all discrete features, so we directly construct a feature matrix W_i ($i \in \{1, 2, \cdots, N\}$, N is the number of basic features) for each basic feature, where $W_i = \{w_{i1}, w_{i2}, \cdots, w_{is}\}$, s is the number of all possible values for this feature. Those feature matrices are randomly initialized and learned during the training process.

However, it is more complicated for stylistic features. As mentioned, stylistic information is mainly textual features, which is of large amount. For instance, a single landing page may have tens of thousands of characters. Thus, a feature extraction model is trained to distill useful information from the landing pages, previous titles, and previous queries.

Model. Our feature extraction model uses GPT2 framework [8], which can be treated as a black box. Input a text sequence x, it outputs an embedding $f(x)$. And the training task is to do binary classification (as shown in Fig. 3). For convenience, we assume that a title (or landing page or query) of a specific ad is closer to all other titles (or landing pages or queries) of the same ad that it is to any titles (or landing pages or queries) of any other ad. To this end, we employ triplet loss [9] that directly reflects what we want to achieve in the feature extraction.

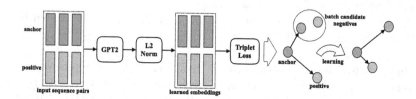

Fig. 3. Illustration of feature extraction model.

Instead of the offline generation of whole triplets, we construct anchor-positive pairs offline while selecting hard negative exemplars online from within a mini-batch. Namely, take titles, for example, we traverse titles from the same ads and pair them up to form anchor-positive pairs. At each step of training, a mini-batch of anchor-positive pairs are fed into the model, a hard negative x_i^n is selected for each anchor-positive such that $\text{argmin}_{x_i^n} ||f(x_i^a) - f(x_i^n)||^2$. Besides, we append a special type token at the beginning of each input sequence to help the model distinguish query, title, and landing page.

Thus, the loss can be formulated as:

$$L = \sum_i^N [||f(x_i^a) - f(x_i^p)||_2^2 - ||f(x_i^a) - f(x_i^n)||_2^2 + \alpha]_+ \tag{2}$$

where N is the cardinality of the training set, and (x_i^a, x_i^p, x_i^n) is a triplet in the training set, referring to anchor, positive and negative sample respectively.

Feature. Since there are thousands of previous ads available, we simply take the most recent k previous titles, landing pages, and queries to construct stylistic features. Take titles of the same customer, for example, the most recent k titles are fed into the pre-trained feature extraction model, and we get $f(x_i), i \in \{1, 2, \cdots, k\}$. Then, a simple summation is performed to get one feature vector for previous titles of same customer, namely, $f(x_{ct}) = \sum_i^k f(x_i)$. The same technique is used on all stylistic features. Besides, the current landing page also contains many texts so, for convenience, we directly use the feature embedding $f(x_{cu})$ learned from this model to represent it.

2.3 Model Framework

The transformer model [11] is adapted for this ad title generation task. The input embedding is replaced by our record embedding to better incorporate the rich related information.

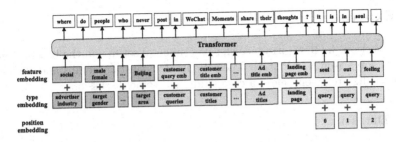

Fig. 4. Illustration of Transformer-based title generation model.

Record Embedding. The input of ad title generation model is a sequence of records. Each record is a tuple of twenty features as constructed in Sect. 2.1. Inspired by previous work [6], we embed features into vectors (as illustrated in Sect. 2.2), and use the concatenation of feature embeddings as the embedding of record. Additionally, to help the model better distinguish different types of features, we add type embedding (a feature matrix T randomly initialized and learned through training) to each feature embedding. The current query shares the same vocabulary with the target title, and to make the model aware of its sequential information, positional encoding [11] is added to query embedding (as shown in Fig. 4). So, the final embedding of each record is formulated as follows:

$$e_i = [e_{i,1}; e_{i,2}; e_{i,3}; \cdots; e_{i,20}] \tag{3}$$

$$e_{ij} = e_{\text{feature}_j} + e_{\text{type}_j}, j \in \{1, 2, \cdots, 19\} \tag{4}$$

$$e_{ij} = e_{\text{feature}_j} + e_{\text{type}_j} + e_{\text{pos}}, j = 20 \tag{5}$$

where $e_i \in \mathcal{R}^{dim}$ is the ith record embedding in the input sequence and $e_{i,j} \in \mathcal{R}^{dim}$ is the jth feature embedding in ith record.

Title Generation. We simply use the whole Transformer model to do title generation. The encoder first maps the input record embedding to a context embedding H that integrates all the input information as encoder output. At each decoding step, the decoder computes the output probability over the whole vocabulary conditioned on the context embedding and previous output words.

$$H = \text{Transformer-Encoder}(e) \tag{6}$$

$$P(y_t|x, y_{\leq t}) = \text{Transformer-Decoder}(y_{\leq t}, H) \tag{7}$$

where e is the record embedding for one sample, H is the encoder output, y is the ground-truth title.

Therefore, the model's training objective is to minimize the negative log-likelihood of the training data concerning all the parameters, as denoted as θ.

$$\mathcal{L} = -\sum_{i=1}^{N} \sum_{j=1}^{|y^{(i)}|} \log P(y_j^{(i)}|x^{(i)}, y_{\leq j}^{(i)}; \theta) \tag{8}$$

where N is the cardinality of the training set, (i) refers to the ith training sample, $|y^{(i)}|$ is the length of the ith ground-truth title.

3   Experiments

In this section, we first introduce the experimental setup and the baseline methods in comparison to APTGN. Then the implementation details and the evaluation results are given before a case study and some discussions.

3.1   Dataset

All the experiments are conducted on a real dataset, which is collected by the data collection platform in this company. The detailed information of each ad is well recorded, including its creation time, customer, advertiser, advertiser's industry, targeting industry, age, gender, platform, landing page, input query and finally selected titles, and so on. We use 30,000,000 records for training, each record contains a title and 20 features as constructed in Sect. 2.1.

3.2   Baselines

Ad title generation methods commonly used in the industry are selected as our baselines, including traditional retrieval model and transformer-based text generation models.

Retrieval Model. This is a traditional search-based method. Previous ad titles and relevant text snippets are collected in ElasticSearch, which is a distributed retrieval framework. Each text is represented by the TF-IDF feature. When recommending titles, the cosine similarity between the input query of the advertiser and texts in the title database is considered, and titles with the higher similarity scores are retrieved and recommended to the advertisers. When there is no query available, we will use keywords mined from texts on the current landing page and previous titles instead.

Query-to-Title. This is a transformer-based generation model. We collect advertisers' input queries and titles for final ads as training data. Same as retrieval model, when there is no query, keywords mined from texts on the current landing page and previous titles will be used as query. With the queries as input, this model is trained to generate suitable titles for the corresponding queries to minimize the cross-entropy loss. The transformer model used here is the same as that in APTGN, which is also pre-trained.

Bidirectional Continuation. This model is the same as Query-to-Title, but with different training data. Instead of directly using input query, here we construct multiple query-title pairs with one title by traversing all the possible N-grams. Each N-gram is paired with the title as a training sample.

3.3   Implementation Details

We pre-train the transformer module used in our title generation model and GPT2 module used in the feature extraction model with 3T Chinese news corpus separately. The size of word embedding is set to 768 and the embeddings are updated during fine-tuning. The size of feature embedding is also set to 768. Once the feature extraction model finishes training, the feature embedding for certain feature remains fixed. For the feature extraction model, the threshold for triplet loss α is set to 0.5 and the threshold for most recent k records is set to 500. For the title generation model, all titles and queries are padded to a maximum sentence length of 65. We perform a mini-batch cross-entropy loss training with a batch size of 64 records for 5 epochs. We use Adam optimizer, and the learning rate is specified as $5e-5$, where warm-up is performed for the first 20000 steps. We adopted nucleus sampling with a probability of 0.75 to promote the diversity of generated titles.

3.4 Automatic Evaluation

The evaluation package released by Chen et al. [3] is used to do the automatic evaluation for the generated titles, including BLEU, METEOR and ROUGE scores. We randomly select 1000 records as a test set. Experimental results on the test set are shown in Table 1. From this table, we can see that our proposed APTGN achieves the best performance on these metrics.

3.5 Online A/B Testing

To further verify the effectiveness of our proposed model, we test our method in the real-world advertising platform using A/B testing. During online A/B testing, advertisers are split equally into two groups according to their unique ID and are directed into a baseline bucket and an experimental bucket. For advertisers in the baseline bucket, ad titles are generated by three baseline models, where the generated titles of the three models are randomly ranked and merged according to equal probability. For advertisers in the experimental bucket, ad titles are generated by all the baseline models plus APTGN, where the generated titles of all the four models are also randomly ranked and merged, but with different probabilities, our model with probability $\frac{1}{2}$, and each baseline model with probability $\frac{1}{6}$. All conditions in the two buckets are identical except for the available title generation models. We apply the adoption rate to measure the performance of models.

$$\text{Adoption Rate} = \frac{\text{Number of adoption}}{\text{Number of request}} \tag{9}$$

the event that one advertiser comes to the platform and selects titles is regarded as one request, and the action that he finally selects titles from our recommended titles (no matter modified or not) is viewed as one adoption.

We deployed A/B testing for 14 days. The overall adoption rate of recommended titles in the baseline is 18.58% while that in the experimental bucket is 23.64%, with a 27.22% improvement compared to baseline. Table 2 shows the adoption rate of different models in the baseline bucket and experimental bucket separately. In the experimental bucket, the adoption rate of our proposed model outperforms the other three models. Compared to the baselines, when our proposed model is added, the adoption rates of the baseline models experience a great decline. This reflects that titles generated by our proposed model are of higher quality in terms of the advertisers' favor.

Table 1. Auto metric scores on test set of different model.

	BLEU1	BLEU2	BLEU3	BLEU4	METEOR	ROUGE_L
Retrieval	23.64	14.52	9.50	6.42	10.88	19.13
Query-to-Title	28.12	18.54	13.17	9.67	13.19	23.70
BiContinuation	29.25	19.45	13.90	10.26	13.62	24.13
APTGN	**38.05**	**27.86**	**21.38**	**16.82**	**18.30**	**32.64**

Table 2. Adoption rate of different models in baseline and experimental buckets.

Models	Experiment			Baseline		
	Request	Adoption	Adoption rate	Request	Adoption	Adoption rate
Retrieval	22738	554	2.44%	22593	1693	7.49%
Query-to-Title	22902	561	2.45%	23397	2014	8.61%
BiContinuation	20228	585	2.89%	20727	1960	9.46%
APTGN	22843	4722	**20.67%**			

Table 3. Sample titles generated by different models.

Query	听评书(listen to storytelling)	
Ground-Truth	经典评书【岳飞传】，正版资源在线听，正版资源免费听！(Classic storytelling [Biography of Yue Fei], online listening to genuine resources for free!)	超简单手账简笔画奶茶，夏日来临，为自己画杯奶茶吧(Super simple planner, summer is coming, draw a cup of milk tea for yourself.)
Retrieval	最近发现了一款神器，里面的评书资源真是太好听了(Recently discover an artifact, the storytelling resources inside are really nice)	还在外面喝奶精奶茶？回家榨果汁吧！网红榨汁机限量抢(Still drinking creamer tea? Go home and squeeze the juice! Net red juicer limited grab)
Query-to-Title	单田芳经典评书全集，更新快，正版资源在线听(Complete works of Shan Tianfang's classic storytelling, fast update, online access to genuine resources)	奶茶也太便宜了吧！这家奶茶店居然才卖这个价！(Milk tea is too cheap! Milk tea is selling at this price at this shop!)
BiContinuation	看小说太累？来喜马拉雅听评书，不伤眼，随时随地听(Tired of reading novels? Come to the Himalaya to listen to the storytelling)	网红爆款泡茶杯，每天只需几块钱，不用再去店里买奶茶啦！(Net red burst teacups, only a few dollars a day, no longer need to go to the store to buy milk tea!)
APTGN	经典评书《岳飞传》，正版资源免费听，更新快，不伤眼！(The classic storytelling "Biography of Yue Fei", the genuine resource is free to listen to, the update is fast and does not hurt the eyes!)	超萌手绘的手账简笔画，各种手绘风格，快来下载吧 (Super cute hand-painted planner, various hand-painted styles, come and download)

3.6 Case Study

We present some titles generated by different models to have an intuitive understanding of title quality. Two general situations are considered, one is with a query and the other without query.

From the second column, we can see that with a query, generally, all the models can generate proper titles. But titles generated by our model can be more detailed and specific. For example, when the query is "听评书 (listen to storytelling)", titles generated by baseline models seem more general while title generated by our model can directly give the detail about the subject "岳飞传 (Biography of Yue Fei)" of the ad.

When there is no query available, the results can show the advantage of our model. From the last column in Table 3, we can see that titles generated by baseline models deviate from the ground-truth to a certain extent, with a wrong subject for example. However, the title generated by our model is very similar to the ground truth. This shows that the traditional way of using previous ads by directly mining keywords from them is not efficient. Our model can make full use of all the available information, thus generate more correct and specific titles. In a real-world application, one goal the platform wants to achieve is to recommend satisfactory titles for advertisers once they come to the platform without bothering to input anything. In this sense, our model has great practical value.

3.7 Discussion

Experiments have shown that our proposed model outperforms all the baseline models. In this section, we make further analysis of those models.

In the case where we already have an extensive title database, the retrieval method can be a quick and effective way to generate titles. However, the long-term use of this method is not appropriate. Duplicate titles would reduce the effectiveness of advertising. A simple keyword-based generation model combined with a nucleus sampling decoding strategy can increase the diversity of generated titles to a certain extent. But due to the limited input information, the relevance and correctness cannot be promised. Besides, all these methods have higher requirements on the input query, which requires the advertisers to provide query keywords as accurately as possible. This violates the motivation of the title recommendation system. Once there is no input query from the advertiser available, the quality of titles generated by keywords mined from landing page and previous ads has shown a relatively large decline. When generating titles for a new ad, it's crucial to know the current interest, basic background, and expression preferences of the advertiser. Current ad material only contains limited information, failing to provide enough details to the generation model. APTGN model leverages massive related advertising data to give a comprehensive description of each ad, thus, the generated titles can be correct and specific.

4   Conclusion

Recommending satisfactory titles for ads is of great significance in real-world applications. Considering the dilemma that limited current information of ad is available while lots of related previous advertisement data wasted, we propose an Ad-Profile-based model to help generate titles for ads automatically. A profile is constructed for each ad with twenty features, which covers three dimensions, basic, current, and stylistic information. Also, a feature extraction model is designed to give a representation vector of each feature. The transformer model is adapted to do title generation taking all the constructed features as input. Automatic metrics and online A/B testing both show that our proposed model significantly outperforms the baseline models which have been commonly used in the advertising platform. In the future, we will further explore how to integrate all the features into the model in a better way.

Acknowledgments. This work is supported in part by grants JC2017005, HW2020004, 20AZD085 and 2020A1515010949. We thank the anonymous reviewers for their valuable comments.

References

1. Ahmed, A.: Ad campaigns generation using deep learning approach (2019)
2. Bartz, K., Barr, C., Aijaz, A.: Natural language generation for sponsored-search advertisements. In: Proceedings of the 9th ACM Conference on Electronic Commerce, pp. 1–9 (2008)
3. Chen, X., et al.: Microsoft coco captions: data collection and evaluation server. arXiv preprint arXiv:1504.00325 (2015)
4. Dennis, W.L., Erwin, A., Galinium, M.: Data mining approach for user profile generation on advertisement serving. In: 2016 8th International Conference on Information Technology and Electrical Engineering (ICITEE), pp. 1–6. IEEE (2016)
5. Hughes, J.W., Chang, K.H., Zhang, R.: Generating better search engine text advertisements with deep reinforcement learning. In: Proceedings of the 25th ACM SIGKDD International Conference on Knowledge Discovery & Data Mining, pp. 2269–2277 (2019)
6. Li, G., Crego, J.M., Senellart, J.: Enhanced transformer model for data-to-text generation. In: Proceedings of the 3rd Workshop on Neural Generation and Translation, pp. 148–156 (2019)
7. Liu, P.: Computational Advertising: The Market and Technology of Internet Commercial Monetization. People Post Press, New York City (2015)
8. Radford, A., Wu, J., Child, R., Luan, D., Amodei, D., Sutskever, I.: Language models are unsupervised multitask learners. OpenAI Blog **1**(8), 9 (2019)
9. Schroff, F., Kalenichenko, D., Philbin, J.: FaceNet: a unified embedding for face recognition and clustering. In: Proceedings of the IEEE Conference on Computer Vision and Pattern Recognition, pp. 815–823 (2015)
10. Simsek, A., Karagoz, P.: Wikipedia enriched advertisement recommendation for microblogs by using sentiment enhanced user profiles. J. Intell. Inf. Syst. **54**(2), 245–269 (2018). https://doi.org/10.1007/s10844-018-0540-5
11. Vaswani, A., et al.: Attention is all you need. In: Advances in Neural Information Processing Systems, pp. 5998–6008 (2017)

Parasitic Network: Zero-Shot Relation Extraction for Knowledge Graph Populating

Shengbin Jia[1,2(✉)], E. Shijia[2], Ling Ding[1], Xiaojun Chen[1], LingLing Yao[2], and Yang Xiang[1]

[1] Tongji University, Shanghai 201804, China
{shengbinjia,dling,xiaojunchen,shxiangyang}@tongji.edu.cn
[2] Tencent, Shanghai 200233, China
{shengbinjia,allene,vincentyao}@tencent.com

Abstract. The relation tuple is the basic unit of the knowledge graph. Conventional relation extraction methods can only identify limited relation classes and not recognize the unseen relation types that have no pre-labeled training data. In this paper, we explore the zero-shot relation extraction to overcome the challenge. The only requisite information about an unseen type is the label name. We propose a Parasitic Neural Network (PNN), where unseen types are parasitic on seen types to get automatic annotation and training. The model learns a mapping between the feature representations of text samples and the distributions of unseen types in a shared semantic space. Experiment results show that our model significantly outperforms others on the unseen relation extraction task and achieves effect improvement of more than 20% when there are not any manual annotations or additional resources. This model, with good performance and fast implementation, can support the industrial knowledge graph populating.

Keywords: Relation extraction · Zero-shot learning · Neural network · Knowledge graph

1 Introduction

The relation extraction (RE) task aims to determine relational facts from the unstructured text. A relational fact usually comprises of a resourceful relation and two entities, which expresses a certain semantic connection between entities, e.g., (Michael Jordan, place_of_birth, New York). RE is a key link in the construction of knowledge graphs (KGs), especially to populate knowledge bases in an automatic manner.

The conventional methods (including one/few -shot learning) [5,14] cannot meet the practical needs of the KG populating in industrial applications. Generally, there are massive fine-grained types of relations in the real KGs. However, these methods are often to distinguish the limited relational taxonomy, where

© Springer Nature Switzerland AG 2021
C. S. Jensen et al. (Eds.): DASFAA 2021, LNCS 12683, pp. 541–552, 2021.
https://doi.org/10.1007/978-3-030-73200-4_38

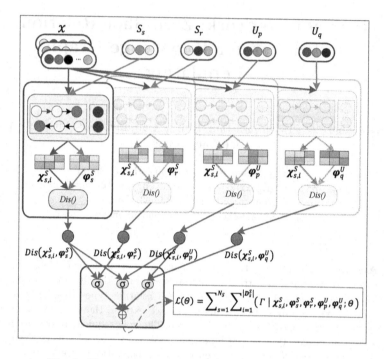

Fig. 1. The architecture of the parasitic neural network.

the relation types are seen and each type must have a certain number of pre-labeled samples. They are unable to generalize to new (unseen) relations (i.e., they will break down when predicting a type that has no training examples). Collecting sufficient labeled instances for training on all expected categories is almost impossible, in contrast with the limited number of relation types covered by existing datasets.

To address the challenge, we develop a zero-shot relation extraction (ZRE), which is under the restriction that the extractor should identify facts of new relation types after learning from limited labeled instances of seen types. The ZRE is a promising learning paradigm by reducing annotation costs and improving application efficiency. However, it is immature and has received limited attention. The existing popular methods address the ZRE task to develop specific transfer learning procedures by reading comprehension [11], textual entailment [19], and so on. We consider these methods to be **indirect-trick**. They need much unnatural descriptive information to improve the understandability of relation types. Annotation costs severely decrease their applicability to new types. In this paper, we are committed to the **direct-trick** method. It does not need any manual intervention to pre-describe relation types. Instead, it just uses the name of type labels that is a natural expression of relation semantics.

Furthermore, we raise a zero-shot learning framework that learns the mapping between the text feature representations and the relation type embeddings (prototypes) in a shared semantic space. To prevent over-fitting seen types and successfully adapt to unseen types, the model requires to solve a principal problem: How to understand the distributions of unseen relation types in the shared space. And to this end, we propose the parasitic neural network model.

In summary, our key contributions are presented as follows. (1) We develop a general zero-shot learning framework for unseen relation extraction by the direct-trick. It emphasizes the non-use of manual annotations or external knowledge. (2) Based on the Parasitism, we propose the PNN that leverages the association of the relation types in a shared semantic space to learn the distributions of unseen types automatically. (3) Our experiment results achieve significant improvement than other methods, both in the direct- and indirect-trick cases.

2   Related Work

The conventional relation extraction was usually regarded as a standard classification task with the pre-fixed relational taxonomy [3,14]. The classifiers were usually involved in minimizing the softmax cross-entropy loss function. They heavily depended on time-consuming and labor-intensive annotated data. Distant-supervision was a primary way to annotate adequate amounts of training data by heuristically aligning knowledge bases and text [17]. However, the distant-supervision could produce a large number of mislabeling. Besides, many researchers formulated relation extraction as the few/one-shot learning tasks which aimed to learn extraction models from one, or only a few, training samples [5,28]. However, a shortcoming common to the above methods was that they would break down when predicting a type that had no training examples. Thus zero-shot extraction was irreplaceable because the training set could not contain such numerous relation categories at once.

Most of the works of zero-shot learning were focused on the area of computer vision, such as the tasks of object recognition [10], image segmentation [12, 18], image retrieval [29], and so on. In the area of natural language processing, the applications of zero-shot learning have been emerging in recent years, such as machine translation [8,30], entity typing [20,21], event extraction [2,6], and knowledge graph completion [4,23].

As for zero-shot relation extraction, it was immature and had received limited attention. By analyzing linguistics, old-fashioned approaches developed unsupervised models (e.g., clustering) based on combinations of manual features, patterns, or corpus-level resources [9,16,27]. They tended to be inefficient and consumed a lot of manpower. The recent methods were to transfer other tasks to produce relations. Levy et al. [11] formulated relation types as various parametrized natural-language questions, then used a reading comprehension model to process the questions to obtain relation facts. By considering the text and the relation description as the premise and hypothesis respectively, Obamuyide et al. [19] transformed the extraction task to determine the truthfulness of the hypothesis by a textual entailment model. It was expensive to manually formulate reading comprehension questions or entailment rules. In addition, transfer-based methods were constrained by the capability of indirect tasks whose errors or defects could be cascaded into relation extraction.

In this paper, we take a universal and all-inclusive manner, which is to model the mapping between text instances and relation type prototypes [25,26]. More importantly, we explore the ZRE via the direct-trick. Because of the extremely scarce relation type information, we set up the parasitic learning.

In addition, it's worth noting that the ZRE is different from the open relation extraction task [1,7]. While open extraction systems need no relation-specific training data, such open relation is rough and non-standardized and should align from different relation phrasings. Here, we hope to extract the canonical relations independent of how the original text is phrased.

3   Methodology

3.1   Parasitism Thought

Let $S = \{S_s \mid s = 1, \cdots, N_S\}$ denotes a set of seen relation types and $U = \{U_u \mid u = 1, \cdots, N_U\}$ unseen types, with $S \cap U = \emptyset$. Suppose that the dataset $\mathbb{D} = \mathbb{D}^S \cap \mathbb{D}^U$ is a collection of text instances. The $\mathbb{D}_s^S = \{x_{s,i}^S \mid y_s^S = S_s\}$ is as the set of labeled training instances belonging to seen types S_s. The $\mathbb{D}^U = \{x_j^U\}$ is as the set of testing instances, meanwhile, $y_j^U \in U$ is to be predicted as the corresponding type labels for x_j^U. In semantic embedding space $\mathbb{R}^z$, the instance x will be embedded to χ and it is assumed to belong to one category. The types will be vectorized as type prototypes $\varphi = \{\varphi^S, \varphi^U\}$. Overall, the ZRE task is defined as: Given $\mathbb{D}^S$, the ZRE system learns the mapping $f(\cdot) : \chi \to \varphi$, which can classify testing instances $\mathbb{D}^U$ (i.e., to predict y^U).

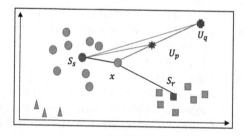

Fig. 2. The distributed associations of a set of relation types $\{S_s, S_r, U_p, U_q\}$. The definition of these symbols is consistent with the Algorithm 1.

The instances with the same relation type will cluster around a single prototype in the shared semantic space, whereas they are far away from other type prototypes. Meanwhile, the more similar types are distributed closer in the space [25,26]. Therefore, we determine the semantic distance $Dis(\cdot)$ between the feature representations χ and the type prototypes φ. Here, the semantic distance

---

Algorithm 1. Parasitic neural network training algorithm.

Require: S, $\mathbb{D}^S$, φ^S, U, φ^U.

1: **Initialize**

2:    Calculate the semantic distances of seen types S to unseen types U, as,
$$D(S_s, U_u) = Dis(\varphi_s^S, \varphi_u^U) \mid s = 1, ..., N_S, \ u = 1, ..., N_U.$$

3:    Obtain the array R by ranking the $D(S, U)$ (from small to large),
$$\forall\{ \ s = 1, ..., N_S, \ m = 1, ..., N_U-1\} \text{ s.t.}$$
$$R[s][m] \in U \wedge D(S_s, R[s][m]) \leqslant D(S_s, R[s][m+1]).$$

4: **for** S_s (as *Host*) in S **do**

5:    **for** $x_{s,i}^S$ in $\mathbb{D}_s^S$ **do**

6:       Select S_r from S randomly;

7:       Select any $U_p = R[s][m']$ from $R[s][1 : N_U-1]$ as *Parasite*;

8:       Select any $U_q = R[s][m'']$ from $R[s][m' : N_U]$ as *Parasite*.

9:       Construct four sets of inputs for PNN sub-networks, as $(x_{s,i}^S, S_s), (x_{s,i}^S, S_r), (x_{s,i}^S, U_p), (x_{s,i}^S, U_q)$.

10:      Run PNN to

11:         learn the $\chi_{s,i}^S$ of instance $x_{s,i}^S$;

12:         obtain the corresponding prototypes φ_s^S, φ_r^S, φ_p^U, φ_q^U;

13:         calculate $Dis(\chi_{s,i}^S, \varphi_s^S)$, $Dis(\chi_{s,i}^S, \varphi_r^S)$, $Dis(\chi_{s,i}^S, \varphi_p^U)$, $Dis(\chi_{s,i}^S, \varphi_q^U)$.

14:      Minimize the Joint energy function in Eq. 2.

15:    **end for**

16: **end for**

---

is a quantification of the mapping $f(\cdot)$. The smaller the distance, the better the mapping fit.

Furthermore, we can establish the following assumptions of the **premise**: (1) Given any relation type R_1 and a corresponding instance x, it should be sure that the semantic distance between x and R_1 is the smallest (or even 0), compared with the distance between x and any other types. (2) For arbitrarily given type R_2 and type R_3 ($R_1 \neq R_2 \neq R_3$), if the semantic distance between R_2 and R_1 is smaller than that between R_3 and R_1, the semantic distance between R_2 and x should be smaller than that between R_3 and x.

The above premises imply the association among the relation types in the shared semantic space. According to this correlation among seen types $\{S_s, S_r\}$ and unseen types $\{U_p, U_q\}$, as vivid shown in Fig. 2, we can create annotations for unseen types (*Parasite*) by considering the instances of seen types as *Host*, just like "Parasitism". The Algorithm 1 (lines 1 to 8) shows the process of data creation. Then, we train the PNN model to learn the distributions of unseen types.

Joint Energy Function. As described in Algorithm 1, each sub-network produces a semantic distance metric. They interact with each other and then joint together. In detail, we establish a series of trunks, shaped like $\{Branch_1, Branch_2\}$, including $\{Dis(\chi_{s,i}^S, \varphi_s^S), Dis(\chi_{s,i}^S, \varphi_r^S)\}$, $\{Dis(\chi_{s,i}^S, \varphi_s^S), Dis(\chi_{s,i}^S, \varphi_p^U)\}$, and $\{Dis(\chi_{s,i}^S, \varphi_p^U), Dis(\chi_{s,i}^S, \varphi_q^U)\}$. According to the premises mentioned above, we will compare the semantic distance between the two in each trunk.

Motivated by the triplet loss [22], we set the σ function to ensure that the $Branch_1$ is smaller than the $Branch_2$ by at least a margin m, as,

$$\sigma(Branch_1, Branch_2, m) = max(Branch_1 - Branch_2 + m, 0). \qquad (1)$$

Thus, the joint energy function is defined as,

$$\mathcal{L}(\Theta) = \sum_{s=1}^{N_S} \sum_{i=1}^{|\mathbb{D}_s^S|} (\Gamma \mid \chi_{s,i}^S, \varphi_s^S, \varphi_r^S, \varphi_p^U, \varphi_q^U; \Theta), \qquad (2)$$

$$\Gamma = \beta \begin{array}{l} \sigma(Dis(\chi_{s,i}^S, \varphi_s^S), \ Dis(\chi_{s,i}^S, \varphi_r^S), \ m_1) + \\ \sigma(Dis(\chi_{s,i}^S, \varphi_s^S), \ Dis(\chi_{s,i}^S, \varphi_p^U), \ m_2) + \\ \gamma \ \sigma(Dis(\chi_{s,i}^S, \varphi_p^U), \ Dis(\chi_{s,i}^S, \varphi_q^U), \ m_3) \end{array} \qquad (3)$$

where we employ the cosine distance (within the range $[0, 2]$), β and γ are the trade-off parameters.

3.2 Network Architecture

As shown in Fig. 1, the PNN consists of four sub-networks that accept distinct inputs but are then joined by a joint energy function.

The parameters between the sub-networks are tied, that is, each network computes the same metric on a shared workbench (shared by *Host* and *Parasite* that seems to be a parasitic energy community). Tying guarantees that two inputs of an identical class cannot be mapped by their respective networks to very different locations, in the semantic space, and each sub-network can also distinguish inputs of varied types.

Text Embedding. The sub-network takes as input one piece of text and a relation label, and the text contains pre-identified head and tail entities. We transform the text instance x into its distributed representation $\boldsymbol{x}$ by adopting triple embeddings $\{\boldsymbol{x}^w, \boldsymbol{x}^c, \boldsymbol{x}^p\}$. The $\boldsymbol{x}^w$ denotes the word embedding. To deal with unregistered words, we use a convolutional neural network to encode character embedding $\boldsymbol{x}^c$ of each word, as [15] doing. The $\boldsymbol{x}^p$ represents the position embedding to specify entity pairs. Similarly to [13], it is defined as the combination of the relative distances from the current word to head or tail entities.

Relation Type Prototype. We achieve the prototype φ with the word embeddings of type labels' names. Word embeddings capture distributional similarities from a large text corpus. Semantically similar words are embedded as nearby vectors, while semantically dissimilar words are presented as vectors far apart. Therefore, word embeddings can effectively reflect the semantic distance between relation types. Each prototype is an average of word embeddings of the core words (i.e., nouns, adjectives, etc., except prepositions, conjunctions) in its label name. We can fine-tune these embeddings along with training.[1]

---

[1] We do not fine-tune these embeddings during training.

Learning Feature Representation from Text. The sample text has latent feature information that is category-invariant and implied in content, syntax, etc. We feed the text embeddings into the bidirectional ordered neurons long short-term memory network (ONLSTM) [24] to encode feature representation χ. The ONLSTM performs tree-like syntactic structure composition operations on a sentence without destroying its sequence form. It can learn context temporal semantics, meanwhile, capture potential syntactic information involved in natural language.

Notably, this syntactic information is critical to the relation extraction task [7]. There are strict semantic associations and formal constraints that the head argument is the agent of the relation and the tail argument is the object of the relation. Integrating syntax-semantics into a neural network can encode better representations of natural language sentences.

Based on the new activation function cumulative softmax (cumax()), the ONLSTM promotes differentiation of the life cycle of information stored inside each neuron: high-ranking neurons will store long-term information, while low-ranking neurons will store short-term information. In detail,

$$f_t = sigmoid\left(W_f x_t + U_f h_{t-1} + b_f\right) \tag{4}$$

$$i_t = sigmoid\left(W_i x_t + U_i h_{t-1} + b_i\right) \tag{5}$$

$$\hat{c}_t = tanh\left(W_c x_t + U_c h_{t-1} + b_c\right) \tag{6}$$

$$\tilde{f}_t = cumax(W_{\tilde{f}} x_t + U_{\tilde{f}} h_{t-1} + b_{\tilde{f}}) \tag{7}$$

$$\tilde{i}_t = cumax(W_{\tilde{i}} x_t + U_{\tilde{i}} h_{t-1} + b_{\tilde{i}}) \tag{8}$$

$$\omega_t = \tilde{f}_t \circ \tilde{i}_t \tag{9}$$

$$c_t = \omega_t \circ \left(f_t \circ c_{t-1} + i_t \circ \hat{c}_t\right) + \left(\tilde{f}_t - \omega_t\right) \circ c_{t-1} + \left(\tilde{i}_t - \omega_t\right) \circ \hat{c}_t \tag{10}$$

$$o_t = \sigma\left(W_o x_t + U_o h_{t-1} + b_o\right) \tag{11}$$

$$h_t = o_t \circ tanh\left(c_t\right). \tag{12}$$

In the above formulas, f_t, i_t, o_t, $\tilde{f}_t$, $\tilde{i}_t$, c_t, h_t represent the forget gate, input gate, output gate, master forget gate, master input gate, cell memory and hidden state, respectively. The W and U are the trainable parameter matrixes, b is the bias. The $\circ$ denotes the Hadamard product.

Extracting Unseen Relations. Once the model is optimized, we determine the possible relation that a test instance x_j^U may represent, if any. The top ranked prediction from the candidate predicted types U, denoted as $C(x_j^U, 1)$, is given by:

$$C(x_j^U, 1) = argmin\ Dis(\chi_j^U, \varphi_u^U),\ \ u = 1, 2, \dots, N_U \tag{13}$$

Moreover, $C(x_j^U, K)$ denotes the Kth most probable relation type predicted for x_j^U.

4   Experiments

4.1   Settings

Dataset. We evaluate models using the zero-shot relation extraction dataset of [11]. It consists of 120 relation types that are from the knowledge base Wikidata. We use the positive labeled relation instances in this dataset. There are 225,060 samples. By applying a similar process to [11] and [19], we randomly select 24 classes as a testing set, 10 classes as the dev set, and the rest as the training set. The results reported for each experiment are the average taken over five runs with independent random initializations. Given different thresholds regarding distance, we can measure the precision (P), recall (R), and F1 of the results. We report the optimal values.

Hyperparameters. We implement the neural network by the Keras. The word embedding is from the GloVe[2] with 100 dimensions. The character embedding is initialized randomly as 50 dimensions. The size of the ONLSTM unit is 100. Parameter optimization is performed with Adam optimizer. To mitigate overfitting, we apply the dropout and early-stopping methods. Besides, we empirically set $m_1 = 0.1$, $m_2 = 0.1$, $m_3 = 0.08$, $\beta = \gamma = 1$.

Comparison Systems. We examine several major components in our model. (1.1) We test the influence of word embedding on prototypes, increasing noise by randomly zeroing its value in varied proportions. (1.2) We compare the bidirectional ONLSTM to the bidirectional LSTM. (1.3) We verify the choice of distance, including logistic regression probability (LR), euclidean distance (EU), and cosine distance (COS).   We compare our PNN-based systems to external systems. (2.1) *120-Softmax* is a conventional 120-dimensional softmax classifier, but we only use seen types to train it. (2.2) *NaiveMAP* learns the mapping between the samples and seen types, by using the single mapping distance as loss directly (*Single*) [23], or by adopting a tied network with triplet loss (*Triplet*) [6]. (2.3) Model of Levy et al. [11][3] is via reading comprehension, by using different descriptions for relation types (i.e., *NL* - the label's name, *SQ* - only a single question template per relation type, *MQ* - multiple questions, and *QE* - an ensemble learning way). (2.4) Model of Obamuyide et al. [19] is based on textual entailment, where *TE* transforms external entailment corpus for training, and *MD* represents training with manual annotations.

4.2   Results and Analysis

Ablation Study. The upper part of Table 1 shows our PNN-based models with different factors. The relation prototype is crucial to the model, and it depends

---

[2] https://nlp.stanford.edu/projects/glove/.

[3] Notably, the methods of Levy et al. input an instance with head entity and relation (question) to predict tail entity (answer), however, our model inputs an instance with head and tail entities to predict relation. But from the perspective that we all aim to obtain fact triples, their results are valuable for reference.

on the quality of the embedding. Fortunately, just using the usual embedding GloVe, we have achieved the F1 value of 58%. Compared with the LSTM, the ONLSTM improves model performance by 7%. It shows that the potential syntactic information captured by the ONLSTM is pretty useful for relation extraction. However, the LSTM explicitly imposes a chain structure on a sentence that is contrary to the potential hierarchical structure of language, resulting in a loss of syntactic information [24]. The choice of distance metric is also important, where the cosine distance (COS) can well improve the effectiveness of a PNN.

Comparison with Other Methods. The middle part of Table 1 presents the results of several direct-trick models. Our PNN (with ONLSTM and COS) remarkably outperforms others. As expected, the conventional classifier 120-Softmax has almost no effect and is at the level of random guessing. The NaiveMAP+Single is insufficient in a zero-shot setting since it cannot capture the association information between types. Furthermore, the NaiveMAP+Triplet has an F1 improvement of more than 40% than NaiveMAP+Single, which proves that the configuration of the tied network structure is appropriate. However, the NaiveMAP+Triplet tends to over-fit seen types. Our model can alleviate this over-fitting effectively by learning the semantic distributions of unseen relation types explicitly, resulting in more than 5% F1 improvement. Besides, our model achieves effect improvement of more than 20% than the model of Levy et al. when there are no manual annotations or additional resources.

Table 1. The performance of the PNN-based models (ablation study) and external systems (for comparison). * indicates the model via the indirect-trick.

Models				P	R	F1
PNN	10%Noise	COS	ONLSTM	50.91	44.97	47.75
	5%Noise	COS	ONLSTM	57.68	49.48	53.26
	0%Noise	COS	LSTM	58.47	45.45	51.13
		COS	ONLSTM	**63.40**	**53.79**	**58.20**
		LR	ONLSTM	57.96	43.28	49.55
		EU	ONLSTM	61.75	52.32	56.64
120-Softmax (with ONLSTM)				3.30	3.30	3.30
NaiveMAP (with COS and ONLSTM)			Single	10.12	8.97	9.51
			Triplet	55.15	50.93	52.95
Levy et al. [11]			NL	40.50	28.56	33.40
			SQ *	37.18	31.24	33.90
			MQ *	43.61	36.45	39.61
			QE *	45.85	37.44	41.11
Obamuyide et al. [19]			TE *	–	–	44.38
			MD *	–	–	64.78
PNN (with ONLSTM)*				**75.15**	**72.83**	**73.98**

As shown in the bottom of Table 1, by introducing manual guidance[4], the performance improvement of PNN is significantly, and other methods by the indirect-trick are inferior to our model. These indirect-trick methods are constrained by extra annotation effort. The less quantity and lower quality of annotation information, the worse the models will perform.

Analyze the Impact of Training Set Size. Figure 3 shows the results of our model after being trained with varying proportions of seen types. As the seen types in the training set increasing, the performance of unseen relation extraction will become better. The reason may be that the diversity of training set reduces the tendency of the model to over-fit seen types. In addition, most of the correct extractions appear in the front part (i.e., top $K \leqslant 5$) of the candidate type ranks. It proves that our model can understand the distributions of unseen relation types in the shared space, where the semantic distance between each sample and its corresponding prototype tends to be minimal.

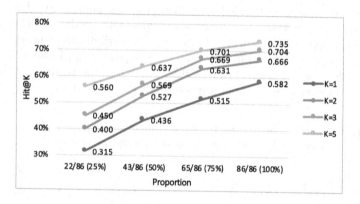

Fig. 3. The effects of our model trained with varying number of seen relation types. The hits@K represents the F1 of correct extractions ranked in the top K in Eq. 13.

Case Study. We sample an unseen relation type "father" and its corresponding instances from the test set. The 1st row of Table 2 presents several unseen types and their respective semantic distance from the target type "father". The 3rd and 5th rows of Table 2 show the semantic distances between each text instance and the relation types. The distance between each instance and the target type is minimum. Besides, the smaller the semantic distance between a relation type and the target type, the smaller the semantic distance between it and the instance corresponding to the target type can be. Therefore, the case demonstrates that our model can learn the category-invariant semantics of text to match the distribution of unseen type prototypes.

---

[4] We manually annotate a sample for each relation type, and use the way in Sect. 3.2 to learn the expression of the text, it is as the relation type prototype.

Table 2. Examples of unseen relation type "father".

father (0)	named_after (0.361)	employer (0.644)	chairperson (0.889)
[Samuel Dirksz van Hoogstraten]$_{entity}$ trained first with his father [Dirk van Hoogstraten]$_{entity}$ and stayed in Dordrecht until about 1640			
0.403	0.562	0.726	1.119
[Bertrade de Montfort]$_{entity}$ was the daughter of [Simon I de Montfort]$_{entity}$ and Agnes, Countess of Evreux			
0.362	0.531	0.728	1.122

5   Conclusion

Knowledge graphs have been widely used in the industry. However, the acquisition of large-scale knowledge makes it challenging for relation extraction. To this end, ZRE is evolving to identify unseen relations. Previous solutions for ZRE rely on expensive annotations that are hard to get, and we suggest an approach to overcomes such limitations. Via the direct-trick, we propose a general zero-shot relation extraction framework. Furthermore, we develop the parasitic neural network. Inspired by parasitism, it owns a tied network structure and expands annotations automatically for unseen relation types to learn their distributions. The experimental results are excellent. In practice, our model has been able to assist workers in populating KGs, which has improved productivity.

Acknowledgments. This work is supported by the National Natural Science Foundation of China under Grant 72071145 and the 2020 Tencent Rhino-Bird Elite Training Program.

References

1. Banko, M., Cafarella, M.J., Soderland, S., Broadhead, M., Etzioni, O.: Open information extraction from the web. In: IJCAI, pp. 2670–2676 (2007)
2. Bronstein, O., Dagan, I., Li, Q., Ji, H., Frank, A.: Seed-based event trigger labeling: how far can event descriptions get us? In: ACL, pp. 372–376 (2015)
3. Doddington, G.R., Mitchell, A., Przybocki, M.A., Ramshaw, L.A., Strassel, S.M., Weischedel, R.M.: The automatic content extraction (ACE) program-tasks, data, and evaluation. In: LREC, vol. 2, p. 1 (2004)
4. Goldstein, O.: Zero-shot relation extraction from word embeddings. Ph.D. thesis, UCLA (2018)
5. Han, X., et al.: FewRel: a large-scale supervised few-shot relation classification dataset with state-of-the-art evaluation. In: EMNLP (2018)
6. Huang, L., Ji, H., Cho, K., Dagan, I., Riedel, S., Voss, C.: Zero-shot transfer learning for event extraction. In: ACL, pp. 2160–2170 (2018)
7. Jia, S., Li, M., Xiang, Y., et al.: Chinese open relation extraction and knowledge base establishment. ACM TALLIP **17**(3), 15 (2018)
8. Johnson, M., et al.: Google's multilingual neural machine translation system: enabling zero-shot translation. TACL **5**, 339–351 (2017)

9. Kok, S., Domingos, P.: Extracting semantic networks from text via relational clustering. In: ECML, pp. 624–639 (2008)

10. Lampert, C.H., Nickisch, H., Harmeling, S.: Learning to detect unseen object classes by between-class attribute transfer. In: CVPR, pp. 951–958 (2009)

11. Levy, O., Seo, M., Choi, E., Zettlemoyer, L.: Zero-shot relation extraction via reading comprehension. In: CoNLL, pp. 333–342 (2017)

12. Li, Z., Gavves, E., Mensink, T., Snoek, C.G.M.: Attributes make sense on segmented objects. In: Fleet, D., Pajdla, T., Schiele, B., Tuytelaars, T. (eds.) ECCV 2014. LNCS, vol. 8694, pp. 350–365. Springer, Cham (2014). https://doi.org/10.1007/978-3-319-10599-4_23

13. Lin, Y., Shen, S., Liu, Z., Luan, H., Sun, M.: Neural relation extraction with selective attention over instances. In: ACL, pp. 2124–2133 (2016)

14. Liu, K.: A survey on neural relation extraction. Sci. China Technol. Sci. **63**, 1971–1989 (2020). https://doi.org/10.1007/s11431-020-1673-6

15. Ma, X., Hovy, E.: End-to-end sequence labeling via bi-directional LSTM-CNNs-CRF. In: ACL, pp. 1064–1074 (2016)

16. Min, B., Shi, S., Grishman, R., Lin, C.Y.: Ensemble semantics for large-scale unsupervised relation extraction. In: EMNLP, pp. 1027–1037 (2012)

17. Mintz, M., Bills, S., Snow, R., Jurafsky, D.: Distant supervision for relation extraction without labeled data. In: ACL, pp. 1003–1011 (2009)

18. Naha, S., Wang, Y.: Object figure-ground segmentation using zero-shot learning. In: ICPR, pp. 2842–2847 (2016)

19. Obamuyide, A., Vlachos, A.: Zero-shot relation classification as textual entailment. In: EMNLP Workshop on FEVER, pp. 72–78 (2018)

20. Obeidat, R., Fern, X., Shahbazi, H., Tadepalli, P.: Description-based zero-shot fine-grained entity typing. In: NAACL, pp. 807–814 (2019)

21. Pasupat, P., Liang, P.: Zero-shot entity extraction from web pages. In: ACL, pp. 391–401 (2014)

22. Schroff, F., Kalenichenko, D., Philbin, J.: FaceNet: a unified embedding for face recognition and clustering. In: CVPR, pp. 815–823 (2015)

23. Shah, H., Villmow, J., Ulges, A., Schwanecke, U., Shafait, F.: An open-world extension to knowledge graph completion models. In: AAAI (2019)

24. Shen, Y., Tan, S., Sordoni, A., Courville, A.: Ordered neurons: integrating tree structures into recurrent neural networks. In: ICLR (2019)

25. Snell, J., Swersky, K., Zemel, R.: Prototypical networks for few-shot learning. In: NeurIPS, pp. 4077–4087 (2017)

26. Socher, R., Ganjoo, M., Manning, C.D., Ng, A.: Zero-shot learning through cross-modal transfer. In: NeurIPS, pp. 935–943 (2013)

27. Yan, Y., Okazaki, N., Matsuo, Y., Yang, Z., Ishizuka, M.: Unsupervised relation extraction by mining Wikipedia texts using information from the web. In: ACL, pp. 1021–1029 (2009)

28. Ye, Z.X., Ling, Z.H.: Multi-level matching and aggregation network for few-shot relation classification. In: ACL (2019)

29. Zhang, Z., Saligrama, V.: Zero-shot learning via joint latent similarity embedding. In: CVPR, pp. 6034–6042 (2016)

30. Zheng, H., Cheng, Y., Liu, Y.: Maximum expected likelihood estimation for zero-resource neural machine translation. In: IJCAI, pp. 4251–4257 (2017)

Graph Attention Networks for New Product Sales Forecasting in E-Commerce

Chuanyu Xu[1], Xiuchong Wang[1], Binbin Hu[2], Da Zhou[3(✉)], Yu Dong[1], Chengfu Huo[1], and Weijun Ren[1]

[1] Alibaba Group, Hangzhou, China
{tracy.xcy,xiuchong.wxc,dongyu.dy,
chengfu.huocf,afei}@alibaba-inc.com
[2] Ant Group, Hangzhou, China
bin.hbb@antfin.com
[3] Xiamen University, Xiamen, China
zhouda@xmu.edu.cn

Abstract. Aiming to discover competitive new products, sales forecasting has been playing an increasingly important role in real-world E-Commerce systems. Current methods either only utilize historical sales records with time series based models, or train powerful classifiers (*e.g.,* DNN and GBDT) with subtle feature engineering. Despite effectiveness, they have limited abilities to make prediction for new products due to the sparsity of product-related features. With the observation on real-world data, we find that some additional time series features (*e.g.,* brand and category) implying product characteristics also play vital roles in new product sales forecasting. Hence, we organize them as a new kind of dense feature called CPV (Category-Property-Value) and propose a Time Series aware Heterogeneous Graph (TSHG) to integrate CPVs and products based time series into a unified framework for fine-grained interaction. Furthermore, we propose a novel Graph Attention Networks based new product Sales Forecasting model (GASF) that jointly exploits high-order structure and time series features derived from THSG for new product sales forecasting with graph attention networks. Moreover, a multi trend attention (MTA) mechanism is also proposed to solve temporal shifting and spatial inconsistency between the time series of products and CPVs. Extensive experiments on an industrial dataset and online system demonstrate the effectiveness of our proposed approaches.

Keywords: Time series aware Heterogeneous Graph · Graph attention networks · Multi trend attention

1 Introduction

With the development of E-Commerce, the user scale on E-Commerce platform is constantly expanding and the consumer demand is increasingly diversified.

C. Xu and X. Wang—Contribute equally to this work.

C. S. Jensen et al. (Eds.): DASFAA 2021, LNCS 12683, pp. 553–565, 2021.
https://doi.org/10.1007/978-3-030-73200-4_39

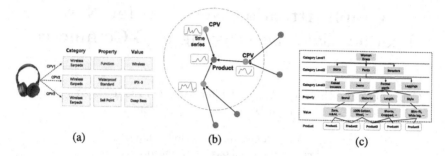

Fig. 1. (a) Illustration of CPVs for wireless headphone. (b) A toy example of TSHG in our scenario. (c) Illustration of CPV structure.

Discovering and supplying high-quality new products has been becoming the core technology to meet diversified needs of customers. According to the statistical data in Alibaba[1], millions of new products are released everyday and they contribute 40% of GMV (Gross Merchandise Volume) in the platform. Therefore, new product sales forecasting, which aims to discover competitive new products, has played a fundamental role in enhancing the productivity of E-Commerce and satisfying user experience.

Intuitively, the new product sales forecasting in E-Commerce can be formulated as a time series prediction problem, which is well explored in numerous studies. Naturally, conventional solutions propose to adopt time series models (e.g., Auto-Regressive Integrated Moving Average (ARIMA) [2] and Long Short Term Memory network (LSTM) [7]) for prediction, which only utilize historical sales records. However, the sparsity and instability of historical records of new products may harm the performance of these models. Besides, a series of feature based methods are proposed to perform subtle feature engineering for each product, and then train a powerful classifier (e.g., Gradient Boosting Decision Tree (GBDT) [4] and Deep Neural Networks [3]) for prediction. Due to the powerful ability of feature learning, these methods have achieved a considerable improvement on sales forecasting task. Nevertheless, we argue that the existing methods have three major limitations for new product sales forecasting.

- **L1:** They commonly utilize product-related features (e.g., user behavior features and product static information) to make prediction for product sales in future. Unfortunately, these features are sparse or even absent for new products in real-world application, which seriously hinders the forecasting performance.
- **L2:** These approaches treat new products and its additional features (e.g., category and brand) separately, which ignores the interaction between them, resulting in sub-optimal performance in the complex scenario.
- **L3:** With the analysis of real-world data, we find the temporal shifting and spatial inconsistency (we will revisit temporal shifting and spatial inconsis-

---

[1] https://www.alibaba.com/.

tency in latter sections) between the time series of products and additional features in our business scenario, which is poorly explored in the existing approaches.

To address these issues, we aim to comprehensively explore and exploit abundant product-related features and time series features in a more proper way, and propose a novel Graph Attention network based Sales Forecasting approach, called GASF shortly.

Inspired from daily business experience that the sales of new products are mainly determined by whether their characteristics meet the market trend, besides original time series features, we propose to construct *Category - Property - Value* (called CPV shortly) features to characterize the trends of new products more comprehensively. Figure 1(a) shows an example. Wireless headphone is described by the CPV "*Wireless Earpads - Function - Wireless*", which denotes that it belongs to the "*Wireless Earpads*" category and has the "*Wireless*" "*Function*". Note that a product can be described by multiple CPVs and a CPV can also be used to describe multiple products. In contrast with historical sales records, the proposed CPV features capture the sales trends in the macro level, especially for new products with limited features or interactions (**L1**). Therefore, it is quite likely to take full advantage of time series features for improving the performance on sales forecasting task by integrating above two aspects of information together.

On the other hand, graph has been proposed as a general approach to model various types of objects. In order to jointly consider products and extracted CPVs together, we propose **T**ime **S**eries aware **H**eterogeneous **G**raph (TSHG for short) to effectively capture underlying specialities of new products for sales forecasting. As shown in Fig. 1(b), products and their CPVs are connected and objects (products or CPVs) in TSHG contains time series. With the help of recently emerging graph neural networks [15], high-order structure derived from TSHG and time series features in products and CPVs can be naturally explored in an unified framework (**L2**).

With the observation of real data in our E-commerce scenario, we find the temporal shifting and spatial inconsistency between the time series of products and CPVs, which means the response speed (*i.e.*, temporal) and intensity (*i.e.*, spatial) of products and CPVs are quite different for the hot spot in the market. To fill this gap, we proposed a novel Multi Trend Attention (MTA) mechanism in GASF, which (1) shifts the trend of the product over multiple time units on the time axis to get multiple distances of the product and CPV trends (spatial inconsistency), and (2) gets the time series trend by taking first-order derivative and ensures the trend in the same space (spatial inconsistency) (**L3**). With MTA, our model is expected to learn fine-grained interaction of time series between products and CPVs beyond topological structure.

To sum up, we make the following contributions:

- Inspired from daily business experience, we construct a new kind of heterogeneous feature called CPV in E-Commerce to overcome the sparsity of new products. Moreover, we propose to frame the new product sales forecasting problem in the setting of TSHG, which integrates the products, CPVs and time series in an unified framework.

- We propose GASF, an end-to-end approach to simultaneously extract time series and structural information in TSHG. To our best knowledge, it is the first attempt to introduce deep graph learning with attention mechanism for sales forecasting task, which provides a new perspective to capture fine-grained interaction between products and other objects in real-world E-commerce scenarios.
- With the analysis of real data, the temporal shifting and spatial inconsistency between the time series of products and CPVs is uncovered and a novel multi trend attention mechanism is designed in GASF model to solve it.
- We perform extensive experiments on an Alibaba dataset for product sales forecasting. The results demonstrate that our model consistently and significantly outperform various state-of-the-arts. Moreover, our model also achieves significant performance improvement on online system.

2   Preliminaries

In real-world E-Commerce systems, a new product is associated with a series of basic information (*i.e.*, category, property and value) when it is released. In order to comprehensively characterize new products for sales forecasting, we proposed to extract category-property-value (CPV) features with normalization as follows:

- **Category:** we apply the clustering technology and calculate frequencies for category names in each cluster, the name with highest frequency will be selected as the standard category and other category names are mapped to it.
- **Property:** We summarize properties on category and normalize them via Word2vec [11,12]. After representing each property as embedding, we follow the same process flow of category to obtain standard properties with clustering and mapping.
- **Value:** We summarize values on category-property. These values can be normalized by synonyms and Word2vec algorithm.

We show an example of well-established CPV structure in Fig. 1(c). CPV is defined as a kind of property and value under a category in E-Commerce. For example, we can observe that "Product2" can be describe as "Woman Dress - Brand - Zara" or "Woman Dress - Material - Wool" in Fig. 1(c).

In this paper, we aim at leveraging above CPV features and time series to effectively capture underlying specialities of new products for sales forecasting. Hence, we frame our task in the setting of time series aware heterogeneous graph, which considers the CPV features and time series in an unified framework.

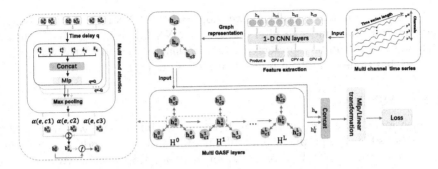

Fig. 2. The overall architecture of the proposed GASF approach.

Definition 1 *Time series aware heterogeneous graph (TSHG). A TSHG is defined as a directed graph $\mathcal{G} = \{\mathcal{V}, \mathcal{E}, \mathcal{T}\}$, where $\mathcal{V} = \mathcal{V}^{CPV} \bigcup \mathcal{V}^{Pro}$ consists of the CPV node set $\mathcal{V}^{CPV}$ and the product node set $\mathcal{V}^{Pro}$ and $\mathcal{E}$ contains edges connecting products and corresponding CPVs. $\mathcal{T} = \{t_v | \in \mathcal{V}\}$ is the set of times series on nodes (products or CPVs). Moreover, $t_v = \{\tau^1, ..., \tau^M\}$, where τ^i is a fixed-length time series and indicates that t_v is a M-channel time series composed of M different single-dimensional time series.*

The TSHG provides a flexible way to model various complex interactions between products and CPVs in an unified framework, which could be used to enhance sales forecasting. Given the above preliminaries, we are ready to formulate our task.

Definition 2 *TSHG enhanced sales forecasting. Given an time series aware heterogeneous graph $\mathcal{G}$, for each product $p \in \mathcal{P}$, we aim to learn prediction functions $\mathcal{F}^C(p|\mathcal{G}; \Theta^C)$ and $\mathcal{F}^R(p|\mathcal{G}; \Theta^R)$ to estimate whether product p will be sold in the future (classification) and its total sales over a time period (regression), respectively. Here Θ^C and Θ^R represent the parameters of the prediction function $\mathcal{F}^C$ and $\mathcal{F}^R$, respectively.*

3 Proposed GASF Model

In this section, we present GASF, an unified model to leverage CPV features and time series for new products sales forecasting with graph attention network. We show the framework of our proposed model in Fig. 2.

3.1 Feature Extraction

Since the original time series for products are multi-channel time series, we firstly present how to extract features to characterize products and CPVs, as show in Fig. 2. Following the well-established technology in previous work [13, 18], we set up multiple convolution neural network (CNN) layers for dimension reduction and feature extraction. Specifically, we adopt two 1-D CNN layers with "VALID" padding and the number of filters of second CNN layer is set to be 1. Hence, each node $v \in \mathcal{V}$ can be represented as a fixed-length vector $\mathbf{h}_v$. It is worthwhile to note that node v can be a product or a CPV in TSHG.

3.2   GASF Layer

As mentioned above, we propose a TSHG to flexibly model various complex interactions between products and CPVs in an unified framework. We now build upon the architecture of graph attention network [15] to recursively capture structural information in TSHG. Distinct from previous works [8,9,15], we propose a multi trend attention (MTA): $\mathbb{R}^d \times \mathbb{R}^d \to \mathbb{R}$ to generate attentive weights between nodes, which overcomes the temporal shifting and spatial inconsistency between products and CPVs in our scenario. Here we start with the description of a single GASF layer, consisting of information propagation and information aggregation, followed by the stack of multiple GASF layers.

Information Propagation. Intuitively, a certain node (a product or a CPV) in TSHG can be easily influenced by its neighbors. In order to capture such fine-grained interactions, we perform information propagation between a target node and its neighbors. Formally, given a node v, we use $\mathcal{N}_v$ to denote its neighbor set. We use linear weighed combination to characterize the local structural information for node v:

$$\mathbf{h}_{\mathcal{N}_v} = \sum_{c \in \mathcal{N}_v} \alpha(v, c) \mathbf{h}_c, \tag{1}$$

where $\alpha(v, c)$ weighs the importance of each propagation on edge $v \leftarrow c$. We implement $\alpha(v, c)$ via multi trend attention, which will be introduce later.

Information Aggregation. Next, we aggregate the node representation $\mathbf{h}_v$ and its neighborhood representation $\mathbf{h}_{\mathcal{N}_v}$ to enhance the expressive ability. We simply takes summation of target node and its neighbor representation as follows:

$$f_{Sum} = \text{ReLU}(\mathbf{h}_v + \mathbf{h}_{\mathcal{N}_v}). \tag{2}$$

High-Order Propagation. Since a single GASF may be inadequate in capturing complex interactions between products and CPVs in TSHG, we further stack multiple GASF layers to explore the information propagated from high-order neighbors. Formally, given a node v, we recursively obtain its representation through information propagation and aggregation in the l-th step as follows:

$$\mathbf{h}_v^{(l)} = f(\mathbf{h}^{(l-1)}, \sum_{c \in \mathcal{N}_v} \alpha(v, c) \mathbf{h}_c^{(l-1)}). \tag{3}$$

Here, we recursively propagate the representations from a target node's neighbors to refine the node's representation in THSG. Moreover, we set $\mathbf{h}_v^{(0)} = \mathbf{h}_v$ at the initial information propagation iteration.

3.3 Multi Trend Attention

The key idea of attention mechanism is to learn a weighted representation across target node and its neighbors, which aims to propagate more informative features from neighbors to target node. Hence, attention mechanism is naturally implemented to learn the similarity between time series of products and CPVs in our well-established TSHG. The more similar the time series trends of the two nodes are, the more relevant they are. This implementation is based on the assumption that products and CPVs have the similar response to the hot spots in market, but temporal shifting and spatial inconsistency between the time series of products and CPVs are widely existed in our scenario.

- **Temporal shifting** means products and CPVs have different response times to market trends, which may lead to a gap between the time series of products and CPVs on the temporal view. In Fig. 3(a) we observe that the trend of product and CPV2 are more similar, which indicates that they are more related to each other. However, CPV2 responds to the market more slowly than the product, resulting in the temporal shifting of time series between between them. This phenomenon is very common in E-Commerce. For example, the release of *"Iphone"* may subsequently lead to the growth of CPV *"Iphone protective cover - Style- Cartoon"* over a period of time.
- **Spatial inconsistency** indicates the different response intensity of time series of products and CPVs for hot spots in market. As shown in Fig. 3(a), it is clear that the euclidean distance of product between CPV1 is smaller than that of product between CPV2, even though the time series trend of CPV2 is more similar to the product. In Fig. 3(b) we notice that these time series trends are comparable in a space where the euclidean distance of the product and CPV2 becomes smaller and more reasonable. It shows that time series trend reflects the similarity between time series in a better way.

Temporal shifting and spatial inconsistency between the time series of products and CPVs reveal that products and CPVs have different response speed and intensity for hot spots in market, which cannot be captured by traditional attention mechanism. Hence, we propose a novel multi trend attention mechanism to overcome this issue, aiming to calculate the relevance of products and CPVs. For convenience, we denote the target product and CPV node in TSHG as e and c, respectively. For each product, we move q times unit for the time series of product e, where $q > 0$ means move forward and $q < 0$ means move backward. Note that we retain the time series where product e and CPV c overlap on the time axis, and thus missing values before and after the retained time series are filled with the first and last value, respectively. Subsequently, we can get fixed-length vectors for product e and CPV c, and denote them as $\mathbf{h}_e^q$ and $\mathbf{h}_c^q$. Now we can get their trends $\mathbf{t}_e^q$ and $\mathbf{t}_c^q$ as follow:

$$\mathbf{t}_e^q = \frac{\mathbf{h}_e^q[1:d] - \mathbf{h}_e^q[0:d-1]}{\mathbf{h}_e^q[0:d-1] + \lambda}, \mathbf{t}_c^q = \frac{\mathbf{h}_c^q[1:d] - \mathbf{h}_c^q[0:d-1]}{\mathbf{h}_c^q[0:d-1] + \lambda}, \quad (4)$$

(a) Time Series   (b) Time Series Trend

Fig. 3. (a) represents sales volume of product, cpv1 and cpv2. (b) represents the approximate 1st-order right partial derivative of these time series. Temporal shifting represents the time difference between the reaction of the product and CPV to the market. Spatial inconsistency represents magnitude of the reaction of the product and CPV to the market, i.e. there exist dimensional inconsistency between the time series of the product and CPV.

where $\mathbf{h}_e^q[0 : d-1]$ and $\mathbf{h}_e^q[1 : d]$ denotes the first and last $(d-1)$ -dimension features of $\mathbf{h}_e^q$, respectively ($\mathbf{h}_c^q[0 : d-1]$ and $\mathbf{h}_c^q[1 : d]$ are similar). λ is a smoothing parameter. In addition, we apply Eq. (4) to $\dot{\mathbf{t}}_e^q$ and $\dot{\mathbf{t}}_c^q$ to get their trend $\ddot{\mathbf{t}}_e^q$ and $\ddot{\mathbf{t}}_c^q$, which can reveal the speed of time series trend change.

Next, we calculate the similarity between the trend of the product e and the CPV c (i.e., $\dot{\mathbf{t}}_e^q$ and $\dot{\mathbf{t}}_c^q$) as well as the speed of their trend changes with q time units interval, which is defined as:

$$\dot{s}^q(e, c) = g(\dot{\mathbf{t}}_e^q, \dot{\mathbf{t}}_c^q), \ddot{s}^q(e, c) = g(\ddot{\mathbf{t}}_e^q, \ddot{\mathbf{t}}_c^q), \tag{5}$$

where $g(\cdot, \cdot)$ measures the similarity of two vectors, which is set as the inverse of euclidean distance in our paper.

By integrating above information together, we are ready to formulate the attention score of product e and CPV c with q time units interval as follows:

$$\alpha^q(e, c) = \text{ReLU}(\mathbf{W}[\dot{\mathbf{t}}_e^q \parallel \dot{\mathbf{t}}_c^q \parallel \ddot{\mathbf{t}}_e^q \parallel \ddot{\mathbf{t}}_c^q \parallel \dot{s}^q(e, c) \parallel \ddot{s}^q(e, c)] + b), \tag{6}$$

where $\mathbf{W}$ and b is the weight matrix and bias, respectively. And $\parallel$ is the concatenation operation.

Since the time unit interval for each product-CPV pair is different from each other, we choose the maximum interval of Q time units to obtain the final attention score as follows:

$$\alpha(e, c) = \max_{-Q \le q \le Q} \alpha^q(e, c) \tag{7}$$

3.4 Model Learning

After L-th propagation, we denote $h_v^{(L)}$ as the final representation, which captures both time series and structural information in TSHG. Inspired by [6,19], we concatenate the initial vector (i.e., $\mathbf{h}_v$) and high-order representation (i.e., $h_v^{(L)}$) for later prediction.

$$\mathbf{h}_v^f = \tanh(\mathbf{W}_f[\mathbf{h}_v^0 \| \mathbf{h}_v^{(L)}] + \mathbf{b}_f), \tag{8}$$

where $\mathbf{W}_f$ and $\mathbf{b}_f$ is the weight matrix and bias vector, respectively. And $\|$ denotes the concatenation operation.

As mentioned above, we aim to predict whether a product will be sold in the future (classification) and its total sales over a time period (regression), respectively. In our work, we feed $\mathbf{h}_f$ into MLP module for classification in order to implement a nonlinear function for feature interaction, while a linear transformation is adopted for regression. To guide the learning progress, we choose the cross entropy function with negative sampling for classification [14] and mean squared error function for regression [1].

4 Experiments

In this section, we conduct comprehensive experimental studies to verify the effectiveness of our method by answering the following three questions:

RQ1 Does our proposed GASF model outperform other state-of-the-art methods on both classification and regression tasks?
RQ2 How does the proposed GASF perform for new products sales forecasting at different released times ?
RQ3 How sensitive is the proposed GASF model to the hyper-parameters ?

4.1 Experimental Settings

Datasets. To demonstrate the effectiveness of the proposed approach, we conduct experiments on an Alibaba[2] real dataset. The dataset contains 8428378 new products and 1765293 CPVs, new products are released over the time period from 01-05-2019 to 30-10-2019. Also, products and CPVs are 7-channel time series composed of 7 different single-dimensional time series. These time series are extracted from online traffic logs and represent seven different user behaviors such as exposure page views (PV), exposure unique visitor (UV), click PV, click UV, add to cart UV, pay UV and order UV. The length of each behavior time series is 60, representing the number of such behavior in the past 60 days. The labels of each product are whether it would be sold in the next 30 days (classification) and its total sales in the next 30 days (regression). In our experiment, the training set is taken over 08-21-2019 to 08-30-2019. The testing set is taken in 09-30-2019.

Evaluation Protocol. We evaluate the performance of the proposed model on two main tasks, namely binary classification and regression. Two metrics are used here for binary classification evaluation: 1) Area Under Curve (AUC) to evaluate the model's ranking performance; and 2) Precision at Top-N (P@N) to evaluate the model's ability of distinguishing top products. Regression task aims to predict the total sales for a product. We introduce two classical metrics

---

[2] https://www.1688.com.

[13] for the performance evaluation: weighted Mean Absolute Percentage Error (wMAPE) and Mean Absolute Error (MAE).

Baselines. We compare our model with the following methods:**Historical Average (HA)** is a heuristic-based baseline. **Lasso** takes the historical sales records as input for Logistic Regression/Linear Regression with L_1 regularization. **Gradient Boosting Decision Tree (GBDT)** is a common used technique for both classification and forecasting regression problem in industry, we carefully design 70 features from these 9 log indicators. **DNN** is a simple neural network architecture with 3 fully connection layers and a linear regression layer. **GBDT-CPV** adds CPV features on the basis of GBDT. We apply a pooling (average, maximum, median) to all neighbor heterogeneous features to improve the acquirement of information. **CNN-WD** [20] is a convolutional neural network based model for sales forecasting in E-Commerce.

Parameter Settings. For all approaches, we tune the model parameters by grid search and report the performance on the testing dataset. For GBDT models, we take these parameters: num_rounds = 200, max_depth = 6, subsample = 0.8 and learning rate = 0.1. For DNN model, the dimensions for fully connected layers are [256, 128], a dropout with p = 0.2 is applied to the output of last fully connected layer and learning rate is set to 0.001. For GASF model, the kernel sizes and number of filters for 1-D CNNs are [3, 7] and [20, 1], the dimensions for fully connected layers are [256, 128] and $\lambda = 0.00001$, we use Adam as optimizer [10] with a learning rate 0.0001.

4.2 Performance Comparison (RQ1)

Now we compare the performance of our GASF with the baselines. The comparison results are shown in Table 1. The main observations are summarized as follows:

1. GASF achieves the best performance on both classification and regression tasks. One-sample paired t-test shows that all the improvements are statistically significant ($p < 0.005$). We think that the outperformance of our GASF would benefit from the design of GATs in capturing the spatial and temporal features from the input jointly.
2. Among all the baselines, HA and ARIMA typically underperform machine learning and deep learning based models, mainly because they only rely on historical sales records of new products. GBDT-CPV model outperforms original GBDT by absolute 3 points in AUC, indicating the significance of CPVs.
3. Thanks to the Multi Trend Attention (MTA), GASF-MTA outperforms GASF for both tasks. This shows the effectiveness of the proposed MTA and also indicates the importance of taking temporal shifting and spatial inconsistency of the time series into account.

Table 1. Overall performance comparison. The best performance of each setting is highlighted as bold font.

Methods	Binary classification				Regression	
	AUC	P@30K	P@300K	P@1M	MAE	wMAPE
HA	0.6834	0.5597	0.1044	0.0541	4.0135	0.8891
Lasso	0.7854	0.5785	0.1045	0.0539	2.9324	0.6495
GBDT	0.8286	0.6492	0.1529	0.0589	2.6822	0.5941
GBDT-CPV	0.8591	0.6566	0.1566	0.0591	2.6746	0.5914
DNN	0.8441	0.6531	0.1584	0.0578	2.5137	0.5570
CNN-WD	0.8475	0.5825	0.1577	0.0581	2.5006	0.5541
GASF	0.8669	**0.6827**	0.1650	0.0592	2.4354	0.5396
GASF-MTA	**0.8750**	0.6821	**0.1661**	**0.0594**	**2.4056**	**0.5329**

Table 2. Performance comparison over different released period. The best performance of each setting is highlighted as bold font.

Methods	AUC			
	1 day	2–10 days	11–30 days	31–90 days
HA	0.5237	0.5967	0.7114	0.7913
GBDT	0.7403	0.7963	0.8547	0.8860
GBDT-CPV	0.7618	0.8117	0.8664	0.8914
CNN-WD	0.6745	0.7815	0.8611	0.8977
GASF-MTA	**0.7807**	**0.8324**	**0.8732**	**0.8995**

4.3 Performance for Different Released Times (RQ2)

In real business, the released period of new products ranges from 1 day to 90 days. The shorter the released period of a new product, the less information the forecasting model can obtain. Additional heterogeneous features are particular useful to alleviate such a "cold-start" problem in new product forecasting. Here, we study the forecasting performance *w.r.t.* the released periods, which varies in the set of {1 day, 2–10 days, 10–30 days, 30–90 days}. We show the AUC comparison results on the classification task in Table 2. The main observations are summarized as follows:

1. With the increase of the released time, all models has achieved a better performance. This is due to the enhancement of new product historical information.
2. For all released period, our proposed GASF-MTA model shows a significant outperformance than other models. Improvements increase as release time decrease, this shows the effectiveness of our proposed model for new product sales forecasting.
3. Among all the GBDT approaches, GBDT-CPV outperforms GBDT for all released period. We should note that the improvements stem from the CPV. This shows the effectiveness of our proposed CPV for new product sales forecasting.

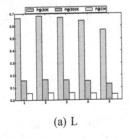

(a) L

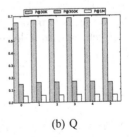

(b) Q

Fig. 4. The impact of key hyper-parameters for GASF (L: the number of the GASF layers; Q: the maximum interval of time units).

4.4   Hyper-Parameter Study (RQ3)

In this section, we explore the impact of key hyper-parameters for GASF and how L and Q empirically influence the learning effect of GASF.

Figure 4(a) shows the performance of GASF with respect to L. We can see that $L = 2$ is better than $L = 1$; however, increasing L beyond 2 gives marginal returns in performance. This makes sense because larger L includes information of further nodes and thus needs deeper GASF to learn, which is more difficult to optimize. In this case, it may be easily over-fitting with more layers [5].

Figure 4(b) shows the performance of GASF with respect to Q. We can see that with the increasing of Q, the performance is improved to a maximum, and then decrease. This indicates the positive effect of using a larger interval for temporal shifting. However, too large interval may introduce noise and compromise the performance.

4.5   Comparison with Online System

We have successfully deployed our proposed GASF in our real promotion business scenario of Alibaba, and compare it with the best online baseline model (*i.e.,* GBDT-CPV) on Dec 2020. The experimental results shows that has a relative gain by **3.4%** and **7.6%** than on GMV and number of deduplicated buyers (BYR) respectively. This observation demonstrates the effectiveness and business value of our proposed approach in E-Commerce.

5   Conclusion and Future Work

In this paper, a novel GAT architecture model is presented for new product sales forecasting in E-Commerce, named GASF. GASF models products and their CPVs by a general graph-structured time series and extracts spatial and temporal features simultaneously. The experiments on two real-world tasks and online system demonstrate a significant outperformance of our proposed model. To the best of our knowledge, this is the first time to apply GATs for sales forecasting. For further improvements, we will pursue two directions. The first is to explore more heterogeneous relations such as completing relation between substitutable products, which may enhance the representation ability of our model. The second is to incorporate side information [16,17] into our multi trend attention, which can provide more flexibility to learn the weights of different product-CPV pair.

References

1. Chai, T., Draxler, R.R.: Root mean square error (RMSE) or mean absolute error (MAE)?-Arguments against avoiding RMSE in the literature. Geosci. Model Dev. 7(3), 1247–1250 (2014)
2. Chatfield, C.: The analysis of time series: an introduction (2003)

3. Chen, C., et al.: How much can a retailer sell? Sales forecasting on Tmall. In: Yang, Q., Zhou, Z.-H., Gong, Z., Zhang, M.-L., Huang, S.-J. (eds.) PAKDD 2019. LNCS (LNAI), vol. 11440, pp. 204–216. Springer, Cham (2019). https://doi.org/10.1007/978-3-030-16145-3_16

4. Friedman, J.H.: Greedy function approximation: a gradient boosting machine. Ann. Stat. **29**, 1189–1232 (2001)

5. Glorot, X., Bengio, Y.: Understanding the difficulty of training deep feedforward neural networks. In: AISTATS, pp. 249–256 (2010)

6. He, K., Zhang, X., Ren, S., Sun, J.: Deep residual learning for image recognition. In: CVPR, pp. 770–778 (2016)

7. Hochreiter, S., Schmidhuber, J.: Long short-term memory. Neural Comput. **9**(8), 1735–1780 (1997)

8. Hu, B., Shi, C., Zhao, W.X., Yu, P.S.: Leveraging meta-path based context for top-n recommendation with a neural co-attention model. In: SIGKDD, pp. 1531–1540 (2018)

9. Hu, B., Zhang, Z., Shi, C., Zhou, J., Li, X., Qi, Y.: Cash-out user detection based on attributed heterogeneous information network with a hierarchical attention mechanism. In: AAAI, pp. 946–953 (2019)

10. Kingma, D.P., Ba, J.: Adam: a method for stochastic optimization. arXiv preprint arXiv:1412.6980 (2014)

11. Mikolov, T., Chen, K., Corrado, G., Dean, J.: Efficient estimation of word representations in vector space. In: ICLR Workshop (2013)

12. Mikolov, T., Sutskever, I., Chen, K., Corrado, G.S., Dean, J.: Distributed representations of words and phrases and their compositionality. In: NIPS, pp. 3111–3119 (2013)

13. Qi, Y., Li, C., Deng, H., Cai, M., Qi, Y., Deng, Y.: A deep neural framework for sales forecasting in e-commerce. In: CIKM, pp. 299–308 (2019)

14. Tang, J., Qu, M., Wang, M., Zhang, M., Yan, J., Mei, Q.: Line: large-scale information network embedding. In: WWW, pp. 1067–1077 (2015)

15. Veličković, P., Cucurull, G., Casanova, A., Romero, A., Lio, P., Bengio, Y.: Graph attention networks. ICLR (2017)

16. Wang, H., Zhang, F., Hou, M., Xie, X., Guo, M., Liu, Q.: Shine: signed heterogeneous information network embedding for sentiment link prediction. In: WSDM, pp. 592–600 (2018)

17. Wang, H., et al.: Knowledge-aware graph neural networks with label smoothness regularization for recommender systems. In: SIGKDD, pp. 968–977 (2019)

18. Wang, J., Sun, T., Liu, B., Cao, Y., Zhu, H.: CLVSA: a convolutional LSTM based variational sequence-to-sequence model with attention for predicting trends of financial markets. In: IJCAI, pp. 3705–3711 (2019)

19. Xu, K., Li, C., Tian, Y., Sonobe, T., Kawarabayashi, K.i., Jegelka, S.: Representation learning on graphs with jumping knowledge networks. arXiv preprint arXiv:1806.03536 (2018)

20. Zhao, K., Wang, C.: Sales forecast in e-commerce using convolutional neural network. arXiv preprint arXiv:1708.07946 (2017)

Transportation Recommendation with Fairness Consideration

Ding Zhou[1], Hao Liu[2(✉)], Tong Xu[1], Le Zhang[1], Rui Zha[1], and Hui Xiong[3(✉)]

[1] School of Computer Science and Technology, University of Science and Technology of China, Hefei, China
{zhouding,laughing,zr990210}@mail.ustc.edu.cn, tongxu@ustc.edu.cn
[2] Business Intelligence Lab, Baidu Research, Beijing, China
liuhao30@baidu.com
[3] Rutgers University, New Brunswick, USA
hxiong@rutgers.edu

Abstract. Recent years have witnessed the widespread use of online map services to recommend transportation routes involving multiple transport modes, such as bus, subway, and taxi. However, existing transportation recommendation services mainly focus on improving the overall user click-through rate that is dominated by mainstream user groups, and thus may result in unsatisfactory recommendations for users with diversified travel needs. In other words, different users may receive unequal services. To this end, in this paper, we first identify two types of unfairness in transportation recommendation, (*i*) the *under-estimate* unfairness which reflects lower recommendation accuracy (*i.e.*, the quality), and (*ii*) the *under-recommend* unfairness which indicates lower recommendation volume (*i.e.*, the quantity) for users who travel in certain regions and during certain time periods. Then, we propose the **F**airness-**A**ware **S**patiotemporal **T**ransportation **R**ecommendation (**FASTR**) framework to mitigate the transportation recommendation bias. In particular, based on a multi-task wide and deep learning model, we propose the dual-focal mechanism for under-estimate mitigation and tailor-designed spatiotemporal fairness metrics and regularizers for under-recommend mitigation. Finally, extensive experiments on two real-world datasets verify the effectiveness of our approach to handle these two types of unfairness.

Keywords: Transportation recommendation · Personalized recommendation · Fairness machine learning

1 Introduction

Transportation recommendation is a one-stop routing service, which aims to help users find the most proper transport mode (*e.g.*, bus, subway, and taxi) and combinations, by given the Origin-Destination pair of users. As an emerging map service in various online navigation applications (*e.g.*, Baidu Maps, Google Maps), transportation recommendation has deeply penetrated the citizens' daily

C. S. Jensen et al. (Eds.): DASFAA 2021, LNCS 12683, pp. 566–578, 2021.
https://doi.org/10.1007/978-3-030-73200-4_40

lives. For instance, the transportation recommendation service on Baidu Maps is answering over ten million queries made by millions of users in China per day. Due to the practicality of transportation recommendation, there has been an increasing attention to this field from both academia and industry. Recently, different strategies are proposed to recommend transport modes for users, such as historical trajectories based strategy [20], shortest distance based strategy [10] and city graph based strategy [13,14,19]. Although existing works can achieve good performance in transportation recommendation, they overlook two types of unfairness that we observe from large-scale historical recommendation log. One is under-estimate unfairness, which may lead to lower recommendation accuracy on minorities. Since the majority loss functions minimize the overall error of model that benefits mainstream groups, this under-estimate unfairness (*e.g.*, big performance gap between different transport modes) is becoming increasingly prevalent. The other is the under-recommend unfairness, which may lead to lower recommendation volume for minorities' transportation needs. For instance, bus and subway that concentrated in the center of the city are the protagonists during rush hour, which may greatly squeeze the recommendation volume of other transport modes like taxi. In other words, users who live in suburban and need taxi at that time can not be recommended and satisfied. Furthermore, these two types of unfairness may increase homogeneity and decrease utility [5] of the recommender services.

Recently, the machine learning fairness community primarily focuses on fairness in classification and has proposed various definitions of fairness [3,17], such as group fairness [4,8] that restricts any two groups to having equal probability of being assigned to the positive predicted class, and equality of opportunity [12] that restricts any two groups to having equal false negative rate. For an unbiased recommendation, [1] and [2] focus on fairness in pointwise and pairwise accuracy of learning to rank, respectively. However, these fairness metrics can not satisfy the spatiotemporal settings in transportation recommendation. Therefore, a more comprehensive solution is still urgently required for these challenges.

To that end, we propose the Fairness-Aware Spatiotemporal Transportation Recommendation (FASTR) framework for effective and fair transportation recommendation. Specifically, we first introduce a wide and deep learning model [7] modified with multi-task mechanism for capturing feature co-occurrence and high-order interaction relationships. Besides, we propose a dual-focal mechanism to mitigate under-estimate unfairness, which consists of task-level focal loss for enhancing the prediction of each individual task and relation-level focal loss for mitigating performance gap between tasks. Then we propose multiple well-designed spatiotemporal fairness metrics to quantify the under-recommend unfairness in certain regions and time periods. Furthermore, with the help of the proposed fairness metrics, a series of tailor-designed regularizers are proposed to guide the optimization for mitigating the under-recommend unfairness.

Overall, the major contributions of our work can be summarized: 1) To the best of our knowledge, our FASTR model is among the first product-level intelligent transportation recommender that focuses on mitigating under-estimate and under-recommend unfairness, 2) We utilize multi-task wide and deep model with the well-designed dual-focal loss for under-estimate unfairness mitigation, and we propose tailor-designed spatiotemporal fairness metrics and regularizers to mitigate under-recommend unfairness, 3) Extensive experiments on real-world datasets verify the effectiveness of our approach on handling under-estimate and under-recommend unfairness.

2   Data Description and Analysis

In this section, we first introduce the datasets and the constructed features used in our work, and we analyze how unfairness appears in transportation recommendation subsequently. Specifically, we collected our datasets from Baidu Maps, a large-scale navigation application, from July 2019 to September 2019 in Beijing and Shanghai. And according to user interaction loop, our source data $\mathcal{D}$ can be further categorized into query records, click records and the corresponding context features. In short, for each sample in our datasets $\mathcal{D}$, its query record represents one transportation search (*e.g.*, Origin-Destination pair) from a user on Baidu Maps, and its click record indicates the user's feedback on different recommendations (*e.g.*, a user may click on specific transportation recommendation for him/her). Meanwhile, the corresponding context features for each sample consist of spatial features, temporal features, meteorological features, user features and transport mode features, where the details are shown in Table 1. Totally, we have 5,327,897 samples with 1,177,844 clicks in Beijing, as well as 5,120,561 samples with 1,190,813 clicks in Shanghai. And for each sample, we consider 7 transport modes can be recommended in datasets $\mathcal{D}$ (*Bus, Bus + Bicycle, Walk, Bus + Taxi, Bicycle, Taxi* and *Drive*).

Table 1. Corresponding context features for each sample

Feature	Composition
Spatial features	*District category, Point-Of-Interest (POI) category, POI count, Transport Mode Click count*
Temporal features	*Hour, Minute, Day of Week, Day of Month, Workday*
Meteorological features	*Weather, Temperature, Air Quality Index, Wind Speed, Wind Direction*
User features	*Demographic Attribute, Social Attribute, User Historical Transport Mode Distribution*
Transport mode features	*Price, Time, Distance*

To further understand the unfairness phenomenon in transportation recommendation, we analyze our datasets from under-estimate and under-recommend aspects in Beijing with our original model [15]. Note that we have similar observations in Shanghai. Since user clicks can proxy the recommendation accuracy

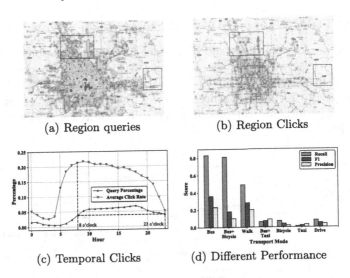

(a) Region queries   (b) Region Clicks

(c) Temporal Clicks   (d) Different Performance

Fig. 1. Distribution and performance of Beijing dataset. (a) queries distribution in region aspect; (b) clicks distribution in region aspect; (c) temporal distribution of average query percentage and clicks rate per hour; (d) the precision, f1-score and precision performance on different transport modes in Beijing.

and volume, we first calculate the distribution of click rate in different regions and time periods to reveal the unfairness. As shown in Fig. 1(a) and Fig. 1(b) that depict the region distribution of queries and clicks respectively in Beijing, the click rate in rectangles of Fig. 1(a) and Fig. 1(b) is much lower or even close to zero compared with other regions, which indicates under-estimate and under-recommend unfairness happened in certain regions. In Fig. 1(c), we can see that the average click rate at 23 o'clock is much lower than 8 o'clock even though they have the same query volume, which shows that transportation recommendation suffers under-estimate and under-recommend unfairness in certain time periods. Furthermore, as shown in Fig. 1(d), we calculate *recall, precision* and *f1-score* for each transport mode in Beijing by the original model [15]. The results show the original model can not give a balanced quality of services for each transport mode and its users, where under-estimate unfairness happened.

3   FASTR Framework

3.1   Overview

The overall workflow of FASTR is shown in Fig. 2, we first input the features mentioned in Sect. 2 to a multi-task wide and deep model for capturing feature co-occurrence relationships. Then, we propose the dual-focal mechanism and the spatiotemporal fairness metrics as well as regularizers to mitigate under-estimate and under-recommend unfairness respectively. Finally, with these well-designed mechanism, metrics, and regularizers, we can have a more balanced quality of recommendation on transport mode for users.

3.2 Multi-task Wide and Deep Learning Model

To capture the co-occurrence and high-order interaction relationships between different features, we adopt the wide and deep learning model [7] that is widely used in many recommender system, and extend it with the multi-task paradigm [16] to serve as our basic model, where the multi-task mechanism improves the performance on minorities and helps to mitigate the under-estimate disparity [9].

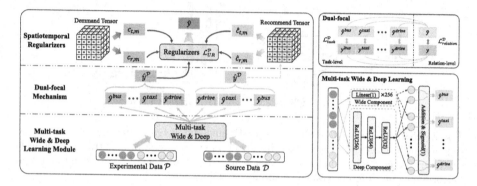

Fig. 2. The overall workflow of FASTR.

Wide and Deep Learning Model. Wide and deep learning consists of a wide component for low-level feature co-occurrence memorization and a deep component for high-level feature co-occurrence generalization. Thus, the wide component with shallow structure is defined as $\hat{y}_i = \mathbf{w}^\top \mathbf{x}_i + b$, where $\mathbf{x}_i$ is the i-th input feature vector, $\mathbf{w}$ is the learnable weighted matrix and b is the bias. The deep component stacks multiple neural network layers to capture higher-order feature representations. Each fully connected layer transform input vector as $\mathbf{z}_{l+1} = \text{ReLU}(\mathbf{w}_l^\top \mathbf{z}_l + b_l)$, where $\mathbf{z}_l$ and $\mathbf{z}_{l+1}$ are the input and output of l-th layer, $\mathbf{w}_l$ and b_l are the weight and bias parameters of layer l. With both wide and deep components, the final prediction of wide and deep learning model can be formulated as $\hat{y}_i = \sigma(\mathbf{w}_w^\top \mathbf{x}_i + \mathbf{w}_d^\top \mathbf{z}_f + b)$, where $\hat{y}_i$ is the final output, σ stands for the activation function, $\mathbf{w}_w$ is the weight parameter of the wide component, and $\mathbf{w}_d$ is the weight parameter of the final output of the deep component $\mathbf{z}_f$.

Multi-task Mechanism. To promote the recommendation performance for users who prefer different transport modes, we follow the settings in [9] who claimed prediction is more accurate when treating the recommendation of each transport mode as an independent task. In particular, we apply the multi-task strategy to divide transportation recommendation into several binary classification tasks that predict whether a user will click on a specific transport mode, where lower-level parameters in the wide and deep components of wide and deep learning model are shared cross all tasks [16]. Notice that we treat the prediction of a transport mode as a binary classification task, where we have 7 tasks (*i.e.*,

7 transport modes) totally in our work. For each transport mode m, the binary classification task of m can be formulated as follows:

$$\hat{y}_i^m = \sigma(\mathbf{w}_w^{m\top}\mathbf{x}_i + \mathbf{w}_d^{m\top}\mathbf{z}_f + b), \tag{1}$$

where $\mathbf{w}_w^m$ and $\mathbf{w}_d^m$ are the task-specific parameters of the wide component and the deep component respectively. And $\hat{y}_i^m \in [0,1]$ indicates the probability of users click on transport mode m.

3.3 Dual-Focal Mechanism for Under-Estimate

As described before, the under-estimate unfairness is usually caused by the performance gap between transport modes in transportation recommendation. Thus, we intuitively need a mechanism that can promote the prediction performance on each transport mode and mitigate the performance gap. In particular, we apply Focal Loss, which has been widely used for computer vision [18] and pays more attention to samples that are more difficult to distinguish for mitigating the under-estimate unfairness. Firstly, we propose task-level focal loss for each binary classification task, which aims to improve the ability of each task on serving the minority users who prefer the task corresponding transport mode but difficult to distinguish. Therefore, we denote task-level focal loss for under-estimate mitigation as the summarization of each binary classification task:

$$\mathcal{L}_{task}^{\mathcal{D}} = -\frac{1}{|\mathcal{D}||\mathcal{M}|}\sum_{i\in\mathcal{D}}\sum_{m\in\mathcal{M}}\big[\alpha_m y_i^m(1-\hat{y}_i^m)^\gamma \log\hat{y}_i^m \\ + (1-\alpha_m)(1-y_i^m)(\hat{y}_i^m)^\gamma \log(1-\hat{y}_i^m)\big], \tag{2}$$

where $\mathcal{M}, \mathcal{D}$ are the set of transport modes and source data respectively. $y_i^m \in \{0,1\}$ is the ground truth that indicates whether user i clicks transport mode m. α_m is the hyperparameter to alleviate binary class imbalance, and γ is the hyperparameter to regulate attentions on the samples that are difficult to distinguish. Taking ground truth y_i^m equals 1 as an example, when the predicted probability $\hat{y}_i^m$ of the i-th sample is nearly 1.0 which means easy to distinguish, the attentions on this sample can be reduced through $(1-\hat{y}_i^m)^\gamma$. And the larger the γ is, the more attention are paid to difficult samples, and vice versa.

Beyond the task-level that focuses on the performance of individual task, we propose relation-level focal loss for mitigating the performance gap between tasks. Specifically, since minority transport modes (*i.e.*, bicycle, taxi and drive) hold less data than the mainstream group, minorities may suffer insufficient training and poor recommendation as described in Sect. 2. Therefore, we apply relation-level focal loss between tasks as follows, where more attention can be paid to minority transport modes.

$$\mathcal{L}_{relation}^{\mathcal{D}} = -\frac{1}{|\mathcal{D}||\mathcal{M}|}\sum_{i\in\mathcal{D}}\sum_{m\in\mathcal{M}}\beta_m y_i^m(1-\hat{y}_i^m)^\gamma \log\hat{y}_i^m, \tag{3}$$

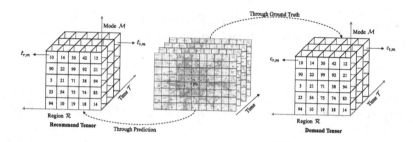

Fig. 3. Demand and recommend tensor construction.

where β_m is the hyperparameter to alleviate multiple class imbalance. With task-level and relation-level focal losses, our dual-focal mechanism can mitigate the under-estimate unfairness by promoting the prediction performance on every transport mode. The overall dual-focal mechanism can be formulated as follows:

$$\mathcal{L}_{UE}^{\mathcal{D}} = \mathcal{L}_{task}^{\mathcal{D}} + \mathcal{L}_{relation}^{\mathcal{D}}. \tag{4}$$

3.4 Spatiotemporal Metrics and Regularizers for Under-Recommend

To mitigate the under-recommend unfairness on recommending lower volume in certain regions and time periods, we first construct demand and recommend tensor in regions and time periods aspects. Then we design a series of spatiotemporal oriented metrics to measure the degree of under-recommend through demand and recommend tensor. And the corresponding regularizers are proposed to mitigate under-recommend unfairness.

Demand and Recommend Tensor Construction. As shown in Fig. 3, we first let $r \in \mathcal{R}$ be the r-th square region of the study area of $\mathcal{R}$, $t \in \mathcal{T}$ be the t-th o'clock in one day, and $m \in \mathcal{M}$ be the m-th transport mode. Then, we calculate the ground truth number of demands to mode m in region r and time t as $c_{r,m}$ and $c_{t,m}$ respectively. Thus we can construct our demand tensor as shown in Fig. 3. What's more, to reflect the under-recommend unfairness of recommender system, we denote $\hat{c}_{r,m}$ and $\hat{c}_{t,m}$ as the recommend volume of recommender system on transport mode m in region r and time t respectively. Then, with the calculated $\hat{c}_{r,m}$ and $\hat{c}_{t,m}$, we formulate recommend tensor in Fig. 3 to proxy recommendation volume.

Region-based Fairness (RF) Metric. Now we formally define our spatial metric RF in the region $\mathcal{R}$ as:

$$RF = P\{(u(r) - u(r')) \leq \epsilon \mid r \neq r', \, r, \, r' \in \mathcal{R}\}, \tag{5}$$

where $u(r)$ denotes the degree of under-recommend in region r, and lower $u(r)$ indicates lower under-recommend. And RF can be interpreted as for any two regions, the differences between $u(r)$ and $u(r')$ are not greater than ϵ. To be more direct and remove the interference on selecting ϵ, we modify our RF metric as follows to measure the degree of under-recommend unfairness in the spatial aspect, which is same to Formula 5.

$$RF = \max_{r \in \mathcal{R}}(u(r)) - \min_{r \in \mathcal{R}}(u(r)). \tag{6}$$

Intuitively, we design under-recommend degree of transport mode m in region r as the scaled differences between demands $c_{r,m}$ and recommend volume $\hat{c}_{r,m}$:

$$u(r, m) = \text{ReLU}\left(\frac{c_{r,m} - \hat{c}_{r,m}}{c_{r,m}}\right), \tag{7}$$

where function ReLU($\cdot$) is utilized to filter out the transport mode that not under-recommend. With $u(r, m)$, $u(r)$ can be calculated as follows:

$$u(r) = \frac{\sum_{m \in \mathcal{M}} u(r, m)}{\sum_{m \in \mathcal{M}} \text{sign}(u(r, m))}, \tag{8}$$

where sign($\cdot$) is a function that treats positive numbers as 1 and 0 for others.

Temporal-based Fairness (TF) Metric. Similar to RF, we define temporal metric TF as follows:

$$TF = \max_{t \in \mathcal{T}}(u(t)) - \min_{t \in \mathcal{T}}(u(t)), \tag{9}$$

where TF measures how different the degree of under-recommend over different transport modes from the perspective of time periods. And $u(t)$ can be calculated as follows:

$$u(t) = \frac{\sum_{m \in \mathcal{M}} u(t, m)}{\sum_{m \in \mathcal{M}} \text{sign}(u(t, m))},$$
$$u(t, m) = \text{ReLU}\left(\frac{c_{t,m} - \hat{c}_{t,m}}{c_{t,m}}\right). \tag{10}$$

Region-Temporal Fairness (RTF) Metric. To measure the whole degree of under-recommend from both spatial and temporal perspective, we formally define RTF, the overall degree of under-recommend, as follows:

$$RTF = \frac{\sum_{r \in \mathcal{R}} u(r)}{|\mathcal{R}|} = \frac{\sum_{t \in \mathcal{T}} u(t)}{|\mathcal{T}|}. \tag{11}$$

Spatiotemporal Regularizer for Under-recommend. Since $\hat{c}_{r,m}$ and $\hat{c}_{r,t}$ can not be calculated directly during model training, we follow the randomized experiments in [1] and collect an experimental data $\mathcal{P}$. Specifically, we rebalance $\mathcal{P}$ to have approximately the same transportation demand distribution at arbitrary regions and time periods. Note that further data restrictions can be applied for alternative goals, and $\mathcal{P}$ is independent of the source data $\mathcal{D}$. Based on RF, TF and experimental data $\mathcal{P}$, we define our spatiotemporal oriented regularizer to constrict under-recommend unfairness:

$$\mathcal{L}_{UR}^{\mathcal{P}} = \lambda_{\mathcal{R}} \frac{\sum_{r \in \mathcal{R}} u(r)}{|\mathcal{R}|} + \lambda_{\mathcal{T}} \frac{\sum_{t \in \mathcal{T}} u(t)}{|\mathcal{T}|}, \tag{12}$$

where $\lambda_{\mathcal{R}}$ and $\lambda_{\mathcal{T}}$ are the weight terms. And the ultimate goal of our mission can be summarized as follows:

$$\mathcal{L} = \mathcal{L}_{UR}^{\mathcal{P}} + \mathcal{L}_{UE}^{\mathcal{D}}. \tag{13}$$

4   Experiments

In this section, we evaluate the performance of our FASTR framework on two real-world datasets described before by the transportation recommendation task.

4.1   Experimental Settings

A) Evaluation Metrics. As described in Sect. 3.4, RF, TF and RTF metrics are utilized to measure the under-recommend degree of our transportation recommender system in the perspective of spatial, temporal and spatiotemporal respectively. Note that lower in RF, TF, and RTF, better in fairness. Besides, we choose to apply macro-recall, variance-recall and maxmin-recall to reveal the performance on mitigating under-estimate unfairness of FASTR in recommending different transport modes. Specifically, the weight of macro-recall for every transport mode is the same, which leads to a fairer evaluation of models. And variance-recall are calculated to measure the differences of performance on predicting different transport modes, where larger the variance is, larger the degree of under-estimate. We also propose the maxmin-recall to describe the difference between the maximum and the minimum recall of transport modes.

B) Baselines & Variants. We compare our approach with four learning-based methods, which are widely used and recognized in the industry, and three variants of our FASTR. Specifically, Logistic Regression (LR) and XGBoost [6] as the most representative models for classification tasks are compared, and the inputs of LR and XGBoost are as same as FASTR. Wide&Deep [7] and DeepFM [11] are two widely acclaimed models for recommendation, who incorporate both shallow and deep relationships between features. Here, we also use the same input as FASTR for them. The ablation study is conducted with three variants defined as follows, 1) FASTR-MR masks dual-focal loss and spatiotemporal regularizers of our FASTR, and we utilize cross-entropy loss for each binary classier, 2) FASTR-MD replaces dual-focal loss with cross-entropy loss, and 3) FASTR-MM masks the multi-task mechanism of our FASTR.

C) Implementation Details. In the implementation phase, we constructed our FASTR by PaddlePaddle[1], which supports a variety of AI-empowered products in Baidu. Specifically, we first transformed categorical features into 32-dimensional embedding vectors, and concatenated them with all other continuous features as the input vector. Then, we fed the input vector into multi-task wide and deep learning model, where the deep component consists of four fully connected layers with 400, 256, 64 and 32 hidden units respectively. And we

---

[1] https://www.paddlepaddle.org.cn/.

chose to use Sigmoid as our activation function. When implementing our fairness constricts, the class weight α_m and β_m for dual-focal loss were set through balance strategy[2], and the hyperparameter γ was set to 3.0. For $\mathcal{L}_{UR}^{\mathcal{P}}$, we set $\lambda_{\mathcal{R}}$ and $\lambda_{\mathcal{T}}$ both equaled to 0.5. Finally, we set the batch size to 128, learning rate to 5e-4 and trained them with Adam optimizer [21].

4.2 Quantitative Evaluations of FASTR

Performance on Mitigating Under-estimate Unfairness. Figure 4(a), 4(b), and 4(c) show the overall performance on mitigating under-estimate of our FASTR and other methods. And we find three observations through these results. Firstly, as shown in Fig. 4, the macro-recall of our FASTR and its variants are better than other methods, and FASTR achieves much lower variance-recall and maxmin-recall, which means we can provide unbiased transportation recommendation for users without causing too much harm to the overall quality (*i.e.,* macro-recall) compared to other methods. Secondly, FASTR consistently outperforms Wide&Deep and FASTR-MM in terms of macro-recall, variance-recall and maxmin-recall metrics, which proves the effectiveness of multi-task mechanism on mitigating under-estimate unfairness. Thirdly, comparing FASTR and FASTR-MD, the former has better performance on variance-recall and maxmin-recall than the later, which demonstrates the effectiveness of our task-level and relation-level focal loss on mitigating under-estimate unfairness.

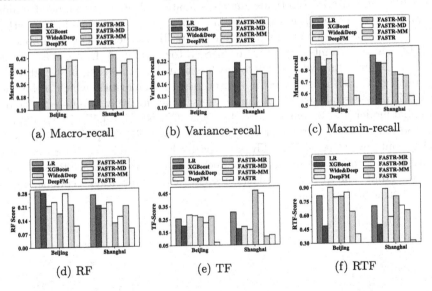

Fig. 4. Overall performance on transportation recommendation.

Performance on Mitigating Under-recommend Unfairness. Figure 4(d), 4(e), and 4(f) depict the ability of baselines, variants and our FASTR on mitigat-

---

[2] https://scikit-learn.org.

ing under-recommend unfairness. And we have three observations through these results. Firstly, our FASTR framework achieves better performance than other methods on RF, TF and RTF metrics, which demonstrates the effectiveness of our spatiotemporal regularizer $\mathcal{L}_{UR}^{P}$ on mitigating under-recommend unfairness in both region and temporal perspective. Specifically, our FASTR framework beats other methods on RF, TF and RTF up to 18.12%, 21.23% and 50.62% in Beijing, and 17.54%, 17.39% and 56.22% in Shanghai respectively. Secondly, DeepFM and Wide&Deep have similar performance on RF and TF metrics but DeepFM's RTF degree is much higher than Wide&Deep. Since DeepFM has a better fitting ability than Wide&Deep, the bias in datasets may cause this gap in RTF. Comparing FASTR with FASTR-MR and FASTR-MD, we find both spatiotemporal regularizer and dual-focal mechanism are useful to mitigate under-recommend unfairness, where the specially designed spatiotemporal regularizer plays better. Thirdly, we compare FASTR with the best baseline XGBoost and draw Fig. 5 by calculating the distribution of RTF in Beijing. We find that our FARSTR framework recommends more densely than XGBoost, as shown in the black box in Fig. 5(b) and Fig. 5(c), which indicates our FASTR suffers lower under-recommend unfairness.

4.3   The Cost of Fairness in Transportation Recommendation

In this paper, we propose to use dual-focal mechanism with spatial and temporal oriented regularizers to mitigate under-estimate and under-recommend unfairness. However, as shown in Fig. 4, the big improvement in fairness brings performance degradation for the mainstream group. To further reveal the impact of our fairness constraints, we apply our FASTR to an online A/B test in November 2019 in Beijing. And we find there has less than 1% decreasing in overall click-through rate with more than 6% improvement in minorities, which means our FASTR provides fair services for more users. Quantitatively, compared with our original model [15], we have 11.3%, 40.1%, 30.6%, 60.8%, 74.5%, 18.8% improving on macro-recall, variance-recall, maxmin-recall, RF, TF, RTF respectively. The results show that our FASTR is acceptable because of its big improvement in fairness for minority groups but little degradation on performance for the mainstream group, and we can provide fairness-aware transportation recommendation for a better user experience.

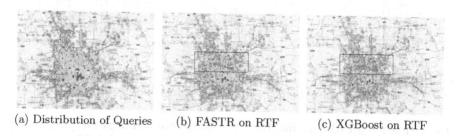

(a) Distribution of Queries   (b) FASTR on RTF   (c) XGBoost on RTF

Fig. 5. Distribution of RTF. (a) queries distribution in Beijing. (b) RTF distribution of FASTR. (c) RTF distribution of XGBoost. Notice that the redder region means lower under-recommend unfairness.

5   Conclusion

In this paper, we investigated the fairness problem in transportation recommendation by mitigating the under-estimate and under-recommend unfairness for users with different travel needs. Specifically, we proposed a Fairness-Aware Spatiotemporal Transportation Recommendation framework (FASTR), which consists of multi-task wide and deep model with dual-focal mechanism for under-estimate unfairness mitigation and tailor-designed spatiotemporal metrics and regularizers for under-recommend unfairness mitigation. Extensive evaluations on real-world datasets validated the effectiveness of our FASTR on mitigating these two types of unfairness, which lead to an unbiased transportation recommendation for users. Besides, through the urban scale A/B test, we confirmed the practicability of our FASTR framework.

Acknowledgement. This research was partially supported by grants from the National Key Research and Development Program of China (Grant No. 2018YFB1402600), and the National Natural Science Foundation of China (Grant No. 91746301, 62072423). And the work was done when the first author interned in Baidu Research.

References

1. Beutel, A., et al.: Fairness in recommendation ranking through pairwise comparisons. In: Proceedings of the 25th ACM SIGKDD International Conference on Knowledge Discovery & Data Mining, pp. 2212–2220 (2019)
2. Beutel, A., Chi, E.H., Cheng, Z., Pham, H., Anderson, J.: Beyond globally optimal: focused learning for improved recommendations. In: Proceedings of the 26th International Conference on World Wide Web, pp. 203–212 (2017)
3. Biega, A.J., Gummadi, K.P., Weikum, G.: Equity of attention: amortizing individual fairness in rankings. In: The 41st International ACM SIGIR Conference on Research & Development in Information Retrieval, pp. 405–414 (2018)
4. Calders, T., Verwer, S.: Three Naive Bayes approaches for discrimination-free classification. Data Min. Knowl. Disc. **21**(2), 277–292 (2010). https://doi.org/10.1007/s10618-010-0190-x
5. Chaney, A.J.B., Stewart, B.M., Engelhardt, B.E.: How algorithmic confounding in recommendation systems increases homogeneity and decreases utility. In: Proceedings of the 12th ACM Conference on Recommender Systems - RecSys 2018 (2018)
6. Chen, T., He, T., Benesty, M., Khotilovich, V., Tang, Y.: Xgboost: extreme gradient boosting. R package version 0.4-2, pp. 1–4 (2015)
7. Cheng, H.T., Koc, L., et al.: Wide & deep learning for recommender systems. In: Proceedings of the 1st Workshop on Deep Learning for Recommender systems (2016)
8. Crowson, C.S., Atkinson, E.J., Therneau, T.M.: Assessing calibration of prognostic risk scores. Stat. Methods Med. Res. **25**(4), 1692–1706 (2016)
9. Das, A., Dantcheva, A., Bremond, F.: Mitigating bias in gender, age and ethnicity classification: a multi-task convolution neural network approach. In: Leal-Taixé, L., Roth, S. (eds.) ECCV 2018. LNCS, vol. 11129, pp. 573–585. Springer, Cham (2019). https://doi.org/10.1007/978-3-030-11009-3_35

10. Fu, L., Sun, D., Rilett, L.R.: Heuristic shortest path algorithms for transportation applications: state of the art. Comput. Oper. Res. **33**, 3324–3343 (2006)
11. Guo, H., Tang, R., Ye, Y., Li, Z., He, X.: DeepFM: a factorization-machine based neural network for CTR prediction. arXiv preprint arXiv:1703.04247 (2017)
12. Hardt, M., Price, E., Srebro, N.: Equality of opportunity in supervised learning. In: Advances in Neural Information Processing Systems, pp. 3315–3323 (2016)
13. Liu, H., Han, J., Fu, Y., Zhou, J., Lu, X., Xiong, H.: Multi-modal transportation recommendation with unified route representation learning. Proc. VLDB Endow. **14**(3), 342–350 (2021)
14. Liu, H., Tong, Y., Han, J., Zhang, P., Lu, X., Xiong, H.: Incorporating multi-source urban data for personalized and context-aware multi-modal transportation recommendation. IEEE Trans. Knowl. Data Eng. (2020)
15. Liu, H., Tong, Y., Zhang, P., Lu, X., Duan, J., Xiong, H.: Hydra: a personalized and context-aware multi-modal transportation recommendation system. In: Proceedings of the 25th ACM SIGKDD (2019)
16. Ruder, S.: An overview of multi-task learning in deep neural networks. arXiv preprint arXiv:1706.05098 (2017)
17. Singh, A., Joachims, T.: Fairness of exposure in rankings. In: Proceedings of the 24th ACM SIGKDD (2018)
18. Wang, Z., She, Q., Ward, T.E.: Generative adversarial networks in computer vision: a survey and taxonomy. arXiv preprint arXiv:1906.01529 (2019)
19. Xu, T., Zhu, H., Xiong, H., Zhong, H., Chen, E.: Exploring the social learning of taxi drivers in latent vehicle-to-vehicle networks. IEEE TMC **19**, 1804–1817 (2019)
20. Zheng, Y.: Trajectory data mining: an overview. ACM Trans. Intell. Syst. Technol. (TIST) **6**(3), 1–41 (2015)
21. Zhong, H., et al.: Adam revisited: a weighted past gradients perspective. Front. Comput. Sci. **14**(5), 1–16 (2020). https://doi.org/10.1007/s11704-019-8457-x

Constraint-Adaptive Rule Mining in Large Databases

Meng Li, Ya-Lin Zhang, Qitao Shi, Xinxing Yang, Qing Cui, Longfei Li, and Jun Zhou$^{(\boxtimes)}$

Ant Group, Hangzhou, China
{lm168260,lyn.zyl,qitao.sqt,xinxing.yangxx,
cuiqing.cq,longyao.llf,jun.zhoujun}@antgroup.com

Abstract. Decision rules are widely used due to their interpretability, efficiency, and stability in various applications, especially for financial tasks, such as fraud detection and loan assessment. In many scenarios, it is highly demanded to generate decision rules under some specific constraints. However, the performance, efficiency, and adaptivity of previous methods, which take no consideration of these constraints, is far from satisfactory in these scenarios, especially when the constraints are relatively tight. In this paper, to deal with this problem, we propose a constraint-adaptive rule mining algorithm named CARM (**C**onstraint **A**daptive **R**ule **M**ining), which is a novel decision tree based model. To provide a practical balance between purity and constraint fitness when building the trees, an adaptive criterion is designed and applied to better meet the constraints. Besides, a rule extraction and pruning process is applied to satisfy the constraints and further alleviate the overfitting problem. In addition, to improve the coverage, an iterative covering framework is proposed in this paper. Experiments on both public and business data sets show that the proposed method is able to achieve better performance, competitive efficiency, as well as low rule complexity when comparing with other methods.

Keywords: Rule induction · Confidence constraint · Adaptive criteria

1 Introduction

The rule-based classification model is one of the most straightforward and interpretable predictive models. A rule-based classification model consists of a set or a list of IF-THEN decision rules, which consist of several attribute conditions and a label as a prediction. In contrast with other models, the rule-based classification model has a multitude of advantages. On one hand, the IF-THEN structure of rules is straightforward to understand. On the other hand, the inference of rule is fast, since only a few binary statements need to be checked to find out which rules are satisfied.

Due to the aforementioned advantages, decision rules are widely employed in industry, especially in financial fields [5,6], and methods for rule mining are

© Springer Nature Switzerland AG 2021
C. S. Jensen et al. (Eds.): DASFAA 2021, LNCS 12683, pp. 579–591, 2021.
https://doi.org/10.1007/978-3-030-73200-4_41

widely studied and improved in recent years [3,14]. Fraud detection, loan assessment, and many other applications have a high demand for interpretability, and tree-based methods are widely applied [16,17] when building machine learning models. Furthermore, these tasks always utilize decision rules to solve binary classification or concept learning problems due to business requirements.

In these real-world tasks, various requirements may arise, which can be formalized as constraints in the rule learning problem. Confidence and coverage are two representatives of these constraints. Confidence is defined as the relative frequency of target (positive) samples in the samples covered by rules, i.e., the precision of the rules. Coverage is defined as the relative frequency of covered target (positive) samples in the whole target (positive) samples, i.e., the recall of the rules. In industry fields, problems of rule mining under specified constraints, especially confidence constraints, are very common and highly demanded.

Nevertheless, the problem of rule mining under specific constraints faces crucial challenges, and few studies have been performed to handle them. The performance of traditional rule mining methods, which take no consideration of confidence constraints, may be far from satisfactory, especially when the constraints are tight. Rules induced by these methods may violate the constraints or have a low coverage on target samples within the constraints. Therefore, it is a crucial challenge to mine constraint-adaptive decision rules, especially for a relatively high confidence constraint on class-imbalanced data. Besides, for industrial tasks, the amount of real business data is usually huge, which introduces exceptionally high requirements for space and time complexity.

To solve the decision rule induction problem with confidence constraint, we propose a constraint-adaptive rule mining algorithm named CARM, which ensures adaptivity, computational efficiency, and the other requirements mentioned above. First of all, we propose a new decision tree algorithm to extract decision rules, together with a new criterion for the selection of the best split points in the tree building process. The new criterion considers both impurity and constraint fitness of the rules related to the split choice (the feature and corresponding threshold) when the tree node is split. If the fitness score is far from satisfying the confidence constraint, the constraint fitness is punished. Besides, a rule extraction method is employed to obtain the rules that satisfy the constraint, and we further apply a pruning process to alleviate the overfitting problem. What's more, we propose an iterative framework to improve coverage of the resulting rules. The proposed method can efficiently improve coverage while satisfying the confidence constraint at the same time.

2   Related Work

Previous work on rule mining is extensive. There are numerous proposals for greedy or heuristic rule mining [2,10–13], especially tree-based methods. These tree-based approaches convert decision tree to rules. For instance, CART is one of the decision trees that have been widely used. The criteria used for splitting in CART is the Gini index. For each candidate split, the impurity calculated by the Gini index of all the sub-partitions is summed, and the split that leads to the maximum reduction in impurity is chosen. Other decision tree methods like ID3 [11] and C4.5 [12] use different kinds of greedy splitting criteria and pruning techniques. An attempt to directly combine the advantages of decision-tree and rule-based learning is the PART [4] algorithm, which does not learn the next rule in isolation but learns a global model in form of a decision tree. However, the splitting criteria of these tree-based methods only measure the impurity but take no consideration of any constraints.

Another specific rule mining approach is called Bayesian Rule Lists [8] (BRL for short). BRL uses Bayesian statistics to learn decision lists from frequent patterns that are pre-mined with the FP-tree algorithm. The goal of the BRL algorithm is to learn an accurate decision list using a selection of the pre-mined conditions while prioritizing lists with few rules and short conditions. SBRL [15], which is later proposed for large data sets, is two orders of magnitude faster than previous work. Unfortunately, it is claimed that it takes about 2.5 h to train a model with one million samples. The efficiency is still not acceptable for industry tasks which are always with tens of millions of samples. What's more, the constraints are not considered in these methods.

Constraint-based rule methods are also proposed by some previous works. Apriori [1] and CARS [9] only exploit the minimum support and the minimum confidence constraint for frequent rule filtering. These two methods face the effective problem on large databases due to a combinatorial explosion of frequent itemsets and are not easy to deal with continuous features. Dense-Miner [7] is designed to exploit constraints such as minimum confidence and a new constraint called minimum improvement during the mining phase. The minimum improvement constraint prunes any rule that does not offer a significant predictive advantage, which increases the efficiency of the algorithm. But the enumeration search of Dense-Miner still costs too much training time, which will hinder its application to large scale tasks.

To apply rule mining techniques to solve the confidence-constraint problem in real-world applications, especially for industrial tasks, the adaptivity, performance, and efficiency of the methods as mentioned earlier are still far from satisfactory. Methods with these above-mentioned characters are in high demand, and in this paper, a novel tree-based method is proposed towards this direction.

3   Problem Statement

Rule learning is considered as a concept learning problem in many industry applications, especially in financial tasks. The concept learning task is to learn a set of rules that describe a single target class. We are given a set of positive samples, for which we know they belong to the target concept, and a set of negative samples, for which we know they do not belong to the target concept, as training information. In this case, it is typically sufficient to learn theory for the target class only. All instances that are not covered by any of the learned rules will be classified as negative.

In concept learning, samples are either positive or negative samples of a given target class, and they are covered (predicted positive) or not covered (predicted negative) by a rule R or a set of rules RS. A perfect set of rules is one that covers all positive samples and covers no negative samples. Confidence of a single rule R or a set of rules RS is defined as the relative frequency of positive samples in the covered samples by rule or rules. This measure or metric is known as several different names, including confidence, precision, and rule accuracy. For simplicity, we call it confidence in the later statement. Coverage of a single rule R or a set of rules RS is defined as the relative frequency of covered positive examples in the positive examples.

In this paper, we consider the confidence-constrained rule learning task, in which confidence constraint is appended upon the traditional rule learning task, and this task can be regarded as a particular case of the concept learning task. The same as concept learning, the goal of confidence-constrained rule learning is to learn rules for the target class such as high-risk applicators in a loan assessment task, fraud applications for insurance, and other practical targets in industrial tasks. Furthermore, the rules must satisfy the confidence constraint, which means the confidence of the learned rules must exceed a threshold θ, which is set by users in advance of the training process. Additionally, another crucial goal of the confidence-constrained rule learning task is to achieve high coverage. More formally, the objective of this problem is to obtain a rule set RS^* that each rule of this rule set satisfies the confidence constraint and maximize coverage of the target class, which is shown in Eq. 1:

$$RS^* = \underset{Conf(R)>=\theta}{argmax} \ (Cov(RS)) \tag{1}$$

In addition, high efficiency is required so that the solution can be deployed in large scale applications. The performance of rules should be stable on training and validation data and the overfitting problem should be avoided.

4 Proposed Method

4.1 Overview

We propose a constraint-adaptive rule mining algorithm (CARM) to solve the confidence-constrained rule learning problem. Concretely, we design a novel tree-based method to learn the rules and design a new criterion to build the trees. A measure called constraint fitness is introduced to describe the fitness level of the splitting result under the constraints, and we take both the impurity and the constraint fitness into consideration while selecting the best split points in the tree building process. Besides, our rule extraction process is designed to satisfy the confidence constraint, and we further apply a rule pruning to alleviate the overfitting problem. Since the coverage and performance may not be satisfactory with rules obtained from one single tree, we propose to utilize an iterative covering framework to expand the rule set, and improve the coverage.

With our proposal, not only the confidence constraint can be satisfied but also the coverage and performance can be improved efficiently. Also, the proposed method can be deployed in distributed mode to satisfy the industrial settings.

4.2 Adaptive Decision Tree

Basically, all decision tree algorithms require a splitting criterion that splits a node to form a tree. In most cases, the criterion depends on a single attribute. There are various splitting criteria based on impurity of a node. The main aim of these splitting criteria is to reduce the impurity of a node. These splitting measures are defined in terms of the class distribution of the samples before and after splitting. We denote $p = (p_1, ..., p_j)$ as the sample proportions of each class in a node D, and $\phi(D)$ as the impurity function. $\phi(D)$ has a maximum when all p_j are equal and reaches the minimum when one of the p_j equals 1 and the others equals 0. Suppose the samples will be divided into left child node D_L and right child node D_R after splitting with feature s and threshold t, then the impurity-based goodness of this split can be defined as the $\Phi(s,t)$, which measures the difference between the weighted impurity score of parent node and two child nodes, as shown in Eq. 2.

$$\Phi(s,t) = \phi(D) - \frac{|D_L|}{|D|}\phi(D_L) - \frac{|D_R|}{|D|}\phi(D_R) \tag{2}$$

in which $|D|$ denotes the sample size in node D. The most commonly encountered impurity functions are Gini index and information entropy. However, these two criteria only consider the impurity of the splitting. In confidence-constrained rule learning tasks, only considering impurity is far from enough.

In this paper, we design a new criterion for the generation of decision trees. A measure called constraint fitness is proposed to describe the fitness level of the splitting result with regard to the constraints, and the criterion takes both impurity and constraint fitness of the corresponding rule into consideration when performing the splitting process.

Concretely, we first define the constraint fitness for a node. The samples in a node D can be regarded as the covered (predicted positive) samples of the corresponding rule R (the path from the root to this node), so we define the confidence $Conf(D)$ of node D as the confidence $Conf(R)$ of the corresponding rule R, i.e., $Conf(D) = Conf(R)$. The constraint fitness $\psi(D)$ of the node is calculated by considering the fitness of the confidence with regard to the constraint, which is shown in Eq. 3. The fitness score will be 0 if the confidence $Conf(D)$ in node D satisfies the constraint, i.e., $Conf(D) \geq \theta$, and will get a punishment $\theta - Conf(D)$ if the confidence $Conf(D)$ in node D doesn't satisfy the constraint, i.e., $Conf(D) < \theta$.

$$\psi(D) = \min(0, Conf(D) - \theta) \tag{3}$$

and the constraint fitness of this split (with feature s and threshold t) is defined as the maximum fitness of the resulting child nodes after the split, which is shown in Eq. 4:

$$\Psi(s,t) = \max\left(\psi(D_L), \psi(D_R)\right) \tag{4}$$

in which D_L and D_R are the resulting child nodes after the split. The intuition is that if the split results in a child node (no matter which child) that satisfies the constraint, which means the corresponding rules from the root to this child node satisfy the constraint, the split choice turns out to be feasible.

We use WSM (weighted sum model) to concatenates the impurity-based goodness and the constraint fitness with a parameter ω, and the impurity-based goodness is divided by the number of samples for normalization. The adaptive goodness $\rho(s,t)$ of the split (with feature s and threshold t) is defined as:

$$\rho(s,t) = \frac{\Phi(s,t)}{|D|} + \omega * \Psi(s,t) \tag{5}$$

Notably, the value of ω is set to infinity, if the maximum allowable depth is reached. In this situation, the constraint should be a rigid constraint.

4.3   Rule Extraction and Pruning

Rules can be easily extracted from the adaptive decision tree. In conventional practices, the path from the root node to the leaf node corresponds to a single rule with the splitting condition of each non-leaf node in the path joint with "AND". Furthermore, since confidence constraint is required in our setting, rules should be filtered from these extracted rules into a rule pool RS_{pool}. The process of extraction and selection is illustrated in Fig. 1, with the confidence (only the confidence on training set for simplicity) of each node annotated. The rules corresponding to leaf nodes D, E, G, H and I are firstly extracted, and if we set the threshold as $\theta = 0.5$, then the rules corresponding to nodes D, E and H are selected into the rule pool RS_{pool} since the confidence of these nodes satisfies the constraint.

However, the rules extracted and selected with the confidence constraint (i.e., the rules in RS_{pool}) are likely to overfit if the length of these rules is relatively long. In order to solve this problem, a constraint-adaptive post-pruning is proposed as shown in Algorithm 1 and Fig. 1. For every rule R^j in rule pool RS_{pool}, we try to prune it and its brother rule $R^j_{brother}$, and replace them with their parent rule R^j_{parent}. The parent rule and brother rule of a specific rule are defined as the rule extracted with the path from the root node to its parent node and brother node (the other child node which shares the same parent node) in a decision tree. As an example, in Fig. 1, Node F is the parent node for H, while I is its corresponding brother node, and the rules related to these two nodes are the parent rule and brother rule for the rule related to node H. If the confidence of the parent rule satisfies the confidence constraint (i.e., $Conf_{train}(R^j_{parent}) \geq \theta$ and $Conf_{val}(R^j_{parent}) \geq \theta$) when applying to training and validation data sets, the rule and its brother rule are pruned and replaced with their parent rule. The post-pruning algorithm traverses all the rules in the rule pool until no rules can be pruned and replaced to reduce the complexity of the rule set, then the pruning

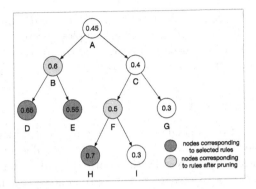

Fig. 1. An illustration of rule extraction, selection and pruning. The numbers in the node represent the confidence of the rule (node).

process can be terminated. For the example in Fig. 1, the rules corresponding to nodes D, E and H are pruned and replaced with the rules corresponding to nodes B and F, which will be collected to make up the pruned rule set.

Algorithm 1. Pseudo-code of Pruning

Input: The raw data set D_{train} and D_{valid}; confidence constraint $CC(Conf(R) >= \theta)$; a set of previously extracted rules RS_{pool}

Output: the pruned rule set RS_{pruned}

1: $RS_{pruned} \leftarrow \emptyset$
2: **while** $RS_{pool} \neq \emptyset$ **do**
3:     Pick a non-pruned rule $R^j \in RS_{pool}$
4:     Find the parent rule R^j_{parent} and the brother rule $R^j_{brother}$ of R^j
5:     Calculate the confidence $Conf_{train}(R^j_{parent})$ and $Conf_{val}(R^j_{parent})$ of the parent rule R^j_{parent} applied on D_{train} and D_{valid}
6:     **if** $Conf_{train}(R^j_{parent}) \geq \theta$ and $Conf_{val}(R^j_{parent}) \geq \theta$ **then**
7:         $RS_{pool} = RS_{pool} \setminus R^j$
8:         **if** $R^j_{brother} \in RS_{pool}$ **then**
9:             $RS_{pool} = RS_{pool} \setminus R^j_{brother}$
10:        **end if**
11:        $RS_{pool} = RS_{pool} \cup R^j_{parent}$
12:    **else**
13:        $RS_{pruned} = RS_{pruned} \cup R^j$
14:    **end if**
15: **end while**
16: **return** RS_{pruned}

In most cases, this post pruning method described above can not only avoid overfitting and reduce the complexity of the rule set, but also increase the coverage obviously when a rule is pruned and replace by its parent rule while his brother rule is not in the rule pool (i.e., the brother rule does not satisfy the confidence constraint). Due to the post-pruning, the target (positive) samples covered by its brother rule can be involved in the covered samples. For concrete illustration in Fig. 1, the target (positive) samples covered by the rule corresponding to node I are recalled to improve the coverage of the final rule set, without a violation of the confidence constraint.

4.4 Iterative Covering Framework

The rules generated from a single adaptive tree may not perform satisfactorily. Generally, a decision tree with a high depth is not intelligible and likely to overfit, so it is a common practice to restrict the depth of the decision tree. However, if the tree is trained with a strictly restricted depth, followed by the rule extraction and pruning process as described in the last subsection, only a small quantity of rules will be finally obtained. Maybe only a small amount of target samples will be covered, which means that the coverage of the resulted rule set may be far from satisfactory.

In this paper, we propose to employ an iterative framework to alleviate this problem. Note that there may be many target samples that are not covered by the learned rules, and one possible reason is that the building process of the adaptive tree is dominated to cover other target samples, which may have very different behavior when compared to the uncovered target samples. To handle this, in the subsequent process, the samples that have been covered will be removed so that they will not influence the learning process of the subsequent adaptive trees. Concretely speaking, an adaptive decision tree is firstly built, and rules are extracted and pruned as mentioned in the last two subsections. Then, each sample that covered by the rules is removed, no matter it is positive or not. After that, a new adaptive decision tree is learned with the remaining samples, and the obtained rules are combined with the preceding ones. This process is repeated until no sample is left or other stop condition encounters.

In each iteration, several rules are obtained, and the size of the remaining samples will gradually reduce so that the efficiency is acceptable. By employing this iterative covering framework, the coverage can be obviously improved, and the effectiveness of the resulting rules can be enormously enhanced.

5   Experiment

5.1   Experiments on Benchmark Data Sets

We first conduct experiments on eight benchmark data sets, with different sample and feature size. All these data are available on the OpenML database[1]. All experiments are performed on a 4-core computer with 16 GB of RAM.

---

[1] https://www.openml.org/.

Table 1. Classification performance (coverage, %) on benchmark data sets (**Bold:** Best among interpretable models; *Italics*: Best overall.)

Dataset	Constraint	Interpretable					Non-interpretable		
		RIPPER	PART	CART	SBRL	**CARM**	LR	RF	SVM
adult	conf_0.3	54.3	69.1	80.3	84.2	*91.9*	78.4	75.2	64.4
positive rate = 0.24	conf_0.5	54.3	54.2	50.4	48.0	*67.7*	58.8	62.5	54.6
	conf_0.7	–	36.1	30.4	30.2	*54.1*	37.8	48.3	43.8
	conf_0.9	–	19.1	17.4	–	*21.3*	15.6	–	15.1
bank-mkt	conf_0.3	38.8	50.2	54.3	44.7	*77.2*	98.8	**99.3**	49.0
positive rate = 0.12	conf_0.5	38.8	43.9	39.1	36.2	*55.0*	**97.5**	97.3	35.4
	conf_0.7	12.8	*20.4*	–	–	–	**94.8**	93.2	–
	conf_0.9	–	*8.5*	–	–	–	81.0	**83.5**	–
banknote	conf_0.5	96.6	*98.3*	97.4	*98.3*	*98.3*	98.6	99.3	**100.0**
positive rate = 0.46	conf_0.7	96.6	*98.3*	97.4	88.8	96.6	97.3	97.9	**100.0**
	conf_0.9	96.6	*98.3*	97.4	79.3	96.6	91.8	94.5	**100.0**
ionosphere	conf_0.5	*86.7*	*86.7*	*86.7*	60.0	*86.7*	66.7	**86.7**	**86.7**
positive rate = 0.37	conf_0.7	–	*13.3*	–	–	–	60.0	**86.7**	**86.7**
	conf_0.9	–	*13.3*	–	–	–	–	**60.0**	–
magic	conf_0.5	*100.0*	*100.0*	*100.0*	87.1	*100.0*	99.6	**100.0**	99.2
positive rate = 0.35	conf_0.7	*100.0*	*100.0*	*100.0*	85.3	*100.0*	97.6	**100.0**	98.1
	conf_0.9	*100.0*	*100.0*	*100.0*	83.8	*100.0*	91.5	99.9	96.3
tic-tac-toe	conf_0.5	97.1	98.5	95.6	66.2	*100.0*	97.1	89.7	98.5
positive rate = 0.34	conf_0.7	97.1	*98.5*	95.6	66.2	97.1	97.1	61.8	94.1
	conf_0.9	*97.1*	77.9	95.6	–	94.1	64.7	17.6	73.5
transfusion	conf_0.3	34.2	39.5	57.9	68.4	*78.9*	**100.0**	97.3	21.1
positive rate = 0.24	conf_0.5	34.2	39.5	36.8	–	*52.6*	**100.0**	92.9	7.9
	conf_0.7	–	–	–	–	–	76.1	**89.4**	2.6
WDBC	conf_0.5	84.6	*89.7*	84.6	66.7	87.2	**98.4**	96.8	94.9
positive rate = 0.39	conf_0.7	84.6	*87.2*	84.6	66.7	*87.2*	**96.8**	93.7	94.9
	conf_0.9	–	69.2	–	66.7	*84.6*	88.9	**90.5**	84.6
Average		54.0	61.9	57.8	47.2	*66.4*	80.2	**81.3**	64.5

Table 2. Rule complexity (count/length) on benchmark data sets

Dataset	RIPPER	PART	CART	SBRL	**CARM**
adult	7/3.3	943/5.4	53/5.8	22/2.0	12.8/4.8
bank-mkt	17/3.3	895/4.4	60/5.9	18/2.6	18.0/5.2
banknote	8/2.3	8/1.6	20/4.9	13/1.4	6.5/3.1
ionosphere	3/1.0	7/2.6	13/4.6	3/1.3	6.3/2.7
magic	2/1.0	2/1.0	2/1.0	14/1.5	1.0/1.0
tic-tac-toe	9/2.8	26/2.8	33/5.4	8/1.6	12.0/3.6
transfusion	3/2.0	7/1.4	35/5.5	5/1.0	22.3/5.0
WDBC	5/1.6	7/1.7	15/4.5	3/1.7	5.5/2.9
Average	6.8/2.2	236.9/2.6	59.1/5.5	10.8/1.6	12.7/3.8

Algorithms for Comparison. We compare CARM with many other rule-based methods and a set of "non-interpretable" methods. For rule-based methods, RIPPER, PART, SBRL and CART are chosen as baselines. Support vector machine (SVM), logistic regression (LR), random forests (RF) are chosen to represent "non-interpretable" methods. RIPPER and PART are implemented in Weka and SBRL is available as R package in Python. CART, SVM, LR and RF are implemented in Python using the scikit-learn package. We use the confidence and coverage of the induced rule list/set that satisfies the constraint as the evaluation metric. The coverage on each data set are obtained from 5-fold cross-validations and all the methods are tested with default setting. Gini index is used as the impurity function while building the adaptive trees. Since the parameter of maximum rule length has a great influence on rule-base methods, we set this parameter of these rule-based methods to 6. The encoding of categorical features and normalization of numerical features are conducted if necessary.

Rule Performance. We evaluate the confidence and coverage of these methods with different constraint setting. Only high-confidence constraint problem are tested because the low-confidence constraint problem can be transformed to high-confidence constraint problem by reverse the problem target. On every benchmark data set, we set several discrete confidence constraints, which are all higher than the proportion of positive samples.

Table 1 lists the classification performances on benchmark data sets. The number in the table represents the coverage of rules that satisfies the specific confidence constraint. If there are no rules generated, the cell is filled with '–'. It can be seen that the proposed method CARM has a competitive performance among the "interpretable" methods. CARM wins in 16 of 26 cases with an average confidence of 66.4%. It is 3.5% percentages ahead of PART which is in the second place in terms of performance.

Influence of Pruning and Iterative Covering Framework. We then validate the influence of the pruning process and iteration covering framework. We set the iteration round to 5, and run experiments with/without the pruning process. Figure 2 shows the result on data set bank-mkt. As we can see, with the pruning process, the coverage of rule set is consistently improved. What's more, as the iteration proceeds, the performance can be evidently boosted, which validates the effect of the iteration covering framework.

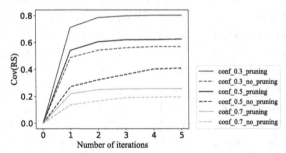

Fig. 2. The performance on different iterations (bank-mkt)

Rule Complexity. Table 2 lists the rule complexity of different methods on benchmark data sets. We evaluate the number of rules and the average length of rules. It can be seen the rules generated by CARM are relatively shorter and less among the "interpretable" methods especially on large data sets (i.e., adult and bank-mkt). The rule count of CARM is significantly less than PART which is in second place in terms of performance. This good property can further provide better efficiency in the deployment stage.

5.2   Experiments on Business Data Sets

Experiments on extra-large scale business data sets are further conducted to verify the effectiveness and scalability of the proposed method on real industrial tasks. The data sets come from the tasks for three different areas. Data set A is from a task of loan assessment, which aims at finding the most at-risk customers with higher overdue probability, so that the system can reduce their credit limit. Data set B is from a task of fraud detection in the insurance area, which aims at detecting fraud applications for insurance, so that the system can reject these applications to avoid the economic losses. Data set C is from the task of intelligent self-service. It is for discovering the potential users of a telephone App who need a specific self-service and the App can push the self-service automatically. Table 3 presents the detailed information of these data sets. As shown, the number of samples or features is extremely large (e.g., up to 10 million samples for data set A), and these data sets are class-imbalanced. All parameters of the evaluated models are set as the default values as above.

Table 3. The information of the business data sets

Dataset	#Train	#Valid	#Test	#Dim
Data A	6000000	2000000	2000000	19
Data B	3000000	1000000	1000000	72
Data C	5175057	1725019	1725019	12

The results are shown in Table 4. SBRL is not compared due to its out-of-memory problem for extremely large data sets. As we can see, the CARM achieves better performance, which validates the effectiveness of the proposed method when applying it in real industrial tasks and make it a choice for performing constraint-based rule induction adaptively for extremely large scale industrial data sets. Actually, this work has been deployed in our system, providing help for many different real-world tasks.

Table 4. Classification performance (coverage, %) on business data sets (**Bold**: Best among interpretable models)

Dataset	Constraint	RIPPER	PART	CART	**CARM**
Data A Positive rate = 0.003	conf_0.1	1.0	4.2	4.2	**9.4**
	conf_0.2	1.0	0.3	-	**2.9**
Data B Positive rate = 0.006	conf_0.1	60.1	87.3	65.2	**95.6**
	conf_0.2	60.1	86.6	40.1	**89.7**
	conf_0.3	59.6	75.4	31.0	**80.9**
	conf_0.4	55.2	61.8	22.1	**74.5**
Data C Positive rate = 0.008	conf_0.1	6.4	53.0	52.6	**68.6**
	conf_0.2	6.4	35.7	38.1	**49.9**
	conf_0.3	6.4	18.5	20.3	**33.1**
	conf_0.4	6.4	11.6	10.4	**20.4**

6 Conclusion

In this paper, we propose a constraint-adaptive rule mining algorithm to deal with the constraint-based rule induction problem. A novel decision tree model with an adaptive criterion is designed to solve this specific problem effectively, and an iterative covering framework is proposed to increase the coverage. We perform experiments on both benchmark and extra-large scale business data sets, which validate the effectiveness of the proposed method.

References

1. Agrawal, R., Mannila, H., Srikant, R., Toivonen, H., Verkamo, A.I.: Fast discovery of association rules. In: Advances in Knowledge Discovery and Data Mining, pp. 307–328. AAAI/MIT Press (1996)
2. Clark, P., Niblett, T.: The CN2 induction algorithm. Mach. Learn. **3**, 261–283 (1989). https://doi.org/10.1023/A:1022641700528
3. Dash, S., Günlük, O., Wei, D.: Boolean decision rules via column generation. In: NIPS, pp. 4660–4670 (2018)
4. Frank, E., Witten, I.H.: Generating accurate rule sets without global optimization. In: ICML, pp. 144–151. Morgan Kaufmann (1998)
5. Gao, Q., Xu, D.: An empirical study on the application of the evidential reasoning rule to decision making in financial investment. Knowl.-Based Syst. **164**, 226–234 (2019)
6. Hajek, P.: Interpretable fuzzy rule-based systems for detecting financial statement fraud. In: MacIntyre, J., Maglogiannis, I., Iliadis, L., Pimenidis, E. (eds.) AIAI 2019. IAICT, vol. 559, pp. 425–436. Springer, Cham (2019). https://doi.org/10.1007/978-3-030-19823-7_36
7. Bayardo Jr., R.J., Agrawal, R., Gunopulos, D.: Constraint-based rule mining in large, dense databases. In: ICDE, pp. 188–197 (1999)

8. Letham, B., Rudin, C., McCormick, T.H., Madigan, D.: Interpretable classifiers using rules and Bayesian analysis: building a better stroke prediction model. Ann. Appl. Stat. **9**(3), 1350–1371 (2015)
9. Liu, B., Hsu, W., Ma, Y.: Integrating classification and association rule mining. In: KDD, pp. 80–86 (1998)
10. Marchand, M., Sokolova, M.: Learning with decision lists of data-dependent features. JMLR **6**, 427–451 (2005)
11. Quinlan, J.R.: Induction of decision trees. Mach. Learn. **1**(1), 81–106 (1986). https://doi.org/10.1007/BF00116251
12. Quinlan, J.R.: C4.5: Programs for Machine Learning. Morgan Kaufmann, Burlington (1993)
13. Rudin, C., Letham, B., Madigan, D.: Learning theory analysis for association rules and sequential event prediction. JMLR **14**(1), 3441–3492 (2013)
14. Wang, T., Rudin, C., Doshi-Velez, F., Liu, Y., Klampfl, E., MacNeille, P.: A Bayesian framework for learning rule sets for interpretable classification. JMLR **18**, 70:1–70:37 (2017)
15. Yang, H., Rudin, C., Seltzer, M.: Scalable Bayesian rule lists. In: ICML, vol. 70, pp. 3921–3930. PMLR (2017)
16. Zhang, Y.L., Li, L.: Interpretable MTL from heterogeneous domains using boosted tree. In: CIKM, pp. 2053–2056 (2019)
17. Zhang, Y., et al.: Distributed deep forest and its application to automatic detection of cash-out fraud. TIST **10**(5), 55:1–55:19 (2019)

Demo Papers

FedTopK: Top-K Queries Optimization over Federated RDF Systems

Ningchao Ge[1], Zheng Qin[1(✉)], Peng Peng[1], and Lei Zou[2]

[1] Hunan University, Changsha, China
{ningchaoge,zqin,hnu16pp}@hnu.edu.cn
[2] Peking University, Beijing, China
zoulei@pku.edu.cn

Abstract. Recently, how to evaluate SPARQL queries over federated RDF systems has become a hot research topic. However, most existing studies mainly focus on implementing and optimizing the basic queries over federated SPARQL systems, and few of them discuss top-k queries. To remedy this defect, this demo designs a system named *FedTopK* that can support top-k queries over federated RDF systems. FedTopK employs a cost-based optimal query plan generation algorithm and a query plan execution optimization strategy to minimize the top-k query cost. In addition, FedTopK uses a query decomposition optimization scheme which allow merge triple patterns with the same multi-sources into one subquery to reduce the remote access times. Experimental studies over real federated RDF datasets show that the demo is efficient.

1 Introduction

In recent years, *R*esource *D*escription *F*ramework (*RDF*) has been widely used in many applications. Many data providers publish their datasets using the RDF model at their own sites, and provide the SPARQL interfaces to support users to submit SPARQL queries. In this paper, an autonomous site with a SPARQL interface is called an *RDF source*. To integrate multiple RDF sources, federated RDF systems have been proposed [2–4].

Right now, practitioners are showing a growing interest in top-k queries, which impose an order on the result set and limit the number of results. Top-k queries can be expressed in SPARQL by including the ORDER BY and LIMIT clauses. However, existing federated RDF systems can only support to alter the sequence of solution mappings after the full evaluation of the graph pattern in the WHERE clause. Therefore, this paper implement a federated RDF system, named FedTopK, which optimize evaluation of top-k queries over federated RDF systems. In summary, FedTopK has the following unique features:

– FedTopK have an incremental query execution strategy in accordance with the characteristics of top-k queries, which can greatly improve the query efficiency by terminating the execution as soon as the requested number of final results has been obtained.

© Springer Nature Switzerland AG 2021
C. S. Jensen et al. (Eds.): DASFAA 2021, LNCS 12683, pp. 595–599, 2021.
https://doi.org/10.1007/978-3-030-73200-4_42

- FedTopK can minimize query cost by a cost-based optimal query plan generation algorithm, which can optimize the join order of subqueries.
- FedTopK can reduce the remote access times effectively by a query decomposition scheme, which allows merge triple patterns with the same multi-sources into one subquery.

2   System Architecture and Key Techniques

Figure 1 shows the system architecture of our proposed federated RDF system FedTopK. It consists of a control site and some RDF sources. We assume that queries are submitted to the control site. The control site decomposes the query into several subqueries on relevant sources and generate a query plan. Then, the decomposed subqueries are sent to their relevant sources and executed. Last, matches of subqueries are returned to the control site and joined to form complete matches according to the query plan. In summary, there are three steps during the query processing of FedTopK: *query decomposition and source selection, cost-based query plan generation* and *query execution.*

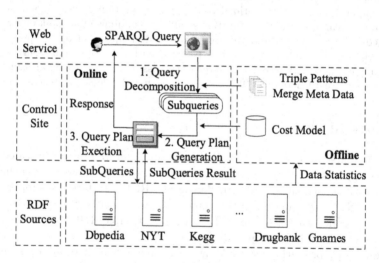

Fig. 1. Scheme for query processing in FedTopK

Query Decomposition and Source Selection. When an user submit a top-k query Q online, the query Q is decomposed into a set of subqueries, $Q = \{q_1@S_1, q_2@S_2, ..., q_n@S_n\}$, where S_i is the set of relevant sources for q_i. FedTopK can merge triple patterns with the same multi-sources into one subquery by maintaining the triple patterns merge conditions from RDF sources offline. It can reduce the communication overhead effectively by reducing the number of subqueries. For example, Fig. 2 shows an example query decomposition and source selection result.

Fig. 2. Example query decomposition and source selection result

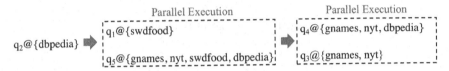

Fig. 3. Example query plan

Cost-Based Optimal Query Plan Generation. A query plan represent a join order of subqueries $Q = \{q_1@S_1, q_2@S_2, ..., q_n@S_n\}$. Different query plans have different query costs. FedTopK designs a cost model to calculate the query cost and join cost of subqueries in accordance with the statistics data maintained from RDF sources offline. On this basic, the optimal query plan can be obtained by a optimal query plan generation algorithm. For example, Fig. 3 shows an example query plan for the query decomposition and source selection result, and we assume this query plan is the optimal one.

Query Execution. The query plan determines the execution order and execution mode (serial and parallel) of subqueries. For query plan in Fig. 3, subquery $q_2@\{dbpedia\}$ is executed firstly. Then, subqueries $q_1@\{swdfood\}$ and $q_5@\{gnames, dbpedia, swdfood, nyt\}$ can be executed in parallel, and so on. Among that, we propose an optimization in accordance with the characteristics of top-k query. During query execution, when a subquery containing the top-k constraint is executed, its results are sorted and incrementally used to generate the final results in order. The execution can stops as soon as the requested number of final results has been obtained.

3 Demonstration

In this demo, we use two famous comprehensive RDF benchmark suites, LargeRDFBench [5] and WatDiv [1], to show the demonstration of FedTopK. The federated RDF system FedTopK can efficiently support both SPARQL basic queries and top-k queries. More demonstrations can be referred with http://47.111.92.242:8080/FedTopK/Demo/index.html.

Fig. 4. Query page of FedTopK

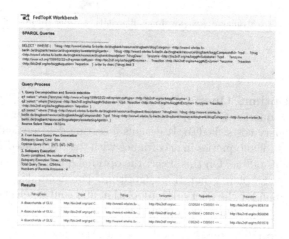

Fig. 5. Query result page of FedTopK

Figure 4 and Fig. 5 demonstrate the two main pages of FedTopK. Users can enter a SPARQL top-k query or select a query statement from the query sample list in Fig. 4. In the top of Fig. 5, FedTopK shows the detail query process including current SPARQL query statement, the set of subqueries after query decomposition, the optimal query plan and the value of query performance indicators. Finally, the query results can be found in the bottom of Fig. 5.

4   Conclusion

FedTopK is a federated RDF system that can support top-k SPARQL queries. It can improve query performance by a cost-based optimal query plan generation algorithm and a query plan execution optimization strategy. It also reduces the remote requests by a query decomposition optimization.

Acknowledgment. This work was supported by the National Natural Science Foundation of China under Grant (No. U20A20174, 61772191), Science and Technology Key Projects of Hunan Province (2019WK2072, 2018TP3001, 2018TP2023,), ChangSha Science and Technology Project (kq2006029), National Key Research and Development Program of China under grant 2019YFB1406401 and Key Research and Development Program of Hubei Province (No. 2020BAB026).

References

1. Aluç, G., Hartig, O., Özsu, M.T., Daudjee, K.: Diversified stress testing of RDF data management systems. In: Mika, P., et al. (eds.) ISWC 2014. LNCS, vol. 8796, pp. 197–212. Springer, Cham (2014). https://doi.org/10.1007/978-3-319-11964-9_13
2. Montoya, G., Skaf-Molli, H., Hose, K.: The *Odyssey* approach for optimizing federated SPARQL queries. In: d'Amato, C., et al. (eds.) ISWC 2017. LNCS, vol. 10587, pp. 471–489. Springer, Cham (2017). https://doi.org/10.1007/978-3-319-68288-4_28
3. Peng, P., Ge, Q., Zou, L., Özsu, M.T., Xu, Z., Zhao, D.: Optimizing multi-query evaluation in federated RDF systems. TKDE (2019)
4. Quilitz, B., Leser, U.: Querying distributed RDF data sources with SPARQL. In: Bechhofer, S., Hauswirth, M., Hoffmann, J., Koubarakis, M. (eds.) ESWC 2008. LNCS, vol. 5021, pp. 524–538. Springer, Heidelberg (2008). https://doi.org/10.1007/978-3-540-68234-9_39
5. Saleem, M., Hasnain, A., Ngomo, A.N.: LargeRDFBench: a billion triples benchmark for SPARQL endpoint federation. J. Web Semant. **48**, 85–125 (2018)

Shopping Around: *CoSurvey* Helps You Make a Wise Choice

Qinhui Chen[1], Liping Hua[1], Junjie Wei[1], Hui Zhao[1,2](✉), and Gang Zhao[3]

[1] Software Engineering Institute, East China Normal University, Shanghai, China
[2] Shanghai Key Laboratory of Trustworthy Computing, Shanghai, China
`hzhao@sei.ecnu.edu.cn`
[3] Microsoft, Beijing, China
`gang.zhao@microsoft.com`

Abstract. When shopping online, customers usually compare commodities with each other before making their purchase decision. In addition to the product price, they also concern the word-of-mouth. However, marketing strategies from various e-commerce platforms, along with the diverse online commodities, make it difficult for customers to distinguish the most cost-effective products. Present cross-platform commodity comparison applications merely focus on product prices, without jointly concerning the reviews. In this demonstration, we developed a web-based application, *CoSurvey*, which matches commodities from various e-commerce platforms and analyzes product comment sentiment on the base of the proposed Attention-BiLSTM-CNN Model. The model uses an attention-based Bi-LSTM network to learn sentence sequence information, uses a CNN to learn sentence structure information, and uses a multilayer perceptron (MLP) to learn meta-information. The meta-information in the comment sentiment analysis task includes comment's *like number, reviewer level, additional image, deliver time, and sentence length*. Besides the keyword query, *CoSurvey* provides customers a survey of cross-platform products price changing trends and comment sentiment evolutions. The high concurrency requirements and load balance are also concerned.

Keywords: Sentiment analysis · Entity resolution · E-commerce · Multiple neural network · Attention mechanism

1 Introduction

With the permeation of online shopping, customers usually shop around on different e-commerce platforms. Besides the price and the brand, product reviews play a decisive role in the final purchase decision making as they can reflect on customer's preferences for the product. However, the inconsistency of cross-platform product descriptions, along with massive product reviews, bring about overwhelming information overload. During shopping festivals such as Singles Day (11.11) and Black Friday, this situation aggravates further.

© Springer Nature Switzerland AG 2021
C. S. Jensen et al. (Eds.): DASFAA 2021, LNCS 12683, pp. 600–603, 2021.
https://doi.org/10.1007/978-3-030-73200-4_43

Although there exist some cross-platform commodity comparison applications, such as Kelkoo[1], and Goggle Product Search[2], these applications only focus on the price comparison and overlook the product review analysis jointly. Therefore, we develop *CoSurvey*, which surveys the product information from different e-commerce platforms to help the customer make a wise shopping decision. The application meets two challenges: (1) Product matching, which aims to align the same product of different e-commerce platforms; (2) Product review sentiment analysis, which concerns not only comment text but also its valuable meta information, such as when the review is delivered, how many consumers agree with the review, etc. Since different platforms have different comment meta information, we normalize them by extracting comments' *like number, reviewer level, additional image, deliver time, and sentence length.*

Our implementations can be summarized as follows:

- We propose and train a deep fusion neural network - Attention-BiLSTM-CNN model – which takes both comment and its meta-information to classify the sentiment polarity. The experimental results demonstrate that our model's precision achieves 94.48%, recall is 94.29%, F1 value is 94.38%.
- We also train Attention-BiLSTM-CNN model to calculate the product pairs' match possibility. The blocking strategy[2] is applied to decrease complexity.
- The system concerns high concurrency. Linux Virtual Server and multi-Nginx servers are employed to implement the load balance.

2   *CoSurvey* System Overview and Key Techniques

As shown in Fig. 1, *CoSurvey* system consists of five layers: data layer, NLP layer, business layer, gateway layer, and visual layer. NLP layer provides key techniques for product matching and review sentiment analysis task, which mainly includes:

Data Pre-Process. Data pre-processing steps include filtering out missing values, normalizing comment meta information and product names, etc.

Sentiment Classification. The Attention-BiLSTM-CNN model is applied to predict product review sentiment. The model will be detailed in Sect. 2.1.

Product Matching. An Attention-BiLSTM-CNN model is trained to match products from different platforms. The input of the model is a pair of product descriptions from different platforms. The output is the possibility of whether the descriptions identify the same product. The blocking strategy is used to divide the whole dataset into several subsets by the product brand. The best-matched pairs are stored in the database.

CoSurvey crawls different platforms' commodity information and stores them in MongoDB. Elasticsearch[3] and Redis[4] are used to moderate database pressure. *CoSurvey* meets high concurrency requirements. We employ multiple LVS and Nginx servers in Openresty[5] to implement the load balance. *CoSurvey* also pro-

---

[1] https://www.kelkoo.co.uk/.
[2] https://shopping.google.com/?nord=1.
[3] https://www.elastic.co/cn/elasticsearch/.
[4] https://redis.io/.
[5] http://openresty.org/cn/.

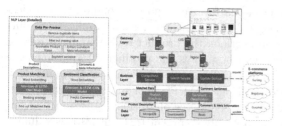
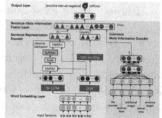

Fig. 1. The framework of *CoSurvey*   **Fig. 2.** Attention-BiLSTM-CNN model architecture

vides customers an interactive web-based interface to browse all commodities or search for a specific commodity using keywords or product characteristics. For both query modes, *CoSurvey* presents all the selling links of the commodity and gives out a detail comparison of its review and price. According to these comparisons, customers can obtain the latent relationship between promotion and product feedback.

2.1 Attention-BiLSTM-CNN Model

Model Structure. The model consists word embedding layer, sentence representation encoder (SRE), comment meta information encoder (CMIE), sentence-meta information fusion layer, and output layer, as is shown in Fig. 2.

The word embeddings are initialized by ERNIE [6] model, which has demonstrated outperform BERT in Chinese corpus. The word embedding was fine-tuned during the training process. In SRE, we fuse CNN and BiLSTM via attention mechanism [3] to fully utilize sentence structural and sequential information. Specifically, Bi-LSTM output R and CNN output C are used to calculate the attention result $H = softmax(\frac{K^T Q}{\sqrt{d}} V)$ (where $K, V = C, Q = R$). C is also fed into an sqrt-pooling layer to obtain the pooling result $C_{pooling}$. In CMIE, we obtain high-dimensional meta-information representation E through MLP. Then, the fusion layer concatenates the outputs of SRE and CMIE, forming a fusing representation $V^{HCE} = [H; \ C_{pooling}; \ E]$. Finally, V^{HCE} is fed into a fully-connected layer with $softmax$ to obtain the sentiment polarity.

Experiments. We compare our model to CNN [1] model, LSTM [4] model, and AT-LSTM [5] model. The experiment results show that our model reaches 94.48% in precision and outperforms other models by 1.65%. We also perform an ablation study, which shows that meta-information makes an improvement by 0.4% in precision in the sentiment classification task.

3   System Demonstration

We provide customers a highly interactive demonstration of our system. Figure 3 shows the main scenarios of the demo. (1) Customers can shop around commodities from different e-commerce platforms and search for products using keywords or product characteristics. (2) Customers can overview the price, the comment number, and the feedback rate distribution of the products. (3) Customers can browse the detailed comparison information for a specific product, like the current lowest price, the word cloud, and the price trends of this product. (4) Customers can view the product sentiment evolution of the product, where the static surveyed emotion, the dynamic emotional tendency, and the supported or inconsistent feedback rates are presented.

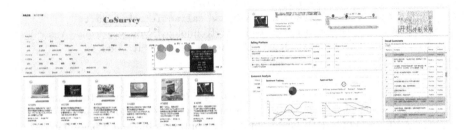

Fig. 3. System Demonstration

4   Conclusion

In this demonstration, we develop a distributed application called *CoSurvey* to survey the product information across different e-commerce platforms. We apply Attention-BiLSTM-CNN Model to implement both the sentiment analysis task and the product matching task. *CoSurvey* provides cross-platform product information survey service to help customers make a wise purchase decision. The application also provides operators insightful feedback to improve the production and the marketing strategy.

Acknowledgements. This work is supported by National Key Research and Development Program (2019YFB2102600).

References

1. Johnson, R., Zhang, T.: Effective use of word order for text categorization with convolutional neural networks. In: NAACL, pp. 103–112 (2015)
2. O'Hare, K., Jurek-Loughrey, A., de Campos, C.: An unsupervised blocking technique for more efficient record linkage. Data Knowl. Eng. **122**, 181–195 (2019)
3. Vaswani, A., et al.: Attention is all you need. In: NIPS, pp. 5998–6008 (2017)
4. Wang, X., Liu, Y., Sun, C.J., Wang, B., Wang, X.: Predicting polarities of tweets by composing word embeddings with long short-term memory. In: IJCNLP, pp. 1343–1353 (2015)
5. Wang, Y., Huang, M., Zhu, X., Zhao, L.: Attention-based LSTM for aspect-level sentiment classification. In: EMNLP, pp. 606–615 (2016)
6. Zhang, Z., Han, X., Liu, Z., Jiang, X., Sun, M., Liu, Q.: ERNIE: enhanced language representation with informative entities. In: ACL, pp. 1441–1451 (2019)

IntRoute: An Integer Programming Based Approach for Best Bus Route Discovery

Chang-Wei Sung[1], Xinghao Yang[2]([✉]), Chung-Shou Liao[1], and Wei Liu[2]

[1] Department of Industrial Engineering, National Tsing Hua University,
Hsinchu, Taiwan
sung103034036@gapp.nthu.edu.tw, csliao@ie.nthu.edu.tw
[2] School of Computer Science, University of Technology Sydney, Ultimo, Australia
xinghao.yang@student.uts.edu.au, wei.liu@uts.edu.au

Abstract. An efficient data-driven public transportation system can improve urban potency. In this research, we propose IntRoute, an Integer Programming (IP) based approach to optimize bus route planning. Specifically, IntRoute first contracts bus stops via clustering and then derives a new bus route via a mixed integer linear program (ILP). This two-phase strategy brings three major merits, i.e., a single bus route without any transfer, the minimal total time consuming, and an efficient optimization algorithm for large-scale problems. Experimental results show that our IntRoute significantly reduces the traditional commuting time in Sydney from 31.53 min down to 18.06 min on average.

1 Introduction

In this study, we consider an important data-driven public transportation problem: finding the best bus route that minimizes passengers' overall commuting time. Given a bus transportation system as well as the requests of specific passengers who commute from different starting locations to a fixed destination, the goal of the problem is designing a new bus route to satisfy the passengers' demand without any transfer.

Previous studies on bus transfer problems were mostly not data-driven [2] due to data skewness problems [3]. In this work, we proposed a new integer-programming based method, which we call IntRoute, to find the route that minimizes the total time cost of the targeted passengers. Specifically, our IntRoute contains two main phases, i.e., the contraction of bus stops via K-means clustering and the derivation of new bus route via a mixed integer linear programming (ILP). The major contributions of this research are listed below.

- We design a single bus route in which all passengers with an identical destination are delivered without any transfers.
- We present a two-phase framework to minimize the total time expense of the specific passengers.
- We develop a genetic algorithm (GA) to solve the integer linear programming (ILP) for large-scale instances.

© Springer Nature Switzerland AG 2021
C. S. Jensen et al. (Eds.): DASFAA 2021, LNCS 12683, pp. 604–607, 2021.
https://doi.org/10.1007/978-3-030-73200-4_44

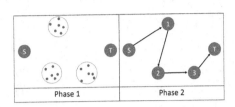

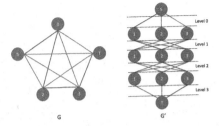

Fig. 1. The framework of the IntRoute.   **Fig. 2.** Graph transformation.

2   Methodology

The main framework of our two-phase IntRoute method are shown in Fig. 1.

Phase 1: Contraction of Bus Stops. In the first phase of IntRoute, we contract multiple requests into a super node by using clustering approaches and determine a pickup bus stop for all the requests inside the super node. In each super node, passengers are asked to walk to the pickup bus stop and wait for buses. The walking time is counted into the total commuting time. We employ K-means clustering, as it consumes the least walking time compared with hierarchical clustering and density-peaks clustering. Then we exploit silhouette method and elbow method to determine the number of pickup stops, i.e., the cluster number n. Both methods indicate the best clustering number should be $n = 20$. Therefore, this phase finds 20 pickup stations that minimizes the passengers total walking time.

Phase 2: Design of One Alternate Route. In the real-world transportation network, an arbitrary node pair (i, j) may be connected (Fig. 2 left). To solve the problem via mathematical IP model, we introduce the *multi-level graph* G' (Fig. 2 right), where each level represents a possible sub-route from one stop to the next one. The red paths in G and G' are equal. The graph G' indicates the order of visiting the pick-up stops directly by the levels.

Modelling via Integer Programming. We denote the node set as $N = \{1, 2, \ldots, n\} \cup \{S, T\}$, where S and T represents the source and destination, respectively. The arc set is $A = \{(i, j) | i \in N, \ j \in N, \ \text{and } i \neq j\}$. The travel time from node i to node j is denoted as c_{ij}, and r_i is the number of passengers who want to go to the destination. $x_{ij}^l \in \{0, 1\}$ represents a binary variable, indicating whether the bus goes from node i to node j at level l on G'. $y_{ijk}^l \in \{0, 1\}$ represents a binary variable which denotes if request k travels through arc (i, j) at level l. $v_k^l \in \{0, 1\}$ represents a binary variable, indicating if request k is served at the level l. Formally, the IP model is formulated as:

$$\min \sum_l \sum_k \sum_{(i,j) \in A} y_{ijk}^l c_{ij} r_k, \quad s.t.$$

$$\sum_{(i,j) \in A} x_{ij}^l = 1, for \ l = 0, 1, \ldots, n \tag{1}$$

$$\sum_{(j,i) \in A} x_{ji}^l = \sum_{(i,k) \in A} x_{ik}^{l+1}, \forall i; \forall l \tag{2}$$

$$\sum_{(S,i)\in A} x_{Si}^0 = \sum_{(i,j)\in A} x_{ij}^1 \tag{3}$$

$$\sum_{(i,j)\in A} x_{ij}^{n-1} = \sum_{(j,T)\in A} x_{jT}^n \tag{4}$$

$$\sum_i \sum_l x_{ij}^l \leq 1, for \quad j = 1, 2, \ldots, n, T \tag{5}$$

$$v_k^l \leq v_k^{(l+1)}, \forall k, \forall l \tag{6}$$

$$\sum_k v_k^l = l, for \ l = 1, 2, \ldots, n \tag{7}$$

$$x_{ij}^l \leq v_i^l, for \ (i,j) \in A, l = 1, 2, \ldots, n \tag{8}$$

$$y_{ijk}^l \leq (x_{ij}^l + v_k^l)/2, \ \forall (i,j) \in A, \forall k, \forall l \tag{9}$$

$$y_{ijk}^l \geq (x_{ij}^l + v_k^l) - 1, \ \forall (i,j) \in A, \forall k, \forall l \tag{10}$$

$$x_{ij}^l \in \{0,1\}; \ v_k^l \in \{0,1\}; \ y_{ijk}^l \in \{0,1\} \tag{11}$$

where (1) ensures that each level is exactly passed once. (2)–(4) ensure that flow conservation of the graph. (5) ensures that each node can be entered at most once. (6) ensures that request k must be served at level $l + 1$ if it is served at level l. (7) ensures that l requests are on the bus when the bus is serving level l. (8) ensures that if arc (i,j) is picked at level l, request i should be served at level l. (9) and (10) ensure that the request k is served at level l on the arc (i,j) if both x_{ij}^l and v_k^l are equal to one. (11) shows x_{ij}^l, v_k^l, and y_{ijk}^l are all binaries.

Optimization via Genetic Algorithm. We design a genetic algorithm (GA) to solve this IP problem. Specifically, the *chromosome* represents possible sequence of the node set $\{1, 2, \ldots, n\}$, and the *population* is a set of chromosomes.

As shown in Algorithm 1, a new chromosome can be generated by the *crossover* between two parent bus routes with a probability of crossover rate R_c. The *mutation* is defined as the position exchange between two randomly selected near-by bus stoops with a probability of R_m. There are two steps in our GA: (1) the randomly population initialization with a given size M, and (2) the population evolution for T generations by crossover and mutation according to a fitness function. We adopt a 2-OPT technique to avoid the cross sub-paths. Besides, we propose a decomposition technique that clusters the optimal route into three sub-route via k-means. We concatenate the clusters in a reverse order, i.e., from the destination node to the start node. This strategy greatly reduces the travel time.

Algorithm 1: Genetic Algorithm for Solving the IP Problem

Input: $T = 1000$, $R_c = 0.1$, $R_m = 0.05$
Output: The chromosome with the best fitness function

1 Initialize the population with size $M = 500$;
2 **for** $i = 1$ **to** T **do**
3   Select new population P_i from P_{i-1};
4   **for** *individual* $p \in P_i$ **do**
5   *offspring* $\leftarrow$ *Crossover*(p, R_c);
6   *offspring* $\leftarrow$ *Mutate*(*offspring*, R_m);
7          $p \leftarrow$ *offspring*;
8   **end**
9 **end**
10 **return** The best chromosome.

Table 1. Routing time before optimization

Bus route	C'	C
2153142, 2155252, 2148445, 2153226, 2121125, 2074117, 212225, 211220, 211118, 2137134, 212746, 2150145, 2145561, 2150302, 2190145, 2135206, 213447, 203833, 200721, 201635, CBD	33.62 min	31.53 min

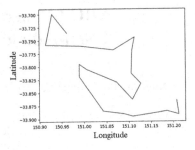

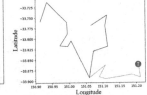

Fig. 3. Bus route without transfer

Table 2. Commute time

Methods	Time
Original route	31.53
Greedy algorithm [1]	53.41
GA	31.27
GA + 2-OPT	33.62
GA + decomposition	18.06

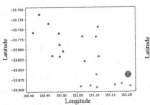

Fig. 4. Bus routes after optimization

3 Experiments and Analysis

Data. The experiment is performed based on publicly available real-world commuting data, retrieved from the card-based transit payment system[1] in Sydney, Australia, including approximately three million trips.

Results. The routing time before optimization are listed in Table 1, where routes are represented by the IDs of the bus stops. Here C denotes the time cost without transfer and without optimization (also demoed in Fig. 3), and C' denotes that of the original commute with transfers. The new route after our optimization is shown in Fig. 4, while the time costs of the bus routes optimized by different methods are listed in Table 2.

Conclusions. Our IntRoute method greatly reduces the time expense for passengers from 31.53 min to 18.06 min on average, saving about 43% of commute time. In future, we plan to investigate more optimization methods for further improving our solutions.

References

1. Li, L., Fu, Z.: The school bus routing problem: a case study. J. Oper. Res. Soc. **53**(5), 552–558 (2002). https://doi.org/10.1057/palgrave.jors.2601341
2. Liu, J., Mao, J., Du, Y.T., Zhao, L., Zhang, Z.: Dynamic bus route adjustment based on hot bus stop pair extraction. In: Li, G., Yang, J., Gama, J., Natwichai, J., Tong, Y. (eds.) DASFAA 2019. LNCS, vol. 11448, pp. 562–566. Springer, Cham (2019). https://doi.org/10.1007/978-3-030-18590-9_87
3. Liu, W., Chawla, S.: A quadratic mean based supervised learning model for managing data skewness. In: Proceedings of SIAM SDM Conference (2011)

---

[1] https://opendata.transport.nsw.gov.au/dataset/opal-tap-on-and-tap-off.

NRCP-Miner: Towards the Discovery of Non-redundant Co-location Patterns

Xuguang Bao[1], Jinjie Lu[1], Tianlong Gu[1], Liang Chang[1(✉)], and Lizhen Wang[2]

[1] Guangxi Key Laboratory of Trusted Software, Guilin University of Electronic Technology,
Guilin 541004, China
changl@guet.edu.cn
[2] Yunnan University, Kunming 650091, China

Abstract. Co-location pattern mining, which refers to discovering neighboring spatial features in geographic space, is an interesting and important task in spatial data mining. However, in practice, the usefulness of prevalent (interesting) co-location patterns generated by traditional frameworks is strongly limited by their huge amount, which may affect the user's following decisions. To address this issue, in this demonstration, we present a novel schema, named NRCP-Miner, aiming at the redundancy reduction for prevalent co-location patterns, i.e., discovering non-redundant co-location patterns by utilizing the spatial distribution information of co-location instances. NRCP-Miner can effectively remove the redundant patterns contained in prevalent co-location patterns, thus further assists the user to make the following decisions. We evaluated the efficiency of NRCP-Miner compared with related state-of-the-art approaches.

Keywords: Spatial data mining · Co-location pattern mining · Prevalent co-location patterns · Redundancy reduction · Decision-making system

1 Introduction

The explosive growth of the spatial data results in significant demand for spatial data mining. Co-location pattern mining, as an important spatial data mining task, has been extensively studied for discovering neighboring relationships of spatial features. A spatial co-location pattern commonly demonstrates neighboring relationships of spatial features. Spatial co-location patterns may yield important insights in many applications, including Earth Science, public health, biology, transportation, etc.

To measure how interesting a co-location pattern is, the PI (Participation Index) value proposed by Huang et al. [1] is commonly used. Given a user-specified minimum prevalence threshold min_prev, for a co-location pattern c, if $PI(c) \geq min\_prev$ satisfies, c is called a prevalent co-location pattern (PCP). As a PCP is a set of spatial features, given a spatial dataset containing m spatial features, the number of generated PCPs can reach as much as 2^m. Furthermore, the PI measure satisfies the anti-monotonicity property [1], i.e., if a PCP c is prevalent, all its subsets are also prevalent. However, most of its subsets are redundant by considering their prevalences or PI values, which

© Springer Nature Switzerland AG 2021
C. S. Jensen et al. (Eds.): DASFAA 2021, LNCS 12683, pp. 608–611, 2021.
https://doi.org/10.1007/978-3-030-73200-4_45

may affect the decisions of the user. Thus, it is crucial to reduce the number of PCPs by redundancy reduction.

To reduce the number of PCPs, two classic condensed representations have been proposed—maximal co-location patterns [2] (MCPs) and closed co-location patterns [3] (CCPs), respectively. However, MCPs are considered as a lossy representation because they ignore the PI values of co-location patterns. Although CCPs are lossless representations considering both prevalences and PI values of co-location patterns, they contain redundancies. Thus, Wang et al. [4] proposed an algorithm called RRClosed to select non-redundant co-location patterns from CCPs. Later, they introduced a new lossless and non-redundant representation called SPI-closed (Super Participation Index-closed) co-location patterns (SCPs), and proposed a method called SPI-Miner [5] to efficiently discover SCPs.

In this demonstration, we present a novel and efficient system, named NRCP-Miner, to discover SCPs. Instead of RRClosed or SPI-Miner, we adopt a clique-based approach [6] to discover PCPs, and then furtherly select SCPs. Because the clique-based approach constructs a hash structure that can be stored permanently and is independent of the prevalence threshold, our proposed system performs more efficiently than RRClosed and SPI-Miner, especially when the system needs to be executed multiple times. Besides, as SCPs are subsets of PCPs, our proposed NRCP-Miner can be applied to domains of PCPs. For example, the mobile service provider may be interested in mobile service SCPs frequently requested by geographical neighboring users. Botanists may be interested in SCPs consisting of symbiotic plant species.

2 System Overview

NRCP-Miner undergoes six steps to generate SCPs, as shown in Fig. 1.

Step 1: *Materialization of the inputted spatial data.* This step first gathers all neighboring relationships of each instance by considering a user-given distance threshold min_dist, and then groups the neighboring relationships as a neighbor list.

Step 2: *Generation of complete cliques.* This step aims to generate complete cliques using the neighbor list. As the enumeration of maximal cliques is considered as an NP-hard problem, we adopt a linear method [6] to generate complete cliques.

Step 3: *Compression of the complete cliques.* As the calculation of the PI value of a co-location c is only based on the instances participating in c, thus, the complete cliques can be compressed into a hash structure.

Step 4: *Generation of PCPs.* Given the instance hash, the PI value of any co-location pattern can be efficiently calculated by considering the user-specified prevalence threshold min_prev.

Step 5: *Selection of CCPs.* As the CCPs are subsets of PCPs, thus, all CCPs can be selected from PCPs by the definition of CCPs, i.e., removing the PCP whose PI value equals the PI value of one of its supersets.

Step 6: *Generation of SCPs.* To generate the SCPs from CCPs, we adopt the latter part of the RRClosed method [4], which generates SCPs from CCPs by designing a NET structure and a lemma for pruning.

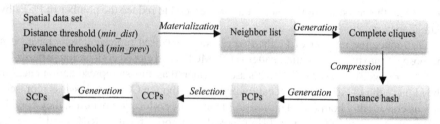

Fig. 1. System description

3 Demonstration Scenarios

NRCP-Miner is well encapsulated with a friendly interface, what the user faces is only a simple user interface. In this demonstration, we use part of the data set from points of interests (POI data) in Beijing to show the demonstration and efficiency of NRCP-Miner. The selected POI data set contains 5,000 POIs (spatial instances).

Demonstration. Figure 2 shows the main interface of NRCP-Miner. Figure 2(a) gives the original spatial instances read from a file or a database, each instance is represented as <feature name, location <x, y>>. The detailed distribution of instances described in Fig. 2(a) is drawn in Fig. 2(b). The parameters with their specified values are listed in Fig. 2(c). Figure 2(d) shows the generated SCPs based on the settings in Fig. 2(c) from the spatial data shown in Fig. 2(a), as well as the number of per-size SCPs and removed CCPs.

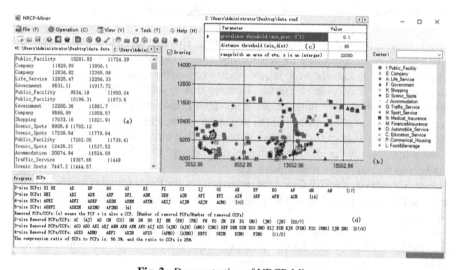

Fig. 2. Demonstration of NRCP-Miner

Efficiency Evaluations. We evaluated the efficiency of NRCP-Miner from two aspects: the compression ratio to CCPs and the running time compared with RRClosed and SPI-Miner. As shown in Fig. 2(d), NRCP-Miner removes 56.3% of PCPs, and 25% of CCPs, and also runs faster than RRClosed and SPI-Miner with the change of the prevalence threshold *min_prev*, as shown in Fig. 3, this is because the hash structure generated by NRCP-Miner is independent of *min_prev*, while the other algorithms have to restart their mining processes with the change of *min_prev*.

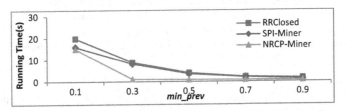

Fig. 3. Efficiency comparison with related literature

4 Conclusion

This demonstration presents a novel and efficient system called NRCP-Miner to discover a newly proposed lossless condensed representation of prevalent co-locations SPI-closed co-locations. Unlike similar approaches mainly focusing on pruning strategies for reducing the number of candidates by using the prevalence threshold, NRCP-Miner gets rid of the constraint of the prevalence threshold. Thus, it can effectively assist the user to find a satisfying prevalence threshold within much less time, and furtherly can well support the decision-making of the user.

Acknowledgements. This work was supported in part by grants (No. 62006057, No. U1811264, No. U1711263, No. 61966009, No. 61762027) from the National Natural Science Foundation of China, in part by National Science Foundation of Guangxi Province (No. 2019GXNSFBA245059), in part by the Key Research and Development Program of Guangxi (No. AD19245011).

References

1. Huang, Y., Shekhar, S., Xiong, H.: Discovering co-location patterns from spatial data sets: a general approach. IEEE Trans. Knowl. Data Eng **16**(12), 1472–1485 (2004)
2. Wang, L., Zhou, L., Lu, J., et al.: An order-clique-based approach for mining maximal co-locations. Inf. Sci. **179**(2009), 3370–3382 (2009)
3. Yoo, J.S., Bow, M.: Mining top-k closed co-location patterns. In: IEEE International Conference on Spatial Data Mining and Geographical Knowledge Services, pp. 100–105 (2011)
4. Wang, L., Bao, X., Zhou, L.: Redundancy reduction for prevalent co-location patterns. IEEE Trans. Knowl. Data Eng. **30**(1), 142–155 (2018)
5. Wang, L., Bao, X., Chen, H., Cao, L.: Effective lossless condensed representation and discovery of spatial co-location patterns. Inf. Sci. **436–437**, 197–213 (2018)
6. Bao, X., Wang, L.: A clique-based approach for co-location pattern mining. Inf. Sci. **490**, 244–264 (2019)

ARCA: A Tool for Area Calculation Based on GPS Data

Sujing Song[1], Jie Sun[2], and Jianqiu Xu[1(✉)]

[1] College of Computer Science and Technology,
Nanjing University of Aeronautics and Astronautics,
Nanjing, People's Republic of China
{SongSuJing,jianqiu}@nuaa.edu.cn
[2] Jiangsu Sea Level Data Technology Co. Ltd., Nanjing, People's Republic of China

Abstract. In this paper, we develop a tool to efficiently and effectively calculate agricultural machinery's working area based on farming machinery's GPS data. The tool works as follows. First, we pre-process GPS data by removing duplicate data, abnormal data and invalid data. Data projection is performed using Gauss-Kruger and the minimum value after projection is used for data transforming and shifting. Second, the tool operates farming machinery trajectory fitting. Finally, an algorithm of area calculation is developed to form the farming machinery's area based on trajectory data produced in the first two steps. The algorithm achieves an error rate 0.29%, and takes 0.03 s to process about 60 GPS records collected in one minute.

Keywords: Farming machinery · GPS data preprocessing · Acreage calculation

1 Introduction

Due to the widespread use of GPS-enabled devices such as smartphones and vehicles, the recording of position data has become very easy, and huge amounts of such data are collected. In the field of agriculture, farming machinery is progressively equipped with positioning equipment. GPS data collected by those devices record agricultural machinery's trajectory in the agricultural field. Agricultural machinery plays an important role in the agricultural field. The common mechanical way of farming machinery is rotary tiller and tractor supporting the operation. In Jiangsu province, hundreds or thousands of rotary tillers are sold per year. Due to the wide application of agricultural machinery used on the farmland, it is an important issue to compute the cultivated land area. In daily life, manual records can calculate the size of cultivated land but cause some problems. To obtain high wages, workers deliberately exaggerated the area of arable land, causing enterprise losses. It is possible to estimate the area of arable land based on the actual land, but the actual farming land may be an irregular shape. Consequently, it is not easy to calculate the area. Even if the farming land is

© Springer Nature Switzerland AG 2021
C. S. Jensen et al. (Eds.): DASFAA 2021, LNCS 12683, pp. 612–616, 2021.
https://doi.org/10.1007/978-3-030-73200-4_46

regular which is easy to calculate, the actual farming land area is not equivalent to the farming land area because the farming task maybe not complete.

GPS data sampling of agricultural machinery is affected by a number of factors such as equipment, sampling frequency and storage mode. Raw GPS data usually contain noise data [2]. Furthermore, latitude and longitude values are not appropriate for area calculation due to numeric problems. Therefore, data transforming and shifting is performed.

To support a fast and accurate calculation of cultivated land area, a software tool named ARCA (Area Calculation) is developed based on GPS data. The input is raw GPS records. The calculation task is achieved by the following steps: data preprocessing, trajectory fitting, region formation by expanding the trajectory, area calculation, and operations of calculating the cultivated land area. The developed tool, (i) achieves an error rate 0.29% and (ii) takes 0.03 s on average to process about 60 GPS records collected in one minute.

2   ARCA

2.1   An Overview

We outline the tool ARCA in Fig. 1. There are three layers: (i) data layer, (ii) functional layer and (iii) display layer. The data layer defines data types *mpoint*, *line*, and *region*. The functional layer includes preprocessing modules and data conversion modules. The display layer shows the cultivated area and the shape of the cultivated land by agricultural machinery.

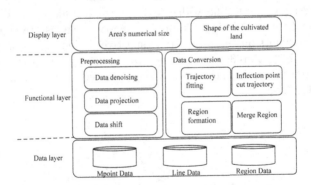

Fig. 1. ARCA structure

2.2   Data Preprocessing

Each GPS record contains a list of attributes among which the following ones have primary information: ID, time, longitude, and latitude. The procedure of removing data noise mainly solves the problems of data duplication, invalidity, and abnormality. Agricultural machinery is likely to collect repeated data which refers to the points that are (i) from consequent GPS records of the same ID and (ii) the distance between each other is zero or close to zero. We use the SNM [1] algorithm performing the sort-detect-merge and eliminating to remove duplicate data. Invalid data refers to the data whose longitude or latitude is zero. Abnormal data refers to data exceeding the defined range.

Latitude and longitude very little in one city, increasing the difficulty of area calculation. To avoid the numeric problem, we use Gauss-Kruger [4] to project latitude and longitude into a rectangular coordinate system, and take the minimum x and y axes from all GPS data to transform x and y axes.

2.3   Area Calculation

There are three alternative solutions for area calculation based on points. Convex hull [5] is a popular algorithm, but will calculate the area of unfinished arable land in the middle, leading to the result larger than the actual value. The square area method takes the GPS point as a center point and expands along the moving direction to form a square area. Farming machinery's width is set as the square side length. All GPS data are used, and all the created squares are involved to calculate the area. The disadvantage of this method is that sparse data points lead to incomplete area calculations. We propose a solution that solves the problems in the two methods.

Our algorithm performs the area calculation in four steps, as demonstrated in Fig. 2. Step 1, GPS data is connected in chronological order to form a trajectory l_i. Step 2, two new trajectories are formed by shifting the trajectory l_i with distance $w/2$ (w refers to the width of farming machinery.). l_i^l is on the left side of l_i and l_i^r is on the right side of l_i. The two regions are constructed by connecting endpoints on the same side respectively. The reg_l is on the left and reg_r is on the right. Step 3, reg_l and reg_r are merged to form the last cultivated area reg_1. Step 4, reg_1 is merged with reg_2 (formed in steps 1–3) to form the final cultivated area.

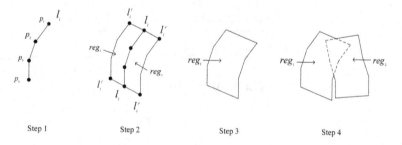

Step 1   Step 2   Step 3   Step 4

Fig. 2. Area calculation procedure

3 Demonstration

The tool is developed in an extensible database system SECONDO [3], running Intel(R) Core I3-2120 3.3 GHz, 4 GB RAM, Ubuntu14.04 64 bit operating system.

Table 1. Experimental statistics

	Group A	Group B	Group C
Data size	15,458	9,433	14,378
Duplicate values	1,914	1,181	1,787
Preproccess(s)	0.97	0.60	0.91
Data conversion(s)	6.58	3.85	6.93
average value(s)	**0.029**	**0.028**	**0.032**

Table 2. Error rate

	Group A	Group B	Group C
Reference	35867.59	24523.94	19181.69
Convexhull Results	33092.46	23306.14	17628.41
Test Results	35975.01	24628.82	19211.17
Error Rate	**0.30%**	**0.43%**	**0.15%**

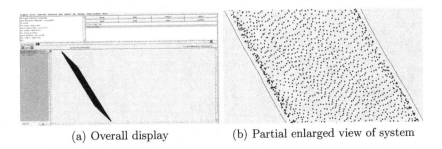

(a) Overall display   (b) Partial enlarged view of system

Fig. 3. Demo display

The experimental records are shown in Table 1. Three groups of three agricultural machinery data are collected for experimental testing. The preprocessing includes denoising, projection and translation and the processed data is to be trajectory fitting. The tool cuts the trajectory at the inflection point, translates the trajectory to form the cultivated land boundary, forms the area, and performs the merging. Finally, the area calculation is executed in SECONDO. This tool takes about 0.03 s to process 60 GPS data records collected in one minute. The calculated error rate is shown in Table 2. The average error rate is 0.29%. The width of farming machinery is set 4 m. The frequency of GPS records is 1 s. At present, we do not have the ground truth of the area of arable land. Therefore, the square area method is selected as the alternative method to obtain the area's reference value. The results of the convex hull solution are shown also. The unit of area is square meters. The system demonstration is shown in Fig. 3. The blue dots represent GPS data after preprocessing and the area within the black frame is the area of cultivated land.

Acknowledgement. This work is supported by National Natural Science Foundation of China (61972198), and Natural Science Foundation of Jiangsu Province of China (BK20191273).

References

1. Draisbach, U., Naumann, F., Szott, S., Wonneberg, O.: Adaptive windows for duplicate detection. In: ICDE (2012)
2. Gao, Q., Zhang, F.L., Wang, R.J., Zhou, F.: Trajectory big data: a review of key technologies in data processing. J. Softw. **28**(4), 959–992 (2017)
3. Güting, R.H., Behr, T., Düntgen, C.: SECONDO: a platform for moving objects database research and for publishing and integrating research implementations. IEEE Data Eng. Bull. **33**(2), 56–63 (2010)
4. Kong, X.Y., Guo, J.M., Liu, Z.Q.: Foundation of Geodesy. Wuhan University Press (2010)
5. de Berg, M., Cheong, O., van Kreveld, M., Overmars, M.: Computational Geometry: Algorithms and Applications. Springer, Heidelberg (2008). https://doi.org/10.1007/978-3-540-77974-2

LSTM Based Sentiment Analysis
for Cryptocurrency Prediction

Xin Huang[1]([✉]), Wenbin Zhang[1], Xuejiao Tang[2], Mingli Zhang[3],
Jayachander Surbiryala[4], Vasileios Iosifidis[2], Zhen Liu[5], and Ji Zhang[6]

[1] University of Maryland, Baltimore County, Baltimore, USA
{xinh1,wenbinzhang}@umbc.edu
[2] Leibniz University Hannover, Hanover, Germany
{xuejiao.tang,iosifidis}@stud.uni-hannover.de
[3] McGill University, Montreal, Canada
mingli.zhang@mcgill.ca
[4] University of Stavanger, Stavanger, Norway
jayachander.surbiryala@uis.no
[5] Guangdong Pharmaceutical University, Guangzhou, China
liu.zhen@gdpu.edu.cn
[6] University of Southern Queensland, Toowoomba, Australia
ji.zhang@usq.edu.au

Abstract. Recent studies in big data analytics and natural language
processing develop automatic techniques in analyzing sentiment in the
social media information. In addition, the growing user base of social
media and the high volume of posts also provide valuable sentiment
information to predict the price fluctuation of the cryptocurrency. This
research is directed to predicting the volatile price movement of cryptocurrency by analyzing the sentiment in social media and finding the
correlation between them. While previous work has been developed to
analyze sentiment in English social media posts, we propose a method
to identify the sentiment of the Chinese social media posts from the
most popular Chinese social media platform Sina-Weibo. We develop
the pipeline to capture Weibo posts, describe the creation of the crypto-
specific sentiment dictionary, and propose a long short-term memory
(LSTM) based recurrent neural network along with the historical cryptocurrency price movement to predict the price trend for future time
frames. The conducted experiments demonstrate the proposed approach
outperforms the state of the art auto regressive based model by 18.5%
in precision and 15.4% in recall.

1 Introduction

Since the birth of Bitcoin, there has been an enormous rise and interest in
the cryptocurrency, a decentralized digital asset developed by the blockchain
technology. This digital currency draws a lot of attention due to its volatility
which provides the opportunity for digital trading with high return. The total

© Springer Nature Switzerland AG 2021
C. S. Jensen et al. (Eds.): DASFAA 2021, LNCS 12683, pp. 617–621, 2021.
https://doi.org/10.1007/978-3-030-73200-4_47

market capitalization of cryptocurrencies has increased from 1 billion dollars to 400 billion dollars in the past decade, with the number still increasing.

On the other hand, the emergence of social media such as Twitter, Reddit and Facebook also makes the latest news and social media posts about financial markets widely accessible. Investors have therefore been utilizing such a variety of digital resources to make trading decisions. Previous studies discovered evidence of such correlation between stock price movement and social media [1]. Sentiment on cryptocurrency social media content with negative emotions, e.g., fear and sadness, neutral emotions, e.g., calm and not sure, or positive emotions, e.g., trust and happiness, can be used to predict cryptocurrency price fluctuations and further to assist the investment decision making. This paper focuses on this trending theme, proposing a recurrent neural network with long short-term memory (LSTM) by utilizing the sentiment analysis of social media to predict the real time price movement of the digital currency.

2   Related Work

Over the past decades, a variety of machine learning techniques have been developed to predict the price movement for the stock market using social media, such as opinion analysis of twitter feeds [3]. [1] used neural networks and daily Twitter feeds as extra predictors to forecast the daily up and down changes in the closing values of the Dow Jones Industrial Average. Moreover, [7] found the Long Short-Term Memory (LSTM) combined with a Twitter sentiment analysis outperforms other machine learning models such as Support Vector Machine in predicting the stock price.

Recent studies have also successfully applied sentiment analysis in various applications, such as predicting the movie revenues [6], analysing sentiment towards US presidential candidates in 2012 [8]. In the English sentiment analysis, Valence Aware Dictionary and sEntiment Reasoner (VADER) [5] is used to classify sentiment in tweets. In this paper, we designed a crypto sentiment dictionary that is customized to cryptocurrency and Chinese Weibo posts. We use LSTM [4] as the neural network learning layer and combine it with the sentiment analysis method to develop a crypto sentiment analyzer that can predict the price movement of cryptocurrency.

3   Methodology

Figure 1 shows the end to end architecture of the LSTM based sentiment deep learning model in predicting the real time price fluctuations. We crawled user posts from China's most popular social media platforms Sina-Weibo, and created a crypto-specific sentiment dictionary with domain-expert knowledge, and then the LSTM recurrent neural network was used to model the sentiment information and make real time prediction for the price trend.

3.1 Data Collection

Chinese investors exchange crypto information via news articles and social media platforms, especially using Sina-Weibo, Wechat and QQ groups. We collect a large-scale Weibo corpus from crawling Chinese microblogs on Sina-Weibo with the cryptocurrency keyword, in particular, Bitcoin, ETH or XPR. The number of crawled cryptocurrency tweets from Weibo is 24,000, as well as 70,000 comments to them, from the most recent 8 days.

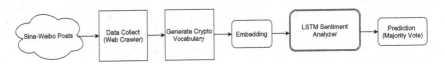

Fig. 1. Architecture of LSTM based cryptocurrency sentiment analysis and price movement prediction. Crypto sentiment dictionary is created to generate the crypto word embedding, LSTM is to learn the sentiment information, and the majority voting on the output of LSTM sentiment analyzer is used to predict the price going up or down.

3.2 Crypto Sentiment Dictionary

The general sentiment dictionary created by natural language processing (NLP) is not applicable in the crypto domain. We introduce a novel way to build a crypto specific sentiment dictionary that can capture the unique characteristics of the crypto social communities. Table 1 shows an example dictionary that is particular to the Chinese crypto words in Sina-Weibo. The first step of generating a crypto sentiment dictionary is to create the vocabulary of the crypto words. We manually label the crawled Weibo posts with ranking, and use seed sentiment words selected by crypto domain experts to do bootstrapping, which adds high frequency new words in highly positive/negative weibo posts into the crypto corpus.

After the crypto corpus is generated, we then create an index mapping dictionary in such a way that the frequently occurring crypto words are assigned lower indexes, similar to the traditional natural language processing. Finally we generate a crypto word encoding for each individual post and use that encoding vector as the training data for the RNN model in sentiment analysis.

Table 1. Crypto specific sentiment dictionary building for Chinese words.

Chinese	Informal Translation	Implied Sentiment
韭菜	Bag Holder	Investors have profit loss
新高	New High	Investors are excited
下车	Abandon Ship	Investors rush to sell off

3.3 LSTM Based Sentiment Analyzer

We develop a long short-term memory network (LSTM) based sentiment analyzer for crypto social media posts. LSTM enables the network to learn long-term relation, by utilizing forget and remember gates that allow the cell to decide which information to block or transmit based on its strength and importance.

The social media post is first tokenized according to crypto word vocabulary and fed into the embedding layer, which converts the word token into the crypto word embedding. The LSTM based recurrent network is trained by taking the sequence of the embedding feature vector. A fully connected layer is used to transform the output of the LSTM later and activated with sigmoid to output the prediction. The labels of the posts used in training were manually labeled and encoded with positive (1), neutral (0) and negative (−1).

4 Evaluation

We used the most recent 7 days' Sina-Weibo posts from top 100 crypto investors accounts as training data and the next 1 day's posts as testing. We use Precision and Recall to measure the performance of our LSTM sentiment predictor. Precision measures the model's ability to return only relevant instances and recall measures the model's ability to classify all relevant instances.

We compare our method with the time series auto regression (AR) approach [2] to evaluate the performance. As Table 2 shows, our approach outperforms the AR approach by 18.5% in precision and 15.4% in recall, exemplifies the effectiveness of the LSTM in analyzing the sentiment of social media content.

Table 2. Precision and recall on evaluating LSTM sentiment analyzer and AR.

Method	Precision	Recall
Auto regression	73.4%	80.2%
LSTM sentiment analyzer	87.0%	92.5%

5 Conclusion

In this paper we introduce a crypto sentiment analyzer by utilizing the recurrent neural network to model the social media sentiment. The model is developed using LSTM and achieves higher precision and recall than the traditional auto regressive approach. The current sentiment analyzer can be used to predict the price fluctuation of the cryptocurrency and integrated to an autonomous trading system to assist the buying or selling of digital assets.

References

1. Bollen, J., Mao, H., Zeng, X.: Twitter mood predicts the stock market. J. Comput. Sci. **2**, 1–8 (2011)
2. Box, G., Jenkins, G., Reinsel, G., Ljung, G.: Time Series Analysis: Forecasting and Control (2016). ISBN 1-118-67502-9. OCLC 915507780
3. Chen, R., Lazer, M.: Sentiment analysis of twitter feeds for the prediction of stock market movement. Stanford Computer Science 229 (2011)
4. Hochreiter, S., Schmidhuber, J.: LSTM can solve hard long time lag problems. In: Proceedings of the 9th International Conference on Neural Information Processing Systems, pp. 473–479 (1996)
5. Hutto, C.J., Gilbert, E.: Vader: a parsimonious rule-based model for sentiment analysis of social media text. In: Eighth International AAAI Conference on Weblogs and Social Media, pp. 216–255 (2014)
6. Joshi, M., Das, D., Gimpel, K., Smith, N.A.: Movie reviews and revenues: an experiment in text regression. In: Human Language Technologies: The 2010 Annual Conference of the North American Chapter of the Association for Computational Linguistics, pp. 293–296 (2010)
7. Pimprikar, R., Ramachadran, S., Senthilkumar, K.: Use of machine learning algorithms and Twitter sentiment analysis for stock market prediction. Int. J. Pure Appl. Math. **115**, 521–526 (2017)
8. Wang, H., Can, D., Kazemzadeh, A., Bar, F., Narayanan, S.: Hand gesture recognition using Fourier descriptors. In: Proceedings of the 50th Annual Meeting of the Association for Computational Linguistics, pp. 115–120 (2012)

SQL-Middleware: Enabling the Blockchain with SQL

Xing Tong[1], Haibo Tang[1], Nan Jiang[1], Wei Fan[1], Yichen Gao[1], Sijia Deng[1],
Zhao Zhang[1(✉)], Cheqing Jin[1], Yingjie Yang[2], and Gang Qin[2]

[1] East China Normal University, Shanghai, China
{xtong,haibtang,njiang,wfan,ycgao,sjdeng}@stu.ecnu.edu.cn,
{zhzhang,cqjin}@dase.ecnu.edu.cn
[2] Ouyeel International Co., Ltd., Shanghai, China
{yangyingjie,qingang}@ouyeel.com

Abstract. With the development of blockchain, blockchain has a broad prospect as a new type of data management system. However, limited to the data modeling method of blockchain, the usability of blockchain is restricted; In addition, every blockchain system has its own native but naive interfaces, when developing based on the different blockchain systems, which will leads to low development efficiency and high development costs. In this study, we construct a SQL-Middleware for blockchain system to solve these problems. The SQL-Middleware first performs relational modeling of blockchain data, mapping the blockchain data into a relational table; On the basis of modeling the blockchain data, SQL-Middleware encapsulates a set of SQL interfaces for blockchain system, thus realizing the unification of interface access methods of different blockchain systems. At last, we implement the SQL-Middleware based on the open source blockchain system CITA. Demonstration shows that the SQL-Middleware greatly improves the data management capabilities of blockchain and simplifies the blockchain access steps.

Keywords: Blockchain · Middleware · Data modeling · SQL

1 Introduction

Blockchain, as a distributed ledger formed by multi-party consensus, can build a credible interactive platform for multiple parties who do not trust each other. As a new type of data management system, blockchain has many drawbacks: First, the expressive capability of blockchain data is weak [1], all data is modeled in a unified transaction format, which limits the potential value of data; Second, from the perspective of evolution of the data management system: from SQL to NoSQL, and then NewSQL, the continuous changes in data management methods have made us aware of the irreplaceable role of SQL in data management, however, generally, blockchain can be classified as a NoSQL system and only

Shanghai Engineering Research Center of Big Data Management.

supports RPC-based naive interfaces. These features restrict the evolution of the blockchain to be an excellent emerging data management system.

In addition, it is worth mentioning that we observe that transactions are structured data, which provides a basis for relational modeling. Based on this observation, we design a middleware called SQL-Middleware for blockchain system to abstract blockchain system as a relational data management system. The middleware we designed has nothing to do with the blockchain consensus and execution process. The reason for adopting the form of middleware is that middleware can transplant to other systems easily and not just limited to a specific blockchain system [2]. However, there are two challenges when building SQL-Middleware for blockchain system: *i*) How to convert a monotonous transaction-based model into different models towards different scenarios; *ii*) How to provide developers with efficient query interfaces. To solve these challenges, we use the **Data** in the transaction as an application-related field and encapsulate the application data into **Data** uniformly, and then we establish a schema for **Data** so that the data in the blockchain system has rich semantic models; Then, we implement part of SQL interfaces for blockchain system, application developers can interact with blockchain system through these SQL interfaces; Finally, we implement a **Terminal** based on the SQL-Middleware which we implemented.

In related research works, there are existing research works, but these works cannot be well compatible with different blockchain systems [1], and cannot support efficient query operations [2,3] or will bring relatively large storage overhead. On the contrary, SQL-Middleware solves these problems very well.

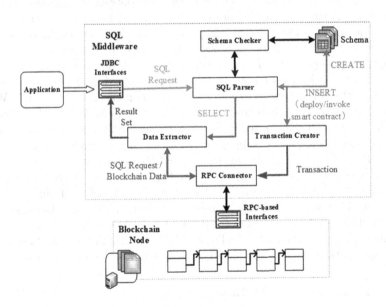

Fig. 1. System overview.

2   System Overview and Key Designs

The overall architecture of the system is shown in Fig. 1. Using SQL-Middleware, we abstract the underlying blockchain system into a SQL-based data management system. On the one hand, the SQL-Middleware model the blockchain system as a relational data management system, which enhance the expressive ability of blockchain data. On the other hand, SQL-Middleware encapsulates the RPC-based interfaces provided by the blockchain system into SQL interfaces, which makes the blockchain system behave like a database. The modules of SQL-Middleware include: *i*) JDBC interface module, applications can transmit SQL statements and result set through this module; *ii*) SQL parser module, which is responsible for parsing the SQL statements transmitted by the application and checking the validity of SQL statements based on locally maintained schema information; *iii*) Transaction Creator and Data Extractor, which are responsible for constructing transactions based on the results of SQL parser, and querying the required data based on the obtained blockchain data; *iv*) RPC Connector module, this module directly connects with the blockchain node by RPC, sends the constructed transaction to the blockchain node for consensus, and pulls the on-chain data from blockchain node.

2.1   Data Modeling and SQL Interfaces

Taking the popular blockchain systems as examples (e.g., Ethereum or CITA all transactions have unified structure. In transactions that used for invoking smart contracts, all parameters needed to call the contract are included in **Data** (Transaction.**Data** = {param1, param2, et al.}), that is, all application-related data is included in **Data**, resulting in a monotonous blockchain transaction model, which limits the data expression capability of blockchain data. To solve this problem, we parse the **Data** in transaction and model it into a table, and each parameter in **Data** is mapped into a field of table. Specifically, we map each function in the smart contract that can be called externally into a table. When smart contract is called, it is equivalent inserting an item into the database.

SQL query is an efficient query method, through SQL query interfaces, complicated query logic can be supported, which is an important support in data mining. However, as far as we can tell, only RPC-based naive interfaces are provided in blockchain system. Compared with SQL, native interfaces of blockchain only support simpler query logic and cannot support rich and efficient query operations. In order to solve this problem, we use the RPC-based interfaces provided by the blockchain system to encapsulate a set of SQL interfaces. We mainly implement three interfaces: CREATE, SELECT, and INSERT. The functions of these SQL statements that supported by SQL-Middleware are shown in Table 1.

3   Demonstrations Details

We develope a SQL-Middleware in C++ using CITA as a case. Based on SQL-Middleware, we implement a **Terminal**. Using this terminal, developers can per-

Table 1. Supported SQL statements and their functions.

SQL	Function	Processing flow
CREATE	Create schemas for on-chain data	Establish schema info to model on-chain data and use schema info to describe on-chain data, then persist it in schema file
INSERT	Insert data into the blockchain	Extract the field info in the INSERT statement, construct it into a transaction format and forward it to blockchain node
SELECT	Read on-chain data that meets the requirements	Pull data on the blockchain node and extract required data according to the schema information

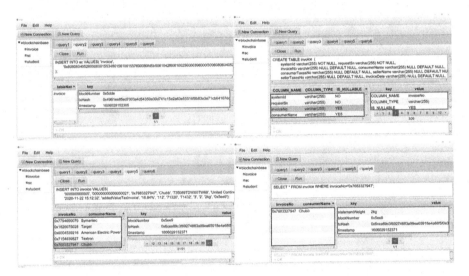

Fig. 2. Terminal panel when performing different operations. (Color figure online)

form assisted work on the blockchain, including designing or viewing the structures or contents of the tables in blockchain. Figure 2 shows the **Terminal** panel when performing different operations. The yellow-marked field represents the locally established schema information. In the green-marked field, we can input SQL statements and send them to our SQL-Middleware, then SQL-Middleware executes the SQL statements and displays the results in the blue-marked field, and red-marked attributes are blockchain attributes of the data. In addition, we develop a blockchain front-end layer based on SQL-Middleware, other systems can access the blockchain system through blockchain front-end layer by **POST** Request, which greatly expands the usage scenarios of blockchain system and deeply excavates the value of blockchain system as a data management system.

Acknowledgments. This work is partially supported by National Science Foundation of China (61972152, U1811264 and U1911203).

References

1. Zhu, Y., Zhang, Z., Jin, C., Zhou, A., Yan, Y.: SEBDB: semantics empowered blockchain database. In: ICDE, pp. 1820–1831. IEEE (2019)
2. El-Hindi, M., Binnig, C., Arasu, A., Kossmann, D., Ramamurthy, R.: BlockchainDB: a shared database on blockchains. In: VLDB Endowment, pp. 1597–1609 (2019)
3. Ji, Y., Chai, Y., Zhou, X., Ren, L., Qin, Y.: Smart intra-query fault tolerance for massive parallel processing databases. Data Sci. Eng. **5**(1), 65–79 (2019). https://doi.org/10.1007/s41019-019-00114-z

Loupe: A Visualization Tool for High-Level Execution Plans in SystemDS

Zhizhen Xu[1], Zihao Chen[1], and Chen Xu[1,2(✉)]

[1] School of Data Science and Engineering,
Shanghai Engineering Research Center of Big Data Management,
East China Normal University, Shanghai, China
{zhizhxu,zhchen}@stu.ecnu.edu.cn, cxu@dase.ecnu.edu.cn
[2] Science and Technology on Parallel and Distributed Processing Laboratory (PDL),
Changsha, China

Abstract. The declarative programming language in SystemDS simplifies users to implement machine learning algorithms. It is able to generate execution jobs on different data processing engines including MapReduce and Spark. The GUI in data processing engines typical visualizes the *low-level* execution process (e.g., RDD transformation in Spark). However, the low-level description in Spark GUI does not show the relationship between DML operations and RDD primitives. In this work, we propose *Loupe*, a tool to visualize *high-level* execution plans in SystemDS to ease users to understand the execution process. This paper introduces the design of the tool and demonstrates a visualization case.

Keywords: SystemDS · Execution plan · Visualization

1 Introduction

With the growing trend of the magnitude of datasets spurred by applications such as search engines, distributed data processing engines like MapReduce [3] and Spark [5], etc., have emerged. It is non-trivial for users to implement sophisticated algorithms by low-level operators in aforementioned data processing systems. To bridge the gap between users' familiar high-level programming languages and complex low-level implementations, SystemDS [2] which originates from SystemML [1,4] has been developed by data management community. It allows users to implement the data science algorithms in DML, a declarative machine learning language. SystemDS compiles and optimizes these algorithms, so as to generate runtime jobs on MapReduce or Spark.

Execution plan in SystemDS describes how an algorithm would be compiled and optimized. Hence, this plan plays an important role in the execution process of an algorithm. For example, SystemDS would optimize multiple matrix multiplications by rearranging the order of operators. It is necessary for users to

© Springer Nature Switzerland AG 2021
C. S. Jensen et al. (Eds.): DASFAA 2021, LNCS 12683, pp. 627–630, 2021.
https://doi.org/10.1007/978-3-030-73200-4_49

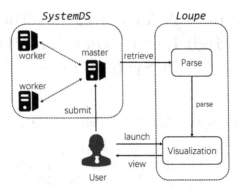

Fig. 1. The architecture of Loupe.

clearly understand execution plans, so that they might be able to identify the performance bottleneck and then improve their implementations. SystemDS can generate different jobs running on a single node, MapReduce, and Spark (i.e., in local mode, MapReduce mode, and Spark mode, respectively). For example, in Spark mode, existing GUI in Spark provides a visualization tool for execution plans in the form of DAGs (Direct Acyclic Graphs) to represent RDD trans-formations. However, the visualized DAG depicts a *low-level* RDD transforma-tion, which does not show the relationship between DML operations and RDD transformations. Consequently, the DAG is difficult for users to understand the execution process of programs in DML. In local mode, there is even no tool to visualize the execution process.

In this work, we propose *Loupe*, a tool to visualize the *high-level* execution plans in SystemDS. Loupe provides a clear description on high-level execution plans, which supplements the low-level DAG visualization in Spark. Moreover, Loupe is able to visualize the execution process on a single node, since it does not tightened with the dedicated data processing engine. In the rest of this paper, Sect. 2 gives an overview of the architecture of Loupe, and Sect. 3 demonstrates the visualization.

2   Architecture

Figure 1 provides the architecture of Loupe. It is composed of two parts: the parse module and the visualization module. The parse module is responsible to retrieve execution plans from SystemDS and transform them into hierarchical blocks, while the visualization module will get these blocks and draw an interactive graph on a web page.

Parse Module. Given the execution plan in plain text format, the parse module is in charge of extracting operators and variables information from it. The parse module splits the whole plan into blocks, which represent logical rela-tions between different parts of an algorithm. There are two types of blocks:

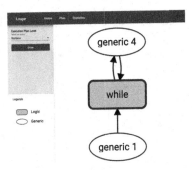

Fig. 2. The block graph.

logic blocks and generic blocks. The logic blocks show the control flow of an algorithm such as *If* and *For*, while the generic blocks describe how operators and variables are arranged in the part of programs. After variables and operators have been attached to the corresponding block, the parse module organizes these blocks in a hierarchical structure according to the logic relations.

Visualization Module. The visualization module retrieves hierarchical blocks with operators and variables information from the parse module. It then provides users with an user-friendly graph with interactive features on a web page.

3 Demonstration

In this section, we will briefly demonstrate how Loupe works. The demonstration uses PageRank, a classical algorithm employed by the search engine.

3.1 Configuration and Submission

Attendees are supposed to configure the log path first so that the parse module could retrieve execution plans correctly. Then, after submitting the PageRank algorithm written in DML script to SystemDS, attendees could view the graph through a web page.

3.2 Visualization

Visiting the web page, attendees first see a block graph, which denotes logical relations between different parts of the algorithm. For example, Fig. 2 shows the block graph of PageRank. Generic block 1 contains initial steps like handling input data. The while block is a logic block indicating that its child blocks are in a *While* loop. Therefore, attendees are able to sort out that generic block 4 inside the while block represents the iterative process of PageRank, which is executed multiple times until convergence.

Upon clicking one block, attendees are able to explore details of a part of the execution plan, visualize the plan and observe how SystemDS compiled and optimized the given algorithm. For example, Fig. 3 shows visualizations of execution

630   Z. Xu et al.

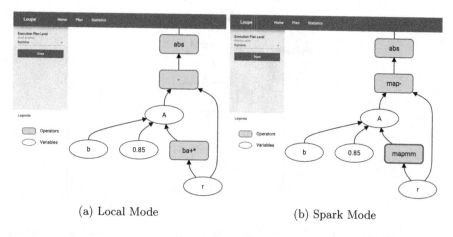

(a) Local Mode   (b) Spark Mode

Fig. 3. Visualizing execution plan for PageRank

plans for PageRank on varying execution mode. Figure 3a depicts the execution plan for PageRank running on local mode. SystemDS compiled the matrix multiplication to a *ba+** operator, the multiplication operator in local mode. Figure 3b depicts the execution plan for PageRank running in Spark mode. The red board block shows that SystemDS compiles the matrix multiplication to a *mapmm* operator which broadcasts one of the matrices to compute products in Spark. This is one kind of the implementations of matrix multiplication in Spark. Via the visualization graph, users learn which operator SystemDS employs to implement the matrix multiplication and whether this operator becomes the performance bottleneck. If so, users are supposed to modify their code or change the execution mode to achieve a better performance. Hence, users would gain a better understanding on the execution plan in SystemDS via Loupe, especially about those compilations and optimizations.

Acknowledgments. This work was supported by the National Natural Science Foundation of China (No. 61902128), Shanghai Sailing Program (No. 19YF1414200).

References

1. Boehm, M., et al.: SystemML: declarative machine learning on spark. PVLDB **9**(13), 1425–1436 (2016)
2. Boehm, M., et al.: SystemDS: a declarative machine learning system for the end-to-end data science lifecycle. In: CIDR (2020)
3. Dean, J., et al.: MapReduce: simplified data processing on large clusters. Commun. ACM **51**(1), 107–113 (2008)
4. Ghoting, A., et al.: SystemML: declarative machine learning on MapReduce. In: ICDE, pp. 231–242 (2011)
5. Zaharia, M., et al.: Resilient distributed datasets: a fault-tolerant abstraction for in-memory cluster computing. In: NSDI, pp. 15–28 (2012)

Ph.D Consortium

Algorithm Fairness Through Data Inclusion, Participation, and Reciprocity

Olalekan J. Akintande$^{(\boxtimes)}$ (iD)

University of Ibadan, Ibadan 20005, Oyo, Nigeria
https://scholar.google.com/citations?hl=en&user=Wn3hpPsAAAAJ

Abstract. Learning algorithms have become the basis of decision making and the modern tool of assessment in all spares of human endeavours. Consequently, several competing arguments about the reliability of learning algorithm remain at AI global debate due to concerns about arguable algorithm biases such as data inclusiveness bias, homogeneity assumption in data structuring, coding bias etc., resulting from human imposed bias, and variance among many others. Recent pieces of evidence (computer vision - misclassification of people of colour, face recognition, among many others) have shown that there is indeed a need for concerns. Evidence suggests that algorithm bias typically can be introduced to learning algorithm during the assemblage of a dataset; such as how the data is collected, digitized, structured, adapted, and entered into a database according to human-designed cataloguing criteria. Therefore, addressing algorithm fairness, bias and variance in artificial intelligence imply addressing the training set bias. We propose a framework of data inclusiveness, participation and reciprocity.

Keywords: AI fairness · Inclusion · Participation · Cross-validation

1 Introduction

Artificial Intelligence; essentially learning algorithms has become a compelling tool for making systematic decisions, based on a large amount of data and facts. This exhilarating field of research has gained wide application and acceptance owing to its peculiar physiognomy and surpassing edge of minimizing space of precariousness and ambivalence. Government and private companies globally rely laboriously on the output of learning algorithms before taking any decision. We have seen the application of learning algorithms in crime detection, computer vision, pattern recognition, weather forecast, robotic and face recognition, economic and financial forecasting, and budgeting etc.

Therefore, learning modelling/algorithms have become the basis of decision making and the modern tool of assessment in all spares of human-daily activities. Consequently, several competing arguments about the reliability of learning

Supervisor: Prof. Olusanya E. Olubusoye.

algorithm remain at various AI global debate due to concerns about arguable algorithm biases such as data inconsistent (non-inclusiveness) bias, diversity bias, unrepresentative, coding bias resulting from human imposed bias, and variance among many others.

Cross-validation is arguably the simplest and widely used model assessment and selection among many other procedures. The importance of cross-validation has resulted in diverse innovative approaches and methods of the cross-validation. Hence, several cross-validation algorithms have been proposed in the literature and with various degrees of pros and cons. This study aims to modify an existing (or propose a new) cross-validation approach to update the existing gap in the literature to address the algorithm bias, data inclusiveness, improve algorithm validation (participation), reliability and bias assessment.

1.1   Problem Statements and Motivation

Artificial intelligence (such as machine learning) remain the game-changer in the new world order of computational, decision and analytic science. It has gained global acceptance and becomes the driven force of decision-making tools in all spheres of human endeavours. However, algorithm bias in artificial intelligence continues to be a major constraint and concern for many particularly the under-represented groups.

Algorithm bias happens when an active learning algorithm under-performed in real-life application. That is, it describes systematic and repeatable errors in a computer system that create unfair outcomes, such as privileging one arbitrary group of users over others. Bias can emerge due to many factors, including but not limited to the design of the algorithm or the unintended or unanticipated use or decisions relating to the way data is coded, collected, selected or used to train the algorithm, Cormen et al. (2009). As algorithms expand their ability to organize society, politics, institutions, and behaviour; sociologists have become concerned with how unexpected output and misrepresentation of data can impact our world. Because algorithms are often rated to be neutral and unbiased, they can erroneously project greater authority than human expertise, and progressively, reliance on algorithms can displace human responsibility for their outcomes.

Recent evidence (such as in; crime and arrest AI algorithm, computer vision - misclassification of people of colour, see Fig. 1 and 2; face recognition, among many others) have shown that there is indeed need for concerns. Weizenbaum (1976) suggests that bias could arise from the data used in training, but also from the way a program is coded. Hence addressing bias and variance in artificial intelligence implies addressing the training set bias.

Thus, to address algorithm fairness, transparency and accountability, we must address the issue of data inclusiveness, and equal representation (in training) and participation in all algorithm designs. We cannot continue to pretend as if gender, colour, race, ethnicity, and origin are insignificant in our societies any longer. They are the nucleus of every community and should also be the

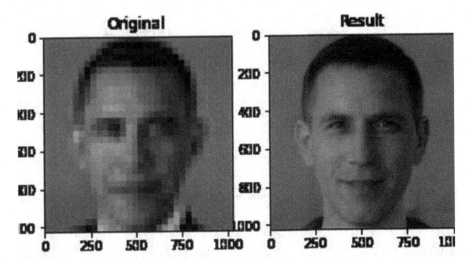

Fig. 1. President Barrack Obama misclassify as Whiteman. Credit: Robert Osazuwa Ness @osazuwa 2020.

nucleus of all algorithmic frameworks. This work aims to address the issue of AI algorithm bias from a data-driven/processes point of view.

Bias and Variance. According to Gillespie et al. (2014), bias can be introduced to an algorithm in several ways. During the assemblage of a dataset, data may be collected, digitized, adapted, and entered into a database according to human-designed cataloguing criteria. Although, modellers play a crucial role in the direction and functionality of algorithm which could be subject to individual bias, preference or judgment. That is, programmers can assign priorities (personal bias), or hierarchies, for how a program assesses (e.g., program design) and sorts the data. This requires human decisions about how data is categorized, and which data is included or discarded, Gillespie et al. (2014) and Diakopoulos (2017).

While this is true, the work of modelling typically follows some standard statistical procedures and are often taken very seriously by the modellers. According to Alake (2020), it could be said that learning models cannot be directly biased by design, and any emergence or cause of bias is external to the architecture and design of the learning framework. Thus, modellers often time have limited jurisdiction over the dataset made available for training and model validation, and criteria set by sponsors perhaps intentionally or unintentionally (environmental bias) limited or under-representative dataset to pose or place some particular race, or people above or below others.

More so, the goal of such a project can influence the programmer priority leaving behind the data (inclusion) ethics. This is the point where algorithm bias plants the seed of unexpected output leading to inequality. Hence, non-inclusive

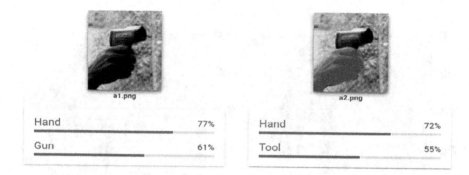

Fig. 2. Bias profiling of people of colour against white. Credit: Nicolas Kayser-Bril @nicolaskb 2020.

or under-representative data promote algorithm bias and result in the source of most biases in model estimation.

Cross-validation approach is an essential tool of all learning algorithms. The goal of cross-validation is to test the model's ability to predict new data that was not used in estimating it, to flag problems like overfitting or selection bias, Cawley and Talbot (2010). In statistics and machine learning, the bias-variance tradeoff is the expression of a set of predictive models whereby models with a lower bias in parameter estimation have a higher variance of the parameter estimates across samples, and vice versa. The bias-variance dilemma or bias-variance problem is the conflict in trying to simultaneously minimize these two sources of error (bias and variance) that prevent supervised learning algorithms from generalizing beyond their training set, Kohavi and Wolpert (1996) and Luxburg and Scholkopf (2011).

Nadkarni (2016) made a data bias case against CV. He noted that even though CV remains a robust technique regardless of variation within the dataset, CV does not have any robust performance against a systematic bias in the choice of data that is adopted for training set particularly when there is built-in bias. More so, the homogeneous assumption in the distribution of training data is associated with the majority of the existing CV. To address the inherent issues of algorithm bias, a cross-validation approach; which is a data-driven method is quite suitable if it accounts for the inherent heterogeneity in training data and proposes a data inclusion approach.

1.2   Aim and Objectives

This study aims to examine the existing cross-validation approaches and propose a modified approach to address the inherent homogeneous assumption leading to algorithm bias and variance associated with the existing CV methods in the model assessment and validation. Hence, the goal of this thesis is to address the issue of algorithm bias and variance from its natural source: data.

Objectives

1. To modify an existing V–fold and Distributed Balanced Stratified (DBS) CV approach leading to the confirmatory CV approach.
2. To assess the performance of the proposed CV on Hold-out, regular CV and DBS CV.
3. In light of 2, to examine in an application, the relevance of the new approach over the existing V-fold, DBS, and other CV approaches.

Essentially, this work aims to propose a CV method that will have a minimal loss, accounts for the inherent heterogeneity in training data, and ensure data inclusion to address under-representation and misclassification. We expect that this procedure will have a robust performance over the existing (random and deterministic) CVs and address bias in AI algorithms due to lack of data inclusion.

References

Cawley, G., Talbot, N.: On over-fitting in model selection & subsequent selection bias in performance evaluation. J. Mach. Learn. Res. **11**, 2079–2107 (2010)

Nadkarni, P.: Core technologies: machine learning and natural language processing. In: Clinical Research Computing (2016)

Alake, R.: Algorithm Bias in Artificial Intelligence Needs to be Discussed (And addressed). https://towardsdatascience.com/algorithm-bias-in-artificial-intelligence-needs-to-be-discussed-and-addressed-8d369d675a70. Accessed 20 May 2020

Cormen, T.H., Leiserson, C.E., Rivest, R.L., Stein, C.: Introduction to Algorithms, 3rd edn., no. 5. MIT Press, Cambridge (2009)

Weizenbaum, J.: Computer Power and Human Reason: From Judgment to Calculation. W.H. Freeman and Company, San Francisco (1976)

Diakopoulos, N.: Enabling accountability of algorithmic media: transparency as a constructive and critical lens. In: Cerquitelli, T., Quercia, D., Pasquale, F. (eds.) Transparent Data Mining for Big and Small Data. SBD, vol. 11, pp. 25–43. Springer, Cham (2017). https://doi.org/10.1007/978-3-319-54024-5_2

Gillespie, T., Boczkowski, P., Foot, K.: Media Technologies, pp. 1–30. MIT Press, Cambridge (2014)

Kohavi, R., Wolpert, D.H.: Bias Plus Variance Decomposition for Zero-One Loss Function. In: ICML (1996)

Luxburg, U.V., Schölkopf, B.: Statistical Learning Theory: Models, Concepts, and Results. Handbook of the History of Logic (2011)

Performance Issues in Scheduling of Real-Time Transactions

Sarvesh Pandey[✉] and Udai Shanker

Computer Science and Engineering Department, Madan Mohan Malaviya
University of Technology, Gorakhpur, India
{sarveshrs,uscs}@mmmut.ac.in

Abstract. The multi-site real-time transactional data-analysis based applications
and the underlying research efforts to improve the performance of such applications have got renewed attention by researchers in the last four years. It reveals
that the current scenario possesses numerous unanswered and truly relevant issues
and challenges requiring a multi-disciplinary research approach to work on and
solve the core database transaction processing related issues. Our focus is to cover
most of the issues and challenges with transaction scheduling algorithms in one
place to put out the current research status. At a high level, the domains covered
are—real-time priority assignment heuristics, real-time concurrency control protocols, and real-time commit processing. The article indeed guides towards the
immediate-future directions requiring actions/ efforts by the modern data-driven
research community.

Keywords: Data analysis · End-to-end application · Priority inversion ·
Real-time applications · Resource conflicts

1 Introduction

These days, a distributed set of external devices are required to perform complex real-
time analysis and the decision-making task. To formulate and drive such application
requirements, the study of a distributed real-time database system (DRTDBS) is of
utmost importance. It is a collection of time-constrained distributed data items with the
transaction as only the medium of update [1]. The DRTDBS should be designed in a way,
so that, it can ensure that the timing requirements are satisfied while the data collection/
update.

The transactions have been categorized into three types based on the consequences
of deadline misses—Soft deadline transactions, Firm deadline transactions, and Hard
deadline transactions [2]. Violations of soft, firm, and hard deadlines result in completely
different consequences. The outcome has some value even after the soft deadline issue
(though lesser) (though lesser), no value if the deadline is firm, and a negative value
(catastrophic consequences) when the deadline is hard [3].

The paper structure is noted in this paragraph. Section 2 discusses various existing
research problems important to DRTDBS performance. Section 3 describes the relation

© Springer Nature Switzerland AG 2021
C. S. Jensen et al. (Eds.): DASFAA 2021, LNCS 12683, pp. 638–642, 2021.
https://doi.org/10.1007/978-3-030-73200-4_51

of the work to the state of the art in DRTDBS transaction management. The proposed protocols and their novelty with respect to the problems addressed are discussed in Sect. 4. The scope of future contributions to various issues and research problems in distributed computing systems for the real-time environment are discussed in Sect. 5. At the end, conclusions are drawn in Sect. 6 (conclusion section).

2 DRTDBS Research Problems

The single most desirable objective in the DRTDBS performance study is the transaction deadline miss percent metric. The lesser the miss-percent, the greater is the acceptability of the firm DRTDBS application. On discussing one level down in the transaction processing hierarchy, the above transaction miss percent metric is mainly affected by the selection of the following set of real-time transaction processing algorithms—priority assignment heuristic, concurrency control protocol [4, 5], and commit processing [6, 7]. When being at the last level of the hierarchy, some of the known and unanswered/inadequately answered problems associated with the above algorithms are deadline computation (the deadline of transaction and/or data), data conflicts, CPU scheduling, the multi-read single-write problem, priority inversion, the pseudo priority inversion, the slow priority inheritance mechanism, distributed deadlock, resource wastage, cyclic restart, the interaction between the lender-borrower and priority inheritance approach, etc.

The above set of problems, associated with real-time transaction processing in a multi-site data-intensive system, can be a challenge as well as a driving force for today's database researchers. The solution to these problems requires a collective coordinated effort across the globe.

3 Literature Work

The DRTDBS design concerns revolve around the scheduling of transactions. Though one can see mature transaction scheduling related research literature available, the constantly changing requirements make the existing algorithms inefficient and in some use cases unacceptable as well [8]. Transaction scheduling, in a broad sense, includes both data scheduling as well as CPU scheduling. The scheduling (of data and CPU) is done in accordance with assigned transaction priorities. Therefore, the role of all three sub transaction scheduling modules—priority assignment, concurrency control protocol and commit processing, is crucial as they are major high-level functions in deciding the system performance.

The design of a priority assignment heuristic is highly dependent on the fact that data required to be utilized by the transaction are known before its initiation or not. The growing complexities—priority inversion and data contention, need to be addressed soon. Moving further with transactions having priorities, the transactions experience conflicts while concurrently executing and trying to access the same data item. The conflict converts into an even bigger problem named priority inversion if the transaction holding the conflicting data has a priority lower than the transaction requesting it. In such cases, honoring the priority of the transaction should be considered in designing a protocol.

However, it is not possible to honor the transaction priority if the conflicting lock-holding low priority sub-transaction is in a PREPARED state—such sub-transactions cannot unilaterally decide as a part of atomic commit processing requirements. A fresh perspective along these lines may be the next step. To get a broader insight into work done in the past, one can refer to [8]. It presents a survey of key transaction scheduling protocols, and various unresolved issues in the context of developing modern time-driven transaction processing systems.

4 Thesis Contributions

The main contributions of this thesis are as follows.

1. To address the issue of high data contention, the Most Dependent Transaction First (MDTF) heuristic has been proposed [9]. The MDTF injects the size of dependent transactions of all directly competing transactions in the computation of their priority value. Performance studies have shown that MDTF provides a trade-off between the NL and EQS heuristics from the performance perspective due to its better handling of data contention. As an extension to the MDTF, we further proposed the contention-aware equal slack (CA-EQS) policy [10]. The CA-EQS is advanced version of MDTF—it performs better than MDTF.
2. We proposed a "Reduction of lengthy transactions starvation effect, Avoidance of deadlock & pseudo priority inversion, and Conditional-restart for an Efficient Resource Utilization" (RACE) concurrency control protocol [11]. The key contributions of this protocol are—reduced wastage of the system resources by eliminating the deadlock, avoiding unnecessary aborts. The protocol also ensures to some extent that long global transactions are not starved.
3. We proposed a "Sophisticated Time and message utilization centred Priority inheritance" (STEP) concomitant protocol [12] that reduces the priority inherit message dissemination time up to half. It also eliminates the priority de-boosting process, since it happens towards the end of execution of a low priority transaction, and therefore, it is needless.
4. The RAPID protocol [13] is the first protocol that addresses the multi-read single-write sequence of events resulting in the write transaction starvation problem. The above problem may result in a bigger performance/ architectural challenge when it occurs hand in hand with the priority inversion problem. RAPID suggested to go with the controlled multi-read environment—it seems to be a win-win tradeoff as switching off multiple-read feature badly hurts concurrency while uncontrolled read may result in the severe high priority write transaction starvation problem.
5. The Early Data Lending Based Distributed Real-time Commit (EDRC) protocol optimistically utilized the Lender-Borrower approach [14]. It re-categorized the data conflict as a locking-locking conflict, locking-processing conflict, and locking-committing conflict. The novel feature of this protocol is that it facilitates to initiate of data lending just after the completion of the data processing task of the cohort.

5   Future Research Challenges

With set goals mainly increased concurrency, stricter consistency (no compromises with data quality), firm time-bound transaction processing requirements, and continuous change in application requirements, it became more challenging to come up with the solutions and their enhancements on a regular basis. As of today, the following research open questions require attention.

1. A widely used performance metric for the firm DRTDBS is the 'transaction miss percent'. However, we are of the view that several new metrics, for instance, resource utilization, conflict percent, rollback percent, etc. should be utilized/developed to get a transparent view on the acceptability of the transaction scheduling algorithms [15, 16].
2. The discussion of priority inversion problem can be widened to nested DRTDBS, mobile DRTDBS, replicated DRTDBS [17, 18], active DRTDBS, energy-aware RTDBS query processing, and the mobile ad-hoc network (MANET) databases.
3. There is one other data-processing research area that we almost stopped looking into as most of the researchers chosen to work on time-constraint databases which is fascinating and in demand [19, 20]. It is a conventional transaction scheduling algorithm. As a lot of work is happening with the focus on the advanced database model, it will be really rewarding if we simply look back at our conventional algorithm with a view that how we can include good features of some of the real-time transaction scheduling algorithms [21].

6   Conclusions

The role of priority inversion in multi-site real-time transactional data analysis is a critical topic that needs to be investigated with a broader perspective. More specifically, the method used to address the above issue in one scenario cannot be applied in the other scenario without careful scientific study and findings because of the increasing complexity and versatility of applications involved.

Acknowledgment. The financial support, during this research, by the Council of Scientific and Industrial Research, New Delhi, India under grant number 1061461137 is acknowledged.

References

1. Pandey, S., Shanker, U.: Priority inversion in DRTDBS: challenges and resolutions. In: Proceedings of the ACM India Joint International Conference on Data Science and Management of Data (CoDS-COMAD 2018), pp. 305–309 (2018)
2. Pandey, S., Shanker, U.: Causes, effects, and consequences of priority inversion in transaction processing. In: Handling Priority Inversion in Time-Constrained Distributed Databases. IGI Global (2020)
3. Shanker, U., Misra, M., Sarje, A.K.: Distributed real time database systems: background and literature review. Int. J. Distrib. Parallel Databases **23**(02), 127–149 (2008)

4. Pandey, S., Shanker, U.: On using priority inheritance-based distributed static two-phase locking protocol. In: Kolhe, M.L., Trivedi, M.C., Tiwari, S., Singh, V.K. (eds.) Advances in Data and Information Sciences. LNNS, vol. 38, pp. 179–188. Springer, Singapore (2018). https://doi.org/10.1007/978-981-10-8360-0_17

5. Pandey, S., Shanker, U.: CART: a real-time concurrency control protocol. In: Desai, B.C., Hong, J., McClatchey, R. (eds.) 22nd International Database Engineering & Applications Symposium (IDEAS 2018). ACM, New York, 18–20 June 2018

6. Pandey, S., Shanker, U.: A one phase priority inheritance commit protocol. In: Negi, A., Bhatnagar, R., Parida, L. (eds.) ICDCIT 2018. LNCS, vol. 10722, pp. 288–294. Springer, Cham (2018). https://doi.org/10.1007/978-3-319-72344-0_24

7. Pandey, S., Shanker, U.: IDRC: a distributed real-time commit protocol. Proc. Comput. Sci. **125**, 290–296 (2018)

8. Pandey, S., Shanker, U.: Transaction scheduling protocols for controlling priority inversion: a review. Comput. Sci. Rev. **35**, 100215 (2020)

9. Pandey, S., Shanker, U.: MDTF: a contention aware priority assignment policy for cohorts in DRTDBS. In: Khosrow-Pour, D.B.A. (ed.) Encyclopedia of Organizational Knowledge, Administration, and Technologies, 1st edn., pp. 742–756. IGI Global (2020)

10. Pandey, S., Shanker, U.: CA-EQS: a contention aware distributed priority assignment heuristic. J. Supercomput. (2020)

11. Pandey, S., Shanker, U.: RACE: a concurrency control protocol for time-constrained transactions. Arab. J. Sci. Eng. (2020)

12. Pandey, S., Shanker, U.: STEP: a concomitant protocol for real time applications. Wirel. Pers. Commun. (2021, under review)

13. Pandey, S., Shanker, U.: RAPID: a real time commit protocol. J. King Saud Univ. – Comput. Inf. Sci. (2020)

14. Pandey, S., Shanker, U.: EDRC: an early data lending based real-time commit protocol. In: Encyclopedia of Information Science and Technology, 5 edn., pp. 800–814 (2021)

15. Haritsa, J., Carey, M., Livny, M.: Value-based scheduling in real-time database systems. VLDB J.—Int. J. Very Large Data **2**(2), 117–152 (1993)

16. Chauhan, N.R., Tripathi, S.P.: Optimal admission control policy based on memetic algorithm in distributed real time database system. Wirel. Pers. Commun. **117**(2), 1123–1141 (2020). https://doi.org/10.1007/s11277-020-07914-x

17. Srivastava, A., Shankar, U., Tiwari, S.K.: A protocol for concurrency control in real-time replicated databases system. IRACST—Int. J. Comput. Netw. Wirel. Commun. (IJCNWC) **2**(3) (2012)

18. Arun, A., Pandey, S., Shanker, U.: A multi-replica centered commit protocol for distributed real-time and embedded applications. Int. J. Syst. Dyn. Appl. (IJSDA) (2021, under revision)

19. Pandey, A.K., Pandey, S., Shanker, U.: LIFT- a new linear two-phase commit protocol. In: Proceedings of 25th Annual International Conference on Advanced Computing and Communications (ADCOM 2019) at IIIT Bangalore (2019)

20. Pandey, S., Pandey, A., Shanker, U.: SP-LIFT: a serial parallel linear and fast-paced recovery-centered transaction commit protocol. SN Comput. Sci. **1**(3), 1–10 (2020)

21. Singh, R.K., Pandey, S., Shanker, U.: A non-database operations aware priority ceiling protocol for hard real-time database systems. In: the Proceedings of 10th International Conference on Computing Communication and Networking Technologies, IIT, Kanpur, India, 6–8 July 2019 (2019)

Semantic Integration of Heterogeneous and Complex Spreadsheet Tables

Sara Bonfitto[✉]

Computer Science Department, Università di Milano, Via Celoria 18, Milan, Italy
sara.bonfitto@unimi.it

Abstract. A great number of companies and institutions use spreadsheets for managing, publishing and sharing their data. Though effective, spreadsheets are mainly designed for being interpreted by humans, and the automatic extraction of their content and interpretation is a complex task. The task becomes even harder when tables present different kinds of mistakes and their layout is complex. In this paper, we outline the approach that we wish to develop during the PhD for answering the research question "how to semi-automatically extract coherent semantic information from heterogeneous and complex spreadsheets?".

Keywords: Heterogeneous spreadsheet tables · Semantic table interpretation · User interfaces · Machine learning

1 Introduction

Recently, our research group was involved in the problem of integration of heterogeneous spreadsheet files that a debt collection agency daily receives from local authorities (e.g. municipalities, tax agency) containing batches of thousand of invoices to be rescued. These spreadsheets are big, heterogeneous and do not follow any standard format or notation (Fig. 1 shows an example). The first row reports the column headers, however, the access keys are not always present and do not follow any specific format. Data occurring in the same column sometimes adhere to different types. For example, the column SSN/VAT contains different data types (actually expressing that invoices can be titled to individual citizens or to companies). Sometimes columns present strings from which different kinds of information can be extracted, as in the case of the "address" column, where different alternative patterns represent the street name, and street/apartment number. Blanks and semi-blank rows can occur in the main table. Semi-blank rows usually contain totals or aggregated data. Blank rows are sometimes used for aesthetic reasons while others for separating rows representing correlated invoices. Indeed, the information about an invoice is not always contained in a single table row. For example, in Fig. 1 there is a correlation among two rows (the fourth row contains the reference to the legal representative associated with

PhD Advisor Prof. Marco Mesiti.

© Springer Nature Switzerland AG 2021
C. S. Jensen et al. (Eds.): DASFAA 2021, LNCS 12683, pp. 643–646, 2021.
https://doi.org/10.1007/978-3-030-73200-4_52

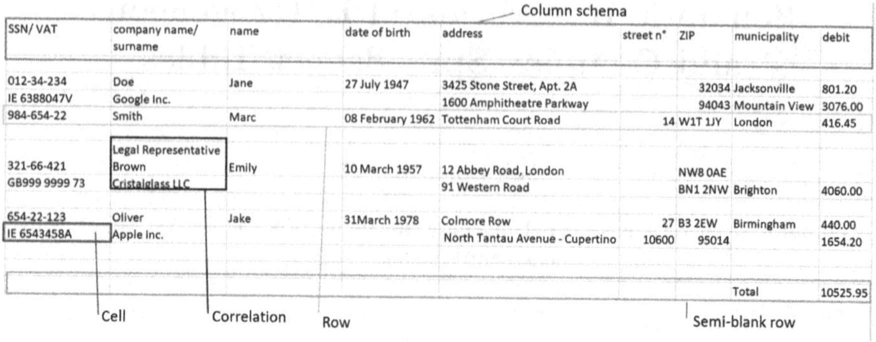

SSN/ VAT	company name/ surname	name	date of birth	address	street n°	ZIP	municipality	debit
012-34-234	Doe	Jane	27 July 1947	3425 Stone Street, Apt. 2A		32034	Jacksonville	801.20
IE 6388047V	Google Inc.			1600 Amphitheatre Parkway		94043	Mountain View	3076.00
984-654-22	Smith	Marc	08 February 1962	Tottenham Court Road	14	W1T 1JY	London	416.45
321-66-421	Legal Representative Brown Cristalglass LLC	Emily	10 March 1957	12 Abbey Road, London		NW8 0AE		
GB999 9999 73				91 Western Road		BN1 2NW	Brighton	4060.00
654-22-123	Oliver	Jake	31March 1978	Colmore Row	27	B3 2EW	Birmingham	440.00
IE 6543458A	Apple Inc.			North Tantau Avenue - Cupertino	10600	95014		1654.20
							Total	10525.95

Cell   Correlation   Row   Semi-blank row

Fig. 1. A spreadsheet example

the invoice in the fifth row). This kind of correlation can be expressed by following different patterns. Last but not least, the information contained in these spreadsheets can contain different kinds of typographical, grammatical and miscalculation errors. The variability of organizations of these spreadsheets prevent the use of well studied approaches for table understanding (e.g. [3]), data repairs and extraction (e.g. [5]), data transformation (e.g. [6]), programming by example (e.g. [4]), and semantic characterization of the information (e.g. [8]). Standard approaches for NLP cannot be applied on short texts like the one that can occur in spreadsheets for extracting patterns. We believe that a completely automatic approach that exploits sophisticated machine learning (ML) techniques cannot properly be used in this context. A semi-automatic approach can be devised to support the user during the process of data cleaning, transformation and semantic characterization that involve the user in the loop in order to tune the prediction system depending on the feedback obtained while processing new spreadsheets. Users need to be supported by easy-to-use graphical interfaces for correcting mistakes and improve the overall performance of the system.

In order to reach this goal, we propose the adoption of a three-phase approach. Phase I is responsible for the spreadsheet cleaning, the identification of the column types and the synthetic error correction. Phase II aims to create a semantic characterization of the table content to be extracted from the spreadsheet and relies on the use of a domain Ontology and, when possible, a Knowledge Base. Phase III relies on the identification of semantic constraints and assertions that need to be checked and maintained on the considered Ontology. The purpose of this phase is to point out semantic mistakes that can be fixed on the RDF representation of spreadsheet tables.

2   The Three-Phase Approach

Phase I: Table Identification and Cleaning. The main purpose of this phase is to correct syntax errors occurring in the data, identifying the correlations existing among table rows, and identifying basic types of each column.

For identifying correlation among table rows, we wish to adopt a declarative pattern-based language for specifying when a correlation exists. Moreover, we wish to develop interfaces for further supporting the users in their manual identification and thus learning new patterns for the interaction. Moreover, we wish to develop a multi-label classification approach for the identification of the cell and column types. Several basic types, domain-specific types and also pattern-based types will be supported. Patterns will be exploited for extracting values from complex strings like for example the address "12 Abbey Road, London" in Fig. 1. A multi-label approach is considered for facing situations like the column "company name/surname" that contains both the company name or the citizen surname. The automatically identified types, however, can contain errors due to the occurrence of mistakes in the data. Therefore specific interfaces should be developed for their easy correction. Moreover, the large amount of invoices to be processed requires the adoption of solutions that apply a single correction to many invoices at the same time.

Phase II: Semantic Characterization of Table Content. The aim of Phase II is to provide a *semantic meta-model* description of the spreadsheet tables by means of annotations w.r.t. a Domain Ontology. Even if many approaches have been proposed so far for this problem, in our research we wish to face the variability of data types identified in the first phase that usually is not considered. Moreover, the semantic meta-model should be coupled with a graphical representation that makes easier to the user checking the automatically generated model and correcting mistakes when needed. Moreover, a ML algorithm will be applied for learning annotations relying on previously established mappings. The user can also change manually the annotations, these modifications should be exploited for tuning the predictions. Our semantic meta-model is inspired by the one used in Karma [7] but differs from it because it is created starting from the types identified in the first phase and allows the extraction of several data from a single column (while Karma only makes a 1:1 correspondence from the data in the spreadsheet to the Ontological concepts).

Phase III: Verification of Semantic Constraints and Assertions. The semantic meta-model is finally used for the automatic extraction and transformation of data in an RDF format according to the domain Ontology. In this phase, we wish to use the semantic constraints and assertions identified on the Ontology to point out semantic mistakes. For example, the total debt amount in Fig. 1 is correct and corresponds to the sum of the single debt imports, the zip code of an address can be validated against the municipality. These are semantic constraints w.r.t. the syntactic constraints identified in Phase I.

3   Concluding Remarks

In this paper, we outlined our main research question and the related problems that should be faced in the next two years of the PhD program.

At current stage, we have started working on a survey on related works in the context of table understanding and semantic interpretation of tables [2]. In this survey, we have outlined the different phases in which the table understanding problem can be organized (localization, segmentation, functional and structural analysis and integration) and presented the main approaches proposed in the last fifteen years. Moreover, an initial solution for the first phase is proposed in [1] by introducing a methodology for determining the value/column types contained in CSV tables that exploits a multi-label prediction algorithm that has been trained on a simulation of typical data available in the considered domain that takes into account the errors occurring in data. This automatic approach has been combined with graphical user interfaces with which the user can check the predicted types and modify them when needed. The modifications can be applied at type-level, thus many values can be modified by a single specification. This initial activity needs to be further enhanced for identifying correlated rows and cells containing an aggregation of other cells and also functional relationships that need to be preserved on data (e.g. the occurrence of an SSN requires the presence of name and surname of an individual).

We are currently working on the second phase of the approach by specifying the semantic-description of a spreadsheet table and its graphical representation. Moreover, an approach for the semi-automatic construction of the model is evolving that takes into account the previously specified meta-models and similarity measures for evaluating their adequateness to the new scenario.

References

1. Bonfitto, S., Cappelletti, L., Trovato, F., Valentini, G., Mesiti, M.: Semi-automatic column type inference for CSV table understanding. In: Bureš, T., et al. (eds.) SOFSEM 2021. LNCS, vol. 12607, pp. 535–549. Springer, Cham (2021). https://doi.org/10.1007/978-3-030-67731-2_39
2. Bonfitto, S., Casiraghi, E., Mesiti, M.: Table understanding approaches for extracting knowledge from heterogeneous tables. WIREs Data Min. Knowl. Disc. (2020, to appear)
3. Holeček, M., Hoskovec, A., Baudiš, P., Klinger, P.: Table understanding in structured documents. In: Proceedings of International Conference on Document Analysis and Recognition Workshops (ICDARW), vol. 5, pp. 158–164 (2019)
4. Jin, Z., Anderson, M.R., Cafarella, M., Jagadish, H.V.: Foofah: transforming data by example. In: Proceedings of ACM SIGMOD, pp. 683–698 (2017). https://doi.org/10.1145/3035918.3064034
5. Kandel, S., Paepcke, A., Hellerstein, J., Heer, J.: Wrangler: interactive visual specification of data transformation scripts. In: ACM Human Factors in Computing Systems (CHI), pp. 3363–3372 (2011). https://doi.org/10.1145/1978942.1979444
6. Shigarov, A., Khristyuk, V., Mikhailov, A., Paramonov, V.: TabbyXL: rule-based spreadsheet data extraction and transformation. In: Damaševičius, R., Vasiljevienė, G. (eds.) ICIST 2019. CCIS, vol. 1078, pp. 59–75. Springer, Cham (2019). https://doi.org/10.1007/978-3-030-30275-7_6
7. Taheriyan, M., Knoblock, C.A., Szekely, P., Ambite, J.L.: Learning the semantics of structured data sources. J. Web Semant. 37, 152–169 (2016)
8. Zhang, Z.: Effective and efficient semantic table interpretation using tableminer^{+}. Semant. Web 8(6), 921–957 (2017). https://doi.org/10.3233/SW-160242

Abstract Model for Multi-model Data

Pavel Čontoš[(✉)] ⓘ

Faculty of Mathematics and Physics, Charles University, Prague, Czech Republic
contos@ksi.mff.cuni.cz

Abstract. In recent years, many so-called multi-model database management systems have emerged, mainly as extensions of the existing
single-model systems, regardless they used to be relational or NoSQL.
These new database systems make new demands on their users. From
the point of view of the conceptual and logical representation, the so
far widely used approaches, especially ER and UML, prove not to be
sufficient enough in many aspects due to the specific properties of multi-
model data. In addition, it is also difficult to query data that is represented in various and often overlapping data models at the logical level.

Keywords: Multi-model data · Abstract model · Transformations ·
Querying · Evolution management · Category theory

1 Introduction

Currently, besides the traditional single-model NoSQL systems, the family of so-
called multi-model database management systems (MMDBMSs) [10] emerged,
allowing us to work with multiple data models at once. These systems are often
based on various proprietary approaches, yet they have no unifying formal background. This may bring many issues and challenges:

- Traditional representation of data at the conceptual layer becomes insufficient. The modeling languages like ER [4] and UML [11] are suitable for the
  relational/object model rather than a combination of multiple data models.
- The level of support of multiple data models in MMDBMSs varies greatly [5],
  e.g., these systems offer different ability to query across various models.
- There are no unified approaches, nor generally applicable methods, allowing
  us to work with multi-model data.

Example 1. Figure 1 illustrates an example of an ER schema and the corresponding data represented by multiple logical models that are, moreover, overlapping
each other. In other words, we illustrate a typical scenario for a multi-model
database.

This work was supported by Czech Science Foundation project 20-22276S.

C. S. Jensen et al. (Eds.): DASFAA 2021, LNCS 12683, pp. 647–651, 2021.
https://doi.org/10.1007/978-3-030-73200-4_53

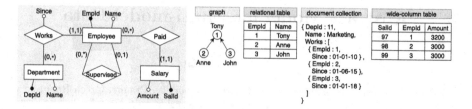

Fig. 1. Example of multi-model data

Although there is a variety of database systems, the usage of multiple single model databases (a polystore) or one multi-model database is a challenging task. If nothing else, the users must then be aware of all the particular involved models and query languages they are working with. In addition, having a query expression spanning over multiple models, external model transformations are required due to joins of data represented by different models at the logical layer.

We believe that category theory [2] allows us to formally describe and represent multi-model data more straightforwardly than the widely used approaches. Therefore, we will exploit the category theory to 1) design an abstract model for the purpose of schema and data instance representation in multi-model database systems, 2) provide internal and external transformations within the abstract model and between the abstract model and particular logical data models, and 3) propose a formal description of multi-model data conceptual querying.

2   Related Work

The idea of an abstract model is not new. Tuijn et al. [14] exploited category theory to describe an approach partially involving the relational and object-relational model. Spivak et al. proceeded with the idea and applied category theory on the relational model, also focusing on querying over the relational model using a proprietary language [12], and model transformations, namely from relational data to RDF triples [13]. Lately, an approach proposed by Liu et al. [9], inspired by Tuijn and Spivak, attempts to transform relational data to document-oriented data, e.g., JSON documents. All the approaches are based on the idea of a category backed by a directed multigraph called *typegraph* [14], unfortunately bringing many limitations, e.g., assumption of the existence of only one and only simple entity identifier.

Considering the logical layer of multiple data models, a unified view of key/value, wide-column, and document model provides the *NoSQL Abstract Model* (NoAM) [1]. The data is represented using named collections of blocks that consist of key/value pairs. The *Tensor Data Model* (TDM) [8] represents the data using tensors (multidimensional matrices). Similarly, associative arrays [6] represent data in tables. However, none of the mentioned approaches can handle graph logical models in a natural way that would allow efficient implementation of data manipulation operations. Finally, there are also recent approaches [3,7] dealing with a similar problem in the area of polystores.

Fig. 2. The concept of a proposed approach

3   Proposed Approach and Methodology

The idea of our approach is illustrated in Fig. 2. We will begin with a proposal of a novel abstract model (i.e., its schema and instance) that would allow us to connect a real-world conceptual schema with the logical layer of a MMDBMS. Having the model, we will then provide algorithms for internal and external model-to-model transformations. Finally, we intend to use the transformations to the evolution management of multi-model data, various data migrations, as well as a foundation for a unified multi-model query processing.

Exploiting category theory, we will start with a translation of the ER model into a proposed abstract model. We believe that such a categorical model based on name/value pairs would potentially be better applicable even on not yet existing data models when compared to the existing and limited models based on entity and relationship types (e.g., tables, documents). Without further details, a concept of the intended abstract model corresponding to the ER schema from Example 1 is presented in Fig. 3.

Next, we plan to use the categorical approach, namely functors, to describe transformations between the models. Figure 4 illustrates its application, e.g., a data migration. In this particular case, a functor is used to migrate a model to another model, e.g., by splitting the **Name** attribute into **Firstname** and **Lastname** attributes, yet representing the same data at the conceptual layer.

These model-to-model transformations can then be directly exploited in multi-model evolution management and query processing. In the former case, such transformations may express changes of the data, i.e., the evolution. In the latter one, the ability of transformations of data in different models is essential for the evaluation of join operations.

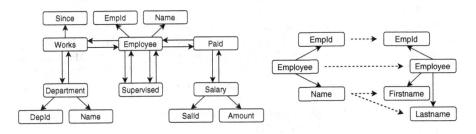

Fig. 3. Abstract model concept   Fig. 4. Data migration

4   Conclusion

To conclude this paper, we provide a list of our near-future plans we believe should contribute to the current state-of-the-art.

- We will attempt to provide a unification layer for the data models supported in MMDBMS.
- We will propose a formal background for multi-model data, as well as mutual transformations of the abstract model and particular logical models.
- We will introduce a unified multi-model query processing.
- We will describe transformations that allow for a correct propagation of changes to data, schemas, and queries induced by the evolution management.
- Within the design of the abstract model, we will consider model extensibility, i.e., the ability to add new models that currently do not exist.

References

1. Atzeni, P., Bugiotti, F., Cabibbo, L., Torlone, R.: Data modeling in the NoSQL world. Comput. Stand. Interfaces **67**, 103149 (2020)
2. Barr, M., Wells, C.: Category Theory for Computing Science, vol. 49. Prentice Hall, New York (1990)
3. Basciani, F., Di Rocco, J., Di Ruscio, D., Pierantonio, A., Iovino, L.: TyphonML: a modeling environment to develop hybrid polystores. In: MODELS 2020, New York, NY, USA (2020)
4. Chen, P.: The entity-relationship model - toward a unified view of data. ACM Trans. Database Syst. (1976)
5. Holubová, I., Svoboda, M., Lu, J.: Unified management of multi-model data. In: Laender, A.H.F., Pernici, B., Lim, E.-P., de Oliveira, J.P.M. (eds.) ER 2019. LNCS, vol. 11788, pp. 439–447. Springer, Cham (2019). https://doi.org/10.1007/978-3-030-33223-5_36
6. Kepner, J., et al.: Associative array model of SQL, NoSQL, and NewSQL databases. In: HPEC 2016, pp. 1–9. IEEE (2016)
7. Kolonko, M., Müllenbach, S.: Polyglot persistence in conceptual modeling for information analysis. In: ACIT 2020 (2020)
8. Leclercq, E., Savonnet, M.: TDM: a tensor data model for logical data independence in polystore systems. In: Gadepally, V., Mattson, T., Stonebraker, M., Wang, F., Luo, G., Teodoro, G. (eds.) DMAH/Poly-2018. LNCS, vol. 11470, pp. 39–56. Springer, Cham (2019). https://doi.org/10.1007/978-3-030-14177-6_4
9. Liu, Z.H., Lu, J., Gawlick, D., Helskyaho, H., Pogossiants, G., Wu, Z.: Multi-model database management systems - a look forward. In: Gadepally, V., Mattson, T., Stonebraker, M., Wang, F., Luo, G., Teodoro, G. (eds.) DMAH/Poly-2018. LNCS, vol. 11470, pp. 16–29. Springer, Cham (2019). https://doi.org/10.1007/978-3-030-14177-6_2
10. Lu, J., Holubová, I.: Multi-model data management: what's new and what's next? In: EDBT 2017, pp. 602–605 (2017)

11. Object Management Group: OMG Unified Modeling Language (OMG UML), version 2.5 (2015). http://www.omg.org/spec/UML/2.5/

12. Spivak, D.I.: Functorial data migration. Inf. Comput. **217**, 31–51 (2012)

13. Spivak, D.I., Wisnesky, R.: Relational foundations for functorial data migration. In: DBPL. ACM (2015)

14. Tuijn, C., Gyssens, M.: CGOOD, a categorical graph-oriented object data model. Theoret. Comput. Sci. **160**(1–2), 217–239 (1996)

User Preference Translation Model for Next Top-k Items Recommendation with Social Relations

Hao-Shang Ma$^{(\boxtimes)}$ and Jen-Wei Huang

Institute of Computer and Communication Engineering,
Department of Electrical Engineering, National Cheng Kung University,
No. 1, University Road, Tainan City, Taiwan
jwhuang@mail.ncku.edu.tw

Abstract. Recommendation systems are used to predict the interests of users through the analysis of historical preferences. Collaborative filtering-based approaches usually ignore the sequential information and sequential recommendation usually focus on the next item prediction. In this work, we would like to determine the next top-k recommendation problem. We propose User Preference Translation Model (UPTM) with item influence embedding and social relations between users. In addition, we will also solve the cold start problem in UPTM.

Keywords: Next top-k recommendation · Influence diffusion embedding · Social recommendation · Cold-start problem

1 Introduction

Recommendation systems are trying to learn the low dimensional representation of users and items. Many features can be adopted in recommendation systems, for example, user-item interactions, user features, item features, and other information such as the temporal factor.

Collaborative filtering-based approaches focus on learning users' preference and predict the items which users will have interest. The sequential relations in users behavior are usually ignored in collaborative filtering. In our opinion, the trigger relations between items are important for users behavior. For example, people usually watch a series of related movies after they watch one of the movie in this series. We would like to discover the target items which can trigger users to buy as much related items as possible. Therefore, we adopt the social influence propagation concept to model the trigger relations between items. In social network, people spread influence to their neighbors and receive influence from their neighbors at the same time. People usually be activated by their friends, family, and followees. Same as the social network, assume an item-item network is formed from users' sequential interaction behavior with item. The propagation of item influence can be used to indicate the likelihood of a user interacting with a related item based on their interaction with the target item.

© Springer Nature Switzerland AG 2021
C. S. Jensen et al. (Eds.): DASFAA 2021, LNCS 12683, pp. 652–655, 2021.
https://doi.org/10.1007/978-3-030-73200-4_54

In sequential recommendation research, we consider the sequential information in users behavior. Given the users' interaction behavior, the problem can be defined as predicting the next item which user will interact. However, we would like to recommend the next k items which users will interact in the future. First, we address the recommendation problem as the next top-k recommendation problem. We propose a User Preference Translation Model with item influence embedding, abbrev. as UPTM, to solve this problem. In the future, we would like to join the social relations between user and propose the social recommendation with UPTM. In addition, the cold start problem is usually ignored in the sequential recommendation systems. We will propose a new scheme to deal with the cold start problem in UPTM.

2   Related Works

2.1   Collaborative Filtering with Deep Neural Network

Collaborative filtering solves the recommendation problem by assuming that users with similar behaviors exhibit similar preferences for items. He et al. [2] propose NeuMF to combine the multi-layer perceptron and matrix factorization to learn the user and item embedding. Wang et al. [4] propose NGCF which is based on the graph neural network. They encodes the collaborative signal which represents the high-order connectivities by performing embedding propagation. Another kind of Generative Adversarial Networks-based methods try to apply GAN to recommendation. Chae et al. [1] suggest a new direction of vector-wise adversarial training and propose the GAN-based CF framework.

2.2   Social Recommendation Systems

On social network, people spread influence to their neighbors and receive influence from their neighbors at the same time. A user is activated by another user since they have same opinion tendency. Social recommendation systems that examine the propagation of influence among network members end up with users who share similar interests connected along diffusion paths. Wu et al. [5] propose DiffNet to model the recursive social influence propagation process and learn the user and item representation in social recommendation. Zhu et al. [6] propose Social Collaborative Mutual Learning Model to combine the item-based collaborative filtering and social collaborative filtering.

3   Methodology

3.1   Problem Definitions

In the current study, recommendations are generated by using previous behavior patterns to predict the items that are likely to interest the user in the future. In the following, $U = \{u_1, u_2, ...u_m\}$ denotes the user set and $I = \{i_1, i_2, ..., i_n\}$

denotes the item set. Each user u has preference record $P_u = (i_{u,1}, i_{u,2}, ...i_{u,t})$, where t is the interactive order. Based on a given P_u, our objective is to recommend the next k items $R_u = \{i_{u,t+1}, i_{u,t+2}, ..., i_{u,t+k}\}$ with which user u is likely to interact.

3.2   User Preference Translation Model

The proposed translation-based recommendation model is illustrated in Fig. 1. The model includes a simulation of item embedding and a translation of user preferences. The proposed scheme is based on the assumption that users are likely to interact with items that are associated with other items that they already possess. We propose using influence diffusion to learn the relationships among items. We first generate item influence diffusion paths from social influence paths sampled from the item-item relation graph from which UPTM learns the item influence embedding by which to encode user preferences. UPTM then learns the parameters in the hidden layer to output the item embedding and generate a recommendation list from the decoder of the translation module, to which is applied a softmax function and top-k sampling.

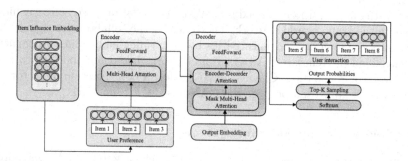

Fig. 1. The framwork of user preference translation model with item influence diffusion embedding

3.3   User Preference Translation Model for Social Recommendation

We would like to apply UPTM on social recommendation by adding the users' social relation in the item embeddings. Each user are encoded into a user embedding to represent the user's social relations in the social network. Then, we add the user embedding into the items to translate the user preference.

4   Experimental Results and Future Works

For evaluating the performance of proposed model, we compare UPTM model with other existing model on real several datasets such as Movielen, Amazon, and Yahoo datasets. We use the precision, recall and NDCG metrics to evaluate the performance for UPTM. In Table 1, we only show the results of NDCG. For overall performance comparison, UPTM outperforms other comparing methods on three metrics for the next k item recommendation problem. All experimental results are shown in our UPTM works [3]. In addition, we will propose the User Preference Translation Model with social recommendation. The social recommendation consider the users' preference and users' social relation simultaneously.

Table 1. Overall performance comparison

Methods	Movielens 1M			Movielens 20M			Amazon Book			Yahoo E-commerce		
	ND@5	ND@10	ND@20	ND@5	ND@10	ND@20	ND@5	ND@10	ND@20	ND@5	ND@10	ND@20
BPR-MF	0.086	0.0812	0.0901	0.0786	0.0864	0.0916	0.0118	0.0107	0.0129	0.0138	0.0141	0.0153
CFGAN	0.1066	0.0962	0.0916	0.0856	0.0986	0.09	0.0166	0.0152	0.0188	0.0152	0.0154	0.0178
NeuMF	0.0945	0.1033	0.1131	0.1193	0.1168	0.1231	0.0135	0.0141	0.0166	0.0144	0.0164	0.0177
NGCF	0.1168	0.1157	0.1235	0.0931	0.0921	0.098	0.0191	0.0176	0.0184	0.0156	0.0187	**0.0215**
UPTM	**0.1888**	**0.1996**	**0.1946**	**0.2001**	**0.2185**	**0.2327**	**0.0268**	**0.0283**	**0.0245**	**0.0189**	**0.0201**	0.0197

However, most sequential recommendation cannot deal the cold start problem since the new items are not shown in the training process. The sequential recommendation usually ignore this problem. In previous recommendation systems, they usually use the popularity of items or the similarity of users' preference to recommend. We would like use the users' social relation and give the initial item embedding for cold start items. We will join this idea to solve the cold start problem in UPTM.

References

1. Chae, D.-K., Kang, J., Kang, J.-S., Kim, S.-W., Lee, J., Lee, J.-T.: CFGAN: a generic collaborative filtering framework based on generative adversarial networks. In: Proceedings of the 27th ACM CIKM, pp. 137–146 (2018)
2. He, X., Liao, L., Zhang, H., Nie, L., Hu, X., Chua, T.-S.: Neural collaborative filtering. In: Proceedings of WWW, pp. 173–182 (2017)
3. Ma, H.-S., Huang, J.-W.: User preference translation model for recommendation system with item influence diffusion embedding. In: Proceedings of IEEE/ACM International Conference on ASONAM, pp. 50–54 (2020)
4. Wang, X., He, X., Wang, M., Feng, F., Chua, T.-S.: Neural graph collaborative filtering. In: Proceedings of the 42nd International ACM SIGIR, pp. 165–174 (2019)
5. Wu, L., Sun, P., Fu, Y., Hong, R., Wang, X., Wang, M.: A neural influence diffusion model for social recommendation. In: Proceedings of the 42nd International ACM SIGIR, pp. 235–244 (2019)
6. Zhu, T., Liu, G., Chen, G.: Social collaborative mutual learning for item recommendation. ACM Trans. Knowl. Discov. Data **14**, 1–19 (2020)

Tutorials

Multi-model Data, Query Languages and Processing Paradigms

Qingsong Guo[1,2(✉)], Jiaheng Lu[1], Chao Zhang[1], and Shuxun Zhang[1]

[1] University of Helsinki, Helsinki, Finland
{qingsong.guo,jiaheng.lu,chao.z.zhang,shuxun.zhang}@helsinki.fi
[2] North University of China, Taiyuan, China

A critical issue in big data management is to address data variety. Data come from disparate sources and may be presented in various models – structured, semi-structured, or unstructured. The increasing availability of multi-model data has triggered the development of Multi-Model DataBase (MMDB) systems [6]. We have found 86 DBMSs[1] (361 systems in total) claimed that they support multi-model data. These MMDBs typically integrate multiple data stores together to accommodate data in the formats that fit the sources best, e.g., key/value pairs, relational tables, graphs, or XML/JSON documents. They also provide unified query languages, which allow users to retrieve data of different models in a single query, i.e., *cross-model query*. Specifying users' interests with a formal query language is a typically challenging task, which becomes even harder in the context of multi-model data. It usually lacks a unified schema to help the users issuing queries, or has an incomplete schema as data come from disparate sources. Similarly, these challenges also incur extra complexity on query evaluation and optimization.

Scope of the Tutorial. In the past decades, many data models have been proposed for practical purposes. Some of them are widely adopted by database systems, i.e., the relational model [3] and its extensions [4], graph model [2], and semi-structured model [1]. Dozens of query languages have been implemented for retrieving multi-model data, such as AsterixDB's SQL++ and ArangoDB's AQL. This tutorial is to offer a comprehensive investigation on these languages and to make a comparative study on their processing paradigms. Multi-model query languages can be classified as four types, i.e., SQL-extensions, document-based language extensions, graph-based language extensions, and native multi-model query languages. We will discuss these languages from the 4 perspectives and will make in-depth comparisons of them from three related aspects: (1) the processing paradigms for cross-model query, (2) their essential semantics, and (3) the strategies for query optimization. We will also discuss the open problems in cross-model query processing of MMDB systems and provide insights on the research challenges and directions for future work. In addition, Finally, we will demonstrate how cross-model queries are processed in MMDB systems such as ArangoDB and OrientDB.

---

[1] DB-Engines Ranking: https://db-engines.com/en/ranking.

© Springer Nature Switzerland AG 2021
C. S. Jensen et al. (Eds.): DASFAA 2021, LNCS 12683, pp. 659–661, 2021.
https://doi.org/10.1007/978-3-030-73200-4

Difference with Our Previous Tutorial. Part of the content in this tutorial was presented in CIKM 2020 [5], which is the first tutorial to discuss state-of-the-art research works and industrial trends on multi-model data query languages. In tutorial [5], we mainly focused on the syntax of query languages and their semantic difference. Furthermore, we invited the participants to write and run cross-model queries[2] with ArangoDB AQL and UniBench [7]. In this tutorial, we will discuss these languages in 4 groups and concentrate on the paradigms for evaluating cross-model queries. We will also provide a hands-on section to demonstrate the cross-model query processing in MMDB systems such as ArangoDB and OrientDB. We will ensure this tutorial has a significant amount of new content(more than 50%) comparing to the previous tutorial.

Tutorial Organization. The tutorial is divided into 5 parts:
Part I: Introduction to multi-model data query languages (10 min)
Part II: SQL-based multi-model query extensions (30 min)
Part III: Document-based multi-model query extensions (20 min)
Part IV: Graph-based multi-model query extensions (15 min)
Part V: Native multi-model query languages (20 min)
Part VI: Demonstration (20 min)
Part VII: Open challenges and future directions (5 min)

Short Bibliographies.

Qingsong Guo is a Postdoctoral Researcher at the University of Helsinki. His research interests include multi-model data management and automatic management of big data with deep learning algorithms.

Jiaheng Lu is a Professor at the University of Helsinki. His main research interests lie in the Big Data management and database systems. He has published more than one hundred journal and conference papers.

Chao Zhang is a senior Ph.D. candidate at the University of Helsinki. His research topic lies in multi-model database benchmarking.

Shuxun Zhang is a Ph.D. candidate at the University of Helsinki. His research topic lies in multi-model database.

References

1. Abiteboul, S., Buneman, P., Suciu, D.: Data on the Web: From Relations to Semistructured Data and XML. Morgan Kaufmann Publishers Inc., San Francisco (1999)

---

[2] https://version.helsinki.fi/chzhang/cikm-2020-hands-on-session-for-multi-model-queries/-/blob/master/hands-on.ipynb.

2. Angles, R., Gutierrez, C.: Survey of graph database models. ACM Comput. Surv. **40**(1), 1–39 (2008)
3. Codd, E.F.: A relational model of data for large shared data banks. Commun. ACM **13**, 377–387 (1970)
4. Codd, E.F.: Extending the database relational model to capture more meaning. ACM Trans. Database Syst. (TODS) **4**(4), 397–434 (1979)
5. Guo, Q., Lu, J., Zhang, C., Sun, C., Yuan, S.: Multi-model data query languages and processing paradigms. In: CIKM 2020, New York (2020)
6. Lu, J., Holubova, I.: Multi-model databases: a new journey to handle the variety of data. ACM Comput. Surv. **52**, 38 (2019)
7. Zhang, C., Lu, J., Xu, P., Chen, Y.: UniBench: a benchmark for multi-model database management systems. In: Nambiar, R., Poess, M. (eds.) TPCTC 2018. LNCS, vol. 11135, pp. 7–23. Springer, Cham (2019). https://doi.org/10.1007/978-3-030-11404-6_2

Lightweight Deep Learning with Model Compression

U. Kang[(✉)]

Seoul National University, Seoul, South Korea
ukang@snu.ac.kr

1 Introduction

Aims & Learning Objectives. How can we perform deep learning efficiently? Deep learning is one of the most widely used machine learning techniques, and is a key driving force of the 4th industrial revolution. Deep learning outperforms many existing algorithms and even humans especially for many difficult tasks including speech recognition, go, language translation, game, etc. One crucial challenge of deep learning, however, is its efficiency both in training and inference. Deep learning requires a lot of parameters which need huge amount of time and space for storage and running. The problem becomes worse in mobile devices like smart phone since they have a limited amount of storage and computing power. It is necessary to design an efficient method for learning and inference in deep learning, which is exactly the goal of this tutorial.

We start with a very brief background of deep learning, including its history, application, and popular models including feedforward neural network, convolutional neural network, and recurrent neural network. Then we describe how to compress deep learning models using techniques including pruning [1], weight sharing [4], quantization [3], approximation, and knowledge distillation [2]. The audience is expected to gain substantial knowledge about reducing time and space in using deep learning.

Outline. Here is the outline of the tutorial.

- Brief overview of deep learning
- Pruning technique
- Weight sharing
- Quantization
- Low-rank approximation
- Distillation

Previous Presentation. The tutorial was presented in IEEE BigComp 2019 conference, and had attracted significant interests.

Length. We plan to deliver a *1.5 h* tutorial. We will spend 1/3 of the time on overview of deep learning, and remaining 2/3 of the time on model compression techniques.

© Springer Nature Switzerland AG 2021
C. S. Jensen et al. (Eds.): DASFAA 2021, LNCS 12683, pp. 662–663, 2021.
https://doi.org/10.1007/978-3-030-73200-4

Target Audience. The target audience consists of data mining professionals who wish to have a comprehensive understandings on model compression. The audience will learn recent developments on model compression and how they could utilize these tools for real-world problems that they are facing with in the wild.

2 About the Instructors

U. Kang is an associate professor in the Department of Computer Science and Engineering of Seoul National University. He received Ph.D. in Computer Science at Carnegie Mellon University, after receiving B.S. in Computer Science and Engineering at Seoul National University. He won 2013 SIGKDD Doctoral Dissertation Award, 2013 New Faculty Award from Microsoft Research Asia, 2016 Korean Young Information Scientist Award, and four "best paper" awards including 2018 ICDM 10-year best paper award. He has published over 90 refereed articles in major data mining, database, and machine learning venues. He holds four U.S. patents. His research interests include big data mining, deep learning, and machine learning.

References

1. Gordon, M.A., Duh, K., Andrews, N.: Compressing BERT: studying the e ects of weight pruning on transfer learning. In: Proceedings of the 5th Workshop on Representation Learning for NLP, RepL4NLP@ACL 2020, Online, 9 July 2020, pp. 143–155 (2020)
2. Hinton, G.E., Vinyals, O., Dean, J.: Distilling the knowledge in a neural network. CoRR, abs/1503.02531 (2015)
3. Hubara, I., Courbariaux, M., Soudry, D., El-Yaniv, R., Bengio, Y.: Binarized neural networks. In: NIPS, pp. 4107–4115 (2016)
4. Lan, Z., Chen, M., Goodman, S., Gimpel, K., Sharma, P., Soricut, R.: ALBERT: A lite BERT for self-supervised learning of language representations. In: ICLR. OpenReview.net (2020)

Discovering Communities over Large Graphs: Algorithms, Applications, and Opportunities

Chaokun Wang[1]([⊠]), Junchao Zhu[1], Zhuo Wang[2], Yunkai Lou[1], Gaoyang Guo[1], and Binbin Wang[1]

[1] School of Software, Tsinghua University, Beijing 100084, China
{chaokun,zhu-jc17,louyk18,ggy16,wbb18}@mails.tsinghua.edu.cn
[2] University of Chinese Academy of Sciences, Beijing, China
wangzhuo@iie.ac.cn

1 Motivation

In the past decades, community discovery has attracted great attention in both academia and industry. This tutorial highlights significant ideas and focuses on the typical techniques for community discovery and community-related research problems arising from urgent practical needs. It consists of the following three main parts: community detection, community search, and applications of communities. The community detection part presents some classical methods and a procedure-oriented benchmark for community detection. The part of community search (a.k.a. local community detection) reviews the representative methods of community search, and brings forward several extensions of this task, including community search with spatial or temporal information, community focusing, forbidden nodes aware community search, and so on. In the part of applications of communities, three interesting studies are demonstrated. The tutorial is never presented anywhere else. The intended length of the tutorial is two hours.

2 Outline

Part I. Introduction and background
  - The concept of community and the importance of community structures.
Part II. Community detection
  - Review of community detection [12].
  - Benchmarks and tools of community detection [9].
Part III. Community search
  - Review of community search [1, 2, 14].
  - Community search extensions [3, 5, 8, 10, 11, 13].
Part IV. Applications of communities
  - Subgraph matching [6].
  - Algorithm recommendation for community detection [4].
  - Synthetic graph generator [7].
Part V. Challenges and Opportunities

This work is supported in part by the National Natural Science Foundation of China (No. 61872207).

C. S. Jensen et al. (Eds.): DASFAA 2021, LNCS 12683, pp. 664–666, 2021.
https://doi.org/10.1007/978-3-030-73200-4

3   Biography

Chaokun Wang is a tenured associate professor at the School of Software, Tsinghua University. He has published over 100 refereed papers in major conferences and journals, including TKDE, SIGMOD, VLDB, ICDE, SIGIR, WWW, KDD, and AAAI. His current research interests include social network analysis, graph data management, and big data systems.

Junchao Zhu is a Ph.D. candidate in the School of Software, Tsinghua University. His main interests include social networks and community search.

Zhuo Wang receives his Ph.D. degree in the University of Chinese Academy of Sciences. His interests include social networks and graph algorithms.

Yunkai Lou is a Ph.D. candidate at Tsinghua University. Yunkai's main research interests include graph algorithms and graph database systems.

Gaoyang Guo is a Ph.D. candidate at Tsinghua University. He is interested in the combination of probabilistic graphical models and deep learning models.

Binbin Wang is currently working toward the M.Eng. degree at Tsinghua University. He is interested in social network generation and graph analysis systems.

References

1. Chen, L., Liu, C., Liao, K., Li, J., Zhou, R.: Contextual community search over large social networks. In: ICDE, pp. 88–99 (2019)
2. Fang, Y., et al.: A survey of community search over big graphs. VLDB J. **29**, 353–392 (2019). https://doi.org/10.1007/s00778-019-00556-x
3. Fang, Y., et al.: On spatial-aware community search. TKDE **31**(4), 783–798 (2019)
4. Guo, G., Wang, C., Ying, X.: Which algorithm performs best: algorithm selection for community detection. In: WWW (Companion Volume), pp. 27–28 (2018)
5. Li, R., Su, J., Qin, L., Yu, J.X., Dai, Q.: Persistent community search in temporal networks. In: ICDE, pp. 797–808 (2018)
6. Lou, Y., Wang, C.: Osmac: optimizing subgraph matching algorithms with community structure. In: ICDE, pp. 1750–1753 (2019)
7. Wang, C., Wang, B., Huang, B., Song, S., Li, Z.: FastSGG: efficient social graph generation using a degree distribution generation model. In: ICDE (2021)
8. Wang, C., Zhu, J.: Forbidden nodes aware community search. In: AAAI, pp. 758–765 (2019)
9. Wang, M., Wang, C., Yu, J.X., Zhang, J.: Community detection in social networks: an in-depth benchmarking study with a procedure-oriented framework. PVLDB **8**(10), 998–1009 (2015)
10. Wang, Z., Wang, C., Wang, W., Gu, X., Li, B., Meng, D.: Adaptive relation discovery from focusing seeds on large networks. In: ICDE, pp. 217–228 (2020)
11. Wang, Z., Wang, W., Wang, C., Gu, X., Li, B., Meng, D.: Community focusing: Yet another query-dependent community detection. In: AAAI, pp. 329–337 (2019)

12. Wu, Y., Jin, R., Li, J., Zhang, X.: Robust local community detection: on free rider effect and its elimination. PVLDB **8**(7), 798–809 (2015)
13. Zhang, F., Lin, X., Zhang, Y., Qin, L., Zhang, W.: Efficient community discovery with user engagement and similarity. VLDB J. **28**(6), 987–1012 (2019). https://doi.org/10.1007/s00778-019-00579-4
14. Zhang, Z., Huang, X., Xu, J., Choi, B., Shang, Z.: Keyword-centric community search. In: ICDE, pp. 422–433 (2019)

AI Governance: Advanced Urban Computing on Informatics Forecasting and Route Planning

Hsun-Ping Hsieh[1,2(⊠)] and Fandel Lin[2]

[1] Department of Electrical Engineering, National Cheng Kung University,
Tainan, Taiwan
hphsieh@mail.ncku.edu.tw
[2] Institute of Computer and Communication Engineering, National Cheng Kung
University, Tainan, Taiwan
q36084028@mail.ncku.edu.tw

Abstract. Urban computing is an interdisciplinary field that combines computing technologies, such as wireless networks, sensors, computational power, and data analytics to improve quality of life in urban areas. Recently, urban computing draws from the evolution of computer science, internet of things (IoT), data science, and artificial intelligence to improve the dealing of immediate citywide events. Two important urban computing issues will be introduced in this work: one is citywide informatics forecasting, while the other one is route planning in urban spaces. These two topics are highly-correlated with city governance.

Keywords: Urban computing · Artificial Intelligence · Spatial-temporal prediction · Route planning · Urban governance

1 Introduction

Urban informatics can be generally defined as signals collected in different locations with time information. For example, stations are deployed to monitor the air quality in different areas for every hour. Or, a travel company may analyze users' check-in behaviors to learn about their preferences and come up with new business strategies. Forecasting urban informatics can help the government or enterprises prepare for emergencies well in advance, and even predict the future of a city to reap profit. For example, the work [4] predicts the possible number of illegal-parking events in big urban spaces. Based on the results, the government can distribute more patrol force to illegal parking hot spots in advance. However, it is challenging to forecast urban informatics, since various features should be considered jointly. Some features (e.g., hourly weather, traffic volume, check-ins) are dynamic, while others (e.g., road network, points-of-interest, population) are rarely updated. In this work, we would like to explore how advanced data engineering technologies and spatial-temporal AI models effectively use heterogeneous urban big data to make predictions and improve urban governance.

© Springer Nature Switzerland AG 2021
C. S. Jensen et al. (Eds.): DASFAA 2021, LNCS 12683, pp. 667–669, 2021.
https://doi.org/10.1007/978-3-030-73200-4

Research focused on routing algorithm in urban computing has developed swiftly over years [2]. With the rapid growth of metropolis and conurbation, the construction and modification of the transportation system have turned out to be one of the most crucial factors to facilitate quality of life [6]. When it comes to transportation in cities, most people rely on either private vehicle, which uses road networks, or mass transit, which uses public transportation networks. Despite the differences, both networks can be naturally transformed into directed graphs [1]. Meanwhile, several methods utilizing various strategies are also proposed to deal with multi-criteria route planning in urban places [3]. On the other hand, with the flourish of hybrid computational intelligent systems that synergize learning-based inference models in recent years, studies that concentrate on targets including recommendation, planning, scheduling according to first-stage inference results are proposed [5]. Despite the promising results achieved by using hybrid computational intelligent systems, several challenges of route planning in urban space still exist and are worth discussing.

2 Biographies of the Presenters

Hsun-Ping Hsieh is an Associate Professor at Department of Electrical Engineering, National Cheng Kung University, Taiwan. H.P.'s research interests include Urban Science, Big Data Mining, and Urban and Geo-social Computing. He had published a series of papers in top conferences and journals, including, SIGKDD, Web conference, SIGIR, CIKM, SIGSPATIAL, Multimedia, ICDM, ECML-PKDD, TIST, and KAIS. H.P.'s academic recognitions include: 2020 MediaTek Social Innovation Special Award, 2019 MOST Einstein Young Scholar Fellowship, 2013 and 2014 Garmin Research Fellowship, and 2013 Excellent Stars of Tomorrow of Microsoft Research Asia.

Fandel Lin is a M.S. student in Institute of Computer and Communication Engineering at National Cheng Kung University, Taiwan. His research interests lie in Urban Computing, Combinatorial Optimization, and Geographic Information Science. Meanwhile, his researches are recognized by the first place award of Student Research Competition in ACM SIGSPATIAL'18 and the second place award of ACM Student Research Competition Grand Final in 2018–2019. Recently, he has published several papers in peer-reviewed journals and conferences, including KAIS, ICDM, PKDD, SIGSPATIAL, and Web conference. He is currently a member of Urban Science and Computing Lab (UCLAB) and supervised by Prof. Hsun-Ping Hsieh.

References

1. Bast, H.: Car or public transport – two worlds Effic. Algorithms **5760**, 355–367 (2009)
2. Bast, H.: Route planning in transportation networks. In: Kliemann, l., Sanders, p., (eds.) Algorithm Engineering Selected Results and Surveys. LNCS, vol. 9220, pp. 19–80. Springer Cham (2016)https://doi.org/10.1007/978-3-319-49487-6_2
3. Delling, D., Sanders, P., Schultes, D., Wagner, D.: Engineering route planning algorithms. Algorithmics Large Complex Netw. **5515**, 117–139 (2009)

4. Jiang, J., Chen, Y.-C., Hsieh, H.-P.: Detection of illegal parking events using spatial-temporal features. In: ACM SIGSPATIAL (2020)
5. Mourad, A., Puchinger, J., Chu.: A survey of models and algorithms for optimizing shared mobility. Transp. Res. Part B: Methodol. **123**, 323–346 (2019)
6. Steg, 1., Gifford, R.: Sustainable transportation and quality of life. J. Transp. Geogr. **13**(1), 59–69 (2005)

Author Index

Printed in the United States
by Baker & Taylor Publisher Services